THE U.S. ATLAS OF NUCLEAR FALLOUT

1951-1970

VOLUME II. COUNTY COMPARISONS

by

RICHARD L. MILLER

THE U.S. ATLAS OF NUCLEAR FALLOUT

1951-1970

VOLUME II. COUNTY COMPARISONS

by

RICHARD L. MILLER

LEGIS BOOKS
THE WOODLANDS, TEXAS

LEGIS BOOKS
Division of Legis Corp.
P.O. Box 7888
The Woodlands, Texas 77387-7888

Printed in the United States of America

Printing number

1 2 3 4 5 6 7 8 9 10

Library of Congress Cataloging-in-Publication Data

Miller, Richard L. (Richard Lee)
The U.S. Atlas of Nuclear Fallout 1951-1970 Volume II: County Comparisons
1. Nuclear weapons–Testing.
2. Radioactive fallout–United States.
3. United States–History.
4. United States–Geography.
5. Science Reference.

ISBN: 978-1-881043-29-4

For Chrissy McHenry Bain

U.S. ATLAS OF NUCLEAR FALLOUT 1951-1970

Volumes in the Series

Vol I: Total Fallout: General Reader Edition
ISBN 978-1-881043-13-3
Vol II: County Comparisons: ISBN 978-1181043-29-4
Vol III: Total Fallout Eastern U.S. ISBN 978-1-881043-28-7
Vol IV: Total Fallout Western U.S ISBN 978-1881043-27-0
Vol V: Calculations ISBN 978-1-881043-19-5 and
ISBN 978-1-881043-31-7 (2nd Edition)

Original Vol I: Total Fallout ISBN 1-881043-12-6 was replaced by the Total Fallout General Reader Edition ISBN 978-1-881043-313; however, this original volume is still available by special order from TwoSixty Press.

Total fallout data values in the US Atlas of Nuclear Fallout are expressed in microCuries/sq meter down to a minimum value of 0.8 microCuries or 800 nanoCuries. For more detailed information regarding county fallout levels, you may want to consider volumes in the following expanded series, which include county fallout values as low as 20 nanoCuries:

Nuclear Fallout from the Nevada Test Site 1951-1970

Publication date: 2nd quarter 2006:

Volumes in the Series:

Southern U.S. (ISBN: 978-1881043-33-1) Fallout data for 786 counties in Alabama, Florida, Georgia, Louisiana, Mississippi, North Carolina and South Carolina.

East Central U.S. (ISBN: 978-1-881043-32-4) Fallout data for 764 counties in Kentucky, Tennessee, Virginia and West Virginia.

Eastern U.S. (ISBN: 978-1881043-34-8) Fallout data for 576 counties in Connecticut, District of Columbia, Delaware, Massachusetts, Maryland, Maine, New Hampshire, New Jersey, New York, Ohio, Pennsylvania, Rhode Island and Vermont.

West Central U.S. (ISBN: 978-1881043-35-5) Fallout data for 812 counties in Kansas, Missouri, Nebraska and Oklahoma.

North Central U.S. (ISBN: 978-1-881043-36-2) Fallout data for 652 counties in Iowa, Illinois, Michigan and Wisconsin.

Western U.S. (ISBN: 978-1881043-37-9) Fallout data for 681 counties in Arizona, California, Colorado, Idaho, Montana, North Dakota, New Mexico, Nevada, Oregon, South Dakota, Utah, Washington and Wyoming.

South Central U.S. (ISBN: 978-1881043-38-6) Fallout data for 578 counties in Arkansas and Texas.

Other Publications by LEGIS BOOKS include:

U.S. Cancer Mortality Rates and Trends 1950-1979 Volumes I-IV

Reprint of an EPA-NCI document originally published in 1983 detailing county-by-county rates for 32 cancers categorized by decade: 1950-59, 1960-69 and 1970-79.

Vol I: ISBN: 978-1-881043-22-5
Vol II: ISBN: 978-1-881043-23-2
Vol III: ISBN: 978-1-881043-25-6
Vol IV: ISBN: 978-1-88-1043-26-3
Four-Volume Set: 978-1-881043-24-9

Additional information may be found on the web at:
http://*www.260press.com/publications.htm*
and
http://*www.legis.com*

TABLE OF CONTENTS

ACKNOWLEDGEMENTS

This book would have been impossible without the dedication and commitment of the scientists of the National Cancer Institute, who produced the original I-131 deposition data. Theirs was the most difficult task of all, and they have my deepest appreciation and respect.

I would also like to thank the following software companies for donating or providing discounted software to support my research: Palisades Corporation (@Risk and Risk Optimizer); Attar Software (Xpert Rule); Decisioneering (Crystal Ball Predictor); Spotfire (Spotfire Pro); MapInfo (MapInfo Pro); Golden Software (Didger); Cytel Software (EGRET); Salford Systems (CART and MARS); and SPSS (Systat). And, of course, a tip of the hat to Microsoft, maker of the only software that should be classified as heavy equipment: Excel 97. Additionally, I would like to express thanks to the makers of Spreadsheet Assistant™ for Excel 97 and 2000. Without it, the work would have taken four times as long to complete.

Thanks to Lynn Anspaugh of the University of Utah for his help and suggestions regarding the Hicks Tables, and to Owen Hoffman of SENES, Oak Ridge, Tennessee, for his assistance, suggestions and patience regarding geometric means, geometric standard deviations, arithmetic means of log normal distributions, and Monte Carlo simulations.

I would especially like to thank my friends (and nuclear specialists) Jim Horner and Karl Johannson, who spent long hours checking the material for accuracy. Couldn't have done it without you guys!

It should be noted that Cliff Honicker performed the original investigations on rainouts in both South Dakota and Minnesota, and also kindly provided me with the AEC reports of investigations. The device specs were provided by the indefatigable nuclear weapons researcher, the late Chuck Hansen.

Additionally, I would like to thank the many librarians who spent time with the advance reviews and made suggestions for enhanced readability. They include: Patty Gross, Council Valley (ID) Library, Wendy Bossie, Caribou (ME) Public Library, Kathy Adam, Shenandehowa (NY) Public Library, Bev Russell, Scottsbluff (NE) Public Library, Arlene Ott, Cody (WY) Public Library and Symma Zerkow, Houston (TX) Public Library

Saving the most important for last, , I would like to thank my mother, Ruth E. Miller, my daughter Trace and my friend Kim Shaw for their continued interest and encouragement and support throughout the long days and nights working on this project. These books belong to them.

PREFACE

The era of aboveground nuclear testing at the Nevada Test Site began on January 27, 1951 with the detonation of Ranger ABLE. Starting in 1951 and continuing until 1962, the federal government detonated well over a hundred nuclear weapons in aboveground tests. It is well known that radioactive fallout fell on Nevada, Utah, Arizona, Colorado and Idaho. What is not generally known, however, is that the prevailing winds carried massive quantities of radioactive fallout to all of the remaining states, even those as far away as Maine and Florida. No county in any of the continental United States escaped fallout from the Nevada Test Site.

What did the fallout contain? How harmful were the radioactive elements in the fallout? What are the long-term health effects of exposure to radioactive fallout? How can an individual citizen know if he or she has been exposed to fallout—and what is the significance of the exposure?

Until recently, there have been few, if any, answers to these questions. Now, however, new data has been made available that sheds light on these questions. In fact, most of the above questions have now been answered.

This book uses recently published government data to show which counties in the United States had the heaviest dose of fallout and when that fallout came down. I hope the research in this book is a starting point for further research into what may potentially be a significant health risk.

I also hope the average reader will be able to use this data to learn where he or she was at the time the fallout came down.

For over ten years significant quantities of radioactive materials from nuclear testing were deposited across the country, making their way into the food chain and exposing several generations of Americans to radiation. While the Atomic Energy Commission was aware that a thunderstorm could "wash" nearly all the radioactive material from a nuclear cloud and send it to earth, there was no effort to evaluate hot spots and rainouts downwind of the Test Site. Additionally, there were relatively few attempts to evaluate potential links between nuclear fallout and disease rates.

In 1981 Harry G. Hicks published tables that allowed researchers to theoretically determine deposition rates of over a hundred fallout components. Three years later, Congress authorized the National Cancer Institute to conduct a study to determine if thyroid cancer in the population could have been caused by radioactive iodine in fallout.

The NCI was given one year to complete the study.

On October 4, 1997, after nearly 15 years of work, the National Cancer Institute announced that the radioactive iodine report was finally ready.
Data for each nuclear test, by fallout day was published for each and every county in the United States. Some counties in Utah, Nevada, Arizona and California were even divided into sub-counties based on geographical differences.

The report in its entirety ran to over 120,000 pages. Perhaps for that reason, the authors of the study decided against publishing the results in book form. Rather, the information was posted on the Internet on the National Cancer Institute Website under the heading "What's New."

Some of the pages had a rushed quality to them. Some map titles were written in pencil and the cloud trajectories appeared more rudimentary than the originals from 1953 and earlier. County lists included some places, such as Washabaugh County, SD, that no longer existed—or, in the case of Columbus County GA, *never* existed (the NCI undoubtedly meant Muscogee County, which is the county surrounding the city of Columbus.)

Worse, for anyone trying to analyze the data, the alphabetical order for counties seemed to vary with the shot. In some tables Arizona came before Arkansas; in others the order was reversed. For many of the tests, the NCI tracked days of fallout beginning with day one (the detonation date) to day ten. But for one notoriously fallout-intensive shot, Buster-Jangle Easy, the fallout tracking began on day *three.* Others, like the subsurface test Buster-Jangle Uncle, were apparently only tracked for one day.

Several nuclear tests were combined, even though the events surrounding the detonation and subsequent cloud paths didn't really warrant it. One important nuclear test, shot Stokes in 1957, was listed as having produced *no* fallout, even though internal Atomic Energy Commission memorandums strongly suggested otherwise.

There were problems in deciphering the fallout deposition rates. The data included obscure and poorly documented statistical terms (more about that later), was difficult to download and once downloaded, nearly impossible to format and evaluate.

From a larger perspective, the study had an enormous shortcoming: it included data for only *one* component of fallout, a form of radioactive iodine, I-131. This short-lived component of fallout accounted for only ***two percent*** of the total fallout deposited across the country.

But even with all those shortcomings, it was still an extraordinary piece of work. In addition to county deposition of radioactive iodine (I-131), the authors included information about population exposure based on milk consumption. A reader (or more appropriately, an Internet surfer) had only to type in her birth date, county data and milk consumption information to determine the estimated dose of radiation her thyroid likely received from fallout. Unfortunately, the most important piece of information was left out: the National Cancer Institute never discussed whether exposures to radioactive materials in fallout could cause cancer.

WHY FALLOUT IN THE 1950S MATTERS IN THE 21ST CENTURY

Because of fallout from the Nevada Test Site, radioactive elements called radionuclides were deposited in every county of the United States. Many of these radionuclides became incorporated into living cells. It is known that atoms of strontium-90 and ruthenium-103 made their way to bones and teeth. Other radionuclides such as iron-59, copper-64 and rubidium-88 may have been incorporated into important enzymes.

While all of these radionuclides emitted radiation of one form or another, many, such as arsenic, beryllium, cadmium, curium, cobalt, rhodium, tin, uranium and yttrium are either suspected or recognized carcinogens.

Others, such as indium, lead, manganese, molybdenum, selenium, tellurium, and tungsten are suspected teratogens, capable of causing birth defects.

All of these elements, in their radioactive form, were significant components of fallout.

Moreover, measurable quantities of many radionuclides can remain in the environment in significant quantities *fifty years* after detonation of the device that produced them. These radionuclides include manganese-54, cobalt-60, americium-241, curium-242, strontium-90, yttrium-90, cesium-137, promethium-147 and europium-155. In other words, radionuclides deposited in 1957 will still be with us in 2007. As a somewhat extreme example, consider the toxic fallout component americium-241: The amount of this element remaining after 50 years from just one shot—Teapot Apple II in 1955—is fully 99.49 per cent of the amount produced *at the time of detonation.*

This influx of radionuclides into the environment from the Nevada Test Site continued for over 12 years. While global fallout, due in part to dilution effects, painted the environment in broad brush strokes, Nevada Test Site fallout was anything but uniform. Some counties received thousands of times the fallout as other counties. In addition, there is strong evidence that rainstorms were instrumental in concentrating the fallout in extremely small areas *within* county boundaries.

Despite this, as of the time of this writing (March 2002), there has *never* been a government-sponsored scientific study published evaluating cancer rates or birth defects in the entire United States against total fallout deposition. Moreover, there has never been an epidemiological investigation of a cancer cluster outside of Utah and Nevada that included nuclear fallout as a potential factor.

Did fallout in the 1950's and 1960's cause cancer? Given the extremely long latency period associated with many cancers, perhaps a better question is: did fallout in the 1950's and 1960's cause cancers in the 1990's?

Hopefully this series of books will help answer that question.

INTRODUCTION

In January 1983, the 97th Congress directed the National Institute of Health to evaluate the extent and effects of the fallout component iodine-131 (I-131) fallout resulting from the Nevada nuclear tests conducted in the 1950s and early 1960s[1].

The task was finally completed and published in 1997. The report consisted of I-131 fallout deposition rates from nearly 60 aboveground and subsurface nuclear tests—for each of 3098 counties. The finished product included data for up to 20 fallout days subsequent to the nuclear test. All together, the report was equivalent to more than 120,000 pages of data.

Officials at the National Cancer Institute decided that publication of such a mass of data would be feasible only over the Internet. Interested citizens could download data for their own home counties and learn just how much I-131 was deposited from each test, and when. Not only that, they could enter their birth date and estimated milk consumption and read an estimate of the radiation exposure to their thyroid.

There were some problems, however. While some areas had a dense network of detection sites—there were 65 in New York state shortly after the Ranger detonations—the fixed radioactivity detection sites numbered only 97. Thus, there was a measure of uncertainty regarding fallout amounts when the county was a significant distance from the detection site. So, instead of a single value for fallout, say, 50 units of fallout per square foot (or in this case, square meter), the data came out more like this: "the amount was somewhere between 5 units of fallout per square foot and 100 units of fallout per square foot."

One way of presenting this information would have been using minimums and maximums: For example, the above information might have read:

Minimum: 5 units of fallout per square foot.

Maximum: 100 units of fallout per square foot.

The explanation would have sounded something like this: "Because back during the 1950s the fallout was collected at a site that was some distance from your county, we don't really know exactly how much fallout came down in ***your county.*** But we're 95 percent sure that the number was somewhere between 5 units per square foot and 100 units per square foot."

The authors didn't do that. Instead, the information presented over the Internet was given in terms of ***GM*** and ***GSD*** with no explanation of what ***GM*** and ***GSD*** actually meant. In practical terms, the information was limited to those individuals with (1) access to the Internet, (2) an ability to download and translate the data, (3) an education or experience that included high-level statistics, and (4) access to a spreadsheet and an ability to perform the complicated mathematics necessary to make sense of the information.

The GM actually meant "***Geometric Mean***" and the GSD referred to "***Geometric Standard Deviation***", statistical terms that are rarely encountered even in college-level statistics courses. Essentially, the GM was a measure of the average fallout deposition, while the GSD was a measure of the uncertainty surrounding that average. In order to work with the data, I needed a single, conservative value. I consulted Dr. Owen Hoffman, an expert on statistics and radiation, who suggested I take the GM and GSD and convert them to a single value using the Arithmetic Mean of the Lognormal Distribution. Here is the calculation:

$$AM = GM\left(\exp\left(\frac{(\ln GSD)^2}{2}\right)\right)$$

Where: AM = Arithmetic mean of the lognormal distribution;
GM = Geometric mean of the lognormal distribution;
GSD = Geometric standard deviation of the lognormal distribution
exp = exponent e = 2.7183
ln = natural logarithm

After obtaining a single conservative value for the range of values for I-131, I then had to convert this value to total fallout. Each nuclear test is associated with a unique ratio between I-131 and total fallout (which includes all the radionuclides produced at the time of detonation.) Fortunately , this information is found in declassified government documents formally titled: *Results of Calculations of External Gamma Radiation Exposure Rates from Fallout and The Related Radionuclide Compositions.* The are more commonly known as the "Hicks Tables," named for their author Dr. Harry Hicks.

Using the Hicks Tables, I calculated the I-131/total fallout ratios for each nuclear test. After applying the appropriate multiplier to each derived I-131 value, I arrived at a total fallout value for each deposition event. Unless indicated, all values in this book and in the series are given as total fallout, expressed either as nanoCuries/square meter or microCuries/square meter.

I-131 AND TOTAL FALLOUT

The published NCI data included values for only a single radioactive component of fallout—one of the radioactive isotopes of the element iodine. This particular radioactive element (called a ***radioisotope*** or ***radionuclide),*** makes up only two percent of the total fallout.

> **Fact:** Fallout from nuclear tests is made up of 150-200 radioactive elements called ***radionuclides.*** Some of these radionuclides include radioactive versions of common elements such as iron, copper, cobalt and sodium. One variety of radioactive iodine, I-131 typically makes up two percent of the total fallout.

Since radioactive iodine 131 makes up two percent of the total, one might think it possible to calculate total fallout simply by multiplying the I-131 amounts by 50. Unfortunately, it is not as easy as that. For one thing, each atomic test produced its own mix of fallout. One atomic test might produce very low levels of the radioactive element americium-24 1, while another test might produce much greater amounts.

Worse, radioactive versions of elements have a unique characteristic that sets them apart from ordinary non-radioactive elements: they change, over time, to *other* elements. The radioactive element beryllium can change to the element lithium. Another radioactive element, lanthanum-144 changes or decays to another element, cerium-144, which in turn decays to yet a different element, praseodymium-144. Each of these elements has unique chemical and physical characteristics.

What was needed was some way to determine what ***kinds*** of radioactive elements were produced for *each particular atomic test.* Not only that, it was essential to know how much of *each* radioactive element was produced in each of the 100 aboveground atomic tests.

Interestingly, that work had actually been done—and published— by a Livermore physicist named Harry G. Hicks in 1981. Through experimentation and long hours on a mainframe computer, Harry Hicks calculated ratios for more than 125 radioactive elements produced by *every* aboveground test ever detonated at the Nevada Test Site.

The work came to be called the Hicks Tables and, importantly, I-131 was among the radionuclides listed. Now, with knowledge of the radioactive iodine-131 levels for each atomic test—and a little mathematics—it was possible to finally determine how much fallout each atomic test actually deposited on each of 3098 U.S. counties.

FACT: The number following the element symbol (I-***131***, for example) is called the ***mass number.*** The mass number is the *sum of the number of protons and the number of neutrons in the element.* This ***mass* number** is unique for each radioactive element. All atoms of iodine-131 (I-131) for example, decay at a different rate and produce *different* amounts of radiation than do all atoms of iodine-***130*** (I-130).

FACT: Each nuclear test is unique in the **type** and **amount** of radioactive elements found in its fallout.

NAMING CONVENTIONS

In the 1950s they were called "atomic tests". In the 1960s and later, they were referred to as "nuclear tests." Some in the weapons community simply referred to them as "shots" or "events." All are synonymous with a detonation of a device that produces radioactive fallout. In most of the tables, this detonation is referred to simply as **shot,** followed by the name of the test or "event."

It is important to remember that all of the nuclear tests associated with this book refer to detonations that occurred at or very near the Nevada Nuclear Test Site in Nye County, Nevada.

Each series of nuclear tests (or atomic tests, or shots—again, they all refer to the same thing) was given a name. The first series of tests in Nevada was termed ***Ranger.*** The first nuclear test in the ranger series was named ***Able.*** It was detonated on January 27, 1951. The second test in the Ranger series was ***Baker.*** Those familiar with military code designations for alphabet (Able, Baker, Charlie, Dog, Echo, Fox, Golf, etc) would probably have expected the third shot in the Ranger series to be named "Charlie." Instead, the third test in the Ranger series was named ***Baker-2.***

FACT: There were 6 major series of aboveground nuclear tests at the Nevada Test Site. They were:

1. Ranger	Jan 27-Feb 6 1951
2. Buster-Jangle	Oct 22-Nov 29 1951
3. Tumbler-Snapper	Apr 1-Jun 5, 1952
4. Upshot-Knothole	Mar 17-Jun 4 1953
5. Teapot	Feb 18-May 15, 1955
6. Plumbbob	May 28-Oct 7, 1957

The final nuclear test covered by this book is shot ***SEDAN,*** an underground test detonated on July 6, 1962.

Through the first, second and third series, the detonation names generally followed the A, B, C, and D—designation. The eighth test in the Tumbler-Snapper series in 1952, shot ***How,*** had a name that began with H—the eighth letter of the alphabet.

By 1953 the nuclear test names had become more personalized: **Annie, Nancy, Dixie** and **Simon** were all names of nuclear tests in this series. The 1955 nuclear test series included shot names such as **Turk, Hornet, Wasp, Bee** and **Ess.**

Shot **Ess** apparently was so-named because it was detonated some sixty feet below the *surface* of the desert.
The final official aboveground series included nuclear tests named for scientists important in the history of physics. There were tests named **Boltzmann, Doppler, Kepler, Stokes, Fizeau,** and **Newton.**

Some shots were named for both scientists and geological formations. Shot **Wheeler** could have easily been named for physicist John Wheeler, there also happened to be a mountain in New Mexico named Wheeler Peak. As coincidence would have it, the nuclear cloud from shot Wheeler eventually drifted over the town of Wheeler, Texas.

In this book, the nuclear tests are generally referred to first by the series, then by the specific name in **all caps,** as in Upshot-Knothole **HARRY.** Abbreviated, the shot may be listed with the series name abbreviated **(R** =**R**anger, **BJ** = **B**uster-**J**angle; **TS** = **T**umbler **S**napper; **UK** = **U**pshot-**K**nothole; **TP** = **T**ea**p**ot, **PB** = **P**lumb**b**ob) followed by the sequence number of the shot in that series.

For example, shot HARRY was the ninth detonation listed by the National Cancer Institute's I-131 survey in the Upshot-Knothole Series. Thus, this nuclear test can also be designated simply as ***UK9.*** In some volumes in this series, I have included an additional bit of information—the day the fallout occurred relative to the day of the actual detonation.

In this case, the symbol **UK9-3** would refer to the *third* day of fallout resulting from the *ninth* shot in the **U**pshot-**K**nothole series, as listed by the National Cancer Institute.

Where space permits, shots are listed by the series name or abbreviation, followed by the shot name. Importantly some nuclear tests were *not* included in the NCI study. Others, such as Teapot BEE and Teapot ESS were combined into one. Also, some tests, such as Ranger-FOX (R-3) and Plumbbob STOKES (PB-7), which were associated with no fallout (according to the 1997 NCI study), were excluded from some lists and maps. A complete list of abbreviations used in the book is found at the end of this section.

MORE ON SHOT DAYS

The term "shot day" refers to a particular day after detonation. For example, TS1-1 refers to the *day of detonation* of the ***first*** shot in the **T**umbler-**S**napper series. That test was formally named Tumbler-Snapper ABIn this case, the day that TS-1 refers to is April 1, 192. The term **TS1-2** refers to one day *after* detonation of Tumbler-Snapper 1—or April 2, 192. TS1 -3 refers to the second day after the detonation. Another way to look at it:The ***2*** in TS1-*2* refers to ***the number of days the fallout has been in the air, including the day of detonation.*** In this case, since TS1 refers to shot Tumbler-Snapper ABand since that detonation occurred on April 1, 192 (see above) TS1-3 refers to April 3, 192

Example: TS3-5

TS refers to the name of the test series. In this case, Tumbler-Snapper, which occurred in the spring of 1952.

The 3 indicates the third nuclear test in the Tumbler-Snapper series.

The **-5** refers to the days after the initial detonation. In this example, **TS3-5** refers to the **fifth day** after the detonation of the third nuclear test in the Tumbler-Snapper series. The different shot series discussed in the volumes in this series, along with their abbreviations include:
Ranger (R); Buster-Jangle (BJ); Tumbler-Snapper (TS); Upshot-Knothole (UpK); Teapot (TP), Plumbbob (PB); Hardtack (HT), Underground (UE or ue).

WHAT *IS* NUCLEAR FALLOUT?

A nuclear device is essentially a ball of fissionable, radioactive material surrounded by explosives. This fissionable material is usually some form of plutonium. Sometimes, another radioactive material, tritium, is added to the mix to make the blast more powerful. Other elements in nuclear devices might include americium-24 1 as part of the neutron source and beryllium as part of the *neutron reflector,* used to bounce neutrons coming from the plutonium *back* to the plutonium. Additionally, there are various structural components such as carbon, copper and iron. When the plutonium sphere fissions, the plutonium atoms may fracture into many different kinds of radioactive elements, from beryllium to cobalt to iron. In all, the resultant soup may include 200 different radioactive types of elements. Other structures that happened to be placed near the device will become radioactive through a process called *neutron capture.* These structures could include the shot cabin or "shot cab," the copper electrical cables, the sodium and magnesium in the sand beneath the fireball, and of course, the iron shot tower—which often weighed between 500 and 700 tons. Most of these objects will be vaporized by the fireball, only to re-form as tiny spheres ranging in from a micron to maybe several thousand microns. A micron equals about 0.0000000254 inches.

FACT: Fallout particles that are carried by the wind from the site of a nuclear detonation may be as large as several thousand microns, to as small as 1 micron in diameter. By way of comparison: One inch is equal to 39,370,079 microns. The droplets making up a summer cumulus cloud have an average diameter of 11.2 microns.

Interestingly, the fallout particles resulting from a high altitude burst are generally spherical, while those from tower or low airbursts are both spherical and irregular-shaped. Those detonated underground produce fallout particles that are generally irregular in shape.[3]

Fallout particles large enough to *have* color can be colorless, gold-yellow, orange, red, brown, green or black. Some may be magnetic.[4]

Close-up, a fallout particle may look like a glossy, brown marble, or a jagged, irregular piece of rock.

FIG 3.0 FALLOUT FROM A NEVADA TEST SITE TOWER SHOT

Figure 3.0 shows a photo of a 0.5 mm (about two-hundredths of an inch in diameter) sphere of fallout taken from one of the tower shots. Note the smaller spheres sticking to its surface.

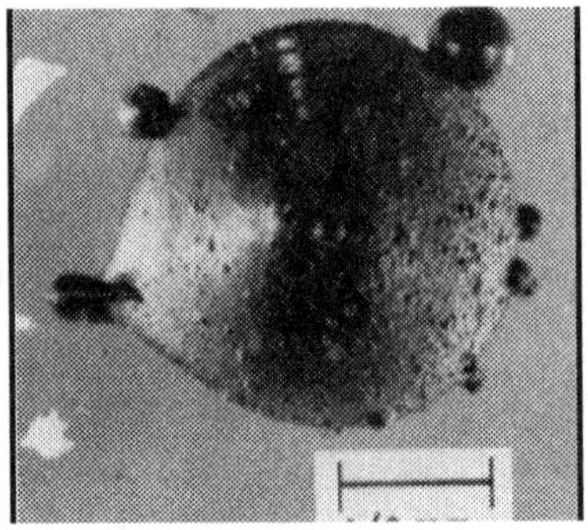

The photo below, Fig 4.0 is of a different kind of fallout particle with a brilliant, glossy surface. It was also the result of a tower shot.

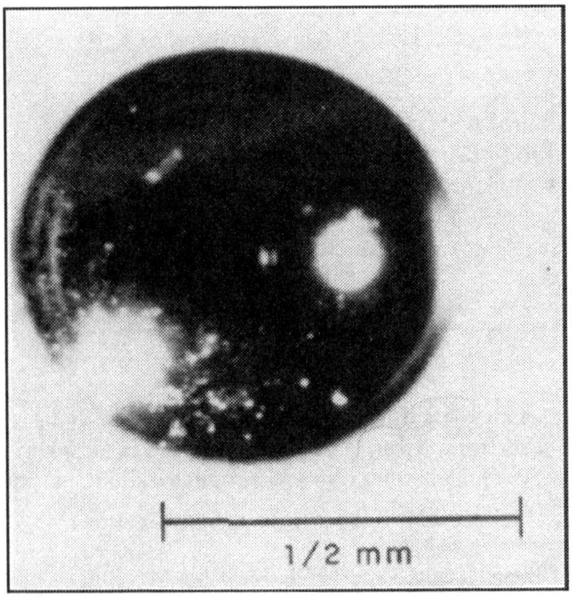

FIG 4.0
FALLOUT PARTICLE FROM AN NTS TOWER SHOT.[5]

WHY IS FALLOUT RADIOACTIVE?

Fallout is radioactive because the elements that make up the fallout emit radiation. This radiation may be in the form of alpha particles, beta particles, low-speed electrons, gamma rays or x-rays. Alpha particles consist of packets of two protons and two neutrons each. Beta particles are simply high-speed electrons. X-rays are very energetic radio waves and gamma rays are similar to X-rays, only more energetic[6].

> **FACT:** The frequency associated with normal household current is 60 cycles per second. That is, it reverses polarity sixty times a second. By comparison, the frequency of a **gamma ray** or **X-ray** can be as high as 10,000,000,000,000,000,000 cycles per second.

Each time radiation—either alpha, beta, gamma, low-energy electrons or x - rays—is emitted by a radioactive atom, that radiation emission is a sign that the atom has changed from one element to another. For example, a radioactive atom of iron may emit radiation and, in the process, become another element, radioactive cobalt. The cobalt atom might then emit radiation and become yet another element, radioactive nickel.

It's important to remember that each radioactive element is unique in the ***rate*** the radiation is emitted. Radioactive iron-55, for example, has a half-life of 2.6 **years,** while radioactive cobalt-55 (which iron-55 becomes upon emitting radiation) has a half-life of only 18 **hours.** That means that radiation is emitted with greater *frequency* from the radioactive cobalt-55 than from the radioactive iron-55.

Each radioactive element is also unique in the ***type and amount*** of radiation it emits. The radioactive element americium-241 produces an alpha particle, while strontium-90 emits a low-energy beta particle. Cobalt-60 produces beta particles, gamma rays *and* X-rays.

Since each nuclear test produces a unique blend of radioactive elements, calculating the total radiation from such a changing mixture of elements is not an easy task.

ATOMIC MASS UNITS

In the above example we used symbols followed by numbers to represent the various radioactive elements. In science, each element is represented by a specific symbol. For example, cobalt is represented by the symbol ***Co,*** manganese by ***Mn*** , and nickel by ***Ni.***

You may remember that each atom is made up primarily of three objects—neutrons, protons and electrons. The standard atom is often visualized as a sphere, with the neutrons and protons packed together at the center of the sphere—at the core—and with the electrons forming the outer layers. The configuration of the electrons determines the chemical characteristics of the atom—that is, how it combines with other atoms to form molecules.

The configuration of the inner core of neutrons and protons, the ***nucleus,*** determines how the atom acts physically—that is, whether it will emit radiation and change to another element.

Scientists have long known that the number of protons determines the type of element. The element carbon has six protons in its nucleus, and therefore is assigned a specific ***atomic number*** on the chart of the elements: 6. Nitrogen has seven protons and is thus assigned the number 7 on the chart of the elements. The more protons an atom has, the heavier it is. Iron has 54 protons while lead has 82 protons. Plutonium, an extremely heavy atom, has an atomic number of 94.

As it turns out, protons carry a positive charge, neutrons carry no charge, and electrons carry a negative charge. **In order to balance out the charges in an atom, the number of electrons should equal the number of protons.** Since the configuration of electrons determine the chemical characteristics of the atom, and since the number of protons actually determines the number of electrons that the atom has available, *it is really the number of protons that determines how an atom will act chemically.*

Scientists have long known that in the single, isolated atom the number of electrons equaled the number of protons. And while the number of electrons might change from molecule to molecule, the number of protons remained fixed. This was why the ancient alchemists had such poor luck in "transmuting" lead to gold. They would have to somehow have to remove three protons from each atom of lead (82 protons per atom) to get atoms of gold (79 protons per atom.)

Scientists examining atoms originally surmised that the number of neutrons—the subatomic objects with no charge—equaled the number of protons. So, if an atom had, say 6 protons (which would make it the element carbon), then it also probably had 6 neutrons.

But early in the 20th century, scientists discovered that **neutrons varied considerably among atoms of the same element.** Since a neutron has no charge, a change in the number of neutrons wouldn't affect the number of electrons and thus the chemical characteristics of the atom. But the fact remained that, due to the extra neutrons, atoms of a given element can vary in weight or ***mass.*** Atoms with the same number of protons but differing numbers of neutrons are ***isotopes*** of each other.

To differentiate between the atoms of an element, scientists decided to create the term **atomic mass unit.** The concept was simple: simply assume that the protons and neutrons **each weigh one atomic mass unit** or a.m.u. Thus, if the sum of neutrons and protons in an atom of carbon, equaled, say, 14, then the isotope was assigned the name *carbon-14.*

As another example, the element cobalt has an atomic number of 27—which means simply that it has twenty-seven (27) protons in its nucleus. **There can be no other element with twenty-seven protons in its nucleus. If there are 27 protons in the atom then it must be an atom of cobalt.** Now, if someone were to count the number of ***neutrons*** in an atom of non-radioactive cobalt, he would likely find the number to be thirty-two. For reasons that require more detailed discussions of nuclear physics, the cobalt atom is most stable with this combination of neutrons and protons. Put another way, cobalt is considered to be *most* stable when the mass number equals 59—that is, when there are 32 neutrons combined with 27 protons.

But there are also cobalt atoms with *more* than 32 neutrons. Some of the cobalt produced as a result of a nuclear detonation might have ***33*** neutrons. For reasons that go beyond the scope of this book, an atom in which the number of neutrons is significantly greater or less than the number of protons is considered *unstable.* That is, an unstable isotope of a particular element tends to change to another isotope—in fact, to another element entirely. And in doing so, emits radiation of one form or another.

Returning to our second cobalt example, we sum the atomic mass units for the neutrons and protons and find the mass equals **60.** In other words, when the atom of cobalt has 27 protons and 33 neutrons, then scientists define it as cobalt-60. Using symbols, the term would be written **Co60.**

The use of Atomic Mass U n its ***is one way for scientists to differentiate between atoms of the same element that have differing numbers of neutrons.***

For example, iodine-130 and iodine-131 both have the **same** number of protons (which is why they are both iodine.) However, iodine-131 has ***one more*** neutron than iodine-130, and as a result behaves differently with regard to the type of radiation emitted. Both iodine-130 and iodine-131 have the exact same chemical properties.

The number following the alphabetical symbol (such as the ***60*** in Co60) refers to the sum of the neutrons and protons in atomic mass units. The important thing to remember is that the radioisotope we are talking about is defined by the symbol (i.e. **Co** for *cobalt)* plus the mass unit number (for example **60**—which means there are 27 protons and 33 neutrons in the nucleus). Taken together, **Co + 60 = Co60.**

Atoms with unstable configurations of neutrons and protons eventually change to another, more stable element or radioisotope. This process is called ***decay.*** It's not the same as what we normally speak of when we talk about decay of organic materials. But it does mean that the atom has lost energy and is transforming itself from an atom of ***lesser*** stability to one of ***greater*** stability. Eventually, in fact, the atom will reach a point where no more radiation is emitted. At that point it is considered ***stable*** and thus no longer radioactive.

Given two atoms, if the symbols are the same, then each of the two atoms are the same chemically. That is, they react with other chemicals in the same way. Both **iodine-130** and **iodine-131** are iodine. Both behave exactly the same chemically. *But because they have different numbers of neutrons* **iodine-130** and **iodine-131** are defined as ***different isotopes*** of iodine. *And each isotope emits its own type and quantity of radiation.*

CURIES AND NANOCURIES

A radioactive element ***(radionuclide, radioisotope)*** that decays rapidly and thus emits more bursts of radiation per second is said to have a ***higher activity*** than a substance that decays slowly and emits fewer bursts of radiation per second.

Scientists have invented a number of units to compare activities of radionuclides. One such unit cited in the 1997 National Cancer Institute Fallout study—and the one used in this book—is the Curie. The Curie was named after Mademoiselle Marie Curie, the French scientist who made numerous important discoveries about radiation. A radioactive substance is considered to be associated with Curie of activity if the radioactive atoms in the substances are decaying at a rate of 37,000,000,000 atoms per second. For those readers familiar with scientific notation, that's 3.7×10^{10} atoms per second (3.7 followed by ten zeroes is equal to the number **37** followed by ***nine*** zeros.)

Scientific notation is occasionally used in this book. Scientific notation is a shorthand way of writing very large or very small numbers. Suppose we wanted to write 37,123,456,789 in scientific notation.

1. First we round the 37,123,456,789 to whatever place we would like. Here, we will round down to simply **37,000,000,000.**
2. Now, we express the first two numbers as a value between 1 and 9.9, here **3.7.**
3. Finally, we divide 37,000,000,000 by 3.7 and get 10,000,000,000. In other words, 3.7 x 10,000,000,000 = 37,000,000,000.
4. Count the zeroes: 10.
5. Now we can write the **second** part of the number,the value 10 plus the all-important ***exponent***—the small number in superscript just to the right of the 10. Here, the 10 along with the exponent is written $\mathbf{10^{10}}$. This value tells us by how many factors of ten we must multiply the base number (here, 3.7) to make it equal to the original number (37,000,000,000).
6. The answer is 10, or 3.7×10^{10}. The same value can be written as 3.7E+10.
7. If the exponent is preceded by a negative sign, as in 3.7×10^{-10} or 3.7E-10, this means the value is very small. In decimals, 3.7×10^{-10} is equal to **0.00000000037.**

It must be noted that one Curie—about 37,000,000,000 disintegrations per section—represents *significant* activity. Anyone recording 37,000,000,000 clicks per second from a vial of fallout would also likely to be exposed to a lethal amount of radiation.

While the aboveground nuclear tests in the 1950s released fallout into the atmosphere equal to *millions* of Curies of activity, the fallout was dispersed so that by the time the tiny spheres of fallout reached the ground the overall activity per square foot was on the order of a ***billionth*** of a Curie. When something is a billionth of something else, scientists describe it using the term ***nano.*** Thus, a billionth of a Curie—equal to 37 bursts of radiation per second—is called a **nanoCurie.** The abbreviation for nano Curie is **nCi.**

The term for ***one-millionth*** of a Curie is a *micro*-Curie, abbreviated **μCi.** A **micro Curie** is equal to a thousand nano-Curies and represents 37,000 bursts of radiation per second. On individual fallout days most counties in the United States received fallout in the **nanoCurie** amounts. Summed over the years from 1951 to 1962, the amounts can generally be easily expressed in terms of **microCuries.** Thus, for some tables, you will see fallout given in nCi/sq meter ***(billionths*** of a Curie per square meter) and for other tables you will see fallout given in μCi/sq meter ***(millionths*** of a Curie per square meter.)

The important thing to remember is that ***dispersion of a 1000-ft diameter nuclear cloud over thousands of cubic miles of air will result in a significant dilution of activity.*** As a consequence, the radioactivity of the fallout, as it finally falls to earth, is measured not in Curies, but in millionths and billionths of Curies: microCuries (μCi/sq meter) and nanoCuries per square meter (nCi/sq meter.)

The fallout values used for the maps in **Sections 1-7** and for the Table in Section 9 are given in *nanoCuries.* A nanoCurie represents an activity level equal to one thousandth of a microCurie. **NanoCuries are represented in this book by the symbol nCi, and represent an activity equal to 37 disintegrations per second.**

Another unit of radioactivity is a *Becquerel (Beq),* named after the French scientist Henri Becquerel, who discovered radioactivity. One *Beq* is equal to one (1) disintegration per second. Thirty-seven Beq is equal to one nanoCurie. . .or 37 disintegrations (and concurrent emissions of radiation) per second.

In **Section 8,** the fallout values for counties with the highest levels of fallout are expressed in terms of **microCuries.** You will see the term microCurie represented by the symbol **μCi. *One microCurie is defined as radiological decay equal to 37,000 disintegrations per second.*** (37,000,000,000 divided by 1,000,000).

SQUARE METERS AND SQUARE YARDS

The activity discussed earlier is given in terms of square meters. A square meter, abbreviated either ***sq meter*** or m_2 simply refers to a square that is 39.37 inches on a side. In other words, a square meter is slightly bigger (about 1.2 times) than a square yard.

So, of you happen to see the fallout level given as 10 nCi/sq meter (10 nanoCuries per square meter) you will know that this means there were 10 x 37 = 370 bursts of radiation per second from a plot of land about 1.2 times as large as a square yard.

FACT: A square meter is slightly larger in area than a square yard, and a little over ten times as large as a square foot.

1 square meter = 1.196 square yards = 10.76 square feet

The 1997 National Cancer Institute study gave fallout levels in terms of nanoCuries per square meter (nCi/sq meter). Again, a square meter is equal to 10.76 square feet, or an area 3 ft. 3 inches on a side. Thus, if you read that fallout from a particular shot produced an average of 1 nCi/sq meter on a given county on a particular day, you may reasonably conclude that for each square plot of land 3.3 ft on a side, the fallout produced 37 **units** of radiation.

Why just units of radiation? Can't we identify the type of radiation—that is, whether the radiation consists of x-rays, gamma or beta particles?

Unfortunately, no. At least not yet anyway. Because, while we know how fast the various mix of elements is changing (the activity level expressed in fractions of a Curie) ***we still don't know what kinds of radioactive elements are in the fallout.*** And since we still don't know what radioactive elements are there, we are not able to say what forms of radiation are produced. Thanks to some mathematics, and a closer look at some important research by government scientists, we will eventually have that information, and that will be discussed in a later volume dealing specifically with radionuclides.

RELATIVE FALLOUT VALUES

A primary purpose of this book is to show ***relative*** fallout levels for each county from each nuclear test. To do that, arithmetic means were calculated from the geometric means and geometric standard deviations given in the NCI tables. These arithmetic means were then used as central values to evaluate relative fallout levels. Where tables or maps refer to Averages, the term specifically refers to the ***arithmetic mean of the log-normal distribution.***

NUCLEAR TESTS AND TEST SERIES ABBREVIATIONS

Nuclear tests and test series were assigned names by those in charge of the detonations. Some of these names, like ***Buster-Jangle, Tumbler-Snapper and Upshot-Knothole*** were quite long and involved. To save space in the tables, these names were abbreviated to ***BJ, TS*** and ***UK.*** To ensure consistency, shot names are spelled out, but represented in all caps, such as shot CLIMAX in the UK (Upshot-Knothole) series.

The following list includes the nuclear tests that are included in this volume. The first column lists the abbreviation of the **shot** (nuclear test) or **series** (group of shots.) The second column gives the ***full name*** of the shot or series. The third column lists the date of the shot or series, and the fourth column tells whether that shot or series was included in the National Cancer Institute's 1997 radioactive iodine study. Not all tests were included in the study. Some smaller detonations, such as **BJ Able** were considered too small to be included in the analysis.

MAJOR NUCLEAR TESTS **Conducted at the Nevada Test Site 1951 -1970** **This list does *not* include all nuclear tests conducted at the NTS between 1951 and 1970**				
ABBRE-VIATION	**NUCLEAR TEST OR TEST SERIES**	**DATES**	**INCLUDED IN NCI STUDY?**	**Kilotons**
R	Ranger Series	27 Jan -6 Feb 1951	Yes	---
BJ	Buster-Jangle Series	22 Oct-29 Nov, 1951	Yes	---
TS	Tumbler-Snapper Series	15 Apr-5 June, 1952	Yes	---
UpK	Upshot-Knothole Series	17 Mar-4 Jun 1953	Yes	---
TP	Teapot Series	18 Feb-15 May, 1955	Yes	---
PB	Plumbbob Series	28 May-7 Oct, 1957	Yes	---
HT	Hardtack II Series	12 Sep-30 Oct, 1958	Yes	---

UE	Underground Series	Roughly, 1961-Sep 1992	Up to 1970	---
(Ranger Able)	Ranger Able	27 Jan 1951	No	1
R1	Ranger Baker	28 Jan 1951	Yes	8
(Ranger Easy)	Ranger Easy	1 Feb 1951	No	1
R2	Ranger Baker-2	2 Feb 1951	Yes	8
R3	Ranger Fox	6 Feb 1951	No	22
(Buster Able)	Buster Able	22 Oct 1951	No	low
BJ1	Buster Baker	28 Oct 1951	Yes	3.5
BJ2	Buster Charlie	30 Oct 1951	Yes	14
BJ3	Buster Dog	1 Nov 1951	Yes	21
BJ4	Buster Easy	5 Nov 1951	Yes	31
BJ5	Jangle Sugar	19 Nov 1951	Yes	1.2
BJ6	Jangle Uncle	29 Nov 1951	Yes	1.2
TS1	Tumbler-Snapper Able	1 Apr 1952	Yes	1
TS2	Tumbler-Snapper Baker	15 Apr 1952	Yes	1
TS3	Tumbler-Snapper Charlie	22 Apr 1952	Yes	31
TS4	Tumbler-Snapper Dog	1 May 1952	Yes	19
TS5	Tumbler-Snapper Easy	7 May 1952	Yes	12
TS6	Tumbler-Snapper Fox	25 May 1952	Yes	11

TS7	Tumbler-Snapper George	1 Jun 1952	Yes	15
TS8	Tumbler-Snapper How	5 Jun 1952	Yes	14
UpK1	Upshot-Knothole Annie	17 Mar 1953	Yes	16
UpK2	Upshot-Knothole Nancy	24 Mar 1953	Yes	24
UpK3	Upshot-Knothole Ruth	31 Mar 1953	Yes	0.2
UpK4	Upshot-Knothole Dixie	6 Apr 1953	Yes	11
UpK5	Upshot-Knothole Ray	11 Apr 1953	Yes	0.2
UpK6	Upshot-Knothole Badger	18 Apr 1953	Yes	23
UpK7	Upshot-Knothole Simon	25 Apr 1953	Yes	43
UpK8	Upshot-Knothole Encore	8 May 1953	Yes	27
UpK9	Upshot-Knothole Harry	19 May 1953	Yes	32
UpK10	Upshot-Knothole Grable	25 May 1953	Yes	15
UpK11	Upshot-Knothole Climax	4 Jun 1953	Yes	61
TP1	Teapot Wasp	18 Feb 1955	Yes	1
TP2	Teapot Moth	22 Feb 1955	Yes	2
TP3	Teapot Tesla	1 Mar 1955	Yes	7
TP4	Teapot Turk	7 Mar 1955	Yes	43

TP5	Teapot Hornet	12 Mar 1955	Yes	4
TP6	Teapot Bee	22 Mar 1955	Combined with Teapot Ess	8
TP6	Teapot Ess	23 Mar 1955	Combined with Teapot Bee	1
TP7	Teapot Apple-1	29 Mar 1955	Yes	14
(Teapot Wasp Prime)	Teapot Wasp Prime	29 March 1955	Combined with Teapot Apple-1	3
(Teapot HA)	Teapot High Altitude test	6 Apr 1955	No	3
TP8	Teapot Post	9 Apr 1955	Yes	2
TP9	Teapot Met	19 Apr 1955	Yes	22
TP10	Teapot Apple-2	5 May 1955	Yes	29
TP11	Teapot Zucchini	15 May 1955	Yes	28
PB1	Plumbbob Boltzmann	28 May 1957	Yes	12
PB1	Plumbbob Franklin	2 Jun 1957	Combined with Plumbbob Boltzmann	0.14
PB2	Plumbbob Wilson	18 June 1957	Yes	10
PB3	Plumbbob Priscilla	24 Jun 1957	Yes	37
PB4	Plumbbob Hood	5 Jul 1957	Yes	74
PB5	Plumbbob Diablo	15 Jul 1957	Yes	17
(Plumbbob John)	Plumbbob John	19 Jul 1957	No	2
PB6	Plumbbob Kepler	24 Jul 1957	Combined with PB Owens	10

PB6	Plumbbob Owens	25 Jul 1957	Combined with PB Kepler	9.7
PB7	Plumbbob Stokes	7 Aug 1957	Yes (however, no fallout listed)	19
PB8	Plumbbob Shasta	18 Aug 1957	Yes	17
PB9	Plumbbob Doppler	23 Aug 1957	Yes	11
PB10	Plumbbob Franklin Prime	30 Aug 1957	Yes (however, no fallout listed)	4.7
PB11	Plumbbob Smoky	31 Aug 1957	Yes	44
PB12	Plumbbob Galileo	2 Sep 1957	Yes	11
PB13	Plumbbob Wheeler	6 Sep 1957	Combined with Coulomb B and LaPlace	0.19
PB13	Plumbbob Coulomb B	6 Sep 1957	Combined with Wheeler and LaPlace	0.3
PB13	Plumbbob LaPlace	8 Sep 1957	Combined with Wheeler and Coulomb B	1
PB14	Plumbbob Fizeau	14 Sep 1957	Yes	11
PB15	Plumbbob Newton	16 Sep 1957	Yes	12
PB16	Plumbbob Whitney	23 Sep 1957	Yes	19
PB17	Plumbbob Charleston	28 Sep 1957	Yes	12
PB18	Plumbbob Morgan	7 Oct 1957	Yes	8
HT1	Hardtack II Eddy	19 Sep 1958	Yes	0.08

HT2	Hardtack II Hidalgo	5 Oct 1958	Yes	0.07
HT3	Hardtack II Quay	10 Oct 1958	Yes	0.07
HT4	Hardtack II Lea	13 Oct 1958	Yes	1.4
HT5	Hardtack II Vesta	17 Oct 1958	Yes	0.02
HT6	Hardtack II Rio Arriba	18 Oct 1958	Yes	0.09
HT7	Hardtack II Socorro	22 Oct 1958	Yes	6.0
HT8	Hardtack II Wrangall	22 Oct 1958	Yes	0.11
HT9	Hardtack II Sanford	26 Oct 1958	Yes	4.9
UE1*	Antler	15 Sep 1961	Yes	2.6
UE2	Danny Boy	5 Mar 1962	Yes	0.43
UE3*	Platte	14 Apr 1962	Yes	1.85
UE4*	Eel	19 May 1962	Yes	4.5
UE5	Des Moines	13 Jun 1962	Yes	2.9
S (UE5)	Sedan	6 Jul 1962	Yes	104
UE7	Johnny Boy	11 Jul 1962	Yes	0.5
UE8	Small Boy	14 Jul 1962	Yes	Low
UE9*	Bandicoot	19 Oct 1962	Yes	12.5
UE10*	Pike	13 Mar 64	Yes	<20
UE11	Sulky	18 Dec 1964	Yes	0.09
UE12	Palanquin	14 Apr 1965	Yes	4.3
UE13	Pinstripe	25 Apr 1966	Yes	<20
UE14	Cabriolet	26 Jan 1968	Yes	2.3
UE15	Buggy	12 Mar 1968	Yes	5.4
UE16	Schooner	8 Dec 1968	Yes	30
UE17	Baneberry	18 Dec 1970	Yes	10

***Fallout values for shots UE-1, -3, -4, -9 and -10 are for I-131.**

EXTRA COUNTIES

In the 1997 National Cancer Institute I-131 report fourteen counties near the Nevada Test Site were subdivided based upon geography and different fallout patterns. These counties are shown in the following illustrations. Table A2.1 of the NCI report defines the parameters of each of the separate subdivisions. The following illustrations were taken from that report: *Estimated Exposures and Thyroid Doses Received by the American People from Iodine-I-131 in Fallout Following Nevada Atmospheric Nuclear Bomb Tests.* Appendices to the Report from the National Cancer Institute. U.S. Department of Health and Human Services: Oct 1997.) In this book, the subdivided counties are differentiated using numerals, as, for example, AZ, Coconino1, Coconino2, Coconino3, etc.

ARIZONA COUNTIES SUBDIVIDED

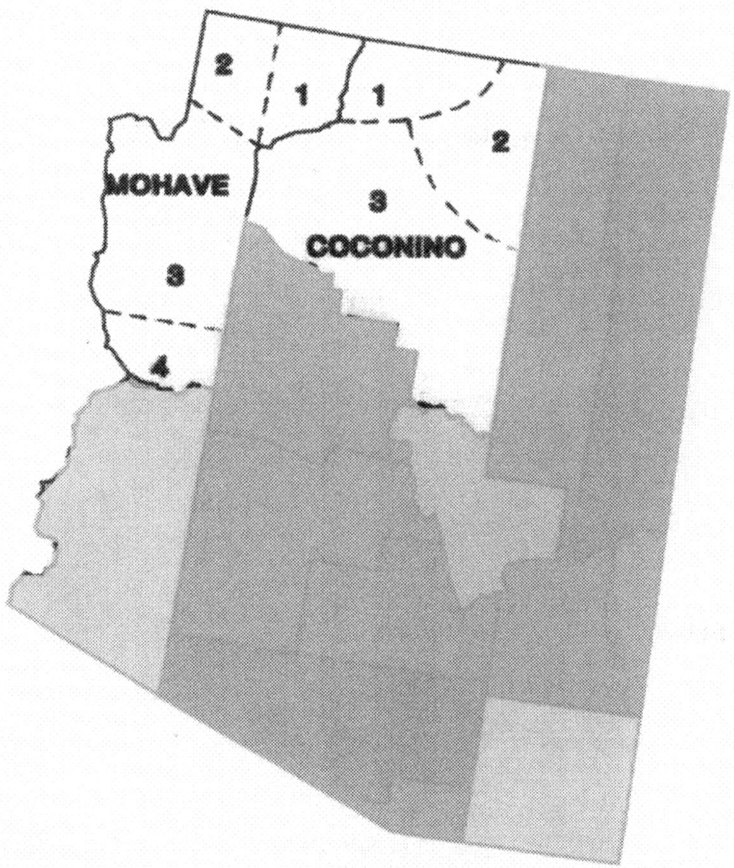

CALIFORNIA COUNTY SUBDIVIDED

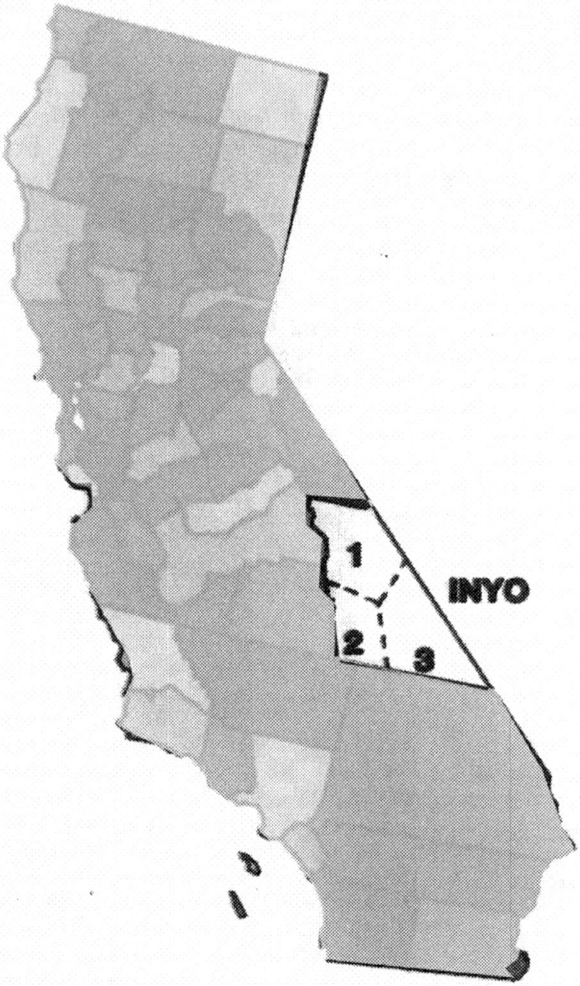

NEVADA COUNTY SUBDIVIDED

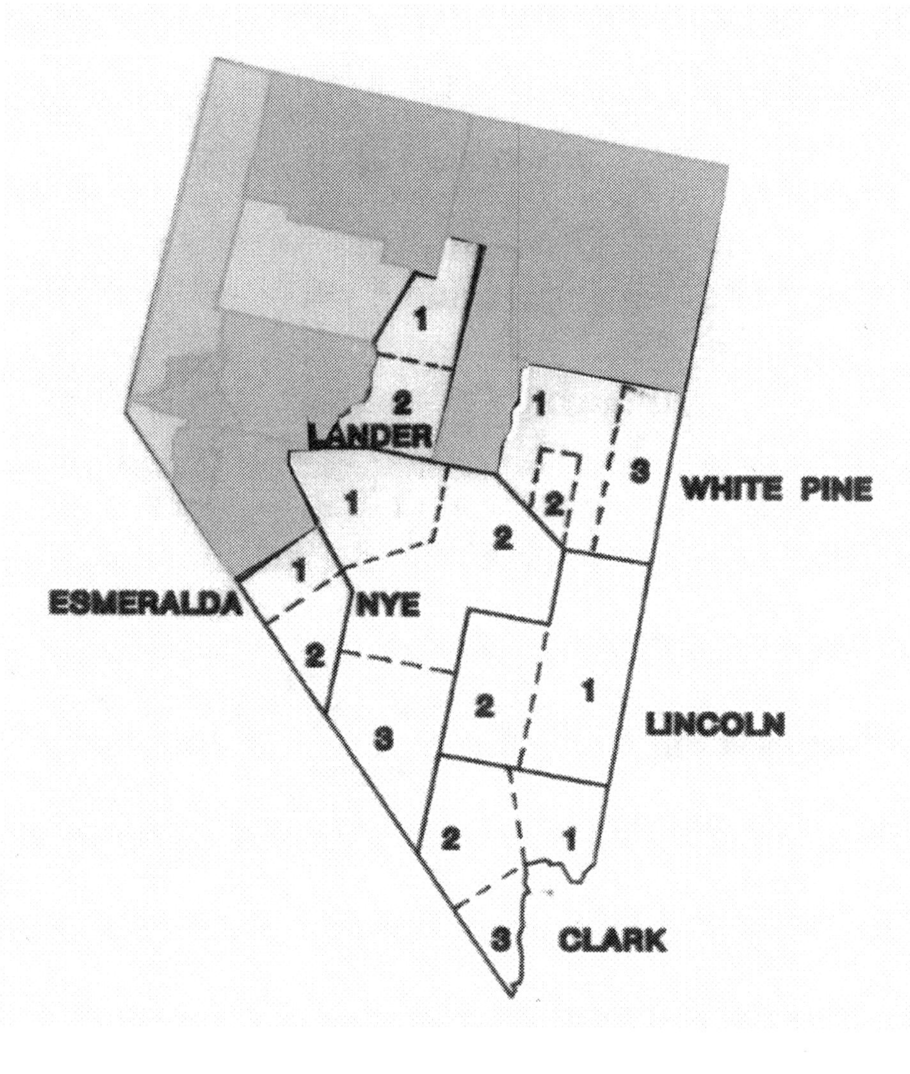

NEVADA COUNTY SUBDIVIDED

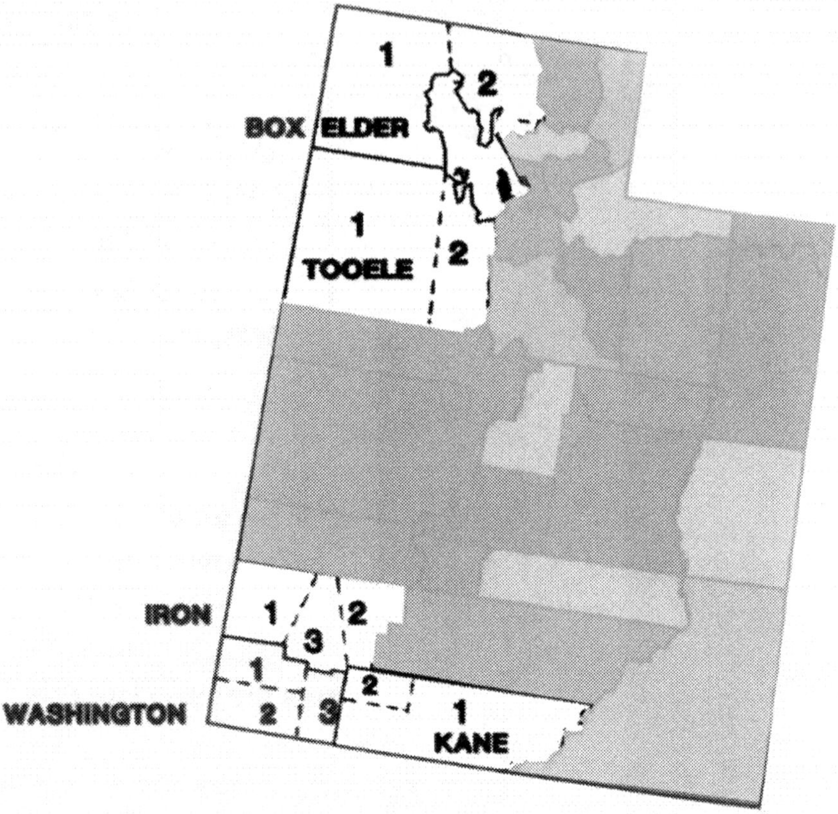

COUNTY COMPARISONS

Changes in the 2005 Version

The original version of the U.S. ATLAS OF NUCLEAR FALLOUT involved material divided into six volumes. Volume I, Total Fallout, Volume II: Radioisotopes, Volume III: County Comparisons, Volume IV Trajectories and Volume V Calculations.

Materials for Volumes II and IV were compiled, but never published.

Careful examination of the material disclosed insert errors in total fallout values for four nuclear tests: Ranger Baker-2, Tumbler-Snapper Easy, Tumbler-Snapper Fox and Tumbler-Snapper How. Since radioisotope values were calculated based upon Hicks Table ratios, the errors in the total fallout values did not show up in radioisotope estimates. While the errors were minor and were within the +/- 5 percent of the total amount, in March, 2005 it was decided to remove the books from publication until the errors had been corrected.

Since that time we viewed this as an opportunity to carefully examine the information for redundancy and ease of use. As a result, tables of information have been reorganized for ease of understanding. Counties are arranged alphabetically and the confusing "county index numbers" have been replaced by nationwide standings. For example, if a county is rated as 96 nationwide, it means that it is the 96th hottest county in terms of fallout for that particular category.

ORGANIZATION

Volume I and II include *TOTAL FALLOUT* data for the Western and Eastern states, respectively. Volume III as before, involves *COUNTY COMPARISONS* in terms of fallout date, and Volume IV *ABOVEGROUND NEVADA TESTS* discusses the major aboveground nuclear tests at the Nevada Test Site. Volume V, *CALCULATIONS*, is unchanged from the earlier series.

This edition of *COUNTY COMPARISONS* includes a table of rankings by specific fallout date. If a county is listed as, for example, 4 for a given date, that means it received the 4th highest fallout amount among all United States counties for that date. In many instances, however, several counties received the same amount of fallout; thus several counties may share the same ranking for a given date.

Values in the COUNTY COMPARISON volume are given in microCuries/square meters. Counties are listed in the main section only if they were among the top ten highest fallout counties for that particular shot day. Those wishing more precise information regarding fallout values may wish to consult the 7-volume series: NUCLEAR FALLOUT FROM THE NEVADA TEST SITE 1951-1970 (LEGIS BOOKS 2005).

NOTES

1. Eisler, Peter and Steve Sternberg. "Fallout: did it Harm? Study shows contaminants fell far from Nevada Test Site." USA Today. 26 July, 1997.
2. Section 7(a) of Public Law 97-414 directs the Secretary of Health and Human Services to :

(1) conduct scientific research and prepare analysis necessary to develop valid and credible assessments of the risks of thyroid cancer that are associated with thyroid doses of Iodine 131;

(2) conduct scientific research and prepare analysis necessary to develop valid and credible methods to estimate the thyroid doses of Iodine 131 that are received by individuals from nuclear bomb fallout; and,

(3) conduct scientific research and prepare analysis necessary to develop valid and credible assessments of the exposure to Iodine 131 that the American people received from Nevada atmospheric nuclear bomb tests."

RADIONUCLIDE ESTIMATION PROCEDURE

Step 1: Construct exponential curve based on Hicks Tables values for the radionuclide and specific shot (in this case, 24Na and shot Hood.)

Step 2: Determine equation that best fits the data. For radionuclides this is usually an exponential curve.

Step 3: With X = day post shot, solve for y to obtain radionuclide values (μ/sq meter) for days 1-10.

Step 4: Adjust estimated radionuclide values by inserting Hicks Values for days 1,2,5 and 10.

Step 5: Perform the same procedure for 131I values.

Step 6: Calculate radionuclide/131I ratios for each day post shot.

Step 7: Multiply the ratios times NCI 131I values specific to each shot to obtain radionuclide deposition estimates for each day post shot.

HICKS AND ADJUSTED VALUES FOR 24Na, SHOT HOOD

DAY	HICKS VALUES 24Na	CALCULATED VALUES 24Na	ADJUSTED VALUES 24Na
1	1.45E-02	1.45E-02	1.45E-02
2	4.78E-03	4.78E-03	4.78E-03
3	---	1.58E-03	1.58E-03
4	---	5.20E-04	5.20E-04
5	1.71E-04	1.72E-04	1.71E-04
6	---	5.66E-05	5.66E-05
7	---	1.87E-05	1.87E-05
8	---	6.16E-06	6.16E-06
9	---	2.03E-06	2.03E-06
10	6.71E-07	6.70E-07	6.71E-07

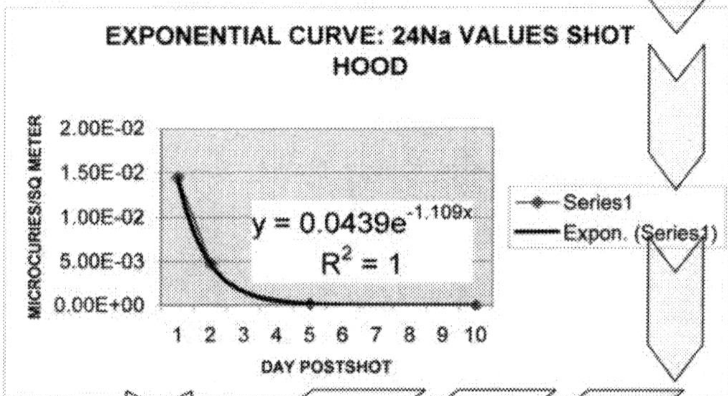

ADJUSTED 24Na/131I RATIOS

DAY	ADJUSTED VALUES 24Na	ADJUSTED 131I VALUES	RATIO 24Na Value/131I Value
1	1.45E-02	5.81E-01	2.49E-02
2	4.78E-03	5.49E-01	8.70E-03
3	1.58E-03	5.50E-01	2.87E-03
4	5.20E-04	4.67E-01	1.11E-03
5	1.71E-04	4.42E-01	3.88E-04
6	5.66E-05	4.00E-01	1.41E-04
7	1.87E-05	3.70E-01	5.04E-05
8	6.16E-06	3.43E-01	1.80E-05
9	2.03E-06	3.17E-01	6.41E-06
10	6.71E-07	2.91E-01	2.30E-06

NCI 131I ESTIMATE:
123.45 nCi/sq meter

A. DAY POST SHOT	B. NCI 131I VALUE (nCi/sq meter)	C. RATIO SPECIFIC FOR SHOT AND DAY POST SHOT	D. FINAL RADIONUCLIDE ESTIMATE (B X C)
8	**123.45 x**	**1.80E-05 =**	**2.22E-03**

COUNTY	uCi/ sqm	SHOT DAY	DATE	RANK FOR DAY	TIMES IN TOP 10
AL AUTAUGA	1.91	TP6-4	25-Mar-55	9	2
AL AUTAUGA	3.14	TP6-5	26-Mar-55	2	2
AL BALDWIN	0.81	UpK3-6	5-Apr-53	9	5
AL BALDWIN	3.14	TP6-5	26-Mar-55	2	5
AL BALDWIN	22.97	PB3-3	26-Jun-57	4	5
AL BALDWIN	1.40	PB6-8	31-Jul-57	8	5
AL BALDWIN	0.98	PB9-7	29-Aug-57	5	5
AL BARBOUR	5.99	PB17-3	30-Sep-57	7	1
AL BIBB	1.62	TS4-6	6-May-52	7	6
AL BIBB	5.11	TS7-8	8-Jun-52	3	6
AL BIBB	1.91	TP6-4	25-Mar-55	9	6
AL BIBB	3.14	TP6-5	26-Mar-55	2	6
AL BIBB	3.96	PB17-3	30-Sep-57	9	6
AL BLOUNT	7.52	TS5-3	9-May-52	10	2
AL BLOUNT	2.87	TP6-4	25-Mar-55	4	2
AL BULLOCK	5.99	PB17-3	30-Sep-57	7	1
AL BUTLER	1.00	TS5-5	11-May-52	7	3
AL BUTLER	3.14	TP6-5	26-Mar-55	2	3
AL BUTLER	0.92	PB6-9	1-Aug-57	6	3
AL CALHOUN	7.66	BJ3-5	5-Nov-51	3	2
AL CALHOUN	1.55	PB2-5	22-Jun-57	6	2
AL CHAMBERS	1.82	PB2-4	21-Jun-57	5	1
AL CHEROKEE	8.10	BJ3-6	6-Nov-51	5	3
AL CHEROKEE	1.91	TP6-4	25-Mar-55	9	3
AL CHEROKEE	3.96	PB17-3	30-Sep-57	9	3
AL CHILTON	2.47	TP10-4	8-May-55	10	1
AL CHOCTAW	1.47	TP6-2	23-Mar-55	9	5
AL CHOCTAW	4.11	TP6-3	24-Mar-55	4	5
AL CHOCTAW	3.14	TP6-5	26-Mar-55	2	5
AL CHOCTAW	11.44	PB3-3	26-Jun-57	6	5
AL CHOCTAW	1.99	PB6-7	30-Jul-57	6	5
AL CLARKE	0.81	UpK3-6	5-Apr-53	9	5
AL CLARKE	1.47	TP6-2	23-Mar-55	9	5
AL CLARKE	4.11	TP6-3	24-Mar-55	4	5
AL CLARKE	3.14	TP6-5	26-Mar-55	2	5

COUNTY	uCi/ sqm	SHOT DAY	DATE	RANK FOR DAY	TIMES IN TOP 10
AL CLARKE	11.44	PB3-3	26-Jun-57	6	5
AL CLAY	1.91	TP6-4	25-Mar-55	9	1
AL CLEBURNE	1.91	TP6-4	25-Mar-55	9	2
AL CLEBURNE	3.96	PB17-3	30-Sep-57	9	2
AL COFFEE	3.14	TP6-5	26-Mar-55	2	2
AL COFFEE	3.96	PB17-3	30-Sep-57	9	2
AL COLBERT	1.62	TS4-6	6-May-52	7	4
AL COLBERT	1.62	TS4-6	6-May-52	7	4
AL COLBERT	0.88	UpK3-6	5-Apr-53	8	4
AL COLBERT	2.78	PB2-5	22-Jun-57	4	4
AL CONECUH	1.47	TP6-2	23-Mar-55	9	5
AL CONECUH	4.11	TP6-3	24-Mar-55	4	5
AL CONECUH	3.14	TP6-5	26-Mar-55	2	5
AL CONECUH	45.85	PB3-3	26-Jun-57	3	5
AL CONECUH	0.98	PB9-7	29-Aug-57	5	5
AL COOSA	2.67	TP10-4	8-May-55	4	1
AL COVINGTON	0.99	UpK5-5	15-Apr-53	2	3
AL COVINGTON	3.14	TP6-5	26-Mar-55	2	3
AL COVINGTON	1.16	PB1-9	5-Jun-57	5	3
AL CRENSHAW	3.14	TP6-5	26-Mar-55	2	1
AL CULLMAN	4.86	BJ3-6	6-Nov-51	7	4
AL CULLMAN	1.91	TP6-4	25-Mar-55	9	4
AL CULLMAN	1.16	PB1-9	5-Jun-57	5	4
AL CULLMAN	3.96	PB17-3	30-Sep-57	9	4
AL DALE	3.14	TP6-5	26-Mar-55	2	2
AL DALE	3.96	PB17-3	30-Sep-57	9	2
AL DALLAS	1.91	TP6-4	25-Mar-55	9	3
AL DALLAS	3.14	TP6-5	26-Mar-55	2	3
AL DALLAS	2.73	PB2-4	21-Jun-57	2	3
AL DE KALB	4.86	BJ3-6	6-Nov-51	7	6
AL DE KALB	1.62	TS4-6	6-May-52	7	6
AL DE KALB	1.62	TS4-6	6-May-52	7	6
AL DE KALB	1.91	TP6-4	25-Mar-55	9	6
AL DE KALB	1.16	PB1-9	5-Jun-57	5	6
AL DE KALB	5.99	PB17-3	30-Sep-57	7	6

COUNTY	uCi/ sqm	SHOT DAY	DATE	RANK FOR DAY	TIMES IN TOP 10
AL ESCAMBIA	3.14	TP6-5	26-Mar-55	2	3
AL ESCAMBIA	1.03	TP8-5	13-Apr-55	2	3
AL ESCAMBIA	22.97	PB3-3	26-Jun-57	4	3
AL ETOWAH	8.10	BJ3-6	6-Nov-51	5	3
AL ETOWAH	2.87	TP6-4	25-Mar-55	4	3
AL ETOWAH	5.99	PB17-3	30-Sep-57	7	3
AL FAYETTE	2.78	PB2-5	22-Jun-57	4	1
AL FRANKLIN	1.82	PB2-4	21-Jun-57	5	2
AL FRANKLIN	2.78	PB2-5	22-Jun-57	4	2
AL GENEVA	3.14	TP6-5	26-Mar-55	2	3
AL GENEVA	0.92	PB6-9	1-Aug-57	6	3
AL GENEVA	3.96	PB17-3	30-Sep-57	9	3
AL GREENE	1.47	TP6-2	23-Mar-55	9	5
AL GREENE	4.11	TP6-3	24-Mar-55	4	5
AL GREENE	2.87	TP6-4	25-Mar-55	4	5
AL GREENE	3.14	TP6-5	26-Mar-55	2	5
AL GREENE	1.89	PB6-7	30-Jul-57	7	5
AL HALE	1.47	TP6-2	23-Mar-55	9	6
AL HALE	4.11	TP6-3	24-Mar-55	4	6
AL HALE	2.87	TP6-4	25-Mar-55	4	6
AL HALE	3.14	TP6-5	26-Mar-55	2	6
AL HALE	2.78	PB2-5	22-Jun-57	4	6
AL HALE	0.83	PB3-9	2-Jul-57	4	6
AL HENRY	3.96	PB17-3	30-Sep-57	9	1
AL HOUSTON	6.40	TS3-7	28-Apr-52	2	6
AL HOUSTON	6.40	TS3-7	28-Apr-52	2	6
AL HOUSTON	1.65	TS3-8	29-Apr-52	5	6
AL HOUSTON	1.65	TS3-8	29-Apr-52	5	6
AL HOUSTON	1.16	PB1-9	5-Jun-57	5	6
AL HOUSTON	5.99	PB17-3	30-Sep-57	7	6
AL JACKSON	4.86	BJ3-6	6-Nov-51	7	3
AL JACKSON	2.87	TP6-4	25-Mar-55	4	3
AL JACKSON	5.99	PB17-3	30-Sep-57	7	3
AL JEFFERSON	1.91	TP6-4	25-Mar-55	9	2
AL JEFFERSON	1.12	PB6-12	4-Aug-57	1	2

COUNTY	uCi/ sqm	SHOT DAY	DATE	RANK FOR DAY	TIMES IN TOP 10
AL LAMAR	4.55	PB2-4	21-Jun-57	1	3
AL LAMAR	2.78	PB2-5	22-Jun-57	4	3
AL LAMAR	0.83	PB3-9	2-Jul-57	4	3
AL LAUDERDALE	0.88	UpK3-6	5-Apr-53	8	4
AL LAUDERDALE	0.89	PB1-9	5-Jun-57	7	4
AL LAUDERDALE	1.82	PB2-4	21-Jun-57	5	4
AL LAUDERDALE	2.78	PB2-5	22-Jun-57	4	4
AL LAWRENCE	1.25	PB3-8	1-Jul-57	3	1
AL LEE	0.99	UpK5-5	15-Apr-53	2	3
AL LEE	3.96	PB17-3	30-Sep-57	9	3
AL LEE	4.73	PB17-6	3-Oct-57	10	3
AL LIMESTONE	3.23	TS4-6	6-May-52	2	4
AL LIMESTONE	3.23	TS4-6	6-May-52	2	4
AL LIMESTONE	2.73	PB2-4	21-Jun-57	2	4
AL LIMESTONE	3.96	PB17-3	30-Sep-57	9	4
AL LOWNDES	3.14	TP6-5	26-Mar-55	2	1
AL MACON	3.96	PB17-3	30-Sep-57	9	1
AL MADISON	4.86	BJ3-6	6-Nov-51	7	6
AL MADISON	3.23	TS4-6	6-May-52	2	6
AL MADISON	3.23	TS4-6	6-May-52	2	6
AL MADISON	2.87	TP6-4	25-Mar-55	4	6
AL MADISON	1.82	PB2-4	21-Jun-57	5	6
AL MADISON	5.99	PB17-3	30-Sep-57	7	6
AL MARENGO	1.47	TP6-2	23-Mar-55	9	5
AL MARENGO	4.11	TP6-3	24-Mar-55	4	5
AL MARENGO	4.30	TP6-4	25-Mar-55	2	5
AL MARENGO	3.14	TP6-5	26-Mar-55	2	5
AL MARENGO	68.82	PB3-3	26-Jun-57	2	5
AL MARION	0.89	PB1-9	5-Jun-57	7	3
AL MARION	4.64	PB2-5	22-Jun-57	2	3
AL MARION	0.83	PB3-9	2-Jul-57	4	3
AL MARSHALL	8.10	BJ3-6	6-Nov-51	5	4
AL MARSHALL	1.91	TP6-4	25-Mar-55	9	4
AL MARSHALL	1.16	PB1-9	5-Jun-57	5	4
AL MARSHALL	5.99	PB17-3	30-Sep-57	7	4

COUNTY	uCi/ sqm	SHOT DAY	DATE	RANK FOR DAY	TIMES IN TOP 10
AL MOBILE	38.63	UpK6-2	19-Apr-53	2	5
AL MOBILE	1.30	TP6-5	26-Mar-55	6	5
AL MOBILE	68.82	PB3-3	26-Jun-57	2	5
AL MOBILE	2.34	PB6-8	31-Jul-57	3	5
AL MOBILE	2.73	PB9-2	24-Aug-57	6	5
AL MONROE	0.81	UpK3-6	5-Apr-53	9	8
AL MONROE	1.23	UpK3-7	6-Apr-53	10	8
AL MONROE	1.47	TP6-2	23-Mar-55	9	8
AL MONROE	4.11	TP6-3	24-Mar-55	4	8
AL MONROE	3.14	TP6-5	26-Mar-55	2	8
AL MONROE	1.03	TP8-5	13-Apr-55	2	8
AL MONROE	45.85	PB3-3	26-Jun-57	3	8
AL MONROE	1.40	PB6-8	31-Jul-57	8	8
AL MONTGOMERY	3.11	TP10-4	8-May-55	2	1
AL MORGAN	4.86	BJ3-6	6-Nov-51	7	4
AL MORGAN	1.91	TP6-4	25-Mar-55	9	4
AL MORGAN	2.73	PB2-4	21-Jun-57	2	4
AL MORGAN	3.96	PB17-3	30-Sep-57	9	4
AL PERRY	5.11	TS7-8	8-Jun-52	3	5
AL PERRY	1.47	TP6-2	23-Mar-55	9	5
AL PERRY	4.11	TP6-3	24-Mar-55	4	5
AL PERRY	3.14	TP6-5	26-Mar-55	2	5
AL PERRY	1.82	PB2-4	21-Jun-57	5	5
AL PICKENS	1.47	TP6-2	23-Mar-55	9	5
AL PICKENS	4.11	TP6-3	24-Mar-55	4	5
AL PICKENS	2.87	TP6-4	25-Mar-55	4	5
AL PICKENS	2.78	PB2-5	22-Jun-57	4	5
AL PICKENS	0.83	PB3-9	2-Jul-57	4	5
AL PIKE	3.14	TP6-5	26-Mar-55	2	4
AL PIKE	1.16	PB1-9	5-Jun-57	5	4
AL PIKE	1.82	PB2-4	21-Jun-57	5	4
AL PIKE	3.96	PB17-3	30-Sep-57	9	4
AL RANDOLPH	1.29	BJ1-2	29-Oct-51	3	3
AL RANDOLPH	5.99	PB17-3	30-Sep-57	7	3
AL RANDOLPH	1.32	UE8-6	19-Jul-62	9	3

COUNTY	uCi/ sqm	SHOT DAY	DATE	RANK FOR DAY	TIMES IN TOP 10
AL RUSSELL	3.96	PB17-3	30-Sep-57	9	1
AL SHELBY	1.82	PB2-4	21-Jun-57	5	2
AL SHELBY	3.96	PB17-3	30-Sep-57	9	2
AL ST CLAIR	1.91	TP6-4	25-Mar-55	9	1
AL SUMTER	1.47	TP6-2	23-Mar-55	9	5
AL SUMTER	4.11	TP6-3	24-Mar-55	4	5
AL SUMTER	3.14	TP6-5	26-Mar-55	2	5
AL SUMTER	1.30	PB2-4	21-Jun-57	9	5
AL SUMTER	4.64	PB2-5	22-Jun-57	2	5
AL TALLADEGA	1.91	TP6-4	25-Mar-55	9	2
AL TALLADEGA	1.55	PB2-5	22-Jun-57	6	2
AL TALLAPOOSA	2.53	TP10-4	8-May-55	7	2
AL TALLAPOOSA	3.96	PB17-3	30-Sep-57	9	2
AL TUSCALOOSA	1.47	TP6-2	23-Mar-55	9	6
AL TUSCALOOSA	8.23	TP6-3	24-Mar-55	2	6
AL TUSCALOOSA	3.14	TP6-5	26-Mar-55	2	6
AL TUSCALOOSA	1.82	PB2-4	21-Jun-57	5	6
AL TUSCALOOSA	2.78	PB2-5	22-Jun-57	4	6
AL TUSCALOOSA	1.27	PB3-9	2-Jul-57	2	6
AL WALKER	1.91	TP6-4	25-Mar-55	9	1
AL WASHINGTON	0.81	UpK3-6	5-Apr-53	9	7
AL WASHINGTON	1.47	TP6-2	23-Mar-55	9	7
AL WASHINGTON	4.11	TP6-3	24-Mar-55	4	7
AL WASHINGTON	3.14	TP6-5	26-Mar-55	2	7
AL WASHINGTON	22.97	PB3-3	26-Jun-57	4	7
AL WASHINGTON	1.40	PB6-8	31-Jul-57	8	7
AL WASHINGTON	2.73	PB9-2	24-Aug-57	6	7
AL WILCOX	1.47	TP6-2	23-Mar-55	9	4
AL WILCOX	4.11	TP6-3	24-Mar-55	4	4
AL WILCOX	3.14	TP6-5	26-Mar-55	2	4
AL WILCOX	22.97	PB3-3	26-Jun-57	4	4
AL WINSTON	1.86	PB2-5	22-Jun-57	5	2
AL WINSTON	1.25	PB3-8	1-Jul-57	3	2

COUNTY	uCi/ sqm	SHOT DAY	DATE	RANK FOR DAY	TIMES IN TOP 10
AR ARKANSAS	12.19	UpK2-7	30-Mar-53	4	9
AR ARKANSAS	14.03	UpK2-8	31-Mar-53	8	9
AR ARKANSAS	3.67	UpK3-6	5-Apr-53	1	9
AR ARKANSAS	6.18	UpK8-3	10-May-53	2	9
AR ARKANSAS	1.52	UpK8-4	11-May-53	6	9
AR ARKANSAS	0.89	UpK8-5	12-May-53	6	9
AR ARKANSAS	1.77	PB1-6	2-Jun-57	10	9
AR ARKANSAS	5.46	PB9-2	24-Aug-57	4	9
AR ARKANSAS	1.25	PB18-9	15-Oct-57	3	9
AR ASHLEY	0.93	UpK8-9	16-May-53	2	3
AR ASHLEY	1.47	TP6-2	23-Mar-55	9	3
AR ASHLEY	4.11	TP6-3	24-Mar-55	4	3
AR BAXTER	18.26	UpK2-7	30-Mar-53	2	6
AR BAXTER	4.10	UpK3-5	4-Apr-53	5	6
AR BAXTER	1.64	PB3-7	30-Jun-57	7	6
AR BAXTER	1.48	PB3-8	1-Jul-57	2	6
AR BAXTER	5.46	PB9-2	24-Aug-57	4	6
AR BAXTER	1.93	PB18-8	14-Oct-57	9	6
AR BENTON	2.23	UpK2-10	2-Apr-53	3	5
AR BENTON	2.10	UpK3-3	2-Apr-53	4	5
AR BENTON	6.83	UpK3-5	4-Apr-53	3	5
AR BENTON	2.45	PB3-8	1-Jul-57	1	5
AR BENTON	10.65	PB4-5	9-Jul-57	10	5
AR BOONE	1.47	UpK2-10	2-Apr-53	5	6
AR BOONE	1.40	UpK3-3	2-Apr-53	6	6
AR BOONE	4.10	UpK3-5	4-Apr-53	5	6
AR BOONE	1.83	UpK8-4	11-May-53	4	6
AR BOONE	1.75	PB3-7	30-Jun-57	4	6
AR BOONE	2.45	PB3-8	1-Jul-57	1	6
AR BRADLEY	4.02	BJ3-5	5-Nov-51	8	7
AR BRADLEY	1.22	UpK3-6	5-Apr-53	6	7
AR BRADLEY	3.91	UpK5-4	14-Apr-53	1	7
AR BRADLEY	1.83	UpK8-4	11-May-53	4	7
AR BRADLEY	1.47	TP6-2	23-Mar-55	9	7
AR BRADLEY	4.11	TP6-3	24-Mar-55	4	7

COUNTY	uCi/ sqm	SHOT DAY	DATE	RANK FOR DAY	TIMES IN TOP 10
AR BRADLEY	1.93	PB18-8	14-Oct-57	9	7
AR CALHOUN	1.22	UpK3-6	5-Apr-53	6	7
AR CALHOUN	3.91	UpK5-4	14-Apr-53	1	7
AR CALHOUN	1.83	UpK8-4	11-May-53	4	7
AR CALHOUN	1.47	TP6-2	23-Mar-55	9	7
AR CALHOUN	4.11	TP6-3	24-Mar-55	4	7
AR CALHOUN	2.03	PB1-6	2-Jun-57	5	7
AR CALHOUN	4.26	PB11-2	1-Sep-57	10	7
AR CARROLL	4.10	UpK3-5	4-Apr-53	5	3
AR CARROLL	2.45	PB3-8	1-Jul-57	1	3
AR CARROLL	1.93	PB18-8	14-Oct-57	9	3
AR CHICOT	1.56	UpK8-9	16-May-53	1	4
AR CHICOT	1.47	TP6-2	23-Mar-55	9	4
AR CHICOT	4.11	TP6-3	24-Mar-55	4	4
AR CHICOT	1.25	PB18-9	15-Oct-57	3	4
AR CLARK	4.02	BJ3-5	5-Nov-51	8	7
AR CLARK	1.22	UpK3-6	5-Apr-53	6	7
AR CLARK	3.91	UpK5-4	14-Apr-53	1	7
AR CLARK	1.32	UpK8-3	10-May-53	9	7
AR CLARK	1.83	UpK8-4	11-May-53	4	7
AR CLARK	3.21	PB18-8	14-Oct-57	6	7
AR CLARK	1.25	PB18-9	15-Oct-57	3	7
AR CLAY	1.22	BJ5-6	24-Nov-51	6	10
AR CLAY	1.22	BJ5-6	24-Nov-51	6	10
AR CLAY	12.19	UpK2-7	30-Mar-53	4	10
AR CLAY	1.49	UpK3-6	5-Apr-53	5	10
AR CLAY	3.68	UpK8-3	10-May-53	4	10
AR CLAY	1.09	TP4-5	11-Mar-55	1	10
AR CLAY	2.78	PB2-5	22-Jun-57	4	10
AR CLAY	2.10	PB3-9	2-Jul-57	1	10
AR CLAY	5.46	PB9-2	24-Aug-57	4	10
AR CLAY	13.72	PB9-3	25-Aug-57	6	10
AR CLEBURNE	12.19	UpK2-7	30-Mar-53	4	10
AR CLEBURNE	21.07	UpK2-8	31-Mar-53	3	10
AR CLEBURNE	3.67	UpK3-6	5-Apr-53	1	10

COUNTY	uCi/ sqm	SHOT DAY	DATE	RANK FOR DAY	TIMES IN TOP 10
AR CLEBURNE	6.18	UpK8-3	10-May-53	2	10
AR CLEBURNE	1.52	UpK8-4	11-May-53	6	10
AR CLEBURNE	0.89	UpK8-5	12-May-53	6	10
AR CLEBURNE	2.94	PB6-6	29-Jul-57	6	10
AR CLEBURNE	8.19	PB9-2	24-Aug-57	3	10
AR CLEBURNE	4.26	PB11-2	1-Sep-57	10	10
AR CLEBURNE	1.93	PB18-8	14-Oct-57	9	10
AR CLEVELAND	1.22	UpK3-6	5-Apr-53	6	3
AR CLEVELAND	3.91	UpK5-4	14-Apr-53	1	3
AR CLEVELAND	1.83	UpK8-4	11-May-53	4	3
AR COLUMBIA	4.02	BJ3-5	5-Nov-51	8	8
AR COLUMBIA	3.91	UpK5-4	14-Apr-53	1	8
AR COLUMBIA	1.83	UpK8-4	11-May-53	4	8
AR COLUMBIA	1.47	TP6-2	23-Mar-55	9	8
AR COLUMBIA	4.11	TP6-3	24-Mar-55	4	8
AR COLUMBIA	1.04	PB18-7	13-Oct-57	7	8
AR COLUMBIA	1.93	PB18-8	14-Oct-57	9	8
AR COLUMBIA	1.25	PB18-9	15-Oct-57	3	8
AR CONWAY	4.02	BJ3-5	5-Nov-51	8	6
AR CONWAY	0.81	UpK3-6	5-Apr-53	9	6
AR CONWAY	1.07	UpK5-4	14-Apr-53	8	6
AR CONWAY	1.83	UpK8-4	11-May-53	4	6
AR CONWAY	1.16	UpK8-5	12-May-53	5	6
AR CONWAY	2.73	PB9-2	24-Aug-57	6	6
AR CRAIGHEAD	17.41	UpK2-3	26-Mar-53	3	10
AR CRAIGHEAD	18.26	UpK2-7	30-Mar-53	2	10
AR CRAIGHEAD	21.07	UpK2-8	31-Mar-53	3	10
AR CRAIGHEAD	2.17	UpK3-6	5-Apr-53	2	10
AR CRAIGHEAD	6.18	UpK8-3	10-May-53	2	10
AR CRAIGHEAD	1.21	PB1-5	1-Jun-57	10	10
AR CRAIGHEAD	4.64	PB2-5	22-Jun-57	2	10
AR CRAIGHEAD	0.83	PB3-9	2-Jul-57	4	10
AR CRAIGHEAD	5.46	PB9-2	24-Aug-57	4	10
AR CRAIGHEAD	13.72	PB9-3	25-Aug-57	6	10
AR CRAWFORD	2.23	UpK2-10	2-Apr-53	3	6

COUNTY	uCi/ sqm	SHOT DAY	DATE	RANK FOR DAY	TIMES IN TOP 10
AR CRAWFORD	2.10	UpK3-3	2-Apr-53	4	6
AR CRAWFORD	4.10	UpK3-5	4-Apr-53	5	6
AR CRAWFORD	1.07	UpK5-4	14-Apr-53	8	6
AR CRAWFORD	1.83	UpK8-4	11-May-53	4	6
AR CRAWFORD	4.15	PB13-9	14-Sep-57	7	6
AR CRITTENDEN	12.28	UpK2-3	26-Mar-53	9	13
AR CRITTENDEN	12.19	UpK2-7	30-Mar-53	4	13
AR CRITTENDEN	21.07	UpK2-8	31-Mar-53	3	13
AR CRITTENDEN	3.67	UpK3-6	5-Apr-53	1	13
AR CRITTENDEN	6.18	UpK8-3	10-May-53	2	13
AR CRITTENDEN	1.52	UpK8-4	11-May-53	6	13
AR CRITTENDEN	0.89	UpK8-5	12-May-53	6	13
AR CRITTENDEN	3.22	TP7-4	1-Apr-55	3	13
AR CRITTENDEN	1.21	PB1-5	1-Jun-57	10	13
AR CRITTENDEN	2.78	PB2-5	22-Jun-57	4	13
AR CRITTENDEN	1.25	PB3-8	1-Jul-57	3	13
AR CRITTENDEN	1.27	PB3-9	2-Jul-57	2	13
AR CRITTENDEN	2.73	PB9-2	24-Aug-57	6	13
AR CROSS	12.51	UpK2-3	26-Mar-53	8	9
AR CROSS	18.26	UpK2-7	30-Mar-53	2	9
AR CROSS	2.17	UpK3-6	5-Apr-53	2	9
AR CROSS	6.18	UpK8-3	10-May-53	2	9
AR CROSS	1.52	UpK8-4	11-May-53	6	9
AR CROSS	1.52	UpK8-5	12-May-53	2	9
AR CROSS	3.02	TP7-4	1-Apr-55	7	9
AR CROSS	2.10	PB3-9	2-Jul-57	1	9
AR CROSS	1.00	PB6-11	3-Aug-57	3	9
AR DALLAS	4.02	BJ3-5	5-Nov-51	8	7
AR DALLAS	1.22	UpK3-6	5-Apr-53	6	7
AR DALLAS	3.91	UpK5-4	14-Apr-53	1	7
AR DALLAS	1.83	UpK8-4	11-May-53	4	7
AR DALLAS	3.04	PB6-6	29-Jul-57	4	7
AR DALLAS	3.21	PB18-8	14-Oct-57	6	7
AR DALLAS	1.25	PB18-9	15-Oct-57	3	7
AR DESHA	12.19	UpK2-7	30-Mar-53	4	5

COUNTY	uCi/ sqm	SHOT DAY	DATE	RANK FOR DAY	TIMES IN TOP 10
AR DESHA	3.67	UpK3-6	5-Apr-53	1	5
AR DESHA	3.68	UpK8-3	10-May-53	4	5
AR DESHA	2.73	PB9-2	24-Aug-57	6	5
AR DESHA	1.93	PB18-8	14-Oct-57	9	5
AR DREW	36.18	UpK7-4	28-Apr-53	10	6
AR DREW	1.56	UpK8-9	16-May-53	1	6
AR DREW	1.47	TP6-2	23-Mar-55	9	6
AR DREW	4.11	TP6-3	24-Mar-55	4	6
AR DREW	1.81	PB1-6	2-Jun-57	7	6
AR DREW	1.93	PB18-8	14-Oct-57	9	6
AR FAULKNER	18.26	UpK2-7	30-Mar-53	2	10
AR FAULKNER	1.47	UpK2-10	2-Apr-53	5	10
AR FAULKNER	1.40	UpK3-3	2-Apr-53	6	10
AR FAULKNER	4.10	UpK3-5	4-Apr-53	5	10
AR FAULKNER	0.81	UpK3-6	5-Apr-53	9	10
AR FAULKNER	1.07	UpK5-4	14-Apr-53	8	10
AR FAULKNER	1.52	UpK8-4	11-May-53	6	10
AR FAULKNER	1.92	PB3-7	30-Jun-57	2	10
AR FAULKNER	3.18	PB6-6	29-Jul-57	3	10
AR FAULKNER	1.93	PB18-8	14-Oct-57	9	10
AR FRANKLIN	1.47	UpK2-10	2-Apr-53	5	9
AR FRANKLIN	1.40	UpK3-3	2-Apr-53	6	9
AR FRANKLIN	4.10	UpK3-5	4-Apr-53	5	9
AR FRANKLIN	1.07	UpK5-4	14-Apr-53	8	9
AR FRANKLIN	1.83	UpK8-4	11-May-53	4	9
AR FRANKLIN	1.16	UpK8-5	12-May-53	5	9
AR FRANKLIN	10.05	TP11-6	20-May-55	5	9
AR FRANKLIN	2.45	PB3-8	1-Jul-57	1	9
AR FRANKLIN	1.93	PB18-8	14-Oct-57	9	9
AR FULTON	24.64	UpK2-3	26-Mar-53	1	7
AR FULTON	18.26	UpK2-7	30-Mar-53	2	7
AR FULTON	14.03	UpK2-8	31-Mar-53	8	7
AR FULTON	2.17	UpK3-6	5-Apr-53	2	7
AR FULTON	2.50	UpK8-3	10-May-53	6	7
AR FULTON	5.46	PB9-2	24-Aug-57	4	7

COUNTY	uCi/ sqm	SHOT DAY	DATE	RANK FOR DAY	TIMES IN TOP 10
AR FULTON	13.72	PB9-3	25-Aug-57	6	7
AR GARLAND	4.02	BJ3-5	5-Nov-51	8	6
AR GARLAND	1.22	UpK3-6	5-Apr-53	6	6
AR GARLAND	3.91	UpK5-4	14-Apr-53	1	6
AR GARLAND	1.32	UpK8-3	10-May-53	9	6
AR GARLAND	1.83	UpK8-4	11-May-53	4	6
AR GARLAND	3.21	PB18-8	14-Oct-57	6	6
AR GRANT	1.22	UpK3-6	5-Apr-53	6	7
AR GRANT	3.91	UpK5-4	14-Apr-53	1	7
AR GRANT	1.32	UpK8-3	10-May-53	9	7
AR GRANT	1.83	UpK8-4	11-May-53	4	7
AR GRANT	1.79	PB1-8	4-Jun-57	1	7
AR GRANT	3.21	PB18-8	14-Oct-57	6	7
AR GRANT	1.25	PB18-9	15-Oct-57	3	7
AR GREENE	1.22	BJ5-6	24-Nov-51	6	8
AR GREENE	1.22	BJ5-6	24-Nov-51	6	8
AR GREENE	18.26	UpK2-7	30-Mar-53	2	8
AR GREENE	2.17	UpK3-6	5-Apr-53	2	8
AR GREENE	3.68	UpK8-3	10-May-53	4	8
AR GREENE	1.21	PB1-5	1-Jun-57	10	8
AR GREENE	1.27	PB3-9	2-Jul-57	2	8
AR GREENE	13.72	PB9-3	25-Aug-57	6	8
AR HEMPSTEAD	1.22	UpK3-6	5-Apr-53	6	7
AR HEMPSTEAD	3.91	UpK5-4	14-Apr-53	1	7
AR HEMPSTEAD	1.83	UpK8-4	11-May-53	4	7
AR HEMPSTEAD	12.96	TP6-2	23-Mar-55	2	7
AR HEMPSTEAD	1.56	PB18-7	13-Oct-57	4	7
AR HEMPSTEAD	3.21	PB18-8	14-Oct-57	6	7
AR HEMPSTEAD	1.25	PB18-9	15-Oct-57	3	7
AR HOT SPRING	4.02	BJ3-5	5-Nov-51	8	7
AR HOT SPRING	1.22	UpK3-6	5-Apr-53	6	7
AR HOT SPRING	3.91	UpK5-4	14-Apr-53	1	7
AR HOT SPRING	1.32	UpK8-3	10-May-53	9	7
AR HOT SPRING	1.83	UpK8-4	11-May-53	4	7
AR HOT SPRING	4.26	PB11-2	1-Sep-57	10	7

COUNTY	uCi/ sqm	SHOT DAY	DATE	RANK FOR DAY	TIMES IN TOP 10
AR HOT SPRING	3.21	PB18-8	14-Oct-57	6	7
AR HOWARD	4.02	BJ3-5	5-Nov-51	8	14
AR HOWARD	1.22	UpK3-6	5-Apr-53	6	14
AR HOWARD	3.91	UpK5-4	14-Apr-53	1	14
AR HOWARD	1.32	UpK8-3	10-May-53	9	14
AR HOWARD	1.83	UpK8-4	11-May-53	4	14
AR HOWARD	1.16	UpK8-5	12-May-53	5	14
AR HOWARD	12.96	TP6-2	23-Mar-55	2	14
AR HOWARD	1.86	PB1-6	2-Jun-57	6	14
AR HOWARD	1.58	PB1-8	4-Jun-57	2	14
AR HOWARD	3.80	PB5-4	18-Jul-57	10	14
AR HOWARD	1.64	PB13-9	14-Sep-57	10	14
AR HOWARD	1.56	PB18-7	13-Oct-57	4	14
AR HOWARD	3.21	PB18-8	14-Oct-57	6	14
AR HOWARD	1.25	PB18-9	15-Oct-57	3	14
AR INDEPENDENCE	2.53	UpK1-4	20-Mar-53	9	9
AR INDEPENDENCE	18.26	UpK2-7	30-Mar-53	2	9
AR INDEPENDENCE	14.03	UpK2-8	31-Mar-53	8	9
AR INDEPENDENCE	2.17	UpK3-6	5-Apr-53	2	9
AR INDEPENDENCE	6.18	UpK8-3	10-May-53	2	9
AR INDEPENDENCE	0.89	UpK8-5	12-May-53	6	9
AR INDEPENDENCE	8.19	PB9-2	24-Aug-57	3	9
AR INDEPENDENCE	13.72	PB9-3	25-Aug-57	6	9
AR INDEPENDENCE	1.25	PB18-9	15-Oct-57	3	9
AR IZARD	12.19	UpK2-7	30-Mar-53	4	8
AR IZARD	14.03	UpK2-8	31-Mar-53	8	8
AR IZARD	2.17	UpK3-6	5-Apr-53	2	8
AR IZARD	3.68	UpK8-3	10-May-53	4	8
AR IZARD	1.52	UpK8-4	11-May-53	6	8
AR IZARD	0.89	UpK8-5	12-May-53	6	8
AR IZARD	8.19	PB9-2	24-Aug-57	3	8
AR IZARD	13.72	PB9-3	25-Aug-57	6	8
AR JACKSON	13.35	UpK2-3	26-Mar-53	7	9
AR JACKSON	18.26	UpK2-7	30-Mar-53	2	9
AR JACKSON	14.03	UpK2-8	31-Mar-53	8	9

COUNTY	uCi/ sqm	SHOT DAY	DATE	RANK FOR DAY	TIMES IN TOP 10
AR JACKSON	3.67	UpK3-6	5-Apr-53	1	9
AR JACKSON	6.18	UpK8-3	10-May-53	2	9
AR JACKSON	0.89	UpK8-5	12-May-53	6	9
AR JACKSON	1.27	PB3-9	2-Jul-57	2	9
AR JACKSON	5.46	PB9-2	24-Aug-57	4	9
AR JACKSON	13.72	PB9-3	25-Aug-57	6	9
AR JEFFERSON	3.77	UpK1-4	20-Mar-53	3	8
AR JEFFERSON	12.19	UpK2-7	30-Mar-53	4	8
AR JEFFERSON	3.67	UpK3-6	5-Apr-53	1	8
AR JEFFERSON	6.18	UpK8-3	10-May-53	2	8
AR JEFFERSON	1.52	UpK8-4	11-May-53	6	8
AR JEFFERSON	0.89	UpK8-5	12-May-53	6	8
AR JEFFERSON	2.99	PB6-6	29-Jul-57	5	8
AR JEFFERSON	1.93	PB18-8	14-Oct-57	9	8
AR JOHNSON	1.47	UpK2-10	2-Apr-53	5	8
AR JOHNSON	1.40	UpK3-3	2-Apr-53	6	8
AR JOHNSON	0.81	UpK3-6	5-Apr-53	9	8
AR JOHNSON	1.07	UpK5-4	14-Apr-53	8	8
AR JOHNSON	1.83	UpK8-4	11-May-53	4	8
AR JOHNSON	1.71	PB3-7	30-Jun-57	6	8
AR JOHNSON	2.45	PB3-8	1-Jul-57	1	8
AR JOHNSON	1.93	PB18-8	14-Oct-57	9	8
AR LAFAYETTE	4.02	BJ3-5	5-Nov-51	8	8
AR LAFAYETTE	1.22	UpK3-6	5-Apr-53	6	8
AR LAFAYETTE	3.91	UpK5-4	14-Apr-53	1	8
AR LAFAYETTE	1.83	UpK8-4	11-May-53	4	8
AR LAFAYETTE	12.96	TP6-2	23-Mar-55	2	8
AR LAFAYETTE	1.56	PB18-7	13-Oct-57	4	8
AR LAFAYETTE	3.21	PB18-8	14-Oct-57	6	8
AR LAFAYETTE	1.25	PB18-9	15-Oct-57	3	8
AR LAWRENCE	1.22	BJ5-6	24-Nov-51	6	10
AR LAWRENCE	1.22	BJ5-6	24-Nov-51	6	10
AR LAWRENCE	18.26	UpK2-7	30-Mar-53	2	10
AR LAWRENCE	14.03	UpK2-8	31-Mar-53	8	10
AR LAWRENCE	1.49	UpK3-6	5-Apr-53	5	10

COUNTY	uCi/ sqm	SHOT DAY	DATE	RANK FOR DAY	TIMES IN TOP 10
AR LAWRENCE	6.18	UpK8-3	10-May-53	2	10
AR LAWRENCE	1.86	PB2-5	22-Jun-57	5	10
AR LAWRENCE	1.27	PB3-9	2-Jul-57	2	10
AR LAWRENCE	5.46	PB9-2	24-Aug-57	4	10
AR LAWRENCE	13.72	PB9-3	25-Aug-57	6	10
AR LEE	12.19	UpK2-7	30-Mar-53	4	7
AR LEE	21.07	UpK2-8	31-Mar-53	3	7
AR LEE	3.67	UpK3-6	5-Apr-53	1	7
AR LEE	6.18	UpK8-3	10-May-53	2	7
AR LEE	0.89	UpK8-5	12-May-53	6	7
AR LEE	1.27	PB3-9	2-Jul-57	2	7
AR LEE	4.26	PB11-2	1-Sep-57	10	7
AR LINCOLN	6.07	UpK2-7	30-Mar-53	7	7
AR LINCOLN	3.67	UpK3-6	5-Apr-53	1	7
AR LINCOLN	3.68	UpK8-3	10-May-53	4	7
AR LINCOLN	1.52	UpK8-4	11-May-53	6	7
AR LINCOLN	0.89	UpK8-5	12-May-53	6	7
AR LINCOLN	2.85	PB6-6	29-Jul-57	8	7
AR LINCOLN	1.93	PB18-8	14-Oct-57	9	7
AR LITTLE RIVER	1.22	UpK3-6	5-Apr-53	6	9
AR LITTLE RIVER	3.91	UpK5-4	14-Apr-53	1	9
AR LITTLE RIVER	1.32	UpK8-3	10-May-53	9	9
AR LITTLE RIVER	1.83	UpK8-4	11-May-53	4	9
AR LITTLE RIVER	12.96	TP6-2	23-Mar-55	2	9
AR LITTLE RIVER	1.47	PB1-8	4-Jun-57	4	9
AR LITTLE RIVER	2.50	PB13-9	14-Sep-57	8	9
AR LITTLE RIVER	2.60	PB18-7	13-Oct-57	1	9
AR LITTLE RIVER	3.21	PB18-8	14-Oct-57	6	9
AR LOGAN	4.10	UpK3-5	4-Apr-53	5	8
AR LOGAN	0.81	UpK3-6	5-Apr-53	9	8
AR LOGAN	1.07	UpK5-4	14-Apr-53	8	8
AR LOGAN	1.83	UpK8-4	11-May-53	4	8
AR LOGAN	10.29	TP11-6	20-May-55	4	8
AR LOGAN	0.97	PB3-8	1-Jul-57	4	8
AR LOGAN	2.11	PB9-2	24-Aug-57	8	8

COUNTY	uCi/ sqm	SHOT DAY	DATE	RANK FOR DAY	TIMES IN TOP 10
AR LOGAN	1.93	PB18-8	14-Oct-57	9	8
AR LONOKE	18.26	UpK2-7	30-Mar-53	2	6
AR LONOKE	3.67	UpK3-6	5-Apr-53	1	6
AR LONOKE	6.18	UpK8-3	10-May-53	2	6
AR LONOKE	1.74	PB3-7	30-Jun-57	5	6
AR LONOKE	2.74	PB6-6	29-Jul-57	10	6
AR LONOKE	3.21	PB18-8	14-Oct-57	6	6
AR MADISON	2.23	UpK2-10	2-Apr-53	3	5
AR MADISON	2.10	UpK3-3	2-Apr-53	4	5
AR MADISON	6.83	UpK3-5	4-Apr-53	3	5
AR MADISON	1.83	UpK8-4	11-May-53	4	5
AR MADISON	2.73	PB9-2	24-Aug-57	6	5
AR MARION	18.26	UpK2-7	30-Mar-53	2	6
AR MARION	4.10	UpK3-5	4-Apr-53	5	6
AR MARION	0.89	UpK8-5	12-May-53	6	6
AR MARION	1.61	PB3-7	30-Jun-57	8	6
AR MARION	1.48	PB3-8	1-Jul-57	2	6
AR MARION	7.99	PB8-3	20-Aug-57	9	6
AR MILLER	1.62	UpK5-4	14-Apr-53	5	8
AR MILLER	12.96	TP6-2	23-Mar-55	2	8
AR MILLER	1.79	PB1-6	2-Jun-57	8	8
AR MILLER	3.80	PB5-4	18-Jul-57	10	8
AR MILLER	1.64	PB13-9	14-Sep-57	10	8
AR MILLER	2.60	PB18-7	13-Oct-57	1	8
AR MILLER	3.21	PB18-8	14-Oct-57	6	8
AR MILLER	1.25	PB18-9	15-Oct-57	3	8
AR MISSISSIPPI	12.19	UpK2-7	30-Mar-53	4	8
AR MISSISSIPPI	14.03	UpK2-8	31-Mar-53	8	8
AR MISSISSIPPI	2.17	UpK3-6	5-Apr-53	2	8
AR MISSISSIPPI	6.18	UpK8-3	10-May-53	2	8
AR MISSISSIPPI	2.92	TP7-4	1-Apr-55	10	8
AR MISSISSIPPI	2.78	PB2-5	22-Jun-57	4	8
AR MISSISSIPPI	0.83	PB3-9	2-Jul-57	4	8
AR MISSISSIPPI	2.73	PB9-2	24-Aug-57	6	8
AR MONROE	18.26	UpK2-7	30-Mar-53	2	10

COUNTY	uCi/ sqm	SHOT DAY	DATE	RANK FOR DAY	TIMES IN TOP 10
AR MONROE	21.07	UpK2-8	31-Mar-53	3	10
AR MONROE	3.67	UpK3-6	5-Apr-53	1	10
AR MONROE	6.18	UpK8-3	10-May-53	2	10
AR MONROE	1.52	UpK8-4	11-May-53	6	10
AR MONROE	0.89	UpK8-5	12-May-53	6	10
AR MONROE	1.16	TP10-7	11-May-55	10	10
AR MONROE	1.25	PB3-8	1-Jul-57	3	10
AR MONROE	5.46	PB9-2	24-Aug-57	4	10
AR MONROE	1.25	PB18-9	15-Oct-57	3	10
AR MONTGOMERY	4.02	BJ3-5	5-Nov-51	8	12
AR MONTGOMERY	1.22	UpK3-6	5-Apr-53	6	12
AR MONTGOMERY	3.91	UpK5-4	14-Apr-53	1	12
AR MONTGOMERY	1.32	UpK8-3	10-May-53	9	12
AR MONTGOMERY	1.83	UpK8-4	11-May-53	4	12
AR MONTGOMERY	1.16	UpK8-5	12-May-53	5	12
AR MONTGOMERY	12.96	TP6-2	23-Mar-55	2	12
AR MONTGOMERY	9.85	TP11-6	20-May-55	6	12
AR MONTGOMERY	3.80	PB5-4	18-Jul-57	10	12
AR MONTGOMERY	2.73	PB9-2	24-Aug-57	6	12
AR MONTGOMERY	1.64	PB13-9	14-Sep-57	10	12
AR MONTGOMERY	3.21	PB18-8	14-Oct-57	6	12
AR NEVADA	4.02	BJ3-5	5-Nov-51	8	11
AR NEVADA	1.22	UpK3-6	5-Apr-53	6	11
AR NEVADA	3.91	UpK5-4	14-Apr-53	1	11
AR NEVADA	1.83	UpK8-4	11-May-53	4	11
AR NEVADA	1.47	TP6-2	23-Mar-55	9	11
AR NEVADA	4.11	TP6-3	24-Mar-55	4	11
AR NEVADA	0.92	TP9-13	27-Apr-55	4	11
AR NEVADA	1.45	PB1-8	4-Jun-57	5	11
AR NEVADA	1.56	PB18-7	13-Oct-57	4	11
AR NEVADA	3.21	PB18-8	14-Oct-57	6	11
AR NEVADA	1.25	PB18-9	15-Oct-57	3	11
AR NEWTON	4.02	BJ3-5	5-Nov-51	8	7
AR NEWTON	1.83	UpK8-4	11-May-53	4	7
AR NEWTON	1.16	UpK8-5	12-May-53	5	7

COUNTY	uCi/ sqm	SHOT DAY	DATE	RANK FOR DAY	TIMES IN TOP 10
AR NEWTON	9.12	TP11-6	20-May-55	9	7
AR NEWTON	1.48	PB3-8	1-Jul-57	2	7
AR NEWTON	2.73	PB9-2	24-Aug-57	6	7
AR NEWTON	1.93	PB18-8	14-Oct-57	9	7
AR OUACHITA	4.02	BJ3-5	5-Nov-51	8	9
AR OUACHITA	1.22	UpK3-6	5-Apr-53	6	9
AR OUACHITA	1.83	UpK8-4	11-May-53	4	9
AR OUACHITA	1.47	TP6-2	23-Mar-55	9	9
AR OUACHITA	4.11	TP6-3	24-Mar-55	4	9
AR OUACHITA	3.52	PB1-6	2-Jun-57	3	9
AR OUACHITA	10.30	PB12-6	7-Sep-57	10	9
AR OUACHITA	1.93	PB18-8	14-Oct-57	9	9
AR OUACHITA	1.25	PB18-9	15-Oct-57	3	9
AR PERRY	0.81	UpK3-6	5-Apr-53	9	7
AR PERRY	1.07	UpK5-4	14-Apr-53	8	7
AR PERRY	1.12	UpK8-4	11-May-53	9	7
AR PERRY	1.16	UpK8-5	12-May-53	5	7
AR PERRY	55.98	TP10-3	7-May-55	7	7
AR PERRY	2.73	PB9-2	24-Aug-57	6	7
AR PERRY	1.93	PB18-8	14-Oct-57	9	7
AR PHILLIPS	18.26	UpK2-7	30-Mar-53	2	9
AR PHILLIPS	3.67	UpK3-6	5-Apr-53	1	9
AR PHILLIPS	6.18	UpK8-3	10-May-53	2	9
AR PHILLIPS	1.52	UpK8-4	11-May-53	6	9
AR PHILLIPS	0.89	UpK8-5	12-May-53	6	9
AR PHILLIPS	2.78	PB2-5	22-Jun-57	4	9
AR PHILLIPS	1.25	PB3-8	1-Jul-57	3	9
AR PHILLIPS	1.27	PB3-9	2-Jul-57	2	9
AR PHILLIPS	5.46	PB9-2	24-Aug-57	4	9
AR PIKE	4.02	BJ3-5	5-Nov-51	8	13
AR PIKE	1.22	UpK3-6	5-Apr-53	6	13
AR PIKE	3.91	UpK5-4	14-Apr-53	1	13
AR PIKE	1.32	UpK8-3	10-May-53	9	13
AR PIKE	1.83	UpK8-4	11-May-53	4	13
AR PIKE	1.16	UpK8-5	12-May-53	5	13

COUNTY	uCi/ sqm	SHOT DAY	DATE	RANK FOR DAY	TIMES IN TOP 10
AR PIKE	12.96	TP6-2	23-Mar-55	2	13
AR PIKE	10.76	TP11-6	20-May-55	2	13
AR PIKE	1.40	PB1-8	4-Jun-57	7	13
AR PIKE	1.64	PB13-9	14-Sep-57	10	13
AR PIKE	1.56	PB18-7	13-Oct-57	4	13
AR PIKE	3.21	PB18-8	14-Oct-57	6	13
AR PIKE	1.25	PB18-9	15-Oct-57	3	13
AR POINSETT	15.89	UpK2-3	26-Mar-53	4	12
AR POINSETT	6.07	UpK2-7	30-Mar-53	7	12
AR POINSETT	21.07	UpK2-8	31-Mar-53	3	12
AR POINSETT	2.17	UpK3-6	5-Apr-53	2	12
AR POINSETT	6.18	UpK8-3	10-May-53	2	12
AR POINSETT	0.89	UpK8-5	12-May-53	6	12
AR POINSETT	2.39	UpK9-6	24-May-53	10	12
AR POINSETT	2.99	TP7-4	1-Apr-55	8	12
AR POINSETT	1.21	PB1-5	1-Jun-57	10	12
AR POINSETT	1.25	PB3-8	1-Jul-57	3	12
AR POINSETT	2.10	PB3-9	2-Jul-57	1	12
AR POINSETT	5.46	PB9-2	24-Aug-57	4	12
AR POLK	4.02	BJ3-5	5-Nov-51	8	11
AR POLK	2.73	UpK3-5	4-Apr-53	8	11
AR POLK	0.81	UpK3-6	5-Apr-53	9	11
AR POLK	1.07	UpK5-4	14-Apr-53	8	11
AR POLK	1.83	UpK8-4	11-May-53	4	11
AR POLK	1.16	UpK8-5	12-May-53	5	11
AR POLK	12.96	TP6-2	23-Mar-55	2	11
AR POLK	1.77	PB1-6	2-Jun-57	10	11
AR POLK	2.50	PB13-9	14-Sep-57	8	11
AR POLK	1.04	PB18-7	13-Oct-57	7	11
AR POLK	1.93	PB18-8	14-Oct-57	9	11
AR POPE	4.02	BJ3-5	5-Nov-51	8	7
AR POPE	1.78	UpK5-4	14-Apr-53	4	7
AR POPE	1.83	UpK8-4	11-May-53	4	7
AR POPE	1.16	UpK8-5	12-May-53	5	7
AR POPE	1.31	PB1-8	4-Jun-57	10	7

COUNTY	uCi/ sqm	SHOT DAY	DATE	RANK FOR DAY	TIMES IN TOP 10
AR POPE	2.06	PB3-7	30-Jun-57	1	7
AR POPE	1.93	PB18-8	14-Oct-57	9	7
AR PRAIRIE	12.19	UpK2-7	30-Mar-53	4	11
AR PRAIRIE	35.10	UpK2-8	31-Mar-53	1	11
AR PRAIRIE	3.67	UpK3-6	5-Apr-53	1	11
AR PRAIRIE	6.18	UpK8-3	10-May-53	2	11
AR PRAIRIE	1.52	UpK8-4	11-May-53	6	11
AR PRAIRIE	0.89	UpK8-5	12-May-53	6	11
AR PRAIRIE	4.02	PB6-6	29-Jul-57	1	11
AR PRAIRIE	13.65	PB9-2	24-Aug-57	1	11
AR PRAIRIE	1.51	PB8-7	24-Aug-57	10	11
AR PRAIRIE	4.26	PB11-2	1-Sep-57	10	11
AR PRAIRIE	1.25	PB18-9	15-Oct-57	3	11
AR PULASKI	3.96	UpK1-4	20-Mar-53	2	8
AR PULASKI	1.22	UpK3-6	5-Apr-53	6	8
AR PULASKI	3.91	UpK5-4	14-Apr-53	1	8
AR PULASKI	1.32	UpK8-3	10-May-53	9	8
AR PULASKI	1.12	UpK8-4	11-May-53	9	8
AR PULASKI	1.82	PB3-7	30-Jun-57	3	8
AR PULASKI	2.78	PB6-6	29-Jul-57	9	8
AR PULASKI	2.73	PB9-2	24-Aug-57	6	8
AR RANDOLPH	1.22	BJ5-6	24-Nov-51	6	12
AR RANDOLPH	1.22	BJ5-6	24-Nov-51	6	12
AR RANDOLPH	18.26	UpK2-7	30-Mar-53	2	12
AR RANDOLPH	14.03	UpK2-8	31-Mar-53	8	12
AR RANDOLPH	1.49	UpK3-6	5-Apr-53	5	12
AR RANDOLPH	3.68	UpK8-3	10-May-53	4	12
AR RANDOLPH	1.36	TP10-7	11-May-55	5	12
AR RANDOLPH	1.86	PB2-5	22-Jun-57	5	12
AR RANDOLPH	0.83	PB3-9	2-Jul-57	4	12
AR RANDOLPH	8.19	PB9-2	24-Aug-57	3	12
AR RANDOLPH	13.72	PB9-3	25-Aug-57	6	12
AR RANDOLPH	4.26	PB11-2	1-Sep-57	10	12
AR SALINE	4.02	BJ3-5	5-Nov-51	8	8
AR SALINE	1.22	UpK3-6	5-Apr-53	6	8

COUNTY	uCi/ sqm	SHOT DAY	DATE	RANK FOR DAY	TIMES IN TOP 10
AR SALINE	3.91	UpK5-4	14-Apr-53	1	8
AR SALINE	1.32	UpK8-3	10-May-53	9	8
AR SALINE	1.83	UpK8-4	11-May-53	4	8
AR SALINE	2.73	PB9-2	24-Aug-57	6	8
AR SALINE	4.26	PB11-2	1-Sep-57	10	8
AR SALINE	3.21	PB18-8	14-Oct-57	6	8
AR SCOTT	0.81	UpK3-6	5-Apr-53	9	7
AR SCOTT	1.07	UpK5-4	14-Apr-53	8	7
AR SCOTT	1.83	UpK8-4	11-May-53	4	7
AR SCOTT	12.96	TP6-2	23-Mar-55	2	7
AR SCOTT	3.80	PB5-4	18-Jul-57	10	7
AR SCOTT	2.73	PB9-2	24-Aug-57	6	7
AR SCOTT	4.15	PB13-9	14-Sep-57	7	7
AR SEARCY	4.02	BJ3-5	5-Nov-51	8	6
AR SEARCY	1.83	UpK8-4	11-May-53	4	6
AR SEARCY	2.06	PB3-7	30-Jun-57	1	6
AR SEARCY	2.45	PB3-8	1-Jul-57	1	6
AR SEARCY	8.19	PB9-2	24-Aug-57	3	6
AR SEARCY	1.93	PB18-8	14-Oct-57	9	6
AR SEBASTIAN	0.93	UpK2-10	2-Apr-53	9	7
AR SEBASTIAN	0.87	UpK3-3	2-Apr-53	10	7
AR SEBASTIAN	1.83	UpK8-4	11-May-53	4	7
AR SEBASTIAN	1.16	UpK8-5	12-May-53	5	7
AR SEBASTIAN	12.96	TP6-2	23-Mar-55	2	7
AR SEBASTIAN	9.82	TP11-6	20-May-55	7	7
AR SEBASTIAN	4.15	PB13-9	14-Sep-57	7	7
AR SEVIER	4.02	BJ3-5	5-Nov-51	8	10
AR SEVIER	1.22	UpK3-6	5-Apr-53	6	10
AR SEVIER	3.91	UpK5-4	14-Apr-53	1	10
AR SEVIER	1.32	UpK8-3	10-May-53	9	10
AR SEVIER	1.83	UpK8-4	11-May-53	4	10
AR SEVIER	12.96	TP6-2	23-Mar-55	2	10
AR SEVIER	2.73	PB9-2	24-Aug-57	6	10
AR SEVIER	1.56	PB18-7	13-Oct-57	4	10
AR SEVIER	3.21	PB18-8	14-Oct-57	6	10

COUNTY	uCi/ sqm	SHOT DAY	DATE	RANK FOR DAY	TIMES IN TOP 10
AR SEVIER	1.25	PB18-9	15-Oct-57	3	10
AR SHARP	1.22	BJ5-6	24-Nov-51	6	11
AR SHARP	1.22	BJ5-6	24-Nov-51	6	11
AR SHARP	18.26	UpK2-7	30-Mar-53	2	11
AR SHARP	14.03	UpK2-8	31-Mar-53	8	11
AR SHARP	1.49	UpK3-6	5-Apr-53	5	11
AR SHARP	6.18	UpK8-3	10-May-53	2	11
AR SHARP	0.89	UpK8-5	12-May-53	6	11
AR SHARP	10.70	TP11-6	20-May-55	3	11
AR SHARP	5.46	PB9-2	24-Aug-57	4	11
AR SHARP	13.72	PB9-3	25-Aug-57	6	11
AR SHARP	1.25	PB18-9	15-Oct-57	3	11
AR ST FRANCIS	12.19	UpK2-7	30-Mar-53	4	10
AR ST FRANCIS	21.07	UpK2-8	31-Mar-53	3	10
AR ST FRANCIS	3.67	UpK3-6	5-Apr-53	1	10
AR ST FRANCIS	6.18	UpK8-3	10-May-53	2	10
AR ST FRANCIS	1.52	UpK8-4	11-May-53	6	10
AR ST FRANCIS	0.89	UpK8-5	12-May-53	6	10
AR ST FRANCIS	1.21	PB1-5	1-Jun-57	10	10
AR ST FRANCIS	2.78	PB2-5	22-Jun-57	4	10
AR ST FRANCIS	1.27	PB3-9	2-Jul-57	2	10
AR ST FRANCIS	2.73	PB9-2	24-Aug-57	6	10
AR STONE	12.19	UpK2-7	30-Mar-53	4	7
AR STONE	14.03	UpK2-8	31-Mar-53	8	7
AR STONE	2.17	UpK3-6	5-Apr-53	2	7
AR STONE	3.68	UpK8-3	10-May-53	4	7
AR STONE	0.89	UpK8-5	12-May-53	6	7
AR STONE	1.54	PB3-7	30-Jun-57	10	7
AR STONE	5.46	PB9-2	24-Aug-57	4	7
AR UNION	2.58	UpK5-4	14-Apr-53	2	6
AR UNION	1.83	UpK8-4	11-May-53	4	6
AR UNION	1.47	TP6-2	23-Mar-55	9	6
AR UNION	4.11	TP6-3	24-Mar-55	4	6
AR UNION	1.89	PB6-7	30-Jul-57	7	6
AR UNION	1.93	PB18-8	14-Oct-57	9	6

COUNTY	uCi/ sqm	SHOT DAY	DATE	RANK FOR DAY	TIMES IN TOP 10
AR VAN BUREN	12.19	UpK2-7	30-Mar-53	4	9
AR VAN BUREN	4.10	UpK3-5	4-Apr-53	5	9
AR VAN BUREN	1.07	UpK5-4	14-Apr-53	8	9
AR VAN BUREN	1.52	UpK8-5	12-May-53	2	9
AR VAN BUREN	8.70	TP11-6	20-May-55	10	9
AR VAN BUREN	2.88	PB6-6	29-Jul-57	7	9
AR VAN BUREN	8.19	PB9-2	24-Aug-57	3	9
AR VAN BUREN	4.26	PB11-2	1-Sep-57	10	9
AR VAN BUREN	1.93	PB18-8	14-Oct-57	9	9
AR WASHINGTON	2.23	UpK2-10	2-Apr-53	3	5
AR WASHINGTON	2.10	UpK3-3	2-Apr-53	4	5
AR WASHINGTON	6.83	UpK3-5	4-Apr-53	3	5
AR WASHINGTON	1.83	UpK8-4	11-May-53	4	5
AR WASHINGTON	1.48	PB3-8	1-Jul-57	2	5
AR WHITE	12.19	UpK2-7	30-Mar-53	4	10
AR WHITE	21.07	UpK2-8	31-Mar-53	3	10
AR WHITE	3.67	UpK3-6	5-Apr-53	1	10
AR WHITE	6.18	UpK8-3	10-May-53	2	10
AR WHITE	1.52	UpK8-4	11-May-53	6	10
AR WHITE	0.89	UpK8-5	12-May-53	6	10
AR WHITE	1.60	PB3-7	30-Jun-57	9	10
AR WHITE	3.96	PB6-6	29-Jul-57	2	10
AR WHITE	5.46	PB9-2	24-Aug-57	4	10
AR WHITE	1.25	PB18-9	15-Oct-57	3	10
AR WOODRUFF	3.06	UpK2-7	30-Mar-53	9	7
AR WOODRUFF	21.07	UpK2-8	31-Mar-53	3	7
AR WOODRUFF	3.67	UpK3-6	5-Apr-53	1	7
AR WOODRUFF	6.18	UpK8-3	10-May-53	2	7
AR WOODRUFF	0.89	UpK8-5	12-May-53	6	7
AR WOODRUFF	11.58	TP11-6	20-May-55	1	7
AR WOODRUFF	13.72	PB9-3	25-Aug-57	6	7
AR YELL	2.73	UpK3-5	4-Apr-53	8	8
AR YELL	0.81	UpK3-6	5-Apr-53	9	8
AR YELL	1.07	UpK5-4	14-Apr-53	8	8
AR YELL	1.12	UpK8-4	11-May-53	9	8

COUNTY	uCi/ sqm	SHOT DAY	DATE	RANK FOR DAY	TIMES IN TOP 10
AR YELL	1.16	UpK8-5	12-May-53	5	8
AR YELL	9.54	TP11-6	20-May-55	8	8
AR YELL	2.73	PB9-2	24-Aug-57	6	8
AR YELL	1.93	PB18-8	14-Oct-57	9	8

COUNTY	uCi/ sqm	SHOT DAY	DATE	RANK FOR DAY	TIMES IN TOP 10
AZ APACHE	4.43	TS4-1	1-May-52	8	10
AZ APACHE	4.43	TS4-1	1-May-52	8	10
AZ APACHE	27.72	UpK1-1	17-Mar-53	9	10
AZ APACHE	306.51	UpK6-1	18-Apr-53	4	10
AZ APACHE	515.70	UpK7-1	25-Apr-53	3	10
AZ APACHE	33.12	UpK11-1	4-Jun-53	9	10
AZ APACHE	26.77	TP5-1	12-Mar-55	10	10
AZ APACHE	32.07	PB3-1	24-Jun-57	7	10
AZ APACHE	9.40	PB13-3	8-Sep-57	6	10
AZ APACHE	9.94	PB18-1	7-Oct-57	9	10
AZ COCONINO1	4.86	TS4-1	1-May-52	6	8
AZ COCONINO1	4.86	TS4-1	1-May-52	6	8
AZ COCONINO1	1522.25	UpK9-1	19-May-53	9	8
AZ COCONINO1	44.82	UpK11-1	4-Jun-53	4	8
AZ COCONINO1	44.11	TP5-1	12-Mar-55	4	8
AZ COCONINO1	47.30	TP6-1	22-Mar-55	5	8
AZ COCONINO1	33.75	TP7-1	29-Mar-55	10	8
AZ COCONINO1	4.97	UE13-1	25-Apr-66	9	8
AZ COCONINO2	5.86	TS4-1	1-May-52	5	8
AZ COCONINO2	5.86	TS4-1	1-May-52	5	8
AZ COCONINO2	10.24	UpK3-1	31-Mar-53	10	8
AZ COCONINO2	315.49	UpK6-1	18-Apr-53	1	8
AZ COCONINO2	875.09	UpK7-1	25-Apr-53	2	8
AZ COCONINO2	115.24	UpK11-1	4-Jun-53	1	8
AZ COCONINO2	45.25	TP5-1	12-Mar-55	2	8
AZ COCONINO2	25.64	PB13-3	8-Sep-57	2	8
AZ COCONINO3	4.31	TS4-1	1-May-52	9	5
AZ COCONINO3	4.31	TS4-1	1-May-52	9	5
AZ COCONINO3	5.12	UpK5-1	11-Apr-53	10	5
AZ COCONINO3	42.48	TP5-1	12-Mar-55	5	5
AZ COCONINO3	25.64	PB13-3	8-Sep-57	2	5
AZ GILA	18.78	UpK3-1	31-Mar-53	4	1
AZ MARICOPA	25.62	UpK3-1	31-Mar-53	2	2
AZ MARICOPA	17.51	TP6-1	22-Mar-55	10	2
AZ MOHAVE1	6.25	TS4-1	1-May-52	4	6

COUNTY	uCi/ sqm	SHOT DAY	DATE	RANK FOR DAY	TIMES IN TOP 10
AZ MOHAVE1	6.25	TS4-1	1-May-52	4	6
AZ MOHAVE1	29.87	TP3-1	1-Mar-55	10	6
AZ MOHAVE1	44.62	TP5-1	12-Mar-55	3	6
AZ MOHAVE1	63.42	TP6-1	22-Mar-55	3	6
AZ MOHAVE1	33.75	TP7-1	29-Mar-55	10	6
AZ MOHAVE2	6.25	TS4-1	1-May-52	4	10
AZ MOHAVE2	6.25	TS4-1	1-May-52	4	10
AZ MOHAVE2	15.93	UpK3-1	31-Mar-53	5	10
AZ MOHAVE2	1986.84	UpK9-1	19-May-53	6	10
AZ MOHAVE2	34.93	UpK11-1	4-Jun-53	7	10
AZ MOHAVE2	30.07	TP3-1	1-Mar-55	9	10
AZ MOHAVE2	114.49	TP6-1	22-Mar-55	2	10
AZ MOHAVE2	207.52	TP11-1	15-May-55	3	10
AZ MOHAVE2	230.98	PB11-1	31-Aug-57	10	10
AZ MOHAVE2	6.48	PB13-3	8-Sep-57	9	10
AZ MOHAVE3	6.41	TS4-1	1-May-52	3	7
AZ MOHAVE3	6.41	TS4-1	1-May-52	3	7
AZ MOHAVE3	11.23	UpK3-1	31-Mar-53	9	7
AZ MOHAVE3	10.24	UpK5-1	11-Apr-53	3	7
AZ MOHAVE3	21.31	TP6-1	22-Mar-55	8	7
AZ MOHAVE3	88.60	TP11-1	15-May-55	8	7
AZ MOHAVE3	6.48	PB13-3	8-Sep-57	10	7
AZ MOHAVE4	6.41	TS4-1	1-May-52	3	7
AZ MOHAVE4	6.41	TS4-1	1-May-52	3	7
AZ MOHAVE4	11.23	UpK3-1	31-Mar-53	9	7
AZ MOHAVE4	10.24	UpK5-1	11-Apr-53	3	7
AZ MOHAVE4	21.31	TP6-1	22-Mar-55	9	7
AZ MOHAVE4	88.60	TP11-1	15-May-55	8	7
AZ MOHAVE4	6.48	PB13-3	8-Sep-57	10	7
AZ NAVAJO	4.57	TS4-1	1-May-52	7	8
AZ NAVAJO	4.57	TS4-1	1-May-52	7	8
AZ NAVAJO	308.72	UpK6-1	18-Apr-53	3	8
AZ NAVAJO	348.78	UpK7-1	25-Apr-53	6	8
AZ NAVAJO	38.55	UpK11-1	4-Jun-53	5	8
AZ NAVAJO	35.80	TP5-1	12-Mar-55	6	8

COUNTY	uCi/ sqm	SHOT DAY	DATE	RANK FOR DAY	TIMES IN TOP 10
AZ NAVAJO	12.68	PB13-3	8-Sep-57	5	8
AZ NAVAJO	9.94	PB18-1	7-Oct-57	9	8
AZ PIMA	0.85	TP1-1	18-Feb-55	4	3
AZ PIMA	148.41	PB4-1	5-Jul-57	1	3
AZ PIMA	148.41	PB4-1	5-Jul-57	1	3
AZ PINAL	14.08	UpK3-1	31-Mar-53	7	1
AZ SANTA CRUZ	0.90	TP1-1	18-Feb-55	3	3
AZ SANTA CRUZ	98.28	PB4-1	5-Jul-57	2	3
AZ SANTA CRUZ	98.28	PB4-1	5-Jul-57	2	3
AZ YAVAPAI	14.51	UpK3-1	31-Mar-53	6	3
AZ YAVAPAI	5.56	UpK5-1	11-Apr-53	6	3
AZ YAVAPAI	23.82	TP6-1	22-Mar-55	7	3
AZ YUMA	46.04	UpK5-1	11-Apr-53	1	2
AZ YUMA	68.08	TP1-1	18-Feb-55	2	2

COUNTY	uCi/ sqm	SHOT DAY	DATE	RANK FOR DAY	TIMES IN TOP 10
CA ALPINE	1.59	TS7-2	2-Jun-52	8	4
CA ALPINE	14.49	UpK6-2	19-Apr-53	7	4
CA ALPINE	1.15	TP6-6	27-Mar-55	4	4
CA ALPINE	1.36	BN-2	19-Dec-70	4	4
CA AMADOR	1.49	TS7-2	2-Jun-52	9	3
CA AMADOR	1.71	UpK6-3	20-Apr-53	7	3
CA AMADOR	1.15	TP6-6	27-Mar-55	4	3
CA BUTTE	1.15	TP6-6	27-Mar-55	4	2
CA BUTTE	1.14	BN-2	19-Dec-70	5	2
CA CALAVERAS	1.15	TP6-6	27-Mar-55	4	1
CA DEL NORTE	1.24	TP6-6	27-Mar-55	3	3
CA DEL NORTE	2.47	PB16-3	25-Sep-57	2	3
CA DEL NORTE	1.29	PB16-4	26-Sep-57	9	3
CA EL DORADO	1.59	TS7-2	2-Jun-52	8	2
CA EL DORADO	1.78	TP6-6	27-Mar-55	1	2
CA FRESNO	12.11	BJ1-1	28-Oct-51	1	2
CA FRESNO	2.19	BJ2-1	30-Oct-51	3	2
CA HUMBOLDT	2.47	PB16-3	25-Sep-57	2	2
CA HUMBOLDT	1.29	PB16-4	26-Sep-57	9	2
CA INYO1	31.32	PB2-1	18-Jun-57	3	2
CA INYO1	88.07	PB6-1	24-Jul-57	5	2
CA INYO2	31.32	PB2-1	18-Jun-57	3	2
CA INYO2	88.07	PB6-1	24-Jul-57	5	2
CA INYO3	72.80	BJ2-1	30-Oct-51	1	5
CA INYO3	100.72	TS7-1	1-Jun-52	6	5
CA INYO3	5.42	UpK5-1	11-Apr-53	8	5
CA INYO3	355.04	TP4-1	7-Mar-55	1	5
CA INYO3	898.07	PB11-1	31-Aug-57	2	5
CA KERN	12.11	BJ1-1	28-Oct-51	1	2
CA KERN	2.19	BJ2-1	30-Oct-51	3	2
CA KINGS	12.11	BJ1-1	28-Oct-51	1	2
CA KINGS	2.19	BJ2-1	30-Oct-51	3	2
CA LAKE	1.15	TP6-6	27-Mar-55	4	1
CA LASSEN	6.98	BN-2	19-Dec-70	1	1
CA LOS ANGELES	71.77	BJ2-1	30-Oct-51	2	1

COUNTY	uCi/ sqm	SHOT DAY	DATE	RANK FOR DAY	TIMES IN TOP 10
CA MODOC	0.98	TP6-6	27-Mar-55	5	1
CA MONO	46.65	PB6-1	24-Jul-57	8	4
CA MONO	9.07	PB13-1	6-Sep-57	10	4
CA MONO	23.62	PB14-1	14-Sep-57	3	4
CA MONO	48.97	PB16-1	23-Sep-57	7	4
CA MONTEREY	12.11	BJ1-1	28-Oct-51	1	2
CA MONTEREY	2.19	BJ2-1	30-Oct-51	3	2
CA NEVADA	1.31	BJ5-2	20-Nov-51	5	5
CA NEVADA	1.49	TS7-2	2-Jun-52	9	5
CA NEVADA	9.66	UpK6-2	19-Apr-53	8	5
CA NEVADA	1.78	TP6-6	27-Mar-55	1	5
CA NEVADA	4.52	BN-2	19-Dec-70	2	5
CA PLACER	1.59	TS7-2	2-Jun-52	8	4
CA PLACER	1.71	UpK6-3	20-Apr-53	7	4
CA PLACER	1.78	TP6-6	27-Mar-55	1	4
CA PLACER	4.52	BN-2	19-Dec-70	2	4
CA PLUMAS	1.78	TP6-6	27-Mar-55	1	1
CA SACRAMENTO	1.42	TS7-2	2-Jun-52	10	1
CA SAN BENITO	12.11	BJ1-1	28-Oct-51	1	2
CA SAN BENITO	2.19	BJ2-1	30-Oct-51	3	2
CA SAN BERNADIN	11.23	UpK3-1	31-Mar-53	9	3
CA SAN BERNADIN	22.71	UpK5-1	11-Apr-53	2	3
CA SAN BERNADIN	107.56	TP1-1	18-Feb-55	1	3
CA SAN LUIS OBI	12.11	BJ1-1	28-Oct-51	1	2
CA SAN LUIS OBI	2.19	BJ2-1	30-Oct-51	3	2
CA SANTA BARBAR	5.03	BJ1-1	28-Oct-51	2	2
CA SANTA BARBAR	0.91	BJ2-1	30-Oct-51	4	2
CA SHASTA	1.42	TP6-6	27-Mar-55	2	3
CA SHASTA	0.99	PB16-3	25-Sep-57	4	3
CA SHASTA	1.29	PB16-4	26-Sep-57	9	3
CA SIERRA	1.59	TS7-2	2-Jun-52	8	4
CA SIERRA	14.49	UpK6-2	19-Apr-53	7	4
CA SIERRA	1.78	TP6-6	27-Mar-55	1	4
CA SIERRA	6.98	BN-2	19-Dec-70	1	4
CA SISKIYOU	0.98	TP6-6	27-Mar-55	5	3

COUNTY	uCi/ sqm	SHOT DAY	DATE	RANK FOR DAY	TIMES IN TOP 10
CA SISKIYOU	1.48	PB16-3	25-Sep-57	3	3
CA SISKIYOU	1.29	PB16-4	26-Sep-57	9	3
CA TEHAMA	1.78	TP6-6	27-Mar-55	1	1
CA TRINITY	1.48	PB16-3	25-Sep-57	3	2
CA TRINITY	1.29	PB16-4	26-Sep-57	9	2
CA TULARE	12.11	BJ1-1	28-Oct-51	1	2
CA TULARE	2.19	BJ2-1	30-Oct-51	3	2
CA VENTURA	12.11	BJ1-1	28-Oct-51	1	2
CA VENTURA	2.19	BJ2-1	30-Oct-51	3	2
CA YUBA	1.78	TP6-6	27-Mar-55	1	2
CA YUBA	1.64	BN-2	19-Dec-70	3	2

COUNTY	uCi/ sqm	SHOT DAY	DATE	RANK FOR DAY	TIMES IN TOP 10
CO ADAMS	1.01	TS6-5	29-May-52	7	15
CO ADAMS	40.29	TP3-2	2-Mar-55	1	15
CO ADAMS	8.90	TP3-3	3-Mar-55	1	15
CO ADAMS	150.71	TP4-2	8-Mar-55	1	15
CO ADAMS	53.28	TP10-2	6-May-55	3	15
CO ADAMS	2.37	PB4-2	6-Jul-57	6	15
CO ADAMS	4.92	PB8-4	21-Aug-57	7	15
CO ADAMS	4.92	PB8-4	21-Aug-57	7	15
CO ADAMS	4.79	PB11-2	1-Sep-57	8	15
CO ADAMS	17.86	PB12-3	4-Sep-57	9	15
CO ADAMS	2.20	PB13-8	13-Sep-57	3	15
CO ADAMS	3.51	PB18-2	8-Oct-57	4	15
CO ADAMS	2.37	PB18-3	9-Oct-57	3	15
CO ADAMS	1.70	PB18-4	10-Oct-57	5	15
CO ADAMS	0.86	PB18-6	12-Oct-57	7	15
CO ALAMOSA	1.09	TS5-12	18-May-52	2	9
CO ALAMOSA	4.00	TP3-2	2-Mar-55	5	9
CO ALAMOSA	22.88	TP4-2	8-Mar-55	6	9
CO ALAMOSA	59.98	TP10-2	6-May-55	3	9
CO ALAMOSA	13.03	TP11-3	17-May-55	8	9
CO ALAMOSA	1.84	PB4-2	6-Jul-57	9	9
CO ALAMOSA	1.73	PB4-10	14-Jul-57	9	9
CO ALAMOSA	4.92	PB8-4	21-Aug-57	7	9
CO ALAMOSA	4.92	PB8-4	21-Aug-57	7	9
CO ARAPAHOE	40.29	TP3-2	2-Mar-55	1	13
CO ARAPAHOE	4.45	TP3-3	3-Mar-55	3	13
CO ARAPAHOE	150.71	TP4-2	8-Mar-55	1	13
CO ARAPAHOE	53.28	TP10-2	6-May-55	3	13
CO ARAPAHOE	9.77	PB8-4	21-Aug-57	3	13
CO ARAPAHOE	9.77	PB8-4	21-Aug-57	3	13
CO ARAPAHOE	4.79	PB11-2	1-Sep-57	8	13
CO ARAPAHOE	3.31	PB13-8	13-Sep-57	2	13
CO ARAPAHOE	4.60	PB18-2	8-Oct-57	2	13
CO ARAPAHOE	2.37	PB18-3	9-Oct-57	3	13
CO ARAPAHOE	1.70	PB18-4	10-Oct-57	5	13

COUNTY	uCi/ sqm	SHOT DAY	DATE	RANK FOR DAY	TIMES IN TOP 10
CO ARAPAHOE	0.86	PB18-6	12-Oct-57	7	13
CO ARAPAHOE	0.10	UE2-7	11-Mar-62	3	13
CO ARCHULETA	20.32	R-1-1	28-Jan-51	3	14
CO ARCHULETA	1.09	TS5-12	18-May-52	2	14
CO ARCHULETA	18.14	TS6-2	26-May-52	5	14
CO ARCHULETA	9.16	TS6-4	28-May-52	1	14
CO ARCHULETA	1.05	UpK3-3	2-Apr-53	7	14
CO ARCHULETA	1.65	UpK4-1	6-Apr-53	6	14
CO ARCHULETA	185.53	UpK6-1	18-Apr-53	8	14
CO ARCHULETA	20.82	UpK6-2	19-Apr-53	5	14
CO ARCHULETA	2.16	UpK6-5	22-Apr-53	1	14
CO ARCHULETA	64.30	UpK7-3	27-Apr-53	2	14
CO ARCHULETA	1.68	TP6-3	24-Mar-55	9	14
CO ARCHULETA	11.19	PB3-2	25-Jun-57	2	14
CO ARCHULETA	9.99	PB13-4	9-Sep-57	1	14
CO ARCHULETA	0.88	PB18-5	11-Oct-57	4	14
CO BACA	0.87	TS4-3	3-May-52	1	14
CO BACA	0.87	TS4-3	3-May-52	1	14
CO BACA	20.82	UpK6-2	19-Apr-53	5	14
CO BACA	7.78	TP3-2	2-Mar-55	3	14
CO BACA	49.44	TP4-2	8-Mar-55	4	14
CO BACA	8.01	TP7-2	30-Mar-55	7	14
CO BACA	32.33	TP10-2	6-May-55	5	14
CO BACA	0.98	PB1-12	8-Jun-57	1	14
CO BACA	4.79	PB11-2	1-Sep-57	8	14
CO BACA	2.20	PB13-8	13-Sep-57	3	14
CO BACA	4.60	PB18-2	8-Oct-57	2	14
CO BACA	4.73	PB18-3	9-Oct-57	1	14
CO BACA	1.70	PB18-4	10-Oct-57	5	14
CO BACA	0.10	UE2-7	11-Mar-62	4	14
CO BENT	0.87	TS4-3	3-May-52	1	13
CO BENT	0.87	TS4-3	3-May-52	1	13
CO BENT	4.00	TP3-2	2-Mar-55	5	13
CO BENT	22.88	TP4-2	8-Mar-55	6	13
CO BENT	59.98	TP10-2	6-May-55	3	13

COUNTY	uCi/ sqm	SHOT DAY	DATE	RANK FOR DAY	TIMES IN TOP 10
CO BENT	1.18	PB1-10	6-Jun-57	2	13
CO BENT	0.89	PB1-11	7-Jun-57	4	13
CO BENT	4.79	PB11-2	1-Sep-57	8	13
CO BENT	2.20	PB13-8	13-Sep-57	3	13
CO BENT	2.30	PB18-2	8-Oct-57	5	13
CO BENT	2.37	PB18-3	9-Oct-57	3	13
CO BENT	1.70	PB18-4	10-Oct-57	5	13
CO BENT	0.86	PB18-6	12-Oct-57	7	13
CO BOULDER	40.29	TP3-2	2-Mar-55	1	17
CO BOULDER	4.45	TP3-3	3-Mar-55	3	17
CO BOULDER	150.71	TP4-2	8-Mar-55	1	17
CO BOULDER	53.28	TP10-2	6-May-55	3	17
CO BOULDER	15.86	TP11-2	16-May-55	6	17
CO BOULDER	14.60	TP11-3	17-May-55	6	17
CO BOULDER	2.37	PB4-2	6-Jul-57	6	17
CO BOULDER	4.92	PB8-4	21-Aug-57	7	17
CO BOULDER	4.92	PB8-4	21-Aug-57	7	17
CO BOULDER	4.79	PB11-2	1-Sep-57	8	17
CO BOULDER	3.31	PB13-8	13-Sep-57	2	17
CO BOULDER	4.60	PB18-2	8-Oct-57	2	17
CO BOULDER	2.37	PB18-3	9-Oct-57	3	17
CO BOULDER	1.70	PB18-4	10-Oct-57	5	17
CO BOULDER	2.54	PB18-6	12-Oct-57	2	17
CO BOULDER	0.10	UE2-7	11-Mar-62	3	17
CO BOULDER	0.82	BN-2	19-Dec-70	7	17
CO CHAFFEE	17.18	TS6-2	26-May-52	9	11
CO CHAFFEE	4.94	TP3-2	2-Mar-55	4	11
CO CHAFFEE	42.87	TP4-2	8-Mar-55	5	11
CO CHAFFEE	39.24	TP10-2	6-May-55	4	11
CO CHAFFEE	11.97	TP11-3	17-May-55	9	11
CO CHAFFEE	4.79	PB11-2	1-Sep-57	8	11
CO CHAFFEE	1.12	PB13-8	13-Sep-57	7	11
CO CHAFFEE	1.14	PB15-4	19-Sep-57	6	11
CO CHAFFEE	2.30	PB18-2	8-Oct-57	5	11
CO CHAFFEE	2.37	PB18-3	9-Oct-57	3	11

COUNTY	uCi/ sqm	SHOT DAY	DATE	RANK FOR DAY	TIMES IN TOP 10
CO CHAFFEE	1.70	PB18-4	10-Oct-57	5	11
CO CHEYENNE	0.87	TS4-3	3-May-52	1	14
CO CHEYENNE	0.87	TS4-3	3-May-52	1	14
CO CHEYENNE	1.31	UpK6-3	20-Apr-53	9	14
CO CHEYENNE	1.32	UpK8-3	10-May-53	9	14
CO CHEYENNE	7.78	TP3-2	2-Mar-55	3	14
CO CHEYENNE	49.44	TP4-2	8-Mar-55	4	14
CO CHEYENNE	32.33	TP10-2	6-May-55	5	14
CO CHEYENNE	0.83	PB3-8	1-Jul-57	5	14
CO CHEYENNE	4.79	PB11-2	1-Sep-57	8	14
CO CHEYENNE	2.20	PB13-8	13-Sep-57	3	14
CO CHEYENNE	3.51	PB18-2	8-Oct-57	4	14
CO CHEYENNE	4.73	PB18-3	9-Oct-57	1	14
CO CHEYENNE	3.41	PB18-4	10-Oct-57	1	14
CO CHEYENNE	1.68	PB18-6	12-Oct-57	4	14
CO CLEAR CREEK	40.29	TP3-2	2-Mar-55	1	18
CO CLEAR CREEK	4.45	TP3-3	3-Mar-55	3	18
CO CLEAR CREEK	150.71	TP4-2	8-Mar-55	1	18
CO CLEAR CREEK	53.28	TP10-2	6-May-55	3	18
CO CLEAR CREEK	15.86	TP11-2	16-May-55	6	18
CO CLEAR CREEK	10.05	PB2-3	20-Jun-57	10	18
CO CLEAR CREEK	2.39	PB4-9	13-Jul-57	3	18
CO CLEAR CREEK	4.56	PB5-4	18-Jul-57	9	18
CO CLEAR CREEK	4.92	PB8-4	21-Aug-57	7	18
CO CLEAR CREEK	4.92	PB8-4	21-Aug-57	7	18
CO CLEAR CREEK	0.82	PB9-8	30-Aug-57	3	18
CO CLEAR CREEK	4.79	PB11-2	1-Sep-57	8	18
CO CLEAR CREEK	3.31	PB13-8	13-Sep-57	2	18
CO CLEAR CREEK	2.30	PB18-2	8-Oct-57	5	18
CO CLEAR CREEK	2.37	PB18-3	9-Oct-57	3	18
CO CLEAR CREEK	1.70	PB18-4	10-Oct-57	5	18
CO CLEAR CREEK	2.54	PB18-6	12-Oct-57	2	18
CO CLEAR CREEK	0.23	UE2-5	9-Mar-62	6	18
CO CONEJOS	20.32	R-1-1	28-Jan-51	3	11
CO CONEJOS	1.09	TS5-12	18-May-52	2	11

COUNTY	uCi/ sqm	SHOT DAY	DATE	RANK FOR DAY	TIMES IN TOP 10
CO CONEJOS	20.82	UpK6-2	19-Apr-53	5	11
CO CONEJOS	1437.92	UpK9-1	19-May-53	10	11
CO CONEJOS	4.00	TP3-2	2-Mar-55	5	11
CO CONEJOS	34.32	TP4-2	8-Mar-55	6	11
CO CONEJOS	89.98	TP10-2	6-May-55	2	11
CO CONEJOS	11.19	PB3-2	25-Jun-57	2	11
CO CONEJOS	1.73	PB4-10	14-Jul-57	9	11
CO CONEJOS	9.99	PB13-4	9-Sep-57	1	11
CO CONEJOS	0.88	PB18-5	11-Oct-57	4	11
CO COSTILLA	1.09	TS5-12	18-May-52	2	7
CO COSTILLA	20.82	UpK6-2	19-Apr-53	5	7
CO COSTILLA	4.00	TP3-2	2-Mar-55	5	7
CO COSTILLA	22.88	TP4-2	8-Mar-55	6	7
CO COSTILLA	119.97	TP10-2	6-May-55	1	7
CO COSTILLA	11.19	PB3-2	25-Jun-57	2	7
CO COSTILLA	9.99	PB13-4	9-Sep-57	1	7
CO CROWLEY	0.87	TS4-3	3-May-52	1	13
CO CROWLEY	0.87	TS4-3	3-May-52	1	13
CO CROWLEY	4.00	TP3-2	2-Mar-55	5	13
CO CROWLEY	22.88	TP4-2	8-Mar-55	6	13
CO CROWLEY	59.98	TP10-2	6-May-55	3	13
CO CROWLEY	1.77	PB1-10	6-Jun-57	1	13
CO CROWLEY	4.79	PB11-2	1-Sep-57	8	13
CO CROWLEY	3.31	PB13-8	13-Sep-57	2	13
CO CROWLEY	2.30	PB18-2	8-Oct-57	5	13
CO CROWLEY	2.37	PB18-3	9-Oct-57	3	13
CO CROWLEY	1.70	PB18-4	10-Oct-57	5	13
CO CROWLEY	0.86	PB18-6	12-Oct-57	7	13
CO CROWLEY	0.10	UE2-7	11-Mar-62	4	13
CO CUSTER	0.83	TS5-11	17-May-52	4	13
CO CUSTER	1.09	TS5-12	18-May-52	2	13
CO CUSTER	16.42	TS6-2	26-May-52	10	13
CO CUSTER	4.00	TP3-2	2-Mar-55	5	13
CO CUSTER	22.88	TP4-2	8-Mar-55	6	13
CO CUSTER	59.98	TP10-2	6-May-55	3	13

COUNTY	uCi/ sqm	SHOT DAY	DATE	RANK FOR DAY	TIMES IN TOP 10
CO CUSTER	19.31	TP11-3	17-May-55	2	13
CO CUSTER	3.46	PB4-10	14-Jul-57	2	13
CO CUSTER	4.79	PB11-2	1-Sep-57	8	13
CO CUSTER	1.12	PB13-8	13-Sep-57	7	13
CO CUSTER	4.60	PB18-2	8-Oct-57	2	13
CO CUSTER	2.37	PB18-3	9-Oct-57	3	13
CO CUSTER	1.70	PB18-4	10-Oct-57	5	13
CO DELTA	33.45	UpK10-1	25-May-53	5	2
CO DELTA	334.56	TP9-1	15-Apr-55	5	2
CO DENVER	16.73	TP3-2	2-Mar-55	2	12
CO DENVER	1.85	TP3-3	3-Mar-55	7	12
CO DENVER	62.59	TP4-2	8-Mar-55	3	12
CO DENVER	22.13	TP10-2	6-May-55	10	12
CO DENVER	13.17	TP11-2	16-May-55	9	12
CO DENVER	6.10	PB8-4	21-Aug-57	5	12
CO DENVER	6.10	PB8-4	21-Aug-57	5	12
CO DENVER	21.92	PB12-3	4-Sep-57	3	12
CO DENVER	1.38	PB13-8	13-Sep-57	5	12
CO DENVER	3.82	PB18-2	8-Oct-57	3	12
CO DENVER	0.98	PB18-3	9-Oct-57	4	12
CO DENVER	0.86	PB18-6	12-Oct-57	7	12
CO DOUGLAS	40.29	TP3-2	2-Mar-55	1	16
CO DOUGLAS	4.45	TP3-3	3-Mar-55	3	16
CO DOUGLAS	150.71	TP4-2	8-Mar-55	1	16
CO DOUGLAS	53.28	TP10-2	6-May-55	3	16
CO DOUGLAS	13.10	TP11-3	17-May-55	7	16
CO DOUGLAS	2.98	PB4-9	13-Jul-57	2	16
CO DOUGLAS	4.92	PB8-4	21-Aug-57	7	16
CO DOUGLAS	4.92	PB8-4	21-Aug-57	7	16
CO DOUGLAS	4.79	PB11-2	1-Sep-57	8	16
CO DOUGLAS	18.16	PB12-3	4-Sep-57	8	16
CO DOUGLAS	3.31	PB13-8	13-Sep-57	2	16
CO DOUGLAS	2.30	PB18-2	8-Oct-57	5	16
CO DOUGLAS	2.37	PB18-3	9-Oct-57	3	16
CO DOUGLAS	1.70	PB18-4	10-Oct-57	5	16

COUNTY	uCi/ sqm	SHOT DAY	DATE	RANK FOR DAY	TIMES IN TOP 10
CO DOUGLAS	0.86	PB18-6	12-Oct-57	7	16
CO DOUGLAS	0.10	UE2-7	11-Mar-62	3	16
CO EAGLE	40.29	TP3-2	2-Mar-55	1	14
CO EAGLE	4.45	TP3-3	3-Mar-55	3	14
CO EAGLE	150.71	TP4-2	8-Mar-55	1	14
CO EAGLE	53.28	TP10-2	6-May-55	3	14
CO EAGLE	15.86	TP11-2	16-May-55	6	14
CO EAGLE	0.94	PB3-10	3-Jul-57	1	14
CO EAGLE	4.73	PB4-2	6-Jul-57	2	14
CO EAGLE	4.92	PB8-4	21-Aug-57	7	14
CO EAGLE	4.92	PB8-4	21-Aug-57	7	14
CO EAGLE	4.79	PB11-2	1-Sep-57	8	14
CO EAGLE	4.43	PB15-4	19-Sep-57	1	14
CO EAGLE	2.30	PB18-2	8-Oct-57	5	14
CO EAGLE	2.37	PB18-3	9-Oct-57	3	14
CO EAGLE	1.70	PB18-4	10-Oct-57	5	14
CO EL PASO	0.87	TS4-3	3-May-52	1	16
CO EL PASO	0.87	TS4-3	3-May-52	1	16
CO EL PASO	2.05	TP3-2	2-Mar-55	8	16
CO EL PASO	17.81	TP4-2	8-Mar-55	8	16
CO EL PASO	1.84	PB4-2	6-Jul-57	9	16
CO EL PASO	1.73	PB4-10	14-Jul-57	9	16
CO EL PASO	4.92	PB8-4	21-Aug-57	7	16
CO EL PASO	4.92	PB8-4	21-Aug-57	7	16
CO EL PASO	4.79	PB11-2	1-Sep-57	8	16
CO EL PASO	19.16	PB12-3	4-Sep-57	5	16
CO EL PASO	3.31	PB13-8	13-Sep-57	2	16
CO EL PASO	9.19	PB18-2	8-Oct-57	1	16
CO EL PASO	4.73	PB18-3	9-Oct-57	1	16
CO EL PASO	1.70	PB18-4	10-Oct-57	5	16
CO EL PASO	1.68	PB18-6	12-Oct-57	4	16
CO EL PASO	0.10	UE2-7	11-Mar-62	4	16
CO ELBERT	0.87	TS4-3	3-May-52	1	16
CO ELBERT	0.87	TS4-3	3-May-52	1	16
CO ELBERT	4.94	TP3-2	2-Mar-55	4	16

COUNTY	uCi/ sqm	SHOT DAY	DATE	RANK FOR DAY	TIMES IN TOP 10
CO ELBERT	42.87	TP4-2	8-Mar-55	5	16
CO ELBERT	39.24	TP10-2	6-May-55	4	16
CO ELBERT	2.37	PB4-2	6-Jul-57	6	16
CO ELBERT	1.73	PB4-10	14-Jul-57	9	16
CO ELBERT	9.77	PB8-4	21-Aug-57	3	16
CO ELBERT	9.77	PB8-4	21-Aug-57	3	16
CO ELBERT	4.79	PB11-2	1-Sep-57	8	16
CO ELBERT	3.31	PB13-8	13-Sep-57	2	16
CO ELBERT	2.30	PB18-2	8-Oct-57	5	16
CO ELBERT	2.37	PB18-3	9-Oct-57	3	16
CO ELBERT	1.70	PB18-4	10-Oct-57	5	16
CO ELBERT	0.86	PB18-6	12-Oct-57	7	16
CO ELBERT	0.10	UE2-7	11-Mar-62	3	16
CO FREMONT	4.94	TP3-2	2-Mar-55	4	12
CO FREMONT	42.87	TP4-2	8-Mar-55	5	12
CO FREMONT	39.24	TP10-2	6-May-55	4	12
CO FREMONT	14.93	TP11-3	17-May-55	5	12
CO FREMONT	1.18	PB1-10	6-Jun-57	2	12
CO FREMONT	5.20	PB4-10	14-Jul-57	1	12
CO FREMONT	4.79	PB11-2	1-Sep-57	8	12
CO FREMONT	2.20	PB13-8	13-Sep-57	3	12
CO FREMONT	4.60	PB18-2	8-Oct-57	2	12
CO FREMONT	2.37	PB18-3	9-Oct-57	3	12
CO FREMONT	1.70	PB18-4	10-Oct-57	5	12
CO FREMONT	1.68	PB18-6	12-Oct-57	4	12
CO GARFIELD	334.56	TP9-1	15-Apr-55	5	3
CO GARFIELD	81.66	PB4-1	5-Jul-57	3	3
CO GARFIELD	81.66	PB4-1	5-Jul-57	3	3
CO GILPIN	40.29	TP3-2	2-Mar-55	1	14
CO GILPIN	4.45	TP3-3	3-Mar-55	3	14
CO GILPIN	150.71	TP4-2	8-Mar-55	1	14
CO GILPIN	53.28	TP10-2	6-May-55	3	14
CO GILPIN	15.86	TP11-2	16-May-55	6	14
CO GILPIN	2.37	PB4-2	6-Jul-57	6	14
CO GILPIN	1.64	PB4-9	13-Jul-57	10	14

COUNTY	uCi/ sqm	SHOT DAY	DATE	RANK FOR DAY	TIMES IN TOP 10
CO GILPIN	4.56	PB5-4	18-Jul-57	9	14
CO GILPIN	4.79	PB11-2	1-Sep-57	8	14
CO GILPIN	3.31	PB13-8	13-Sep-57	2	14
CO GILPIN	2.30	PB18-2	8-Oct-57	5	14
CO GILPIN	2.37	PB18-3	9-Oct-57	3	14
CO GILPIN	1.70	PB18-4	10-Oct-57	5	14
CO GILPIN	2.54	PB18-6	12-Oct-57	2	14
CO GRAND	7.88	UpK11-2	5-Jun-53	5	24
CO GRAND	40.29	TP3-2	2-Mar-55	1	24
CO GRAND	4.45	TP3-3	3-Mar-55	3	24
CO GRAND	150.71	TP4-2	8-Mar-55	1	24
CO GRAND	6.57	TP4-3	9-Mar-55	10	24
CO GRAND	53.28	TP10-2	6-May-55	3	24
CO GRAND	31.71	TP11-2	16-May-55	3	24
CO GRAND	10.05	PB2-3	20-Jun-57	10	24
CO GRAND	4.56	PB5-4	18-Jul-57	9	24
CO GRAND	9.77	PB8-4	21-Aug-57	3	24
CO GRAND	9.77	PB8-4	21-Aug-57	3	24
CO GRAND	4.79	PB11-2	1-Sep-57	8	24
CO GRAND	3.31	PB13-8	13-Sep-57	2	24
CO GRAND	2.30	PB18-2	8-Oct-57	5	24
CO GRAND	2.37	PB18-3	9-Oct-57	3	24
CO GRAND	1.70	PB18-4	10-Oct-57	5	24
CO GRAND	0.86	PB18-6	12-Oct-57	7	24
CO GRAND	0.65	UE2-4	8-Mar-62	5	24
CO GRAND	0.22	UE2-5	9-Mar-62	7	24
CO GRAND	0.11	UE2-6	10-Mar-62	6	24
CO GRAND	0.10	UE2-7	11-Mar-62	4	24
CO GRAND	18.01	UE8-1	14-Jul-62	4	24
CO GRAND	8.30	UE8-3	16-Jul-62	10	24
CO GRAND	0.82	BN-2	19-Dec-70	7	24
CO GUNNISON	24.69	TS6-2	26-May-52	2	15
CO GUNNISON	1.36	UpK5-3	13-Apr-53	2	15
CO GUNNISON	1570.04	UpK9-1	19-May-53	8	15
CO GUNNISON	11.26	UpK9-3	21-May-53	6	15

COUNTY	uCi/ sqm	SHOT DAY	DATE	RANK FOR DAY	TIMES IN TOP 10
CO GUNNISON	4.00	TP3-2	2-Mar-55	5	15
CO GUNNISON	64.56	TP4-2	8-Mar-55	2	15
CO GUNNISON	6.69	TP4-3	9-Mar-55	9	15
CO GUNNISON	3.05	TP6-3	24-Mar-55	5	15
CO GUNNISON	1.89	TP8-2	10-Apr-55	8	15
CO GUNNISON	2.90	TP10-10	14-May-55	1	15
CO GUNNISON	1.71	PB4-2	6-Jul-57	10	15
CO GUNNISON	10.34	PB5-3	17-Jul-57	2	15
CO GUNNISON	2.21	PB15-4	19-Sep-57	2	15
CO GUNNISON	1.33	PB18-2	8-Oct-57	6	15
CO GUNNISON	0.10	UE2-7	11-Mar-62	3	15
CO HINSDALE	30.27	TS6-2	26-May-52	1	12
CO HINSDALE	67.75	TS6-3	27-May-52	2	12
CO HINSDALE	1.92	UpK4-1	6-Apr-53	3	12
CO HINSDALE	12.12	UpK9-3	21-May-53	2	12
CO HINSDALE	2.44	UpK10-3	27-May-53	2	12
CO HINSDALE	4.00	TP3-2	2-Mar-55	5	12
CO HINSDALE	64.56	TP4-2	8-Mar-55	2	12
CO HINSDALE	2.90	TP10-10	14-May-55	1	12
CO HINSDALE	17.19	TP11-3	17-May-55	3	12
CO HINSDALE	11.39	PB4-5	9-Jul-57	8	12
CO HINSDALE	2.21	PB15-4	19-Sep-57	2	12
CO HINSDALE	1.33	PB18-2	8-Oct-57	6	12
CO HUERFANO	3.27	TS5-12	18-May-52	1	12
CO HUERFANO	17.61	TS6-2	26-May-52	8	12
CO HUERFANO	1.06	TS6-5	29-May-52	5	12
CO HUERFANO	4.00	TP3-2	2-Mar-55	5	12
CO HUERFANO	22.88	TP4-2	8-Mar-55	6	12
CO HUERFANO	59.98	TP10-2	6-May-55	3	12
CO HUERFANO	4.56	PB5-4	18-Jul-57	9	12
CO HUERFANO	4.79	PB11-2	1-Sep-57	8	12
CO HUERFANO	1.12	PB13-8	13-Sep-57	7	12
CO HUERFANO	4.60	PB18-2	8-Oct-57	2	12
CO HUERFANO	2.37	PB18-3	9-Oct-57	3	12
CO HUERFANO	1.70	PB18-4	10-Oct-57	5	12

COUNTY	uCi/ sqm	SHOT DAY	DATE	RANK FOR DAY	TIMES IN TOP 10
CO JACKSON	10.78	UpK11-2	5-Jun-53	2	13
CO JACKSON	3.16	TP3-2	2-Mar-55	7	13
CO JACKSON	5.32	TP8-2	10-Apr-55	1	13
CO JACKSON	24.89	TP10-2	6-May-55	9	13
CO JACKSON	7.30	PB2-2	19-Jun-57	3	13
CO JACKSON	0.83	PB8-2	19-Aug-57	8	13
CO JACKSON	3.07	PB8-4	21-Aug-57	10	13
CO JACKSON	3.07	PB8-4	21-Aug-57	10	13
CO JACKSON	2.37	PB18-3	9-Oct-57	3	13
CO JACKSON	1.44	PB18-4	10-Oct-57	6	13
CO JACKSON	0.86	PB18-6	12-Oct-57	7	13
CO JACKSON	0.21	UE2-6	10-Mar-62	3	13
CO JACKSON	0.91	BN-2	19-Dec-70	6	13
CO JEFFERSON	6.73	UpK11-2	5-Jun-53	10	18
CO JEFFERSON	40.29	TP3-2	2-Mar-55	1	18
CO JEFFERSON	4.45	TP3-3	3-Mar-55	3	18
CO JEFFERSON	150.71	TP4-2	8-Mar-55	1	18
CO JEFFERSON	53.28	TP10-2	6-May-55	3	18
CO JEFFERSON	15.02	TP11-3	17-May-55	4	18
CO JEFFERSON	2.37	PB4-2	6-Jul-57	6	18
CO JEFFERSON	4.92	PB8-4	21-Aug-57	7	18
CO JEFFERSON	4.92	PB8-4	21-Aug-57	7	18
CO JEFFERSON	4.79	PB11-2	1-Sep-57	8	18
CO JEFFERSON	3.31	PB13-8	13-Sep-57	2	18
CO JEFFERSON	3.51	PB18-2	8-Oct-57	4	18
CO JEFFERSON	2.37	PB18-3	9-Oct-57	3	18
CO JEFFERSON	1.70	PB18-4	10-Oct-57	5	18
CO JEFFERSON	1.25	PB18-5	11-Oct-57	2	18
CO JEFFERSON	0.86	PB18-6	12-Oct-57	7	18
CO JEFFERSON	0.10	UE2-7	11-Mar-62	3	18
CO JEFFERSON	8.30	UE8-3	16-Jul-62	10	18
CO KIOWA	0.87	TS4-3	3-May-52	1	17
CO KIOWA	0.87	TS4-3	3-May-52	1	17
CO KIOWA	1.31	UpK6-3	20-Apr-53	9	17
CO KIOWA	1.90	UpK6-4	21-Apr-53	2	17

COUNTY	uCi/ sqm	SHOT DAY	DATE	RANK FOR DAY	TIMES IN TOP 10
CO KIOWA	68.64	UpK7-3	27-Apr-53	1	17
CO KIOWA	7.78	TP3-2	2-Mar-55	3	17
CO KIOWA	49.44	TP4-2	8-Mar-55	4	17
CO KIOWA	32.33	TP10-2	6-May-55	5	17
CO KIOWA	1.33	PB1-11	7-Jun-57	3	17
CO KIOWA	10.05	PB2-3	20-Jun-57	10	17
CO KIOWA	1.05	PB3-9	2-Jul-57	3	17
CO KIOWA	4.79	PB11-2	1-Sep-57	8	17
CO KIOWA	3.31	PB13-8	13-Sep-57	2	17
CO KIOWA	2.30	PB18-2	8-Oct-57	5	17
CO KIOWA	2.37	PB18-3	9-Oct-57	3	17
CO KIOWA	1.70	PB18-4	10-Oct-57	5	17
CO KIOWA	0.86	PB18-6	12-Oct-57	7	17
CO KIT CARSON	1.11	BJ5-6	24-Nov-51	7	20
CO KIT CARSON	1.11	BJ5-6	24-Nov-51	7	20
CO KIT CARSON	0.87	TS4-3	3-May-52	1	20
CO KIT CARSON	0.87	TS4-3	3-May-52	1	20
CO KIT CARSON	1.31	UpK6-3	20-Apr-53	9	20
CO KIT CARSON	7.78	TP3-2	2-Mar-55	3	20
CO KIT CARSON	49.44	TP4-2	8-Mar-55	4	20
CO KIT CARSON	5.38	TP7-2	30-Mar-55	10	20
CO KIT CARSON	32.33	TP10-2	6-May-55	5	20
CO KIT CARSON	0.89	PB1-11	7-Jun-57	4	20
CO KIT CARSON	0.83	PB3-8	1-Jul-57	5	20
CO KIT CARSON	1.73	PB4-10	14-Jul-57	9	20
CO KIT CARSON	1.96	PB5-5	19-Jul-57	9	20
CO KIT CARSON	9.58	PB11-2	1-Sep-57	4	20
CO KIT CARSON	3.31	PB13-8	13-Sep-57	2	20
CO KIT CARSON	3.51	PB18-2	8-Oct-57	4	20
CO KIT CARSON	4.73	PB18-3	9-Oct-57	1	20
CO KIT CARSON	2.56	PB18-4	10-Oct-57	3	20
CO KIT CARSON	1.68	PB18-6	12-Oct-57	4	20
CO KIT CARSON	18.17	UE8-3	16-Jul-62	3	20
CO LA PLATA	179.46	R-1-1	28-Jan-51	1	4
CO LA PLATA	89.12	TS7-1	1-Jun-52	10	4

COUNTY	uCi/ sqm	SHOT DAY	DATE	RANK FOR DAY	TIMES IN TOP 10
CO LA PLATA	45.73	UpK1-1	17-Mar-53	8	4
CO LA PLATA	10.06	PB18-1	7-Oct-57	8	4
CO LAKE	40.29	TP3-2	2-Mar-55	1	16
CO LAKE	4.45	TP3-3	3-Mar-55	3	16
CO LAKE	150.71	TP4-2	8-Mar-55	1	16
CO LAKE	10.18	TP4-3	9-Mar-55	3	16
CO LAKE	53.28	TP10-2	6-May-55	3	16
CO LAKE	15.86	TP11-2	16-May-55	6	16
CO LAKE	1.73	PB4-10	14-Jul-57	9	16
CO LAKE	4.92	PB8-4	21-Aug-57	7	16
CO LAKE	4.92	PB8-4	21-Aug-57	7	16
CO LAKE	4.79	PB11-2	1-Sep-57	8	16
CO LAKE	2.20	PB13-8	13-Sep-57	3	16
CO LAKE	2.21	PB15-4	19-Sep-57	2	16
CO LAKE	2.30	PB18-2	8-Oct-57	5	16
CO LAKE	2.37	PB18-3	9-Oct-57	3	16
CO LAKE	1.70	PB18-4	10-Oct-57	5	16
CO LAKE	0.10	UE2-7	11-Mar-62	3	16
CO LARIMER	9.64	UpK11-2	5-Jun-53	3	14
CO LARIMER	3.16	TP3-2	2-Mar-55	7	14
CO LARIMER	24.89	TP10-2	6-May-55	9	14
CO LARIMER	23.13	TP11-2	16-May-55	4	14
CO LARIMER	7.30	PB2-2	19-Jun-57	3	14
CO LARIMER	1.65	PB8-2	19-Aug-57	6	14
CO LARIMER	7.76	PB9-3	25-Aug-57	9	14
CO LARIMER	3.68	PB9-4	26-Aug-57	9	14
CO LARIMER	0.83	PB9-6	28-Aug-57	8	14
CO LARIMER	2.37	PB18-3	9-Oct-57	3	14
CO LARIMER	1.44	PB18-4	10-Oct-57	6	14
CO LARIMER	0.86	PB18-6	12-Oct-57	7	14
CO LARIMER	0.11	UE2-6	10-Mar-62	6	14
CO LARIMER	0.10	UE2-7	11-Mar-62	3	14
CO LAS ANIMAS	1.02	TS4-4	4-May-52	1	15
CO LAS ANIMAS	1.02	TS4-4	4-May-52	1	15
CO LAS ANIMAS	1.30	TS4-6	6-May-52	10	15

COUNTY	uCi/ sqm	SHOT DAY	DATE	RANK FOR DAY	TIMES IN TOP 10
CO LAS ANIMAS	1.30	TS4-6	6-May-52	10	15
CO LAS ANIMAS	20.82	UpK6-2	19-Apr-53	5	15
CO LAS ANIMAS	4.00	TP3-2	2-Mar-55	5	15
CO LAS ANIMAS	22.88	TP4-2	8-Mar-55	6	15
CO LAS ANIMAS	7.27	TP7-2	30-Mar-55	8	15
CO LAS ANIMAS	59.98	TP10-2	6-May-55	3	15
CO LAS ANIMAS	1.73	PB4-10	14-Jul-57	9	15
CO LAS ANIMAS	4.56	PB5-4	18-Jul-57	9	15
CO LAS ANIMAS	4.92	PB8-4	21-Aug-57	7	15
CO LAS ANIMAS	4.92	PB8-4	21-Aug-57	7	15
CO LAS ANIMAS	1.12	PB13-8	13-Sep-57	7	15
CO LAS ANIMAS	0.88	PB18-5	11-Oct-57	4	15
CO LINCOLN	0.87	TS4-3	3-May-52	1	17
CO LINCOLN	0.87	TS4-3	3-May-52	1	17
CO LINCOLN	4.00	TP3-2	2-Mar-55	5	17
CO LINCOLN	22.88	TP4-2	8-Mar-55	6	17
CO LINCOLN	59.98	TP10-2	6-May-55	3	17
CO LINCOLN	1.19	PB1-9	5-Jun-57	4	17
CO LINCOLN	1.65	PB4-9	13-Jul-57	9	17
CO LINCOLN	1.73	PB4-10	14-Jul-57	9	17
CO LINCOLN	4.56	PB5-4	18-Jul-57	9	17
CO LINCOLN	9.77	PB8-4	21-Aug-57	3	17
CO LINCOLN	9.77	PB8-4	21-Aug-57	3	17
CO LINCOLN	4.79	PB11-2	1-Sep-57	8	17
CO LINCOLN	3.31	PB13-8	13-Sep-57	2	17
CO LINCOLN	4.60	PB18-2	8-Oct-57	2	17
CO LINCOLN	3.55	PB18-3	9-Oct-57	2	17
CO LINCOLN	1.70	PB18-4	10-Oct-57	5	17
CO LINCOLN	0.86	PB18-6	12-Oct-57	7	17
CO LOGAN	0.82	TS5-9	15-May-52	6	19
CO LOGAN	54.38	TS6-3	27-May-52	4	19
CO LOGAN	9.92	UpK2-7	30-Mar-53	5	19
CO LOGAN	4.00	TP3-2	2-Mar-55	5	19
CO LOGAN	14.24	TP11-2	16-May-55	7	19
CO LOGAN	6.07	PB5-2	16-Jul-57	10	19

COUNTY	uCi/ sqm	SHOT DAY	DATE	RANK FOR DAY	TIMES IN TOP 10
CO LOGAN	6.79	PB5-3	17-Jul-57	7	19
CO LOGAN	1.96	PB5-5	19-Jul-57	9	19
CO LOGAN	0.83	PB8-2	19-Aug-57	8	19
CO LOGAN	3.07	PB8-4	21-Aug-57	10	19
CO LOGAN	3.07	PB8-4	21-Aug-57	10	19
CO LOGAN	7.76	PB9-3	25-Aug-57	9	19
CO LOGAN	3.68	PB9-4	26-Aug-57	9	19
CO LOGAN	0.83	PB9-6	28-Aug-57	8	19
CO LOGAN	0.97	PB18-2	8-Oct-57	7	19
CO LOGAN	4.73	PB18-3	9-Oct-57	1	19
CO LOGAN	1.44	PB18-4	10-Oct-57	6	19
CO LOGAN	2.54	PB18-6	12-Oct-57	2	19
CO LOGAN	0.10	UE2-7	11-Mar-62	3	19
CO MESA	29.36	UpK10-1	25-May-53	8	2
CO MESA	395.91	TP9-1	15-Apr-55	3	2
CO MINERAL	24.26	TS6-2	26-May-52	3	11
CO MINERAL	1.65	UpK4-1	6-Apr-53	7	11
CO MINERAL	1.36	UpK5-3	13-Apr-53	2	11
CO MINERAL	41.14	UpK7-3	27-Apr-53	9	11
CO MINERAL	4.00	TP3-2	2-Mar-55	5	11
CO MINERAL	64.56	TP4-2	8-Mar-55	2	11
CO MINERAL	2.90	TP10-10	14-May-55	1	11
CO MINERAL	1.71	PB4-2	6-Jul-57	10	11
CO MINERAL	1.14	PB15-4	19-Sep-57	6	11
CO MINERAL	1.33	PB18-2	8-Oct-57	6	11
CO MINERAL	0.10	UE2-7	11-Mar-62	4	11
CO MOFFAT	240.97	TP10-1	5-May-55	7	1
CO MONTEZUMA	47.60	UpK1-1	17-Mar-53	7	2
CO MONTEZUMA	10.06	PB18-1	7-Oct-57	8	2
CO MONTROSE	133.62	TS7-1	1-Jun-52	5	2
CO MONTROSE	88.97	TP9-1	15-Apr-55	9	2
CO MORGAN	0.82	TS5-9	15-May-52	6	16
CO MORGAN	40.29	TP3-2	2-Mar-55	1	16
CO MORGAN	4.45	TP3-3	3-Mar-55	3	16
CO MORGAN	150.71	TP4-2	8-Mar-55	1	16

COUNTY	uCi/ sqm	SHOT DAY	DATE	RANK FOR DAY	TIMES IN TOP 10
CO MORGAN	53.28	TP10-2	6-May-55	3	16
CO MORGAN	47.57	TP11-2	16-May-55	1	16
CO MORGAN	4.56	PB5-4	18-Jul-57	9	16
CO MORGAN	24.46	PB8-4	21-Aug-57	1	16
CO MORGAN	24.46	PB8-4	21-Aug-57	1	16
CO MORGAN	0.88	PB9-6	28-Aug-57	7	16
CO MORGAN	4.79	PB11-2	1-Sep-57	8	16
CO MORGAN	2.20	PB13-8	13-Sep-57	3	16
CO MORGAN	2.30	PB18-2	8-Oct-57	5	16
CO MORGAN	2.37	PB18-3	9-Oct-57	3	16
CO MORGAN	1.70	PB18-4	10-Oct-57	5	16
CO MORGAN	0.86	PB18-6	12-Oct-57	7	16
CO OTERO	0.87	TS4-3	3-May-52	1	12
CO OTERO	0.87	TS4-3	3-May-52	1	12
CO OTERO	4.00	TP3-2	2-Mar-55	5	12
CO OTERO	22.88	TP4-2	8-Mar-55	6	12
CO OTERO	59.98	TP10-2	6-May-55	3	12
CO OTERO	1.19	PB1-9	5-Jun-57	4	12
CO OTERO	4.79	PB11-2	1-Sep-57	8	12
CO OTERO	2.20	PB13-8	13-Sep-57	3	12
CO OTERO	2.30	PB18-2	8-Oct-57	5	12
CO OTERO	2.37	PB18-3	9-Oct-57	3	12
CO OTERO	1.70	PB18-4	10-Oct-57	5	12
CO OTERO	0.10	UE2-7	11-Mar-62	3	12
CO PARK	4.94	TP3-2	2-Mar-55	4	10
CO PARK	42.87	TP4-2	8-Mar-55	5	10
CO PARK	39.24	TP10-2	6-May-55	4	10
CO PARK	2.37	PB4-2	6-Jul-57	6	10
CO PARK	4.79	PB11-2	1-Sep-57	8	10
CO PARK	3.31	PB13-8	13-Sep-57	2	10
CO PARK	2.30	PB18-2	8-Oct-57	5	10
CO PARK	2.37	PB18-3	9-Oct-57	3	10
CO PARK	1.70	PB18-4	10-Oct-57	5	10
CO PARK	2.54	PB18-6	12-Oct-57	2	10
CO PHILLIPS	0.87	TS4-3	3-May-52	1	17

COUNTY	uCi/ sqm	SHOT DAY	DATE	RANK FOR DAY	TIMES IN TOP 10
CO PHILLIPS	0.87	TS4-3	3-May-52	1	17
CO PHILLIPS	98.44	TS6-3	27-May-52	1	17
CO PHILLIPS	1.31	UpK6-3	20-Apr-53	9	17
CO PHILLIPS	7.78	TP3-2	2-Mar-55	3	17
CO PHILLIPS	49.44	TP4-2	8-Mar-55	4	17
CO PHILLIPS	32.33	TP10-2	6-May-55	5	17
CO PHILLIPS	1.71	PB4-2	6-Jul-57	10	17
CO PHILLIPS	0.83	PB8-2	19-Aug-57	8	17
CO PHILLIPS	23.29	PB9-3	25-Aug-57	3	17
CO PHILLIPS	11.05	PB9-4	26-Aug-57	6	17
CO PHILLIPS	1.62	PB9-6	28-Aug-57	2	17
CO PHILLIPS	4.73	PB18-3	9-Oct-57	1	17
CO PHILLIPS	1.44	PB18-4	10-Oct-57	6	17
CO PHILLIPS	2.54	PB18-6	12-Oct-57	2	17
CO PHILLIPS	0.10	UE2-7	11-Mar-62	3	17
CO PHILLIPS	18.17	UE8-3	16-Jul-62	3	17
CO PITKIN	17.71	TS6-2	26-May-52	7	18
CO PITKIN	2.72	UpK5-3	13-Apr-53	1	18
CO PITKIN	45.32	UpK7-3	27-Apr-53	8	18
CO PITKIN	10.85	UpK9-3	21-May-53	9	18
CO PITKIN	4.00	TP3-2	2-Mar-55	5	18
CO PITKIN	64.56	TP4-2	8-Mar-55	2	18
CO PITKIN	8.05	TP4-3	9-Mar-55	5	18
CO PITKIN	3.05	TP6-3	24-Mar-55	5	18
CO PITKIN	9.28	TP7-2	30-Mar-55	6	18
CO PITKIN	3.83	TP8-2	10-Apr-55	3	18
CO PITKIN	2.67	TP8-3	11-Apr-55	3	18
CO PITKIN	2.90	TP10-10	14-May-55	1	18
CO PITKIN	20.50	TP11-3	17-May-55	1	18
CO PITKIN	0.88	PB9-6	28-Aug-57	7	18
CO PITKIN	4.43	PB15-4	19-Sep-57	1	18
CO PITKIN	1.33	PB18-2	8-Oct-57	6	18
CO PITKIN	0.59	UE2-4	8-Mar-62	7	18
CO PITKIN	0.10	UE2-7	11-Mar-62	3	18
CO PROWERS	0.87	TS4-3	3-May-52	1	15

COUNTY	uCi/ sqm	SHOT DAY	DATE	RANK FOR DAY	TIMES IN TOP 10
CO PROWERS	0.87	TS4-3	3-May-52	1	15
CO PROWERS	1.31	UpK6-3	20-Apr-53	9	15
CO PROWERS	7.78	TP3-2	2-Mar-55	3	15
CO PROWERS	49.44	TP4-2	8-Mar-55	4	15
CO PROWERS	32.33	TP10-2	6-May-55	5	15
CO PROWERS	2.48	TP10-4	8-May-55	9	15
CO PROWERS	0.89	PB1-11	7-Jun-57	4	15
CO PROWERS	0.88	PB9-6	28-Aug-57	7	15
CO PROWERS	4.79	PB11-2	1-Sep-57	8	15
CO PROWERS	3.31	PB13-8	13-Sep-57	2	15
CO PROWERS	2.30	PB18-2	8-Oct-57	5	15
CO PROWERS	2.37	PB18-3	9-Oct-57	3	15
CO PROWERS	1.70	PB18-4	10-Oct-57	5	15
CO PROWERS	0.86	PB18-6	12-Oct-57	7	15
CO PUEBLO	0.87	TS4-3	3-May-52	1	10
CO PUEBLO	0.87	TS4-3	3-May-52	1	10
CO PUEBLO	1.66	TP3-2	2-Mar-55	10	10
CO PUEBLO	9.50	TP4-2	8-Mar-55	10	10
CO PUEBLO	24.91	TP10-2	6-May-55	8	10
CO PUEBLO	4.79	PB11-2	1-Sep-57	8	10
CO PUEBLO	1.12	PB13-8	13-Sep-57	7	10
CO PUEBLO	4.60	PB18-2	8-Oct-57	2	10
CO PUEBLO	3.55	PB18-3	9-Oct-57	2	10
CO PUEBLO	1.70	PB18-4	10-Oct-57	5	10
CO RIO BLANCO	165.45	TP9-1	15-Apr-55	8	4
CO RIO BLANCO	240.07	TP10-1	5-May-55	8	4
CO RIO BLANCO	81.28	PB4-1	5-Jul-57	3	4
CO RIO BLANCO	81.28	PB4-1	5-Jul-57	3	4
CO RIO GRANDE	1.09	TS5-12	18-May-52	2	8
CO RIO GRANDE	18.04	TS6-2	26-May-52	6	8
CO RIO GRANDE	4.00	TP3-2	2-Mar-55	5	8
CO RIO GRANDE	22.88	TP4-2	8-Mar-55	6	8
CO RIO GRANDE	59.98	TP10-2	6-May-55	3	8
CO RIO GRANDE	11.19	PB3-2	25-Jun-57	2	8
CO RIO GRANDE	0.85	PB4-3	7-Jul-57	10	8

COUNTY	uCi/ sqm	SHOT DAY	DATE	RANK FOR DAY	TIMES IN TOP 10
CO RIO GRANDE	9.99	PB13-4	9-Sep-57	1	8
CO ROUTT	2.72	UpK5-3	13-Apr-53	1	18
CO ROUTT	1.07	UpK5-4	14-Apr-53	8	18
CO ROUTT	1.58	UpK8-1	8-May-53	8	18
CO ROUTT	17.89	UpK11-2	5-Jun-53	1	18
CO ROUTT	4.00	TP3-2	2-Mar-55	5	18
CO ROUTT	1.72	TP3-3	3-Mar-55	8	18
CO ROUTT	64.56	TP4-2	8-Mar-55	2	18
CO ROUTT	1.52	TP6-3	24-Mar-55	10	18
CO ROUTT	6.22	TP7-2	30-Mar-55	9	18
CO ROUTT	2.22	TP8-2	10-Apr-55	6	18
CO ROUTT	2.90	TP10-10	14-May-55	1	18
CO ROUTT	3.42	PB4-2	6-Jul-57	4	18
CO ROUTT	6.87	PB5-3	17-Jul-57	6	18
CO ROUTT	4.43	PB15-4	19-Sep-57	1	18
CO ROUTT	1.33	PB18-2	8-Oct-57	6	18
CO ROUTT	0.98	UE2-4	8-Mar-62	2	18
CO ROUTT	0.10	UE2-7	11-Mar-62	4	18
CO ROUTT	0.91	BN-2	19-Dec-70	6	18
CO SAGUACHE	19.00	TS6-2	26-May-52	4	13
CO SAGUACHE	4.94	TP3-2	2-Mar-55	4	13
CO SAGUACHE	42.87	TP4-2	8-Mar-55	5	13
CO SAGUACHE	39.24	TP10-2	6-May-55	4	13
CO SAGUACHE	11.88	TP11-3	17-May-55	10	13
CO SAGUACHE	4.73	PB4-2	6-Jul-57	2	13
CO SAGUACHE	9.77	PB8-4	21-Aug-57	3	13
CO SAGUACHE	9.77	PB8-4	21-Aug-57	3	13
CO SAGUACHE	4.79	PB11-2	1-Sep-57	8	13
CO SAGUACHE	1.14	PB15-4	19-Sep-57	6	13
CO SAGUACHE	2.30	PB18-2	8-Oct-57	5	13
CO SAGUACHE	2.37	PB18-3	9-Oct-57	3	13
CO SAGUACHE	1.70	PB18-4	10-Oct-57	5	13
CO SEDGWICK	1.35	TS5-9	15-May-52	3	15
CO SEDGWICK	14.85	UpK2-7	30-Mar-53	3	15
CO SEDGWICK	4.00	TP3-2	2-Mar-55	5	15

COUNTY	uCi/ sqm	SHOT DAY	DATE	RANK FOR DAY	TIMES IN TOP 10
CO SEDGWICK	14.08	PB2-3	20-Jun-57	6	15
CO SEDGWICK	4.87	PB5-4	18-Jul-57	8	15
CO SEDGWICK	3.26	PB5-5	19-Jul-57	5	15
CO SEDGWICK	1.30	PB8-2	19-Aug-57	7	15
CO SEDGWICK	23.29	PB9-3	25-Aug-57	3	15
CO SEDGWICK	11.05	PB9-4	26-Aug-57	6	15
CO SEDGWICK	0.83	PB9-6	28-Aug-57	8	15
CO SEDGWICK	4.73	PB18-3	9-Oct-57	1	15
CO SEDGWICK	1.44	PB18-4	10-Oct-57	6	15
CO SEDGWICK	1.68	PB18-6	12-Oct-57	4	15
CO SEDGWICK	0.10	UE2-7	11-Mar-62	3	15
CO SEDGWICK	18.17	UE8-3	16-Jul-62	3	15
CO SUMMIT	40.29	TP3-2	2-Mar-55	1	20
CO SUMMIT	4.45	TP3-3	3-Mar-55	3	20
CO SUMMIT	150.71	TP4-2	8-Mar-55	1	20
CO SUMMIT	7.87	TP4-3	9-Mar-55	6	20
CO SUMMIT	2.35	TP8-2	10-Apr-55	5	20
CO SUMMIT	53.28	TP10-2	6-May-55	3	20
CO SUMMIT	15.86	TP11-2	16-May-55	6	20
CO SUMMIT	10.05	PB2-3	20-Jun-57	10	20
CO SUMMIT	4.92	PB8-4	21-Aug-57	7	20
CO SUMMIT	4.92	PB8-4	21-Aug-57	7	20
CO SUMMIT	4.79	PB11-2	1-Sep-57	8	20
CO SUMMIT	3.31	PB13-8	13-Sep-57	2	20
CO SUMMIT	2.30	PB18-2	8-Oct-57	5	20
CO SUMMIT	2.37	PB18-3	9-Oct-57	3	20
CO SUMMIT	1.70	PB18-4	10-Oct-57	5	20
CO SUMMIT	0.86	PB18-6	12-Oct-57	7	20
CO SUMMIT	0.33	UE2-4	8-Mar-62	10	20
CO SUMMIT	0.12	UE2-5	9-Mar-62	8	20
CO SUMMIT	0.10	UE2-7	11-Mar-62	3	20
CO SUMMIT	8.30	UE8-3	16-Jul-62	10	20
CO TELLER	4.94	TP3-2	2-Mar-55	4	11
CO TELLER	42.87	TP4-2	8-Mar-55	5	11
CO TELLER	39.24	TP10-2	6-May-55	4	11

COUNTY	uCi/ sqm	SHOT DAY	DATE	RANK FOR DAY	TIMES IN TOP 10
CO TELLER	2.37	PB4-2	6-Jul-57	6	11
CO TELLER	4.79	PB11-2	1-Sep-57	8	11
CO TELLER	3.31	PB13-8	13-Sep-57	2	11
CO TELLER	2.30	PB18-2	8-Oct-57	5	11
CO TELLER	2.37	PB18-3	9-Oct-57	3	11
CO TELLER	1.70	PB18-4	10-Oct-57	5	11
CO TELLER	2.54	PB18-6	12-Oct-57	2	11
CO TELLER	0.10	UE2-7	11-Mar-62	4	11
CO WASHINGTON	0.87	TS4-3	3-May-52	1	18
CO WASHINGTON	0.87	TS4-3	3-May-52	1	18
CO WASHINGTON	1.31	UpK6-3	20-Apr-53	9	18
CO WASHINGTON	7.78	TP3-2	2-Mar-55	3	18
CO WASHINGTON	49.44	TP4-2	8-Mar-55	4	18
CO WASHINGTON	5.38	TP7-2	30-Mar-55	10	18
CO WASHINGTON	32.33	TP10-2	6-May-55	5	18
CO WASHINGTON	1.18	PB1-10	6-Jun-57	2	18
CO WASHINGTON	4.73	PB4-2	6-Jul-57	2	18
CO WASHINGTON	1.96	PB5-5	19-Jul-57	9	18
CO WASHINGTON	9.77	PB8-4	21-Aug-57	3	18
CO WASHINGTON	9.77	PB8-4	21-Aug-57	3	18
CO WASHINGTON	4.79	PB11-2	1-Sep-57	8	18
CO WASHINGTON	1.12	PB13-8	13-Sep-57	7	18
CO WASHINGTON	4.60	PB18-2	8-Oct-57	2	18
CO WASHINGTON	2.37	PB18-3	9-Oct-57	3	18
CO WASHINGTON	1.70	PB18-4	10-Oct-57	5	18
CO WASHINGTON	2.54	PB18-6	12-Oct-57	2	18
CO WELD	3.16	TP3-2	2-Mar-55	7	12
CO WELD	24.89	TP10-2	6-May-55	9	12
CO WELD	14.59	PB2-2	19-Jun-57	2	12
CO WELD	0.84	PB5-8	22-Jul-57	8	12
CO WELD	0.83	PB8-2	19-Aug-57	8	12
CO WELD	3.68	PB9-4	26-Aug-57	9	12
CO WELD	1.62	PB9-6	28-Aug-57	2	12
CO WELD	0.97	PB18-2	8-Oct-57	7	12
CO WELD	2.37	PB18-3	9-Oct-57	3	12

COUNTY	uCi/ sqm	SHOT DAY	DATE	RANK FOR DAY	TIMES IN TOP 10
CO WELD	1.44	PB18-4	10-Oct-57	6	12
CO WELD	0.86	PB18-6	12-Oct-57	7	12
CO WELD	10.77	UE8-3	16-Jul-62	7	12
CO YUMA	0.87	TS4-3	3-May-52	1	14
CO YUMA	0.87	TS4-3	3-May-52	1	14
CO YUMA	41.63	TS6-3	27-May-52	8	14
CO YUMA	1.31	UpK6-3	20-Apr-53	9	14
CO YUMA	7.78	TP3-2	2-Mar-55	3	14
CO YUMA	49.44	TP4-2	8-Mar-55	4	14
CO YUMA	32.33	TP10-2	6-May-55	5	14
CO YUMA	1.73	PB4-10	14-Jul-57	9	14
CO YUMA	4.87	PB5-4	18-Jul-57	8	14
CO YUMA	1.96	PB5-5	19-Jul-57	9	14
CO YUMA	3.51	PB18-2	8-Oct-57	4	14
CO YUMA	4.73	PB18-3	9-Oct-57	1	14
CO YUMA	1.70	PB18-4	10-Oct-57	5	14
CO YUMA	2.54	PB18-6	12-Oct-57	2	14

COUNTY	uCi/ sqm	SHOT DAY	DATE	RANK FOR DAY	TIMES IN TOP 10
CT FAIRFIELD	11.62	BJ3-7	7-Nov-51	3	3
CT FAIRFIELD	2.71	TP11-9	23-May-55	10	3
CT FAIRFIELD	2.99	PB5-8	22-Jul-57	4	3
CT HARTFORD	1.16	TP7-6	3-Apr-55	6	2
CT HARTFORD	5.97	PB5-8	22-Jul-57	3	2
CT LITCHFIELD	30.19	R1-2	29-Jan-51	8	2
CT LITCHFIELD	5.97	PB5-8	22-Jul-57	3	2
CT MIDDLESEX	93.65	UpK4-2	7-Apr-53	8	2
CT MIDDLESEX	8.96	PB5-8	22-Jul-57	2	2
CT NEW HAVEN	1.21	TP11-8	22-May-55	10	3
CT NEW HAVEN	3.60	TP11-9	23-May-55	7	3
CT NEW HAVEN	5.97	PB5-8	22-Jul-57	3	3
CT NEW LONDON	1.00	UpK4-3	8-Apr-53	10	4
CT NEW LONDON	0.99	UpK5-5	15-Apr-53	2	4
CT NEW LONDON	1.83	UpK8-4	11-May-53	4	4
CT NEW LONDON	5.97	PB5-8	22-Jul-57	3	4
CT TOLLAND	30.19	R1-2	29-Jan-51	8	2
CT TOLLAND	2.99	PB5-8	22-Jul-57	4	2
CT WINDHAM	1.00	UpK4-3	8-Apr-53	10	5
CT WINDHAM	0.99	UpK5-5	15-Apr-53	2	5
CT WINDHAM	1.33	UpK7-6	30-Apr-53	7	5
CT WINDHAM	1.83	UpK8-4	11-May-53	4	5
CT WINDHAM	2.99	PB5-8	22-Jul-57	4	5

COUNTY	uCi/ sqm	SHOT DAY	DATE	RANK FOR DAY	TIMES IN TOP 10
DC WASHINGTON	19.54	BJ2-3	1-Nov-51	7	4
DC WASHINGTON	0.96	UpK8-6	13-May-53	3	4
DC WASHINGTON	5.97	PB5-8	22-Jul-57	3	4
DC WASHINGTON	0.94	PB5-9	23-Jul-57	4	4

COUNTY	uCi/ sqm	SHOT DAY	DATE	RANK FOR DAY	TIMES IN TOP 10
DE KENT	3.98	BJ4-6	10-Nov-51	1	2
DE KENT	2.99	PB5-8	22-Jul-57	4	2
DE NEW CASTLE	3.98	BJ4-6	10-Nov-51	1	4
DE NEW CASTLE	1.42	UpK3-7	6-Apr-53	8	4
DE NEW CASTLE	2.03	TP6-4	25-Mar-55	8	4
DE NEW CASTLE	8.96	PB5-8	22-Jul-57	2	4
DE SUSSEX	3.98	BJ4-6	10-Nov-51	1	7
DE SUSSEX	1.42	UpK3-7	6-Apr-53	8	7
DE SUSSEX	2.03	TP6-4	25-Mar-55	8	7
DE SUSSEX	0.89	PB1-9	5-Jun-57	7	7
DE SUSSEX	1.37	PB5-7	21-Jul-57	10	7
DE SUSSEX	5.97	PB5-8	22-Jul-57	3	7
DE SUSSEX	0.94	PB5-9	23-Jul-57	4	7

COUNTY	uCi/ sqm	SHOT DAY	DATE	RANK FOR DAY	TIMES IN TOP 10
FL ALACHUA	5.73	TS3-8	29-Apr-52	2	9
FL ALACHUA	5.73	TS3-8	29-Apr-52	2	9
FL ALACHUA	1.30	UpK3-7	6-Apr-53	9	9
FL ALACHUA	3.89	UpK4-3	8-Apr-53	4	9
FL ALACHUA	1.46	TP6-7	28-Mar-55	6	9
FL ALACHUA	2.73	PB2-4	21-Jun-57	2	9
FL ALACHUA	0.92	PB6-9	1-Aug-57	6	9
FL ALACHUA	2.19	PB8-5	22-Aug-57	8	9
FL ALACHUA	11.66	PB11-4	3-Sep-57	5	9
FL BAKER	5.73	TS3-8	29-Apr-52	2	9
FL BAKER	5.73	TS3-8	29-Apr-52	2	9
FL BAKER	1.30	UpK3-7	6-Apr-53	9	9
FL BAKER	3.89	UpK4-3	8-Apr-53	4	9
FL BAKER	1.46	TP6-7	28-Mar-55	6	9
FL BAKER	0.85	TP10-10	14-May-55	2	9
FL BAKER	1.75	PB6-7	30-Jul-57	8	9
FL BAKER	2.19	PB8-5	22-Aug-57	8	9
FL BAKER	11.66	PB11-4	3-Sep-57	5	9
FL BAY	6.40	TS3-7	28-Apr-52	2	6
FL BAY	6.40	TS3-7	28-Apr-52	2	6
FL BAY	1.65	TS3-8	29-Apr-52	5	6
FL BAY	1.65	TS3-8	29-Apr-52	5	6
FL BAY	3.14	TP6-5	26-Mar-55	2	6
FL BAY	3.96	PB17-3	30-Sep-57	9	6
FL BRADFORD	5.73	TS3-8	29-Apr-52	2	7
FL BRADFORD	5.73	TS3-8	29-Apr-52	2	7
FL BRADFORD	1.30	UpK3-7	6-Apr-53	9	7
FL BRADFORD	3.89	UpK4-3	8-Apr-53	4	7
FL BRADFORD	1.82	PB2-4	21-Jun-57	5	7
FL BRADFORD	5.05	PB4-14	18-Jul-57	6	7
FL BRADFORD	1.75	PB6-7	30-Jul-57	8	7
FL BREVARD	5.73	TS3-8	29-Apr-52	2	7
FL BREVARD	5.73	TS3-8	29-Apr-52	2	7
FL BREVARD	3.89	UpK4-3	8-Apr-53	4	7
FL BREVARD	1.46	TP6-7	28-Mar-55	6	7

COUNTY	uCi/ sqm	SHOT DAY	DATE	RANK FOR DAY	TIMES IN TOP 10
FL BREVARD	3.61	PB8-5	22-Aug-57	5	7
FL BREVARD	19.43	PB11-4	3-Sep-57	2	7
FL BREVARD	1.74	UE8-7	20-Jul-62	3	7
FL BROWARD	1.16	R-2-4	5-Feb-51	2	5
FL BROWARD	5.73	TS3-8	29-Apr-52	2	5
FL BROWARD	5.73	TS3-8	29-Apr-52	2	5
FL BROWARD	3.12	UpK6-3	20-Apr-53	3	5
FL BROWARD	2.19	PB8-5	22-Aug-57	8	5
FL CALHOUN	6.40	TS3-7	28-Apr-52	2	8
FL CALHOUN	6.40	TS3-7	28-Apr-52	2	8
FL CALHOUN	1.65	TS3-8	29-Apr-52	5	8
FL CALHOUN	1.65	TS3-8	29-Apr-52	5	8
FL CALHOUN	3.14	TP6-5	26-Mar-55	2	8
FL CALHOUN	3.24	PB12-8	9-Sep-57	7	8
FL CALHOUN	5.99	PB17-3	30-Sep-57	7	8
FL CALHOUN	1.25	UE8-6	19-Jul-62	10	8
FL CHARLOTTE	5.73	TS3-8	29-Apr-52	2	5
FL CHARLOTTE	5.73	TS3-8	29-Apr-52	2	5
FL CHARLOTTE	3.12	UpK6-3	20-Apr-53	3	5
FL CHARLOTTE	2.73	PB2-4	21-Jun-57	2	5
FL CHARLOTTE	3.61	PB8-5	22-Aug-57	5	5
FL CITRUS	5.73	TS3-8	29-Apr-52	2	10
FL CITRUS	5.73	TS3-8	29-Apr-52	2	10
FL CITRUS	1.30	UpK3-7	6-Apr-53	9	10
FL CITRUS	3.89	UpK4-3	8-Apr-53	4	10
FL CITRUS	2.73	PB2-4	21-Jun-57	2	10
FL CITRUS	1.75	PB6-7	30-Jul-57	8	10
FL CITRUS	0.92	PB6-9	1-Aug-57	6	10
FL CITRUS	2.19	PB8-5	22-Aug-57	8	10
FL CITRUS	11.66	PB11-4	3-Sep-57	5	10
FL CITRUS	1.88	UE8-7	20-Jul-62	2	10
FL CLAY	5.73	TS3-8	29-Apr-52	2	9
FL CLAY	5.73	TS3-8	29-Apr-52	2	9
FL CLAY	1.30	UpK3-7	6-Apr-53	9	9
FL CLAY	3.89	UpK4-3	8-Apr-53	4	9

COUNTY	uCi/ sqm	SHOT DAY	DATE	RANK FOR DAY	TIMES IN TOP 10
FL CLAY	1.46	TP6-7	28-Mar-55	6	9
FL CLAY	1.82	PB2-4	21-Jun-57	5	9
FL CLAY	5.05	PB4-14	18-Jul-57	6	9
FL CLAY	1.75	PB6-7	30-Jul-57	8	9
FL CLAY	1.37	UE8-7	20-Jul-62	4	9
FL COLLIER	1.11	R-2-4	5-Feb-51	3	6
FL COLLIER	5.73	TS3-8	29-Apr-52	2	6
FL COLLIER	5.73	TS3-8	29-Apr-52	2	6
FL COLLIER	3.12	UpK6-3	20-Apr-53	3	6
FL COLLIER	2.73	PB2-4	21-Jun-57	2	6
FL COLLIER	3.61	PB8-5	22-Aug-57	5	6
FL COLUMBIA	5.73	TS3-8	29-Apr-52	2	7
FL COLUMBIA	5.73	TS3-8	29-Apr-52	2	7
FL COLUMBIA	1.30	UpK3-7	6-Apr-53	9	7
FL COLUMBIA	3.89	UpK4-3	8-Apr-53	4	7
FL COLUMBIA	1.75	PB6-7	30-Jul-57	8	7
FL COLUMBIA	2.19	PB8-5	22-Aug-57	8	7
FL COLUMBIA	11.66	PB11-4	3-Sep-57	5	7
FL DADE	2.77	R-2-4	5-Feb-51	1	4
FL DADE	5.73	TS3-8	29-Apr-52	2	4
FL DADE	5.73	TS3-8	29-Apr-52	2	4
FL DADE	1.96	UpK6-3	20-Apr-53	5	4
FL DE SOTO	5.73	TS3-8	29-Apr-52	2	3
FL DE SOTO	5.73	TS3-8	29-Apr-52	2	3
FL DE SOTO	3.12	UpK6-3	20-Apr-53	3	3
FL DIXIE	6.40	TS3-7	28-Apr-52	2	10
FL DIXIE	6.40	TS3-7	28-Apr-52	2	10
FL DIXIE	1.65	TS3-8	29-Apr-52	5	10
FL DIXIE	1.65	TS3-8	29-Apr-52	5	10
FL DIXIE	7.78	UpK4-3	8-Apr-53	2	10
FL DIXIE	0.85	TP10-10	14-May-55	2	10
FL DIXIE	1.82	PB2-4	21-Jun-57	5	10
FL DIXIE	2.19	PB8-5	22-Aug-57	8	10
FL DIXIE	11.66	PB11-4	3-Sep-57	5	10
FL DIXIE	1.92	UE8-6	19-Jul-62	6	10

COUNTY	uCi/ sqm	SHOT DAY	DATE	RANK FOR DAY	TIMES IN TOP 10
FL DUVAL	2.38	TS3-8	29-Apr-52	4	7
FL DUVAL	2.38	TS3-8	29-Apr-52	4	7
FL DUVAL	1.62	UpK4-3	8-Apr-53	9	7
FL DUVAL	1.01	TP6-7	28-Mar-55	7	7
FL DUVAL	1.82	PB2-4	21-Jun-57	5	7
FL DUVAL	1.51	PB5-8	22-Jul-57	7	7
FL DUVAL	0.94	PB5-9	23-Jul-57	4	7
FL ESCAMBIA	3.14	TP6-5	26-Mar-55	2	3
FL ESCAMBIA	22.97	PB3-3	26-Jun-57	4	3
FL ESCAMBIA	0.98	PB9-7	29-Aug-57	5	3
FL FLAGLER	5.73	TS3-8	29-Apr-52	2	9
FL FLAGLER	5.73	TS3-8	29-Apr-52	2	9
FL FLAGLER	1.30	UpK3-7	6-Apr-53	9	9
FL FLAGLER	3.89	UpK4-3	8-Apr-53	4	9
FL FLAGLER	5.05	PB4-14	18-Jul-57	6	9
FL FLAGLER	2.19	PB8-5	22-Aug-57	8	9
FL FLAGLER	19.43	PB11-4	3-Sep-57	2	9
FL FLAGLER	1.73	PB18-7	13-Oct-57	3	9
FL FLAGLER	1.37	UE8-7	20-Jul-62	4	9
FL FRANKLIN	0.95	R-2-5	6-Feb-51	2	9
FL FRANKLIN	6.40	TS3-7	28-Apr-52	2	9
FL FRANKLIN	6.40	TS3-7	28-Apr-52	2	9
FL FRANKLIN	1.65	TS3-8	29-Apr-52	5	9
FL FRANKLIN	1.65	TS3-8	29-Apr-52	5	9
FL FRANKLIN	1.55	PB2-5	22-Jun-57	6	9
FL FRANKLIN	2.19	PB8-5	22-Aug-57	8	9
FL FRANKLIN	11.66	PB11-4	3-Sep-57	5	9
FL FRANKLIN	0.80	PB13-7	12-Sep-57	7	9
FL GADSDEN	6.40	TS3-7	28-Apr-52	2	5
FL GADSDEN	6.40	TS3-7	28-Apr-52	2	5
FL GADSDEN	1.65	TS3-8	29-Apr-52	5	5
FL GADSDEN	1.65	TS3-8	29-Apr-52	5	5
FL GADSDEN	1.82	PB2-4	21-Jun-57	5	5
FL GILCHRIST	5.73	TS3-8	29-Apr-52	2	8
FL GILCHRIST	5.73	TS3-8	29-Apr-52	2	8

COUNTY	uCi/ sqm	SHOT DAY	DATE	RANK FOR DAY	TIMES IN TOP 10
FL GILCHRIST	7.78	UpK4-3	8-Apr-53	2	8
FL GILCHRIST	0.85	TP10-10	14-May-55	2	8
FL GILCHRIST	1.82	PB2-4	21-Jun-57	5	8
FL GILCHRIST	1.75	PB6-7	30-Jul-57	8	8
FL GILCHRIST	2.19	PB8-5	22-Aug-57	8	8
FL GILCHRIST	11.66	PB11-4	3-Sep-57	5	8
FL GLADES	5.73	TS3-8	29-Apr-52	2	4
FL GLADES	5.73	TS3-8	29-Apr-52	2	4
FL GLADES	3.12	UpK6-3	20-Apr-53	3	4
FL GLADES	3.61	PB8-5	22-Aug-57	5	4
FL GULF	6.40	TS3-7	28-Apr-52	2	6
FL GULF	6.40	TS3-7	28-Apr-52	2	6
FL GULF	1.65	TS3-8	29-Apr-52	5	6
FL GULF	1.65	TS3-8	29-Apr-52	5	6
FL GULF	3.14	TP6-5	26-Mar-55	2	6
FL GULF	1.82	PB2-4	21-Jun-57	5	6
FL HAMILTON	5.73	TS3-8	29-Apr-52	2	5
FL HAMILTON	5.73	TS3-8	29-Apr-52	2	5
FL HAMILTON	1.30	UpK3-7	6-Apr-53	9	5
FL HAMILTON	7.78	UpK4-3	8-Apr-53	2	5
FL HAMILTON	2.19	PB8-5	22-Aug-57	8	5
FL HARDEE	5.73	TS3-8	29-Apr-52	2	3
FL HARDEE	5.73	TS3-8	29-Apr-52	2	3
FL HARDEE	3.12	UpK6-3	20-Apr-53	3	3
FL HENDRY	5.73	TS3-8	29-Apr-52	2	6
FL HENDRY	5.73	TS3-8	29-Apr-52	2	6
FL HENDRY	3.12	UpK6-3	20-Apr-53	3	6
FL HENDRY	0.82	TP10-10	14-May-55	3	6
FL HENDRY	1.82	PB2-4	21-Jun-57	5	6
FL HENDRY	2.19	PB8-5	22-Aug-57	8	6
FL HERNANDO	5.73	TS3-8	29-Apr-52	2	11
FL HERNANDO	5.73	TS3-8	29-Apr-52	2	11
FL HERNANDO	1.30	UpK3-7	6-Apr-53	9	11
FL HERNANDO	3.89	UpK4-3	8-Apr-53	4	11
FL HERNANDO	0.97	TP6-7	28-Mar-55	8	11

COUNTY	uCi/ sqm	SHOT DAY	DATE	RANK FOR DAY	TIMES IN TOP 10
FL HERNANDO	2.73	PB2-4	21-Jun-57	2	11
FL HERNANDO	1.75	PB6-7	30-Jul-57	8	11
FL HERNANDO	0.92	PB6-9	1-Aug-57	6	11
FL HERNANDO	19.43	PB11-4	3-Sep-57	2	11
FL HERNANDO	0.80	PB13-7	12-Sep-57	7	11
FL HERNANDO	1.06	UE8-7	20-Jul-62	6	11
FL HIGHLANDS	5.73	TS3-8	29-Apr-52	2	5
FL HIGHLANDS	5.73	TS3-8	29-Apr-52	2	5
FL HIGHLANDS	3.12	UpK6-3	20-Apr-53	3	5
FL HIGHLANDS	1.82	PB2-4	21-Jun-57	5	5
FL HIGHLANDS	0.93	PB6-10	2-Aug-57	5	5
FL HILLSBOROUGH	1.18	R-2-5	6-Feb-51	1	9
FL HILLSBOROUGH	22.91	TS3-8	29-Apr-52	1	9
FL HILLSBOROUGH	22.91	TS3-8	29-Apr-52	1	9
FL HILLSBOROUGH	3.89	UpK4-3	8-Apr-53	4	9
FL HILLSBOROUGH	0.97	TP6-7	28-Mar-55	8	9
FL HILLSBOROUGH	2.73	PB2-4	21-Jun-57	2	9
FL HILLSBOROUGH	2.19	PB8-5	22-Aug-57	8	9
FL HILLSBOROUGH	7.77	PB11-4	3-Sep-57	8	9
FL HILLSBOROUGH	0.68	UE8-7	20-Jul-62	10	9
FL HOLMES	3.14	TP6-5	26-Mar-55	2	5
FL HOLMES	1.16	PB1-9	5-Jun-57	5	5
FL HOLMES	0.92	PB6-9	1-Aug-57	6	5
FL HOLMES	0.80	PB13-7	12-Sep-57	7	5
FL HOLMES	3.96	PB17-3	30-Sep-57	9	5
FL INDIAN RIVER	5.73	TS3-8	29-Apr-52	2	7
FL INDIAN RIVER	5.73	TS3-8	29-Apr-52	2	7
FL INDIAN RIVER	3.12	UpK6-3	20-Apr-53	3	7
FL INDIAN RIVER	0.82	TP10-10	14-May-55	3	7
FL INDIAN RIVER	2.83	PB5-9	23-Jul-57	2	7
FL INDIAN RIVER	0.93	PB6-10	2-Aug-57	5	7
FL INDIAN RIVER	0.27	UE8-8	21-Jul-62	4	7
FL JACKSON	6.40	TS3-7	28-Apr-52	2	6
FL JACKSON	6.40	TS3-7	28-Apr-52	2	6
FL JACKSON	1.65	TS3-8	29-Apr-52	5	6

COUNTY	uCi/ sqm	SHOT DAY	DATE	RANK FOR DAY	TIMES IN TOP 10
FL JACKSON	1.65	TS3-8	29-Apr-52	5	6
FL JACKSON	3.14	TP6-5	26-Mar-55	2	6
FL JACKSON	1.16	PB1-9	5-Jun-57	5	6
FL JEFFERSON	6.40	TS3-7	28-Apr-52	2	6
FL JEFFERSON	6.40	TS3-7	28-Apr-52	2	6
FL JEFFERSON	1.65	TS3-8	29-Apr-52	5	6
FL JEFFERSON	1.65	TS3-8	29-Apr-52	5	6
FL JEFFERSON	3.89	UpK4-3	8-Apr-53	4	6
FL JEFFERSON	0.85	TP10-10	14-May-55	2	6
FL LAFAYETTE	6.40	TS3-7	28-Apr-52	2	7
FL LAFAYETTE	6.40	TS3-7	28-Apr-52	2	7
FL LAFAYETTE	1.65	TS3-8	29-Apr-52	5	7
FL LAFAYETTE	1.65	TS3-8	29-Apr-52	5	7
FL LAFAYETTE	1.30	UpK3-7	6-Apr-53	9	7
FL LAFAYETTE	3.89	UpK4-3	8-Apr-53	4	7
FL LAFAYETTE	0.92	PB6-9	1-Aug-57	6	7
FL LAKE	1.18	R-2-5	6-Feb-51	1	12
FL LAKE	5.73	TS3-8	29-Apr-52	2	12
FL LAKE	5.73	TS3-8	29-Apr-52	2	12
FL LAKE	1.30	UpK3-7	6-Apr-53	9	12
FL LAKE	3.89	UpK4-3	8-Apr-53	4	12
FL LAKE	1.46	TP6-7	28-Mar-55	6	12
FL LAKE	2.73	PB2-4	21-Jun-57	2	12
FL LAKE	1.89	PB5-9	23-Jul-57	3	12
FL LAKE	3.61	PB8-5	22-Aug-57	5	12
FL LAKE	19.43	PB11-4	3-Sep-57	2	12
FL LAKE	1.04	PB18-7	13-Oct-57	7	12
FL LAKE	1.37	UE8-7	20-Jul-62	4	12
FL LEE	5.73	TS3-8	29-Apr-52	2	4
FL LEE	5.73	TS3-8	29-Apr-52	2	4
FL LEE	3.12	UpK6-3	20-Apr-53	3	4
FL LEE	3.61	PB8-5	22-Aug-57	5	4
FL LEON	2.66	TS3-7	28-Apr-52	4	5
FL LEON	2.66	TS3-7	28-Apr-52	4	5
FL LEON	15.57	UpK4-3	8-Apr-53	1	5

COUNTY	uCi/ sqm	SHOT DAY	DATE	RANK FOR DAY	TIMES IN TOP 10
FL LEON	2.19	PB8-5	22-Aug-57	8	5
FL LEON	7.77	PB11-4	3-Sep-57	8	5
FL LEVY	5.73	TS3-8	29-Apr-52	2	5
FL LEVY	5.73	TS3-8	29-Apr-52	2	5
FL LEVY	1.30	UpK3-7	6-Apr-53	9	5
FL LEVY	3.89	UpK4-3	8-Apr-53	4	5
FL LEVY	0.92	PB6-9	1-Aug-57	6	5
FL LIBERTY	0.95	R-2-5	6-Feb-51	2	8
FL LIBERTY	6.40	TS3-7	28-Apr-52	2	8
FL LIBERTY	6.40	TS3-7	28-Apr-52	2	8
FL LIBERTY	1.65	TS3-8	29-Apr-52	5	8
FL LIBERTY	1.65	TS3-8	29-Apr-52	5	8
FL LIBERTY	3.14	TP6-5	26-Mar-55	2	8
FL LIBERTY	1.46	TP6-7	28-Mar-55	6	8
FL LIBERTY	2.94	PB12-8	9-Sep-57	8	8
FL MADISON	6.40	TS3-7	28-Apr-52	2	6
FL MADISON	6.40	TS3-7	28-Apr-52	2	6
FL MADISON	1.65	TS3-8	29-Apr-52	5	6
FL MADISON	1.65	TS3-8	29-Apr-52	5	6
FL MADISON	7.78	UpK4-3	8-Apr-53	2	6
FL MADISON	7.77	PB11-4	3-Sep-57	8	6
FL MANATEE	5.73	TS3-8	29-Apr-52	2	7
FL MANATEE	5.73	TS3-8	29-Apr-52	2	7
FL MANATEE	3.12	UpK6-3	20-Apr-53	3	7
FL MANATEE	4.55	PB2-4	21-Jun-57	1	7
FL MANATEE	1.55	PB2-5	22-Jun-57	6	7
FL MANATEE	0.93	PB6-10	2-Aug-57	5	7
FL MANATEE	1.12	PB6-12	4-Aug-57	1	7
FL MARION	5.73	TS3-8	29-Apr-52	2	9
FL MARION	5.73	TS3-8	29-Apr-52	2	9
FL MARION	1.30	UpK3-7	6-Apr-53	9	9
FL MARION	3.89	UpK4-3	8-Apr-53	4	9
FL MARION	1.46	TP6-7	28-Mar-55	6	9
FL MARION	2.75	TP10-6	10-May-55	2	9
FL MARION	2.73	PB2-4	21-Jun-57	2	9

COUNTY	uCi/ sqm	SHOT DAY	DATE	RANK FOR DAY	TIMES IN TOP 10
FL MARION	2.19	PB8-5	22-Aug-57	8	9
FL MARION	7.77	PB11-4	3-Sep-57	8	9
FL MARTIN	5.73	TS3-8	29-Apr-52	2	6
FL MARTIN	5.73	TS3-8	29-Apr-52	2	6
FL MARTIN	3.12	UpK6-3	20-Apr-53	3	6
FL MARTIN	0.82	TP10-10	14-May-55	3	6
FL MARTIN	1.82	PB2-4	21-Jun-57	5	6
FL MARTIN	0.45	UE8-8	21-Jul-62	2	6
FL MONROE	5.73	TS3-8	29-Apr-52	2	5
FL MONROE	5.73	TS3-8	29-Apr-52	2	5
FL MONROE	3.12	UpK6-3	20-Apr-53	3	5
FL MONROE	1.82	PB2-4	21-Jun-57	5	5
FL MONROE	2.19	PB8-5	22-Aug-57	8	5
FL NASSAU	5.73	TS3-8	29-Apr-52	2	8
FL NASSAU	5.73	TS3-8	29-Apr-52	2	8
FL NASSAU	1.30	UpK3-7	6-Apr-53	9	8
FL NASSAU	3.89	UpK4-3	8-Apr-53	4	8
FL NASSAU	0.97	TP6-7	28-Mar-55	8	8
FL NASSAU	1.51	PB5-8	22-Jul-57	7	8
FL NASSAU	11.66	PB11-4	3-Sep-57	5	8
FL NASSAU	1.04	PB18-7	13-Oct-57	7	8
FL OKALOOSA	3.14	TP6-5	26-Mar-55	2	2
FL OKALOOSA	68.82	PB3-3	26-Jun-57	2	2
FL OKEECHOBEE	5.73	TS3-8	29-Apr-52	2	4
FL OKEECHOBEE	5.73	TS3-8	29-Apr-52	2	4
FL OKEECHOBEE	3.12	UpK6-3	20-Apr-53	3	4
FL OKEECHOBEE	0.93	PB6-10	2-Aug-57	5	4
FL ORANGE	5.73	TS3-8	29-Apr-52	2	10
FL ORANGE	5.73	TS3-8	29-Apr-52	2	10
FL ORANGE	1.30	UpK3-7	6-Apr-53	9	10
FL ORANGE	3.89	UpK4-3	8-Apr-53	4	10
FL ORANGE	1.46	TP6-7	28-Mar-55	6	10
FL ORANGE	1.82	PB2-4	21-Jun-57	5	10
FL ORANGE	5.05	PB4-14	18-Jul-57	6	10
FL ORANGE	3.61	PB8-5	22-Aug-57	5	10

COUNTY	uCi/ sqm	SHOT DAY	DATE	RANK FOR DAY	TIMES IN TOP 10
FL ORANGE	11.66	PB11-4	3-Sep-57	5	10
FL ORANGE	1.04	PB18-7	13-Oct-57	7	10
FL OSCEOLA	5.73	TS3-8	29-Apr-52	2	9
FL OSCEOLA	5.73	TS3-8	29-Apr-52	2	9
FL OSCEOLA	1.30	UpK3-7	6-Apr-53	9	9
FL OSCEOLA	3.89	UpK4-3	8-Apr-53	4	9
FL OSCEOLA	1.46	TP6-7	28-Mar-55	6	9
FL OSCEOLA	4.55	PB2-4	21-Jun-57	1	9
FL OSCEOLA	5.05	PB4-14	18-Jul-57	6	9
FL OSCEOLA	1.89	PB5-9	23-Jul-57	3	9
FL OSCEOLA	0.99	UE8-7	20-Jul-62	7	9
FL PALM BEACH	1.11	R-2-4	5-Feb-51	3	6
FL PALM BEACH	5.73	TS3-8	29-Apr-52	2	6
FL PALM BEACH	5.73	TS3-8	29-Apr-52	2	6
FL PALM BEACH	3.12	UpK6-3	20-Apr-53	3	6
FL PALM BEACH	0.82	TP10-10	14-May-55	3	6
FL PALM BEACH	2.19	PB8-5	22-Aug-57	8	6
FL PASCO	5.73	TS3-8	29-Apr-52	2	7
FL PASCO	5.73	TS3-8	29-Apr-52	2	7
FL PASCO	1.30	UpK3-7	6-Apr-53	9	7
FL PASCO	3.89	UpK4-3	8-Apr-53	4	7
FL PASCO	0.97	TP6-7	28-Mar-55	8	7
FL PASCO	1.82	PB2-4	21-Jun-57	5	7
FL PASCO	1.04	PB18-7	13-Oct-57	7	7
FL PINELLAS	5.73	TS3-8	29-Apr-52	2	8
FL PINELLAS	5.73	TS3-8	29-Apr-52	2	8
FL PINELLAS	3.89	UpK4-3	8-Apr-53	4	8
FL PINELLAS	1.46	TP6-7	28-Mar-55	6	8
FL PINELLAS	2.73	PB2-4	21-Jun-57	2	8
FL PINELLAS	1.55	PB2-5	22-Jun-57	6	8
FL PINELLAS	2.19	PB8-5	22-Aug-57	8	8
FL PINELLAS	11.66	PB11-4	3-Sep-57	5	8
FL POLK	5.73	TS3-8	29-Apr-52	2	9
FL POLK	5.73	TS3-8	29-Apr-52	2	9
FL POLK	3.89	UpK4-3	8-Apr-53	4	9

COUNTY	uCi/ sqm	SHOT DAY	DATE	RANK FOR DAY	TIMES IN TOP 10
FL POLK	1.46	TP6-7	28-Mar-55	6	9
FL POLK	2.73	PB2-4	21-Jun-57	2	9
FL POLK	0.94	PB5-9	23-Jul-57	4	9
FL POLK	2.19	PB8-5	22-Aug-57	8	9
FL POLK	11.66	PB11-4	3-Sep-57	5	9
FL POLK	0.99	UE8-7	20-Jul-62	7	9
FL PUTNAM	5.73	TS3-8	29-Apr-52	2	9
FL PUTNAM	5.73	TS3-8	29-Apr-52	2	9
FL PUTNAM	1.30	UpK3-7	6-Apr-53	9	9
FL PUTNAM	3.89	UpK4-3	8-Apr-53	4	9
FL PUTNAM	1.46	TP6-7	28-Mar-55	6	9
FL PUTNAM	0.94	PB5-9	23-Jul-57	4	9
FL PUTNAM	11.66	PB11-4	3-Sep-57	5	9
FL PUTNAM	1.04	PB18-7	13-Oct-57	7	9
FL PUTNAM	1.37	UE8-7	20-Jul-62	4	9
FL SANTA ROSA	0.81	UpK3-6	5-Apr-53	9	4
FL SANTA ROSA	3.14	TP6-5	26-Mar-55	2	4
FL SANTA ROSA	22.97	PB3-3	26-Jun-57	4	4
FL SANTA ROSA	1.40	PB6-8	31-Jul-57	8	4
FL SARASOTA	5.73	TS3-8	29-Apr-52	2	7
FL SARASOTA	5.73	TS3-8	29-Apr-52	2	7
FL SARASOTA	3.12	UpK6-3	20-Apr-53	3	7
FL SARASOTA	2.73	PB2-4	21-Jun-57	2	7
FL SARASOTA	0.93	PB6-10	2-Aug-57	5	7
FL SARASOTA	1.12	PB6-12	4-Aug-57	1	7
FL SARASOTA	2.19	PB8-5	22-Aug-57	8	7
FL SEMINOLE	5.73	TS3-8	29-Apr-52	2	11
FL SEMINOLE	5.73	TS3-8	29-Apr-52	2	11
FL SEMINOLE	1.30	UpK3-7	6-Apr-53	9	11
FL SEMINOLE	3.89	UpK4-3	8-Apr-53	4	11
FL SEMINOLE	1.46	TP6-7	28-Mar-55	6	11
FL SEMINOLE	5.64	TP10-6	10-May-55	1	11
FL SEMINOLE	1.82	PB2-4	21-Jun-57	5	11
FL SEMINOLE	5.05	PB4-14	18-Jul-57	6	11
FL SEMINOLE	3.61	PB8-5	22-Aug-57	5	11

COUNTY	uCi/ sqm	SHOT DAY	DATE	RANK FOR DAY	TIMES IN TOP 10
FL SEMINOLE	11.66	PB11-4	3-Sep-57	5	11
FL SEMINOLE	1.04	PB18-7	13-Oct-57	7	11
FL ST JOHNS	5.73	TS3-8	29-Apr-52	2	8
FL ST JOHNS	5.73	TS3-8	29-Apr-52	2	8
FL ST JOHNS	1.30	UpK3-7	6-Apr-53	9	8
FL ST JOHNS	3.89	UpK4-3	8-Apr-53	4	8
FL ST JOHNS	1.46	TP6-7	28-Mar-55	6	8
FL ST JOHNS	4.72	PB5-9	23-Jul-57	1	8
FL ST JOHNS	19.43	PB11-4	3-Sep-57	2	8
FL ST JOHNS	1.37	UE8-7	20-Jul-62	4	8
FL ST LUCIE	5.73	TS3-8	29-Apr-52	2	4
FL ST LUCIE	5.73	TS3-8	29-Apr-52	2	4
FL ST LUCIE	3.12	UpK6-3	20-Apr-53	3	4
FL ST LUCIE	0.08	UE8-8	21-Jul-62	9	4
FL SUMTER	5.73	TS3-8	29-Apr-52	2	9
FL SUMTER	5.73	TS3-8	29-Apr-52	2	9
FL SUMTER	20.85	UpK2-3	26-Mar-53	2	9
FL SUMTER	1.30	UpK3-7	6-Apr-53	9	9
FL SUMTER	3.89	UpK4-3	8-Apr-53	4	9
FL SUMTER	1.46	TP6-7	28-Mar-55	6	9
FL SUMTER	0.92	PB6-9	1-Aug-57	6	9
FL SUMTER	1.14	PB6-11	3-Aug-57	1	9
FL SUMTER	1.88	UE8-7	20-Jul-62	2	9
FL SUWANNEE	5.73	TS3-8	29-Apr-52	2	5
FL SUWANNEE	5.73	TS3-8	29-Apr-52	2	5
FL SUWANNEE	1.30	UpK3-7	6-Apr-53	9	5
FL SUWANNEE	3.89	UpK4-3	8-Apr-53	4	5
FL SUWANNEE	1.82	PB2-4	21-Jun-57	5	5
FL TAYLOR	6.40	TS3-7	28-Apr-52	2	6
FL TAYLOR	6.40	TS3-7	28-Apr-52	2	6
FL TAYLOR	1.65	TS3-8	29-Apr-52	5	6
FL TAYLOR	1.65	TS3-8	29-Apr-52	5	6
FL TAYLOR	3.89	UpK4-3	8-Apr-53	4	6
FL TAYLOR	0.76	UE8-7	20-Jul-62	8	6
FL UNION	5.73	TS3-8	29-Apr-52	2	7

COUNTY	uCi/ sqm	SHOT DAY	DATE	RANK FOR DAY	TIMES IN TOP 10
FL UNION	5.73	TS3-8	29-Apr-52	2	7
FL UNION	1.30	UpK3-7	6-Apr-53	9	7
FL UNION	3.89	UpK4-3	8-Apr-53	4	7
FL UNION	1.75	PB6-7	30-Jul-57	8	7
FL UNION	2.19	PB8-5	22-Aug-57	8	7
FL UNION	11.66	PB11-4	3-Sep-57	5	7
FL VOLUSIA	5.73	TS3-8	29-Apr-52	2	11
FL VOLUSIA	5.73	TS3-8	29-Apr-52	2	11
FL VOLUSIA	1.30	UpK3-7	6-Apr-53	9	11
FL VOLUSIA	3.89	UpK4-3	8-Apr-53	4	11
FL VOLUSIA	1.46	TP6-7	28-Mar-55	6	11
FL VOLUSIA	5.05	PB4-14	18-Jul-57	6	11
FL VOLUSIA	0.94	PB5-9	23-Jul-57	4	11
FL VOLUSIA	2.19	PB8-5	22-Aug-57	8	11
FL VOLUSIA	19.43	PB11-4	3-Sep-57	2	11
FL VOLUSIA	1.73	PB18-7	13-Oct-57	3	11
FL VOLUSIA	1.37	UE8-7	20-Jul-62	4	11
FL WAKULLA	6.40	TS3-7	28-Apr-52	2	6
FL WAKULLA	6.40	TS3-7	28-Apr-52	2	6
FL WAKULLA	1.65	TS3-8	29-Apr-52	5	6
FL WAKULLA	1.65	TS3-8	29-Apr-52	5	6
FL WAKULLA	1.30	UpK3-7	6-Apr-53	9	6
FL WAKULLA	3.89	UpK4-3	8-Apr-53	4	6
FL WALTON	3.14	TP6-5	26-Mar-55	2	2
FL WALTON	1.93	PB1-9	5-Jun-57	1	2
FL WASHINGTON	6.40	TS3-7	28-Apr-52	2	6
FL WASHINGTON	6.40	TS3-7	28-Apr-52	2	6
FL WASHINGTON	1.65	TS3-8	29-Apr-52	5	6
FL WASHINGTON	1.65	TS3-8	29-Apr-52	5	6
FL WASHINGTON	3.14	TP6-5	26-Mar-55	2	6
FL WASHINGTON	1.16	PB1-9	5-Jun-57	5	6

COUNTY	uCi/ sqm	SHOT DAY	DATE	RANK FOR DAY	TIMES IN TOP 10
GA APPLING	1.35	TS3-7	28-Apr-52	5	6
GA APPLING	1.35	TS3-7	28-Apr-52	5	6
GA APPLING	3.89	UpK4-3	8-Apr-53	4	6
GA APPLING	0.85	TP10-10	14-May-55	2	6
GA APPLING	2.73	PB2-4	21-Jun-57	2	6
GA APPLING	1.51	PB5-8	22-Jul-57	7	6
GA ATKINSON	5.73	TS3-8	29-Apr-52	2	5
GA ATKINSON	5.73	TS3-8	29-Apr-52	2	5
GA ATKINSON	1.30	UpK3-7	6-Apr-53	9	5
GA ATKINSON	3.89	UpK4-3	8-Apr-53	4	5
GA ATKINSON	2.73	PB2-4	21-Jun-57	2	5
GA BACON	1.35	TS3-7	28-Apr-52	5	5
GA BACON	1.35	TS3-7	28-Apr-52	5	5
GA BACON	7.78	UpK4-3	8-Apr-53	2	5
GA BACON	2.73	PB2-4	21-Jun-57	2	5
GA BACON	7.77	PB11-4	3-Sep-57	8	5
GA BAKER	6.40	TS3-7	28-Apr-52	2	5
GA BAKER	6.40	TS3-7	28-Apr-52	2	5
GA BAKER	1.65	TS3-8	29-Apr-52	5	5
GA BAKER	1.65	TS3-8	29-Apr-52	5	5
GA BAKER	5.99	PB17-3	30-Sep-57	7	5
GA BALDWIN	1.29	BJ1-2	29-Oct-51	3	2
GA BALDWIN	5.99	PB17-3	30-Sep-57	7	2
GA BANKS	1.82	PB2-4	21-Jun-57	5	2
GA BANKS	5.99	PB17-3	30-Sep-57	7	2
GA BARROW	1.08	TS5-5	11-May-52	4	2
GA BARROW	5.99	PB17-3	30-Sep-57	7	2
GA BARTOW	0.86	BJ1-2	29-Oct-51	5	3
GA BARTOW	1.91	TP6-4	25-Mar-55	9	3
GA BARTOW	3.96	PB17-3	30-Sep-57	9	3
GA BEN HILL	6.40	TS3-7	28-Apr-52	2	5
GA BEN HILL	6.40	TS3-7	28-Apr-52	2	5
GA BEN HILL	1.65	TS3-8	29-Apr-52	5	5
GA BEN HILL	1.65	TS3-8	29-Apr-52	5	5
GA BEN HILL	3.89	UpK4-3	8-Apr-53	4	5

COUNTY	uCi/ sqm	SHOT DAY	DATE	RANK FOR DAY	TIMES IN TOP 10
GA BERRIEN	6.40	TS3-7	28-Apr-52	2	8
GA BERRIEN	6.40	TS3-7	28-Apr-52	2	8
GA BERRIEN	1.65	TS3-8	29-Apr-52	5	8
GA BERRIEN	1.65	TS3-8	29-Apr-52	5	8
GA BERRIEN	1.30	UpK3-7	6-Apr-53	9	8
GA BERRIEN	3.89	UpK4-3	8-Apr-53	4	8
GA BERRIEN	0.97	TP6-7	28-Mar-55	8	8
GA BERRIEN	2.73	PB2-4	21-Jun-57	2	8
GA BIBB	21.09	PB16-5	27-Sep-57	9	2
GA BIBB	3.96	PB17-3	30-Sep-57	9	2
GA BLECKLEY	1.49	TS4-6	6-May-52	8	5
GA BLECKLEY	1.49	TS4-6	6-May-52	8	5
GA BLECKLEY	1.04	UpK5-2	12-Apr-53	4	5
GA BLECKLEY	0.85	TP10-10	14-May-55	2	5
GA BLECKLEY	3.96	PB17-3	30-Sep-57	9	5
GA BRANTLEY	5.73	TS3-8	29-Apr-52	2	8
GA BRANTLEY	5.73	TS3-8	29-Apr-52	2	8
GA BRANTLEY	1.30	UpK3-7	6-Apr-53	9	8
GA BRANTLEY	3.89	UpK4-3	8-Apr-53	4	8
GA BRANTLEY	5.05	PB4-14	18-Jul-57	6	8
GA BRANTLEY	1.51	PB5-8	22-Jul-57	7	8
GA BRANTLEY	1.81	UE8-6	19-Jul-62	7	8
GA BRANTLEY	1.37	UE8-7	20-Jul-62	4	8
GA BROOKS	6.40	TS3-7	28-Apr-52	2	6
GA BROOKS	6.40	TS3-7	28-Apr-52	2	6
GA BROOKS	1.65	TS3-8	29-Apr-52	5	6
GA BROOKS	1.65	TS3-8	29-Apr-52	5	6
GA BROOKS	1.30	UpK3-7	6-Apr-53	9	6
GA BROOKS	3.89	UpK4-3	8-Apr-53	4	6
GA BRYAN	2.63	TS3-7	28-Apr-52	4	7
GA BRYAN	2.63	TS3-7	28-Apr-52	4	7
GA BRYAN	2.12	TS6-6	30-May-52	9	7
GA BRYAN	5.05	PB4-14	18-Jul-57	6	7
GA BRYAN	1.51	PB5-8	22-Jul-57	7	7
GA BRYAN	0.94	PB5-9	23-Jul-57	4	7

COUNTY	uCi/ sqm	SHOT DAY	DATE	RANK FOR DAY	TIMES IN TOP 10
GA BRYAN	0.27	UE8-8	21-Jul-62	4	7
GA BULLOCH	2.63	TS3-7	28-Apr-52	4	4
GA BULLOCH	2.63	TS3-7	28-Apr-52	4	4
GA BULLOCH	1.51	PB5-8	22-Jul-57	7	4
GA BULLOCH	1.89	PB5-9	23-Jul-57	3	4
GA BURKE	1.35	TS3-7	28-Apr-52	5	3
GA BURKE	1.35	TS3-7	28-Apr-52	5	3
GA BURKE	1.51	PB5-8	22-Jul-57	7	3
GA BUTTS	0.86	BJ1-2	29-Oct-51	5	2
GA BUTTS	5.99	PB17-3	30-Sep-57	7	2
GA CALHOUN	6.40	TS3-7	28-Apr-52	2	5
GA CALHOUN	6.40	TS3-7	28-Apr-52	2	5
GA CALHOUN	1.65	TS3-8	29-Apr-52	5	5
GA CALHOUN	1.65	TS3-8	29-Apr-52	5	5
GA CALHOUN	5.99	PB17-3	30-Sep-57	7	5
GA CAMDEN	5.73	TS3-8	29-Apr-52	2	8
GA CAMDEN	5.73	TS3-8	29-Apr-52	2	8
GA CAMDEN	1.30	UpK3-7	6-Apr-53	9	8
GA CAMDEN	3.89	UpK4-3	8-Apr-53	4	8
GA CAMDEN	0.97	TP6-7	28-Mar-55	8	8
GA CAMDEN	0.85	TP10-10	14-May-55	2	8
GA CAMDEN	1.51	PB5-8	22-Jul-57	7	8
GA CAMDEN	11.66	PB11-4	3-Sep-57	5	8
GA CANDLER	1.35	TS3-7	28-Apr-52	5	4
GA CANDLER	1.35	TS3-7	28-Apr-52	5	4
GA CANDLER	3.89	UpK4-3	8-Apr-53	4	4
GA CANDLER	1.51	PB5-8	22-Jul-57	7	4
GA CARROLL	1.49	TS4-6	6-May-52	8	4
GA CARROLL	1.49	TS4-6	6-May-52	8	4
GA CARROLL	0.95	TS5-5	11-May-52	9	4
GA CARROLL	3.96	PB17-3	30-Sep-57	9	4
GA CATOOSA	8.10	BJ3-6	6-Nov-51	5	3
GA CATOOSA	1.91	TP6-4	25-Mar-55	9	3
GA CATOOSA	5.99	PB17-3	30-Sep-57	7	3
GA CHARLTON	5.73	TS3-8	29-Apr-52	2	7

COUNTY	uCi/ sqm	SHOT DAY	DATE	RANK FOR DAY	TIMES IN TOP 10
GA CHARLTON	5.73	TS3-8	29-Apr-52	2	7
GA CHARLTON	1.30	UpK3-7	6-Apr-53	9	7
GA CHARLTON	3.89	UpK4-3	8-Apr-53	4	7
GA CHARLTON	1.46	TP6-7	28-Mar-55	6	7
GA CHARLTON	11.66	PB11-4	3-Sep-57	5	7
GA CHARLTON	1.81	UE8-6	19-Jul-62	7	7
GA CHATHAM	1.09	TS3-7	28-Apr-52	6	6
GA CHATHAM	1.09	TS3-7	28-Apr-52	6	6
GA CHATHAM	5.05	PB4-14	18-Jul-57	6	6
GA CHATHAM	1.51	PB5-8	22-Jul-57	7	6
GA CHATHAM	0.94	PB5-9	23-Jul-57	4	6
GA CHATHAM	0.27	UE8-8	21-Jul-62	4	6
GA CHATTAHOOCHE	1.16	PB1-9	5-Jun-57	5	2
GA CHATTAHOOCHE	5.99	PB17-3	30-Sep-57	7	2
GA CHATTOOGA	8.10	BJ3-6	6-Nov-51	5	3
GA CHATTOOGA	2.87	TP6-4	25-Mar-55	4	3
GA CHATTOOGA	5.99	PB17-3	30-Sep-57	7	3
GA CHEROKEE	0.86	BJ1-2	29-Oct-51	5	5
GA CHEROKEE	1.91	TP6-4	25-Mar-55	9	5
GA CHEROKEE	1.16	PB1-9	5-Jun-57	5	5
GA CHEROKEE	1.82	PB2-4	21-Jun-57	5	5
GA CHEROKEE	5.99	PB17-3	30-Sep-57	7	5
GA CLARKE	1.29	BJ1-2	29-Oct-51	3	2
GA CLARKE	9.96	PB17-3	30-Sep-57	3	2
GA CLAY	6.40	TS3-7	28-Apr-52	2	6
GA CLAY	6.40	TS3-7	28-Apr-52	2	6
GA CLAY	1.65	TS3-8	29-Apr-52	5	6
GA CLAY	1.65	TS3-8	29-Apr-52	5	6
GA CLAY	1.16	PB1-9	5-Jun-57	5	6
GA CLAY	9.96	PB17-3	30-Sep-57	3	6
GA CLAYTON	26.92	PB16-5	27-Sep-57	2	3
GA CLAYTON	5.99	PB17-3	30-Sep-57	7	3
GA CLAYTON	5.58	PB17-6	3-Oct-57	6	3
GA CLINCH	5.73	TS3-8	29-Apr-52	2	6
GA CLINCH	5.73	TS3-8	29-Apr-52	2	6

COUNTY	uCi/ sqm	SHOT DAY	DATE	RANK FOR DAY	TIMES IN TOP 10
GA CLINCH	1.30	UpK3-7	6-Apr-53	9	6
GA CLINCH	3.89	UpK4-3	8-Apr-53	4	6
GA CLINCH	0.97	TP6-7	28-Mar-55	8	6
GA CLINCH	7.77	PB11-4	3-Sep-57	8	6
GA COBB	3.96	PB17-3	30-Sep-57	9	1
GA COFFEE	5.73	TS3-8	29-Apr-52	2	6
GA COFFEE	5.73	TS3-8	29-Apr-52	2	6
GA COFFEE	1.30	UpK3-7	6-Apr-53	9	6
GA COFFEE	3.89	UpK4-3	8-Apr-53	4	6
GA COFFEE	0.85	TP10-10	14-May-55	2	6
GA COFFEE	7.77	PB11-4	3-Sep-57	8	6
GA COLQUITT	6.40	TS3-7	28-Apr-52	2	7
GA COLQUITT	6.40	TS3-7	28-Apr-52	2	7
GA COLQUITT	1.65	TS3-8	29-Apr-52	5	7
GA COLQUITT	1.65	TS3-8	29-Apr-52	5	7
GA COLQUITT	1.30	UpK3-7	6-Apr-53	9	7
GA COLQUITT	3.89	UpK4-3	8-Apr-53	4	7
GA COLQUITT	11.66	PB11-4	3-Sep-57	5	7
GA COLUMBIA	2.20	PB2-8	25-Jun-57	6	2
GA COLUMBIA	1.51	PB5-8	22-Jul-57	7	2
GA COOK	6.40	TS3-7	28-Apr-52	2	8
GA COOK	6.40	TS3-7	28-Apr-52	2	8
GA COOK	1.65	TS3-8	29-Apr-52	5	8
GA COOK	1.65	TS3-8	29-Apr-52	5	8
GA COOK	1.30	UpK3-7	6-Apr-53	9	8
GA COOK	3.89	UpK4-3	8-Apr-53	4	8
GA COOK	0.97	TP6-7	28-Mar-55	8	8
GA COOK	2.73	PB2-4	21-Jun-57	2	8
GA COWETA	0.97	TS5-5	11-May-52	8	2
GA COWETA	5.99	PB17-3	30-Sep-57	7	2
GA CRAWFORD	22.11	PB16-5	27-Sep-57	8	3
GA CRAWFORD	9.96	PB17-3	30-Sep-57	3	3
GA CRAWFORD	5.12	PB17-6	3-Oct-57	8	3
GA CRISP	6.40	TS3-7	28-Apr-52	2	4
GA CRISP	6.40	TS3-7	28-Apr-52	2	4

COUNTY	uCi/ sqm	SHOT DAY	DATE	RANK FOR DAY	TIMES IN TOP 10
GA CRISP	1.65	TS3-8	29-Apr-52	5	4
GA CRISP	1.65	TS3-8	29-Apr-52	5	4
GA DADE	4.86	BJ3-6	6-Nov-51	7	3
GA DADE	2.87	TP6-4	25-Mar-55	4	3
GA DADE	3.96	PB17-3	30-Sep-57	9	3
GA DAWSON	1.29	BJ1-2	29-Oct-51	3	3
GA DAWSON	2.87	TP6-4	25-Mar-55	4	3
GA DAWSON	5.99	PB17-3	30-Sep-57	7	3
GA DE KALB	0.86	BJ1-2	29-Oct-51	5	3
GA DE KALB	25.90	PB16-5	27-Sep-57	4	3
GA DE KALB	3.96	PB17-3	30-Sep-57	9	3
GA DECATUR	6.40	TS3-7	28-Apr-52	2	7
GA DECATUR	6.40	TS3-7	28-Apr-52	2	7
GA DECATUR	1.65	TS3-8	29-Apr-52	5	7
GA DECATUR	1.65	TS3-8	29-Apr-52	5	7
GA DECATUR	1.30	UpK3-7	6-Apr-53	9	7
GA DECATUR	3.89	UpK4-3	8-Apr-53	4	7
GA DECATUR	1.46	TP6-7	28-Mar-55	6	7
GA DODGE	0.86	BJ1-2	29-Oct-51	5	2
GA DODGE	3.96	PB17-3	30-Sep-57	9	2
GA DOUGHERTY	6.40	TS3-7	28-Apr-52	2	5
GA DOUGHERTY	6.40	TS3-7	28-Apr-52	2	5
GA DOUGHERTY	1.65	TS3-8	29-Apr-52	5	5
GA DOUGHERTY	1.65	TS3-8	29-Apr-52	5	5
GA DOUGHERTY	9.96	PB17-3	30-Sep-57	3	5
GA DOUGLAS	0.86	BJ1-2	29-Oct-51	5	5
GA DOUGLAS	1.91	TP6-4	25-Mar-55	9	5
GA DOUGLAS	23.79	PB16-5	27-Sep-57	6	5
GA DOUGLAS	5.99	PB17-3	30-Sep-57	7	5
GA DOUGLAS	6.15	PB17-6	3-Oct-57	3	5
GA EARLY	6.40	TS3-7	28-Apr-52	2	5
GA EARLY	6.40	TS3-7	28-Apr-52	2	5
GA EARLY	1.65	TS3-8	29-Apr-52	5	5
GA EARLY	1.65	TS3-8	29-Apr-52	5	5
GA EARLY	1.12	PB6-12	4-Aug-57	1	5

COUNTY	uCi/ sqm	SHOT DAY	DATE	RANK FOR DAY	TIMES IN TOP 10
GA ECHOLS	5.73	TS3-8	29-Apr-52	2	4
GA ECHOLS	5.73	TS3-8	29-Apr-52	2	4
GA ECHOLS	3.89	UpK4-3	8-Apr-53	4	4
GA ECHOLS	2.19	PB8-5	22-Aug-57	8	4
GA EFFINGHAM	1.35	TS3-7	28-Apr-52	5	5
GA EFFINGHAM	1.35	TS3-7	28-Apr-52	5	5
GA EFFINGHAM	1.51	PB5-8	22-Jul-57	7	5
GA EFFINGHAM	1.89	PB5-9	23-Jul-57	3	5
GA EFFINGHAM	0.08	UE8-8	21-Jul-62	10	5
GA ELBERT	1.13	TS5-5	11-May-52	3	4
GA ELBERT	1.91	TP6-4	25-Mar-55	9	4
GA ELBERT	1.51	PB5-8	22-Jul-57	7	4
GA ELBERT	5.99	PB17-3	30-Sep-57	7	4
GA EMANUEL	1.35	TS3-7	28-Apr-52	5	4
GA EMANUEL	1.35	TS3-7	28-Apr-52	5	4
GA EMANUEL	1.51	PB5-8	22-Jul-57	7	4
GA EMANUEL	5.99	PB17-3	30-Sep-57	7	4
GA EVANS	1.35	TS3-7	28-Apr-52	5	7
GA EVANS	1.35	TS3-7	28-Apr-52	5	7
GA EVANS	3.89	UpK4-3	8-Apr-53	4	7
GA EVANS	0.85	TP10-10	14-May-55	2	7
GA EVANS	2.73	PB2-4	21-Jun-57	2	7
GA EVANS	1.51	PB5-8	22-Jul-57	7	7
GA EVANS	1.89	PB5-9	23-Jul-57	3	7
GA FANNIN	4.86	BJ3-6	6-Nov-51	7	8
GA FANNIN	1.49	TS4-6	6-May-52	8	8
GA FANNIN	1.49	TS4-6	6-May-52	8	8
GA FANNIN	1.00	TS5-5	11-May-52	7	8
GA FANNIN	2.87	TP6-4	25-Mar-55	4	8
GA FANNIN	1.16	PB1-9	5-Jun-57	5	8
GA FANNIN	1.82	PB2-4	21-Jun-57	5	8
GA FANNIN	3.96	PB17-3	30-Sep-57	9	8
GA FAYETTE	29.19	PB16-5	27-Sep-57	1	3
GA FAYETTE	5.99	PB17-3	30-Sep-57	7	3
GA FAYETTE	6.27	PB17-6	3-Oct-57	2	3

COUNTY	uCi/ sqm	SHOT DAY	DATE	RANK FOR DAY	TIMES IN TOP 10
GA FLOYD	4.59	BJ3-5	5-Nov-51	6	5
GA FLOYD	4.86	BJ3-6	6-Nov-51	7	5
GA FLOYD	1.91	TP6-4	25-Mar-55	9	5
GA FLOYD	1.16	PB1-9	5-Jun-57	5	5
GA FLOYD	3.96	PB17-3	30-Sep-57	9	5
GA FORSYTH	1.91	TP6-4	25-Mar-55	9	2
GA FORSYTH	5.99	PB17-3	30-Sep-57	7	2
GA FRANKLIN	0.86	BJ1-2	29-Oct-51	5	3
GA FRANKLIN	5.97	PB5-8	22-Jul-57	3	3
GA FRANKLIN	5.99	PB17-3	30-Sep-57	7	3
GA FULTON	26.01	PB16-5	27-Sep-57	3	1
GA GILMER	3.24	BJ3-6	6-Nov-51	10	4
GA GILMER	2.87	TP6-4	25-Mar-55	4	4
GA GILMER	1.16	PB1-9	5-Jun-57	5	4
GA GILMER	3.96	PB17-3	30-Sep-57	9	4
GA GLASCOCK	1.51	PB5-8	22-Jul-57	7	3
GA GLASCOCK	9.96	PB17-3	30-Sep-57	3	3
GA GLASCOCK	2.58	UE8-6	19-Jul-62	4	3
GA GLYNN	5.73	TS3-8	29-Apr-52	2	8
GA GLYNN	5.73	TS3-8	29-Apr-52	2	8
GA GLYNN	1.30	UpK3-7	6-Apr-53	9	8
GA GLYNN	3.89	UpK4-3	8-Apr-53	4	8
GA GLYNN	0.97	TP6-7	28-Mar-55	8	8
GA GLYNN	0.85	TP10-10	14-May-55	2	8
GA GLYNN	1.51	PB5-8	22-Jul-57	7	8
GA GLYNN	11.66	PB11-4	3-Sep-57	5	8
GA GORDON	8.10	BJ3-6	6-Nov-51	5	6
GA GORDON	0.95	TS5-5	11-May-52	9	6
GA GORDON	1.91	TP6-4	25-Mar-55	9	6
GA GORDON	1.16	PB1-9	5-Jun-57	5	6
GA GORDON	1.82	PB2-4	21-Jun-57	5	6
GA GORDON	3.96	PB17-3	30-Sep-57	9	6
GA GRADY	6.40	TS3-7	28-Apr-52	2	6
GA GRADY	6.40	TS3-7	28-Apr-52	2	6
GA GRADY	1.65	TS3-8	29-Apr-52	5	6

COUNTY	uCi/ sqm	SHOT DAY	DATE	RANK FOR DAY	TIMES IN TOP 10
GA GRADY	1.65	TS3-8	29-Apr-52	5	6
GA GRADY	0.99	UpK5-5	15-Apr-53	2	6
GA GRADY	0.69	UE8-7	20-Jul-62	9	6
GA GREENE	1.29	BJ1-2	29-Oct-51	3	3
GA GREENE	1.51	PB5-8	22-Jul-57	7	3
GA GREENE	5.99	PB17-3	30-Sep-57	7	3
GA GWINNETT	1.29	BJ1-2	29-Oct-51	3	4
GA GWINNETT	1.91	TP6-4	25-Mar-55	9	4
GA GWINNETT	22.38	PB16-5	27-Sep-57	7	4
GA GWINNETT	5.99	PB17-3	30-Sep-57	7	4
GA HABERSHAM	5.99	PB17-3	30-Sep-57	7	3
GA HABERSHAM	2.11	UE8-5	18-Jul-62	7	3
GA HABERSHAM	1.32	UE8-6	19-Jul-62	9	3
GA HALL	0.86	BJ1-2	29-Oct-51	5	3
GA HALL	1.91	TP6-4	25-Mar-55	9	3
GA HALL	5.99	PB17-3	30-Sep-57	7	3
GA HANCOCK	1.29	BJ1-2	29-Oct-51	3	4
GA HANCOCK	1.16	PB1-9	5-Jun-57	5	4
GA HANCOCK	1.51	PB5-8	22-Jul-57	7	4
GA HANCOCK	5.99	PB17-3	30-Sep-57	7	4
GA HARALSON	1.91	TP6-4	25-Mar-55	9	4
GA HARALSON	2.73	PB2-4	21-Jun-57	2	4
GA HARALSON	1.55	PB2-5	22-Jun-57	6	4
GA HARALSON	5.99	PB17-3	30-Sep-57	7	4
GA HART	1.91	TP6-4	25-Mar-55	9	3
GA HART	1.51	PB5-8	22-Jul-57	7	3
GA HART	5.99	PB17-3	30-Sep-57	7	3
GA HEARD	0.86	BJ1-2	29-Oct-51	5	3
GA HEARD	1.04	UpK5-2	12-Apr-53	4	3
GA HEARD	5.99	PB17-3	30-Sep-57	7	3
GA HENRY	0.86	BJ1-2	29-Oct-51	5	2
GA HENRY	5.99	PB17-3	30-Sep-57	7	2
GA HOUSTON	1.49	TS4-6	6-May-52	8	5
GA HOUSTON	1.49	TS4-6	6-May-52	8	5
GA HOUSTON	1.04	UpK5-2	12-Apr-53	4	5

COUNTY	uCi/ sqm	SHOT DAY	DATE	RANK FOR DAY	TIMES IN TOP 10
GA HOUSTON	1.82	PB2-4	21-Jun-57	5	5
GA HOUSTON	3.96	PB17-3	30-Sep-57	9	5
GA IRWIN	6.40	TS3-7	28-Apr-52	2	5
GA IRWIN	6.40	TS3-7	28-Apr-52	2	5
GA IRWIN	1.65	TS3-8	29-Apr-52	5	5
GA IRWIN	1.65	TS3-8	29-Apr-52	5	5
GA IRWIN	3.89	UpK4-3	8-Apr-53	4	5
GA JACKSON	1.29	BJ1-2	29-Oct-51	3	3
GA JACKSON	1.04	UpK5-2	12-Apr-53	4	3
GA JACKSON	5.99	PB17-3	30-Sep-57	7	3
GA JASPER	9.96	PB17-3	30-Sep-57	3	1
GA JEFF DAVIS	1.35	TS3-7	28-Apr-52	5	5
GA JEFF DAVIS	1.35	TS3-7	28-Apr-52	5	5
GA JEFF DAVIS	7.78	UpK4-3	8-Apr-53	2	5
GA JEFF DAVIS	2.73	PB2-4	21-Jun-57	2	5
GA JEFF DAVIS	7.77	PB11-4	3-Sep-57	8	5
GA JEFFERSON	1.35	TS3-7	28-Apr-52	5	5
GA JEFFERSON	1.35	TS3-7	28-Apr-52	5	5
GA JEFFERSON	5.97	PB5-8	22-Jul-57	3	5
GA JEFFERSON	5.99	PB17-3	30-Sep-57	7	5
GA JEFFERSON	2.58	UE8-6	19-Jul-62	4	5
GA JENKINS	1.35	TS3-7	28-Apr-52	5	4
GA JENKINS	1.35	TS3-7	28-Apr-52	5	4
GA JENKINS	1.51	PB5-8	22-Jul-57	7	4
GA JENKINS	1.89	PB5-9	23-Jul-57	3	4
GA JOHNSON	0.86	BJ1-2	29-Oct-51	5	5
GA JOHNSON	1.35	TS3-7	28-Apr-52	5	5
GA JOHNSON	1.35	TS3-7	28-Apr-52	5	5
GA JOHNSON	5.97	PB5-8	22-Jul-57	3	5
GA JOHNSON	5.99	PB17-3	30-Sep-57	7	5
GA JONES	3.96	PB17-3	30-Sep-57	9	1
GA LAMAR	0.86	BJ1-2	29-Oct-51	5	2
GA LAMAR	5.99	PB17-3	30-Sep-57	7	2
GA LANIER	6.40	TS3-7	28-Apr-52	2	8
GA LANIER	6.40	TS3-7	28-Apr-52	2	8

COUNTY	uCi/ sqm	SHOT DAY	DATE	RANK FOR DAY	TIMES IN TOP 10
GA LANIER	1.65	TS3-8	29-Apr-52	5	8
GA LANIER	1.65	TS3-8	29-Apr-52	5	8
GA LANIER	1.30	UpK3-7	6-Apr-53	9	8
GA LANIER	3.89	UpK4-3	8-Apr-53	4	8
GA LANIER	0.97	TP6-7	28-Mar-55	8	8
GA LANIER	7.77	PB11-4	3-Sep-57	8	8
GA LAURENS	0.86	BJ1-2	29-Oct-51	5	4
GA LAURENS	1.35	TS3-7	28-Apr-52	5	4
GA LAURENS	1.35	TS3-7	28-Apr-52	5	4
GA LAURENS	1.51	PB5-8	22-Jul-57	7	4
GA LEE	6.40	TS3-7	28-Apr-52	2	4
GA LEE	6.40	TS3-7	28-Apr-52	2	4
GA LEE	1.65	TS3-8	29-Apr-52	5	4
GA LEE	1.65	TS3-8	29-Apr-52	5	4
GA LIBERTY	2.63	TS3-7	28-Apr-52	4	9
GA LIBERTY	2.63	TS3-7	28-Apr-52	4	9
GA LIBERTY	3.89	UpK4-3	8-Apr-53	4	9
GA LIBERTY	0.85	TP10-10	14-May-55	2	9
GA LIBERTY	1.51	PB5-8	22-Jul-57	7	9
GA LIBERTY	1.89	PB5-9	23-Jul-57	3	9
GA LIBERTY	3.61	PB8-5	22-Aug-57	5	9
GA LIBERTY	19.43	PB11-4	3-Sep-57	2	9
GA LIBERTY	0.08	UE8-8	21-Jul-62	9	9
GA LINCOLN	3.73	BJ2-5	3-Nov-51	9	3
GA LINCOLN	2.20	PB2-8	25-Jun-57	6	3
GA LINCOLN	1.51	PB5-8	22-Jul-57	7	3
GA LONG	2.63	TS3-7	28-Apr-52	4	7
GA LONG	2.63	TS3-7	28-Apr-52	4	7
GA LONG	3.89	UpK4-3	8-Apr-53	4	7
GA LONG	0.85	TP10-10	14-May-55	2	7
GA LONG	1.51	PB5-8	22-Jul-57	7	7
GA LONG	3.61	PB8-5	22-Aug-57	5	7
GA LONG	19.43	PB11-4	3-Sep-57	2	7
GA LOWNDES	6.40	TS3-7	28-Apr-52	2	6
GA LOWNDES	6.40	TS3-7	28-Apr-52	2	6

COUNTY	uCi/ sqm	SHOT DAY	DATE	RANK FOR DAY	TIMES IN TOP 10
GA LOWNDES	1.65	TS3-8	29-Apr-52	5	6
GA LOWNDES	1.65	TS3-8	29-Apr-52	5	6
GA LOWNDES	3.89	UpK4-3	8-Apr-53	4	6
GA LOWNDES	7.77	PB11-4	3-Sep-57	8	6
GA LUMPKIN	1.91	TP6-4	25-Mar-55	9	3
GA LUMPKIN	1.82	PB2-4	21-Jun-57	5	3
GA LUMPKIN	5.99	PB17-3	30-Sep-57	7	3
GA MADISON	0.86	BJ1-2	29-Oct-51	5	3
GA MADISON	1.51	PB5-8	22-Jul-57	7	3
GA MADISON	5.99	PB17-3	30-Sep-57	7	3
GA MCDUFFIE	2.12	TS6-6	30-May-52	10	4
GA MCDUFFIE	1.51	PB5-8	22-Jul-57	7	4
GA MCDUFFIE	9.96	PB17-3	30-Sep-57	3	4
GA MCDUFFIE	1.37	UE8-7	20-Jul-62	4	4
GA MCINTOSH	3.89	BJ2-2	31-Oct-51	10	7
GA MCINTOSH	1.35	TS3-7	28-Apr-52	5	7
GA MCINTOSH	1.35	TS3-7	28-Apr-52	5	7
GA MCINTOSH	3.89	UpK4-3	8-Apr-53	4	7
GA MCINTOSH	0.97	TP6-7	28-Mar-55	8	7
GA MCINTOSH	0.85	TP10-10	14-May-55	2	7
GA MCINTOSH	1.51	PB5-8	22-Jul-57	7	7
GA MERIWETHER	2.15	BJ1-2	29-Oct-51	1	2
GA MERIWETHER	5.99	PB17-3	30-Sep-57	7	2
GA MILLER	6.40	TS3-7	28-Apr-52	2	5
GA MILLER	6.40	TS3-7	28-Apr-52	2	5
GA MILLER	1.65	TS3-8	29-Apr-52	5	5
GA MILLER	1.65	TS3-8	29-Apr-52	5	5
GA MILLER	1.12	PB6-12	4-Aug-57	1	5
GA MITCHELL	6.40	TS3-7	28-Apr-52	2	7
GA MITCHELL	6.40	TS3-7	28-Apr-52	2	7
GA MITCHELL	1.65	TS3-8	29-Apr-52	5	7
GA MITCHELL	1.65	TS3-8	29-Apr-52	5	7
GA MITCHELL	1.16	PB1-9	5-Jun-57	5	7
GA MITCHELL	9.96	PB17-3	30-Sep-57	3	7
GA MITCHELL	1.81	UE8-6	19-Jul-62	7	7

COUNTY	uCi/ sqm	SHOT DAY	DATE	RANK FOR DAY	TIMES IN TOP 10
GA MONROE	24.06	PB16-5	27-Sep-57	5	2
GA MONROE	3.96	PB17-3	30-Sep-57	9	2
GA MONTGOMERY	1.35	TS3-7	28-Apr-52	5	5
GA MONTGOMERY	1.35	TS3-7	28-Apr-52	5	5
GA MONTGOMERY	1.30	UpK3-7	6-Apr-53	9	5
GA MONTGOMERY	3.89	UpK4-3	8-Apr-53	4	5
GA MONTGOMERY	1.51	PB5-8	22-Jul-57	7	5
GA MORGAN	1.29	BJ1-2	29-Oct-51	3	3
GA MORGAN	5.99	PB17-3	30-Sep-57	7	3
GA MORGAN	1.88	UE8-7	20-Jul-62	2	3
GA MURRAY	8.10	BJ3-6	6-Nov-51	5	3
GA MURRAY	2.87	TP6-4	25-Mar-55	4	3
GA MURRAY	3.96	PB17-3	30-Sep-57	9	3
GA NEWTON	0.86	BJ1-2	29-Oct-51	5	4
GA NEWTON	1.82	PB2-4	21-Jun-57	5	4
GA NEWTON	5.99	PB17-3	30-Sep-57	7	4
GA NEWTON	1.33	UE8-7	20-Jul-62	5	4
GA OCONEE	1.29	BJ1-2	29-Oct-51	3	2
GA OCONEE	3.96	PB17-3	30-Sep-57	9	2
GA OGLETHORPE	0.86	BJ1-2	29-Oct-51	5	4
GA OGLETHORPE	1.51	PB5-8	22-Jul-57	7	4
GA OGLETHORPE	5.99	PB17-3	30-Sep-57	7	4
GA OGLETHORPE	6.73	PB17-6	3-Oct-57	1	4
GA PAULDING	1.29	BJ1-2	29-Oct-51	3	5
GA PAULDING	1.91	TP6-4	25-Mar-55	9	5
GA PAULDING	1.82	PB2-4	21-Jun-57	5	5
GA PAULDING	5.99	PB17-3	30-Sep-57	7	5
GA PAULDING	5.63	PB17-6	3-Oct-57	5	5
GA PEACH	9.96	PB17-3	30-Sep-57	3	1
GA PICKENS	0.86	BJ1-2	29-Oct-51	5	4
GA PICKENS	2.87	TP6-4	25-Mar-55	4	4
GA PICKENS	1.82	PB2-4	21-Jun-57	5	4
GA PICKENS	5.99	PB17-3	30-Sep-57	7	4
GA PIERCE	5.73	TS3-8	29-Apr-52	2	6
GA PIERCE	5.73	TS3-8	29-Apr-52	2	6

COUNTY	uCi/ sqm	SHOT DAY	DATE	RANK FOR DAY	TIMES IN TOP 10
GA PIERCE	1.30	UpK3-7	6-Apr-53	9	6
GA PIERCE	3.89	UpK4-3	8-Apr-53	4	6
GA PIERCE	0.97	TP6-7	28-Mar-55	8	6
GA PIERCE	1.51	PB5-8	22-Jul-57	7	6
GA PIKE	2.15	BJ1-2	29-Oct-51	1	2
GA PIKE	5.99	PB17-3	30-Sep-57	7	2
GA POLK	1.91	TP6-4	25-Mar-55	9	1
GA PULASKI	1.04	UpK5-2	12-Apr-53	4	3
GA PULASKI	2.73	PB2-4	21-Jun-57	2	3
GA PULASKI	9.96	PB17-3	30-Sep-57	3	3
GA PUTNAM	1.16	PB1-9	5-Jun-57	5	2
GA PUTNAM	9.96	PB17-3	30-Sep-57	3	2
GA QUITMAN	1.16	PB1-9	5-Jun-57	5	2
GA QUITMAN	9.96	PB17-3	30-Sep-57	3	2
GA RABUN	0.86	BJ1-2	29-Oct-51	5	7
GA RABUN	3.11	TP6-4	25-Mar-55	3	7
GA RABUN	0.97	TP6-5	26-Mar-55	8	7
GA RABUN	6.96	PB2-5	22-Jun-57	1	7
GA RABUN	5.97	PB5-8	22-Jul-57	3	7
GA RABUN	14.98	PB17-3	30-Sep-57	2	7
GA RABUN	2.11	UE8-5	18-Jul-62	7	7
GA RANDOLPH	6.40	TS3-7	28-Apr-52	2	4
GA RANDOLPH	6.40	TS3-7	28-Apr-52	2	4
GA RANDOLPH	1.65	TS3-8	29-Apr-52	5	4
GA RANDOLPH	1.65	TS3-8	29-Apr-52	5	4
GA RICHMOND	1.51	PB5-8	22-Jul-57	7	1
GA ROCKDALE	0.86	BJ1-2	29-Oct-51	5	4
GA ROCKDALE	1.04	UpK5-2	12-Apr-53	4	4
GA ROCKDALE	5.99	PB17-3	30-Sep-57	7	4
GA ROCKDALE	1.33	UE8-7	20-Jul-62	5	4
GA SCHLEY	2.73	PB2-4	21-Jun-57	2	3
GA SCHLEY	20.96	PB16-5	27-Sep-57	10	3
GA SCHLEY	5.99	PB17-3	30-Sep-57	7	3
GA SCREVEN	2.63	TS3-7	28-Apr-52	4	5
GA SCREVEN	2.63	TS3-7	28-Apr-52	4	5

COUNTY	uCi/ sqm	SHOT DAY	DATE	RANK FOR DAY	TIMES IN TOP 10
GA SCREVEN	1.51	PB5-8	22-Jul-57	7	5
GA SCREVEN	1.89	PB5-9	23-Jul-57	3	5
GA SCREVEN	0.10	UE8-8	21-Jul-62	8	5
GA SEMINOLE	6.40	TS3-7	28-Apr-52	2	4
GA SEMINOLE	6.40	TS3-7	28-Apr-52	2	4
GA SEMINOLE	1.65	TS3-8	29-Apr-52	5	4
GA SEMINOLE	1.65	TS3-8	29-Apr-52	5	4
GA SPALDING	0.86	BJ1-2	29-Oct-51	5	2
GA SPALDING	5.99	PB17-3	30-Sep-57	7	2
GA STEPHENS	0.86	BJ1-2	29-Oct-51	5	4
GA STEPHENS	1.51	PB5-8	22-Jul-57	7	4
GA STEPHENS	5.99	PB17-3	30-Sep-57	7	4
GA STEPHENS	2.11	UE8-5	18-Jul-62	7	4
GA STEWART	1.16	PB1-9	5-Jun-57	5	2
GA STEWART	5.99	PB17-3	30-Sep-57	7	2
GA SUMTER	6.40	TS3-7	28-Apr-52	2	4
GA SUMTER	6.40	TS3-7	28-Apr-52	2	4
GA SUMTER	1.65	TS3-8	29-Apr-52	5	4
GA SUMTER	1.65	TS3-8	29-Apr-52	5	4
GA TALBOT	0.92	TS5-5	11-May-52	10	1
GA TALIAFERRO	1.29	BJ1-2	29-Oct-51	3	5
GA TALIAFERRO	0.95	TS5-5	11-May-52	9	5
GA TALIAFERRO	1.51	PB5-8	22-Jul-57	7	5
GA TALIAFERRO	9.96	PB17-3	30-Sep-57	3	5
GA TALIAFERRO	1.37	UE8-7	20-Jul-62	4	5
GA TATTNALL	1.35	TS3-7	28-Apr-52	5	7
GA TATTNALL	1.35	TS3-7	28-Apr-52	5	7
GA TATTNALL	3.89	UpK4-3	8-Apr-53	4	7
GA TATTNALL	0.85	TP10-10	14-May-55	2	7
GA TATTNALL	2.73	PB2-4	21-Jun-57	2	7
GA TATTNALL	1.51	PB5-8	22-Jul-57	7	7
GA TATTNALL	1.89	PB5-9	23-Jul-57	3	7
GA TAYLOR	5.99	PB17-3	30-Sep-57	7	1
GA TELFAIR	1.35	TS3-7	28-Apr-52	5	5
GA TELFAIR	1.35	TS3-7	28-Apr-52	5	5

COUNTY	uCi/ sqm	SHOT DAY	DATE	RANK FOR DAY	TIMES IN TOP 10
GA TELFAIR	1.30	UpK3-7	6-Apr-53	9	5
GA TELFAIR	3.89	UpK4-3	8-Apr-53	4	5
GA TELFAIR	5.05	PB4-14	18-Jul-57	6	5
GA TERRELL	6.40	TS3-7	28-Apr-52	2	5
GA TERRELL	6.40	TS3-7	28-Apr-52	2	5
GA TERRELL	1.65	TS3-8	29-Apr-52	5	5
GA TERRELL	1.65	TS3-8	29-Apr-52	5	5
GA TERRELL	9.96	PB17-3	30-Sep-57	3	5
GA THOMAS	6.40	TS3-7	28-Apr-52	2	8
GA THOMAS	6.40	TS3-7	28-Apr-52	2	8
GA THOMAS	1.65	TS3-8	29-Apr-52	5	8
GA THOMAS	1.65	TS3-8	29-Apr-52	5	8
GA THOMAS	1.30	UpK3-7	6-Apr-53	9	8
GA THOMAS	3.89	UpK4-3	8-Apr-53	4	8
GA THOMAS	0.97	TP6-7	28-Mar-55	8	8
GA THOMAS	3.17	UE8-6	19-Jul-62	2	8
GA TIFT	6.40	TS3-7	28-Apr-52	2	6
GA TIFT	6.40	TS3-7	28-Apr-52	2	6
GA TIFT	1.65	TS3-8	29-Apr-52	5	6
GA TIFT	1.65	TS3-8	29-Apr-52	5	6
GA TIFT	1.30	UpK3-7	6-Apr-53	9	6
GA TIFT	3.89	UpK4-3	8-Apr-53	4	6
GA TOOMBS	1.35	TS3-7	28-Apr-52	5	5
GA TOOMBS	1.35	TS3-7	28-Apr-52	5	5
GA TOOMBS	1.30	UpK3-7	6-Apr-53	9	5
GA TOOMBS	3.89	UpK4-3	8-Apr-53	4	5
GA TOOMBS	1.51	PB5-8	22-Jul-57	7	5
GA TOWNS	1.55	UpK5-2	12-Apr-53	2	5
GA TOWNS	3.11	TP6-4	25-Mar-55	3	5
GA TOWNS	0.97	TP6-5	26-Mar-55	8	5
GA TOWNS	2.78	PB2-5	22-Jun-57	4	5
GA TOWNS	14.98	PB17-3	30-Sep-57	2	5
GA TREUTLEN	1.35	TS3-7	28-Apr-52	5	3
GA TREUTLEN	1.35	TS3-7	28-Apr-52	5	3
GA TREUTLEN	1.51	PB5-8	22-Jul-57	7	3

COUNTY	uCi/ sqm	SHOT DAY	DATE	RANK FOR DAY	TIMES IN TOP 10
GA TROUP	1.29	BJ1-2	29-Oct-51	3	6
GA TROUP	1.49	TS4-6	6-May-52	8	6
GA TROUP	1.49	TS4-6	6-May-52	8	6
GA TROUP	0.92	TS5-5	11-May-52	10	6
GA TROUP	2.73	PB2-4	21-Jun-57	2	6
GA TROUP	5.99	PB17-3	30-Sep-57	7	6
GA TURNER	6.40	TS3-7	28-Apr-52	2	5
GA TURNER	6.40	TS3-7	28-Apr-52	2	5
GA TURNER	1.65	TS3-8	29-Apr-52	5	5
GA TURNER	1.65	TS3-8	29-Apr-52	5	5
GA TURNER	1.04	UpK5-2	12-Apr-53	4	5
GA TWIGGS	1.49	TS4-6	6-May-52	8	5
GA TWIGGS	1.49	TS4-6	6-May-52	8	5
GA TWIGGS	1.04	UpK5-2	12-Apr-53	4	5
GA TWIGGS	1.82	PB2-4	21-Jun-57	5	5
GA TWIGGS	3.96	PB17-3	30-Sep-57	9	5
GA UNION	0.86	BJ1-2	29-Oct-51	5	5
GA UNION	1.49	TS4-6	6-May-52	8	5
GA UNION	1.49	TS4-6	6-May-52	8	5
GA UNION	2.87	TP6-4	25-Mar-55	4	5
GA UNION	3.96	PB17-3	30-Sep-57	9	5
GA WALKER	8.10	BJ3-6	6-Nov-51	5	3
GA WALKER	1.91	TP6-4	25-Mar-55	9	3
GA WALKER	5.99	PB17-3	30-Sep-57	7	3
GA WALTON	0.86	BJ1-2	29-Oct-51	5	5
GA WALTON	1.16	PB1-9	5-Jun-57	5	5
GA WALTON	1.82	PB2-4	21-Jun-57	5	5
GA WALTON	5.99	PB17-3	30-Sep-57	7	5
GA WALTON	1.88	UE8-7	20-Jul-62	2	5
GA WARE	5.73	TS3-8	29-Apr-52	2	6
GA WARE	5.73	TS3-8	29-Apr-52	2	6
GA WARE	1.30	UpK3-7	6-Apr-53	9	6
GA WARE	3.89	UpK4-3	8-Apr-53	4	6
GA WARE	0.97	TP6-7	28-Mar-55	8	6
GA WARE	1.81	UE8-6	19-Jul-62	7	6

COUNTY	uCi/ sqm	SHOT DAY	DATE	RANK FOR DAY	TIMES IN TOP 10
GA WARREN	1.29	BJ1-2	29-Oct-51	3	4
GA WARREN	1.51	PB5-8	22-Jul-57	7	4
GA WARREN	9.96	PB17-3	30-Sep-57	3	4
GA WARREN	1.37	UE8-7	20-Jul-62	4	4
GA WASHINGTON	0.86	BJ1-2	29-Oct-51	5	3
GA WASHINGTON	5.97	PB5-8	22-Jul-57	3	3
GA WASHINGTON	5.99	PB17-3	30-Sep-57	7	3
GA WAYNE	1.35	TS3-7	28-Apr-52	5	8
GA WAYNE	1.35	TS3-7	28-Apr-52	5	8
GA WAYNE	1.30	UpK3-7	6-Apr-53	9	8
GA WAYNE	3.89	UpK4-3	8-Apr-53	4	8
GA WAYNE	0.97	TP6-7	28-Mar-55	8	8
GA WAYNE	1.51	PB5-8	22-Jul-57	7	8
GA WAYNE	19.43	PB11-4	3-Sep-57	2	8
GA WAYNE	2.58	UE8-6	19-Jul-62	4	8
GA WEBSTER	6.40	TS3-7	28-Apr-52	2	7
GA WEBSTER	6.40	TS3-7	28-Apr-52	2	7
GA WEBSTER	1.65	TS3-8	29-Apr-52	5	7
GA WEBSTER	1.65	TS3-8	29-Apr-52	5	7
GA WEBSTER	1.04	UpK5-2	12-Apr-53	4	7
GA WEBSTER	1.16	PB1-9	5-Jun-57	5	7
GA WEBSTER	5.99	PB17-3	30-Sep-57	7	7
GA WHEELER	1.35	TS3-7	28-Apr-52	5	4
GA WHEELER	1.35	TS3-7	28-Apr-52	5	4
GA WHEELER	1.30	UpK3-7	6-Apr-53	9	4
GA WHEELER	3.89	UpK4-3	8-Apr-53	4	4
GA WHITE	1.91	TP6-4	25-Mar-55	9	3
GA WHITE	2.73	PB2-4	21-Jun-57	2	3
GA WHITE	5.99	PB17-3	30-Sep-57	7	3
GA WHITFIELD	8.10	BJ3-6	6-Nov-51	5	3
GA WHITFIELD	2.87	TP6-4	25-Mar-55	4	3
GA WHITFIELD	3.96	PB17-3	30-Sep-57	9	3
GA WILCOX	6.40	TS3-7	28-Apr-52	2	6
GA WILCOX	6.40	TS3-7	28-Apr-52	2	6
GA WILCOX	1.65	TS3-8	29-Apr-52	5	6

COUNTY	uCi/ sqm	SHOT DAY	DATE	RANK FOR DAY	TIMES IN TOP 10
GA WILCOX	1.65	TS3-8	29-Apr-52	5	6
GA WILCOX	1.04	UpK5-2	12-Apr-53	4	6
GA WILCOX	5.99	PB17-3	30-Sep-57	7	6
GA WILKES	1.51	PB5-8	22-Jul-57	7	2
GA WILKES	5.99	PB17-3	30-Sep-57	7	2
GA WILKINSON	1.29	BJ1-2	29-Oct-51	3	2
GA WILKINSON	5.99	PB17-3	30-Sep-57	7	2
GA WORTH	6.40	TS3-7	28-Apr-52	2	7
GA WORTH	6.40	TS3-7	28-Apr-52	2	7
GA WORTH	1.65	TS3-8	29-Apr-52	5	7
GA WORTH	1.65	TS3-8	29-Apr-52	5	7
GA WORTH	0.80	PB13-7	12-Sep-57	7	7
GA WORTH	5.99	PB17-3	30-Sep-57	7	7
GA WORTH	1.32	UE8-6	19-Jul-62	9	7

COUNTY	uCi/ sqm	SHOT DAY	DATE	RANK FOR DAY	TIMES IN TOP 10
IA ADAIR	2.50	TP10-4	8-May-55	8	7
IA ADAIR	1.86	PB4-10	14-Jul-57	8	7
IA ADAIR	2.48	PB5-7	21-Jul-57	6	7
IA ADAIR	15.90	PB8-3	20-Aug-57	7	7
IA ADAIR	83.35	PB9-4	26-Aug-57	1	7
IA ADAIR	345.56	S-3	8-Jul-62	8	7
IA ADAIR	0.24	UE7-5	15-Jul-62	5	7
IA ADAMS	59.11	TS7-4	4-Jun-52	6	7
IA ADAMS	3.09	PB4-10	14-Jul-57	5	7
IA ADAMS	23.89	PB8-3	20-Aug-57	5	7
IA ADAMS	50.01	PB9-4	26-Aug-57	2	7
IA ADAMS	2.04	PB9-5	27-Aug-57	8	7
IA ADAMS	0.86	PB18-6	12-Oct-57	7	7
IA ADAMS	345.56	S-3	8-Jul-62	8	7
IA ALLAMAKEE	0.99	TS3-8	29-Apr-52	8	12
IA ALLAMAKEE	0.99	TS3-8	29-Apr-52	8	12
IA ALLAMAKEE	1.19	PB4-3	7-Jul-57	6	12
IA ALLAMAKEE	2.48	PB5-7	21-Jul-57	6	12
IA ALLAMAKEE	2.43	PB6-9	1-Aug-57	2	12
IA ALLAMAKEE	13.72	PB9-3	25-Aug-57	6	12
IA ALLAMAKEE	8.37	PB9-4	26-Aug-57	7	12
IA ALLAMAKEE	2.04	PB9-5	27-Aug-57	8	12
IA ALLAMAKEE	3.98	PB18-8	14-Oct-57	4	12
IA ALLAMAKEE	0.08	UE2-7	11-Mar-62	5	12
IA ALLAMAKEE	0.13	UE7-5	15-Jul-62	8	12
IA ALLAMAKEE	9.84	UE12-4	17-Apr-65	4	12
IA APPANOOSE	182.07	TS7-3	3-Jun-52	4	8
IA APPANOOSE	1.67	PB5-7	21-Jul-57	7	8
IA APPANOOSE	15.90	PB8-3	20-Aug-57	7	8
IA APPANOOSE	8.37	PB9-4	26-Aug-57	7	8
IA APPANOOSE	2.04	PB9-5	27-Aug-57	8	8
IA APPANOOSE	0.81	PB15-4	19-Sep-57	7	8
IA APPANOOSE	345.56	S-3	8-Jul-62	8	8
IA APPANOOSE	0.33	UE7-5	15-Jul-62	3	8
IA AUDUBON	0.93	TS8-10	14-Jun-52	1	10

COUNTY	uCi/ sqm	SHOT DAY	DATE	RANK FOR DAY	TIMES IN TOP 10
IA AUDUBON	2.80	UpK1-4	20-Mar-53	7	10
IA AUDUBON	13.52	UpK2-3	26-Mar-53	6	10
IA AUDUBON	5.67	UpK9-5	23-May-53	1	10
IA AUDUBON	1.15	TP7-6	3-Apr-55	7	10
IA AUDUBON	1.86	PB4-10	14-Jul-57	8	10
IA AUDUBON	83.35	PB9-4	26-Aug-57	1	10
IA AUDUBON	0.81	PB15-4	19-Sep-57	7	10
IA AUDUBON	0.86	PB18-6	12-Oct-57	7	10
IA AUDUBON	345.56	S-3	8-Jul-62	8	10
IA BENTON	1.54	PB6-9	1-Aug-57	4	8
IA BENTON	27.43	PB9-3	25-Aug-57	2	8
IA BENTON	50.01	PB9-4	26-Aug-57	2	8
IA BENTON	3.39	PB9-5	27-Aug-57	4	8
IA BENTON	3.98	PB18-8	14-Oct-57	4	8
IA BENTON	0.08	UE2-7	11-Mar-62	5	8
IA BENTON	527.44	S-3	8-Jul-62	5	8
IA BENTON	0.24	UE7-5	15-Jul-62	5	8
IA BLACK HAWK	0.99	TS3-8	29-Apr-52	8	9
IA BLACK HAWK	0.99	TS3-8	29-Apr-52	8	9
IA BLACK HAWK	1.02	PB4-3	7-Jul-57	8	9
IA BLACK HAWK	1.67	PB5-7	21-Jul-57	7	9
IA BLACK HAWK	27.43	PB9-3	25-Aug-57	2	9
IA BLACK HAWK	50.01	PB9-4	26-Aug-57	2	9
IA BLACK HAWK	3.39	PB9-5	27-Aug-57	4	9
IA BLACK HAWK	3.98	PB18-8	14-Oct-57	4	9
IA BLACK HAWK	0.08	UE2-7	11-Mar-62	5	9
IA BOONE	1.38	TP7-6	3-Apr-55	1	6
IA BOONE	2.48	PB5-7	21-Jul-57	6	6
IA BOONE	83.35	PB9-4	26-Aug-57	1	6
IA BOONE	0.08	UE2-7	11-Mar-62	5	6
IA BOONE	527.44	S-3	8-Jul-62	5	6
IA BOONE	0.24	UE7-5	15-Jul-62	5	6
IA BREMER	0.99	TS3-8	29-Apr-52	8	11
IA BREMER	0.99	TS3-8	29-Apr-52	8	11
IA BREMER	1.36	PB4-3	7-Jul-57	4	11

COUNTY	uCi/ sqm	SHOT DAY	DATE	RANK FOR DAY	TIMES IN TOP 10
IA BREMER	3.38	PB6-8	31-Jul-57	1	11
IA BREMER	0.92	PB6-9	1-Aug-57	6	11
IA BREMER	27.43	PB9-3	25-Aug-57	2	11
IA BREMER	50.01	PB9-4	26-Aug-57	2	11
IA BREMER	2.04	PB9-5	27-Aug-57	8	11
IA BREMER	2.65	PB18-8	14-Oct-57	8	11
IA BREMER	0.08	UE2-7	11-Mar-62	5	11
IA BREMER	22.99	UE12-4	17-Apr-65	1	11
IA BUCHANAN	0.99	TS3-8	29-Apr-52	8	12
IA BUCHANAN	0.99	TS3-8	29-Apr-52	8	12
IA BUCHANAN	1.32	UpK8-3	10-May-53	9	12
IA BUCHANAN	1.10	PB4-3	7-Jul-57	7	12
IA BUCHANAN	2.48	PB5-7	21-Jul-57	6	12
IA BUCHANAN	0.92	PB6-9	1-Aug-57	6	12
IA BUCHANAN	13.72	PB9-3	25-Aug-57	6	12
IA BUCHANAN	50.01	PB9-4	26-Aug-57	2	12
IA BUCHANAN	2.04	PB9-5	27-Aug-57	8	12
IA BUCHANAN	3.98	PB18-8	14-Oct-57	4	12
IA BUCHANAN	0.08	UE2-7	11-Mar-62	5	12
IA BUCHANAN	0.24	UE7-5	15-Jul-62	5	12
IA BUENA VISTA	7.03	UpK7-5	29-Apr-53	6	8
IA BUENA VISTA	1.30	PB2-4	21-Jun-57	9	8
IA BUENA VISTA	1.19	PB4-3	7-Jul-57	6	8
IA BUENA VISTA	4.02	PB5-5	19-Jul-57	3	8
IA BUENA VISTA	0.81	PB15-5	20-Sep-57	9	8
IA BUENA VISTA	0.86	PB18-6	12-Oct-57	7	8
IA BUENA VISTA	0.17	UE2-7	11-Mar-62	2	8
IA BUENA VISTA	527.44	S-3	8-Jul-62	5	8
IA BUTLER	0.99	TS3-8	29-Apr-52	8	7
IA BUTLER	0.99	TS3-8	29-Apr-52	8	7
IA BUTLER	1.36	PB4-3	7-Jul-57	4	7
IA BUTLER	16.67	PB9-4	26-Aug-57	4	7
IA BUTLER	2.04	PB9-5	27-Aug-57	8	7
IA BUTLER	0.12	UE7-5	15-Jul-62	9	7
IA BUTLER	9.32	UE12-4	17-Apr-65	6	7

COUNTY	uCi/ sqm	SHOT DAY	DATE	RANK FOR DAY	TIMES IN TOP 10
IA CALHOUN	1.74	TS4-6	6-May-52	5	7
IA CALHOUN	1.74	TS4-6	6-May-52	5	7
IA CALHOUN	1.19	PB4-3	7-Jul-57	6	7
IA CALHOUN	2.01	PB5-5	19-Jul-57	8	7
IA CALHOUN	0.17	UE2-7	11-Mar-62	2	7
IA CALHOUN	527.44	S-3	8-Jul-62	5	7
IA CALHOUN	0.24	UE7-5	15-Jul-62	5	7
IA CARROLL	52.94	TS7-4	4-Jun-52	10	7
IA CARROLL	1.12	TP7-6	3-Apr-55	8	7
IA CARROLL	1.30	PB2-4	21-Jun-57	9	7
IA CARROLL	3.09	PB4-10	14-Jul-57	5	7
IA CARROLL	83.35	PB9-4	26-Aug-57	1	7
IA CARROLL	0.10	UE2-7	11-Mar-62	4	7
IA CARROLL	527.44	S-3	8-Jul-62	5	7
IA CASS	59.65	TS7-4	4-Jun-52	5	7
IA CASS	1.86	PB4-10	14-Jul-57	8	7
IA CASS	7.99	PB8-3	20-Aug-57	9	7
IA CASS	83.35	PB9-4	26-Aug-57	1	7
IA CASS	3.39	PB9-5	27-Aug-57	4	7
IA CASS	0.86	PB18-6	12-Oct-57	7	7
IA CASS	345.56	S-3	8-Jul-62	8	7
IA CEDAR	1.52	PB4-3	7-Jul-57	2	9
IA CEDAR	1.67	PB5-7	21-Jul-57	7	9
IA CEDAR	13.72	PB9-3	25-Aug-57	6	9
IA CEDAR	33.34	PB9-4	26-Aug-57	3	9
IA CEDAR	3.39	PB9-5	27-Aug-57	4	9
IA CEDAR	0.89	PB9-7	29-Aug-57	6	9
IA CEDAR	6.63	PB18-8	14-Oct-57	2	9
IA CEDAR	0.08	UE2-7	11-Mar-62	5	9
IA CEDAR	0.12	UE7-5	15-Jul-62	9	9
IA CERRO GORDO	0.99	TS3-8	29-Apr-52	8	6
IA CERRO GORDO	0.99	TS3-8	29-Apr-52	8	6
IA CERRO GORDO	6.09	PB5-5	19-Jul-57	2	6
IA CERRO GORDO	0.81	PB15-4	19-Sep-57	7	6
IA CERRO GORDO	0.17	UE2-7	11-Mar-62	2	6

COUNTY	uCi/ sqm	SHOT DAY	DATE	RANK FOR DAY	TIMES IN TOP 10
IA CERRO GORDO	9.84	UE12-4	17-Apr-65	4	6
IA CHEROKEE	1.37	TS4-6	6-May-52	9	8
IA CHEROKEE	1.37	TS4-6	6-May-52	9	8
IA CHEROKEE	1.30	PB2-4	21-Jun-57	9	8
IA CHEROKEE	0.85	PB4-3	7-Jul-57	10	8
IA CHEROKEE	2.01	PB5-5	19-Jul-57	8	8
IA CHEROKEE	0.10	UE2-7	11-Mar-62	4	8
IA CHEROKEE	527.44	S-3	8-Jul-62	5	8
IA CHEROKEE	0.22	UE7-5	15-Jul-62	6	8
IA CHICKASAW	0.99	TS3-8	29-Apr-52	8	10
IA CHICKASAW	0.99	TS3-8	29-Apr-52	8	10
IA CHICKASAW	1.02	PB4-3	7-Jul-57	8	10
IA CHICKASAW	3.38	PB6-8	31-Jul-57	1	10
IA CHICKASAW	13.72	PB9-3	25-Aug-57	6	10
IA CHICKASAW	16.67	PB9-4	26-Aug-57	4	10
IA CHICKASAW	2.04	PB9-5	27-Aug-57	8	10
IA CHICKASAW	3.98	PB18-8	14-Oct-57	4	10
IA CHICKASAW	0.24	UE7-5	15-Jul-62	5	10
IA CHICKASAW	9.32	UE12-4	17-Apr-65	6	10
IA CLARKE	54.98	TS7-4	4-Jun-52	9	9
IA CLARKE	1.86	PB4-10	14-Jul-57	8	9
IA CLARKE	1.67	PB5-7	21-Jul-57	7	9
IA CLARKE	15.90	PB8-3	20-Aug-57	7	9
IA CLARKE	50.01	PB9-4	26-Aug-57	2	9
IA CLARKE	2.04	PB9-5	27-Aug-57	8	9
IA CLARKE	0.81	PB15-5	20-Sep-57	9	9
IA CLARKE	345.56	S-3	8-Jul-62	8	9
IA CLARKE	0.24	UE7-5	15-Jul-62	5	9
IA CLAY	1.44	BJ5-3	21-Nov-51	6	8
IA CLAY	7.29	UpK7-5	29-Apr-53	4	8
IA CLAY	1.30	PB2-4	21-Jun-57	9	8
IA CLAY	0.85	PB4-3	7-Jul-57	10	8
IA CLAY	0.17	UE2-7	11-Mar-62	2	8
IA CLAY	527.44	S-3	8-Jul-62	5	8
IA CLAY	0.13	UE7-5	15-Jul-62	8	8

COUNTY	uCi/ sqm	SHOT DAY	DATE	RANK FOR DAY	TIMES IN TOP 10
IA CLAY	11.43	UE12-4	17-Apr-65	3	8
IA CLAYTON	0.99	TS3-8	29-Apr-52	8	13
IA CLAYTON	0.99	TS3-8	29-Apr-52	8	13
IA CLAYTON	1.02	PB4-3	7-Jul-57	8	13
IA CLAYTON	2.48	PB5-7	21-Jul-57	6	13
IA CLAYTON	3.38	PB6-8	31-Jul-57	1	13
IA CLAYTON	0.92	PB6-9	1-Aug-57	6	13
IA CLAYTON	13.72	PB9-3	25-Aug-57	6	13
IA CLAYTON	33.34	PB9-4	26-Aug-57	3	13
IA CLAYTON	2.04	PB9-5	27-Aug-57	8	13
IA CLAYTON	3.98	PB18-8	14-Oct-57	4	13
IA CLAYTON	0.08	UE2-7	11-Mar-62	5	13
IA CLAYTON	0.33	UE7-5	15-Jul-62	3	13
IA CLAYTON	9.84	UE12-4	17-Apr-65	4	13
IA CLINTON	0.93	PB4-3	7-Jul-57	9	8
IA CLINTON	2.48	PB5-7	21-Jul-57	6	8
IA CLINTON	13.72	PB9-3	25-Aug-57	6	8
IA CLINTON	16.67	PB9-4	26-Aug-57	4	8
IA CLINTON	3.39	PB9-5	27-Aug-57	4	8
IA CLINTON	3.98	PB18-8	14-Oct-57	4	8
IA CLINTON	0.08	UE2-7	11-Mar-62	5	8
IA CLINTON	0.12	UE7-5	15-Jul-62	9	8
IA CRAWFORD	0.93	TS8-10	14-Jun-52	1	9
IA CRAWFORD	1.30	PB2-4	21-Jun-57	9	9
IA CRAWFORD	1.19	PB4-3	7-Jul-57	6	9
IA CRAWFORD	6.09	PB5-5	19-Jul-57	2	9
IA CRAWFORD	0.80	PB13-7	12-Sep-57	7	9
IA CRAWFORD	0.81	PB15-4	19-Sep-57	7	9
IA CRAWFORD	0.86	PB18-6	12-Oct-57	7	9
IA CRAWFORD	0.17	UE2-7	11-Mar-62	2	9
IA CRAWFORD	527.44	S-3	8-Jul-62	5	9
IA DALLAS	11.01	UpK2-3	26-Mar-53	10	4
IA DALLAS	1.11	TP7-6	3-Apr-55	9	4
IA DALLAS	83.35	PB9-4	26-Aug-57	1	4
IA DALLAS	0.13	UE7-5	15-Jul-62	8	4

COUNTY	uCi/ sqm	SHOT DAY	DATE	RANK FOR DAY	TIMES IN TOP 10
IA DAVIS	2.48	PB5-7	21-Jul-57	6	8
IA DAVIS	7.99	PB8-3	20-Aug-57	9	8
IA DAVIS	13.72	PB9-3	25-Aug-57	6	8
IA DAVIS	83.35	PB9-4	26-Aug-57	1	8
IA DAVIS	0.98	PB9-6	28-Aug-57	6	8
IA DAVIS	3.98	PB18-8	14-Oct-57	4	8
IA DAVIS	345.56	S-3	8-Jul-62	8	8
IA DAVIS	0.33	UE7-5	15-Jul-62	3	8
IA DECATUR	4.50	UpK11-3	6-Jun-53	4	8
IA DECATUR	3.09	PB4-10	14-Jul-57	5	8
IA DECATUR	4.15	PB5-7	21-Jul-57	2	8
IA DECATUR	39.79	PB8-3	20-Aug-57	3	8
IA DECATUR	33.34	PB9-4	26-Aug-57	3	8
IA DECATUR	1.21	PB15-4	19-Sep-57	5	8
IA DECATUR	345.56	S-3	8-Jul-62	8	8
IA DECATUR	0.24	UE7-5	15-Jul-62	5	8
IA DELAWARE	0.99	TS3-8	29-Apr-52	8	15
IA DELAWARE	0.99	TS3-8	29-Apr-52	8	15
IA DELAWARE	64.44	UpK9-2	20-May-53	8	15
IA DELAWARE	0.99	TP8-5	13-Apr-55	3	15
IA DELAWARE	1.02	PB4-3	7-Jul-57	8	15
IA DELAWARE	1.67	PB5-7	21-Jul-57	7	15
IA DELAWARE	3.38	PB6-8	31-Jul-57	1	15
IA DELAWARE	13.72	PB9-3	25-Aug-57	6	15
IA DELAWARE	50.01	PB9-4	26-Aug-57	2	15
IA DELAWARE	2.04	PB9-5	27-Aug-57	8	15
IA DELAWARE	3.98	PB18-8	14-Oct-57	4	15
IA DELAWARE	0.08	UE2-7	11-Mar-62	5	15
IA DELAWARE	345.56	S-3	8-Jul-62	8	15
IA DELAWARE	0.12	UE7-5	15-Jul-62	9	15
IA DELAWARE	9.84	UE12-4	17-Apr-65	4	15
IA DES MOINES	1.42	UpK5-4	14-Apr-53	6	11
IA DES MOINES	0.87	UpK6-4	21-Apr-53	6	11
IA DES MOINES	0.99	PB5-6	20-Jul-57	9	11
IA DES MOINES	4.15	PB5-7	21-Jul-57	2	11

COUNTY	uCi/ sqm	SHOT DAY	DATE	RANK FOR DAY	TIMES IN TOP 10
IA DES MOINES	13.72	PB9-3	25-Aug-57	6	11
IA DES MOINES	33.34	PB9-4	26-Aug-57	3	11
IA DES MOINES	2.04	PB9-5	27-Aug-57	8	11
IA DES MOINES	2.25	PB18-7	13-Oct-57	2	11
IA DES MOINES	3.98	PB18-8	14-Oct-57	4	11
IA DES MOINES	0.08	UE2-7	11-Mar-62	5	11
IA DES MOINES	0.33	UE7-5	15-Jul-62	3	11
IA DICKINSON	1.15	TS8-11	15-Jun-52	3	10
IA DICKINSON	7.42	UpK7-5	29-Apr-53	3	10
IA DICKINSON	1.20	PB2-2	19-Jun-57	10	10
IA DICKINSON	1.30	PB2-4	21-Jun-57	9	10
IA DICKINSON	1.52	PB4-3	7-Jul-57	2	10
IA DICKINSON	2.01	PB5-5	19-Jul-57	8	10
IA DICKINSON	0.17	UE2-7	11-Mar-62	2	10
IA DICKINSON	527.44	S-3	8-Jul-62	5	10
IA DICKINSON	0.12	UE7-5	15-Jul-62	9	10
IA DICKINSON	9.84	UE12-4	17-Apr-65	4	10
IA DUBUQUE	1.27	PB4-3	7-Jul-57	5	12
IA DUBUQUE	2.48	PB5-7	21-Jul-57	6	12
IA DUBUQUE	3.38	PB6-8	31-Jul-57	1	12
IA DUBUQUE	0.92	PB6-9	1-Aug-57	6	12
IA DUBUQUE	13.72	PB9-3	25-Aug-57	6	12
IA DUBUQUE	33.34	PB9-4	26-Aug-57	3	12
IA DUBUQUE	2.04	PB9-5	27-Aug-57	8	12
IA DUBUQUE	0.89	PB9-7	29-Aug-57	6	12
IA DUBUQUE	3.98	PB18-8	14-Oct-57	4	12
IA DUBUQUE	0.08	UE2-7	11-Mar-62	5	12
IA DUBUQUE	0.12	UE7-5	15-Jul-62	9	12
IA DUBUQUE	9.32	UE12-4	17-Apr-65	6	12
IA EMMET	1.15	TS8-11	15-Jun-52	3	6
IA EMMET	6.93	UpK7-5	29-Apr-53	9	6
IA EMMET	1.30	PB2-4	21-Jun-57	9	6
IA EMMET	1.19	PB4-3	7-Jul-57	6	6
IA EMMET	2.01	PB5-5	19-Jul-57	8	6
IA EMMET	527.44	S-3	8-Jul-62	5	6

COUNTY	uCi/ sqm	SHOT DAY	DATE	RANK FOR DAY	TIMES IN TOP 10
IA FAYETTE	0.99	TS3-8	29-Apr-52	8	12
IA FAYETTE	0.99	TS3-8	29-Apr-52	8	12
IA FAYETTE	1.02	PB4-3	7-Jul-57	8	12
IA FAYETTE	2.48	PB5-7	21-Jul-57	6	12
IA FAYETTE	3.38	PB6-8	31-Jul-57	1	12
IA FAYETTE	13.72	PB9-3	25-Aug-57	6	12
IA FAYETTE	50.01	PB9-4	26-Aug-57	2	12
IA FAYETTE	2.04	PB9-5	27-Aug-57	8	12
IA FAYETTE	3.98	PB18-8	14-Oct-57	4	12
IA FAYETTE	0.08	UE2-7	11-Mar-62	5	12
IA FAYETTE	0.12	UE7-5	15-Jul-62	9	12
IA FAYETTE	9.32	UE12-4	17-Apr-65	6	12
IA FLOYD	0.99	TS3-8	29-Apr-52	8	9
IA FLOYD	0.99	TS3-8	29-Apr-52	8	9
IA FLOYD	1.20	PB2-2	19-Jun-57	10	9
IA FLOYD	1.67	PB5-7	21-Jul-57	7	9
IA FLOYD	8.37	PB9-4	26-Aug-57	7	9
IA FLOYD	2.04	PB9-5	27-Aug-57	8	9
IA FLOYD	0.10	UE2-7	11-Mar-62	4	9
IA FLOYD	527.44	S-3	8-Jul-62	5	9
IA FLOYD	9.84	UE12-4	17-Apr-65	4	9
IA FRANKLIN	1.74	TS4-6	6-May-52	5	10
IA FRANKLIN	1.74	TS4-6	6-May-52	5	10
IA FRANKLIN	1.27	PB4-3	7-Jul-57	5	10
IA FRANKLIN	2.48	PB5-7	21-Jul-57	6	10
IA FRANKLIN	83.35	PB9-4	26-Aug-57	1	10
IA FRANKLIN	2.04	PB9-5	27-Aug-57	8	10
IA FRANKLIN	0.98	PB9-6	28-Aug-57	6	10
IA FRANKLIN	0.17	UE2-7	11-Mar-62	2	10
IA FRANKLIN	0.13	UE7-5	15-Jul-62	8	10
IA FRANKLIN	3.94	UE8-4	17-Jul-62	10	10
IA FREMONT	0.98	UpK11-5	8-Jun-53	3	2
IA FREMONT	0.81	PB15-4	19-Sep-57	7	2
IA GREENE	0.93	PB4-3	7-Jul-57	9	7
IA GREENE	2.22	PB4-9	13-Jul-57	4	7

COUNTY	uCi/ sqm	SHOT DAY	DATE	RANK FOR DAY	TIMES IN TOP 10
IA GREENE	50.01	PB9-4	26-Aug-57	2	7
IA GREENE	2.04	PB9-5	27-Aug-57	8	7
IA GREENE	0.10	UE2-7	11-Mar-62	3	7
IA GREENE	527.44	S-3	8-Jul-62	5	7
IA GREENE	0.24	UE7-5	15-Jul-62	5	7
IA GRUNDY	2.39	PB2-2	19-Jun-57	6	7
IA GRUNDY	4.15	PB5-7	21-Jul-57	2	7
IA GRUNDY	0.92	PB6-9	1-Aug-57	6	7
IA GRUNDY	33.34	PB9-4	26-Aug-57	3	7
IA GRUNDY	3.39	PB9-5	27-Aug-57	4	7
IA GRUNDY	0.89	PB9-7	29-Aug-57	6	7
IA GRUNDY	0.12	UE7-5	15-Jul-62	9	7
IA GUTHRIE	4.73	UpK9-5	23-May-53	4	6
IA GUTHRIE	1.86	PB4-10	14-Jul-57	8	6
IA GUTHRIE	50.01	PB9-4	26-Aug-57	2	6
IA GUTHRIE	2.04	PB9-5	27-Aug-57	8	6
IA GUTHRIE	345.56	S-3	8-Jul-62	8	6
IA GUTHRIE	0.12	UE7-5	15-Jul-62	9	6
IA HAMILTON	1.27	PB4-3	7-Jul-57	5	8
IA HAMILTON	2.48	PB5-7	21-Jul-57	6	8
IA HAMILTON	50.01	PB9-4	26-Aug-57	2	8
IA HAMILTON	2.04	PB9-5	27-Aug-57	8	8
IA HAMILTON	0.10	UE2-7	11-Mar-62	4	8
IA HAMILTON	527.44	S-3	8-Jul-62	5	8
IA HAMILTON	0.13	UE7-5	15-Jul-62	8	8
IA HAMILTON	3.94	UE8-4	17-Jul-62	10	8
IA HANCOCK	2.39	PB2-2	19-Jun-57	7	6
IA HANCOCK	1.27	PB4-3	7-Jul-57	5	6
IA HANCOCK	0.10	UE2-7	11-Mar-62	3	6
IA HANCOCK	527.44	S-3	8-Jul-62	5	6
IA HANCOCK	0.33	UE7-5	15-Jul-62	3	6
IA HANCOCK	9.32	UE12-4	17-Apr-65	6	6
IA HARDIN	1.36	PB4-3	7-Jul-57	4	6
IA HARDIN	2.48	PB5-7	21-Jul-57	6	6
IA HARDIN	50.01	PB9-4	26-Aug-57	2	6

COUNTY	uCi/ sqm	SHOT DAY	DATE	RANK FOR DAY	TIMES IN TOP 10
IA HARDIN	2.04	PB9-5	27-Aug-57	8	6
IA HARDIN	0.08	UE2-7	11-Mar-62	5	6
IA HARDIN	0.12	UE7-5	15-Jul-62	9	6
IA HARRISON	1.30	PB2-4	21-Jun-57	9	6
IA HARRISON	1.10	PB4-3	7-Jul-57	7	6
IA HARRISON	3.09	PB4-10	14-Jul-57	5	6
IA HARRISON	4.02	PB5-5	19-Jul-57	3	6
IA HARRISON	0.10	UE2-7	11-Mar-62	4	6
IA HARRISON	0.82	BN-2	19-Dec-70	7	6
IA HENRY	4.15	PB5-7	21-Jul-57	2	8
IA HENRY	13.72	PB9-3	25-Aug-57	6	8
IA HENRY	50.01	PB9-4	26-Aug-57	2	8
IA HENRY	3.39	PB9-5	27-Aug-57	4	8
IA HENRY	6.63	PB18-8	14-Oct-57	2	8
IA HENRY	0.08	UE2-7	11-Mar-62	5	8
IA HENRY	345.56	S-3	8-Jul-62	8	8
IA HENRY	0.24	UE7-5	15-Jul-62	5	8
IA HOWARD	0.99	TS3-8	29-Apr-52	8	12
IA HOWARD	0.99	TS3-8	29-Apr-52	8	12
IA HOWARD	1.27	PB4-3	7-Jul-57	5	12
IA HOWARD	1.67	PB5-7	21-Jul-57	7	12
IA HOWARD	13.72	PB9-3	25-Aug-57	6	12
IA HOWARD	8.37	PB9-4	26-Aug-57	7	12
IA HOWARD	2.04	PB9-5	27-Aug-57	8	12
IA HOWARD	3.98	PB18-8	14-Oct-57	4	12
IA HOWARD	0.10	UE2-7	11-Mar-62	4	12
IA HOWARD	950.00	S-3	8-Jul-62	1	12
IA HOWARD	0.24	UE7-5	15-Jul-62	5	12
IA HOWARD	9.84	UE12-4	17-Apr-65	4	12
IA HUMBOLDT	1.74	TS4-6	6-May-52	5	6
IA HUMBOLDT	1.74	TS4-6	6-May-52	5	6
IA HUMBOLDT	1.61	PB4-3	7-Jul-57	1	6
IA HUMBOLDT	6.09	PB5-5	19-Jul-57	2	6
IA HUMBOLDT	0.17	UE2-7	11-Mar-62	2	6
IA HUMBOLDT	527.44	S-3	8-Jul-62	5	6

COUNTY	uCi/ sqm	SHOT DAY	DATE	RANK FOR DAY	TIMES IN TOP 10
IA IDA	1.37	TS4-6	6-May-52	9	9
IA IDA	1.37	TS4-6	6-May-52	9	9
IA IDA	1.33	TP7-6	3-Apr-55	3	9
IA IDA	1.10	PB4-3	7-Jul-57	7	9
IA IDA	2.01	PB5-5	19-Jul-57	8	9
IA IDA	0.86	PB18-6	12-Oct-57	7	9
IA IDA	0.17	UE2-7	11-Mar-62	2	9
IA IDA	527.44	S-3	8-Jul-62	5	9
IA IDA	0.17	UE7-5	15-Jul-62	7	9
IA IOWA	2.48	PB5-7	21-Jul-57	6	7
IA IOWA	1.54	PB6-9	1-Aug-57	4	7
IA IOWA	13.72	PB9-3	25-Aug-57	6	7
IA IOWA	33.34	PB9-4	26-Aug-57	3	7
IA IOWA	3.39	PB9-5	27-Aug-57	4	7
IA IOWA	3.98	PB18-8	14-Oct-57	4	7
IA IOWA	527.44	S-3	8-Jul-62	5	7
IA JACKSON	1.27	PB4-3	7-Jul-57	5	8
IA JACKSON	2.48	PB5-7	21-Jul-57	6	8
IA JACKSON	13.72	PB9-3	25-Aug-57	6	8
IA JACKSON	50.01	PB9-4	26-Aug-57	2	8
IA JACKSON	2.04	PB9-5	27-Aug-57	8	8
IA JACKSON	3.98	PB18-8	14-Oct-57	4	8
IA JACKSON	0.08	UE2-7	11-Mar-62	5	8
IA JACKSON	0.13	UE7-5	15-Jul-62	8	8
IA JASPER	3.35	PB4-9	13-Jul-57	1	8
IA JASPER	2.48	PB5-7	21-Jul-57	6	8
IA JASPER	7.99	PB8-3	20-Aug-57	9	8
IA JASPER	83.35	PB9-4	26-Aug-57	1	8
IA JASPER	2.04	PB9-5	27-Aug-57	8	8
IA JASPER	0.81	PB15-5	20-Sep-57	9	8
IA JASPER	527.44	S-3	8-Jul-62	5	8
IA JASPER	0.13	UE7-5	15-Jul-62	8	8
IA JEFFERSON	1.74	TS4-6	6-May-52	5	10
IA JEFFERSON	1.74	TS4-6	6-May-52	5	10
IA JEFFERSON	1.67	PB5-7	21-Jul-57	7	10

COUNTY	uCi/ sqm	SHOT DAY	DATE	RANK FOR DAY	TIMES IN TOP 10
IA JEFFERSON	0.92	PB6-9	1-Aug-57	6	10
IA JEFFERSON	13.72	PB9-3	25-Aug-57	6	10
IA JEFFERSON	8.37	PB9-4	26-Aug-57	7	10
IA JEFFERSON	3.39	PB9-5	27-Aug-57	4	10
IA JEFFERSON	3.98	PB18-8	14-Oct-57	4	10
IA JEFFERSON	621.15	S-3	8-Jul-62	2	10
IA JEFFERSON	0.13	UE7-5	15-Jul-62	8	10
IA JOHNSON	1.74	TS4-6	6-May-52	5	11
IA JOHNSON	1.74	TS4-6	6-May-52	5	11
IA JOHNSON	1.52	PB4-3	7-Jul-57	2	11
IA JOHNSON	0.99	PB5-6	20-Jul-57	9	11
IA JOHNSON	1.40	PB6-8	31-Jul-57	7	11
IA JOHNSON	11.39	PB9-3	25-Aug-57	7	11
IA JOHNSON	6.92	PB9-4	26-Aug-57	7	11
IA JOHNSON	2.75	PB18-8	14-Oct-57	7	11
IA JOHNSON	0.08	UE2-7	11-Mar-62	5	11
IA JOHNSON	345.56	S-3	8-Jul-62	8	11
IA JOHNSON	0.13	UE7-5	15-Jul-62	8	11
IA JONES	1.44	PB4-3	7-Jul-57	3	7
IA JONES	13.72	PB9-3	25-Aug-57	6	7
IA JONES	16.67	PB9-4	26-Aug-57	4	7
IA JONES	3.39	PB9-5	27-Aug-57	4	7
IA JONES	0.89	PB9-7	29-Aug-57	6	7
IA JONES	3.98	PB18-8	14-Oct-57	4	7
IA JONES	0.13	UE7-5	15-Jul-62	8	7
IA KEOKUK	1.27	TS8-9	13-Jun-52	1	9
IA KEOKUK	1.10	PB4-3	7-Jul-57	7	9
IA KEOKUK	1.54	PB6-9	1-Aug-57	4	9
IA KEOKUK	7.99	PB8-3	20-Aug-57	9	9
IA KEOKUK	13.72	PB9-3	25-Aug-57	6	9
IA KEOKUK	50.01	PB9-4	26-Aug-57	2	9
IA KEOKUK	2.04	PB9-5	27-Aug-57	8	9
IA KEOKUK	3.98	PB18-8	14-Oct-57	4	9
IA KEOKUK	345.56	S-3	8-Jul-62	8	9
IA KOSSUTH	1.74	TS4-6	6-May-52	5	8

COUNTY	uCi/ sqm	SHOT DAY	DATE	RANK FOR DAY	TIMES IN TOP 10
IA KOSSUTH	1.74	TS4-6	6-May-52	5	8
IA KOSSUTH	1.15	TS8-11	15-Jun-52	3	8
IA KOSSUTH	1.18	PB5-6	20-Jul-57	7	8
IA KOSSUTH	0.17	UE2-7	11-Mar-62	2	8
IA KOSSUTH	527.44	S-3	8-Jul-62	5	8
IA KOSSUTH	0.13	UE7-5	15-Jul-62	8	8
IA KOSSUTH	9.32	UE12-4	17-Apr-65	6	8
IA LEE	1.83	PB4-9	13-Jul-57	6	9
IA LEE	4.15	PB5-7	21-Jul-57	2	9
IA LEE	13.72	PB9-3	25-Aug-57	6	9
IA LEE	16.67	PB9-4	26-Aug-57	4	9
IA LEE	2.04	PB9-5	27-Aug-57	8	9
IA LEE	6.63	PB18-8	14-Oct-57	2	9
IA LEE	0.08	UE2-7	11-Mar-62	5	9
IA LEE	621.15	S-3	8-Jul-62	2	9
IA LEE	0.33	UE7-5	15-Jul-62	3	9
IA LINN	1.67	PB5-7	21-Jul-57	7	9
IA LINN	2.24	PB6-8	31-Jul-57	4	9
IA LINN	13.72	PB9-3	25-Aug-57	6	9
IA LINN	50.01	PB9-4	26-Aug-57	2	9
IA LINN	3.39	PB9-5	27-Aug-57	4	9
IA LINN	3.98	PB18-8	14-Oct-57	4	9
IA LINN	0.08	UE2-7	11-Mar-62	5	9
IA LINN	0.24	UE7-5	15-Jul-62	5	9
IA LINN	9.84	UE12-4	17-Apr-65	4	9
IA LOUISA	4.15	PB5-7	21-Jul-57	2	8
IA LOUISA	13.72	PB9-3	25-Aug-57	6	8
IA LOUISA	16.67	PB9-4	26-Aug-57	4	8
IA LOUISA	3.39	PB9-5	27-Aug-57	4	8
IA LOUISA	6.63	PB18-8	14-Oct-57	2	8
IA LOUISA	0.08	UE2-7	11-Mar-62	5	8
IA LOUISA	345.56	S-3	8-Jul-62	8	8
IA LOUISA	0.13	UE7-5	15-Jul-62	8	8
IA LUCAS	4.85	PB5-3	17-Jul-57	9	5
IA LUCAS	2.48	PB5-7	21-Jul-57	6	5

COUNTY	uCi/ sqm	SHOT DAY	DATE	RANK FOR DAY	TIMES IN TOP 10
IA LUCAS	23.89	PB8-3	20-Aug-57	5	5
IA LUCAS	50.01	PB9-4	26-Aug-57	2	5
IA LUCAS	0.81	PB15-5	20-Sep-57	9	5
IA LYON	1.44	BJ5-3	21-Nov-51	6	8
IA LYON	1.37	TS4-6	6-May-52	9	8
IA LYON	1.37	TS4-6	6-May-52	9	8
IA LYON	3.97	UpK2-2	25-Mar-53	10	8
IA LYON	6.09	PB5-5	19-Jul-57	2	8
IA LYON	0.10	UE2-7	11-Mar-62	4	8
IA LYON	527.44	S-3	8-Jul-62	5	8
IA LYON	9.84	UE12-4	17-Apr-65	4	8
IA MADISON	14.82	UpK2-3	26-Mar-53	5	6
IA MADISON	1.86	PB4-10	14-Jul-57	8	6
IA MADISON	23.89	PB8-3	20-Aug-57	5	6
IA MADISON	83.35	PB9-4	26-Aug-57	1	6
IA MADISON	345.56	S-3	8-Jul-62	8	6
IA MADISON	0.24	UE7-5	15-Jul-62	5	6
IA MAHASKA	1.67	PB5-7	21-Jul-57	7	7
IA MAHASKA	15.90	PB8-3	20-Aug-57	7	7
IA MAHASKA	33.34	PB9-4	26-Aug-57	3	7
IA MAHASKA	2.04	PB9-5	27-Aug-57	8	7
IA MAHASKA	0.81	PB15-4	19-Sep-57	7	7
IA MAHASKA	0.81	PB15-5	20-Sep-57	9	7
IA MAHASKA	345.56	S-3	8-Jul-62	8	7
IA MARION	0.92	PB6-9	1-Aug-57	6	7
IA MARION	15.90	PB8-3	20-Aug-57	7	7
IA MARION	50.01	PB9-4	26-Aug-57	2	7
IA MARION	3.39	PB9-5	27-Aug-57	4	7
IA MARION	0.81	PB15-5	20-Sep-57	9	7
IA MARION	345.56	S-3	8-Jul-62	8	7
IA MARION	0.24	UE7-5	15-Jul-62	5	7
IA MARSHALL	61.87	UpK9-2	20-May-53	9	9
IA MARSHALL	1.68	PB4-9	13-Jul-57	7	9
IA MARSHALL	4.15	PB5-7	21-Jul-57	2	9
IA MARSHALL	0.92	PB6-9	1-Aug-57	6	9

COUNTY	uCi/ sqm	SHOT DAY	DATE	RANK FOR DAY	TIMES IN TOP 10
IA MARSHALL	50.01	PB9-4	26-Aug-57	2	9
IA MARSHALL	3.39	PB9-5	27-Aug-57	4	9
IA MARSHALL	0.81	PB15-5	20-Sep-57	9	9
IA MARSHALL	527.44	S-3	8-Jul-62	5	9
IA MARSHALL	0.13	UE7-5	15-Jul-62	8	9
IA MILLS	1.67	PB5-7	21-Jul-57	7	4
IA MILLS	23.89	PB8-3	20-Aug-57	5	4
IA MILLS	0.86	PB18-6	12-Oct-57	7	4
IA MILLS	0.24	UE7-5	15-Jul-62	5	4
IA MITCHELL	0.99	TS3-8	29-Apr-52	8	9
IA MITCHELL	0.99	TS3-8	29-Apr-52	8	9
IA MITCHELL	69.62	UpK9-2	20-May-53	4	9
IA MITCHELL	2.54	PB2-4	21-Jun-57	4	9
IA MITCHELL	1.63	PB6-9	1-Aug-57	3	9
IA MITCHELL	1.55	PB15-3	18-Sep-57	10	9
IA MITCHELL	950.00	S-3	8-Jul-62	1	9
IA MITCHELL	0.13	UE7-5	15-Jul-62	8	9
IA MITCHELL	9.32	UE12-4	17-Apr-65	6	9
IA MONONA	2.30	TS4-6	6-May-52	3	9
IA MONONA	2.30	TS4-6	6-May-52	3	9
IA MONONA	1.30	PB2-4	21-Jun-57	9	9
IA MONONA	0.85	PB4-3	7-Jul-57	10	9
IA MONONA	4.02	PB5-5	19-Jul-57	3	9
IA MONONA	0.80	PB13-7	12-Sep-57	7	9
IA MONONA	0.17	UE2-7	11-Mar-62	2	9
IA MONONA	527.44	S-3	8-Jul-62	5	9
IA MONONA	0.17	UE7-5	15-Jul-62	7	9
IA MONROE	155.64	TS7-3	3-Jun-52	6	7
IA MONROE	0.99	TP8-5	13-Apr-55	3	7
IA MONROE	15.90	PB8-3	20-Aug-57	7	7
IA MONROE	16.67	PB9-4	26-Aug-57	4	7
IA MONROE	1.21	PB15-4	19-Sep-57	5	7
IA MONROE	345.56	S-3	8-Jul-62	8	7
IA MONROE	0.24	UE7-5	15-Jul-62	5	7
IA MONTGOMERY	83.35	PB9-4	26-Aug-57	1	2

COUNTY	uCi/ sqm	SHOT DAY	DATE	RANK FOR DAY	TIMES IN TOP 10
IA MONTGOMERY	2.04	PB9-5	27-Aug-57	8	2
IA MUSCATINE	2.48	PB5-7	21-Jul-57	6	7
IA MUSCATINE	13.72	PB9-3	25-Aug-57	6	7
IA MUSCATINE	16.67	PB9-4	26-Aug-57	4	7
IA MUSCATINE	3.39	PB9-5	27-Aug-57	4	7
IA MUSCATINE	3.98	PB18-8	14-Oct-57	4	7
IA MUSCATINE	0.08	UE2-7	11-Mar-62	5	7
IA MUSCATINE	0.12	UE7-5	15-Jul-62	9	7
IA O BRIEN	1.44	BJ5-3	21-Nov-51	6	10
IA O BRIEN	1.37	TS4-6	6-May-52	9	10
IA O BRIEN	1.37	TS4-6	6-May-52	9	10
IA O BRIEN	1.15	TS8-11	15-Jun-52	3	10
IA O BRIEN	1.40	UpK11-4	7-Jun-53	7	10
IA O BRIEN	1.30	PB2-4	21-Jun-57	9	10
IA O BRIEN	0.85	PB4-3	7-Jul-57	10	10
IA O BRIEN	0.17	UE2-7	11-Mar-62	2	10
IA O BRIEN	527.44	S-3	8-Jul-62	5	10
IA O BRIEN	9.84	UE12-4	17-Apr-65	4	10
IA OSCEOLA	1.44	BJ5-3	21-Nov-51	6	12
IA OSCEOLA	1.37	TS4-6	6-May-52	9	12
IA OSCEOLA	1.37	TS4-6	6-May-52	9	12
IA OSCEOLA	0.93	TS8-10	14-Jun-52	1	12
IA OSCEOLA	1.72	TS8-11	15-Jun-52	2	12
IA OSCEOLA	7.03	UpK7-5	29-Apr-53	7	12
IA OSCEOLA	1.30	PB2-4	21-Jun-57	9	12
IA OSCEOLA	1.10	PB4-3	7-Jul-57	7	12
IA OSCEOLA	2.01	PB5-5	19-Jul-57	8	12
IA OSCEOLA	1.18	PB5-6	20-Jul-57	7	12
IA OSCEOLA	0.17	UE2-7	11-Mar-62	2	12
IA OSCEOLA	527.44	S-3	8-Jul-62	5	12
IA PAGE	1.67	PB5-7	21-Jul-57	7	4
IA PAGE	23.89	PB8-3	20-Aug-57	5	4
IA PAGE	0.86	PB18-6	12-Oct-57	7	4
IA PAGE	0.24	UE7-5	15-Jul-62	5	4
IA PALO ALTO	1.74	TS4-6	6-May-52	5	6

COUNTY	uCi/ sqm	SHOT DAY	DATE	RANK FOR DAY	TIMES IN TOP 10
IA PALO ALTO	1.74	TS4-6	6-May-52	5	6
IA PALO ALTO	0.85	PB4-3	7-Jul-57	10	6
IA PALO ALTO	0.30	UE2-7	11-Mar-62	1	6
IA PALO ALTO	527.44	S-3	8-Jul-62	5	6
IA PALO ALTO	0.13	UE7-5	15-Jul-62	8	6
IA PLYMOUTH	1.08	BJ5-3	21-Nov-51	7	8
IA PLYMOUTH	1.37	TS4-6	6-May-52	9	8
IA PLYMOUTH	1.37	TS4-6	6-May-52	9	8
IA PLYMOUTH	3.97	UpK2-2	25-Mar-53	10	8
IA PLYMOUTH	1.44	PB4-3	7-Jul-57	3	8
IA PLYMOUTH	4.02	PB5-5	19-Jul-57	3	8
IA PLYMOUTH	0.10	UE2-7	11-Mar-62	4	8
IA PLYMOUTH	527.44	S-3	8-Jul-62	5	8
IA POCAHONTAS	1.44	BJ5-3	21-Nov-51	6	5
IA POCAHONTAS	1.19	PB4-3	7-Jul-57	6	5
IA POCAHONTAS	0.10	UE2-7	11-Mar-62	4	5
IA POCAHONTAS	527.44	S-3	8-Jul-62	5	5
IA POCAHONTAS	0.13	UE7-5	15-Jul-62	8	5
IA POLK	0.98	TP7-5	2-Apr-55	1	5
IA POLK	83.35	PB9-4	26-Aug-57	1	5
IA POLK	2.04	PB9-5	27-Aug-57	8	5
IA POLK	1.68	PB17-9	6-Oct-57	3	5
IA POLK	0.17	UE7-5	15-Jul-62	7	5
IA POTTAWATTAMI	4.17	UpK9-5	23-May-53	9	3
IA POTTAWATTAMI	83.35	PB9-4	26-Aug-57	1	3
IA POTTAWATTAMI	3.39	PB9-5	27-Aug-57	4	3
IA POWESHIEK	1.10	PB4-3	7-Jul-57	7	9
IA POWESHIEK	2.48	PB5-7	21-Jul-57	6	9
IA POWESHIEK	0.92	PB6-9	1-Aug-57	6	9
IA POWESHIEK	7.99	PB8-3	20-Aug-57	9	9
IA POWESHIEK	54.86	PB9-3	25-Aug-57	1	9
IA POWESHIEK	16.67	PB9-4	26-Aug-57	4	9
IA POWESHIEK	3.39	PB9-5	27-Aug-57	4	9
IA POWESHIEK	3.98	PB18-8	14-Oct-57	4	9
IA POWESHIEK	0.13	UE7-5	15-Jul-62	8	9

COUNTY	uCi/ sqm	SHOT DAY	DATE	RANK FOR DAY	TIMES IN TOP 10
IA RINGGOLD	89.78	TS7-4	4-Jun-52	1	10
IA RINGGOLD	0.99	TP8-5	13-Apr-55	3	10
IA RINGGOLD	2.68	TP10-4	8-May-55	3	10
IA RINGGOLD	3.09	PB4-10	14-Jul-57	5	10
IA RINGGOLD	39.79	PB8-3	20-Aug-57	3	10
IA RINGGOLD	33.34	PB9-4	26-Aug-57	3	10
IA RINGGOLD	1.21	PB15-4	19-Sep-57	5	10
IA RINGGOLD	0.81	PB15-5	20-Sep-57	9	10
IA RINGGOLD	1.44	PB18-6	12-Oct-57	5	10
IA RINGGOLD	345.56	S-3	8-Jul-62	8	10
IA SAC	1.74	TS4-6	6-May-52	5	7
IA SAC	1.74	TS4-6	6-May-52	5	7
IA SAC	1.19	PB4-3	7-Jul-57	6	7
IA SAC	0.81	PB15-4	19-Sep-57	7	7
IA SAC	0.17	UE2-7	11-Mar-62	2	7
IA SAC	527.44	S-3	8-Jul-62	5	7
IA SAC	0.75	BN-2	19-Dec-70	8	7
IA SCOTT	1.67	PB5-7	21-Jul-57	7	7
IA SCOTT	13.72	PB9-3	25-Aug-57	6	7
IA SCOTT	16.67	PB9-4	26-Aug-57	4	7
IA SCOTT	3.39	PB9-5	27-Aug-57	4	7
IA SCOTT	2.65	PB18-8	14-Oct-57	8	7
IA SCOTT	0.08	UE2-7	11-Mar-62	5	7
IA SCOTT	0.33	UE7-5	15-Jul-62	3	7
IA SHELBY	60.05	TS7-4	4-Jun-52	3	10
IA SHELBY	5.67	UpK9-5	23-May-53	1	10
IA SHELBY	1.09	TP7-6	3-Apr-55	10	10
IA SHELBY	0.85	PB4-3	7-Jul-57	10	10
IA SHELBY	1.86	PB4-10	14-Jul-57	8	10
IA SHELBY	83.35	PB9-4	26-Aug-57	1	10
IA SHELBY	2.04	PB9-5	27-Aug-57	8	10
IA SHELBY	0.81	PB15-4	19-Sep-57	7	10
IA SHELBY	0.10	UE2-7	11-Mar-62	4	10
IA SHELBY	345.56	S-3	8-Jul-62	8	10
IA SIOUX	1.44	BJ5-3	21-Nov-51	6	9

COUNTY	uCi/ sqm	SHOT DAY	DATE	RANK FOR DAY	TIMES IN TOP 10
IA SIOUX	1.37	TS4-6	6-May-52	9	9
IA SIOUX	1.37	TS4-6	6-May-52	9	9
IA SIOUX	3.97	UpK2-2	25-Mar-53	10	9
IA SIOUX	1.10	PB4-3	7-Jul-57	7	9
IA SIOUX	2.01	PB5-5	19-Jul-57	8	9
IA SIOUX	0.17	UE2-7	11-Mar-62	2	9
IA SIOUX	527.44	S-3	8-Jul-62	5	9
IA SIOUX	11.43	UE12-4	17-Apr-65	3	9
IA STORY	4.15	PB5-7	21-Jul-57	2	7
IA STORY	50.01	PB9-4	26-Aug-57	2	7
IA STORY	3.39	PB9-5	27-Aug-57	4	7
IA STORY	0.81	PB15-5	20-Sep-57	9	7
IA STORY	0.08	UE2-7	11-Mar-62	5	7
IA STORY	527.44	S-3	8-Jul-62	5	7
IA STORY	0.13	UE7-5	15-Jul-62	8	7
IA TAMA	2.21	PB4-9	13-Jul-57	5	8
IA TAMA	1.67	PB5-7	21-Jul-57	7	8
IA TAMA	0.92	PB6-9	1-Aug-57	6	8
IA TAMA	54.86	PB9-3	25-Aug-57	1	8
IA TAMA	33.34	PB9-4	26-Aug-57	3	8
IA TAMA	3.39	PB9-5	27-Aug-57	4	8
IA TAMA	3.98	PB18-8	14-Oct-57	4	8
IA TAMA	0.13	UE7-5	15-Jul-62	8	8
IA TAYLOR	1.86	PB4-10	14-Jul-57	8	6
IA TAYLOR	39.79	PB8-3	20-Aug-57	3	6
IA TAYLOR	33.34	PB9-4	26-Aug-57	3	6
IA TAYLOR	2.04	PB9-5	27-Aug-57	8	6
IA TAYLOR	1.21	PB15-4	19-Sep-57	5	6
IA TAYLOR	0.12	UE7-5	15-Jul-62	9	6
IA UNION	56.82	TS7-4	4-Jun-52	8	8
IA UNION	1.86	PB4-10	14-Jul-57	8	8
IA UNION	1.67	PB5-7	21-Jul-57	7	8
IA UNION	23.89	PB8-3	20-Aug-57	5	8
IA UNION	50.01	PB9-4	26-Aug-57	2	8
IA UNION	2.04	PB9-5	27-Aug-57	8	8

COUNTY	uCi/ sqm	SHOT DAY	DATE	RANK FOR DAY	TIMES IN TOP 10
IA UNION	0.81	PB15-5	20-Sep-57	9	8
IA UNION	345.56	S-3	8-Jul-62	8	8
IA VAN BUREN	1.67	PB5-7	21-Jul-57	7	8
IA VAN BUREN	27.43	PB9-3	25-Aug-57	2	8
IA VAN BUREN	33.34	PB9-4	26-Aug-57	3	8
IA VAN BUREN	2.04	PB9-5	27-Aug-57	8	8
IA VAN BUREN	6.63	PB18-8	14-Oct-57	2	8
IA VAN BUREN	0.08	UE2-7	11-Mar-62	5	8
IA VAN BUREN	621.15	S-3	8-Jul-62	3	8
IA VAN BUREN	0.12	UE7-5	15-Jul-62	9	8
IA WAPELLO	2.48	PB5-7	21-Jul-57	6	8
IA WAPELLO	7.99	PB8-3	20-Aug-57	9	8
IA WAPELLO	54.86	PB9-3	25-Aug-57	1	8
IA WAPELLO	50.01	PB9-4	26-Aug-57	2	8
IA WAPELLO	2.04	PB9-5	27-Aug-57	8	8
IA WAPELLO	3.98	PB18-8	14-Oct-57	4	8
IA WAPELLO	345.56	S-3	8-Jul-62	8	8
IA WAPELLO	0.13	UE7-5	15-Jul-62	8	8
IA WARREN	1.86	PB4-10	14-Jul-57	8	6
IA WARREN	23.89	PB8-3	20-Aug-57	5	6
IA WARREN	50.01	PB9-4	26-Aug-57	2	6
IA WARREN	2.04	PB9-5	27-Aug-57	8	6
IA WARREN	0.93	PB17-9	6-Oct-57	10	6
IA WARREN	0.24	UE7-5	15-Jul-62	5	6
IA WASHINGTON	2.48	PB5-7	21-Jul-57	6	8
IA WASHINGTON	13.72	PB9-3	25-Aug-57	6	8
IA WASHINGTON	8.37	PB9-4	26-Aug-57	7	8
IA WASHINGTON	2.04	PB9-5	27-Aug-57	8	8
IA WASHINGTON	3.98	PB18-8	14-Oct-57	4	8
IA WASHINGTON	1.25	PB18-9	15-Oct-57	3	8
IA WASHINGTON	0.08	UE2-7	11-Mar-62	5	8
IA WASHINGTON	621.15	S-3	8-Jul-62	3	8
IA WAYNE	2.48	PB5-7	21-Jul-57	6	7
IA WAYNE	15.90	PB8-3	20-Aug-57	7	7
IA WAYNE	16.67	PB9-4	26-Aug-57	4	7

COUNTY	uCi/ sqm	SHOT DAY	DATE	RANK FOR DAY	TIMES IN TOP 10
IA WAYNE	2.04	PB9-5	27-Aug-57	8	7
IA WAYNE	0.81	PB15-4	19-Sep-57	7	7
IA WAYNE	345.56	S-3	8-Jul-62	8	7
IA WAYNE	0.24	UE7-5	15-Jul-62	5	7
IA WEBSTER	1.69	PB4-3	7-Jul-57	1	5
IA WEBSTER	6.09	PB5-5	19-Jul-57	2	5
IA WEBSTER	0.17	UE2-7	11-Mar-62	2	5
IA WEBSTER	0.24	UE7-5	15-Jul-62	5	5
IA WEBSTER	3.94	UE8-4	17-Jul-62	10	5
IA WINNEBAGO	0.99	TS3-8	29-Apr-52	8	9
IA WINNEBAGO	0.99	TS3-8	29-Apr-52	8	9
IA WINNEBAGO	2.54	PB2-4	21-Jun-57	4	9
IA WINNEBAGO	0.93	PB4-3	7-Jul-57	9	9
IA WINNEBAGO	1.55	PB15-3	18-Sep-57	10	9
IA WINNEBAGO	1.16	PB15-5	20-Sep-57	7	9
IA WINNEBAGO	0.10	UE2-7	11-Mar-62	4	9
IA WINNEBAGO	527.44	S-3	8-Jul-62	5	9
IA WINNEBAGO	9.32	UE12-4	17-Apr-65	6	9
IA WINNESHIEK	0.99	TS3-8	29-Apr-52	8	12
IA WINNESHIEK	0.99	TS3-8	29-Apr-52	8	12
IA WINNESHIEK	0.93	PB4-3	7-Jul-57	9	12
IA WINNESHIEK	1.67	PB5-7	21-Jul-57	7	12
IA WINNESHIEK	2.43	PB6-9	1-Aug-57	2	12
IA WINNESHIEK	13.72	PB9-3	25-Aug-57	6	12
IA WINNESHIEK	8.37	PB9-4	26-Aug-57	7	12
IA WINNESHIEK	3.98	PB18-8	14-Oct-57	4	12
IA WINNESHIEK	0.10	UE2-7	11-Mar-62	4	12
IA WINNESHIEK	527.44	S-3	8-Jul-62	5	12
IA WINNESHIEK	0.12	UE7-5	15-Jul-62	9	12
IA WINNESHIEK	9.32	UE12-4	17-Apr-65	6	12
IA WOODBURY	1.26	UpK11-4	7-Jun-53	9	8
IA WOODBURY	0.85	PB4-3	7-Jul-57	10	8
IA WOODBURY	4.02	PB5-5	19-Jul-57	3	8
IA WOODBURY	0.80	PB13-7	12-Sep-57	7	8
IA WOODBURY	0.17	UE2-7	11-Mar-62	2	8

COUNTY	uCi/ sqm	SHOT DAY	DATE	RANK FOR DAY	TIMES IN TOP 10
IA WOODBURY	138.24	S-2	7-Jul-62	7	8
IA WOODBURY	527.44	S-3	8-Jul-62	5	8
IA WOODBURY	0.12	UE7-5	15-Jul-62	9	8
IA WORTH	0.99	TS3-8	29-Apr-52	8	9
IA WORTH	0.99	TS3-8	29-Apr-52	8	9
IA WORTH	2.54	PB2-4	21-Jun-57	4	9
IA WORTH	1.27	PB4-3	7-Jul-57	5	9
IA WORTH	1.97	PB8-5	22-Aug-57	9	9
IA WORTH	1.16	PB15-5	20-Sep-57	7	9
IA WORTH	0.10	UE2-7	11-Mar-62	4	9
IA WORTH	950.00	S-3	8-Jul-62	1	9
IA WORTH	9.32	UE12-4	17-Apr-65	6	9
IA WRIGHT	1.27	TS8-9	13-Jun-52	1	4
IA WRIGHT	1.27	PB4-3	7-Jul-57	5	4
IA WRIGHT	0.10	UE2-7	11-Mar-62	4	4
IA WRIGHT	0.12	UE7-5	15-Jul-62	9	4

COUNTY	uCi/ sqm	SHOT DAY	DATE	RANK FOR DAY	TIMES IN TOP 10
ID ADA	59.69	BJ5-1	19-Nov-51	5	3
ID ADA	13.87	PB6-2	25-Jul-57	9	3
ID ADA	13.00	PB17-1	28-Sep-57	8	3
ID ADAMS	29.58	BJ5-2	20-Nov-51	1	23
ID ADAMS	9.82	BJ5-3	21-Nov-51	4	23
ID ADAMS	2.98	BJ5-4	22-Nov-51	1	23
ID ADAMS	2.98	BJ5-4	22-Nov-51	1	23
ID ADAMS	2.14	BJ5-5	23-Nov-51	1	23
ID ADAMS	2.14	BJ5-5	23-Nov-51	1	23
ID ADAMS	1.06	BJ5-6	24-Nov-51	8	23
ID ADAMS	1.06	BJ5-6	24-Nov-51	8	23
ID ADAMS	1.35	BJ5-7	25-Nov-51	2	23
ID ADAMS	1.35	BJ5-7	25-Nov-51	2	23
ID ADAMS	3.37	BJ5-8	26-Nov-51	4	23
ID ADAMS	3.37	BJ5-8	26-Nov-51	4	23
ID ADAMS	5.30	TS8-4	8-Jun-52	1	23
ID ADAMS	3.01	TS8-5	9-Jun-52	6	23
ID ADAMS	1.40	TS8-8	12-Jun-52	2	23
ID ADAMS	12.28	PB6-3	26-Jul-57	6	23
ID ADAMS	2.67	PB15-3	18-Sep-57	4	23
ID ADAMS	7.58	PB16-3	25-Sep-57	1	23
ID ADAMS	1.29	PB16-4	26-Sep-57	9	23
ID ADAMS	1.23	PB17-4	1-Oct-57	1	23
ID ADAMS	0.85	PB18-4	10-Oct-57	7	23
ID ADAMS	0.65	BN-3	20-Dec-70	8	23
ID ADAMS	0.36	BN-4	21-Dec-70	4	23
ID BANNOCK	17.32	PB17-1	28-Sep-57	5	1
ID BEAR LAKE	17.32	PB17-1	28-Sep-57	5	2
ID BEAR LAKE	3.89	UE13-1	25-Apr-66	10	2
ID BENEWAH	1.94	PB1-5	1-Jun-57	7	8
ID BENEWAH	1.40	PB1-9	5-Jun-57	3	8
ID BENEWAH	12.28	PB6-3	26-Jul-57	6	8
ID BENEWAH	7.58	PB16-3	25-Sep-57	1	8
ID BENEWAH	1.29	PB16-4	26-Sep-57	9	8
ID BENEWAH	6.34	PB17-3	30-Sep-57	6	8

COUNTY	uCi/ sqm	SHOT DAY	DATE	RANK FOR DAY	TIMES IN TOP 10
ID BENEWAH	0.85	PB18-4	10-Oct-57	7	8
ID BENEWAH	0.46	BN-4	21-Dec-70	2	8
ID BINGHAM	17.32	PB17-1	28-Sep-57	5	1
ID BLAINE	1.31	TS5-9	15-May-52	4	11
ID BLAINE	1.59	TS7-2	2-Jun-52	8	11
ID BLAINE	777.08	TS8-1	5-Jun-52	3	11
ID BLAINE	12.28	PB6-3	26-Jul-57	6	11
ID BLAINE	2.67	PB15-3	18-Sep-57	4	11
ID BLAINE	7.58	PB16-3	25-Sep-57	1	11
ID BLAINE	1.29	PB16-4	26-Sep-57	9	11
ID BLAINE	10.80	PB17-5	2-Oct-57	4	11
ID BLAINE	0.85	PB18-4	10-Oct-57	7	11
ID BLAINE	1.14	BN-3	20-Dec-70	3	11
ID BLAINE	0.13	BN-4	21-Dec-70	10	11
ID BOISE	19.72	BJ5-2	20-Nov-51	2	21
ID BOISE	14.77	BJ5-3	21-Nov-51	3	21
ID BOISE	2.98	BJ5-4	22-Nov-51	1	21
ID BOISE	2.98	BJ5-4	22-Nov-51	1	21
ID BOISE	2.14	BJ5-5	23-Nov-51	1	21
ID BOISE	2.14	BJ5-5	23-Nov-51	1	21
ID BOISE	1.01	BJ5-7	25-Nov-51	3	21
ID BOISE	1.01	BJ5-7	25-Nov-51	3	21
ID BOISE	3.37	BJ5-8	26-Nov-51	4	21
ID BOISE	3.37	BJ5-8	26-Nov-51	4	21
ID BOISE	2.74	TS8-5	9-Jun-52	9	21
ID BOISE	0.93	TS8-8	12-Jun-52	3	21
ID BOISE	1.04	UpK7-12	6-May-53	1	21
ID BOISE	12.28	PB6-3	26-Jul-57	6	21
ID BOISE	2.67	PB15-3	18-Sep-57	4	21
ID BOISE	7.58	PB16-3	25-Sep-57	1	21
ID BOISE	1.29	PB16-4	26-Sep-57	9	21
ID BOISE	0.85	PB18-4	10-Oct-57	7	21
ID BOISE	1.18	BN-3	20-Dec-70	2	21
ID BOISE	0.38	BN-4	21-Dec-70	3	21
ID BOISE	0.20	BN-5	22-Dec-70	3	21

COUNTY	uCi/ sqm	SHOT DAY	DATE	RANK FOR DAY	TIMES IN TOP 10
ID BONNER	19.55	UpK10-2	26-May-53	5	9
ID BONNER	0.83	PB1-9	5-Jun-57	8	9
ID BONNER	15.78	PB6-2	25-Jul-57	4	9
ID BONNER	12.28	PB6-3	26-Jul-57	6	9
ID BONNER	2.67	PB15-3	18-Sep-57	4	9
ID BONNER	7.58	PB16-3	25-Sep-57	1	9
ID BONNER	2.58	PB16-4	26-Sep-57	7	9
ID BONNER	0.85	PB18-4	10-Oct-57	7	9
ID BONNER	0.38	BN-4	21-Dec-70	3	9
ID BONNEVILLE	17.32	PB17-1	28-Sep-57	5	3
ID BONNEVILLE	3.89	UE13-1	25-Apr-66	10	3
ID BONNEVILLE	1.84	BN-1	18-Dec-70	8	3
ID BOUNDARY	12.58	UpK10-2	26-May-53	9	5
ID BOUNDARY	12.28	PB6-3	26-Jul-57	6	5
ID BOUNDARY	7.58	PB16-3	25-Sep-57	1	5
ID BOUNDARY	2.58	PB16-4	26-Sep-57	7	5
ID BOUNDARY	0.85	PB18-4	10-Oct-57	7	5
ID BUTTE	1.31	TS5-9	15-May-52	4	11
ID BUTTE	2.66	TS7-2	2-Jun-52	2	11
ID BUTTE	1.72	TS8-6	10-Jun-52	9	11
ID BUTTE	12.28	PB6-3	26-Jul-57	6	11
ID BUTTE	2.67	PB15-3	18-Sep-57	4	11
ID BUTTE	7.58	PB16-3	25-Sep-57	1	11
ID BUTTE	1.29	PB16-4	26-Sep-57	9	11
ID BUTTE	0.85	PB18-4	10-Oct-57	7	11
ID BUTTE	2.59	UE2-2	6-Mar-62	4	11
ID BUTTE	1.47	BN-3	20-Dec-70	1	11
ID BUTTE	0.13	BN-4	21-Dec-70	10	11
ID CAMAS	9.86	BJ5-2	20-Nov-51	3	16
ID CAMAS	9.82	BJ5-3	21-Nov-51	4	16
ID CAMAS	2.98	BJ5-4	22-Nov-51	1	16
ID CAMAS	2.98	BJ5-4	22-Nov-51	1	16
ID CAMAS	2.14	BJ5-5	23-Nov-51	1	16
ID CAMAS	2.14	BJ5-5	23-Nov-51	1	16
ID CAMAS	3.37	BJ5-8	26-Nov-51	4	16

COUNTY	uCi/ sqm	SHOT DAY	DATE	RANK FOR DAY	TIMES IN TOP 10
ID CAMAS	3.37	BJ5-8	26-Nov-51	4	16
ID CAMAS	335.59	TS8-1	5-Jun-52	8	16
ID CAMAS	12.28	PB6-3	26-Jul-57	6	16
ID CAMAS	2.67	PB15-3	18-Sep-57	4	16
ID CAMAS	7.58	PB16-3	25-Sep-57	1	16
ID CAMAS	1.29	PB16-4	26-Sep-57	9	16
ID CAMAS	0.85	PB18-4	10-Oct-57	7	16
ID CAMAS	1.14	BN-3	20-Dec-70	3	16
ID CAMAS	0.13	BN-4	21-Dec-70	10	16
ID CANYON	45.24	BJ5-1	19-Nov-51	8	1
ID CARIBOU	17.32	PB17-1	28-Sep-57	5	2
ID CARIBOU	3.89	UE13-1	25-Apr-66	10	2
ID CASSIA	92.60	TS7-1	1-Jun-52	7	2
ID CASSIA	17.32	PB17-1	28-Sep-57	5	2
ID CLARK	1.31	TS5-9	15-May-52	4	13
ID CLARK	1.59	TS7-2	2-Jun-52	8	13
ID CLARK	1.33	PB1-11	7-Jun-57	3	13
ID CLARK	3.59	PB2-2	19-Jun-57	4	13
ID CLARK	12.01	PB5-2	16-Jul-57	4	13
ID CLARK	29.83	PB8-2	19-Aug-57	2	13
ID CLARK	11.17	PB9-2	24-Aug-57	2	13
ID CLARK	7.62	PB15-1	16-Sep-57	10	13
ID CLARK	3.36	PB15-3	18-Sep-57	3	13
ID CLARK	1.94	PB16-2	24-Sep-57	1	13
ID CLARK	2.07	UE2-2	6-Mar-62	6	13
ID CLARK	0.88	BN-3	20-Dec-70	5	13
ID CLARK	0.23	BN-4	21-Dec-70	8	13
ID CLEARWATER	5.00	BJ5-2	20-Nov-51	4	21
ID CLEARWATER	39.28	BJ5-3	21-Nov-51	1	21
ID CLEARWATER	2.98	BJ5-4	22-Nov-51	1	21
ID CLEARWATER	2.98	BJ5-4	22-Nov-51	1	21
ID CLEARWATER	2.14	BJ5-5	23-Nov-51	1	21
ID CLEARWATER	2.14	BJ5-5	23-Nov-51	1	21
ID CLEARWATER	1.06	BJ5-6	24-Nov-51	8	21
ID CLEARWATER	1.06	BJ5-6	24-Nov-51	8	21

COUNTY	uCi/ sqm	SHOT DAY	DATE	RANK FOR DAY	TIMES IN TOP 10
ID CLEARWATER	1.35	BJ5-7	25-Nov-51	2	21
ID CLEARWATER	1.35	BJ5-7	25-Nov-51	2	21
ID CLEARWATER	1.12	BJ5-8	26-Nov-51	6	21
ID CLEARWATER	1.12	BJ5-8	26-Nov-51	6	21
ID CLEARWATER	2.01	TS8-7	11-Jun-52	5	21
ID CLEARWATER	1.40	PB1-9	5-Jun-57	3	21
ID CLEARWATER	12.28	PB6-3	26-Jul-57	6	21
ID CLEARWATER	7.58	PB16-3	25-Sep-57	1	21
ID CLEARWATER	1.29	PB16-4	26-Sep-57	9	21
ID CLEARWATER	0.93	PB18-3	9-Oct-57	5	21
ID CLEARWATER	0.85	PB18-4	10-Oct-57	7	21
ID CLEARWATER	0.46	BN-4	21-Dec-70	2	21
ID CLEARWATER	0.22	BN-5	22-Dec-70	1	21
ID CUSTER	9.86	BJ5-2	20-Nov-51	3	19
ID CUSTER	14.77	BJ5-3	21-Nov-51	3	19
ID CUSTER	2.98	BJ5-4	22-Nov-51	1	19
ID CUSTER	2.98	BJ5-4	22-Nov-51	1	19
ID CUSTER	2.14	BJ5-5	23-Nov-51	1	19
ID CUSTER	2.14	BJ5-5	23-Nov-51	1	19
ID CUSTER	2.25	BJ5-8	26-Nov-51	5	19
ID CUSTER	2.25	BJ5-8	26-Nov-51	5	19
ID CUSTER	819.83	TS8-1	5-Jun-52	2	19
ID CUSTER	21.00	PB6-2	25-Jul-57	1	19
ID CUSTER	12.28	PB6-3	26-Jul-57	6	19
ID CUSTER	2.67	PB15-3	18-Sep-57	4	19
ID CUSTER	7.58	PB16-3	25-Sep-57	1	19
ID CUSTER	1.29	PB16-4	26-Sep-57	9	19
ID CUSTER	0.85	PB18-4	10-Oct-57	7	19
ID CUSTER	2.20	UE7-2	12-Jul-62	9	19
ID CUSTER	1.02	BN-3	20-Dec-70	4	19
ID CUSTER	0.46	BN-4	21-Dec-70	2	19
ID CUSTER	0.14	BN-5	22-Dec-70	4	19
ID ELMORE	45.35	BJ5-1	19-Nov-51	7	3
ID ELMORE	14.71	PB6-2	25-Jul-57	8	3
ID ELMORE	17.32	PB17-1	28-Sep-57	5	3

COUNTY	uCi/ sqm	SHOT DAY	DATE	RANK FOR DAY	TIMES IN TOP 10
ID FRANKLIN	9.73	PB15-1	16-Sep-57	6	3
ID FRANKLIN	17.32	PB17-1	28-Sep-57	5	3
ID FRANKLIN	3.89	UE13-1	25-Apr-66	10	3
ID FREMONT	1.31	TS5-9	15-May-52	4	12
ID FREMONT	1.91	PB2-2	19-Jun-57	8	12
ID FREMONT	15.87	PB6-2	25-Jul-57	3	12
ID FREMONT	29.83	PB8-2	19-Aug-57	2	12
ID FREMONT	1.86	PB9-2	24-Aug-57	9	12
ID FREMONT	17.28	PB17-5	2-Oct-57	1	12
ID FREMONT	2.18	UE2-2	6-Mar-62	5	12
ID FREMONT	0.59	UE2-3	7-Mar-62	6	12
ID FREMONT	0.11	UE2-6	10-Mar-62	6	12
ID FREMONT	0.77	BN-3	20-Dec-70	7	12
ID FREMONT	0.24	BN-4	21-Dec-70	7	12
ID FREMONT	0.22	BN-5	22-Dec-70	1	12
ID GEM	29.58	BJ5-2	20-Nov-51	1	16
ID GEM	9.82	BJ5-3	21-Nov-51	4	16
ID GEM	2.98	BJ5-4	22-Nov-51	1	16
ID GEM	2.98	BJ5-4	22-Nov-51	1	16
ID GEM	2.14	BJ5-5	23-Nov-51	1	16
ID GEM	2.14	BJ5-5	23-Nov-51	1	16
ID GEM	2.25	BJ5-8	26-Nov-51	5	16
ID GEM	2.25	BJ5-8	26-Nov-51	5	16
ID GEM	862.38	TS8-1	5-Jun-52	1	16
ID GEM	3.68	TS8-5	9-Jun-52	2	16
ID GEM	0.93	TS8-8	12-Jun-52	3	16
ID GEM	12.28	PB6-3	26-Jul-57	6	16
ID GEM	7.58	PB16-3	25-Sep-57	1	16
ID GEM	1.29	PB16-4	26-Sep-57	9	16
ID GEM	0.85	PB18-4	10-Oct-57	7	16
ID GEM	0.46	BN-4	21-Dec-70	2	16
ID GOODING	311.64	TS8-1	5-Jun-52	10	2
ID GOODING	17.32	PB17-1	28-Sep-57	5	2
ID IDAHO	19.72	BJ5-2	20-Nov-51	2	28
ID IDAHO	19.64	BJ5-3	21-Nov-51	2	28

COUNTY	uCi/ sqm	SHOT DAY	DATE	RANK FOR DAY	TIMES IN TOP 10
ID IDAHO	2.98	BJ5-4	22-Nov-51	1	28
ID IDAHO	2.98	BJ5-4	22-Nov-51	1	28
ID IDAHO	2.14	BJ5-5	23-Nov-51	1	28
ID IDAHO	2.14	BJ5-5	23-Nov-51	1	28
ID IDAHO	1.06	BJ5-6	24-Nov-51	8	28
ID IDAHO	1.06	BJ5-6	24-Nov-51	8	28
ID IDAHO	1.35	BJ5-7	25-Nov-51	2	28
ID IDAHO	1.35	BJ5-7	25-Nov-51	2	28
ID IDAHO	1.12	BJ5-8	26-Nov-51	6	28
ID IDAHO	1.12	BJ5-8	26-Nov-51	6	28
ID IDAHO	341.82	TS8-1	5-Jun-52	7	28
ID IDAHO	3.25	TS8-5	9-Jun-52	5	28
ID IDAHO	1.93	TS8-6	10-Jun-52	6	28
ID IDAHO	3.10	TS8-7	11-Jun-52	1	28
ID IDAHO	2.33	TS8-8	12-Jun-52	1	28
ID IDAHO	1.04	UpK7-12	6-May-53	1	28
ID IDAHO	0.83	PB1-9	5-Jun-57	8	28
ID IDAHO	15.78	PB6-2	25-Jul-57	4	28
ID IDAHO	12.28	PB6-3	26-Jul-57	6	28
ID IDAHO	1.98	PB15-3	18-Sep-57	7	28
ID IDAHO	7.58	PB16-3	25-Sep-57	1	28
ID IDAHO	1.94	PB16-4	26-Sep-57	8	28
ID IDAHO	0.85	PB18-4	10-Oct-57	7	28
ID IDAHO	33.38	UE12-3	16-Apr-65	7	28
ID IDAHO	0.46	BN-4	21-Dec-70	2	28
ID IDAHO	0.21	BN-5	22-Dec-70	2	28
ID JEFFERSON	3.08	TS7-2	2-Jun-52	1	11
ID JEFFERSON	2.39	PB2-2	19-Jun-57	7	11
ID JEFFERSON	12.01	PB5-2	16-Jul-57	4	11
ID JEFFERSON	29.83	PB8-2	19-Aug-57	2	11
ID JEFFERSON	11.17	PB9-2	24-Aug-57	2	11
ID JEFFERSON	7.62	PB15-1	16-Sep-57	10	11
ID JEFFERSON	1.64	PB15-3	18-Sep-57	9	11
ID JEFFERSON	1.94	PB16-2	24-Sep-57	1	11
ID JEFFERSON	2.73	UE2-2	6-Mar-62	3	11

COUNTY	uCi/ sqm	SHOT DAY	DATE	RANK FOR DAY	TIMES IN TOP 10
ID JEFFERSON	0.81	BN-3	20-Dec-70	6	11
ID JEFFERSON	0.20	BN-5	22-Dec-70	3	11
ID JEROME	17.32	PB17-1	28-Sep-57	5	1
ID KOOTENAI	13.40	UpK10-2	26-May-53	7	7
ID KOOTENAI	0.83	PB1-9	5-Jun-57	8	7
ID KOOTENAI	12.28	PB6-3	26-Jul-57	6	7
ID KOOTENAI	7.58	PB16-3	25-Sep-57	1	7
ID KOOTENAI	1.29	PB16-4	26-Sep-57	9	7
ID KOOTENAI	0.85	PB18-4	10-Oct-57	7	7
ID KOOTENAI	0.38	BN-4	21-Dec-70	3	7
ID LATAH	12.95	UpK10-2	26-May-53	8	10
ID LATAH	2.91	PB1-5	1-Jun-57	6	10
ID LATAH	0.83	PB1-9	5-Jun-57	8	10
ID LATAH	21.00	PB6-2	25-Jul-57	1	10
ID LATAH	12.28	PB6-3	26-Jul-57	6	10
ID LATAH	7.58	PB16-3	25-Sep-57	1	10
ID LATAH	1.29	PB16-4	26-Sep-57	9	10
ID LATAH	9.52	PB17-3	30-Sep-57	4	10
ID LATAH	0.85	PB18-4	10-Oct-57	7	10
ID LATAH	0.36	BN-4	21-Dec-70	4	10
ID LEMHI	1.15	TS5-8	14-May-52	3	13
ID LEMHI	3.94	TS5-9	15-May-52	1	13
ID LEMHI	557.61	TS8-1	5-Jun-52	4	13
ID LEMHI	1.94	PB1-5	1-Jun-57	7	13
ID LEMHI	3.61	PB1-6	2-Jun-57	2	13
ID LEMHI	21.00	PB6-2	25-Jul-57	1	13
ID LEMHI	12.28	PB6-3	26-Jul-57	6	13
ID LEMHI	5.34	PB15-3	18-Sep-57	1	13
ID LEMHI	7.58	PB16-3	25-Sep-57	1	13
ID LEMHI	1.29	PB16-4	26-Sep-57	9	13
ID LEMHI	0.85	PB18-4	10-Oct-57	7	13
ID LEMHI	2.47	UE7-2	12-Jul-62	6	13
ID LEMHI	33.29	UE12-3	16-Apr-65	8	13
ID LEWIS	2.77	TS8-5	9-Jun-52	8	9
ID LEWIS	1.94	PB1-5	1-Jun-57	7	9

COUNTY	uCi/ sqm	SHOT DAY	DATE	RANK FOR DAY	TIMES IN TOP 10
ID LEWIS	0.83	PB1-9	5-Jun-57	8	9
ID LEWIS	12.28	PB6-3	26-Jul-57	6	9
ID LEWIS	7.58	PB16-3	25-Sep-57	1	9
ID LEWIS	1.29	PB16-4	26-Sep-57	9	9
ID LEWIS	0.85	PB18-4	10-Oct-57	7	9
ID LEWIS	0.46	BN-4	21-Dec-70	2	9
ID LEWIS	0.20	BN-5	22-Dec-70	3	9
ID LINCOLN	17.32	PB17-1	28-Sep-57	5	1
ID MADISON	2.12	TS7-2	2-Jun-52	5	10
ID MADISON	2.39	PB2-2	19-Jun-57	7	10
ID MADISON	12.01	PB5-2	16-Jul-57	4	10
ID MADISON	29.83	PB8-2	19-Aug-57	2	10
ID MADISON	11.17	PB9-2	24-Aug-57	2	10
ID MADISON	7.62	PB15-1	16-Sep-57	10	10
ID MADISON	1.64	PB15-3	18-Sep-57	9	10
ID MADISON	1.94	PB16-2	24-Sep-57	1	10
ID MADISON	4.66	UE2-2	6-Mar-62	2	10
ID MADISON	0.88	BN-3	20-Dec-70	5	10
ID MINIDOKA	17.32	PB17-1	28-Sep-57	5	1
ID NEZ PERCE	0.83	PB1-9	5-Jun-57	8	6
ID NEZ PERCE	12.28	PB6-3	26-Jul-57	6	6
ID NEZ PERCE	7.58	PB16-3	25-Sep-57	1	6
ID NEZ PERCE	5.16	PB16-4	26-Sep-57	6	6
ID NEZ PERCE	6.34	PB17-3	30-Sep-57	6	6
ID NEZ PERCE	0.85	PB18-4	10-Oct-57	7	6
ID ONEIDA	91.94	TS7-1	1-Jun-52	8	3
ID ONEIDA	9.73	PB15-1	16-Sep-57	6	3
ID ONEIDA	17.32	PB17-1	28-Sep-57	5	3
ID OWYHEE	91.04	BJ5-1	19-Nov-51	4	3
ID OWYHEE	14.71	PB6-2	25-Jul-57	8	3
ID OWYHEE	17.32	PB17-1	28-Sep-57	5	3
ID PAYETTE	19.72	BJ5-2	20-Nov-51	2	15
ID PAYETTE	9.82	BJ5-3	21-Nov-51	4	15
ID PAYETTE	2.98	BJ5-4	22-Nov-51	1	15
ID PAYETTE	2.98	BJ5-4	22-Nov-51	1	15

COUNTY	uCi/ sqm	SHOT DAY	DATE	RANK FOR DAY	TIMES IN TOP 10
ID PAYETTE	2.14	BJ5-5	23-Nov-51	1	15
ID PAYETTE	2.14	BJ5-5	23-Nov-51	1	15
ID PAYETTE	5.05	TS8-5	9-Jun-52	1	15
ID PAYETTE	12.28	PB6-3	26-Jul-57	6	15
ID PAYETTE	7.58	PB16-3	25-Sep-57	1	15
ID PAYETTE	1.29	PB16-4	26-Sep-57	9	15
ID PAYETTE	9.52	PB17-3	30-Sep-57	4	15
ID PAYETTE	1.00	PB17-4	1-Oct-57	2	15
ID PAYETTE	0.85	PB18-4	10-Oct-57	7	15
ID PAYETTE	0.62	BN-3	20-Dec-70	9	15
ID PAYETTE	0.38	BN-4	21-Dec-70	3	15
ID POWER	9.73	PB15-1	16-Sep-57	6	2
ID POWER	17.32	PB17-1	28-Sep-57	5	2
ID SHOSHONE	1.94	PB1-5	1-Jun-57	7	7
ID SHOSHONE	0.83	PB1-9	5-Jun-57	8	7
ID SHOSHONE	12.28	PB6-3	26-Jul-57	6	7
ID SHOSHONE	7.58	PB16-3	25-Sep-57	1	7
ID SHOSHONE	2.58	PB16-4	26-Sep-57	7	7
ID SHOSHONE	0.85	PB18-4	10-Oct-57	7	7
ID SHOSHONE	0.38	BN-4	21-Dec-70	3	7
ID TETON	24.44	TS5-2	8-May-52	9	17
ID TETON	2.66	PB1-11	7-Jun-57	1	17
ID TETON	2.39	PB2-2	19-Jun-57	7	17
ID TETON	23.91	PB5-2	16-Jul-57	2	17
ID TETON	15.87	PB6-2	25-Jul-57	3	17
ID TETON	29.83	PB8-2	19-Aug-57	2	17
ID TETON	11.17	PB9-2	24-Aug-57	2	17
ID TETON	7.62	PB15-1	16-Sep-57	10	17
ID TETON	3.36	PB15-3	18-Sep-57	3	17
ID TETON	1.94	PB16-2	24-Sep-57	1	17
ID TETON	4.91	UE2-2	6-Mar-62	1	17
ID TETON	0.62	UE2-3	7-Mar-62	5	17
ID TETON	0.12	UE2-6	10-Mar-62	5	17
ID TETON	2.90	UE7-2	12-Jul-62	4	17
ID TETON	36.05	UE12-3	16-Apr-65	6	17

COUNTY	uCi/ sqm	SHOT DAY	DATE	RANK FOR DAY	TIMES IN TOP 10
ID TETON	9.75	UE12-4	17-Apr-65	5	17
ID TETON	0.20	BN-5	22-Dec-70	3	17
ID TWIN FALLS	91.29	TS7-1	1-Jun-52	9	3
ID TWIN FALLS	14.71	PB6-2	25-Jul-57	8	3
ID TWIN FALLS	17.32	PB17-1	28-Sep-57	5	3
ID VALLEY	19.72	BJ5-2	20-Nov-51	2	25
ID VALLEY	19.64	BJ5-3	21-Nov-51	2	25
ID VALLEY	2.98	BJ5-4	22-Nov-51	1	25
ID VALLEY	2.98	BJ5-4	22-Nov-51	1	25
ID VALLEY	2.14	BJ5-5	23-Nov-51	1	25
ID VALLEY	2.14	BJ5-5	23-Nov-51	1	25
ID VALLEY	1.06	BJ5-6	24-Nov-51	8	25
ID VALLEY	1.06	BJ5-6	24-Nov-51	8	25
ID VALLEY	1.35	BJ5-7	25-Nov-51	2	25
ID VALLEY	1.35	BJ5-7	25-Nov-51	2	25
ID VALLEY	2.25	BJ5-8	26-Nov-51	5	25
ID VALLEY	2.25	BJ5-8	26-Nov-51	5	25
ID VALLEY	376.04	TS8-1	5-Jun-52	6	25
ID VALLEY	3.50	TS8-5	9-Jun-52	3	25
ID VALLEY	0.93	TS8-8	12-Jun-52	3	25
ID VALLEY	21.00	PB6-2	25-Jul-57	1	25
ID VALLEY	12.28	PB6-3	26-Jul-57	6	25
ID VALLEY	5.34	PB15-3	18-Sep-57	1	25
ID VALLEY	7.58	PB16-3	25-Sep-57	1	25
ID VALLEY	1.29	PB16-4	26-Sep-57	9	25
ID VALLEY	0.80	PB17-4	1-Oct-57	4	25
ID VALLEY	0.85	PB18-4	10-Oct-57	7	25
ID VALLEY	0.65	BN-3	20-Dec-70	8	25
ID VALLEY	0.46	BN-4	21-Dec-70	2	25
ID VALLEY	0.20	BN-5	22-Dec-70	3	25
ID WASHINGTON	9.86	BJ5-2	20-Nov-51	3	21
ID WASHINGTON	9.82	BJ5-3	21-Nov-51	4	21
ID WASHINGTON	2.98	BJ5-4	22-Nov-51	1	21
ID WASHINGTON	2.98	BJ5-4	22-Nov-51	1	21
ID WASHINGTON	2.14	BJ5-5	23-Nov-51	1	21

COUNTY	uCi/ sqm	SHOT DAY	DATE	RANK FOR DAY	TIMES IN TOP 10
ID WASHINGTON	2.14	BJ5-5	23-Nov-51	1	21
ID WASHINGTON	3.17	BJ5-6	24-Nov-51	3	21
ID WASHINGTON	3.17	BJ5-6	24-Nov-51	3	21
ID WASHINGTON	4.05	BJ5-7	25-Nov-51	1	21
ID WASHINGTON	4.05	BJ5-7	25-Nov-51	1	21
ID WASHINGTON	2.25	BJ5-8	26-Nov-51	5	21
ID WASHINGTON	2.25	BJ5-8	26-Nov-51	5	21
ID WASHINGTON	1.40	TS8-8	12-Jun-52	2	21
ID WASHINGTON	1.94	PB1-5	1-Jun-57	7	21
ID WASHINGTON	12.28	PB6-3	26-Jul-57	6	21
ID WASHINGTON	7.58	PB16-3	25-Sep-57	1	21
ID WASHINGTON	1.29	PB16-4	26-Sep-57	9	21
ID WASHINGTON	6.34	PB17-3	30-Sep-57	6	21
ID WASHINGTON	0.85	PB18-4	10-Oct-57	7	21
ID WASHINGTON	0.62	BN-3	20-Dec-70	9	21
ID WASHINGTON	0.36	BN-4	21-Dec-70	4	21

COUNTY	uCi/ sqm	SHOT DAY	DATE	RANK FOR DAY	TIMES IN TOP 10
IL ADAMS	1.42	UpK5-4	14-Apr-53	6	7
IL ADAMS	0.87	UpK6-4	21-Apr-53	6	7
IL ADAMS	13.72	PB9-3	25-Aug-57	6	7
IL ADAMS	16.67	PB9-4	26-Aug-57	4	7
IL ADAMS	2.25	PB18-7	13-Oct-57	2	7
IL ADAMS	1.02	PB18-9	15-Oct-57	4	7
IL ADAMS	0.33	UE7-5	15-Jul-62	3	7
IL ALEXANDER	0.87	UpK6-4	21-Apr-53	6	1
IL BOND	1.42	UpK5-4	14-Apr-53	6	3
IL BOND	0.87	UpK6-4	21-Apr-53	6	3
IL BOND	1.94	PB3-3	26-Jun-57	9	3
IL BOONE	1.96	TS8-7	11-Jun-52	6	9
IL BOONE	0.93	PB4-3	7-Jul-57	9	9
IL BOONE	13.72	PB9-3	25-Aug-57	6	9
IL BOONE	16.67	PB9-4	26-Aug-57	4	9
IL BOONE	3.55	PB9-5	27-Aug-57	3	9
IL BOONE	0.89	PB9-7	29-Aug-57	6	9
IL BOONE	0.91	PB18-9	15-Oct-57	5	9
IL BOONE	0.08	UE2-7	11-Mar-62	5	9
IL BOONE	0.12	UE7-5	15-Jul-62	9	9
IL BROWN	4.04	TS8-4	8-Jun-52	7	8
IL BROWN	1.42	UpK5-4	14-Apr-53	6	8
IL BROWN	0.87	UpK6-4	21-Apr-53	6	8
IL BROWN	1.86	PB4-10	14-Jul-57	8	8
IL BROWN	13.72	PB9-3	25-Aug-57	6	8
IL BROWN	33.34	PB9-4	26-Aug-57	3	8
IL BROWN	1.71	PB18-9	15-Oct-57	1	8
IL BROWN	0.12	UE7-5	15-Jul-62	9	8
IL BUREAU	13.72	PB9-3	25-Aug-57	6	6
IL BUREAU	8.37	PB9-4	26-Aug-57	7	6
IL BUREAU	2.15	PB9-5	27-Aug-57	7	6
IL BUREAU	0.98	PB9-6	28-Aug-57	6	6
IL BUREAU	1.02	PB18-9	15-Oct-57	4	6
IL BUREAU	0.12	UE7-5	15-Jul-62	9	6
IL CALHOUN	0.87	UpK6-4	21-Apr-53	6	5

COUNTY	uCi/ sqm	SHOT DAY	DATE	RANK FOR DAY	TIMES IN TOP 10
IL CALHOUN	2.48	PB5-7	21-Jul-57	6	5
IL CALHOUN	13.72	PB9-3	25-Aug-57	6	5
IL CALHOUN	8.37	PB9-4	26-Aug-57	7	5
IL CALHOUN	1.13	PB18-7	13-Oct-57	6	5
IL CARROLL	1.44	PB4-3	7-Jul-57	3	7
IL CARROLL	4.15	PB5-7	21-Jul-57	2	7
IL CARROLL	13.72	PB9-3	25-Aug-57	6	7
IL CARROLL	8.37	PB9-4	26-Aug-57	7	7
IL CARROLL	2.04	PB9-5	27-Aug-57	8	7
IL CARROLL	2.65	PB18-8	14-Oct-57	8	7
IL CARROLL	0.12	UE7-5	15-Jul-62	9	7
IL CASS	1.42	UpK5-4	14-Apr-53	6	7
IL CASS	0.87	UpK6-4	21-Apr-53	6	7
IL CASS	13.72	PB9-3	25-Aug-57	6	7
IL CASS	16.67	PB9-4	26-Aug-57	4	7
IL CASS	0.87	PB18-7	13-Oct-57	8	7
IL CASS	1.02	PB18-9	15-Oct-57	4	7
IL CASS	0.12	UE7-5	15-Jul-62	9	7
IL CHAMPAIGN	1.48	TS4-5	5-May-52	6	7
IL CHAMPAIGN	1.48	TS4-5	5-May-52	6	7
IL CHAMPAIGN	1.42	UpK5-4	14-Apr-53	6	7
IL CHAMPAIGN	0.87	UpK6-4	21-Apr-53	6	7
IL CHAMPAIGN	2.15	PB9-5	27-Aug-57	7	7
IL CHAMPAIGN	1.71	PB18-9	15-Oct-57	1	7
IL CHAMPAIGN	0.53	UE7-5	15-Jul-62	1	7
IL CHRISTIAN	0.87	UpK6-4	21-Apr-53	6	6
IL CHRISTIAN	13.72	PB9-3	25-Aug-57	6	6
IL CHRISTIAN	8.37	PB9-4	26-Aug-57	7	6
IL CHRISTIAN	0.87	PB18-7	13-Oct-57	8	6
IL CHRISTIAN	1.71	PB18-9	15-Oct-57	1	6
IL CHRISTIAN	0.22	UE7-5	15-Jul-62	6	6
IL CLARK	4.52	PB9-5	27-Aug-57	2	2
IL CLARK	0.26	UE7-5	15-Jul-62	4	2
IL CLAY	1.42	UpK5-4	14-Apr-53	6	5
IL CLAY	0.87	UpK6-4	21-Apr-53	6	5

COUNTY	uCi/ sqm	SHOT DAY	DATE	RANK FOR DAY	TIMES IN TOP 10
IL CLAY	1.06	TP7-8	5-Apr-55	5	5
IL CLAY	4.85	PB5-3	17-Jul-57	9	5
IL CLAY	1.02	PB18-9	15-Oct-57	4	5
IL CLINTON	0.87	UpK6-4	21-Apr-53	6	2
IL CLINTON	4.85	PB5-3	17-Jul-57	9	2
IL COLES	1.48	TS4-5	5-May-52	6	7
IL COLES	1.48	TS4-5	5-May-52	6	7
IL COLES	1.42	UpK5-4	14-Apr-53	6	7
IL COLES	0.87	UpK6-4	21-Apr-53	6	7
IL COLES	2.15	PB9-5	27-Aug-57	7	7
IL COLES	1.71	PB18-9	15-Oct-57	1	7
IL COLES	0.11	UE7-5	15-Jul-62	10	7
IL COOK	4.52	PB9-5	27-Aug-57	2	2
IL COOK	0.12	UE7-5	15-Jul-62	9	2
IL CRAWFORD	1.02	UpK3-6	5-Apr-53	7	3
IL CRAWFORD	4.52	PB9-5	27-Aug-57	2	3
IL CRAWFORD	0.11	UE7-5	15-Jul-62	10	3
IL CUMBERLAND	1.42	UpK5-4	14-Apr-53	6	4
IL CUMBERLAND	0.87	UpK6-4	21-Apr-53	6	4
IL CUMBERLAND	10.87	TP4-3	9-Mar-55	2	4
IL CUMBERLAND	1.71	PB18-9	15-Oct-57	1	4
IL DE KALB	1.88	TS8-7	11-Jun-52	8	6
IL DE KALB	13.72	PB9-3	25-Aug-57	6	6
IL DE KALB	33.34	PB9-4	26-Aug-57	3	6
IL DE KALB	2.15	PB9-5	27-Aug-57	7	6
IL DE KALB	0.08	UE2-7	11-Mar-62	5	6
IL DE KALB	0.33	UE7-5	15-Jul-62	3	6
IL DE WITT	1.42	UpK5-4	14-Apr-53	6	5
IL DE WITT	0.87	UpK6-4	21-Apr-53	6	5
IL DE WITT	2.15	PB9-5	27-Aug-57	7	5
IL DE WITT	1.71	PB18-9	15-Oct-57	1	5
IL DE WITT	0.33	UE7-5	15-Jul-62	3	5
IL DOUGLAS	1.42	UpK5-4	14-Apr-53	6	4
IL DOUGLAS	0.87	UpK6-4	21-Apr-53	6	4
IL DOUGLAS	1.71	PB18-9	15-Oct-57	1	4

COUNTY	uCi/ sqm	SHOT DAY	DATE	RANK FOR DAY	TIMES IN TOP 10
IL DOUGLAS	0.12	UE7-5	15-Jul-62	9	4
IL DU PAGE	0.33	UE7-5	15-Jul-62	3	1
IL EDGAR	1.51	PB9-5	27-Aug-57	9	2
IL EDGAR	0.26	UE7-5	15-Jul-62	4	2
IL EDWARDS	1.42	UpK5-4	14-Apr-53	6	3
IL EDWARDS	0.87	UpK6-4	21-Apr-53	6	3
IL EDWARDS	1.71	PB18-9	15-Oct-57	1	3
IL EFFINGHAM	1.42	UpK5-4	14-Apr-53	6	4
IL EFFINGHAM	0.87	UpK6-4	21-Apr-53	6	4
IL EFFINGHAM	7.27	PB5-3	17-Jul-57	5	4
IL EFFINGHAM	1.71	PB18-9	15-Oct-57	1	4
IL FAYETTE	3.97	TS8-4	8-Jun-52	8	5
IL FAYETTE	1.42	UpK5-4	14-Apr-53	6	5
IL FAYETTE	0.87	UpK6-4	21-Apr-53	6	5
IL FAYETTE	1.65	PB4-10	14-Jul-57	10	5
IL FAYETTE	1.71	PB18-9	15-Oct-57	1	5
IL FORD	2.15	PB9-5	27-Aug-57	7	3
IL FORD	1.52	PB18-9	15-Oct-57	2	3
IL FORD	0.53	UE7-5	15-Jul-62	1	3
IL FRANKLIN	0.87	UpK6-4	21-Apr-53	6	2
IL FRANKLIN	8.19	PB9-2	24-Aug-57	3	2
IL FULTON	4.79	TS8-4	8-Jun-52	4	10
IL FULTON	2.13	UpK5-4	14-Apr-53	3	10
IL FULTON	0.87	UpK6-4	21-Apr-53	6	10
IL FULTON	7.76	TP4-3	9-Mar-55	7	10
IL FULTON	13.72	PB9-3	25-Aug-57	6	10
IL FULTON	16.67	PB9-4	26-Aug-57	4	10
IL FULTON	2.15	PB9-5	27-Aug-57	7	10
IL FULTON	1.02	PB18-9	15-Oct-57	4	10
IL FULTON	345.56	S-3	8-Jul-62	8	10
IL FULTON	0.33	UE7-5	15-Jul-62	3	10
IL GALLATIN	0.87	UpK6-4	21-Apr-53	6	2
IL GALLATIN	7.27	PB5-3	17-Jul-57	5	2
IL GREENE	0.87	UpK6-4	21-Apr-53	6	6
IL GREENE	13.97	PB4-4	8-Jul-57	5	6

COUNTY	uCi/ sqm	SHOT DAY	DATE	RANK FOR DAY	TIMES IN TOP 10
IL GREENE	1.67	PB5-7	21-Jul-57	7	6
IL GREENE	13.72	PB9-3	25-Aug-57	6	6
IL GREENE	8.37	PB9-4	26-Aug-57	7	6
IL GREENE	0.13	UE7-5	15-Jul-62	8	6
IL GRUNDY	0.91	PB18-9	15-Oct-57	5	2
IL GRUNDY	0.12	UE7-5	15-Jul-62	9	2
IL HAMILTON	0.87	UpK6-4	21-Apr-53	6	1
IL HANCOCK	1.42	UpK5-4	14-Apr-53	6	9
IL HANCOCK	0.87	UpK6-4	21-Apr-53	6	9
IL HANCOCK	1.67	PB5-7	21-Jul-57	7	9
IL HANCOCK	13.72	PB9-3	25-Aug-57	6	9
IL HANCOCK	33.34	PB9-4	26-Aug-57	3	9
IL HANCOCK	1.13	PB18-7	13-Oct-57	6	9
IL HANCOCK	6.63	PB18-8	14-Oct-57	2	9
IL HANCOCK	621.15	S-3	8-Jul-62	3	9
IL HANCOCK	0.33	UE7-5	15-Jul-62	3	9
IL HARDIN	0.87	UpK6-4	21-Apr-53	6	4
IL HARDIN	1.06	TP7-8	5-Apr-55	5	4
IL HARDIN	1.17	PB2-4	21-Jun-57	10	4
IL HARDIN	0.89	PB13-7	12-Sep-57	5	4
IL HENDERSON	2.13	UpK5-4	14-Apr-53	3	8
IL HENDERSON	0.87	UpK6-4	21-Apr-53	6	8
IL HENDERSON	2.48	PB5-7	21-Jul-57	6	8
IL HENDERSON	13.72	PB9-3	25-Aug-57	6	8
IL HENDERSON	16.67	PB9-4	26-Aug-57	4	8
IL HENDERSON	2.04	PB9-5	27-Aug-57	8	8
IL HENDERSON	3.98	PB18-8	14-Oct-57	4	8
IL HENDERSON	0.33	UE7-5	15-Jul-62	3	8
IL HENRY	2.13	UpK5-4	14-Apr-53	3	8
IL HENRY	0.87	UpK6-4	21-Apr-53	6	8
IL HENRY	2.48	PB5-7	21-Jul-57	6	8
IL HENRY	13.72	PB9-3	25-Aug-57	6	8
IL HENRY	16.67	PB9-4	26-Aug-57	4	8
IL HENRY	2.04	PB9-5	27-Aug-57	8	8
IL HENRY	3.98	PB18-8	14-Oct-57	4	8

COUNTY	uCi/ sqm	SHOT DAY	DATE	RANK FOR DAY	TIMES IN TOP 10
IL HENRY	0.33	UE7-5	15-Jul-62	3	8
IL IROQUOIS	3.01	PB9-5	27-Aug-57	6	3
IL IROQUOIS	1.52	PB18-9	15-Oct-57	2	3
IL IROQUOIS	0.33	UE7-5	15-Jul-62	3	3
IL JACKSON	0.87	UpK6-4	21-Apr-53	6	3
IL JACKSON	5.46	PB9-2	24-Aug-57	4	3
IL JACKSON	1.43	PB15-7	22-Sep-57	3	3
IL JASPER	1.42	UpK5-4	14-Apr-53	6	6
IL JASPER	0.87	UpK6-4	21-Apr-53	6	6
IL JASPER	4.85	PB5-3	17-Jul-57	9	6
IL JASPER	0.99	PB6-10	2-Aug-57	3	6
IL JASPER	2.15	PB9-5	27-Aug-57	7	6
IL JASPER	1.71	PB18-9	15-Oct-57	1	6
IL JEFFERSON	0.87	UpK6-4	21-Apr-53	6	2
IL JEFFERSON	1.06	TP7-8	5-Apr-55	5	2
IL JERSEY	0.87	UpK6-4	21-Apr-53	6	4
IL JERSEY	13.72	PB9-3	25-Aug-57	6	4
IL JERSEY	8.37	PB9-4	26-Aug-57	7	4
IL JERSEY	2.25	PB18-7	13-Oct-57	2	4
IL JO DAVIESS	1.52	PB4-3	7-Jul-57	2	8
IL JO DAVIESS	4.15	PB5-7	21-Jul-57	2	8
IL JO DAVIESS	13.72	PB9-3	25-Aug-57	6	8
IL JO DAVIESS	33.34	PB9-4	26-Aug-57	3	8
IL JO DAVIESS	2.04	PB9-5	27-Aug-57	8	8
IL JO DAVIESS	3.98	PB18-8	14-Oct-57	4	8
IL JO DAVIESS	0.08	UE2-7	11-Mar-62	5	8
IL JO DAVIESS	0.13	UE7-5	15-Jul-62	8	8
IL JOHNSON	0.87	UpK6-4	21-Apr-53	6	3
IL JOHNSON	7.27	PB5-3	17-Jul-57	5	3
IL JOHNSON	2.73	PB9-2	24-Aug-57	6	3
IL KANE	13.72	PB9-3	25-Aug-57	6	6
IL KANE	16.67	PB9-4	26-Aug-57	4	6
IL KANE	2.15	PB9-5	27-Aug-57	7	6
IL KANE	0.91	PB18-9	15-Oct-57	5	6
IL KANE	0.08	UE2-7	11-Mar-62	5	6

COUNTY	uCi/ sqm	SHOT DAY	DATE	RANK FOR DAY	TIMES IN TOP 10
IL KANE	0.12	UE7-5	15-Jul-62	9	6
IL KANKAKEE	1.48	TS4-5	5-May-52	6	5
IL KANKAKEE	1.48	TS4-5	5-May-52	6	5
IL KANKAKEE	3.01	PB9-5	27-Aug-57	6	5
IL KANKAKEE	0.91	PB18-9	15-Oct-57	5	5
IL KANKAKEE	0.33	UE7-5	15-Jul-62	3	5
IL KENDALL	2.15	PB9-5	27-Aug-57	7	2
IL KENDALL	0.12	UE7-5	15-Jul-62	9	2
IL KNOX	2.13	UpK5-4	14-Apr-53	3	8
IL KNOX	0.87	UpK6-4	21-Apr-53	6	8
IL KNOX	13.72	PB9-3	25-Aug-57	6	8
IL KNOX	8.37	PB9-4	26-Aug-57	7	8
IL KNOX	3.55	PB9-5	27-Aug-57	3	8
IL KNOX	1.02	PB18-9	15-Oct-57	4	8
IL KNOX	0.08	UE2-7	11-Mar-62	5	8
IL KNOX	0.33	UE7-5	15-Jul-62	3	8
IL LA SALLE	13.72	PB9-3	25-Aug-57	6	4
IL LA SALLE	16.67	PB9-4	26-Aug-57	4	4
IL LA SALLE	0.91	PB18-9	15-Oct-57	5	4
IL LA SALLE	0.33	UE7-5	15-Jul-62	3	4
IL LAKE	0.99	PB5-6	20-Jul-57	9	5
IL LAKE	2.15	PB9-5	27-Aug-57	7	5
IL LAKE	0.08	UE2-7	11-Mar-62	5	5
IL LAKE	0.12	UE7-5	15-Jul-62	9	5
IL LAKE	14.56	UE12-4	17-Apr-65	2	5
IL LAWRENCE	1.51	PB9-5	27-Aug-57	9	2
IL LAWRENCE	1.12	PB13-7	12-Sep-57	4	2
IL LEE	13.72	PB9-3	25-Aug-57	6	4
IL LEE	8.37	PB9-4	26-Aug-57	7	4
IL LEE	0.91	PB18-9	15-Oct-57	5	4
IL LEE	0.12	UE7-5	15-Jul-62	9	4
IL LIVINGSTON	4.40	TS8-4	8-Jun-52	5	3
IL LIVINGSTON	0.91	PB18-9	15-Oct-57	5	3
IL LIVINGSTON	0.53	UE7-5	15-Jul-62	1	3
IL LOGAN	3.88	TS8-4	8-Jun-52	9	8

COUNTY	uCi/ sqm	SHOT DAY	DATE	RANK FOR DAY	TIMES IN TOP 10
IL LOGAN	1.42	UpK5-4	14-Apr-53	6	8
IL LOGAN	0.87	UpK6-4	21-Apr-53	6	8
IL LOGAN	13.72	PB9-3	25-Aug-57	6	8
IL LOGAN	33.34	PB9-4	26-Aug-57	3	8
IL LOGAN	2.15	PB9-5	27-Aug-57	7	8
IL LOGAN	1.02	PB18-9	15-Oct-57	4	8
IL LOGAN	0.12	UE7-5	15-Jul-62	9	8
IL MACON	1.42	UpK5-4	14-Apr-53	6	5
IL MACON	0.87	UpK6-4	21-Apr-53	6	5
IL MACON	1.65	PB4-10	14-Jul-57	10	5
IL MACON	1.71	PB18-9	15-Oct-57	1	5
IL MACON	0.22	UE7-5	15-Jul-62	6	5
IL MACOUPIN	0.87	UpK6-4	21-Apr-53	6	5
IL MACOUPIN	13.72	PB9-3	25-Aug-57	6	5
IL MACOUPIN	8.37	PB9-4	26-Aug-57	7	5
IL MACOUPIN	1.13	PB18-7	13-Oct-57	6	5
IL MACOUPIN	1.71	PB18-9	15-Oct-57	1	5
IL MADISON	0.87	UpK6-4	21-Apr-53	6	5
IL MADISON	1.06	TP7-8	5-Apr-55	5	5
IL MADISON	1.94	PB3-3	26-Jun-57	9	5
IL MADISON	13.72	PB9-3	25-Aug-57	6	5
IL MADISON	8.37	PB9-4	26-Aug-57	7	5
IL MARION	0.87	UpK6-4	21-Apr-53	6	2
IL MARION	1.65	PB4-10	14-Jul-57	10	2
IL MARSHALL	13.72	PB9-3	25-Aug-57	6	6
IL MARSHALL	16.67	PB9-4	26-Aug-57	4	6
IL MARSHALL	2.15	PB9-5	27-Aug-57	7	6
IL MARSHALL	1.13	PB18-7	13-Oct-57	6	6
IL MARSHALL	1.02	PB18-9	15-Oct-57	4	6
IL MARSHALL	0.53	UE7-5	15-Jul-62	1	6
IL MASON	1.42	UpK5-4	14-Apr-53	6	7
IL MASON	0.87	UpK6-4	21-Apr-53	6	7
IL MASON	13.72	PB9-3	25-Aug-57	6	7
IL MASON	16.67	PB9-4	26-Aug-57	4	7
IL MASON	2.15	PB9-5	27-Aug-57	7	7

COUNTY	uCi/ sqm	SHOT DAY	DATE	RANK FOR DAY	TIMES IN TOP 10
IL MASON	1.02	PB18-9	15-Oct-57	4	7
IL MASON	0.53	UE7-5	15-Jul-62	1	7
IL MASSAC	0.87	UpK6-4	21-Apr-53	6	5
IL MASSAC	1.06	TP7-8	5-Apr-55	5	5
IL MASSAC	7.27	PB5-3	17-Jul-57	5	5
IL MASSAC	2.73	PB9-2	24-Aug-57	6	5
IL MASSAC	0.89	PB13-7	12-Sep-57	5	5
IL MCDONOUGH	1.42	UpK5-4	14-Apr-53	6	8
IL MCDONOUGH	0.87	UpK6-4	21-Apr-53	6	8
IL MCDONOUGH	13.72	PB9-3	25-Aug-57	6	8
IL MCDONOUGH	33.34	PB9-4	26-Aug-57	3	8
IL MCDONOUGH	2.15	PB9-5	27-Aug-57	7	8
IL MCDONOUGH	1.71	PB18-9	15-Oct-57	1	8
IL MCDONOUGH	345.56	S-3	8-Jul-62	8	8
IL MCDONOUGH	0.53	UE7-5	15-Jul-62	1	8
IL MCHENRY	13.72	PB9-3	25-Aug-57	6	7
IL MCHENRY	16.67	PB9-4	26-Aug-57	4	7
IL MCHENRY	2.15	PB9-5	27-Aug-57	7	7
IL MCHENRY	0.89	PB9-7	29-Aug-57	6	7
IL MCHENRY	0.08	UE2-7	11-Mar-62	5	7
IL MCHENRY	0.33	UE7-5	15-Jul-62	3	7
IL MCHENRY	14.56	UE12-4	17-Apr-65	2	7
IL MCLEAN	5.24	TS8-4	8-Jun-52	2	6
IL MCLEAN	1.42	UpK5-4	14-Apr-53	6	6
IL MCLEAN	0.87	UpK6-4	21-Apr-53	6	6
IL MCLEAN	2.15	PB9-5	27-Aug-57	7	6
IL MCLEAN	1.02	PB18-9	15-Oct-57	4	6
IL MCLEAN	0.53	UE7-5	15-Jul-62	1	6
IL MENARD	1.42	UpK5-4	14-Apr-53	6	7
IL MENARD	0.87	UpK6-4	21-Apr-53	6	7
IL MENARD	2.48	PB5-7	21-Jul-57	6	7
IL MENARD	3.38	PB6-8	31-Jul-57	1	7
IL MENARD	13.72	PB9-3	25-Aug-57	6	7
IL MENARD	8.37	PB9-4	26-Aug-57	7	7
IL MENARD	0.12	UE7-5	15-Jul-62	9	7

COUNTY	uCi/ sqm	SHOT DAY	DATE	RANK FOR DAY	TIMES IN TOP 10
IL MERCER	2.13	UpK5-4	14-Apr-53	3	9
IL MERCER	0.87	UpK6-4	21-Apr-53	6	9
IL MERCER	2.48	PB5-7	21-Jul-57	6	9
IL MERCER	13.72	PB9-3	25-Aug-57	6	9
IL MERCER	16.67	PB9-4	26-Aug-57	4	9
IL MERCER	2.04	PB9-5	27-Aug-57	8	9
IL MERCER	3.98	PB18-8	14-Oct-57	4	9
IL MERCER	0.08	UE2-7	11-Mar-62	5	9
IL MERCER	0.12	UE7-5	15-Jul-62	9	9
IL MONROE	0.87	UpK6-4	21-Apr-53	6	3
IL MONROE	1.49	PB2-5	22-Jun-57	7	3
IL MONROE	13.72	PB9-3	25-Aug-57	6	3
IL MONTGOMERY	0.87	UpK6-4	21-Apr-53	6	5
IL MONTGOMERY	1.06	TP7-8	5-Apr-55	5	5
IL MONTGOMERY	13.72	PB9-3	25-Aug-57	6	5
IL MONTGOMERY	8.37	PB9-4	26-Aug-57	7	5
IL MONTGOMERY	1.71	PB18-9	15-Oct-57	1	5
IL MORGAN	1.42	UpK5-4	14-Apr-53	6	6
IL MORGAN	0.87	UpK6-4	21-Apr-53	6	6
IL MORGAN	13.72	PB9-3	25-Aug-57	6	6
IL MORGAN	33.34	PB9-4	26-Aug-57	3	6
IL MORGAN	1.13	PB18-7	13-Oct-57	6	6
IL MORGAN	1.71	PB18-9	15-Oct-57	1	6
IL MOULTRIE	1.42	UpK5-4	14-Apr-53	6	4
IL MOULTRIE	0.87	UpK6-4	21-Apr-53	6	4
IL MOULTRIE	1.71	PB18-9	15-Oct-57	1	4
IL MOULTRIE	0.12	UE7-5	15-Jul-62	9	4
IL OGLE	13.72	PB9-3	25-Aug-57	6	6
IL OGLE	33.34	PB9-4	26-Aug-57	3	6
IL OGLE	2.04	PB9-5	27-Aug-57	8	6
IL OGLE	2.65	PB18-8	14-Oct-57	8	6
IL OGLE	0.91	PB18-9	15-Oct-57	5	6
IL OGLE	0.33	UE7-5	15-Jul-62	3	6
IL PEORIA	2.13	UpK5-4	14-Apr-53	3	10
IL PEORIA	0.87	UpK6-4	21-Apr-53	6	10

COUNTY	uCi/ sqm	SHOT DAY	DATE	RANK FOR DAY	TIMES IN TOP 10
IL PEORIA	7.28	TP4-3	9-Mar-55	8	10
IL PEORIA	1.67	PB5-7	21-Jul-57	7	10
IL PEORIA	13.72	PB9-3	25-Aug-57	6	10
IL PEORIA	16.67	PB9-4	26-Aug-57	4	10
IL PEORIA	2.15	PB9-5	27-Aug-57	7	10
IL PEORIA	0.87	PB18-7	13-Oct-57	8	10
IL PEORIA	1.02	PB18-9	15-Oct-57	4	10
IL PEORIA	0.33	UE7-5	15-Jul-62	3	10
IL PERRY	0.87	UpK6-4	21-Apr-53	6	1
IL PIATT	1.42	UpK5-4	14-Apr-53	6	5
IL PIATT	0.87	UpK6-4	21-Apr-53	6	5
IL PIATT	2.15	PB9-5	27-Aug-57	7	5
IL PIATT	1.71	PB18-9	15-Oct-57	1	5
IL PIATT	0.39	UE7-5	15-Jul-62	2	5
IL PIKE	0.87	UpK6-4	21-Apr-53	6	7
IL PIKE	1.86	PB4-10	14-Jul-57	8	7
IL PIKE	1.67	PB5-7	21-Jul-57	7	7
IL PIKE	27.43	PB9-3	25-Aug-57	2	7
IL PIKE	16.67	PB9-4	26-Aug-57	4	7
IL PIKE	1.13	PB18-7	13-Oct-57	6	7
IL PIKE	0.12	UE7-5	15-Jul-62	9	7
IL POPE	0.87	UpK6-4	21-Apr-53	6	6
IL POPE	1.06	TP7-8	5-Apr-55	5	6
IL POPE	1.49	PB1-9	5-Jun-57	2	6
IL POPE	7.27	PB5-3	17-Jul-57	5	6
IL POPE	2.73	PB9-2	24-Aug-57	6	6
IL POPE	0.89	PB13-7	12-Sep-57	5	6
IL PULASKI	0.87	UpK6-4	21-Apr-53	6	3
IL PULASKI	0.83	PB3-9	2-Jul-57	4	3
IL PULASKI	4.85	PB5-3	17-Jul-57	9	3
IL PUTNAM	4.28	TS8-4	8-Jun-52	6	6
IL PUTNAM	13.72	PB9-3	25-Aug-57	6	6
IL PUTNAM	8.37	PB9-4	26-Aug-57	7	6
IL PUTNAM	2.15	PB9-5	27-Aug-57	7	6
IL PUTNAM	1.02	PB18-9	15-Oct-57	4	6

COUNTY	uCi/ sqm	SHOT DAY	DATE	RANK FOR DAY	TIMES IN TOP 10
IL PUTNAM	0.53	UE7-5	15-Jul-62	1	6
IL RANDOLPH	0.87	UpK6-4	21-Apr-53	6	4
IL RANDOLPH	5.46	PB9-2	24-Aug-57	4	4
IL RANDOLPH	13.72	PB9-3	25-Aug-57	6	4
IL RANDOLPH	1.13	PB18-7	13-Oct-57	6	4
IL RICHLAND	0.87	UpK6-4	21-Apr-53	6	4
IL RICHLAND	1.06	TP7-8	5-Apr-55	5	4
IL RICHLAND	1.65	PB4-10	14-Jul-57	10	4
IL RICHLAND	1.02	PB18-9	15-Oct-57	4	4
IL ROCK ISLAND	1.42	UpK5-4	14-Apr-53	6	11
IL ROCK ISLAND	0.87	UpK6-4	21-Apr-53	6	11
IL ROCK ISLAND	4.15	PB5-7	21-Jul-57	2	11
IL ROCK ISLAND	2.24	PB6-8	31-Jul-57	4	11
IL ROCK ISLAND	13.72	PB9-3	25-Aug-57	6	11
IL ROCK ISLAND	16.67	PB9-4	26-Aug-57	4	11
IL ROCK ISLAND	3.39	PB9-5	27-Aug-57	4	11
IL ROCK ISLAND	1.13	PB18-7	13-Oct-57	6	11
IL ROCK ISLAND	3.98	PB18-8	14-Oct-57	4	11
IL ROCK ISLAND	0.08	UE2-7	11-Mar-62	5	11
IL ROCK ISLAND	0.12	UE7-5	15-Jul-62	9	11
IL SALINE	0.87	UpK6-4	21-Apr-53	6	3
IL SALINE	4.85	PB5-3	17-Jul-57	9	3
IL SALINE	2.73	PB9-2	24-Aug-57	6	3
IL SANGAMON	0.87	UpK6-4	21-Apr-53	6	7
IL SANGAMON	9.53	TP4-3	9-Mar-55	4	7
IL SANGAMON	1.67	PB5-7	21-Jul-57	7	7
IL SANGAMON	13.72	PB9-3	25-Aug-57	6	7
IL SANGAMON	16.67	PB9-4	26-Aug-57	4	7
IL SANGAMON	1.47	PB9-5	27-Aug-57	10	7
IL SANGAMON	0.13	UE7-5	15-Jul-62	8	7
IL SCHUYLER	1.42	UpK5-4	14-Apr-53	6	9
IL SCHUYLER	0.87	UpK6-4	21-Apr-53	6	9
IL SCHUYLER	1.86	PB4-10	14-Jul-57	8	9
IL SCHUYLER	13.72	PB9-3	25-Aug-57	6	9
IL SCHUYLER	8.37	PB9-4	26-Aug-57	7	9

COUNTY	uCi/ sqm	SHOT DAY	DATE	RANK FOR DAY	TIMES IN TOP 10
IL SCHUYLER	2.15	PB9-5	27-Aug-57	7	9
IL SCHUYLER	1.13	PB18-7	13-Oct-57	6	9
IL SCHUYLER	1.02	PB18-9	15-Oct-57	4	9
IL SCHUYLER	0.33	UE7-5	15-Jul-62	3	9
IL SCOTT	0.87	UpK6-4	21-Apr-53	6	5
IL SCOTT	13.72	PB9-3	25-Aug-57	6	5
IL SCOTT	33.34	PB9-4	26-Aug-57	3	5
IL SCOTT	1.13	PB18-7	13-Oct-57	6	5
IL SCOTT	0.12	UE7-5	15-Jul-62	9	5
IL SHELBY	1.42	UpK5-4	14-Apr-53	6	3
IL SHELBY	0.87	UpK6-4	21-Apr-53	6	3
IL SHELBY	1.71	PB18-9	15-Oct-57	1	3
IL ST CLAIR	0.87	UpK6-4	21-Apr-53	6	3
IL ST CLAIR	13.72	PB9-3	25-Aug-57	6	3
IL ST CLAIR	1.13	PB18-7	13-Oct-57	6	3
IL STARK	2.13	UpK5-4	14-Apr-53	3	9
IL STARK	0.87	UpK6-4	21-Apr-53	6	9
IL STARK	4.15	PB5-7	21-Jul-57	2	9
IL STARK	13.72	PB9-3	25-Aug-57	6	9
IL STARK	16.67	PB9-4	26-Aug-57	4	9
IL STARK	2.15	PB9-5	27-Aug-57	7	9
IL STARK	1.13	PB18-7	13-Oct-57	6	9
IL STARK	1.02	PB18-9	15-Oct-57	4	9
IL STARK	0.53	UE7-5	15-Jul-62	1	9
IL STEPHENSON	1.36	PB4-3	7-Jul-57	4	7
IL STEPHENSON	13.72	PB9-3	25-Aug-57	6	7
IL STEPHENSON	16.67	PB9-4	26-Aug-57	4	7
IL STEPHENSON	2.04	PB9-5	27-Aug-57	8	7
IL STEPHENSON	3.98	PB18-8	14-Oct-57	4	7
IL STEPHENSON	0.08	UE2-7	11-Mar-62	5	7
IL STEPHENSON	0.12	UE7-5	15-Jul-62	9	7
IL TAZEWELL	2.13	UpK5-4	14-Apr-53	3	8
IL TAZEWELL	0.87	UpK6-4	21-Apr-53	6	8
IL TAZEWELL	21.82	TP4-3	9-Mar-55	1	8
IL TAZEWELL	13.72	PB9-3	25-Aug-57	6	8

COUNTY	uCi/ sqm	SHOT DAY	DATE	RANK FOR DAY	TIMES IN TOP 10
IL TAZEWELL	8.37	PB9-4	26-Aug-57	7	8
IL TAZEWELL	2.15	PB9-5	27-Aug-57	7	8
IL TAZEWELL	1.02	PB18-9	15-Oct-57	4	8
IL TAZEWELL	0.33	UE7-5	15-Jul-62	3	8
IL UNION	3.84	TS8-4	8-Jun-52	10	2
IL UNION	0.87	UpK6-4	21-Apr-53	6	2
IL VERMILION	4.52	PB9-5	27-Aug-57	2	2
IL VERMILION	0.53	UE7-5	15-Jul-62	1	2
IL WABASH	1.02	UpK3-6	5-Apr-53	7	1
IL WARREN	2.13	UpK5-4	14-Apr-53	3	10
IL WARREN	0.87	UpK6-4	21-Apr-53	6	10
IL WARREN	2.48	PB5-7	21-Jul-57	6	10
IL WARREN	13.72	PB9-3	25-Aug-57	6	10
IL WARREN	33.34	PB9-4	26-Aug-57	3	10
IL WARREN	2.04	PB9-5	27-Aug-57	8	10
IL WARREN	3.98	PB18-8	14-Oct-57	4	10
IL WARREN	0.08	UE2-7	11-Mar-62	5	10
IL WARREN	345.56	S-3	8-Jul-62	8	10
IL WARREN	0.33	UE7-5	15-Jul-62	3	10
IL WASHINGTON	0.87	UpK6-4	21-Apr-53	6	2
IL WASHINGTON	0.89	PB1-9	5-Jun-57	7	2
IL WAYNE	0.87	UpK6-4	21-Apr-53	6	3
IL WAYNE	4.85	PB5-3	17-Jul-57	9	3
IL WAYNE	0.11	UE7-5	15-Jul-62	10	3
IL WHITE	0.87	UpK6-4	21-Apr-53	6	1
IL WHITESIDE	0.85	PB4-3	7-Jul-57	10	7
IL WHITESIDE	2.48	PB5-7	21-Jul-57	6	7
IL WHITESIDE	13.72	PB9-3	25-Aug-57	6	7
IL WHITESIDE	16.67	PB9-4	26-Aug-57	4	7
IL WHITESIDE	3.39	PB9-5	27-Aug-57	4	7
IL WHITESIDE	3.98	PB18-8	14-Oct-57	4	7
IL WHITESIDE	0.12	UE7-5	15-Jul-62	9	7
IL WILL	2.15	PB9-5	27-Aug-57	7	3
IL WILL	0.91	PB18-9	15-Oct-57	5	3
IL WILL	0.33	UE7-5	15-Jul-62	3	3

COUNTY	uCi/ sqm	SHOT DAY	DATE	RANK FOR DAY	TIMES IN TOP 10
IL WILLIAMSON	0.87	UpK6-4	21-Apr-53	6	1
IL WINNEBAGO	2.03	TS8-7	11-Jun-52	4	10
IL WINNEBAGO	1.02	PB4-3	7-Jul-57	8	10
IL WINNEBAGO	13.72	PB9-3	25-Aug-57	6	10
IL WINNEBAGO	16.67	PB9-4	26-Aug-57	4	10
IL WINNEBAGO	3.39	PB9-5	27-Aug-57	4	10
IL WINNEBAGO	0.89	PB9-7	29-Aug-57	6	10
IL WINNEBAGO	2.65	PB18-8	14-Oct-57	8	10
IL WINNEBAGO	0.91	PB18-9	15-Oct-57	5	10
IL WINNEBAGO	0.08	UE2-7	11-Mar-62	5	10
IL WINNEBAGO	0.12	UE7-5	15-Jul-62	9	10
IL WOODFORD	5.30	TS8-4	8-Jun-52	1	8
IL WOODFORD	2.13	UpK5-4	14-Apr-53	3	8
IL WOODFORD	0.87	UpK6-4	21-Apr-53	6	8
IL WOODFORD	13.72	PB9-3	25-Aug-57	6	8
IL WOODFORD	16.67	PB9-4	26-Aug-57	4	8
IL WOODFORD	3.55	PB9-5	27-Aug-57	3	8
IL WOODFORD	1.02	PB18-9	15-Oct-57	4	8
IL WOODFORD	0.53	UE7-5	15-Jul-62	1	8

COUNTY	uCi/ sqm	SHOT DAY	DATE	RANK FOR DAY	TIMES IN TOP 10
IN ADAMS	2.87	TP3-3	3-Mar-55	4	4
IN ADAMS	1.51	PB9-5	27-Aug-57	9	4
IN ADAMS	85.56	S-4	9-Jul-62	9	4
IN ADAMS	0.12	UE7-5	15-Jul-62	9	4
IN ALLEN	0.99	R1-3	30-Jan-51	3	7
IN ALLEN	1.11	TS4-5	5-May-52	10	7
IN ALLEN	1.11	TS4-5	5-May-52	10	7
IN ALLEN	2.87	TP3-3	3-Mar-55	4	7
IN ALLEN	1.51	PB9-5	27-Aug-57	9	7
IN ALLEN	208.78	S-4	9-Jul-62	3	7
IN ALLEN	0.13	UE7-5	15-Jul-62	8	7
IN BARTHOLOMEW	7.53	PB9-5	27-Aug-57	1	3
IN BARTHOLOMEW	85.56	S-4	9-Jul-62	9	3
IN BARTHOLOMEW	0.13	UE7-5	15-Jul-62	8	3
IN BENTON	3.01	PB9-5	27-Aug-57	6	3
IN BENTON	1.52	PB18-9	15-Oct-57	2	3
IN BENTON	0.33	UE7-5	15-Jul-62	3	3
IN BLACKFORD	1.63	TS4-5	5-May-52	5	4
IN BLACKFORD	1.63	TS4-5	5-May-52	5	4
IN BLACKFORD	1.55	UpK11-4	7-Jun-53	4	4
IN BLACKFORD	0.53	UE7-5	15-Jul-62	1	4
IN BOONE	1.48	TS4-5	5-May-52	6	7
IN BOONE	1.48	TS4-5	5-May-52	6	7
IN BOONE	1.02	UpK3-6	5-Apr-53	7	7
IN BOONE	1.55	UpK11-4	7-Jun-53	4	7
IN BOONE	10.82	PB3-4	27-Jun-57	1	7
IN BOONE	4.52	PB9-5	27-Aug-57	2	7
IN BOONE	0.11	UE7-5	15-Jul-62	10	7
IN BROWN	1.18	TS4-5	5-May-52	9	5
IN BROWN	1.18	TS4-5	5-May-52	9	5
IN BROWN	1.02	UpK3-6	5-Apr-53	7	5
IN BROWN	7.58	PB3-4	27-Jun-57	10	5
IN BROWN	0.13	UE7-5	15-Jul-62	8	5
IN CARROLL	3.70	TS4-5	5-May-52	1	5
IN CARROLL	3.70	TS4-5	5-May-52	1	5

COUNTY	uCi/ sqm	SHOT DAY	DATE	RANK FOR DAY	TIMES IN TOP 10
IN CARROLL	1.22	UpK11-4	7-Jun-53	10	5
IN CARROLL	3.01	PB9-5	27-Aug-57	6	5
IN CARROLL	0.12	UE7-5	15-Jul-62	9	5
IN CASS	3.01	PB9-5	27-Aug-57	6	3
IN CASS	103.59	S-4	9-Jul-62	6	3
IN CASS	0.33	UE7-5	15-Jul-62	3	3
IN CLARK	1.56	UpK3-6	5-Apr-53	3	2
IN CLARK	1.86	PB2-5	22-Jun-57	5	2
IN CLAY	1.48	TS4-5	5-May-52	6	5
IN CLAY	1.48	TS4-5	5-May-52	6	5
IN CLAY	1.02	UpK3-6	5-Apr-53	7	5
IN CLAY	3.01	PB9-5	27-Aug-57	6	5
IN CLAY	0.26	UE7-5	15-Jul-62	4	5
IN CLINTON	3.70	TS4-5	5-May-52	1	5
IN CLINTON	3.70	TS4-5	5-May-52	1	5
IN CLINTON	1.44	UpK11-4	7-Jun-53	6	5
IN CLINTON	3.01	PB9-5	27-Aug-57	6	5
IN CLINTON	0.33	UE7-5	15-Jul-62	3	5
IN CRAWFORD	1.02	UpK3-6	5-Apr-53	7	1
IN DAVIESS	1.56	UpK3-6	5-Apr-53	3	3
IN DAVIESS	0.99	PB6-10	2-Aug-57	3	3
IN DAVIESS	1.51	PB9-5	27-Aug-57	9	3
IN DE KALB	1.72	TP3-3	3-Mar-55	8	4
IN DE KALB	3.01	PB9-5	27-Aug-57	6	4
IN DE KALB	0.83	PB18-9	15-Oct-57	6	4
IN DE KALB	0.13	UE7-5	15-Jul-62	8	4
IN DEARBORN	1.02	UpK3-6	5-Apr-53	7	3
IN DEARBORN	39.25	PB6-3	26-Jul-57	1	3
IN DEARBORN	1.34	PB13-7	12-Sep-57	2	3
IN DECATUR	23.55	PB6-3	26-Jul-57	3	3
IN DECATUR	4.52	PB9-5	27-Aug-57	2	3
IN DECATUR	0.13	UE7-5	15-Jul-62	8	3
IN DELAWARE	1.50	UpK11-4	7-Jun-53	5	4
IN DELAWARE	3.01	PB9-5	27-Aug-57	6	4
IN DELAWARE	85.56	S-4	9-Jul-62	9	4

COUNTY	uCi/ sqm	SHOT DAY	DATE	RANK FOR DAY	TIMES IN TOP 10
IN DELAWARE	0.26	UE7-5	15-Jul-62	4	4
IN DUBOIS	1.56	UpK3-6	5-Apr-53	3	1
IN ELKHART	1.51	PB9-5	27-Aug-57	9	2
IN ELKHART	0.13	UE7-5	15-Jul-62	8	2
IN FAYETTE	1.02	UpK3-6	5-Apr-53	7	5
IN FAYETTE	1.17	PB2-4	21-Jun-57	10	5
IN FAYETTE	7.85	PB6-3	26-Jul-57	7	5
IN FAYETTE	0.89	PB13-7	12-Sep-57	5	5
IN FAYETTE	0.11	UE7-5	15-Jul-62	10	5
IN FLOYD	1.02	UpK3-6	5-Apr-53	7	2
IN FLOYD	2.54	PB17-2	29-Sep-57	2	2
IN FOUNTAIN	4.52	PB9-5	27-Aug-57	2	2
IN FOUNTAIN	0.22	UE7-5	15-Jul-62	6	2
IN FRANKLIN	1.02	UpK3-6	5-Apr-53	7	8
IN FRANKLIN	0.89	UpK8-5	12-May-53	6	8
IN FRANKLIN	1.19	PB2-5	22-Jun-57	9	8
IN FRANKLIN	23.55	PB6-3	26-Jul-57	3	8
IN FRANKLIN	4.52	PB9-5	27-Aug-57	2	8
IN FRANKLIN	0.89	PB13-7	12-Sep-57	5	8
IN FRANKLIN	90.29	S-4	9-Jul-62	8	8
IN FRANKLIN	0.11	UE7-5	15-Jul-62	10	8
IN FULTON	1.51	PB9-5	27-Aug-57	9	3
IN FULTON	103.59	S-4	9-Jul-62	6	3
IN FULTON	0.12	UE7-5	15-Jul-62	9	3
IN GIBSON	1.56	UpK3-6	5-Apr-53	3	2
IN GIBSON	1.17	PB2-4	21-Jun-57	10	2
IN GRANT	208.78	S-4	9-Jul-62	3	2
IN GRANT	0.33	UE7-5	15-Jul-62	3	2
IN GREENE	1.02	UpK3-6	5-Apr-53	7	2
IN GREENE	3.01	PB9-5	27-Aug-57	6	2
IN HAMILTON	1.02	UpK3-6	5-Apr-53	7	6
IN HAMILTON	1.78	UpK11-4	7-Jun-53	1	6
IN HAMILTON	9.19	PB3-4	27-Jun-57	4	6
IN HAMILTON	4.52	PB9-5	27-Aug-57	2	6
IN HAMILTON	208.78	S-4	9-Jul-62	3	6

COUNTY	uCi/ sqm	SHOT DAY	DATE	RANK FOR DAY	TIMES IN TOP 10
IN HAMILTON	0.26	UE7-5	15-Jul-62	4	6
IN HANCOCK	1.02	UpK3-6	5-Apr-53	7	4
IN HANCOCK	1.72	UpK11-4	7-Jun-53	2	4
IN HANCOCK	7.53	PB9-5	27-Aug-57	1	4
IN HANCOCK	0.11	UE7-5	15-Jul-62	10	4
IN HARRISON	1.02	UpK3-6	5-Apr-53	7	2
IN HARRISON	2.54	PB17-2	29-Sep-57	2	2
IN HENDRICKS	1.02	UpK3-6	5-Apr-53	7	4
IN HENDRICKS	10.18	PB3-4	27-Jun-57	3	4
IN HENDRICKS	4.52	PB9-5	27-Aug-57	2	4
IN HENDRICKS	0.11	UE7-5	15-Jul-62	10	4
IN HENRY	1.02	UpK3-6	5-Apr-53	7	4
IN HENRY	7.53	PB9-5	27-Aug-57	1	4
IN HENRY	90.29	S-4	9-Jul-62	8	4
IN HENRY	0.26	UE7-5	15-Jul-62	4	4
IN HOWARD	2.67	TS4-5	5-May-52	2	6
IN HOWARD	2.67	TS4-5	5-May-52	2	6
IN HOWARD	1.22	UpK11-4	7-Jun-53	10	6
IN HOWARD	3.01	PB9-5	27-Aug-57	6	6
IN HOWARD	208.78	S-4	9-Jul-62	3	6
IN HOWARD	0.33	UE7-5	15-Jul-62	3	6
IN HUNTINGTON	1.28	UpK11-4	7-Jun-53	8	4
IN HUNTINGTON	2.87	TP3-3	3-Mar-55	4	4
IN HUNTINGTON	3.01	PB9-5	27-Aug-57	6	4
IN HUNTINGTON	0.12	UE7-5	15-Jul-62	9	4
IN JACKSON	1.56	UpK3-6	5-Apr-53	3	3
IN JACKSON	90.29	S-4	9-Jul-62	8	3
IN JACKSON	0.13	UE7-5	15-Jul-62	8	3
IN JASPER	3.01	PB9-5	27-Aug-57	6	2
IN JASPER	1.52	PB18-9	15-Oct-57	2	2
IN JAY	1.51	PB9-5	27-Aug-57	9	2
IN JAY	0.22	UE7-5	15-Jul-62	6	2
IN JEFFERSON	1.56	UpK3-6	5-Apr-53	3	2
IN JEFFERSON	1.86	PB2-5	22-Jun-57	5	2
IN JENNINGS	1.56	UpK3-6	5-Apr-53	3	4

COUNTY	uCi/ sqm	SHOT DAY	DATE	RANK FOR DAY	TIMES IN TOP 10
IN JENNINGS	0.89	UpK8-5	12-May-53	6	4
IN JENNINGS	1.17	PB2-4	21-Jun-57	10	4
IN JENNINGS	0.13	UE7-5	15-Jul-62	8	4
IN JOHNSON	1.02	UpK3-6	5-Apr-53	7	4
IN JOHNSON	1.64	PB6-10	2-Aug-57	1	4
IN JOHNSON	1.51	PB9-5	27-Aug-57	9	4
IN JOHNSON	0.13	UE7-5	15-Jul-62	8	4
IN KNOX	1.56	UpK3-6	5-Apr-53	3	3
IN KNOX	1.51	PB9-5	27-Aug-57	9	3
IN KNOX	1.12	PB13-7	12-Sep-57	4	3
IN KOSCIUSKO	2.87	TP3-3	3-Mar-55	4	2
IN KOSCIUSKO	0.13	UE7-5	15-Jul-62	8	2
IN LA PORTE	2.63	UpK9-6	24-May-53	4	4
IN LA PORTE	4.52	PB9-5	27-Aug-57	2	4
IN LA PORTE	1.52	PB18-9	15-Oct-57	2	4
IN LA PORTE	0.12	UE7-5	15-Jul-62	9	4
IN LAGRANGE	1.42	R1-3	30-Jan-51	1	4
IN LAGRANGE	1.72	TP3-3	3-Mar-55	8	4
IN LAGRANGE	0.89	PB15-6	21-Sep-57	2	4
IN LAGRANGE	0.12	UE7-5	15-Jul-62	9	4
IN LAKE	4.52	PB9-5	27-Aug-57	2	3
IN LAKE	0.91	PB18-9	15-Oct-57	5	3
IN LAKE	0.12	UE7-5	15-Jul-62	9	3
IN LAWRENCE	1.56	UpK3-6	5-Apr-53	3	2
IN LAWRENCE	103.59	S-4	9-Jul-62	6	2
IN MADISON	1.02	UpK3-6	5-Apr-53	7	6
IN MADISON	1.55	UpK11-4	7-Jun-53	4	6
IN MADISON	8.85	PB3-4	27-Jun-57	5	6
IN MADISON	4.52	PB9-5	27-Aug-57	2	6
IN MADISON	90.29	S-4	9-Jul-62	8	6
IN MADISON	0.11	UE7-5	15-Jul-62	10	6
IN MARION	1.48	TS4-5	5-May-52	6	7
IN MARION	1.48	TS4-5	5-May-52	6	7
IN MARION	1.02	UpK3-6	5-Apr-53	7	7
IN MARION	1.61	UpK11-4	7-Jun-53	3	7

COUNTY	uCi/ sqm	SHOT DAY	DATE	RANK FOR DAY	TIMES IN TOP 10
IN MARION	7.65	PB3-4	27-Jun-57	9	7
IN MARION	3.13	PB9-5	27-Aug-57	5	7
IN MARION	0.11	UE7-5	15-Jul-62	10	7
IN MARSHALL	3.01	PB9-5	27-Aug-57	6	3
IN MARSHALL	0.83	PB18-9	15-Oct-57	6	3
IN MARSHALL	0.12	UE7-5	15-Jul-62	9	3
IN MARTIN	1.56	UpK3-6	5-Apr-53	3	2
IN MARTIN	0.99	PB6-10	2-Aug-57	3	2
IN MIAMI	0.12	UE7-5	15-Jul-62	9	1
IN MONROE	1.02	UpK3-6	5-Apr-53	7	3
IN MONROE	3.01	PB9-5	27-Aug-57	6	3
IN MONROE	0.11	UE7-5	15-Jul-62	10	3
IN MONTGOMERY	1.48	TS4-5	5-May-52	6	5
IN MONTGOMERY	1.48	TS4-5	5-May-52	6	5
IN MONTGOMERY	8.85	PB3-4	27-Jun-57	5	5
IN MONTGOMERY	3.01	PB9-5	27-Aug-57	6	5
IN MONTGOMERY	0.22	UE7-5	15-Jul-62	6	5
IN MORGAN	1.02	UpK3-6	5-Apr-53	7	3
IN MORGAN	4.52	PB9-5	27-Aug-57	2	3
IN MORGAN	0.26	UE7-5	15-Jul-62	4	3
IN NEWTON	1.10	PB4-3	7-Jul-57	7	4
IN NEWTON	4.52	PB9-5	27-Aug-57	2	4
IN NEWTON	1.52	PB18-9	15-Oct-57	2	4
IN NEWTON	0.33	UE7-5	15-Jul-62	3	4
IN NOBLE	1.72	TP3-3	3-Mar-55	8	3
IN NOBLE	1.51	PB9-5	27-Aug-57	9	3
IN NOBLE	0.13	UE7-5	15-Jul-62	8	3
IN OHIO	1.56	UpK3-6	5-Apr-53	3	3
IN OHIO	23.55	PB6-3	26-Jul-57	3	3
IN OHIO	2.23	PB13-7	12-Sep-57	1	3
IN ORANGE	1.56	UpK3-6	5-Apr-53	3	1
IN OWEN	1.02	UpK3-6	5-Apr-53	7	3
IN OWEN	4.52	PB9-5	27-Aug-57	2	3
IN OWEN	103.59	S-4	9-Jul-62	6	3
IN PARKE	1.02	UpK3-6	5-Apr-53	7	3

COUNTY	uCi/ sqm	SHOT DAY	DATE	RANK FOR DAY	TIMES IN TOP 10
IN PARKE	8.11	PB3-4	27-Jun-57	6	3
IN PARKE	3.01	PB9-5	27-Aug-57	6	3
IN PERRY	1.02	UpK3-6	5-Apr-53	7	1
IN PIKE	1.56	UpK3-6	5-Apr-53	3	2
IN PIKE	1.86	PB2-5	22-Jun-57	5	2
IN PORTER	3.01	PB9-5	27-Aug-57	6	3
IN PORTER	0.91	PB18-9	15-Oct-57	5	3
IN PORTER	0.33	UE7-5	15-Jul-62	3	3
IN POSEY	1.56	UpK3-6	5-Apr-53	3	4
IN POSEY	1.73	UpK6-5	22-Apr-53	2	4
IN POSEY	7.27	PB5-3	17-Jul-57	5	4
IN POSEY	0.99	PB6-10	2-Aug-57	3	4
IN PULASKI	3.01	PB9-5	27-Aug-57	6	2
IN PULASKI	1.52	PB18-9	15-Oct-57	2	2
IN PUTNAM	1.02	UpK3-6	5-Apr-53	7	4
IN PUTNAM	10.25	PB3-4	27-Jun-57	2	4
IN PUTNAM	4.52	PB9-5	27-Aug-57	2	4
IN PUTNAM	0.11	UE7-5	15-Jul-62	10	4
IN RANDOLPH	0.99	R1-3	30-Jan-51	3	6
IN RANDOLPH	1.02	UpK3-6	5-Apr-53	7	6
IN RANDOLPH	1.44	UpK11-4	7-Jun-53	6	6
IN RANDOLPH	1.51	PB9-5	27-Aug-57	9	6
IN RANDOLPH	85.56	S-4	9-Jul-62	9	6
IN RANDOLPH	0.26	UE7-5	15-Jul-62	4	6
IN RIPLEY	2.37	TS4-5	5-May-52	3	6
IN RIPLEY	2.37	TS4-5	5-May-52	3	6
IN RIPLEY	1.02	UpK3-6	5-Apr-53	7	6
IN RIPLEY	39.25	PB6-3	26-Jul-57	1	6
IN RIPLEY	1.34	PB13-7	12-Sep-57	2	6
IN RIPLEY	0.13	UE7-5	15-Jul-62	8	6
IN RUSH	1.02	UpK3-6	5-Apr-53	7	3
IN RUSH	4.52	PB9-5	27-Aug-57	2	3
IN RUSH	0.11	UE7-5	15-Jul-62	10	3
IN SCOTT	1.56	UpK3-6	5-Apr-53	3	3
IN SCOTT	1.86	PB2-5	22-Jun-57	5	3

COUNTY	uCi/ sqm	SHOT DAY	DATE	RANK FOR DAY	TIMES IN TOP 10
IN SCOTT	85.56	S-4	9-Jul-62	9	3
IN SHELBY	1.02	UpK3-6	5-Apr-53	7	4
IN SHELBY	4.52	PB9-5	27-Aug-57	2	4
IN SHELBY	90.29	S-4	9-Jul-62	8	4
IN SHELBY	0.13	UE7-5	15-Jul-62	8	4
IN SPENCER	1.02	UpK3-6	5-Apr-53	7	1
IN ST JOSEPH	7.53	PB9-5	27-Aug-57	1	3
IN ST JOSEPH	0.83	PB18-9	15-Oct-57	6	3
IN ST JOSEPH	0.12	UE7-5	15-Jul-62	9	3
IN STARKE	3.01	PB9-5	27-Aug-57	6	3
IN STARKE	1.52	PB18-9	15-Oct-57	2	3
IN STARKE	0.12	UE7-5	15-Jul-62	9	3
IN STEUBEN	1.72	TP3-3	3-Mar-55	8	4
IN STEUBEN	0.85	PB13-7	12-Sep-57	6	4
IN STEUBEN	0.89	PB15-6	21-Sep-57	2	4
IN STEUBEN	0.33	UE7-5	15-Jul-62	3	4
IN SULLIVAN	7.76	PB3-4	27-Jun-57	7	2
IN SULLIVAN	4.52	PB9-5	27-Aug-57	2	2
IN SWITZERLAND	1.02	UpK3-6	5-Apr-53	7	3
IN SWITZERLAND	1.47	PB6-7	30-Jul-57	10	3
IN SWITZERLAND	0.89	PB13-7	12-Sep-57	5	3
IN TIPPECANOE	3.70	TS4-5	5-May-52	1	4
IN TIPPECANOE	3.70	TS4-5	5-May-52	1	4
IN TIPPECANOE	4.52	PB9-5	27-Aug-57	2	4
IN TIPPECANOE	0.33	UE7-5	15-Jul-62	3	4
IN TIPTON	2.67	TS4-5	5-May-52	2	6
IN TIPTON	2.67	TS4-5	5-May-52	2	6
IN TIPTON	1.61	UpK11-4	7-Jun-53	3	6
IN TIPTON	1.51	PB9-5	27-Aug-57	9	6
IN TIPTON	208.78	S-4	9-Jul-62	3	6
IN TIPTON	0.53	UE7-5	15-Jul-62	1	6
IN UNION	2.00	TS4-5	5-May-52	4	8
IN UNION	2.00	TS4-5	5-May-52	4	8
IN UNION	1.02	UpK3-6	5-Apr-53	7	8
IN UNION	1.17	PB2-4	21-Jun-57	10	8

COUNTY	uCi/ sqm	SHOT DAY	DATE	RANK FOR DAY	TIMES IN TOP 10
IN UNION	7.85	PB6-3	26-Jul-57	7	8
IN UNION	0.89	PB13-7	12-Sep-57	5	8
IN UNION	85.56	S-4	9-Jul-62	9	8
IN UNION	0.11	UE7-5	15-Jul-62	10	8
IN VANDERBURGH	1.56	UpK3-6	5-Apr-53	3	2
IN VANDERBURGH	1.73	UpK6-5	22-Apr-53	2	2
IN VERMILLION	1.02	UpK3-6	5-Apr-53	7	3
IN VERMILLION	6.69	TP4-3	9-Mar-55	9	3
IN VERMILLION	0.11	UE7-5	15-Jul-62	10	3
IN VIGO	7.71	PB3-4	27-Jun-57	8	2
IN VIGO	4.52	PB9-5	27-Aug-57	2	2
IN WABASH	2.87	TP3-3	3-Mar-55	4	2
IN WABASH	0.12	UE7-5	15-Jul-62	9	2
IN WARREN	1.29	PB6-10	2-Aug-57	2	3
IN WARREN	4.52	PB9-5	27-Aug-57	2	3
IN WARREN	0.33	UE7-5	15-Jul-62	3	3
IN WARRICK	1.56	UpK3-6	5-Apr-53	3	1
IN WASHINGTON	1.56	UpK3-6	5-Apr-53	3	2
IN WASHINGTON	1.17	PB2-4	21-Jun-57	10	2
IN WAYNE	2.00	TS4-5	5-May-52	4	7
IN WAYNE	2.00	TS4-5	5-May-52	4	7
IN WAYNE	1.02	UpK3-6	5-Apr-53	7	7
IN WAYNE	3.01	PB9-5	27-Aug-57	6	7
IN WAYNE	1.34	PB13-7	12-Sep-57	2	7
IN WAYNE	208.78	S-4	9-Jul-62	3	7
IN WAYNE	0.13	UE7-5	15-Jul-62	8	7
IN WELLS	2.87	TP3-3	3-Mar-55	4	4
IN WELLS	1.51	PB9-5	27-Aug-57	9	4
IN WELLS	208.78	S-4	9-Jul-62	3	4
IN WELLS	0.33	UE7-5	15-Jul-62	3	4
IN WHITE	1.51	PB9-5	27-Aug-57	9	2
IN WHITE	0.13	UE7-5	15-Jul-62	8	2
IN WHITLEY	2.87	TP3-3	3-Mar-55	4	3
IN WHITLEY	1.51	PB9-5	27-Aug-57	9	3
IN WHITLEY	0.12	UE7-5	15-Jul-62	9	3

COUNTY	uCi/ sqm	SHOT DAY	DATE	RANK FOR DAY	TIMES IN TOP 10
KS ALLEN	5.84	BJ2-2	31-Oct-51	7	4
KS ALLEN	0.93	PB3-8	1-Jul-57	4	4
KS ALLEN	0.83	PB13-8	13-Sep-57	9	4
KS ALLEN	4.04	PB14-7	20-Sep-57	1	4
KS ANDERSON	6.96	UpK11-2	5-Jun-53	8	5
KS ANDERSON	0.83	PB13-8	13-Sep-57	9	5
KS ANDERSON	4.04	PB14-7	20-Sep-57	1	5
KS ANDERSON	0.08	UE2-7	11-Mar-62	5	5
KS ANDERSON	0.13	UE7-5	15-Jul-62	8	5
KS ATCHISON	12.24	TP4-2	8-Mar-55	9	6
KS ATCHISON	4.15	PB5-7	21-Jul-57	2	6
KS ATCHISON	19.46	PB12-4	5-Sep-57	10	6
KS ATCHISON	1.21	PB15-4	19-Sep-57	5	6
KS ATCHISON	0.81	PB15-5	20-Sep-57	9	6
KS ATCHISON	0.13	UE7-5	15-Jul-62	8	6
KS BARBER	1.22	BJ5-6	24-Nov-51	6	9
KS BARBER	1.22	BJ5-6	24-Nov-51	6	9
KS BARBER	1.30	TS4-6	6-May-52	10	9
KS BARBER	1.30	TS4-6	6-May-52	10	9
KS BARBER	3.88	TS6-7	31-May-52	7	9
KS BARBER	3.78	PB3-2	25-Jun-57	3	9
KS BARBER	2.33	PB3-3	26-Jun-57	8	9
KS BARBER	2.11	PB13-8	13-Sep-57	4	9
KS BARBER	0.81	PB14-7	20-Sep-57	9	9
KS BARTON	1.38	TS3-8	29-Apr-52	6	15
KS BARTON	1.38	TS3-8	29-Apr-52	6	15
KS BARTON	1.12	TS3-9	30-Apr-52	1	15
KS BARTON	1.12	TS3-9	30-Apr-52	1	15
KS BARTON	3.78	TS6-7	31-May-52	9	15
KS BARTON	12.24	TP4-2	8-Mar-55	9	15
KS BARTON	1.09	TP8-4	12-Apr-55	9	15
KS BARTON	2.33	PB3-3	26-Jun-57	8	15
KS BARTON	0.94	PB5-6	20-Jul-57	9	15
KS BARTON	1.63	PB5-7	21-Jul-57	8	15
KS BARTON	19.73	PB8-3	20-Aug-57	6	15

COUNTY	uCi/ sqm	SHOT DAY	DATE	RANK FOR DAY	TIMES IN TOP 10
KS BARTON	3.55	PB13-4	9-Sep-57	5	15
KS BARTON	26.73	PB13-5	10-Sep-57	5	15
KS BARTON	2.11	PB13-8	13-Sep-57	4	15
KS BARTON	14.44	UE8-4	17-Jul-62	1	15
KS BOURBON	5.84	BJ2-2	31-Oct-51	7	3
KS BOURBON	0.93	PB3-8	1-Jul-57	4	3
KS BOURBON	4.04	PB14-7	20-Sep-57	1	3
KS BROWN	12.24	TP4-2	8-Mar-55	9	6
KS BROWN	2.48	PB5-7	21-Jul-57	6	6
KS BROWN	7.99	PB8-3	20-Aug-57	9	6
KS BROWN	1.21	PB15-4	19-Sep-57	5	6
KS BROWN	0.08	UE2-7	11-Mar-62	5	6
KS BROWN	0.33	UE7-5	15-Jul-62	3	6
KS BUTLER	5.84	BJ2-2	31-Oct-51	7	6
KS BUTLER	1.54	PB3-2	25-Jun-57	8	6
KS BUTLER	1.63	PB5-7	21-Jul-57	8	6
KS BUTLER	1.28	PB13-8	13-Sep-57	6	6
KS BUTLER	1.64	PB13-9	14-Sep-57	10	6
KS BUTLER	4.04	PB14-7	20-Sep-57	1	6
KS CHASE	1.28	PB13-8	13-Sep-57	6	3
KS CHASE	1.64	PB13-9	14-Sep-57	10	3
KS CHASE	4.04	PB14-7	20-Sep-57	1	3
KS CHAUTAUQUA	8.76	BJ2-2	31-Oct-51	3	7
KS CHAUTAUQUA	0.89	TP9-13	27-Apr-55	5	7
KS CHAUTAUQUA	2.56	TP10-4	8-May-55	6	7
KS CHAUTAUQUA	1.76	PB2-4	21-Jun-57	6	7
KS CHAUTAUQUA	2.11	PB13-8	13-Sep-57	4	7
KS CHAUTAUQUA	2.50	PB13-9	14-Sep-57	8	7
KS CHAUTAUQUA	1.62	PB14-7	20-Sep-57	4	7
KS CHEROKEE	8.76	BJ2-2	31-Oct-51	3	5
KS CHEROKEE	2.23	UpK2-10	2-Apr-53	3	5
KS CHEROKEE	2.10	UpK3-3	2-Apr-53	4	5
KS CHEROKEE	2.73	UpK3-5	4-Apr-53	8	5
KS CHEROKEE	2.42	PB14-7	20-Sep-57	2	5
KS CHEYENNE	0.87	TS4-3	3-May-52	1	14

COUNTY	uCi/ sqm	SHOT DAY	DATE	RANK FOR DAY	TIMES IN TOP 10
KS CHEYENNE	0.87	TS4-3	3-May-52	1	14
KS CHEYENNE	51.08	TS6-3	27-May-52	5	14
KS CHEYENNE	1.31	UpK6-3	20-Apr-53	9	14
KS CHEYENNE	7.78	TP3-2	2-Mar-55	3	14
KS CHEYENNE	49.44	TP4-2	8-Mar-55	4	14
KS CHEYENNE	3.04	TP8-3	11-Apr-55	2	14
KS CHEYENNE	32.33	TP10-2	6-May-55	5	14
KS CHEYENNE	1.73	PB4-10	14-Jul-57	9	14
KS CHEYENNE	1.96	PB5-5	19-Jul-57	9	14
KS CHEYENNE	4.60	PB18-2	8-Oct-57	2	14
KS CHEYENNE	2.37	PB18-3	9-Oct-57	3	14
KS CHEYENNE	1.70	PB18-4	10-Oct-57	5	14
KS CHEYENNE	0.86	PB18-6	12-Oct-57	7	14
KS CLARK	2.48	BJ5-6	24-Nov-51	4	6
KS CLARK	2.48	BJ5-6	24-Nov-51	4	6
KS CLARK	0.93	PB3-8	1-Jul-57	4	6
KS CLARK	2.11	PB13-8	13-Sep-57	4	6
KS CLARK	0.81	PB14-7	20-Sep-57	9	6
KS CLARK	1.92	PB18-6	12-Oct-57	3	6
KS CLAY	1.68	TS4-6	6-May-52	6	11
KS CLAY	1.68	TS4-6	6-May-52	6	11
KS CLAY	3.89	TS6-7	31-May-52	6	11
KS CLAY	12.24	TP4-2	8-Mar-55	9	11
KS CLAY	2.33	PB3-3	26-Jun-57	8	11
KS CLAY	2.69	PB5-7	21-Jul-57	5	11
KS CLAY	24.70	PB13-5	10-Sep-57	9	11
KS CLAY	1.28	PB13-8	13-Sep-57	6	11
KS CLAY	0.81	PB14-7	20-Sep-57	9	11
KS CLAY	0.13	UE7-5	15-Jul-62	8	11
KS CLAY	5.25	UE8-4	17-Jul-62	6	11
KS CLOUD	1.13	TP8-4	12-Apr-55	7	5
KS CLOUD	5.36	PB13-4	9-Sep-57	3	5
KS CLOUD	1.28	PB13-8	13-Sep-57	6	5
KS CLOUD	0.81	PB14-7	20-Sep-57	9	5
KS CLOUD	4.35	UE8-4	17-Jul-62	8	5

COUNTY	uCi/ sqm	SHOT DAY	DATE	RANK FOR DAY	TIMES IN TOP 10
KS COFFEY	0.83	PB13-8	13-Sep-57	9	3
KS COFFEY	1.64	PB13-9	14-Sep-57	10	3
KS COFFEY	4.04	PB14-7	20-Sep-57	1	3
KS COMANCHE	1.22	BJ5-6	24-Nov-51	6	10
KS COMANCHE	1.22	BJ5-6	24-Nov-51	6	10
KS COMANCHE	4.70	TS6-7	31-May-52	1	10
KS COMANCHE	4.43	UpK9-5	23-May-53	6	10
KS COMANCHE	2.24	PB3-2	25-Jun-57	5	10
KS COMANCHE	0.93	PB3-8	1-Jul-57	4	10
KS COMANCHE	0.86	PB6-11	3-Aug-57	4	10
KS COMANCHE	2.11	PB13-8	13-Sep-57	4	10
KS COMANCHE	1.64	PB13-9	14-Sep-57	10	10
KS COMANCHE	1.92	PB18-6	12-Oct-57	3	10
KS COWLEY	3.70	BJ5-6	24-Nov-51	2	6
KS COWLEY	3.70	BJ5-6	24-Nov-51	2	6
KS COWLEY	2.69	PB5-7	21-Jul-57	5	6
KS COWLEY	2.11	PB13-8	13-Sep-57	4	6
KS COWLEY	2.50	PB13-9	14-Sep-57	8	6
KS COWLEY	4.04	PB14-7	20-Sep-57	1	6
KS CRAWFORD	8.76	BJ2-2	31-Oct-51	3	4
KS CRAWFORD	22.84	PB12-4	5-Sep-57	2	4
KS CRAWFORD	0.83	PB13-8	13-Sep-57	9	4
KS CRAWFORD	2.42	PB14-7	20-Sep-57	2	4
KS DECATUR	0.87	TS4-3	3-May-52	1	15
KS DECATUR	0.87	TS4-3	3-May-52	1	15
KS DECATUR	1.31	UpK6-3	20-Apr-53	9	15
KS DECATUR	1.32	UpK8-3	10-May-53	9	15
KS DECATUR	7.78	TP3-2	2-Mar-55	3	15
KS DECATUR	49.44	TP4-2	8-Mar-55	4	15
KS DECATUR	32.33	TP10-2	6-May-55	5	15
KS DECATUR	3.46	PB4-10	14-Jul-57	3	15
KS DECATUR	4.92	PB8-4	21-Aug-57	7	15
KS DECATUR	4.92	PB8-4	21-Aug-57	7	15
KS DECATUR	5.36	PB13-4	9-Sep-57	3	15
KS DECATUR	2.30	PB18-2	8-Oct-57	5	15

COUNTY	uCi/ sqm	SHOT DAY	DATE	RANK FOR DAY	TIMES IN TOP 10
KS DECATUR	2.37	PB18-3	9-Oct-57	3	15
KS DECATUR	1.70	PB18-4	10-Oct-57	5	15
KS DECATUR	7.94	UE8-4	17-Jul-62	4	15
KS DICKINSON	12.24	TP4-2	8-Mar-55	9	9
KS DICKINSON	1.35	TP8-4	12-Apr-55	3	9
KS DICKINSON	2.33	PB3-3	26-Jun-57	8	9
KS DICKINSON	1.63	PB5-7	21-Jul-57	8	9
KS DICKINSON	24.86	PB13-5	10-Sep-57	8	9
KS DICKINSON	1.28	PB13-8	13-Sep-57	6	9
KS DICKINSON	1.62	PB14-7	20-Sep-57	4	9
KS DICKINSON	0.12	UE7-5	15-Jul-62	9	9
KS DICKINSON	10.54	UE8-4	17-Jul-62	2	9
KS DONIPHAN	4.86	UpK11-3	6-Jun-53	2	7
KS DONIPHAN	1.67	PB5-7	21-Jul-57	7	7
KS DONIPHAN	0.83	PB13-8	13-Sep-57	9	7
KS DONIPHAN	1.21	PB15-4	19-Sep-57	5	7
KS DONIPHAN	0.81	PB15-5	20-Sep-57	9	7
KS DONIPHAN	0.86	PB18-6	12-Oct-57	7	7
KS DONIPHAN	0.12	UE7-5	15-Jul-62	9	7
KS DOUGLAS	1.99	TS4-6	6-May-52	4	4
KS DOUGLAS	1.99	TS4-6	6-May-52	4	4
KS DOUGLAS	4.04	PB14-7	20-Sep-57	1	4
KS DOUGLAS	0.12	UE7-5	15-Jul-62	9	4
KS EDWARDS	2.48	BJ5-6	24-Nov-51	4	8
KS EDWARDS	2.48	BJ5-6	24-Nov-51	4	8
KS EDWARDS	3.81	TS6-7	31-May-52	8	8
KS EDWARDS	3.78	PB3-2	25-Jun-57	3	8
KS EDWARDS	2.33	PB3-3	26-Jun-57	8	8
KS EDWARDS	0.94	PB5-6	20-Jul-57	9	8
KS EDWARDS	2.11	PB13-8	13-Sep-57	4	8
KS EDWARDS	4.55	UE8-4	17-Jul-62	7	8
KS ELK	5.84	BJ2-2	31-Oct-51	7	5
KS ELK	1.76	PB2-4	21-Jun-57	6	5
KS ELK	1.28	PB13-8	13-Sep-57	6	5
KS ELK	2.50	PB13-9	14-Sep-57	8	5

COUNTY	uCi/ sqm	SHOT DAY	DATE	RANK FOR DAY	TIMES IN TOP 10
KS ELK	4.04	PB14-7	20-Sep-57	1	5
KS ELLIS	1.38	TS3-8	29-Apr-52	6	17
KS ELLIS	1.38	TS3-8	29-Apr-52	6	17
KS ELLIS	1.12	TS3-9	30-Apr-52	1	17
KS ELLIS	1.12	TS3-9	30-Apr-52	1	17
KS ELLIS	12.24	TP4-2	8-Mar-55	9	17
KS ELLIS	3.77	TP7-3	31-Mar-55	2	17
KS ELLIS	1.12	TP8-4	12-Apr-55	8	17
KS ELLIS	0.93	PB3-8	1-Jul-57	4	17
KS ELLIS	19.73	PB8-3	20-Aug-57	6	17
KS ELLIS	5.36	PB8-4	21-Aug-57	6	17
KS ELLIS	5.36	PB8-4	21-Aug-57	6	17
KS ELLIS	2.11	PB13-8	13-Sep-57	4	17
KS ELLIS	2.30	PB18-2	8-Oct-57	5	17
KS ELLIS	2.37	PB18-3	9-Oct-57	3	17
KS ELLIS	1.70	PB18-4	10-Oct-57	5	17
KS ELLIS	1.29	PB18-6	12-Oct-57	6	17
KS ELLIS	4.55	UE8-4	17-Jul-62	7	17
KS ELLSWORTH	0.92	TS3-8	29-Apr-52	9	12
KS ELLSWORTH	0.92	TS3-8	29-Apr-52	9	12
KS ELLSWORTH	1.68	TS4-6	6-May-52	6	12
KS ELLSWORTH	1.68	TS4-6	6-May-52	6	12
KS ELLSWORTH	12.24	TP4-2	8-Mar-55	9	12
KS ELLSWORTH	1.27	TP8-4	12-Apr-55	4	12
KS ELLSWORTH	2.33	PB3-3	26-Jun-57	8	12
KS ELLSWORTH	24.93	PB13-5	10-Sep-57	7	12
KS ELLSWORTH	2.11	PB13-8	13-Sep-57	4	12
KS ELLSWORTH	0.81	PB14-7	20-Sep-57	9	12
KS ELLSWORTH	0.13	UE7-5	15-Jul-62	8	12
KS ELLSWORTH	8.02	UE8-4	17-Jul-62	3	12
KS FINNEY	0.87	TS4-3	3-May-52	1	12
KS FINNEY	0.87	TS4-3	3-May-52	1	12
KS FINNEY	1.31	UpK6-3	20-Apr-53	9	12
KS FINNEY	7.78	TP3-2	2-Mar-55	3	12
KS FINNEY	49.44	TP4-2	8-Mar-55	4	12

COUNTY	uCi/ sqm	SHOT DAY	DATE	RANK FOR DAY	TIMES IN TOP 10
KS FINNEY	8.01	TP7-2	30-Mar-55	7	12
KS FINNEY	32.33	TP10-2	6-May-55	5	12
KS FINNEY	1.96	PB5-5	19-Jul-57	9	12
KS FINNEY	2.11	PB13-8	13-Sep-57	4	12
KS FINNEY	2.30	PB18-2	8-Oct-57	5	12
KS FINNEY	2.37	PB18-3	9-Oct-57	3	12
KS FINNEY	1.70	PB18-4	10-Oct-57	5	12
KS FORD	1.22	BJ5-6	24-Nov-51	6	17
KS FORD	1.22	BJ5-6	24-Nov-51	6	17
KS FORD	0.87	TS4-3	3-May-52	1	17
KS FORD	0.87	TS4-3	3-May-52	1	17
KS FORD	1.31	UpK6-3	20-Apr-53	9	17
KS FORD	7.78	TP3-2	2-Mar-55	3	17
KS FORD	49.44	TP4-2	8-Mar-55	4	17
KS FORD	32.33	TP10-2	6-May-55	5	17
KS FORD	0.86	PB6-11	3-Aug-57	4	17
KS FORD	3.55	PB13-4	9-Sep-57	5	17
KS FORD	1.34	PB13-7	12-Sep-57	2	17
KS FORD	1.28	PB13-8	13-Sep-57	6	17
KS FORD	4.60	PB18-2	8-Oct-57	2	17
KS FORD	2.37	PB18-3	9-Oct-57	3	17
KS FORD	1.70	PB18-4	10-Oct-57	5	17
KS FORD	1.92	PB18-6	12-Oct-57	3	17
KS FORD	7.94	UE8-4	17-Jul-62	4	17
KS FRANKLIN	1.30	TS4-6	6-May-52	10	5
KS FRANKLIN	1.30	TS4-6	6-May-52	10	5
KS FRANKLIN	1.54	PB3-2	25-Jun-57	8	5
KS FRANKLIN	4.04	PB14-7	20-Sep-57	1	5
KS FRANKLIN	0.13	UE7-5	15-Jul-62	8	5
KS GEARY	1.68	TS4-6	6-May-52	6	10
KS GEARY	1.68	TS4-6	6-May-52	6	10
KS GEARY	12.24	TP4-2	8-Mar-55	9	10
KS GEARY	1.24	TP8-4	12-Apr-55	5	10
KS GEARY	1.76	PB2-4	21-Jun-57	6	10
KS GEARY	2.69	PB5-7	21-Jul-57	5	10

COUNTY	uCi/ sqm	SHOT DAY	DATE	RANK FOR DAY	TIMES IN TOP 10
KS GEARY	3.55	PB13-4	9-Sep-57	5	10
KS GEARY	1.28	PB13-8	13-Sep-57	6	10
KS GEARY	1.62	PB14-7	20-Sep-57	4	10
KS GEARY	0.22	UE7-5	15-Jul-62	6	10
KS GOVE	0.87	TS4-3	3-May-52	1	14
KS GOVE	0.87	TS4-3	3-May-52	1	14
KS GOVE	1.31	UpK6-3	20-Apr-53	9	14
KS GOVE	7.78	TP3-2	2-Mar-55	3	14
KS GOVE	49.44	TP4-2	8-Mar-55	4	14
KS GOVE	2.91	TP7-3	31-Mar-55	8	14
KS GOVE	32.33	TP10-2	6-May-55	5	14
KS GOVE	1.54	PB3-2	25-Jun-57	8	14
KS GOVE	1.63	PB5-7	21-Jul-57	8	14
KS GOVE	2.30	PB18-2	8-Oct-57	5	14
KS GOVE	2.37	PB18-3	9-Oct-57	3	14
KS GOVE	1.70	PB18-4	10-Oct-57	5	14
KS GOVE	0.10	UE2-7	11-Mar-62	4	14
KS GOVE	4.55	UE8-4	17-Jul-62	7	14
KS GRAHAM	0.87	TS4-3	3-May-52	1	17
KS GRAHAM	0.87	TS4-3	3-May-52	1	17
KS GRAHAM	1.31	UpK6-3	20-Apr-53	9	17
KS GRAHAM	7.78	TP3-2	2-Mar-55	3	17
KS GRAHAM	49.44	TP4-2	8-Mar-55	4	17
KS GRAHAM	32.33	TP10-2	6-May-55	5	17
KS GRAHAM	2.33	PB3-3	26-Jun-57	8	17
KS GRAHAM	1.96	PB5-5	19-Jul-57	9	17
KS GRAHAM	19.73	PB8-3	20-Aug-57	6	17
KS GRAHAM	0.89	PB13-7	12-Sep-57	5	17
KS GRAHAM	1.28	PB13-8	13-Sep-57	6	17
KS GRAHAM	0.81	PB14-7	20-Sep-57	9	17
KS GRAHAM	2.30	PB18-2	8-Oct-57	5	17
KS GRAHAM	2.37	PB18-3	9-Oct-57	3	17
KS GRAHAM	1.70	PB18-4	10-Oct-57	5	17
KS GRAHAM	0.10	UE2-7	11-Mar-62	3	17
KS GRAHAM	4.55	UE8-4	17-Jul-62	7	17

COUNTY	uCi/ sqm	SHOT DAY	DATE	RANK FOR DAY	TIMES IN TOP 10
KS GRANT	0.87	TS4-3	3-May-52	1	12
KS GRANT	0.87	TS4-3	3-May-52	1	12
KS GRANT	1.31	UpK6-3	20-Apr-53	9	12
KS GRANT	7.78	TP3-2	2-Mar-55	3	12
KS GRANT	49.44	TP4-2	8-Mar-55	4	12
KS GRANT	3.04	TP7-3	31-Mar-55	7	12
KS GRANT	2.22	TP8-3	11-Apr-55	10	12
KS GRANT	32.33	TP10-2	6-May-55	5	12
KS GRANT	3.31	PB13-8	13-Sep-57	2	12
KS GRANT	2.30	PB18-2	8-Oct-57	5	12
KS GRANT	2.37	PB18-3	9-Oct-57	3	12
KS GRANT	1.70	PB18-4	10-Oct-57	5	12
KS GRAY	1.22	BJ5-6	24-Nov-51	6	15
KS GRAY	1.22	BJ5-6	24-Nov-51	6	15
KS GRAY	0.87	TS4-3	3-May-52	1	15
KS GRAY	0.87	TS4-3	3-May-52	1	15
KS GRAY	1.31	UpK6-3	20-Apr-53	9	15
KS GRAY	7.78	TP3-2	2-Mar-55	3	15
KS GRAY	49.44	TP4-2	8-Mar-55	4	15
KS GRAY	5.38	TP7-2	30-Mar-55	10	15
KS GRAY	32.33	TP10-2	6-May-55	5	15
KS GRAY	0.86	PB6-11	3-Aug-57	4	15
KS GRAY	2.11	PB13-8	13-Sep-57	4	15
KS GRAY	4.60	PB18-2	8-Oct-57	2	15
KS GRAY	2.37	PB18-3	9-Oct-57	3	15
KS GRAY	1.70	PB18-4	10-Oct-57	5	15
KS GRAY	1.92	PB18-6	12-Oct-57	3	15
KS GREELEY	0.87	TS4-3	3-May-52	1	14
KS GREELEY	0.87	TS4-3	3-May-52	1	14
KS GREELEY	1.31	UpK6-3	20-Apr-53	9	14
KS GREELEY	7.78	TP3-2	2-Mar-55	3	14
KS GREELEY	49.44	TP4-2	8-Mar-55	4	14
KS GREELEY	32.33	TP10-2	6-May-55	5	14
KS GREELEY	1.33	PB1-11	7-Jun-57	3	14
KS GREELEY	9.58	PB11-2	1-Sep-57	4	14

COUNTY	uCi/ sqm	SHOT DAY	DATE	RANK FOR DAY	TIMES IN TOP 10
KS GREELEY	2.20	PB13-8	13-Sep-57	3	14
KS GREELEY	2.30	PB18-2	8-Oct-57	5	14
KS GREELEY	2.37	PB18-3	9-Oct-57	3	14
KS GREELEY	1.70	PB18-4	10-Oct-57	5	14
KS GREELEY	0.86	PB18-6	12-Oct-57	7	14
KS GREELEY	0.10	UE2-7	11-Mar-62	3	14
KS GREENWOOD	1.63	PB5-7	21-Jul-57	8	4
KS GREENWOOD	1.28	PB13-8	13-Sep-57	6	4
KS GREENWOOD	2.50	PB13-9	14-Sep-57	8	4
KS GREENWOOD	4.04	PB14-7	20-Sep-57	1	4
KS HAMILTON	0.87	TS4-3	3-May-52	1	15
KS HAMILTON	0.87	TS4-3	3-May-52	1	15
KS HAMILTON	1.31	UpK6-3	20-Apr-53	9	15
KS HAMILTON	7.78	TP3-2	2-Mar-55	3	15
KS HAMILTON	49.44	TP4-2	8-Mar-55	4	15
KS HAMILTON	32.33	TP10-2	6-May-55	5	15
KS HAMILTON	4.56	PB5-4	18-Jul-57	9	15
KS HAMILTON	1.47	PB9-6	28-Aug-57	3	15
KS HAMILTON	4.79	PB11-2	1-Sep-57	8	15
KS HAMILTON	2.20	PB13-8	13-Sep-57	3	15
KS HAMILTON	2.30	PB18-2	8-Oct-57	5	15
KS HAMILTON	2.37	PB18-3	9-Oct-57	3	15
KS HAMILTON	1.70	PB18-4	10-Oct-57	5	15
KS HAMILTON	0.86	PB18-6	12-Oct-57	7	15
KS HAMILTON	0.10	UE2-7	11-Mar-62	4	15
KS HARPER	2.48	BJ5-6	24-Nov-51	4	7
KS HARPER	2.48	BJ5-6	24-Nov-51	4	7
KS HARPER	2.24	PB3-2	25-Jun-57	5	7
KS HARPER	1.63	PB5-7	21-Jul-57	8	7
KS HARPER	2.11	PB13-8	13-Sep-57	4	7
KS HARPER	4.15	PB13-9	14-Sep-57	7	7
KS HARPER	2.42	PB14-7	20-Sep-57	2	7
KS HARVEY	1.76	PB2-4	21-Jun-57	6	6
KS HARVEY	2.24	PB3-2	25-Jun-57	5	6
KS HARVEY	1.28	PB13-8	13-Sep-57	6	6

COUNTY	uCi/ sqm	SHOT DAY	DATE	RANK FOR DAY	TIMES IN TOP 10
KS HARVEY	2.50	PB13-9	14-Sep-57	8	6
KS HARVEY	2.42	PB14-7	20-Sep-57	2	6
KS HARVEY	5.99	UE8-4	17-Jul-62	5	6
KS HASKELL	1.22	BJ5-6	24-Nov-51	6	13
KS HASKELL	1.22	BJ5-6	24-Nov-51	6	13
KS HASKELL	0.87	TS4-3	3-May-52	1	13
KS HASKELL	0.87	TS4-3	3-May-52	1	13
KS HASKELL	1.31	UpK6-3	20-Apr-53	9	13
KS HASKELL	7.78	TP3-2	2-Mar-55	3	13
KS HASKELL	49.44	TP4-2	8-Mar-55	4	13
KS HASKELL	32.33	TP10-2	6-May-55	5	13
KS HASKELL	1.28	PB13-8	13-Sep-57	6	13
KS HASKELL	2.30	PB18-2	8-Oct-57	5	13
KS HASKELL	2.37	PB18-3	9-Oct-57	3	13
KS HASKELL	1.70	PB18-4	10-Oct-57	5	13
KS HASKELL	1.92	PB18-6	12-Oct-57	3	13
KS HODGEMAN	1.22	BJ5-6	24-Nov-51	6	16
KS HODGEMAN	1.22	BJ5-6	24-Nov-51	6	16
KS HODGEMAN	0.87	TS4-3	3-May-52	1	16
KS HODGEMAN	0.87	TS4-3	3-May-52	1	16
KS HODGEMAN	1.31	UpK6-3	20-Apr-53	9	16
KS HODGEMAN	7.78	TP3-2	2-Mar-55	3	16
KS HODGEMAN	49.44	TP4-2	8-Mar-55	4	16
KS HODGEMAN	5.38	TP7-2	30-Mar-55	10	16
KS HODGEMAN	32.33	TP10-2	6-May-55	5	16
KS HODGEMAN	2.24	PB3-2	25-Jun-57	5	16
KS HODGEMAN	3.26	PB5-5	19-Jul-57	5	16
KS HODGEMAN	2.11	PB13-8	13-Sep-57	4	16
KS HODGEMAN	2.30	PB18-2	8-Oct-57	5	16
KS HODGEMAN	2.37	PB18-3	9-Oct-57	3	16
KS HODGEMAN	1.70	PB18-4	10-Oct-57	5	16
KS HODGEMAN	4.55	UE8-4	17-Jul-62	7	16
KS JACKSON	12.24	TP4-2	8-Mar-55	9	4
KS JACKSON	0.83	PB13-8	13-Sep-57	9	4
KS JACKSON	2.42	PB14-7	20-Sep-57	2	4

COUNTY	uCi/ sqm	SHOT DAY	DATE	RANK FOR DAY	TIMES IN TOP 10
KS JACKSON	0.12	UE7-5	15-Jul-62	9	4
KS JEFFERSON	1.68	TS4-6	6-May-52	6	10
KS JEFFERSON	1.68	TS4-6	6-May-52	6	10
KS JEFFERSON	6.74	UpK11-2	5-Jun-53	9	10
KS JEFFERSON	12.24	TP4-2	8-Mar-55	9	10
KS JEFFERSON	20.34	PB12-4	5-Sep-57	9	10
KS JEFFERSON	3.55	PB13-4	9-Sep-57	5	10
KS JEFFERSON	0.83	PB13-8	13-Sep-57	9	10
KS JEFFERSON	2.42	PB14-7	20-Sep-57	2	10
KS JEFFERSON	0.86	PB18-6	12-Oct-57	7	10
KS JEFFERSON	0.12	UE7-5	15-Jul-62	9	10
KS JEWELL	12.24	TP4-2	8-Mar-55	9	5
KS JEWELL	19.73	PB8-3	20-Aug-57	6	5
KS JEWELL	1.28	PB13-8	13-Sep-57	6	5
KS JEWELL	0.81	PB14-7	20-Sep-57	9	5
KS JEWELL	8.02	UE8-4	17-Jul-62	3	5
KS JOHNSON	0.84	UpK10-7	31-May-53	6	8
KS JOHNSON	0.97	PB3-8	1-Jul-57	4	8
KS JOHNSON	17.72	PB4-5	9-Jul-57	3	8
KS JOHNSON	1.67	PB5-7	21-Jul-57	7	8
KS JOHNSON	7.99	PB8-3	20-Aug-57	9	8
KS JOHNSON	4.04	PB14-7	20-Sep-57	1	8
KS JOHNSON	2.02	PB15-5	20-Sep-57	3	8
KS JOHNSON	5.26	PB18-8	14-Oct-57	3	8
KS KEARNY	0.87	TS4-3	3-May-52	1	18
KS KEARNY	0.87	TS4-3	3-May-52	1	18
KS KEARNY	1.31	UpK6-3	20-Apr-53	9	18
KS KEARNY	7.78	TP3-2	2-Mar-55	3	18
KS KEARNY	49.44	TP4-2	8-Mar-55	4	18
KS KEARNY	5.38	TP7-2	30-Mar-55	10	18
KS KEARNY	32.33	TP10-2	6-May-55	5	18
KS KEARNY	1.76	PB2-4	21-Jun-57	6	18
KS KEARNY	1.73	PB4-10	14-Jul-57	9	18
KS KEARNY	4.56	PB5-4	18-Jul-57	9	18
KS KEARNY	1.96	PB5-5	19-Jul-57	9	18

COUNTY	uCi/ sqm	SHOT DAY	DATE	RANK FOR DAY	TIMES IN TOP 10
KS KEARNY	3.55	PB13-4	9-Sep-57	5	18
KS KEARNY	0.89	PB13-7	12-Sep-57	5	18
KS KEARNY	2.20	PB13-8	13-Sep-57	3	18
KS KEARNY	2.30	PB18-2	8-Oct-57	5	18
KS KEARNY	2.37	PB18-3	9-Oct-57	3	18
KS KEARNY	1.70	PB18-4	10-Oct-57	5	18
KS KEARNY	0.10	UE2-7	11-Mar-62	4	18
KS KINGMAN	2.48	BJ5-6	24-Nov-51	4	8
KS KINGMAN	2.48	BJ5-6	24-Nov-51	4	8
KS KINGMAN	3.78	PB3-2	25-Jun-57	3	8
KS KINGMAN	1.63	PB5-7	21-Jul-57	8	8
KS KINGMAN	17.59	PB12-3	4-Sep-57	10	8
KS KINGMAN	1.28	PB13-8	13-Sep-57	6	8
KS KINGMAN	4.15	PB13-9	14-Sep-57	7	8
KS KINGMAN	1.62	PB14-7	20-Sep-57	4	8
KS KIOWA	2.48	BJ5-6	24-Nov-51	4	6
KS KIOWA	2.48	BJ5-6	24-Nov-51	4	6
KS KIOWA	1.76	PB2-4	21-Jun-57	6	6
KS KIOWA	3.78	PB3-2	25-Jun-57	3	6
KS KIOWA	0.93	PB3-8	1-Jul-57	4	6
KS KIOWA	2.11	PB13-8	13-Sep-57	4	6
KS LABETTE	5.84	BJ2-2	31-Oct-51	7	5
KS LABETTE	5.36	PB13-4	9-Sep-57	3	5
KS LABETTE	1.28	PB13-8	13-Sep-57	6	5
KS LABETTE	2.50	PB13-9	14-Sep-57	8	5
KS LABETTE	1.62	PB14-7	20-Sep-57	4	5
KS LANE	1.22	BJ5-6	24-Nov-51	6	16
KS LANE	1.22	BJ5-6	24-Nov-51	6	16
KS LANE	0.87	TS4-3	3-May-52	1	16
KS LANE	0.87	TS4-3	3-May-52	1	16
KS LANE	1.31	UpK6-3	20-Apr-53	9	16
KS LANE	7.78	TP3-2	2-Mar-55	3	16
KS LANE	49.44	TP4-2	8-Mar-55	4	16
KS LANE	2.82	TP7-3	31-Mar-55	10	16
KS LANE	32.33	TP10-2	6-May-55	5	16

COUNTY	uCi/ sqm	SHOT DAY	DATE	RANK FOR DAY	TIMES IN TOP 10
KS LANE	3.26	PB5-5	19-Jul-57	5	16
KS LANE	2.11	PB13-8	13-Sep-57	4	16
KS LANE	2.30	PB18-2	8-Oct-57	5	16
KS LANE	2.37	PB18-3	9-Oct-57	3	16
KS LANE	1.70	PB18-4	10-Oct-57	5	16
KS LANE	0.10	UE2-7	11-Mar-62	3	16
KS LANE	4.55	UE8-4	17-Jul-62	7	16
KS LEAVENWORTH	1.68	TS4-6	6-May-52	6	12
KS LEAVENWORTH	1.68	TS4-6	6-May-52	6	12
KS LEAVENWORTH	4.58	UpK11-3	6-Jun-53	3	12
KS LEAVENWORTH	12.24	TP4-2	8-Mar-55	9	12
KS LEAVENWORTH	0.99	PB5-6	20-Jul-57	9	12
KS LEAVENWORTH	4.15	PB5-7	21-Jul-57	2	12
KS LEAVENWORTH	7.99	PB8-3	20-Aug-57	9	12
KS LEAVENWORTH	20.41	PB12-4	5-Sep-57	8	12
KS LEAVENWORTH	3.55	PB13-4	9-Sep-57	5	12
KS LEAVENWORTH	2.42	PB14-7	20-Sep-57	2	12
KS LEAVENWORTH	0.86	PB18-6	12-Oct-57	7	12
KS LEAVENWORTH	0.12	UE7-5	15-Jul-62	9	12
KS LINCOLN	0.92	TS3-8	29-Apr-52	9	11
KS LINCOLN	0.92	TS3-8	29-Apr-52	9	11
KS LINCOLN	12.24	TP4-2	8-Mar-55	9	11
KS LINCOLN	3.60	TP7-3	31-Mar-55	5	11
KS LINCOLN	2.33	PB3-3	26-Jun-57	8	11
KS LINCOLN	0.93	PB3-8	1-Jul-57	4	11
KS LINCOLN	19.73	PB8-3	20-Aug-57	6	11
KS LINCOLN	26.87	PB13-5	10-Sep-57	4	11
KS LINCOLN	2.11	PB13-8	13-Sep-57	4	11
KS LINCOLN	0.81	PB14-7	20-Sep-57	9	11
KS LINCOLN	14.44	UE8-4	17-Jul-62	1	11
KS LINN	4.04	PB14-7	20-Sep-57	1	2
KS LINN	0.22	UE7-5	15-Jul-62	6	2
KS LOGAN	0.87	TS4-3	3-May-52	1	13
KS LOGAN	0.87	TS4-3	3-May-52	1	13
KS LOGAN	1.31	UpK6-3	20-Apr-53	9	13

COUNTY	uCi/ sqm	SHOT DAY	DATE	RANK FOR DAY	TIMES IN TOP 10
KS LOGAN	7.78	TP3-2	2-Mar-55	3	13
KS LOGAN	49.44	TP4-2	8-Mar-55	4	13
KS LOGAN	32.33	TP10-2	6-May-55	5	13
KS LOGAN	4.92	PB8-4	21-Aug-57	7	13
KS LOGAN	4.92	PB8-4	21-Aug-57	7	13
KS LOGAN	0.88	PB9-6	28-Aug-57	7	13
KS LOGAN	2.30	PB18-2	8-Oct-57	5	13
KS LOGAN	2.37	PB18-3	9-Oct-57	3	13
KS LOGAN	1.70	PB18-4	10-Oct-57	5	13
KS LOGAN	0.86	PB18-6	12-Oct-57	7	13
KS LYON	3.55	PB13-4	9-Sep-57	5	3
KS LYON	4.04	PB14-7	20-Sep-57	1	3
KS LYON	1.92	PB18-6	12-Oct-57	3	3
KS MARION	3.91	TS6-7	31-May-52	5	7
KS MARION	1.54	PB3-2	25-Jun-57	8	7
KS MARION	3.55	PB13-4	9-Sep-57	5	7
KS MARION	0.83	PB13-8	13-Sep-57	9	7
KS MARION	1.64	PB13-9	14-Sep-57	10	7
KS MARION	2.42	PB14-7	20-Sep-57	2	7
KS MARION	10.54	UE8-4	17-Jul-62	2	7
KS MARSHALL	0.92	TS3-8	29-Apr-52	9	7
KS MARSHALL	0.92	TS3-8	29-Apr-52	9	7
KS MARSHALL	12.24	TP4-2	8-Mar-55	9	7
KS MARSHALL	1.54	PB3-2	25-Jun-57	8	7
KS MARSHALL	3.55	PB13-4	9-Sep-57	5	7
KS MARSHALL	0.81	PB14-7	20-Sep-57	9	7
KS MARSHALL	0.13	UE7-5	15-Jul-62	8	7
KS MCPHERSON	1.15	TP8-4	12-Apr-55	6	9
KS MCPHERSON	2.33	PB3-3	26-Jun-57	8	9
KS MCPHERSON	1.63	PB5-7	21-Jul-57	8	9
KS MCPHERSON	19.06	PB12-3	4-Sep-57	6	9
KS MCPHERSON	30.39	PB13-5	10-Sep-57	1	9
KS MCPHERSON	1.28	PB13-8	13-Sep-57	6	9
KS MCPHERSON	2.42	PB14-7	20-Sep-57	2	9
KS MCPHERSON	0.11	UE7-5	15-Jul-62	10	9

COUNTY	uCi/ sqm	SHOT DAY	DATE	RANK FOR DAY	TIMES IN TOP 10
KS MCPHERSON	10.54	UE8-4	17-Jul-62	2	9
KS MEADE	1.22	BJ5-6	24-Nov-51	6	15
KS MEADE	1.22	BJ5-6	24-Nov-51	6	15
KS MEADE	1.02	TS4-4	4-May-52	1	15
KS MEADE	1.02	TS4-4	4-May-52	1	15
KS MEADE	1.74	TS4-6	6-May-52	5	15
KS MEADE	1.74	TS4-6	6-May-52	5	15
KS MEADE	5.46	UpK3-5	4-Apr-53	4	15
KS MEADE	4.23	UpK6-3	20-Apr-53	2	15
KS MEADE	25.62	TP7-2	30-Mar-55	4	15
KS MEADE	31.48	TP10-2	6-May-55	6	15
KS MEADE	2.11	PB13-8	13-Sep-57	4	15
KS MEADE	4.60	PB18-2	8-Oct-57	2	15
KS MEADE	2.37	PB18-3	9-Oct-57	3	15
KS MEADE	1.70	PB18-4	10-Oct-57	5	15
KS MEADE	1.92	PB18-6	12-Oct-57	3	15
KS MIAMI	7.67	UpK11-2	5-Jun-53	6	3
KS MIAMI	0.89	TP9-13	27-Apr-55	5	3
KS MIAMI	4.04	PB14-7	20-Sep-57	1	3
KS MITCHELL	1.38	TS3-8	29-Apr-52	6	11
KS MITCHELL	1.38	TS3-8	29-Apr-52	6	11
KS MITCHELL	1.12	TS3-9	30-Apr-52	1	11
KS MITCHELL	1.12	TS3-9	30-Apr-52	1	11
KS MITCHELL	12.24	TP4-2	8-Mar-55	9	11
KS MITCHELL	0.93	PB3-8	1-Jul-57	4	11
KS MITCHELL	19.73	PB8-3	20-Aug-57	6	11
KS MITCHELL	12.75	PB13-3	8-Sep-57	4	11
KS MITCHELL	1.28	PB13-8	13-Sep-57	6	11
KS MITCHELL	0.81	PB14-7	20-Sep-57	9	11
KS MITCHELL	14.44	UE8-4	17-Jul-62	1	11
KS MONTGOMERY	5.84	BJ2-2	31-Oct-51	7	7
KS MONTGOMERY	2.60	TP10-4	8-May-55	5	7
KS MONTGOMERY	1.76	PB2-4	21-Jun-57	6	7
KS MONTGOMERY	2.69	PB5-7	21-Jul-57	5	7
KS MONTGOMERY	1.28	PB13-8	13-Sep-57	6	7

COUNTY	uCi/ sqm	SHOT DAY	DATE	RANK FOR DAY	TIMES IN TOP 10
KS MONTGOMERY	1.64	PB13-9	14-Sep-57	10	7
KS MONTGOMERY	0.81	PB14-7	20-Sep-57	9	7
KS MORRIS	1.30	TS4-6	6-May-52	10	7
KS MORRIS	1.30	TS4-6	6-May-52	10	7
KS MORRIS	2.33	PB3-3	26-Jun-57	8	7
KS MORRIS	0.83	PB13-8	13-Sep-57	9	7
KS MORRIS	4.04	PB14-7	20-Sep-57	1	7
KS MORRIS	1.29	PB18-6	12-Oct-57	6	7
KS MORRIS	0.22	UE7-5	15-Jul-62	6	7
KS MORTON	1.02	TS4-4	4-May-52	1	12
KS MORTON	1.02	TS4-4	4-May-52	1	12
KS MORTON	4.23	UpK6-3	20-Apr-53	2	12
KS MORTON	25.62	TP7-2	30-Mar-55	4	12
KS MORTON	31.48	TP10-2	6-May-55	6	12
KS MORTON	0.83	PB3-8	1-Jul-57	5	12
KS MORTON	1.47	PB9-6	28-Aug-57	3	12
KS MORTON	4.79	PB11-2	1-Sep-57	8	12
KS MORTON	5.51	PB13-8	13-Sep-57	1	12
KS MORTON	4.60	PB18-2	8-Oct-57	2	12
KS MORTON	2.37	PB18-3	9-Oct-57	3	12
KS MORTON	1.70	PB18-4	10-Oct-57	5	12
KS NEMAHA	12.24	TP4-2	8-Mar-55	9	5
KS NEMAHA	3.55	PB13-4	9-Sep-57	5	5
KS NEMAHA	0.83	PB13-8	13-Sep-57	9	5
KS NEMAHA	1.62	PB14-7	20-Sep-57	4	5
KS NEMAHA	0.12	UE7-5	15-Jul-62	9	5
KS NEOSHO	5.84	BJ2-2	31-Oct-51	7	6
KS NEOSHO	0.89	TP9-13	27-Apr-55	5	6
KS NEOSHO	2.69	PB5-7	21-Jul-57	5	6
KS NEOSHO	1.28	PB13-8	13-Sep-57	6	6
KS NEOSHO	2.42	PB14-7	20-Sep-57	2	6
KS NEOSHO	0.11	UE7-5	15-Jul-62	10	6
KS NESS	1.22	BJ5-6	24-Nov-51	6	17
KS NESS	1.22	BJ5-6	24-Nov-51	6	17
KS NESS	0.87	TS4-3	3-May-52	1	17

COUNTY	uCi/ sqm	SHOT DAY	DATE	RANK FOR DAY	TIMES IN TOP 10
KS NESS	0.87	TS4-3	3-May-52	1	17
KS NESS	1.31	UpK6-3	20-Apr-53	9	17
KS NESS	7.78	TP3-2	2-Mar-55	3	17
KS NESS	49.44	TP4-2	8-Mar-55	4	17
KS NESS	3.90	TP7-3	31-Mar-55	1	17
KS NESS	32.33	TP10-2	6-May-55	5	17
KS NESS	1.54	PB3-2	25-Jun-57	8	17
KS NESS	3.46	PB4-10	14-Jul-57	4	17
KS NESS	1.96	PB5-5	19-Jul-57	9	17
KS NESS	2.11	PB13-8	13-Sep-57	4	17
KS NESS	2.30	PB18-2	8-Oct-57	5	17
KS NESS	2.37	PB18-3	9-Oct-57	3	17
KS NESS	1.70	PB18-4	10-Oct-57	5	17
KS NESS	4.55	UE8-4	17-Jul-62	7	17
KS NORTON	0.87	TS4-3	3-May-52	1	13
KS NORTON	0.87	TS4-3	3-May-52	1	13
KS NORTON	1.31	UpK6-3	20-Apr-53	9	13
KS NORTON	7.78	TP3-2	2-Mar-55	3	13
KS NORTON	49.44	TP4-2	8-Mar-55	4	13
KS NORTON	32.33	TP10-2	6-May-55	5	13
KS NORTON	19.73	PB8-3	20-Aug-57	6	13
KS NORTON	3.55	PB13-4	9-Sep-57	5	13
KS NORTON	2.30	PB18-2	8-Oct-57	5	13
KS NORTON	2.37	PB18-3	9-Oct-57	3	13
KS NORTON	1.70	PB18-4	10-Oct-57	5	13
KS NORTON	0.10	UE2-7	11-Mar-62	4	13
KS NORTON	4.55	UE8-4	17-Jul-62	7	13
KS OSAGE	1.30	TS4-6	6-May-52	10	6
KS OSAGE	1.30	TS4-6	6-May-52	10	6
KS OSAGE	0.99	PB5-6	20-Jul-57	9	6
KS OSAGE	1.63	PB5-7	21-Jul-57	8	6
KS OSAGE	4.04	PB14-7	20-Sep-57	1	6
KS OSAGE	0.13	UE7-5	15-Jul-62	8	6
KS OSBORNE	1.38	TS3-8	29-Apr-52	6	12
KS OSBORNE	1.38	TS3-8	29-Apr-52	6	12

COUNTY	uCi/ sqm	SHOT DAY	DATE	RANK FOR DAY	TIMES IN TOP 10
KS OSBORNE	1.12	TS3-9	30-Apr-52	1	12
KS OSBORNE	1.12	TS3-9	30-Apr-52	1	12
KS OSBORNE	12.24	TP4-2	8-Mar-55	9	12
KS OSBORNE	3.73	TP7-3	31-Mar-55	3	12
KS OSBORNE	2.33	PB3-3	26-Jun-57	8	12
KS OSBORNE	19.73	PB8-3	20-Aug-57	6	12
KS OSBORNE	2.85	PB8-5	22-Aug-57	7	12
KS OSBORNE	1.28	PB13-8	13-Sep-57	6	12
KS OSBORNE	1.29	PB18-6	12-Oct-57	6	12
KS OSBORNE	14.44	UE8-4	17-Jul-62	1	12
KS OTTAWA	0.92	TS3-8	29-Apr-52	9	13
KS OTTAWA	0.92	TS3-8	29-Apr-52	9	13
KS OTTAWA	4.24	TS6-7	31-May-52	4	13
KS OTTAWA	12.24	TP4-2	8-Mar-55	9	13
KS OTTAWA	2.90	TP7-3	31-Mar-55	9	13
KS OTTAWA	1.40	TP8-4	12-Apr-55	2	13
KS OTTAWA	2.33	PB3-3	26-Jun-57	8	13
KS OTTAWA	0.93	PB3-8	1-Jul-57	4	13
KS OTTAWA	1.63	PB5-7	21-Jul-57	8	13
KS OTTAWA	27.63	PB13-5	10-Sep-57	2	13
KS OTTAWA	2.11	PB13-8	13-Sep-57	4	13
KS OTTAWA	0.81	PB14-7	20-Sep-57	9	13
KS OTTAWA	10.54	UE8-4	17-Jul-62	2	13
KS PAWNEE	1.22	BJ5-6	24-Nov-51	6	9
KS PAWNEE	1.22	BJ5-6	24-Nov-51	6	9
KS PAWNEE	2.24	PB3-2	25-Jun-57	5	9
KS PAWNEE	2.33	PB3-3	26-Jun-57	8	9
KS PAWNEE	0.94	PB5-6	20-Jul-57	9	9
KS PAWNEE	5.36	PB13-4	9-Sep-57	3	9
KS PAWNEE	2.11	PB13-8	13-Sep-57	4	9
KS PAWNEE	2.50	PB13-9	14-Sep-57	8	9
KS PAWNEE	4.55	UE8-4	17-Jul-62	7	9
KS PHILLIPS	1.38	TS3-8	29-Apr-52	6	17
KS PHILLIPS	1.38	TS3-8	29-Apr-52	6	17
KS PHILLIPS	1.12	TS3-9	30-Apr-52	1	17

COUNTY	uCi/ sqm	SHOT DAY	DATE	RANK FOR DAY	TIMES IN TOP 10
KS PHILLIPS	1.12	TS3-9	30-Apr-52	1	17
KS PHILLIPS	12.24	TP4-2	8-Mar-55	9	17
KS PHILLIPS	0.93	PB3-8	1-Jul-57	4	17
KS PHILLIPS	0.83	PB8-2	19-Aug-57	8	17
KS PHILLIPS	19.73	PB8-3	20-Aug-57	6	17
KS PHILLIPS	3.78	PB8-5	22-Aug-57	4	17
KS PHILLIPS	5.36	PB13-4	9-Sep-57	3	17
KS PHILLIPS	0.83	PB13-8	13-Sep-57	9	17
KS PHILLIPS	0.81	PB14-7	20-Sep-57	9	17
KS PHILLIPS	2.37	PB18-3	9-Oct-57	3	17
KS PHILLIPS	1.44	PB18-4	10-Oct-57	6	17
KS PHILLIPS	1.29	PB18-6	12-Oct-57	6	17
KS PHILLIPS	0.10	UE2-7	11-Mar-62	4	17
KS PHILLIPS	14.44	UE8-4	17-Jul-62	1	17
KS POTTAWATOMIE	0.92	TS3-8	29-Apr-52	9	13
KS POTTAWATOMIE	0.92	TS3-8	29-Apr-52	9	13
KS POTTAWATOMIE	1.68	TS4-6	6-May-52	6	13
KS POTTAWATOMIE	1.68	TS4-6	6-May-52	6	13
KS POTTAWATOMIE	12.24	TP4-2	8-Mar-55	9	13
KS POTTAWATOMIE	1.54	PB3-2	25-Jun-57	8	13
KS POTTAWATOMIE	2.33	PB3-3	26-Jun-57	8	13
KS POTTAWATOMIE	0.93	PB3-8	1-Jul-57	4	13
KS POTTAWATOMIE	1.65	PB5-6	20-Jul-57	5	13
KS POTTAWATOMIE	1.63	PB5-7	21-Jul-57	8	13
KS POTTAWATOMIE	1.28	PB13-8	13-Sep-57	6	13
KS POTTAWATOMIE	1.62	PB14-7	20-Sep-57	4	13
KS POTTAWATOMIE	0.33	UE7-5	15-Jul-62	3	13
KS PRATT	2.48	BJ5-6	24-Nov-51	4	7
KS PRATT	2.48	BJ5-6	24-Nov-51	4	7
KS PRATT	1.76	PB2-4	21-Jun-57	6	7
KS PRATT	3.78	PB3-2	25-Jun-57	3	7
KS PRATT	1.28	PB13-8	13-Sep-57	6	7
KS PRATT	2.50	PB13-9	14-Sep-57	8	7
KS PRATT	1.62	PB14-7	20-Sep-57	4	7
KS RAWLINS	0.87	TS4-3	3-May-52	1	12

COUNTY	uCi/ sqm	SHOT DAY	DATE	RANK FOR DAY	TIMES IN TOP 10
KS RAWLINS	0.87	TS4-3	3-May-52	1	12
KS RAWLINS	1.31	UpK6-3	20-Apr-53	9	12
KS RAWLINS	7.78	TP3-2	2-Mar-55	3	12
KS RAWLINS	49.44	TP4-2	8-Mar-55	4	12
KS RAWLINS	32.33	TP10-2	6-May-55	5	12
KS RAWLINS	0.88	PB9-6	28-Aug-57	7	12
KS RAWLINS	2.30	PB18-2	8-Oct-57	5	12
KS RAWLINS	2.37	PB18-3	9-Oct-57	3	12
KS RAWLINS	1.70	PB18-4	10-Oct-57	5	12
KS RAWLINS	0.86	PB18-6	12-Oct-57	7	12
KS RAWLINS	4.55	UE8-4	17-Jul-62	7	12
KS RENO	1.22	BJ5-6	24-Nov-51	6	9
KS RENO	1.22	BJ5-6	24-Nov-51	6	9
KS RENO	46.20	UpK7-3	27-Apr-53	7	9
KS RENO	2.24	PB3-2	25-Jun-57	5	9
KS RENO	1.63	PB5-7	21-Jul-57	8	9
KS RENO	18.20	PB12-3	4-Sep-57	7	9
KS RENO	2.11	PB13-8	13-Sep-57	4	9
KS RENO	2.50	PB13-9	14-Sep-57	8	9
KS RENO	0.81	PB14-7	20-Sep-57	9	9
KS REPUBLIC	12.24	TP4-2	8-Mar-55	9	6
KS REPUBLIC	1.63	PB5-7	21-Jul-57	8	6
KS REPUBLIC	39.46	PB8-3	20-Aug-57	4	6
KS REPUBLIC	0.83	PB13-8	13-Sep-57	9	6
KS REPUBLIC	1.62	PB14-7	20-Sep-57	4	6
KS REPUBLIC	10.54	UE8-4	17-Jul-62	2	6
KS RICE	1.30	TS4-6	6-May-52	10	9
KS RICE	1.30	TS4-6	6-May-52	10	9
KS RICE	1.12	TP8-4	12-Apr-55	8	9
KS RICE	2.33	PB3-3	26-Jun-57	8	9
KS RICE	1.63	PB5-7	21-Jul-57	8	9
KS RICE	26.93	PB13-5	10-Sep-57	3	9
KS RICE	2.11	PB13-8	13-Sep-57	4	9
KS RICE	0.13	UE7-5	15-Jul-62	8	9
KS RICE	8.02	UE8-4	17-Jul-62	3	9

COUNTY	uCi/ sqm	SHOT DAY	DATE	RANK FOR DAY	TIMES IN TOP 10
KS RILEY	12.24	TP4-2	8-Mar-55	9	7
KS RILEY	2.24	PB3-2	25-Jun-57	5	7
KS RILEY	2.33	PB3-3	26-Jun-57	8	7
KS RILEY	1.63	PB5-7	21-Jul-57	8	7
KS RILEY	0.83	PB13-8	13-Sep-57	9	7
KS RILEY	0.81	PB14-7	20-Sep-57	9	7
KS RILEY	0.22	UE7-5	15-Jul-62	6	7
KS ROOKS	1.38	TS3-8	29-Apr-52	6	15
KS ROOKS	1.38	TS3-8	29-Apr-52	6	15
KS ROOKS	1.12	TS3-9	30-Apr-52	1	15
KS ROOKS	1.12	TS3-9	30-Apr-52	1	15
KS ROOKS	12.24	TP4-2	8-Mar-55	9	15
KS ROOKS	2.33	PB3-3	26-Jun-57	8	15
KS ROOKS	1.96	PB5-5	19-Jul-57	9	15
KS ROOKS	19.73	PB8-3	20-Aug-57	6	15
KS ROOKS	1.28	PB13-8	13-Sep-57	6	15
KS ROOKS	2.30	PB18-2	8-Oct-57	5	15
KS ROOKS	2.37	PB18-3	9-Oct-57	3	15
KS ROOKS	1.70	PB18-4	10-Oct-57	5	15
KS ROOKS	1.29	PB18-6	12-Oct-57	6	15
KS ROOKS	0.10	UE2-7	11-Mar-62	4	15
KS ROOKS	14.44	UE8-4	17-Jul-62	1	15
KS RUSH	2.48	BJ5-6	24-Nov-51	4	16
KS RUSH	2.48	BJ5-6	24-Nov-51	4	16
KS RUSH	1.38	TS3-8	29-Apr-52	6	16
KS RUSH	1.38	TS3-8	29-Apr-52	6	16
KS RUSH	1.12	TS3-9	30-Apr-52	1	16
KS RUSH	1.12	TS3-9	30-Apr-52	1	16
KS RUSH	12.24	TP4-2	8-Mar-55	9	16
KS RUSH	3.69	TP7-3	31-Mar-55	4	16
KS RUSH	1.06	TP8-4	12-Apr-55	10	16
KS RUSH	2.33	PB3-3	26-Jun-57	8	16
KS RUSH	1.96	PB5-5	19-Jul-57	9	16
KS RUSH	2.11	PB13-8	13-Sep-57	4	16
KS RUSH	2.30	PB18-2	8-Oct-57	5	16

COUNTY	uCi/ sqm	SHOT DAY	DATE	RANK FOR DAY	TIMES IN TOP 10
KS RUSH	2.37	PB18-3	9-Oct-57	3	16
KS RUSH	1.70	PB18-4	10-Oct-57	5	16
KS RUSH	4.55	UE8-4	17-Jul-62	7	16
KS RUSSELL	1.38	TS3-8	29-Apr-52	6	15
KS RUSSELL	1.38	TS3-8	29-Apr-52	6	15
KS RUSSELL	1.12	TS3-9	30-Apr-52	1	15
KS RUSSELL	1.12	TS3-9	30-Apr-52	1	15
KS RUSSELL	12.24	TP4-2	8-Mar-55	9	15
KS RUSSELL	2.33	PB3-3	26-Jun-57	8	15
KS RUSSELL	0.93	PB3-8	1-Jul-57	4	15
KS RUSSELL	19.73	PB8-3	20-Aug-57	6	15
KS RUSSELL	1.92	PB8-5	22-Aug-57	10	15
KS RUSSELL	3.55	PB13-4	9-Sep-57	5	15
KS RUSSELL	24.61	PB13-5	10-Sep-57	10	15
KS RUSSELL	1.34	PB13-7	12-Sep-57	2	15
KS RUSSELL	2.11	PB13-8	13-Sep-57	4	15
KS RUSSELL	0.81	PB14-7	20-Sep-57	9	15
KS RUSSELL	8.02	UE8-4	17-Jul-62	3	15
KS SALINE	0.92	TS3-8	29-Apr-52	9	13
KS SALINE	0.92	TS3-8	29-Apr-52	9	13
KS SALINE	3.63	TS6-7	31-May-52	10	13
KS SALINE	12.24	TP4-2	8-Mar-55	9	13
KS SALINE	1.41	TP8-4	12-Apr-55	1	13
KS SALINE	2.33	PB3-3	26-Jun-57	8	13
KS SALINE	0.93	PB3-8	1-Jul-57	4	13
KS SALINE	0.94	PB5-6	20-Jul-57	9	13
KS SALINE	3.55	PB13-4	9-Sep-57	5	13
KS SALINE	26.36	PB13-5	10-Sep-57	6	13
KS SALINE	1.28	PB13-8	13-Sep-57	6	13
KS SALINE	0.81	PB14-7	20-Sep-57	9	13
KS SALINE	10.54	UE8-4	17-Jul-62	2	13
KS SCOTT	0.87	TS4-3	3-May-52	1	14
KS SCOTT	0.87	TS4-3	3-May-52	1	14
KS SCOTT	1.31	UpK6-3	20-Apr-53	9	14
KS SCOTT	7.78	TP3-2	2-Mar-55	3	14

COUNTY	uCi/ sqm	SHOT DAY	DATE	RANK FOR DAY	TIMES IN TOP 10
KS SCOTT	49.44	TP4-2	8-Mar-55	4	14
KS SCOTT	32.33	TP10-2	6-May-55	5	14
KS SCOTT	1.96	PB5-5	19-Jul-57	9	14
KS SCOTT	4.92	PB8-4	21-Aug-57	7	14
KS SCOTT	4.92	PB8-4	21-Aug-57	7	14
KS SCOTT	5.51	PB13-8	13-Sep-57	1	14
KS SCOTT	2.30	PB18-2	8-Oct-57	5	14
KS SCOTT	2.37	PB18-3	9-Oct-57	3	14
KS SCOTT	1.70	PB18-4	10-Oct-57	5	14
KS SCOTT	0.10	UE2-7	11-Mar-62	4	14
KS SEDGWICK	2.48	BJ5-6	24-Nov-51	4	6
KS SEDGWICK	2.48	BJ5-6	24-Nov-51	4	6
KS SEDGWICK	5.65	TS8-3	7-Jun-52	2	6
KS SEDGWICK	22.14	PB12-3	4-Sep-57	2	6
KS SEDGWICK	1.28	PB13-8	13-Sep-57	6	6
KS SEDGWICK	2.50	PB13-9	14-Sep-57	8	6
KS SEWARD	2.48	BJ5-6	24-Nov-51	4	15
KS SEWARD	2.48	BJ5-6	24-Nov-51	4	15
KS SEWARD	1.02	TS4-4	4-May-52	1	15
KS SEWARD	1.02	TS4-4	4-May-52	1	15
KS SEWARD	1.74	TS4-6	6-May-52	5	15
KS SEWARD	1.74	TS4-6	6-May-52	5	15
KS SEWARD	5.46	UpK3-5	4-Apr-53	4	15
KS SEWARD	4.23	UpK6-3	20-Apr-53	2	15
KS SEWARD	25.62	TP7-2	30-Mar-55	4	15
KS SEWARD	31.48	TP10-2	6-May-55	6	15
KS SEWARD	1.28	PB13-8	13-Sep-57	6	15
KS SEWARD	2.30	PB18-2	8-Oct-57	5	15
KS SEWARD	2.37	PB18-3	9-Oct-57	3	15
KS SEWARD	1.70	PB18-4	10-Oct-57	5	15
KS SEWARD	1.92	PB18-6	12-Oct-57	3	15
KS SHAWNEE	12.24	TP4-2	8-Mar-55	9	7
KS SHAWNEE	2.33	PB3-3	26-Jun-57	8	7
KS SHAWNEE	20.88	PB12-4	5-Sep-57	6	7
KS SHAWNEE	3.55	PB13-4	9-Sep-57	5	7

COUNTY	uCi/ sqm	SHOT DAY	DATE	RANK FOR DAY	TIMES IN TOP 10
KS SHAWNEE	1.28	PB13-8	13-Sep-57	6	7
KS SHAWNEE	0.86	PB18-6	12-Oct-57	7	7
KS SHAWNEE	0.12	UE7-5	15-Jul-62	9	7
KS SHERIDAN	0.87	TS4-3	3-May-52	1	14
KS SHERIDAN	0.87	TS4-3	3-May-52	1	14
KS SHERIDAN	1.31	UpK6-3	20-Apr-53	9	14
KS SHERIDAN	7.78	TP3-2	2-Mar-55	3	14
KS SHERIDAN	49.44	TP4-2	8-Mar-55	4	14
KS SHERIDAN	32.33	TP10-2	6-May-55	5	14
KS SHERIDAN	1.54	PB3-2	25-Jun-57	8	14
KS SHERIDAN	2.33	PB3-3	26-Jun-57	8	14
KS SHERIDAN	4.87	PB5-4	18-Jul-57	8	14
KS SHERIDAN	2.30	PB18-2	8-Oct-57	5	14
KS SHERIDAN	2.37	PB18-3	9-Oct-57	3	14
KS SHERIDAN	1.70	PB18-4	10-Oct-57	5	14
KS SHERIDAN	0.10	UE2-7	11-Mar-62	3	14
KS SHERIDAN	4.55	UE8-4	17-Jul-62	7	14
KS SHERMAN	3.23	TP3-2	2-Mar-55	6	8
KS SHERMAN	20.53	TP4-2	8-Mar-55	7	8
KS SHERMAN	14.08	PB2-3	20-Jun-57	6	8
KS SHERMAN	1.96	PB5-5	19-Jul-57	9	8
KS SHERMAN	4.60	PB18-2	8-Oct-57	2	8
KS SHERMAN	2.37	PB18-3	9-Oct-57	3	8
KS SHERMAN	1.70	PB18-4	10-Oct-57	5	8
KS SHERMAN	0.86	PB18-6	12-Oct-57	7	8
KS SMITH	0.92	TS3-8	29-Apr-52	9	13
KS SMITH	0.92	TS3-8	29-Apr-52	9	13
KS SMITH	12.24	TP4-2	8-Mar-55	9	13
KS SMITH	0.86	PB6-11	3-Aug-57	4	13
KS SMITH	3.30	PB8-2	19-Aug-57	5	13
KS SMITH	19.73	PB8-3	20-Aug-57	6	13
KS SMITH	3.78	PB8-5	22-Aug-57	4	13
KS SMITH	3.55	PB13-4	9-Sep-57	5	13
KS SMITH	0.83	PB13-8	13-Sep-57	9	13
KS SMITH	0.81	PB14-7	20-Sep-57	9	13

COUNTY	uCi/ sqm	SHOT DAY	DATE	RANK FOR DAY	TIMES IN TOP 10
KS SMITH	1.29	PB18-6	12-Oct-57	6	13
KS SMITH	0.10	UE2-7	11-Mar-62	4	13
KS SMITH	14.44	UE8-4	17-Jul-62	1	13
KS STAFFORD	1.22	BJ5-6	24-Nov-51	6	8
KS STAFFORD	1.22	BJ5-6	24-Nov-51	6	8
KS STAFFORD	4.30	TS6-7	31-May-52	2	8
KS STAFFORD	0.94	PB5-6	20-Jul-57	9	8
KS STAFFORD	27.63	PB13-5	10-Sep-57	2	8
KS STAFFORD	2.11	PB13-8	13-Sep-57	4	8
KS STAFFORD	0.81	PB14-7	20-Sep-57	9	8
KS STAFFORD	8.02	UE8-4	17-Jul-62	3	8
KS STANTON	0.87	TS4-3	3-May-52	1	15
KS STANTON	0.87	TS4-3	3-May-52	1	15
KS STANTON	1.31	UpK6-3	20-Apr-53	9	15
KS STANTON	7.78	TP3-2	2-Mar-55	3	15
KS STANTON	49.44	TP4-2	8-Mar-55	4	15
KS STANTON	3.49	TP8-3	11-Apr-55	1	15
KS STANTON	32.33	TP10-2	6-May-55	5	15
KS STANTON	15.08	PB2-3	20-Jun-57	5	15
KS STANTON	0.88	PB9-6	28-Aug-57	7	15
KS STANTON	4.79	PB11-2	1-Sep-57	8	15
KS STANTON	3.31	PB13-8	13-Sep-57	2	15
KS STANTON	2.30	PB18-2	8-Oct-57	5	15
KS STANTON	2.37	PB18-3	9-Oct-57	3	15
KS STANTON	1.70	PB18-4	10-Oct-57	5	15
KS STANTON	0.10	UE2-7	11-Mar-62	4	15
KS STEVENS	1.02	TS4-4	4-May-52	1	13
KS STEVENS	1.02	TS4-4	4-May-52	1	13
KS STEVENS	2.73	UpK3-5	4-Apr-53	8	13
KS STEVENS	4.23	UpK6-3	20-Apr-53	2	13
KS STEVENS	153.69	TP7-2	30-Mar-55	1	13
KS STEVENS	1.25	TP9-14	28-Apr-55	3	13
KS STEVENS	31.48	TP10-2	6-May-55	6	13
KS STEVENS	5.36	PB13-4	9-Sep-57	3	13
KS STEVENS	2.23	PB13-7	12-Sep-57	1	13

COUNTY	uCi/ sqm	SHOT DAY	DATE	RANK FOR DAY	TIMES IN TOP 10
KS STEVENS	2.30	PB18-2	8-Oct-57	5	13
KS STEVENS	2.37	PB18-3	9-Oct-57	3	13
KS STEVENS	1.70	PB18-4	10-Oct-57	5	13
KS STEVENS	1.92	PB18-6	12-Oct-57	3	13
KS SUMNER	3.70	BJ5-6	24-Nov-51	2	8
KS SUMNER	3.70	BJ5-6	24-Nov-51	2	8
KS SUMNER	2.24	PB3-2	25-Jun-57	5	8
KS SUMNER	2.69	PB5-7	21-Jul-57	5	8
KS SUMNER	22.94	PB12-3	4-Sep-57	1	8
KS SUMNER	1.28	PB13-8	13-Sep-57	6	8
KS SUMNER	4.15	PB13-9	14-Sep-57	7	8
KS SUMNER	2.42	PB14-7	20-Sep-57	2	8
KS THOMAS	0.87	TS4-3	3-May-52	1	12
KS THOMAS	0.87	TS4-3	3-May-52	1	12
KS THOMAS	1.31	UpK6-3	20-Apr-53	9	12
KS THOMAS	7.78	TP3-2	2-Mar-55	3	12
KS THOMAS	49.44	TP4-2	8-Mar-55	4	12
KS THOMAS	32.33	TP10-2	6-May-55	5	12
KS THOMAS	1.73	PB4-10	14-Jul-57	9	12
KS THOMAS	0.88	PB9-6	28-Aug-57	7	12
KS THOMAS	2.30	PB18-2	8-Oct-57	5	12
KS THOMAS	2.37	PB18-3	9-Oct-57	3	12
KS THOMAS	1.70	PB18-4	10-Oct-57	5	12
KS THOMAS	0.86	PB18-6	12-Oct-57	7	12
KS TREGO	0.87	TS4-3	3-May-52	1	13
KS TREGO	0.87	TS4-3	3-May-52	1	13
KS TREGO	1.31	UpK6-3	20-Apr-53	9	13
KS TREGO	7.78	TP3-2	2-Mar-55	3	13
KS TREGO	49.44	TP4-2	8-Mar-55	4	13
KS TREGO	32.33	TP10-2	6-May-55	5	13
KS TREGO	2.33	PB3-3	26-Jun-57	8	13
KS TREGO	2.11	PB13-8	13-Sep-57	4	13
KS TREGO	2.30	PB18-2	8-Oct-57	5	13
KS TREGO	2.37	PB18-3	9-Oct-57	3	13
KS TREGO	1.70	PB18-4	10-Oct-57	5	13

COUNTY	uCi/ sqm	SHOT DAY	DATE	RANK FOR DAY	TIMES IN TOP 10
KS TREGO	0.10	UE2-7	11-Mar-62	4	13
KS TREGO	4.55	UE8-4	17-Jul-62	7	13
KS WABAUNSEE	12.24	TP4-2	8-Mar-55	9	10
KS WABAUNSEE	1.54	PB3-2	25-Jun-57	8	10
KS WABAUNSEE	2.33	PB3-3	26-Jun-57	8	10
KS WABAUNSEE	1.65	PB5-6	20-Jul-57	5	10
KS WABAUNSEE	1.63	PB5-7	21-Jul-57	8	10
KS WABAUNSEE	3.55	PB13-4	9-Sep-57	5	10
KS WABAUNSEE	1.28	PB13-8	13-Sep-57	6	10
KS WABAUNSEE	2.42	PB14-7	20-Sep-57	2	10
KS WABAUNSEE	0.08	UE2-7	11-Mar-62	5	10
KS WABAUNSEE	0.12	UE7-5	15-Jul-62	9	10
KS WALLACE	0.87	TS4-3	3-May-52	1	15
KS WALLACE	0.87	TS4-3	3-May-52	1	15
KS WALLACE	1.31	UpK6-3	20-Apr-53	9	15
KS WALLACE	7.78	TP3-2	2-Mar-55	3	15
KS WALLACE	49.44	TP4-2	8-Mar-55	4	15
KS WALLACE	32.33	TP10-2	6-May-55	5	15
KS WALLACE	2.22	PB1-11	7-Jun-57	2	15
KS WALLACE	0.83	PB3-8	1-Jul-57	5	15
KS WALLACE	4.79	PB11-2	1-Sep-57	8	15
KS WALLACE	2.20	PB13-8	13-Sep-57	3	15
KS WALLACE	3.51	PB18-2	8-Oct-57	4	15
KS WALLACE	2.37	PB18-3	9-Oct-57	3	15
KS WALLACE	1.70	PB18-4	10-Oct-57	5	15
KS WALLACE	1.68	PB18-6	12-Oct-57	4	15
KS WALLACE	0.10	UE2-7	11-Mar-62	4	15
KS WASHINGTON	12.24	TP4-2	8-Mar-55	9	6
KS WASHINGTON	1.09	TP8-4	12-Apr-55	9	6
KS WASHINGTON	7.99	PB8-3	20-Aug-57	9	6
KS WASHINGTON	1.62	PB14-7	20-Sep-57	4	6
KS WASHINGTON	0.24	UE7-5	15-Jul-62	5	6
KS WASHINGTON	4.12	UE8-4	17-Jul-62	9	6
KS WICHITA	0.87	TS4-3	3-May-52	1	16
KS WICHITA	0.87	TS4-3	3-May-52	1	16

COUNTY	uCi/ sqm	SHOT DAY	DATE	RANK FOR DAY	TIMES IN TOP 10
KS WICHITA	1.31	UpK6-3	20-Apr-53	9	16
KS WICHITA	7.78	TP3-2	2-Mar-55	3	16
KS WICHITA	49.44	TP4-2	8-Mar-55	4	16
KS WICHITA	32.33	TP10-2	6-May-55	5	16
KS WICHITA	1.33	PB1-11	7-Jun-57	3	16
KS WICHITA	4.92	PB8-4	21-Aug-57	7	16
KS WICHITA	4.92	PB8-4	21-Aug-57	7	16
KS WICHITA	1.47	PB9-6	28-Aug-57	3	16
KS WICHITA	4.79	PB11-2	1-Sep-57	8	16
KS WICHITA	3.31	PB13-8	13-Sep-57	2	16
KS WICHITA	2.30	PB18-2	8-Oct-57	5	16
KS WICHITA	2.37	PB18-3	9-Oct-57	3	16
KS WICHITA	1.70	PB18-4	10-Oct-57	5	16
KS WICHITA	0.86	PB18-6	12-Oct-57	7	16
KS WILSON	8.76	BJ2-2	31-Oct-51	3	6
KS WILSON	23.58	PB12-4	5-Sep-57	1	6
KS WILSON	0.83	PB13-8	13-Sep-57	9	6
KS WILSON	2.50	PB13-9	14-Sep-57	8	6
KS WILSON	4.04	PB14-7	20-Sep-57	1	6
KS WILSON	0.11	UE7-5	15-Jul-62	10	6
KS WOODSON	5.84	BJ2-2	31-Oct-51	7	5
KS WOODSON	1.28	PB13-8	13-Sep-57	6	5
KS WOODSON	1.64	PB13-9	14-Sep-57	10	5
KS WOODSON	4.04	PB14-7	20-Sep-57	1	5
KS WOODSON	0.11	UE7-5	15-Jul-62	10	5
KS WYANDOTTE	3.86	UpK11-3	6-Jun-53	5	9
KS WYANDOTTE	1.48	PB3-8	1-Jul-57	2	9
KS WYANDOTTE	10.65	PB4-5	9-Jul-57	10	9
KS WYANDOTTE	2.48	PB5-7	21-Jul-57	6	9
KS WYANDOTTE	7.99	PB8-3	20-Aug-57	9	9
KS WYANDOTTE	2.02	PB15-5	20-Sep-57	3	9
KS WYANDOTTE	5.26	PB18-8	14-Oct-57	3	9
KS WYANDOTTE	0.08	UE2-7	11-Mar-62	5	9
KS WYANDOTTE	0.12	UE7-5	15-Jul-62	9	9

COUNTY	uCi/ sqm	SHOT DAY	DATE	RANK FOR DAY	TIMES IN TOP 10
KY ADAIR	2.41	TP10-6	10-May-55	4	3
KY ADAIR	16.75	PB4-5	9-Jul-57	4	3
KY ADAIR	2.58	UE8-6	19-Jul-62	4	3
KY ANDERSON	1.02	UpK3-6	5-Apr-53	7	1
KY BALLARD	0.87	UpK6-4	21-Apr-53	6	2
KY BALLARD	4.85	PB5-3	17-Jul-57	9	2
KY BATH	1.34	PB13-7	12-Sep-57	2	2
KY BATH	208.78	S-4	9-Jul-62	3	2
KY BELL	1.04	UpK5-2	12-Apr-53	4	6
KY BELL	3.11	TP6-4	25-Mar-55	3	6
KY BELL	0.97	TP6-5	26-Mar-55	8	6
KY BELL	4.18	PB2-5	22-Jun-57	3	6
KY BELL	2.23	PB5-8	22-Jul-57	6	6
KY BELL	3.79	PB17-3	30-Sep-57	10	6
KY BOONE	1.02	UpK3-6	5-Apr-53	7	3
KY BOONE	15.70	PB6-3	26-Jul-57	5	3
KY BOONE	90.29	S-4	9-Jul-62	8	3
KY BOURBON	0.89	PB13-7	12-Sep-57	5	2
KY BOURBON	1.08	PB15-7	22-Sep-57	5	2
KY BOYD	2.03	TP6-4	25-Mar-55	8	7
KY BOYD	0.97	TP6-5	26-Mar-55	8	7
KY BOYD	2.99	PB5-8	22-Jul-57	4	7
KY BOYD	23.55	PB6-3	26-Jul-57	3	7
KY BOYD	2.23	PB13-7	12-Sep-57	1	7
KY BOYD	1.08	PB15-7	22-Sep-57	5	7
KY BOYD	208.78	S-4	9-Jul-62	3	7
KY BOYLE	16.75	PB4-5	9-Jul-57	4	2
KY BOYLE	2.54	PB17-2	29-Sep-57	2	2
KY BRACKEN	0.89	PB13-7	12-Sep-57	5	3
KY BRACKEN	1.08	PB15-7	22-Sep-57	5	3
KY BRACKEN	85.56	S-4	9-Jul-62	9	3
KY BREATHITT	0.97	TP6-5	26-Mar-55	8	7
KY BREATHITT	1.39	PB2-5	22-Jun-57	8	7
KY BREATHITT	5.97	PB5-8	22-Jul-57	3	7
KY BREATHITT	23.55	PB6-3	26-Jul-57	3	7

COUNTY	uCi/ sqm	SHOT DAY	DATE	RANK FOR DAY	TIMES IN TOP 10
KY BREATHITT	2.06	PB6-7	30-Jul-57	5	7
KY BREATHITT	5.64	PB17-3	30-Sep-57	8	7
KY BREATHITT	164.78	S-4	9-Jul-62	4	7
KY BRECKINRIDGE	1.02	UpK3-6	5-Apr-53	7	1
KY BULLITT	1.56	UpK3-6	5-Apr-53	3	1
KY CALDWELL	7.27	PB5-3	17-Jul-57	5	3
KY CALDWELL	2.73	PB9-2	24-Aug-57	6	3
KY CALDWELL	0.87	PB18-7	13-Oct-57	8	3
KY CALLOWAY	1.17	PB2-4	21-Jun-57	10	2
KY CALLOWAY	1.27	PB3-9	2-Jul-57	2	2
KY CAMPBELL	1.02	UpK3-6	5-Apr-53	7	5
KY CAMPBELL	7.85	PB6-3	26-Jul-57	7	5
KY CAMPBELL	9.38	PB6-4	27-Jul-57	1	5
KY CAMPBELL	1.08	PB15-7	22-Sep-57	5	5
KY CAMPBELL	208.78	S-4	9-Jul-62	3	5
KY CARLISLE	1.22	BJ5-6	24-Nov-51	6	8
KY CARLISLE	1.22	BJ5-6	24-Nov-51	6	8
KY CARLISLE	12.19	UpK2-7	30-Mar-53	4	8
KY CARLISLE	2.17	UpK3-6	5-Apr-53	2	8
KY CARLISLE	3.68	UpK8-3	10-May-53	4	8
KY CARLISLE	1.09	TP4-5	11-Mar-55	1	8
KY CARLISLE	1.21	PB1-5	1-Jun-57	10	8
KY CARLISLE	0.83	PB3-9	2-Jul-57	4	8
KY CARROLL	208.78	S-4	9-Jul-62	3	1
KY CARTER	0.97	TP6-5	26-Mar-55	8	5
KY CARTER	8.96	PB5-8	22-Jul-57	2	5
KY CARTER	2.23	PB13-7	12-Sep-57	1	5
KY CARTER	1.08	PB15-7	22-Sep-57	5	5
KY CARTER	208.78	S-4	9-Jul-62	3	5
KY CASEY	1.02	UpK3-6	5-Apr-53	7	3
KY CASEY	11.16	PB4-5	9-Jul-57	9	3
KY CASEY	2.54	PB17-2	29-Sep-57	2	3
KY CHRISTIAN	16.75	PB4-5	9-Jul-57	4	2
KY CHRISTIAN	2.73	PB9-2	24-Aug-57	6	2
KY CLARK	1.34	PB13-7	12-Sep-57	2	2

COUNTY	uCi/ sqm	SHOT DAY	DATE	RANK FOR DAY	TIMES IN TOP 10
KY CLARK	149.13	S-4	9-Jul-62	5	2
KY CLAY	3.11	TP6-4	25-Mar-55	3	4
KY CLAY	0.97	TP6-5	26-Mar-55	8	4
KY CLAY	3.79	PB17-3	30-Sep-57	10	4
KY CLAY	1.92	UE8-5	18-Jul-62	9	4
KY CLINTON	2.03	TP6-4	25-Mar-55	8	4
KY CLINTON	0.97	TP6-5	26-Mar-55	8	4
KY CLINTON	4.18	PB2-5	22-Jun-57	3	4
KY CLINTON	3.79	PB17-3	30-Sep-57	10	4
KY CRITTENDEN	1.17	PB2-4	21-Jun-57	10	1
KY CUMBERLAND	1.56	UpK3-6	5-Apr-53	3	2
KY CUMBERLAND	27.91	PB4-5	9-Jul-57	1	2
KY DAVIESS	1.56	UpK3-6	5-Apr-53	3	2
KY DAVIESS	0.99	PB6-10	2-Aug-57	3	2
KY EDMONSON	1.56	UpK3-6	5-Apr-53	3	1
KY ELLIOTT	2.03	TP6-4	25-Mar-55	8	7
KY ELLIOTT	0.97	TP6-5	26-Mar-55	8	7
KY ELLIOTT	1.51	PB5-8	22-Jul-57	7	7
KY ELLIOTT	2.45	PB6-7	30-Jul-57	3	7
KY ELLIOTT	2.23	PB13-7	12-Sep-57	1	7
KY ELLIOTT	1.08	PB15-7	22-Sep-57	5	7
KY ELLIOTT	208.78	S-4	9-Jul-62	3	7
KY ESTILL	0.97	TP6-5	26-Mar-55	8	5
KY ESTILL	23.55	PB6-3	26-Jul-57	3	5
KY ESTILL	2.23	PB13-7	12-Sep-57	1	5
KY ESTILL	1.08	PB15-7	22-Sep-57	5	5
KY ESTILL	94.45	S-4	9-Jul-62	7	5
KY FAYETTE	1.56	UpK3-6	5-Apr-53	3	3
KY FAYETTE	1.17	PB2-4	21-Jun-57	10	3
KY FAYETTE	1.34	PB13-7	12-Sep-57	2	3
KY FLEMING	1.02	UpK3-6	5-Apr-53	7	4
KY FLEMING	1.34	PB13-7	12-Sep-57	2	4
KY FLEMING	1.08	PB15-7	22-Sep-57	5	4
KY FLEMING	208.78	S-4	9-Jul-62	3	4
KY FLOYD	1.04	UpK5-2	12-Apr-53	4	9

COUNTY	uCi/ sqm	SHOT DAY	DATE	RANK FOR DAY	TIMES IN TOP 10
KY FLOYD	19.39	UpK9-4	22-May-53	6	9
KY FLOYD	2.03	TP6-4	25-Mar-55	8	9
KY FLOYD	0.97	TP6-5	26-Mar-55	8	9
KY FLOYD	6.96	PB2-5	22-Jun-57	1	9
KY FLOYD	2.99	PB5-8	22-Jul-57	4	9
KY FLOYD	14.98	PB17-3	30-Sep-57	2	9
KY FLOYD	149.13	S-4	9-Jul-62	5	9
KY FLOYD	1.92	UE8-5	18-Jul-62	9	9
KY FRANKLIN	149.13	S-4	9-Jul-62	5	1
KY FULTON	18.26	UpK2-7	30-Mar-53	2	4
KY FULTON	21.07	UpK2-8	31-Mar-53	3	4
KY FULTON	3.68	UpK8-3	10-May-53	4	4
KY FULTON	2.73	PB9-2	24-Aug-57	6	4
KY GALLATIN	1.47	PB6-7	30-Jul-57	10	3
KY GALLATIN	0.89	PB13-7	12-Sep-57	5	3
KY GALLATIN	208.78	S-4	9-Jul-62	3	3
KY GARRARD	1.02	UpK3-6	5-Apr-53	7	2
KY GARRARD	16.75	PB4-5	9-Jul-57	4	2
KY GRANT	1.56	UpK3-6	5-Apr-53	3	3
KY GRANT	1.79	PB15-7	22-Sep-57	2	3
KY GRANT	85.56	S-4	9-Jul-62	9	3
KY GRAVES	0.83	PB3-9	2-Jul-57	4	2
KY GRAVES	2.73	PB9-2	24-Aug-57	6	2
KY GRAYSON	1.02	UpK3-6	5-Apr-53	7	1
KY GREEN	2.58	UE8-6	19-Jul-62	4	1
KY GREENUP	1.04	UpK5-2	12-Apr-53	4	7
KY GREENUP	3.11	TP6-4	25-Mar-55	3	7
KY GREENUP	0.97	TP6-5	26-Mar-55	8	7
KY GREENUP	2.99	PB5-8	22-Jul-57	4	7
KY GREENUP	1.34	PB13-7	12-Sep-57	2	7
KY GREENUP	1.08	PB15-7	22-Sep-57	5	7
KY GREENUP	359.52	S-4	9-Jul-62	1	7
KY HANCOCK	1.02	UpK3-6	5-Apr-53	7	1
KY HARDIN	1.02	UpK3-6	5-Apr-53	7	1
KY HARLAN	1.55	UpK5-2	12-Apr-53	2	7

COUNTY	uCi/ sqm	SHOT DAY	DATE	RANK FOR DAY	TIMES IN TOP 10
KY HARLAN	3.11	TP6-4	25-Mar-55	3	7
KY HARLAN	1.93	TP6-5	26-Mar-55	4	7
KY HARLAN	4.18	PB2-5	22-Jun-57	3	7
KY HARLAN	1.51	PB5-8	22-Jul-57	7	7
KY HARLAN	3.79	PB17-3	30-Sep-57	10	7
KY HARLAN	77.90	S-4	9-Jul-62	10	7
KY HARRISON	1.19	PB2-5	22-Jun-57	9	4
KY HARRISON	0.89	PB13-7	12-Sep-57	5	4
KY HARRISON	1.08	PB15-7	22-Sep-57	5	4
KY HARRISON	208.78	S-4	9-Jul-62	3	4
KY HART	1.02	UpK3-6	5-Apr-53	7	1
KY HENDERSON	7.27	PB5-3	17-Jul-57	5	1
KY HENRY	1.02	UpK3-6	5-Apr-53	7	2
KY HENRY	208.78	S-4	9-Jul-62	3	2
KY HICKMAN	18.26	UpK2-7	30-Mar-53	2	6
KY HICKMAN	21.07	UpK2-8	31-Mar-53	3	6
KY HICKMAN	1.49	UpK3-6	5-Apr-53	5	6
KY HICKMAN	3.68	UpK8-3	10-May-53	4	6
KY HICKMAN	1.27	PB3-9	2-Jul-57	2	6
KY HICKMAN	2.73	PB9-2	24-Aug-57	6	6
KY HOPKINS	1.65	PB4-10	14-Jul-57	10	2
KY HOPKINS	4.85	PB5-3	17-Jul-57	9	2
KY JACKSON	3.11	TP6-4	25-Mar-55	3	5
KY JACKSON	0.97	TP6-5	26-Mar-55	8	5
KY JACKSON	1.39	PB2-5	22-Jun-57	8	5
KY JACKSON	3.79	PB17-3	30-Sep-57	10	5
KY JACKSON	164.78	S-4	9-Jul-62	4	5
KY JESSAMINE	1.56	UpK3-6	5-Apr-53	3	1
KY JOHNSON	1.04	UpK5-2	12-Apr-53	4	9
KY JOHNSON	3.11	TP6-4	25-Mar-55	3	9
KY JOHNSON	0.97	TP6-5	26-Mar-55	8	9
KY JOHNSON	1.19	PB2-5	22-Jun-57	9	9
KY JOHNSON	1.51	PB5-8	22-Jul-57	7	9
KY JOHNSON	2.23	PB13-7	12-Sep-57	1	9
KY JOHNSON	1.08	PB15-7	22-Sep-57	5	9

COUNTY	uCi/ sqm	SHOT DAY	DATE	RANK FOR DAY	TIMES IN TOP 10
KY JOHNSON	149.13	S-4	9-Jul-62	5	9
KY JOHNSON	2.03	UE8-5	18-Jul-62	8	9
KY KENTON	1.02	UpK3-6	5-Apr-53	7	5
KY KENTON	23.55	PB6-3	26-Jul-57	3	5
KY KENTON	6.39	PB6-4	27-Jul-57	6	5
KY KENTON	1.34	PB13-7	12-Sep-57	2	5
KY KENTON	208.78	S-4	9-Jul-62	3	5
KY KNOTT	3.11	TP6-4	25-Mar-55	3	7
KY KNOTT	0.97	TP6-5	26-Mar-55	8	7
KY KNOTT	4.18	PB2-5	22-Jun-57	3	7
KY KNOTT	1.51	PB5-8	22-Jul-57	7	7
KY KNOTT	7.85	PB6-3	26-Jul-57	7	7
KY KNOTT	3.79	PB17-3	30-Sep-57	10	7
KY KNOTT	94.45	S-4	9-Jul-62	7	7
KY KNOX	3.11	TP6-4	25-Mar-55	3	4
KY KNOX	0.97	TP6-5	26-Mar-55	8	4
KY KNOX	4.18	PB2-5	22-Jun-57	3	4
KY KNOX	3.79	PB17-3	30-Sep-57	10	4
KY LARUE	1.02	UpK3-6	5-Apr-53	7	1
KY LAUREL	1.83	TP6-3	24-Mar-55	7	4
KY LAUREL	0.97	TP6-5	26-Mar-55	8	4
KY LAUREL	6.96	PB2-5	22-Jun-57	1	4
KY LAUREL	3.79	PB17-3	30-Sep-57	10	4
KY LAWRENCE	2.03	TP6-4	25-Mar-55	8	6
KY LAWRENCE	0.97	TP6-5	26-Mar-55	8	6
KY LAWRENCE	2.99	PB5-8	22-Jul-57	4	6
KY LAWRENCE	2.23	PB13-7	12-Sep-57	1	6
KY LAWRENCE	1.08	PB15-7	22-Sep-57	5	6
KY LAWRENCE	359.52	S-4	9-Jul-62	1	6
KY LEE	2.03	TP6-4	25-Mar-55	8	4
KY LEE	0.97	TP6-5	26-Mar-55	8	4
KY LEE	7.85	PB6-3	26-Jul-57	7	4
KY LEE	3.79	PB17-3	30-Sep-57	10	4
KY LESLIE	3.11	TP6-4	25-Mar-55	3	8
KY LESLIE	0.97	TP6-5	26-Mar-55	8	8

COUNTY	uCi/ sqm	SHOT DAY	DATE	RANK FOR DAY	TIMES IN TOP 10
KY LESLIE	4.18	PB2-5	22-Jun-57	3	8
KY LESLIE	14.93	PB5-8	22-Jul-57	1	8
KY LESLIE	3.79	PB17-3	30-Sep-57	10	8
KY LESLIE	77.90	S-4	9-Jul-62	10	8
KY LESLIE	47.25	S-5	10-Jul-62	2	8
KY LESLIE	1.92	UE8-5	18-Jul-62	9	8
KY LETCHER	3.11	TP6-4	25-Mar-55	3	8
KY LETCHER	0.97	TP6-5	26-Mar-55	8	8
KY LETCHER	4.18	PB2-5	22-Jun-57	3	8
KY LETCHER	1.51	PB5-8	22-Jul-57	7	8
KY LETCHER	7.85	PB6-3	26-Jul-57	7	8
KY LETCHER	3.79	PB17-3	30-Sep-57	10	8
KY LETCHER	94.45	S-4	9-Jul-62	7	8
KY LETCHER	1.92	UE8-5	18-Jul-62	9	8
KY LEWIS	1.47	PB6-7	30-Jul-57	10	4
KY LEWIS	1.34	PB13-7	12-Sep-57	2	4
KY LEWIS	1.79	PB15-7	22-Sep-57	2	4
KY LEWIS	359.52	S-4	9-Jul-62	1	4
KY LINCOLN	1.02	UpK3-6	5-Apr-53	7	2
KY LINCOLN	16.75	PB4-5	9-Jul-57	4	2
KY LIVINGSTON	0.87	UpK6-4	21-Apr-53	6	5
KY LIVINGSTON	1.06	TP7-8	5-Apr-55	5	5
KY LIVINGSTON	7.27	PB5-3	17-Jul-57	5	5
KY LIVINGSTON	2.73	PB9-2	24-Aug-57	6	5
KY LIVINGSTON	0.89	PB13-7	12-Sep-57	5	5
KY LYON	11.16	PB4-5	9-Jul-57	9	3
KY LYON	1.65	PB4-10	14-Jul-57	10	3
KY LYON	2.73	PB9-2	24-Aug-57	6	3
KY MADISON	1.02	UpK3-6	5-Apr-53	7	5
KY MADISON	15.70	PB6-3	26-Jul-57	5	5
KY MADISON	2.23	PB13-7	12-Sep-57	1	5
KY MADISON	1.08	PB15-7	22-Sep-57	5	5
KY MADISON	94.45	S-4	9-Jul-62	7	5
KY MAGOFFIN	3.11	TP6-4	25-Mar-55	3	6
KY MAGOFFIN	0.97	TP6-5	26-Mar-55	8	6

COUNTY	uCi/ sqm	SHOT DAY	DATE	RANK FOR DAY	TIMES IN TOP 10
KY MAGOFFIN	4.18	PB2-5	22-Jun-57	3	6
KY MAGOFFIN	8.96	PB5-8	22-Jul-57	2	6
KY MAGOFFIN	3.79	PB17-3	30-Sep-57	10	6
KY MAGOFFIN	149.13	S-4	9-Jul-62	5	6
KY MARSHALL	2.73	PB9-2	24-Aug-57	6	1
KY MARTIN	3.11	TP6-4	25-Mar-55	3	5
KY MARTIN	0.97	TP6-5	26-Mar-55	8	5
KY MARTIN	4.18	PB2-5	22-Jun-57	3	5
KY MARTIN	14.93	PB5-8	22-Jul-57	1	5
KY MARTIN	3.79	PB17-3	30-Sep-57	10	5
KY MASON	1.34	PB13-7	12-Sep-57	2	3
KY MASON	1.08	PB15-7	22-Sep-57	5	3
KY MASON	208.78	S-4	9-Jul-62	3	3
KY MCCRACKEN	0.87	UpK6-4	21-Apr-53	6	4
KY MCCRACKEN	1.61	TP7-8	5-Apr-55	3	4
KY MCCRACKEN	7.27	PB5-3	17-Jul-57	5	4
KY MCCRACKEN	2.73	PB9-2	24-Aug-57	6	4
KY MCCREARY	3.11	TP6-4	25-Mar-55	3	4
KY MCCREARY	0.97	TP6-5	26-Mar-55	8	4
KY MCCREARY	1.39	PB2-5	22-Jun-57	8	4
KY MCCREARY	3.79	PB17-3	30-Sep-57	10	4
KY MCLEAN	7.27	PB5-3	17-Jul-57	5	1
KY MEADE	1.56	UpK3-6	5-Apr-53	3	1
KY MENIFEE	2.03	TP6-4	25-Mar-55	8	6
KY MENIFEE	0.97	TP6-5	26-Mar-55	8	6
KY MENIFEE	39.25	PB6-3	26-Jul-57	1	6
KY MENIFEE	2.23	PB13-7	12-Sep-57	1	6
KY MENIFEE	1.08	PB15-7	22-Sep-57	5	6
KY MENIFEE	149.13	S-4	9-Jul-62	5	6
KY MERCER	1.02	UpK3-6	5-Apr-53	7	2
KY MERCER	16.75	PB4-5	9-Jul-57	4	2
KY METCALFE	1.02	UpK3-6	5-Apr-53	7	2
KY METCALFE	11.16	PB4-5	9-Jul-57	9	2
KY MONROE	16.75	PB4-5	9-Jul-57	4	2
KY MONROE	2.54	PB17-2	29-Sep-57	2	2

COUNTY	uCi/ sqm	SHOT DAY	DATE	RANK FOR DAY	TIMES IN TOP 10
KY MONTGOMERY	1.56	UpK3-6	5-Apr-53	3	2
KY MONTGOMERY	1.34	PB13-7	12-Sep-57	2	2
KY MORGAN	2.03	TP6-4	25-Mar-55	8	7
KY MORGAN	0.97	TP6-5	26-Mar-55	8	7
KY MORGAN	1.51	PB5-8	22-Jul-57	7	7
KY MORGAN	2.45	PB6-7	30-Jul-57	3	7
KY MORGAN	2.23	PB13-7	12-Sep-57	1	7
KY MORGAN	1.08	PB15-7	22-Sep-57	5	7
KY MORGAN	149.13	S-4	9-Jul-62	5	7
KY MUHLENBERG	1.39	TP10-7	11-May-55	4	1
KY NELSON	1.56	UpK3-6	5-Apr-53	3	1
KY NICHOLAS	1.56	UpK3-6	5-Apr-53	3	4
KY NICHOLAS	0.89	PB13-7	12-Sep-57	5	4
KY NICHOLAS	1.08	PB15-7	22-Sep-57	5	4
KY NICHOLAS	208.78	S-4	9-Jul-62	3	4
KY OHIO	1.56	UpK3-6	5-Apr-53	3	1
KY OWEN	1.02	UpK3-6	5-Apr-53	7	4
KY OWEN	1.34	PB13-7	12-Sep-57	2	4
KY OWEN	1.08	PB15-7	22-Sep-57	5	4
KY OWEN	208.78	S-4	9-Jul-62	3	4
KY OWSLEY	2.03	TP6-4	25-Mar-55	8	5
KY OWSLEY	0.97	TP6-5	26-Mar-55	8	5
KY OWSLEY	7.85	PB6-3	26-Jul-57	7	5
KY OWSLEY	3.79	PB17-3	30-Sep-57	10	5
KY OWSLEY	164.78	S-4	9-Jul-62	4	5
KY PENDLETON	1.02	UpK3-6	5-Apr-53	7	6
KY PENDLETON	1.19	PB2-5	22-Jun-57	9	6
KY PENDLETON	7.98	PB6-4	27-Jul-57	2	6
KY PENDLETON	0.89	PB13-7	12-Sep-57	5	6
KY PENDLETON	1.08	PB15-7	22-Sep-57	5	6
KY PENDLETON	85.56	S-4	9-Jul-62	9	6
KY PERRY	1.04	UpK5-2	12-Apr-53	4	10
KY PERRY	17.36	UpK9-4	22-May-53	7	10
KY PERRY	2.03	TP6-4	25-Mar-55	8	10
KY PERRY	1.93	TP6-5	26-Mar-55	4	10

COUNTY	uCi/ sqm	SHOT DAY	DATE	RANK FOR DAY	TIMES IN TOP 10
KY PERRY	2.78	PB2-5	22-Jun-57	4	10
KY PERRY	1.51	PB5-8	22-Jul-57	7	10
KY PERRY	7.85	PB6-3	26-Jul-57	7	10
KY PERRY	3.79	PB17-3	30-Sep-57	10	10
KY PERRY	94.45	S-4	9-Jul-62	7	10
KY PERRY	1.92	UE8-5	18-Jul-62	9	10
KY PIKE	1.04	UpK5-2	12-Apr-53	4	7
KY PIKE	2.03	TP6-4	25-Mar-55	8	7
KY PIKE	0.97	TP6-5	26-Mar-55	8	7
KY PIKE	2.78	PB2-5	22-Jun-57	4	7
KY PIKE	8.96	PB5-8	22-Jul-57	2	7
KY PIKE	7.49	PB17-3	30-Sep-57	5	7
KY PIKE	149.13	S-4	9-Jul-62	5	7
KY POWELL	0.97	TP6-5	26-Mar-55	8	5
KY POWELL	23.55	PB6-3	26-Jul-57	3	5
KY POWELL	2.23	PB13-7	12-Sep-57	1	5
KY POWELL	1.08	PB15-7	22-Sep-57	5	5
KY POWELL	263.65	S-4	9-Jul-62	2	5
KY PULASKI	0.97	TP6-5	26-Mar-55	8	3
KY PULASKI	2.78	PB2-5	22-Jun-57	4	3
KY PULASKI	3.79	PB17-3	30-Sep-57	10	3
KY ROBERTSON	0.89	PB13-7	12-Sep-57	5	3
KY ROBERTSON	1.08	PB15-7	22-Sep-57	5	3
KY ROBERTSON	208.78	S-4	9-Jul-62	3	3
KY ROCKCASTLE	3.11	TP6-4	25-Mar-55	3	6
KY ROCKCASTLE	0.97	TP6-5	26-Mar-55	8	6
KY ROCKCASTLE	1.39	PB2-5	22-Jun-57	8	6
KY ROCKCASTLE	3.79	PB17-3	30-Sep-57	10	6
KY ROCKCASTLE	94.45	S-4	9-Jul-62	7	6
KY ROCKCASTLE	1.92	UE8-5	18-Jul-62	9	6
KY ROWAN	3.11	TP6-4	25-Mar-55	3	6
KY ROWAN	0.97	TP6-5	26-Mar-55	8	6
KY ROWAN	15.70	PB6-3	26-Jul-57	5	6
KY ROWAN	2.23	PB13-7	12-Sep-57	1	6
KY ROWAN	1.08	PB15-7	22-Sep-57	5	6

COUNTY	uCi/ sqm	SHOT DAY	DATE	RANK FOR DAY	TIMES IN TOP 10
KY ROWAN	208.78	S-4	9-Jul-62	3	6
KY RUSSELL	16.75	PB4-5	9-Jul-57	4	1
KY SCOTT	1.02	UpK3-6	5-Apr-53	7	3
KY SCOTT	1.34	PB13-7	12-Sep-57	2	3
KY SCOTT	149.13	S-4	9-Jul-62	5	3
KY SHELBY	1.02	UpK3-6	5-Apr-53	7	2
KY SHELBY	149.13	S-4	9-Jul-62	5	2
KY SIMPSON	0.86	UpK1-6	22-Mar-53	8	2
KY SIMPSON	16.75	PB4-5	9-Jul-57	4	2
KY SPENCER	1.02	UpK3-6	5-Apr-53	7	1
KY TAYLOR	1.56	UpK3-6	5-Apr-53	3	1
KY TODD	16.75	PB4-5	9-Jul-57	4	1
KY TRIGG	2.73	PB9-2	24-Aug-57	6	1
KY TRIMBLE	1.56	UpK3-6	5-Apr-53	3	2
KY TRIMBLE	359.52	S-4	9-Jul-62	1	2
KY UNION	7.27	PB5-3	17-Jul-57	5	1
KY WARREN	2.12	TP10-6	10-May-55	8	3
KY WARREN	1.17	PB2-4	21-Jun-57	10	3
KY WARREN	1.86	PB2-5	22-Jun-57	5	3
KY WASHINGTON	2.21	TP10-6	10-May-55	6	1
KY WAYNE	2.03	TP6-4	25-Mar-55	8	5
KY WAYNE	0.97	TP6-5	26-Mar-55	8	5
KY WAYNE	2.78	PB2-5	22-Jun-57	4	5
KY WAYNE	13.06	PB4-5	9-Jul-57	7	5
KY WAYNE	3.79	PB17-3	30-Sep-57	10	5
KY WHITLEY	1.04	UpK5-2	12-Apr-53	4	5
KY WHITLEY	1.83	TP6-3	24-Mar-55	7	5
KY WHITLEY	3.11	TP6-4	25-Mar-55	3	5
KY WHITLEY	0.97	TP6-5	26-Mar-55	8	5
KY WHITLEY	3.79	PB17-3	30-Sep-57	10	5
KY WOLFE	1.04	UpK5-2	12-Apr-53	4	6
KY WOLFE	0.97	TP6-5	26-Mar-55	8	6
KY WOLFE	23.55	PB6-3	26-Jul-57	3	6
KY WOLFE	1.47	PB6-7	30-Jul-57	10	6
KY WOLFE	2.23	PB13-7	12-Sep-57	1	6

COUNTY	uCi/ sqm	SHOT DAY	DATE	RANK FOR DAY	TIMES IN TOP 10
KY WOLFE	149.13	S-4	9-Jul-62	5	6
KY WOODFORD	1.02	UpK3-6	5-Apr-53	7	3
KY WOODFORD	2.54	PB17-2	29-Sep-57	2	3
KY WOODFORD	149.13	S-4	9-Jul-62	5	3

COUNTY	uCi/ sqm	SHOT DAY	DATE	RANK FOR DAY	TIMES IN TOP 10
LA ACADIA	79.98	UpK6-2	19-Apr-53	1	6
LA ACADIA	2.72	UpK6-3	20-Apr-53	4	6
LA ACADIA	0.81	UpK6-7	24-Apr-53	1	6
LA ACADIA	1.34	UpK8-5	12-May-53	3	6
LA ACADIA	6.40	TP6-3	24-Mar-55	3	6
LA ACADIA	114.67	PB3-3	26-Jun-57	1	6
LA ALLEN	79.98	UpK6-2	19-Apr-53	1	6
LA ALLEN	2.72	UpK6-3	20-Apr-53	4	6
LA ALLEN	0.81	UpK6-7	24-Apr-53	1	6
LA ALLEN	1.34	UpK8-5	12-May-53	3	6
LA ALLEN	6.40	TP6-3	24-Mar-55	3	6
LA ALLEN	68.82	PB3-3	26-Jun-57	2	6
LA ASCENSION	1.01	TS6-5	29-May-52	7	4
LA ASCENSION	38.63	UpK6-2	19-Apr-53	2	4
LA ASCENSION	68.82	PB3-3	26-Jun-57	2	4
LA ASCENSION	1.99	PB6-7	30-Jul-57	6	4
LA ASSUMPTION	38.63	UpK6-2	19-Apr-53	2	2
LA ASSUMPTION	114.67	PB3-3	26-Jun-57	1	2
LA AVOYELLES	0.99	UpK5-5	15-Apr-53	2	7
LA AVOYELLES	1.16	UpK8-5	12-May-53	5	7
LA AVOYELLES	1.47	TP6-2	23-Mar-55	9	7
LA AVOYELLES	4.11	TP6-3	24-Mar-55	4	7
LA AVOYELLES	114.67	PB3-3	26-Jun-57	1	7
LA AVOYELLES	1.99	PB6-7	30-Jul-57	6	7
LA AVOYELLES	2.34	PB6-8	31-Jul-57	3	7
LA BEAUREGARD	79.98	UpK6-2	19-Apr-53	1	5
LA BEAUREGARD	2.72	UpK6-3	20-Apr-53	4	5
LA BEAUREGARD	6.40	TP6-3	24-Mar-55	3	5
LA BEAUREGARD	1.53	PB1-7	3-Jun-57	6	5
LA BEAUREGARD	114.67	PB3-3	26-Jun-57	1	5
LA BIENVILLE	4.02	BJ3-5	5-Nov-51	8	5
LA BIENVILLE	1.16	UpK8-5	12-May-53	5	5
LA BIENVILLE	6.40	TP6-3	24-Mar-55	3	5
LA BIENVILLE	1.43	PB1-8	4-Jun-57	6	5
LA BIENVILLE	3.21	PB18-8	14-Oct-57	6	5

COUNTY	uCi/ sqm	SHOT DAY	DATE	RANK FOR DAY	TIMES IN TOP 10
LA BOSSIER	4.02	BJ3-5	5-Nov-51	8	7
LA BOSSIER	3.91	UpK5-4	14-Apr-53	1	7
LA BOSSIER	1.83	UpK8-4	11-May-53	4	7
LA BOSSIER	6.40	TP6-3	24-Mar-55	3	7
LA BOSSIER	1.45	PB1-8	4-Jun-57	5	7
LA BOSSIER	1.56	PB18-7	13-Oct-57	4	7
LA BOSSIER	3.21	PB18-8	14-Oct-57	6	7
LA CADDO	2.58	UpK5-4	14-Apr-53	2	6
LA CADDO	1.83	UpK8-4	11-May-53	4	6
LA CADDO	1.16	UpK8-5	12-May-53	5	6
LA CADDO	6.40	TP6-3	24-Mar-55	3	6
LA CADDO	1.56	PB18-7	13-Oct-57	4	6
LA CADDO	3.21	PB18-8	14-Oct-57	6	6
LA CALCASIEU	79.98	UpK6-2	19-Apr-53	1	6
LA CALCASIEU	2.72	UpK6-3	20-Apr-53	4	6
LA CALCASIEU	0.81	UpK6-7	24-Apr-53	1	6
LA CALCASIEU	1.34	UpK8-5	12-May-53	3	6
LA CALCASIEU	6.40	TP6-3	24-Mar-55	3	6
LA CALCASIEU	114.67	PB3-3	26-Jun-57	1	6
LA CALDWELL	1.16	UpK8-5	12-May-53	5	6
LA CALDWELL	1.56	UpK8-9	16-May-53	1	6
LA CALDWELL	1.47	TP6-2	23-Mar-55	9	6
LA CALDWELL	4.11	TP6-3	24-Mar-55	4	6
LA CALDWELL	1.36	PB1-8	4-Jun-57	8	6
LA CALDWELL	1.93	PB18-8	14-Oct-57	9	6
LA CAMERON	79.98	UpK6-2	19-Apr-53	1	7
LA CAMERON	2.72	UpK6-3	20-Apr-53	4	7
LA CAMERON	0.81	UpK6-7	24-Apr-53	1	7
LA CAMERON	1.34	UpK8-5	12-May-53	3	7
LA CAMERON	6.40	TP6-3	24-Mar-55	3	7
LA CAMERON	114.67	PB3-3	26-Jun-57	1	7
LA CAMERON	1.07	PB9-8	30-Aug-57	2	7
LA CATAHOULA	1.47	TP6-2	23-Mar-55	9	3
LA CATAHOULA	4.11	TP6-3	24-Mar-55	4	3
LA CATAHOULA	68.82	PB3-3	26-Jun-57	2	3

COUNTY	uCi/ sqm	SHOT DAY	DATE	RANK FOR DAY	TIMES IN TOP 10
LA CLAIBORNE	1.83	UpK8-4	11-May-53	4	7
LA CLAIBORNE	1.47	TP6-2	23-Mar-55	9	7
LA CLAIBORNE	4.11	TP6-3	24-Mar-55	4	7
LA CLAIBORNE	1.58	PB1-8	4-Jun-57	2	7
LA CLAIBORNE	1.89	PB6-7	30-Jul-57	7	7
LA CLAIBORNE	3.21	PB18-8	14-Oct-57	6	7
LA CLAIBORNE	1.25	PB18-9	15-Oct-57	3	7
LA CONCORDIA	1.16	UpK8-5	12-May-53	5	5
LA CONCORDIA	1.56	UpK8-9	16-May-53	1	5
LA CONCORDIA	1.47	TP6-2	23-Mar-55	9	5
LA CONCORDIA	4.11	TP6-3	24-Mar-55	4	5
LA CONCORDIA	68.82	PB3-3	26-Jun-57	2	5
LA DE SOTO	4.02	BJ3-5	5-Nov-51	8	7
LA DE SOTO	1.33	UpK5-4	14-Apr-53	7	7
LA DE SOTO	1.83	UpK8-4	11-May-53	4	7
LA DE SOTO	1.16	UpK8-5	12-May-53	5	7
LA DE SOTO	6.40	TP6-3	24-Mar-55	3	7
LA DE SOTO	3.21	PB18-8	14-Oct-57	6	7
LA DE SOTO	1.25	PB18-9	15-Oct-57	3	7
LA EAST BATON R	38.63	UpK6-2	19-Apr-53	2	4
LA EAST BATON R	114.67	PB3-3	26-Jun-57	1	4
LA EAST BATON R	1.99	PB6-7	30-Jul-57	6	4
LA EAST BATON R	2.34	PB6-8	31-Jul-57	3	4
LA EAST CARROLL	1.16	UpK8-5	12-May-53	5	6
LA EAST CARROLL	1.56	UpK8-9	16-May-53	1	6
LA EAST CARROLL	1.47	TP6-2	23-Mar-55	9	6
LA EAST CARROLL	4.11	TP6-3	24-Mar-55	4	6
LA EAST CARROLL	1.93	PB18-8	14-Oct-57	9	6
LA EAST CARROLL	1.25	PB18-9	15-Oct-57	3	6
LA EAST FELICIA	38.63	UpK6-2	19-Apr-53	2	4
LA EAST FELICIA	0.80	UpK8-5	12-May-53	7	4
LA EAST FELICIA	68.82	PB3-3	26-Jun-57	2	4
LA EAST FELICIA	1.40	PB6-8	31-Jul-57	8	4
LA EVANGELINE	79.98	UpK6-2	19-Apr-53	1	6
LA EVANGELINE	2.72	UpK6-3	20-Apr-53	4	6

COUNTY	uCi/ sqm	SHOT DAY	DATE	RANK FOR DAY	TIMES IN TOP 10
LA EVANGELINE	0.81	UpK6-7	24-Apr-53	1	6
LA EVANGELINE	1.34	UpK8-5	12-May-53	3	6
LA EVANGELINE	6.40	TP6-3	24-Mar-55	3	6
LA EVANGELINE	114.67	PB3-3	26-Jun-57	1	6
LA FRANKLIN	1.16	UpK8-5	12-May-53	5	6
LA FRANKLIN	1.56	UpK8-9	16-May-53	1	6
LA FRANKLIN	1.47	TP6-2	23-Mar-55	9	6
LA FRANKLIN	4.11	TP6-3	24-Mar-55	4	6
LA FRANKLIN	5.46	PB9-2	24-Aug-57	4	6
LA FRANKLIN	1.25	PB18-9	15-Oct-57	3	6
LA GRANT	79.98	UpK6-2	19-Apr-53	1	7
LA GRANT	2.72	UpK6-3	20-Apr-53	4	7
LA GRANT	1.34	UpK8-5	12-May-53	3	7
LA GRANT	6.40	TP6-3	24-Mar-55	3	7
LA GRANT	2.48	TP8-3	11-Apr-55	5	7
LA GRANT	1.30	PB2-4	21-Jun-57	9	7
LA GRANT	114.67	PB3-3	26-Jun-57	1	7
LA IBERIA	38.63	UpK6-2	19-Apr-53	2	4
LA IBERIA	1.34	UpK8-5	12-May-53	3	4
LA IBERIA	1.69	TP10-7	11-May-55	1	4
LA IBERIA	114.67	PB3-3	26-Jun-57	1	4
LA IBERVILLE	38.63	UpK6-2	19-Apr-53	2	5
LA IBERVILLE	1.34	UpK8-5	12-May-53	3	5
LA IBERVILLE	114.67	PB3-3	26-Jun-57	1	5
LA IBERVILLE	1.99	PB6-7	30-Jul-57	6	5
LA IBERVILLE	1.40	PB6-8	31-Jul-57	8	5
LA JACKSON	1.12	UpK8-4	11-May-53	9	5
LA JACKSON	1.47	TP6-2	23-Mar-55	9	5
LA JACKSON	4.11	TP6-3	24-Mar-55	4	5
LA JACKSON	1.40	PB6-8	31-Jul-57	8	5
LA JACKSON	1.93	PB18-8	14-Oct-57	9	5
LA JEFFERSON	38.63	UpK6-2	19-Apr-53	2	4
LA JEFFERSON	22.97	PB3-3	26-Jun-57	4	4
LA JEFFERSON	0.98	PB9-7	29-Aug-57	5	4
LA JEFFERSON	1.07	PB9-8	30-Aug-57	2	4

COUNTY	uCi/ sqm	SHOT DAY	DATE	RANK FOR DAY	TIMES IN TOP 10
LA JEFFERSON DA	79.98	UpK6-2	19-Apr-53	1	5
LA JEFFERSON DA	2.72	UpK6-3	20-Apr-53	4	5
LA JEFFERSON DA	6.40	TP6-3	24-Mar-55	3	5
LA JEFFERSON DA	68.82	PB3-3	26-Jun-57	2	5
LA JEFFERSON DA	1.40	PB6-8	31-Jul-57	8	5
LA LA SALLE	4.02	BJ3-5	5-Nov-51	8	10
LA LA SALLE	40.05	UpK7-4	28-Apr-53	8	10
LA LA SALLE	1.16	UpK8-5	12-May-53	5	10
LA LA SALLE	1.56	UpK8-9	16-May-53	1	10
LA LA SALLE	1.47	TP6-2	23-Mar-55	9	10
LA LA SALLE	4.11	TP6-3	24-Mar-55	4	10
LA LA SALLE	2.39	TP8-3	11-Apr-55	7	10
LA LA SALLE	114.67	PB3-3	26-Jun-57	1	10
LA LA SALLE	1.99	PB6-7	30-Jul-57	6	10
LA LA SALLE	2.34	PB6-8	31-Jul-57	3	10
LA LAFAYETTE	79.98	UpK6-2	19-Apr-53	1	6
LA LAFAYETTE	2.72	UpK6-3	20-Apr-53	4	6
LA LAFAYETTE	1.34	UpK8-5	12-May-53	3	6
LA LAFAYETTE	6.40	TP6-3	24-Mar-55	3	6
LA LAFAYETTE	1.29	TP10-7	11-May-55	6	6
LA LAFAYETTE	114.67	PB3-3	26-Jun-57	1	6
LA LAFOURCHE	38.63	UpK6-2	19-Apr-53	2	3
LA LAFOURCHE	45.85	PB3-3	26-Jun-57	3	3
LA LAFOURCHE	55.66	PB12-6	7-Sep-57	1	3
LA LINCOLN	3.91	UpK5-4	14-Apr-53	1	6
LA LINCOLN	1.12	UpK8-4	11-May-53	9	6
LA LINCOLN	1.47	TP6-2	23-Mar-55	9	6
LA LINCOLN	4.11	TP6-3	24-Mar-55	4	6
LA LINCOLN	1.40	PB6-8	31-Jul-57	8	6
LA LINCOLN	1.93	PB18-8	14-Oct-57	9	6
LA LIVINGSTON	1.10	TS6-5	29-May-52	4	5
LA LIVINGSTON	38.63	UpK6-2	19-Apr-53	2	5
LA LIVINGSTON	68.82	PB3-3	26-Jun-57	2	5
LA LIVINGSTON	1.99	PB6-7	30-Jul-57	6	5
LA LIVINGSTON	2.34	PB6-8	31-Jul-57	3	5

COUNTY	uCi/ sqm	SHOT DAY	DATE	RANK FOR DAY	TIMES IN TOP 10
LA MADISON	44.35	UpK7-4	28-Apr-53	6	7
LA MADISON	1.16	UpK8-5	12-May-53	5	7
LA MADISON	1.56	UpK8-9	16-May-53	1	7
LA MADISON	1.47	TP6-2	23-Mar-55	9	7
LA MADISON	4.11	TP6-3	24-Mar-55	4	7
LA MADISON	10.55	PB12-6	7-Sep-57	8	7
LA MADISON	1.25	PB18-9	15-Oct-57	3	7
LA MOREHOUSE	0.93	UpK8-9	16-May-53	2	4
LA MOREHOUSE	1.47	TP6-2	23-Mar-55	9	4
LA MOREHOUSE	4.11	TP6-3	24-Mar-55	4	4
LA MOREHOUSE	0.89	PB1-9	5-Jun-57	7	4
LA NATCHITOCHES	2.58	UpK5-4	14-Apr-53	2	5
LA NATCHITOCHES	1.83	UpK8-4	11-May-53	4	5
LA NATCHITOCHES	1.16	UpK8-5	12-May-53	5	5
LA NATCHITOCHES	6.40	TP6-3	24-Mar-55	3	5
LA NATCHITOCHES	1.93	PB18-8	14-Oct-57	9	5
LA ORLEANS	16.04	UpK6-2	19-Apr-53	6	2
LA ORLEANS	19.04	PB3-3	26-Jun-57	5	2
LA OUACHITA	1.56	UpK8-9	16-May-53	1	5
LA OUACHITA	1.47	TP6-2	23-Mar-55	9	5
LA OUACHITA	4.11	TP6-3	24-Mar-55	4	5
LA OUACHITA	1.40	PB6-8	31-Jul-57	8	5
LA OUACHITA	5.46	PB9-2	24-Aug-57	4	5
LA PLAQUEMINES	38.63	UpK6-2	19-Apr-53	2	4
LA PLAQUEMINES	68.82	PB3-3	26-Jun-57	2	4
LA PLAQUEMINES	1.99	PB6-7	30-Jul-57	6	4
LA PLAQUEMINES	2.34	PB6-8	31-Jul-57	3	4
LA POINTE COUPE	38.63	UpK6-2	19-Apr-53	2	6
LA POINTE COUPE	1.34	UpK8-5	12-May-53	3	6
LA POINTE COUPE	114.67	PB3-3	26-Jun-57	1	6
LA POINTE COUPE	1.99	PB6-7	30-Jul-57	6	6
LA POINTE COUPE	2.34	PB6-8	31-Jul-57	3	6
LA POINTE COUPE	1.06	PB6-11	3-Aug-57	2	6
LA RAPIDES	79.98	UpK6-2	19-Apr-53	1	8
LA RAPIDES	2.72	UpK6-3	20-Apr-53	4	8

COUNTY	uCi/ sqm	SHOT DAY	DATE	RANK FOR DAY	TIMES IN TOP 10
LA RAPIDES	0.81	UpK6-7	24-Apr-53	1	8
LA RAPIDES	1.34	UpK8-5	12-May-53	3	8
LA RAPIDES	6.40	TP6-3	24-Mar-55	3	8
LA RAPIDES	1.30	PB2-4	21-Jun-57	9	8
LA RAPIDES	114.67	PB3-3	26-Jun-57	1	8
LA RAPIDES	1.40	PB6-8	31-Jul-57	8	8
LA RED RIVER	4.02	BJ3-5	5-Nov-51	8	6
LA RED RIVER	2.58	UpK5-4	14-Apr-53	2	6
LA RED RIVER	1.83	UpK8-4	11-May-53	4	6
LA RED RIVER	1.16	UpK8-5	12-May-53	5	6
LA RED RIVER	6.40	TP6-3	24-Mar-55	3	6
LA RED RIVER	3.21	PB18-8	14-Oct-57	6	6
LA RICHLAND	0.93	UpK8-9	16-May-53	2	4
LA RICHLAND	1.47	TP6-2	23-Mar-55	9	4
LA RICHLAND	4.11	TP6-3	24-Mar-55	4	4
LA RICHLAND	5.46	PB9-2	24-Aug-57	4	4
LA SABINE	79.98	UpK6-2	19-Apr-53	1	7
LA SABINE	2.72	UpK6-3	20-Apr-53	4	7
LA SABINE	0.81	UpK6-7	24-Apr-53	1	7
LA SABINE	1.34	UpK8-5	12-May-53	3	7
LA SABINE	6.40	TP6-3	24-Mar-55	3	7
LA SABINE	1.36	PB1-7	3-Jun-57	8	7
LA SABINE	3.21	PB18-8	14-Oct-57	6	7
LA ST BERNARD	38.63	UpK6-2	19-Apr-53	2	3
LA ST BERNARD	68.82	PB3-3	26-Jun-57	2	3
LA ST BERNARD	0.98	PB9-7	29-Aug-57	5	3
LA ST CHARLES	38.63	UpK6-2	19-Apr-53	2	4
LA ST CHARLES	45.85	PB3-3	26-Jun-57	3	4
LA ST CHARLES	1.40	PB6-8	31-Jul-57	8	4
LA ST CHARLES	52.53	PB12-6	7-Sep-57	2	4
LA ST HELENA	38.63	UpK6-2	19-Apr-53	2	4
LA ST HELENA	114.67	PB3-3	26-Jun-57	1	4
LA ST HELENA	1.99	PB6-7	30-Jul-57	6	4
LA ST HELENA	2.34	PB6-8	31-Jul-57	3	4
LA ST JAMES	38.63	UpK6-2	19-Apr-53	2	4

COUNTY	uCi/ sqm	SHOT DAY	DATE	RANK FOR DAY	TIMES IN TOP 10
LA ST JAMES	1.49	PB1-7	3-Jun-57	7	4
LA ST JAMES	45.85	PB3-3	26-Jun-57	3	4
LA ST JAMES	1.99	PB6-7	30-Jul-57	6	4
LA ST JOHN THE	38.63	UpK6-2	19-Apr-53	2	5
LA ST JOHN THE	1.53	PB1-7	3-Jun-57	5	5
LA ST JOHN THE	45.85	PB3-3	26-Jun-57	3	5
LA ST JOHN THE	1.99	PB6-7	30-Jul-57	6	5
LA ST JOHN THE	2.34	PB6-8	31-Jul-57	3	5
LA ST LANDRY	38.63	UpK6-2	19-Apr-53	2	4
LA ST LANDRY	1.34	UpK8-5	12-May-53	3	4
LA ST LANDRY	68.82	PB3-3	26-Jun-57	2	4
LA ST LANDRY	1.40	PB6-8	31-Jul-57	8	4
LA ST MARTIN	38.63	UpK6-2	19-Apr-53	2	3
LA ST MARTIN	114.67	PB3-3	26-Jun-57	1	3
LA ST MARTIN	1.40	PB6-8	31-Jul-57	8	3
LA ST MARY	38.63	UpK6-2	19-Apr-53	2	3
LA ST MARY	114.67	PB3-3	26-Jun-57	1	3
LA ST MARY	1.06	PB6-11	3-Aug-57	2	3
LA ST TAMMANY	38.63	UpK6-2	19-Apr-53	2	3
LA ST TAMMANY	68.82	PB3-3	26-Jun-57	2	3
LA ST TAMMANY	1.06	PB6-11	3-Aug-57	2	3
LA TANGIPAHOA	38.63	UpK6-2	19-Apr-53	2	3
LA TANGIPAHOA	114.67	PB3-3	26-Jun-57	1	3
LA TANGIPAHOA	1.99	PB6-7	30-Jul-57	6	3
LA TENSAS	43.69	UpK7-4	28-Apr-53	7	7
LA TENSAS	1.16	UpK8-5	12-May-53	5	7
LA TENSAS	1.56	UpK8-9	16-May-53	1	7
LA TENSAS	1.47	TP6-2	23-Mar-55	9	7
LA TENSAS	4.11	TP6-3	24-Mar-55	4	7
LA TENSAS	1.30	PB2-4	21-Jun-57	9	7
LA TENSAS	114.67	PB3-3	26-Jun-57	1	7
LA TERREBONNE	38.63	UpK6-2	19-Apr-53	2	5
LA TERREBONNE	0.80	UpK8-5	12-May-53	7	5
LA TERREBONNE	1.79	PB1-7	3-Jun-57	2	5
LA TERREBONNE	114.67	PB3-3	26-Jun-57	1	5

COUNTY	uCi/ sqm	SHOT DAY	DATE	RANK FOR DAY	TIMES IN TOP 10
LA TERREBONNE	2.34	PB6-8	31-Jul-57	3	5
LA UNION	3.91	UpK5-4	14-Apr-53	1	5
LA UNION	1.12	UpK8-4	11-May-53	9	5
LA UNION	1.47	TP6-2	23-Mar-55	9	5
LA UNION	4.11	TP6-3	24-Mar-55	4	5
LA UNION	1.93	PB18-8	14-Oct-57	9	5
LA VERMILION	1.97	UpK4-1	6-Apr-53	2	9
LA VERMILION	79.98	UpK6-2	19-Apr-53	1	9
LA VERMILION	2.72	UpK6-3	20-Apr-53	4	9
LA VERMILION	0.81	UpK6-7	24-Apr-53	1	9
LA VERMILION	1.34	UpK8-5	12-May-53	3	9
LA VERMILION	6.40	TP6-3	24-Mar-55	3	9
LA VERMILION	1.30	PB2-4	21-Jun-57	9	9
LA VERMILION	68.82	PB3-3	26-Jun-57	2	9
LA VERMILION	1.40	PB6-8	31-Jul-57	8	9
LA VERNON	79.98	UpK6-2	19-Apr-53	1	6
LA VERNON	2.72	UpK6-3	20-Apr-53	4	6
LA VERNON	0.81	UpK6-7	24-Apr-53	1	6
LA VERNON	1.34	UpK8-5	12-May-53	3	6
LA VERNON	6.40	TP6-3	24-Mar-55	3	6
LA VERNON	11.44	PB3-3	26-Jun-57	6	6
LA WASHINGTON	38.63	UpK6-2	19-Apr-53	2	5
LA WASHINGTON	1.29	TP8-5	13-Apr-55	1	5
LA WASHINGTON	68.82	PB3-3	26-Jun-57	2	5
LA WASHINGTON	1.40	PB6-8	31-Jul-57	8	5
LA WASHINGTON	1.07	PB9-8	30-Aug-57	2	5
LA WEBSTER	1.22	UpK3-6	5-Apr-53	6	8
LA WEBSTER	3.91	UpK5-4	14-Apr-53	1	8
LA WEBSTER	1.83	UpK8-4	11-May-53	4	8
LA WEBSTER	6.40	TP6-3	24-Mar-55	3	8
LA WEBSTER	1.49	PB1-8	4-Jun-57	3	8
LA WEBSTER	1.04	PB18-7	13-Oct-57	7	8
LA WEBSTER	3.21	PB18-8	14-Oct-57	6	8
LA WEBSTER	1.25	PB18-9	15-Oct-57	3	8
LA WEST BATON R	38.63	UpK6-2	19-Apr-53	2	5

COUNTY	uCi/ sqm	SHOT DAY	DATE	RANK FOR DAY	TIMES IN TOP 10
LA WEST BATON R	1.34	UpK8-5	12-May-53	3	5
LA WEST BATON R	114.67	PB3-3	26-Jun-57	1	5
LA WEST BATON R	1.99	PB6-7	30-Jul-57	6	5
LA WEST BATON R	2.34	PB6-8	31-Jul-57	3	5
LA WEST CARROLL	1.47	TP6-2	23-Mar-55	9	4
LA WEST CARROLL	4.11	TP6-3	24-Mar-55	4	4
LA WEST CARROLL	1.93	PB18-8	14-Oct-57	9	4
LA WEST CARROLL	1.25	PB18-9	15-Oct-57	3	4
LA WEST FELICIA	38.63	UpK6-2	19-Apr-53	2	4
LA WEST FELICIA	114.67	PB3-3	26-Jun-57	1	4
LA WEST FELICIA	1.99	PB6-7	30-Jul-57	6	4
LA WEST FELICIA	2.34	PB6-8	31-Jul-57	3	4
LA WINN	1.12	UpK8-4	11-May-53	9	5
LA WINN	1.16	UpK8-5	12-May-53	5	5
LA WINN	6.40	TP6-3	24-Mar-55	3	5
LA WINN	1.34	PB1-8	4-Jun-57	9	5
LA WINN	1.40	PB6-8	31-Jul-57	8	5

COUNTY	uCi/ sqm	SHOT DAY	DATE	RANK FOR DAY	TIMES IN TOP 10
MA BARNSTABLE	87.64	UpK4-2	7-Apr-53	10	5
MA BARNSTABLE	3.56	UpK4-3	8-Apr-53	5	5
MA BARNSTABLE	3.03	UpK4-4	9-Apr-53	3	5
MA BARNSTABLE	0.99	UpK5-5	15-Apr-53	2	5
MA BARNSTABLE	2.99	PB5-8	22-Jul-57	4	5
MA BERKSHIRE	1.37	TS4-6	6-May-52	9	8
MA BERKSHIRE	1.37	TS4-6	6-May-52	9	8
MA BERKSHIRE	1.45	UpK7-6	30-Apr-53	6	8
MA BERKSHIRE	5.26	UpK8-3	10-May-53	3	8
MA BERKSHIRE	1.32	UpK8-4	11-May-53	7	8
MA BERKSHIRE	1.36	TP7-6	3-Apr-55	2	8
MA BERKSHIRE	5.97	PB5-8	22-Jul-57	3	8
MA BERKSHIRE	1.57	PB6-7	30-Jul-57	9	8
MA BRISTOL	153.91	UpK4-2	7-Apr-53	3	4
MA BRISTOL	1.89	UpK4-3	8-Apr-53	8	4
MA BRISTOL	1.83	UpK8-4	11-May-53	4	4
MA BRISTOL	5.97	PB5-8	22-Jul-57	3	4
MA DUKES	3.71	BJ3-7	7-Nov-51	10	6
MA DUKES	92.31	UpK4-2	7-Apr-53	9	6
MA DUKES	3.89	UpK4-3	8-Apr-53	4	6
MA DUKES	1.83	UpK8-4	11-May-53	4	6
MA DUKES	1.09	TP6-5	26-Mar-55	7	6
MA DUKES	5.97	PB5-8	22-Jul-57	3	6
MA ESSEX	37.60	R1-2	29-Jan-51	6	7
MA ESSEX	5.16	TS8-3	7-Jun-52	9	7
MA ESSEX	4.67	UpK4-3	8-Apr-53	3	7
MA ESSEX	3.98	UpK4-4	9-Apr-53	2	7
MA ESSEX	0.99	UpK5-5	15-Apr-53	2	7
MA ESSEX	1.09	TP6-5	26-Mar-55	7	7
MA ESSEX	2.99	PB5-8	22-Jul-57	4	7
MA FRANKLIN	3.71	BJ3-7	7-Nov-51	10	6
MA FRANKLIN	1.32	UpK8-3	10-May-53	9	6
MA FRANKLIN	1.72	UpK8-4	11-May-53	5	6
MA FRANKLIN	1.45	TP6-5	26-Mar-55	5	6
MA FRANKLIN	1.36	PB5-6	20-Jul-57	6	6

COUNTY	uCi/ sqm	SHOT DAY	DATE	RANK FOR DAY	TIMES IN TOP 10
MA FRANKLIN	2.99	PB5-8	22-Jul-57	4	6
MA HAMPDEN	30.19	R1-2	29-Jan-51	8	5
MA HAMPDEN	3.71	BJ3-7	7-Nov-51	10	5
MA HAMPDEN	2.68	PB5-6	20-Jul-57	3	5
MA HAMPDEN	2.99	PB5-8	22-Jul-57	4	5
MA HAMPDEN	1.27	PB6-8	31-Jul-57	9	5
MA HAMPSHIRE	1.36	PB5-6	20-Jul-57	6	3
MA HAMPSHIRE	8.96	PB5-8	22-Jul-57	2	3
MA HAMPSHIRE	19.27	PB11-5	4-Sep-57	4	3
MA MIDDLESEX	37.60	R1-2	29-Jan-51	6	9
MA MIDDLESEX	3.71	BJ3-7	7-Nov-51	10	9
MA MIDDLESEX	122.70	UpK4-2	7-Apr-53	5	9
MA MIDDLESEX	3.56	UpK4-3	8-Apr-53	5	9
MA MIDDLESEX	7.95	UpK4-4	9-Apr-53	1	9
MA MIDDLESEX	1.06	UpK4-5	10-Apr-53	1	9
MA MIDDLESEX	0.99	UpK5-5	15-Apr-53	2	9
MA MIDDLESEX	1.09	TP6-5	26-Mar-55	7	9
MA MIDDLESEX	2.99	PB5-8	22-Jul-57	4	9
MA NANTUCKET	2.34	UpK4-3	8-Apr-53	6	4
MA NANTUCKET	7.95	UpK4-4	9-Apr-53	1	4
MA NANTUCKET	1.48	UpK5-5	15-Apr-53	1	4
MA NANTUCKET	2.99	PB5-8	22-Jul-57	4	4
MA NORFOLK	37.60	R1-2	29-Jan-51	6	7
MA NORFOLK	181.57	UpK4-2	7-Apr-53	2	7
MA NORFOLK	4.67	UpK4-3	8-Apr-53	3	7
MA NORFOLK	3.98	UpK4-4	9-Apr-53	2	7
MA NORFOLK	0.99	UpK5-5	15-Apr-53	2	7
MA NORFOLK	1.09	TP6-5	26-Mar-55	7	7
MA NORFOLK	5.97	PB5-8	22-Jul-57	3	7
MA PLYMOUTH	1.61	BJ2-6	4-Nov-51	9	5
MA PLYMOUTH	205.83	UpK4-2	7-Apr-53	1	5
MA PLYMOUTH	3.56	UpK4-3	8-Apr-53	5	5
MA PLYMOUTH	3.03	UpK4-4	9-Apr-53	3	5
MA PLYMOUTH	5.97	PB5-8	22-Jul-57	3	5
MA SUFFOLK	37.60	R1-2	29-Jan-51	6	5

COUNTY	uCi/ sqm	SHOT DAY	DATE	RANK FOR DAY	TIMES IN TOP 10
MA SUFFOLK	123.51	UpK4-2	7-Apr-53	4	5
MA SUFFOLK	1.94	UpK4-3	8-Apr-53	7	5
MA SUFFOLK	0.83	UpK4-4	9-Apr-53	6	5
MA SUFFOLK	2.99	PB5-8	22-Jul-57	4	5
MA WORCESTER	1.89	UpK4-3	8-Apr-53	8	4
MA WORCESTER	1.83	UpK8-4	11-May-53	4	4
MA WORCESTER	1.09	TP6-5	26-Mar-55	7	4
MA WORCESTER	5.97	PB5-8	22-Jul-57	3	4

COUNTY	uCi/ sqm	SHOT DAY	DATE	RANK FOR DAY	TIMES IN TOP 10
MD ALLEGANY	1.29	TP11-8	22-May-55	9	3
MD ALLEGANY	8.96	PB5-8	22-Jul-57	2	3
MD ALLEGANY	2.83	PB5-9	23-Jul-57	2	3
MD ANNE ARUNDEL	8.96	PB5-8	22-Jul-57	2	3
MD ANNE ARUNDEL	2.83	PB5-9	23-Jul-57	2	3
MD ANNE ARUNDEL	2.54	PB8-7	24-Aug-57	7	3
MD BALTIMORE	8.96	PB5-8	22-Jul-57	2	3
MD BALTIMORE	1.89	PB5-9	23-Jul-57	3	3
MD BALTIMORE	2.11	UE8-5	18-Jul-62	7	3
MD CALVERT	1.06	TP9-10	24-Apr-55	1	4
MD CALVERT	8.96	PB5-8	22-Jul-57	2	4
MD CALVERT	2.83	PB5-9	23-Jul-57	2	4
MD CALVERT	2.54	PB8-7	24-Aug-57	7	4
MD CAROLINE	1.67	TP6-4	25-Mar-55	10	3
MD CAROLINE	5.97	PB5-8	22-Jul-57	3	3
MD CAROLINE	1.89	PB5-9	23-Jul-57	3	3
MD CARROLL	2.51	TP6-4	25-Mar-55	5	4
MD CARROLL	1.51	PB5-8	22-Jul-57	7	4
MD CARROLL	2.83	PB5-9	23-Jul-57	2	4
MD CARROLL	2.11	UE8-5	18-Jul-62	7	4
MD CECIL	4.63	BJ3-6	6-Nov-51	9	2
MD CECIL	8.96	PB5-8	22-Jul-57	2	2
MD CHARLES	20.31	BJ2-3	1-Nov-51	6	5
MD CHARLES	1.40	TS5-4	10-May-52	9	5
MD CHARLES	2.99	PB5-8	22-Jul-57	4	5
MD CHARLES	1.89	PB5-9	23-Jul-57	3	5
MD CHARLES	2.54	PB8-7	24-Aug-57	7	5
MD DORCHESTER	5.97	PB5-8	22-Jul-57	3	3
MD DORCHESTER	2.83	PB5-9	23-Jul-57	2	3
MD DORCHESTER	1.51	PB8-7	24-Aug-57	10	3
MD FREDERICK	1.04	UpK4-4	9-Apr-53	5	5
MD FREDERICK	8.96	PB5-8	22-Jul-57	2	5
MD FREDERICK	1.89	PB5-9	23-Jul-57	3	5
MD FREDERICK	1.51	PB8-7	24-Aug-57	10	5
MD FREDERICK	2.11	UE8-5	18-Jul-62	7	5

COUNTY	uCi/ sqm	SHOT DAY	DATE	RANK FOR DAY	TIMES IN TOP 10
MD GARRETT	1.29	TP11-8	22-May-55	9	4
MD GARRETT	8.96	PB5-8	22-Jul-57	2	4
MD GARRETT	2.83	PB5-9	23-Jul-57	2	4
MD GARRETT	103.59	S-4	9-Jul-62	6	4
MD HARFORD	4.63	BJ3-6	6-Nov-51	9	5
MD HARFORD	6.63	UpK10-6	30-May-53	1	5
MD HARFORD	1.67	TP6-4	25-Mar-55	10	5
MD HARFORD	8.96	PB5-8	22-Jul-57	2	5
MD HARFORD	0.94	PB5-9	23-Jul-57	4	5
MD HOWARD	1.67	TP6-4	25-Mar-55	10	4
MD HOWARD	8.96	PB5-8	22-Jul-57	2	4
MD HOWARD	2.83	PB5-9	23-Jul-57	2	4
MD HOWARD	2.54	PB8-7	24-Aug-57	7	4
MD KENT	8.96	PB5-8	22-Jul-57	2	2
MD KENT	0.94	PB5-9	23-Jul-57	4	2
MD MONTGOMERY	1.04	UpK4-4	9-Apr-53	5	5
MD MONTGOMERY	0.96	UpK8-6	13-May-53	3	5
MD MONTGOMERY	8.96	PB5-8	22-Jul-57	2	5
MD MONTGOMERY	0.94	PB5-9	23-Jul-57	4	5
MD MONTGOMERY	1.51	PB8-7	24-Aug-57	10	5
MD PRINCE GEORG	18.02	BJ2-3	1-Nov-51	8	5
MD PRINCE GEORG	1.04	UpK4-4	9-Apr-53	5	5
MD PRINCE GEORG	2.99	PB5-8	22-Jul-57	4	5
MD PRINCE GEORG	2.83	PB5-9	23-Jul-57	2	5
MD PRINCE GEORG	1.51	PB8-7	24-Aug-57	10	5
MD QUEEN ANNES	2.99	PB5-8	22-Jul-57	4	2
MD QUEEN ANNES	1.89	PB5-9	23-Jul-57	3	2
MD SOMERSET	0.90	R-2-6	7-Feb-51	1	7
MD SOMERSET	11.58	BJ3-6	6-Nov-51	3	7
MD SOMERSET	3.43	BJ4-7	11-Nov-51	1	7
MD SOMERSET	1.67	TP6-4	25-Mar-55	10	7
MD SOMERSET	8.96	PB5-8	22-Jul-57	2	7
MD SOMERSET	2.83	PB5-9	23-Jul-57	2	7
MD SOMERSET	1.32	UE8-6	19-Jul-62	9	7
MD ST MARYS	8.96	PB5-8	22-Jul-57	2	3

COUNTY	uCi/ sqm	SHOT DAY	DATE	RANK FOR DAY	TIMES IN TOP 10
MD ST MARYS	2.83	PB5-9	23-Jul-57	2	3
MD ST MARYS	2.54	PB8-7	24-Aug-57	7	3
MD TALBOT	8.96	PB5-8	22-Jul-57	2	2
MD TALBOT	1.89	PB5-9	23-Jul-57	3	2
MD WASHINGTON	1.04	UpK4-4	9-Apr-53	5	4
MD WASHINGTON	8.96	PB5-8	22-Jul-57	2	4
MD WASHINGTON	2.83	PB5-9	23-Jul-57	2	4
MD WASHINGTON	2.11	UE8-5	18-Jul-62	7	4
MD WICOMICO	1.67	TP6-4	25-Mar-55	10	3
MD WICOMICO	8.96	PB5-8	22-Jul-57	2	3
MD WICOMICO	1.89	PB5-9	23-Jul-57	3	3
MD WORCESTER	3.98	BJ4-6	10-Nov-51	1	5
MD WORCESTER	1.67	TP6-4	25-Mar-55	10	5
MD WORCESTER	14.93	PB5-8	22-Jul-57	1	5
MD WORCESTER	1.89	PB5-9	23-Jul-57	3	5
MD WORCESTER	1.32	UE8-6	19-Jul-62	9	5

COUNTY	uCi/ sqm	SHOT DAY	DATE	RANK FOR DAY	TIMES IN TOP 10
ME ANDROSCOGGIN	0.99	R1-3	30-Jan-51	3	7
ME ANDROSCOGGIN	2.34	UpK4-3	8-Apr-53	6	7
ME ANDROSCOGGIN	3.98	UpK4-4	9-Apr-53	2	7
ME ANDROSCOGGIN	1.48	UpK5-5	15-Apr-53	1	7
ME ANDROSCOGGIN	2.68	PB5-6	20-Jul-57	3	7
ME ANDROSCOGGIN	2.99	PB5-8	22-Jul-57	4	7
ME ANDROSCOGGIN	0.24	UE7-5	15-Jul-62	5	7
ME AROOSTOOK	85.07	BJ2-4	2-Nov-51	1	6
ME AROOSTOOK	24.09	UpK2-4	27-Mar-53	1	6
ME AROOSTOOK	1.45	TP6-5	26-Mar-55	5	6
ME AROOSTOOK	1.36	PB5-6	20-Jul-57	6	6
ME AROOSTOOK	8.96	PB5-8	22-Jul-57	2	6
ME AROOSTOOK	0.39	UE7-5	15-Jul-62	2	6
ME CUMBERLAND	12.07	BJ2-2	31-Oct-51	1	7
ME CUMBERLAND	2.34	UpK4-3	8-Apr-53	6	7
ME CUMBERLAND	3.98	UpK4-4	9-Apr-53	2	7
ME CUMBERLAND	1.48	UpK5-5	15-Apr-53	1	7
ME CUMBERLAND	2.68	PB5-6	20-Jul-57	3	7
ME CUMBERLAND	1.51	PB5-8	22-Jul-57	7	7
ME CUMBERLAND	0.12	UE7-5	15-Jul-62	9	7
ME FRANKLIN	7.98	BJ2-2	31-Oct-51	4	9
ME FRANKLIN	5.36	UpK2-4	27-Mar-53	7	9
ME FRANKLIN	2.34	UpK4-3	8-Apr-53	6	9
ME FRANKLIN	1.99	UpK4-4	9-Apr-53	4	9
ME FRANKLIN	0.97	TP6-5	26-Mar-55	8	9
ME FRANKLIN	1.14	PB2-6	23-Jun-57	7	9
ME FRANKLIN	1.36	PB5-6	20-Jul-57	6	9
ME FRANKLIN	1.51	PB5-8	22-Jul-57	7	9
ME FRANKLIN	0.13	UE7-5	15-Jul-62	8	9
ME HANCOCK	1.17	R1-3	30-Jan-51	2	6
ME HANCOCK	14.03	UpK2-4	27-Mar-53	5	6
ME HANCOCK	2.41	TP6-5	26-Mar-55	3	6
ME HANCOCK	1.36	PB5-6	20-Jul-57	6	6
ME HANCOCK	2.99	PB5-8	22-Jul-57	4	6
ME HANCOCK	0.24	UE7-5	15-Jul-62	5	6

COUNTY	uCi/ sqm	SHOT DAY	DATE	RANK FOR DAY	TIMES IN TOP 10
ME KENNEBEC	1.42	R1-3	30-Jan-51	1	8
ME KENNEBEC	2.34	UpK4-3	8-Apr-53	6	8
ME KENNEBEC	3.03	UpK4-4	9-Apr-53	3	8
ME KENNEBEC	0.86	UpK5-6	16-Apr-53	1	8
ME KENNEBEC	0.97	TP6-5	26-Mar-55	8	8
ME KENNEBEC	1.36	PB5-6	20-Jul-57	6	8
ME KENNEBEC	2.99	PB5-8	22-Jul-57	4	8
ME KENNEBEC	0.24	UE7-5	15-Jul-62	5	8
ME KNOX	1.42	R1-3	30-Jan-51	1	5
ME KNOX	3.74	UpK2-4	27-Mar-53	9	5
ME KNOX	1.45	TP6-5	26-Mar-55	5	5
ME KNOX	1.36	PB5-6	20-Jul-57	6	5
ME KNOX	5.97	PB5-8	22-Jul-57	3	5
ME LINCOLN	1.42	R1-3	30-Jan-51	1	7
ME LINCOLN	2.34	UpK4-3	8-Apr-53	6	7
ME LINCOLN	1.99	UpK4-4	9-Apr-53	4	7
ME LINCOLN	0.86	UpK5-6	16-Apr-53	1	7
ME LINCOLN	2.41	TP6-5	26-Mar-55	3	7
ME LINCOLN	1.36	PB5-6	20-Jul-57	6	7
ME LINCOLN	1.51	PB5-8	22-Jul-57	7	7
ME OXFORD	0.99	R1-3	30-Jan-51	3	8
ME OXFORD	12.07	BJ2-2	31-Oct-51	1	8
ME OXFORD	2.34	UpK4-3	8-Apr-53	6	8
ME OXFORD	3.98	UpK4-4	9-Apr-53	2	8
ME OXFORD	1.48	UpK5-5	15-Apr-53	1	8
ME OXFORD	2.68	PB5-6	20-Jul-57	3	8
ME OXFORD	2.99	PB5-8	22-Jul-57	4	8
ME OXFORD	0.33	UE7-5	15-Jul-62	3	8
ME PENOBSCOT	0.99	R1-3	30-Jan-51	3	7
ME PENOBSCOT	45.80	BJ2-4	2-Nov-51	4	7
ME PENOBSCOT	17.23	UpK2-4	27-Mar-53	3	7
ME PENOBSCOT	0.97	TP6-5	26-Mar-55	8	7
ME PENOBSCOT	2.68	PB5-6	20-Jul-57	3	7
ME PENOBSCOT	2.99	PB5-8	22-Jul-57	4	7
ME PENOBSCOT	0.12	UE7-5	15-Jul-62	9	7

COUNTY	uCi/ sqm	SHOT DAY	DATE	RANK FOR DAY	TIMES IN TOP 10
ME PISCATAQUIS	44.31	BJ2-4	2-Nov-51	5	7
ME PISCATAQUIS	14.25	UpK2-4	27-Mar-53	4	7
ME PISCATAQUIS	1.17	UpK11-5	8-Jun-53	2	7
ME PISCATAQUIS	1.45	TP6-5	26-Mar-55	5	7
ME PISCATAQUIS	1.36	PB5-6	20-Jul-57	6	7
ME PISCATAQUIS	2.99	PB5-8	22-Jul-57	4	7
ME PISCATAQUIS	0.12	UE7-5	15-Jul-62	9	7
ME SAGADAHOC	12.07	BJ2-2	31-Oct-51	1	8
ME SAGADAHOC	2.34	UpK4-3	8-Apr-53	6	8
ME SAGADAHOC	3.98	UpK4-4	9-Apr-53	2	8
ME SAGADAHOC	0.99	UpK5-5	15-Apr-53	2	8
ME SAGADAHOC	0.86	UpK5-6	16-Apr-53	1	8
ME SAGADAHOC	2.68	PB5-6	20-Jul-57	3	8
ME SAGADAHOC	1.51	PB5-8	22-Jul-57	7	8
ME SAGADAHOC	0.24	UE7-5	15-Jul-62	5	8
ME SOMERSET	0.99	R1-3	30-Jan-51	3	6
ME SOMERSET	9.56	UpK2-4	27-Mar-53	6	6
ME SOMERSET	1.45	TP6-5	26-Mar-55	5	6
ME SOMERSET	1.36	PB5-6	20-Jul-57	6	6
ME SOMERSET	2.99	PB5-8	22-Jul-57	4	6
ME SOMERSET	0.24	UE7-5	15-Jul-62	5	6
ME WALDO	1.42	R1-3	30-Jan-51	1	6
ME WALDO	4.93	UpK2-4	27-Mar-53	8	6
ME WALDO	1.17	UpK11-5	8-Jun-53	2	6
ME WALDO	1.45	TP6-5	26-Mar-55	5	6
ME WALDO	1.36	PB5-6	20-Jul-57	6	6
ME WALDO	5.97	PB5-8	22-Jul-57	3	6
ME WASHINGTON	1.42	R1-3	30-Jan-51	1	8
ME WASHINGTON	37.84	BJ2-4	2-Nov-51	6	8
ME WASHINGTON	21.42	UpK2-4	27-Mar-53	2	8
ME WASHINGTON	1.45	TP6-5	26-Mar-55	5	8
ME WASHINGTON	1.36	PB5-6	20-Jul-57	6	8
ME WASHINGTON	2.99	PB5-8	22-Jul-57	4	8
ME WASHINGTON	0.96	PB6-10	2-Aug-57	4	8
ME WASHINGTON	0.33	UE7-5	15-Jul-62	3	8

COUNTY	uCi/ sqm	SHOT DAY	DATE	RANK FOR DAY	TIMES IN TOP 10
ME YORK	2.34	UpK4-3	8-Apr-53	6	6
ME YORK	1.99	UpK4-4	9-Apr-53	4	6
ME YORK	0.97	TP6-5	26-Mar-55	8	6
ME YORK	2.68	PB5-6	20-Jul-57	3	6
ME YORK	1.51	PB5-8	22-Jul-57	7	6
ME YORK	0.24	UE7-5	15-Jul-62	5	6

COUNTY	uCi/ sqm	SHOT DAY	DATE	RANK FOR DAY	TIMES IN TOP 10
MI ALCONA	1.52	UpK10-5	29-May-53	7	4
MI ALCONA	0.97	UpK10-10	3-Jun-53	1	4
MI ALCONA	1.51	PB9-5	27-Aug-57	9	4
MI ALCONA	6.71	PB11-4	3-Sep-57	9	4
MI ALGER	1.60	TP9-4	18-Apr-55	1	3
MI ALGER	1.51	PB9-5	27-Aug-57	9	3
MI ALGER	0.89	PB13-7	12-Sep-57	5	3
MI ALLEGAN	1.72	TP3-3	3-Mar-55	8	3
MI ALLEGAN	1.02	PB4-3	7-Jul-57	8	3
MI ALLEGAN	3.01	PB9-5	27-Aug-57	6	3
MI ALPENA	1.53	UpK10-5	29-May-53	6	2
MI ALPENA	3.01	PB9-5	27-Aug-57	6	2
MI ANTRIM	1.45	UpK10-5	29-May-53	9	4
MI ANTRIM	0.97	UpK10-10	3-Jun-53	1	4
MI ANTRIM	3.01	PB9-5	27-Aug-57	6	4
MI ANTRIM	6.71	PB11-4	3-Sep-57	9	4
MI ARENAC	0.97	UpK10-10	3-Jun-53	1	4
MI ARENAC	0.93	PB4-3	7-Jul-57	9	4
MI ARENAC	1.51	PB9-5	27-Aug-57	9	4
MI ARENAC	7.97	UE12-4	17-Apr-65	8	4
MI BARAGA	3.41	UpK10-4	28-May-53	3	4
MI BARAGA	0.93	TP9-16	30-Apr-55	1	4
MI BARAGA	2.01	PB5-5	19-Jul-57	8	4
MI BARAGA	0.89	PB13-7	12-Sep-57	5	4
MI BARRY	1.72	TP3-3	3-Mar-55	8	4
MI BARRY	1.02	PB4-3	7-Jul-57	8	4
MI BARRY	4.52	PB9-5	27-Aug-57	2	4
MI BARRY	0.83	PB18-9	15-Oct-57	6	4
MI BAY	0.97	UpK10-10	3-Jun-53	1	3
MI BAY	1.02	PB4-3	7-Jul-57	8	3
MI BAY	1.51	PB9-5	27-Aug-57	9	3
MI BENZIE	2.95	UpK9-6	24-May-53	1	3
MI BENZIE	4.52	PB9-5	27-Aug-57	2	3
MI BENZIE	7.97	UE12-4	17-Apr-65	8	3
MI BERRIEN	3.01	PB9-5	27-Aug-57	6	3

COUNTY	uCi/ sqm	SHOT DAY	DATE	RANK FOR DAY	TIMES IN TOP 10
MI BERRIEN	0.33	UE7-5	15-Jul-62	3	3
MI BERRIEN	14.56	UE12-4	17-Apr-65	2	3
MI BRANCH	1.72	TP3-3	3-Mar-55	8	5
MI BRANCH	0.93	PB4-3	7-Jul-57	9	5
MI BRANCH	4.52	PB9-5	27-Aug-57	2	5
MI BRANCH	0.85	PB13-7	12-Sep-57	6	5
MI BRANCH	0.33	UE7-5	15-Jul-62	3	5
MI CALHOUN	1.02	PB4-3	7-Jul-57	8	3
MI CALHOUN	3.01	PB9-5	27-Aug-57	6	3
MI CALHOUN	0.13	UE7-5	15-Jul-62	8	3
MI CASS	4.52	PB9-5	27-Aug-57	2	3
MI CASS	0.83	PB18-9	15-Oct-57	6	3
MI CASS	0.33	UE7-5	15-Jul-62	3	3
MI CHARLEVOIX	0.97	UpK10-10	3-Jun-53	1	3
MI CHARLEVOIX	3.01	PB9-5	27-Aug-57	6	3
MI CHARLEVOIX	6.71	PB11-4	3-Sep-57	9	3
MI CHEBOYGAN	0.97	UpK10-10	3-Jun-53	1	3
MI CHEBOYGAN	1.51	PB9-5	27-Aug-57	9	3
MI CHEBOYGAN	6.71	PB11-4	3-Sep-57	9	3
MI CHIPPEWA	5.66	TS8-3	7-Jun-52	1	3
MI CHIPPEWA	1.58	UpK10-5	29-May-53	5	3
MI CHIPPEWA	1.51	PB9-5	27-Aug-57	9	3
MI CLARE	1.36	PB4-3	7-Jul-57	4	2
MI CLARE	1.25	PB13-7	12-Sep-57	3	2
MI CLINTON	1.02	PB4-3	7-Jul-57	8	4
MI CLINTON	4.52	PB9-5	27-Aug-57	2	4
MI CLINTON	0.83	PB18-9	15-Oct-57	6	4
MI CLINTON	7.97	UE12-4	17-Apr-65	8	4
MI CRAWFORD	0.97	UpK10-10	3-Jun-53	1	3
MI CRAWFORD	1.27	PB4-3	7-Jul-57	5	3
MI CRAWFORD	6.71	PB11-4	3-Sep-57	9	3
MI DELTA	3.01	PB9-5	27-Aug-57	6	1
MI DICKINSON	6.09	PB5-5	19-Jul-57	2	2
MI DICKINSON	1.97	PB8-5	22-Aug-57	9	2
MI EATON	1.72	TP3-3	3-Mar-55	8	5

COUNTY	uCi/ sqm	SHOT DAY	DATE	RANK FOR DAY	TIMES IN TOP 10
MI EATON	35.89	TP9-2	16-Apr-55	10	5
MI EATON	1.02	PB4-3	7-Jul-57	8	5
MI EATON	3.01	PB9-5	27-Aug-57	6	5
MI EATON	0.85	PB13-7	12-Sep-57	6	5
MI EMMET	1.29	TP8-5	13-Apr-55	1	4
MI EMMET	1.16	TP9-4	18-Apr-55	9	4
MI EMMET	3.01	PB9-5	27-Aug-57	6	4
MI EMMET	6.71	PB11-4	3-Sep-57	9	4
MI GENESEE	1.36	PB4-3	7-Jul-57	4	3
MI GENESEE	4.52	PB9-5	27-Aug-57	2	3
MI GENESEE	7.97	UE12-4	17-Apr-65	8	3
MI GLADWIN	0.97	UpK10-10	3-Jun-53	1	4
MI GLADWIN	1.02	PB4-3	7-Jul-57	8	4
MI GLADWIN	1.51	PB9-5	27-Aug-57	9	4
MI GLADWIN	7.97	UE12-4	17-Apr-65	8	4
MI GOGEBIC	1.50	PB2-4	21-Jun-57	8	4
MI GOGEBIC	6.09	PB5-5	19-Jul-57	2	4
MI GOGEBIC	1.16	PB15-5	20-Sep-57	7	4
MI GOGEBIC	2.54	PB17-2	29-Sep-57	2	4
MI GRAND TRAVER	0.85	TS4-4	4-May-52	2	5
MI GRAND TRAVER	0.85	TS4-4	4-May-52	2	5
MI GRAND TRAVER	2.60	UpK9-6	24-May-53	6	5
MI GRAND TRAVER	1.36	PB4-3	7-Jul-57	4	5
MI GRAND TRAVER	1.51	PB9-5	27-Aug-57	9	5
MI GRATIOT	1.02	PB4-3	7-Jul-57	8	3
MI GRATIOT	3.01	PB9-5	27-Aug-57	6	3
MI GRATIOT	0.83	PB18-9	15-Oct-57	6	3
MI HILLSDALE	0.93	PB4-3	7-Jul-57	9	3
MI HILLSDALE	3.01	PB9-5	27-Aug-57	6	3
MI HILLSDALE	0.12	UE7-5	15-Jul-62	9	3
MI HOUGHTON	0.93	TP9-16	30-Apr-55	1	5
MI HOUGHTON	1.50	PB2-4	21-Jun-57	8	5
MI HOUGHTON	6.09	PB5-5	19-Jul-57	2	5
MI HOUGHTON	1.16	PB15-5	20-Sep-57	7	5
MI HOUGHTON	1.90	PB17-2	29-Sep-57	9	5

COUNTY	uCi/ sqm	SHOT DAY	DATE	RANK FOR DAY	TIMES IN TOP 10
MI HURON	1.37	BJ4-7	11-Nov-51	3	7
MI HURON	0.95	BJ4-8	12-Nov-51	2	7
MI HURON	2.45	UpK10-6	30-May-53	9	7
MI HURON	0.97	UpK10-10	3-Jun-53	1	7
MI HURON	40.40	TP9-2	16-Apr-55	7	7
MI HURON	1.51	PB9-5	27-Aug-57	9	7
MI HURON	0.85	PB13-7	12-Sep-57	6	7
MI INGHAM	1.02	PB4-3	7-Jul-57	8	2
MI INGHAM	3.01	PB9-5	27-Aug-57	6	2
MI IONIA	1.02	PB4-3	7-Jul-57	8	4
MI IONIA	4.52	PB9-5	27-Aug-57	2	4
MI IONIA	0.83	PB18-9	15-Oct-57	6	4
MI IONIA	14.56	UE12-4	17-Apr-65	2	4
MI IOSCO	0.97	UpK10-10	3-Jun-53	1	3
MI IOSCO	1.51	PB9-5	27-Aug-57	9	3
MI IOSCO	6.71	PB11-4	3-Sep-57	9	3
MI IRON	0.93	TP9-16	30-Apr-55	1	4
MI IRON	6.09	PB5-5	19-Jul-57	2	4
MI IRON	1.97	PB8-5	22-Aug-57	9	4
MI IRON	0.97	PB16-2	24-Sep-57	2	4
MI ISABELLA	2.46	UpK9-6	24-May-53	7	4
MI ISABELLA	1.36	PB4-3	7-Jul-57	4	4
MI ISABELLA	1.51	PB9-5	27-Aug-57	9	4
MI ISABELLA	7.97	UE12-4	17-Apr-65	8	4
MI JACKSON	1.02	PB4-3	7-Jul-57	8	4
MI JACKSON	3.01	PB9-5	27-Aug-57	6	4
MI JACKSON	0.83	PB18-9	15-Oct-57	6	4
MI JACKSON	0.13	UE7-5	15-Jul-62	8	4
MI KALAMAZOO	1.72	TP3-3	3-Mar-55	8	5
MI KALAMAZOO	1.02	PB4-3	7-Jul-57	8	5
MI KALAMAZOO	3.01	PB9-5	27-Aug-57	6	5
MI KALAMAZOO	0.83	PB18-9	15-Oct-57	6	5
MI KALAMAZOO	0.24	UE7-5	15-Jul-62	5	5
MI KALKASKA	0.97	UpK10-10	3-Jun-53	1	2
MI KALKASKA	1.02	PB4-3	7-Jul-57	8	2

COUNTY	uCi/ sqm	SHOT DAY	DATE	RANK FOR DAY	TIMES IN TOP 10
MI KENT	4.52	PB9-5	27-Aug-57	2	2
MI KENT	0.17	UE7-5	15-Jul-62	7	2
MI KEWEENAW	6.09	PB5-5	19-Jul-57	2	1
MI LAKE	2.88	UpK9-6	24-May-53	2	1
MI LAPEER	4.21	BJ3-5	5-Nov-51	7	6
MI LAPEER	1.26	TP9-4	18-Apr-55	5	6
MI LAPEER	1.02	PB4-3	7-Jul-57	8	6
MI LAPEER	3.01	PB9-5	27-Aug-57	6	6
MI LAPEER	0.85	PB13-7	12-Sep-57	6	6
MI LAPEER	7.97	UE12-4	17-Apr-65	8	6
MI LEELANAU	1.26	TP9-4	18-Apr-55	6	1
MI LENAWEE	0.93	PB4-3	7-Jul-57	9	3
MI LENAWEE	4.52	PB9-5	27-Aug-57	2	3
MI LENAWEE	0.24	UE7-5	15-Jul-62	5	3
MI LIVINGSTON	1.02	PB4-3	7-Jul-57	8	3
MI LIVINGSTON	3.01	PB9-5	27-Aug-57	6	3
MI LIVINGSTON	0.85	PB13-7	12-Sep-57	6	3
MI LUCE	5.56	TS8-3	7-Jun-52	4	4
MI LUCE	2.34	UpK10-6	30-May-53	10	4
MI LUCE	0.92	PB6-9	1-Aug-57	6	4
MI LUCE	1.51	PB9-5	27-Aug-57	9	4
MI MACKINAC	5.42	TS8-3	7-Jun-52	5	3
MI MACKINAC	1.21	TP9-4	18-Apr-55	7	3
MI MACKINAC	1.51	PB9-5	27-Aug-57	9	3
MI MACOMB	6.32	BJ3-5	5-Nov-51	4	6
MI MACOMB	0.95	BJ4-8	12-Nov-51	2	6
MI MACOMB	53.93	TP9-2	16-Apr-55	5	6
MI MACOMB	1.02	PB4-3	7-Jul-57	8	6
MI MACOMB	1.51	PB9-5	27-Aug-57	9	6
MI MACOMB	7.97	UE12-4	17-Apr-65	8	6
MI MANISTEE	2.78	UpK9-6	24-May-53	3	4
MI MANISTEE	1.02	PB4-3	7-Jul-57	8	4
MI MANISTEE	1.51	PB9-5	27-Aug-57	9	4
MI MANISTEE	7.97	UE12-4	17-Apr-65	8	4
MI MARQUETTE	4.02	PB5-5	19-Jul-57	3	1

COUNTY	uCi/ sqm	SHOT DAY	DATE	RANK FOR DAY	TIMES IN TOP 10
MI MASON	1.51	PB9-5	27-Aug-57	9	2
MI MASON	7.97	UE12-4	17-Apr-65	8	2
MI MECOSTA	1.02	PB4-3	7-Jul-57	8	4
MI MECOSTA	1.51	PB9-5	27-Aug-57	9	4
MI MECOSTA	0.08	UE2-7	11-Mar-62	5	4
MI MECOSTA	7.97	UE12-4	17-Apr-65	8	4
MI MENOMINEE	3.55	UpK10-4	28-May-53	2	4
MI MENOMINEE	13.72	PB9-3	25-Aug-57	6	4
MI MENOMINEE	8.37	PB9-4	26-Aug-57	7	4
MI MENOMINEE	2.04	PB9-5	27-Aug-57	8	4
MI MIDLAND	1.36	PB4-3	7-Jul-57	4	5
MI MIDLAND	3.01	PB9-5	27-Aug-57	6	5
MI MIDLAND	0.83	PB18-9	15-Oct-57	6	5
MI MIDLAND	0.08	UE2-7	11-Mar-62	5	5
MI MIDLAND	7.97	UE12-4	17-Apr-65	8	5
MI MISSAUKEE	0.85	TS4-4	4-May-52	2	5
MI MISSAUKEE	0.85	TS4-4	4-May-52	2	5
MI MISSAUKEE	1.36	PB4-3	7-Jul-57	4	5
MI MISSAUKEE	1.51	PB9-5	27-Aug-57	9	5
MI MISSAUKEE	6.71	PB11-4	3-Sep-57	9	5
MI MONROE	0.93	PB4-3	7-Jul-57	9	4
MI MONROE	1.51	PB9-5	27-Aug-57	9	4
MI MONROE	1.25	PB13-7	12-Sep-57	3	4
MI MONROE	85.56	S-4	9-Jul-62	9	4
MI MONTCALM	1.02	PB4-3	7-Jul-57	8	2
MI MONTCALM	14.56	UE12-4	17-Apr-65	2	2
MI MONTMORENCY	0.97	UpK10-10	3-Jun-53	1	3
MI MONTMORENCY	1.39	TP9-4	18-Apr-55	2	3
MI MONTMORENCY	3.01	PB9-5	27-Aug-57	6	3
MI MUSKEGON	2.61	UpK9-6	24-May-53	5	3
MI MUSKEGON	3.01	PB9-5	27-Aug-57	6	3
MI MUSKEGON	14.56	UE12-4	17-Apr-65	2	3
MI NEWAYGO	3.01	PB9-5	27-Aug-57	6	2
MI NEWAYGO	14.56	UE12-4	17-Apr-65	2	2
MI OAKLAND	0.95	BJ4-8	12-Nov-51	2	2

COUNTY	uCi/ sqm	SHOT DAY	DATE	RANK FOR DAY	TIMES IN TOP 10
MI OAKLAND	1.10	PB4-3	7-Jul-57	7	2
MI OCEANA	2.43	UpK9-6	24-May-53	8	3
MI OCEANA	3.01	PB9-5	27-Aug-57	6	3
MI OCEANA	14.56	UE12-4	17-Apr-65	2	3
MI OGEMAW	1.42	UpK10-5	29-May-53	10	5
MI OGEMAW	0.97	UpK10-10	3-Jun-53	1	5
MI OGEMAW	1.14	TP9-4	18-Apr-55	10	5
MI OGEMAW	0.85	PB13-7	12-Sep-57	6	5
MI OGEMAW	7.97	UE12-4	17-Apr-65	8	5
MI ONTONAGON	0.93	TP9-16	30-Apr-55	1	3
MI ONTONAGON	1.50	PB2-4	21-Jun-57	8	3
MI ONTONAGON	2.01	PB5-5	19-Jul-57	8	3
MI OSCEOLA	7.97	UE12-4	17-Apr-65	8	1
MI OSCODA	0.97	UpK10-10	3-Jun-53	1	2
MI OSCODA	1.51	PB9-5	27-Aug-57	9	2
MI OTSEGO	0.97	UpK10-10	3-Jun-53	1	4
MI OTSEGO	1.14	TP9-4	18-Apr-55	10	4
MI OTSEGO	1.51	PB9-5	27-Aug-57	9	4
MI OTSEGO	6.71	PB11-4	3-Sep-57	9	4
MI OTTAWA	4.52	PB9-5	27-Aug-57	2	1
MI PRESQUE ISLE	1.69	UpK10-5	29-May-53	3	4
MI PRESQUE ISLE	0.97	UpK10-10	3-Jun-53	1	4
MI PRESQUE ISLE	1.51	PB9-5	27-Aug-57	9	4
MI PRESQUE ISLE	6.71	PB11-4	3-Sep-57	9	4
MI ROSCOMMON	1.27	PB4-3	7-Jul-57	5	3
MI ROSCOMMON	1.51	PB9-5	27-Aug-57	9	3
MI ROSCOMMON	6.71	PB11-4	3-Sep-57	9	3
MI SAGINAW	0.85	TS4-4	4-May-52	2	5
MI SAGINAW	0.85	TS4-4	4-May-52	2	5
MI SAGINAW	1.36	PB4-3	7-Jul-57	4	5
MI SAGINAW	3.01	PB9-5	27-Aug-57	6	5
MI SAGINAW	7.97	UE12-4	17-Apr-65	8	5
MI SANILAC	0.95	BJ4-8	12-Nov-51	2	5
MI SANILAC	1.31	TP9-4	18-Apr-55	3	5
MI SANILAC	0.93	PB4-3	7-Jul-57	9	5

COUNTY	uCi/ sqm	SHOT DAY	DATE	RANK FOR DAY	TIMES IN TOP 10
MI SANILAC	1.51	PB9-5	27-Aug-57	9	5
MI SANILAC	7.97	UE12-4	17-Apr-65	8	5
MI SCHOOLCRAFT	1.51	PB9-5	27-Aug-57	9	1
MI SHIAWASSEE	1.14	TP9-4	18-Apr-55	10	4
MI SHIAWASSEE	1.02	PB4-3	7-Jul-57	8	4
MI SHIAWASSEE	1.25	PB13-7	12-Sep-57	3	4
MI SHIAWASSEE	7.97	UE12-4	17-Apr-65	8	4
MI ST CLAIR	4.21	BJ3-5	5-Nov-51	7	7
MI ST CLAIR	0.95	BJ4-8	12-Nov-51	2	7
MI ST CLAIR	1.31	TP9-4	18-Apr-55	4	7
MI ST CLAIR	0.93	PB4-3	7-Jul-57	9	7
MI ST CLAIR	3.01	PB9-5	27-Aug-57	6	7
MI ST CLAIR	0.85	PB13-7	12-Sep-57	6	7
MI ST CLAIR	7.97	UE12-4	17-Apr-65	8	7
MI ST JOSEPH	1.02	PB4-3	7-Jul-57	8	4
MI ST JOSEPH	3.01	PB9-5	27-Aug-57	6	4
MI ST JOSEPH	0.83	PB18-9	15-Oct-57	6	4
MI ST JOSEPH	0.12	UE7-5	15-Jul-62	9	4
MI TUSCOLA	1.18	TP9-4	18-Apr-55	8	4
MI TUSCOLA	1.02	PB4-3	7-Jul-57	8	4
MI TUSCOLA	3.01	PB9-5	27-Aug-57	6	4
MI TUSCOLA	7.97	UE12-4	17-Apr-65	8	4
MI VAN BUREN	3.01	PB9-5	27-Aug-57	6	2
MI VAN BUREN	0.12	UE7-5	15-Jul-62	9	2
MI WASHTENAW	1.02	PB4-3	7-Jul-57	8	4
MI WASHTENAW	3.01	PB9-5	27-Aug-57	6	4
MI WASHTENAW	0.85	PB13-7	12-Sep-57	6	4
MI WASHTENAW	0.17	UE7-5	15-Jul-62	7	4
MI WAYNE	6.32	BJ3-5	5-Nov-51	4	7
MI WAYNE	1.37	BJ4-7	11-Nov-51	3	7
MI WAYNE	1.43	BJ4-8	12-Nov-51	1	7
MI WAYNE	81.47	TP9-2	16-Apr-55	1	7
MI WAYNE	1.52	PB4-3	7-Jul-57	2	7
MI WAYNE	3.01	PB9-5	27-Aug-57	6	7
MI WAYNE	0.17	UE7-5	15-Jul-62	7	7

COUNTY	uCi/ sqm	SHOT DAY	DATE	RANK FOR DAY	TIMES IN TOP 10
MI WEXFORD	2.42	UpK9-6	24-May-53	9	4
MI WEXFORD	1.36	PB4-3	7-Jul-57	4	4
MI WEXFORD	1.51	PB9-5	27-Aug-57	9	4
MI WEXFORD	7.97	UE12-4	17-Apr-65	8	4

COUNTY	uCi/ sqm	SHOT DAY	DATE	RANK FOR DAY	TIMES IN TOP 10
MN AITKIN	2.88	UpK10-4	28-May-53	5	6
MN AITKIN	1.50	PB2-4	21-Jun-57	8	6
MN AITKIN	10.11	PB5-5	19-Jul-57	1	6
MN AITKIN	4.11	PB15-2	17-Sep-57	9	6
MN AITKIN	1.55	PB15-3	18-Sep-57	10	6
MN AITKIN	0.62	UE2-4	8-Mar-62	6	6
MN ANOKA	2.93	UpK10-4	28-May-53	4	7
MN ANOKA	2.54	PB2-4	21-Jun-57	4	7
MN ANOKA	0.86	PB2-6	23-Jun-57	8	7
MN ANOKA	10.11	PB5-5	19-Jul-57	1	7
MN ANOKA	1.97	PB8-5	22-Aug-57	9	7
MN ANOKA	1.55	PB15-3	18-Sep-57	10	7
MN ANOKA	0.10	UE2-7	11-Mar-62	4	7
MN BECKER	6.09	PB5-5	19-Jul-57	2	5
MN BECKER	1.20	PB6-10	2-Aug-57	2	5
MN BECKER	2.58	PB15-3	18-Sep-57	5	5
MN BECKER	0.98	UE2-4	8-Mar-62	2	5
MN BECKER	0.21	UE2-6	10-Mar-62	3	5
MN BELTRAMI	2.54	PB2-4	21-Jun-57	4	5
MN BELTRAMI	6.09	PB5-5	19-Jul-57	2	5
MN BELTRAMI	4.11	PB15-2	17-Sep-57	9	5
MN BELTRAMI	1.55	PB15-3	18-Sep-57	10	5
MN BELTRAMI	0.93	UE2-4	8-Mar-62	3	5
MN BENTON	2.54	PB2-4	21-Jun-57	4	4
MN BENTON	10.11	PB5-5	19-Jul-57	1	4
MN BENTON	1.97	PB8-5	22-Aug-57	9	4
MN BENTON	1.55	PB15-3	18-Sep-57	10	4
MN BIG STONE	2.01	PB5-5	19-Jul-57	8	5
MN BIG STONE	1.20	PB6-10	2-Aug-57	2	5
MN BIG STONE	2.58	PB15-3	18-Sep-57	5	5
MN BIG STONE	0.98	UE2-4	8-Mar-62	2	5
MN BIG STONE	0.10	UE2-7	11-Mar-62	3	5
MN BLUE EARTH	2.54	PB2-4	21-Jun-57	4	8
MN BLUE EARTH	0.86	PB2-6	23-Jun-57	8	8
MN BLUE EARTH	0.85	PB4-3	7-Jul-57	10	8

COUNTY	uCi/ sqm	SHOT DAY	DATE	RANK FOR DAY	TIMES IN TOP 10
MN BLUE EARTH	6.09	PB5-5	19-Jul-57	2	8
MN BLUE EARTH	1.97	PB8-5	22-Aug-57	9	8
MN BLUE EARTH	0.17	UE2-7	11-Mar-62	2	8
MN BLUE EARTH	325.19	S-3	8-Jul-62	10	8
MN BLUE EARTH	9.32	UE12-4	17-Apr-65	6	8
MN BROWN	7.00	UpK7-5	29-Apr-53	8	11
MN BROWN	2.54	PB2-4	21-Jun-57	4	11
MN BROWN	0.86	PB2-6	23-Jun-57	8	11
MN BROWN	6.09	PB5-5	19-Jul-57	2	11
MN BROWN	1.18	PB5-6	20-Jul-57	7	11
MN BROWN	1.97	PB8-5	22-Aug-57	9	11
MN BROWN	1.55	PB15-3	18-Sep-57	10	11
MN BROWN	0.10	UE2-7	11-Mar-62	4	11
MN BROWN	325.19	S-3	8-Jul-62	10	11
MN BROWN	0.12	UE7-5	15-Jul-62	9	11
MN BROWN	9.84	UE12-4	17-Apr-65	4	11
MN CARLTON	2.81	UpK10-4	28-May-53	6	5
MN CARLTON	2.54	PB2-4	21-Jun-57	4	5
MN CARLTON	10.11	PB5-5	19-Jul-57	1	5
MN CARLTON	4.11	PB15-2	17-Sep-57	9	5
MN CARLTON	1.55	PB15-3	18-Sep-57	10	5
MN CARVER	2.54	PB2-4	21-Jun-57	4	5
MN CARVER	10.11	PB5-5	19-Jul-57	1	5
MN CARVER	1.18	PB5-6	20-Jul-57	7	5
MN CARVER	4.05	PB6-9	1-Aug-57	1	5
MN CARVER	0.22	UE7-5	15-Jul-62	6	5
MN CASS	2.54	PB2-4	21-Jun-57	4	6
MN CASS	6.09	PB5-5	19-Jul-57	2	6
MN CASS	4.11	PB15-2	17-Sep-57	9	6
MN CASS	2.58	PB15-3	18-Sep-57	5	6
MN CASS	0.93	UE2-4	8-Mar-62	3	6
MN CASS	0.10	UE2-7	11-Mar-62	3	6
MN CHIPPEWA	4.05	PB6-9	1-Aug-57	1	5
MN CHIPPEWA	1.20	PB6-10	2-Aug-57	2	5
MN CHIPPEWA	1.97	PB8-5	22-Aug-57	9	5

COUNTY	uCi/ sqm	SHOT DAY	DATE	RANK FOR DAY	TIMES IN TOP 10
MN CHIPPEWA	2.58	PB15-3	18-Sep-57	5	5
MN CHIPPEWA	0.10	UE2-7	11-Mar-62	4	5
MN CHISAGO	2.66	UpK10-4	28-May-53	8	6
MN CHISAGO	1.50	PB2-4	21-Jun-57	8	6
MN CHISAGO	4.02	PB5-5	19-Jul-57	3	6
MN CHISAGO	1.97	PB8-5	22-Aug-57	9	6
MN CHISAGO	1.55	PB15-3	18-Sep-57	10	6
MN CHISAGO	0.17	UE2-7	11-Mar-62	2	6
MN CLAY	1.32	TS3-8	29-Apr-52	7	6
MN CLAY	1.32	TS3-8	29-Apr-52	7	6
MN CLAY	2.54	PB2-4	21-Jun-57	4	6
MN CLAY	1.55	PB15-3	18-Sep-57	10	6
MN CLAY	0.98	UE2-4	8-Mar-62	2	6
MN CLAY	0.10	UE2-7	11-Mar-62	3	6
MN CLEARWATER	2.54	PB2-4	21-Jun-57	4	4
MN CLEARWATER	4.11	PB15-2	17-Sep-57	9	4
MN CLEARWATER	1.55	PB15-3	18-Sep-57	10	4
MN CLEARWATER	0.93	UE2-4	8-Mar-62	3	4
MN COOK	1.50	PB2-4	21-Jun-57	8	3
MN COOK	4.02	PB5-5	19-Jul-57	3	3
MN COOK	1.55	PB15-3	18-Sep-57	10	3
MN COTTONWOOD	7.49	UpK7-5	29-Apr-53	2	9
MN COTTONWOOD	2.54	PB2-4	21-Jun-57	4	9
MN COTTONWOOD	4.02	PB5-5	19-Jul-57	3	9
MN COTTONWOOD	1.18	PB5-6	20-Jul-57	7	9
MN COTTONWOOD	1.97	PB8-5	22-Aug-57	9	9
MN COTTONWOOD	1.55	PB15-3	18-Sep-57	10	9
MN COTTONWOOD	0.17	UE2-7	11-Mar-62	2	9
MN COTTONWOOD	0.12	UE7-5	15-Jul-62	9	9
MN COTTONWOOD	9.84	UE12-4	17-Apr-65	4	9
MN CROW WING	2.37	UpK10-4	28-May-53	10	8
MN CROW WING	1.50	PB2-4	21-Jun-57	8	8
MN CROW WING	6.09	PB5-5	19-Jul-57	2	8
MN CROW WING	4.11	PB15-2	17-Sep-57	9	8
MN CROW WING	1.55	PB15-3	18-Sep-57	10	8

COUNTY	uCi/ sqm	SHOT DAY	DATE	RANK FOR DAY	TIMES IN TOP 10
MN CROW WING	0.98	UE2-4	8-Mar-62	2	8
MN CROW WING	0.12	UE2-5	9-Mar-62	9	8
MN CROW WING	0.10	UE2-7	11-Mar-62	3	8
MN DAKOTA	0.99	TS3-8	29-Apr-52	8	10
MN DAKOTA	0.99	TS3-8	29-Apr-52	8	10
MN DAKOTA	2.54	PB2-4	21-Jun-57	4	10
MN DAKOTA	10.11	PB5-5	19-Jul-57	1	10
MN DAKOTA	1.20	PB6-10	2-Aug-57	2	10
MN DAKOTA	1.97	PB8-5	22-Aug-57	9	10
MN DAKOTA	1.55	PB15-3	18-Sep-57	10	10
MN DAKOTA	1.16	PB15-5	20-Sep-57	7	10
MN DAKOTA	0.17	UE2-7	11-Mar-62	2	10
MN DAKOTA	0.22	UE7-5	15-Jul-62	6	10
MN DODGE	5.92	TS3-8	29-Apr-52	2	8
MN DODGE	5.92	TS3-8	29-Apr-52	2	8
MN DODGE	2.54	PB2-4	21-Jun-57	4	8
MN DODGE	0.93	PB4-3	7-Jul-57	9	8
MN DODGE	1.18	PB5-6	20-Jul-57	7	8
MN DODGE	0.10	UE2-7	11-Mar-62	3	8
MN DODGE	0.12	UE7-5	15-Jul-62	9	8
MN DODGE	22.99	UE12-4	17-Apr-65	1	8
MN DOUGLAS	2.54	PB2-4	21-Jun-57	4	8
MN DOUGLAS	2.01	PB5-5	19-Jul-57	8	8
MN DOUGLAS	4.05	PB6-9	1-Aug-57	1	8
MN DOUGLAS	6.81	PB15-2	17-Sep-57	4	8
MN DOUGLAS	1.55	PB15-3	18-Sep-57	10	8
MN DOUGLAS	1.18	UE2-4	8-Mar-62	1	8
MN DOUGLAS	0.21	UE2-6	10-Mar-62	3	8
MN DOUGLAS	0.10	UE2-7	11-Mar-62	3	8
MN FARIBAULT	0.99	TS3-8	29-Apr-52	8	9
MN FARIBAULT	0.99	TS3-8	29-Apr-52	8	9
MN FARIBAULT	1.50	PB2-4	21-Jun-57	8	9
MN FARIBAULT	1.19	PB4-3	7-Jul-57	6	9
MN FARIBAULT	6.09	PB5-5	19-Jul-57	2	9
MN FARIBAULT	1.55	PB15-3	18-Sep-57	10	9

COUNTY	uCi/ sqm	SHOT DAY	DATE	RANK FOR DAY	TIMES IN TOP 10
MN FARIBAULT	0.10	UE2-7	11-Mar-62	4	9
MN FARIBAULT	584.61	S-3	8-Jul-62	4	9
MN FARIBAULT	0.13	UE7-5	15-Jul-62	8	9
MN FILLMORE	0.99	TS3-8	29-Apr-52	8	10
MN FILLMORE	0.99	TS3-8	29-Apr-52	8	10
MN FILLMORE	1.50	PB2-4	21-Jun-57	8	10
MN FILLMORE	1.19	PB4-3	7-Jul-57	6	10
MN FILLMORE	1.18	PB5-6	20-Jul-57	7	10
MN FILLMORE	1.47	PB6-7	30-Jul-57	10	10
MN FILLMORE	0.17	UE2-7	11-Mar-62	2	10
MN FILLMORE	327.38	S-3	8-Jul-62	9	10
MN FILLMORE	0.24	UE7-5	15-Jul-62	5	10
MN FILLMORE	9.84	UE12-4	17-Apr-65	4	10
MN FREEBORN	0.99	TS3-8	29-Apr-52	8	7
MN FREEBORN	0.99	TS3-8	29-Apr-52	8	7
MN FREEBORN	2.54	PB2-4	21-Jun-57	4	7
MN FREEBORN	1.19	PB4-3	7-Jul-57	6	7
MN FREEBORN	10.11	PB5-5	19-Jul-57	1	7
MN FREEBORN	0.10	UE2-7	11-Mar-62	4	7
MN FREEBORN	9.84	UE12-4	17-Apr-65	4	7
MN GOODHUE	0.99	TS3-8	29-Apr-52	8	8
MN GOODHUE	0.99	TS3-8	29-Apr-52	8	8
MN GOODHUE	2.54	PB2-4	21-Jun-57	4	8
MN GOODHUE	1.61	PB4-3	7-Jul-57	1	8
MN GOODHUE	10.11	PB5-5	19-Jul-57	1	8
MN GOODHUE	0.10	UE2-7	11-Mar-62	4	8
MN GOODHUE	0.22	UE7-5	15-Jul-62	6	8
MN GOODHUE	14.56	UE12-4	17-Apr-65	2	8
MN GRANT	1.50	PB2-4	21-Jun-57	8	7
MN GRANT	2.01	PB5-5	19-Jul-57	8	7
MN GRANT	1.20	PB6-10	2-Aug-57	2	7
MN GRANT	4.11	PB15-2	17-Sep-57	9	7
MN GRANT	1.55	PB15-3	18-Sep-57	10	7
MN GRANT	0.98	UE2-4	8-Mar-62	2	7
MN GRANT	0.10	UE2-7	11-Mar-62	4	7

COUNTY	uCi/ sqm	SHOT DAY	DATE	RANK FOR DAY	TIMES IN TOP 10
MN HENNEPIN	0.93	PB4-3	7-Jul-57	9	5
MN HENNEPIN	4.20	PB5-5	19-Jul-57	3	5
MN HENNEPIN	1.01	PB6-9	1-Aug-57	5	5
MN HENNEPIN	0.17	UE2-7	11-Mar-62	2	5
MN HENNEPIN	0.22	UE7-5	15-Jul-62	6	5
MN HOUSTON	0.99	TS3-8	29-Apr-52	8	11
MN HOUSTON	0.99	TS3-8	29-Apr-52	8	11
MN HOUSTON	1.52	PB4-3	7-Jul-57	2	11
MN HOUSTON	4.15	PB5-7	21-Jul-57	2	11
MN HOUSTON	2.24	PB6-8	31-Jul-57	4	11
MN HOUSTON	13.72	PB9-3	25-Aug-57	6	11
MN HOUSTON	16.67	PB9-4	26-Aug-57	4	11
MN HOUSTON	2.04	PB9-5	27-Aug-57	8	11
MN HOUSTON	3.98	PB18-8	14-Oct-57	4	11
MN HOUSTON	0.17	UE2-7	11-Mar-62	2	11
MN HOUSTON	9.32	UE12-4	17-Apr-65	6	11
MN HUBBARD	1.50	PB2-4	21-Jun-57	8	6
MN HUBBARD	6.09	PB5-5	19-Jul-57	2	6
MN HUBBARD	2.58	PB15-3	18-Sep-57	5	6
MN HUBBARD	0.93	UE2-4	8-Mar-62	3	6
MN HUBBARD	0.44	UE2-5	9-Mar-62	1	6
MN HUBBARD	0.10	UE2-7	11-Mar-62	3	6
MN ISANTI	2.66	UpK10-4	28-May-53	8	7
MN ISANTI	1.50	PB2-4	21-Jun-57	8	7
MN ISANTI	4.02	PB5-5	19-Jul-57	3	7
MN ISANTI	2.43	PB6-9	1-Aug-57	2	7
MN ISANTI	1.97	PB8-5	22-Aug-57	9	7
MN ISANTI	1.55	PB15-3	18-Sep-57	10	7
MN ISANTI	0.17	UE2-7	11-Mar-62	2	7
MN ITASCA	2.54	PB2-4	21-Jun-57	4	5
MN ITASCA	10.11	PB5-5	19-Jul-57	1	5
MN ITASCA	4.11	PB15-2	17-Sep-57	9	5
MN ITASCA	2.58	PB15-3	18-Sep-57	5	5
MN ITASCA	0.65	UE2-4	8-Mar-62	5	5
MN JACKSON	7.52	UpK7-5	29-Apr-53	1	8

COUNTY	uCi/ sqm	SHOT DAY	DATE	RANK FOR DAY	TIMES IN TOP 10
MN JACKSON	2.54	PB2-4	21-Jun-57	4	8
MN JACKSON	1.44	PB4-3	7-Jul-57	3	8
MN JACKSON	2.01	PB5-5	19-Jul-57	8	8
MN JACKSON	1.55	PB15-3	18-Sep-57	10	8
MN JACKSON	0.17	UE2-7	11-Mar-62	2	8
MN JACKSON	0.13	UE7-5	15-Jul-62	8	8
MN JACKSON	11.43	UE12-4	17-Apr-65	3	8
MN KANABEC	2.54	PB2-4	21-Jun-57	4	4
MN KANABEC	6.09	PB5-5	19-Jul-57	2	4
MN KANABEC	2.58	PB15-3	18-Sep-57	5	4
MN KANABEC	0.17	UE2-7	11-Mar-62	2	4
MN KANDIYOHI	2.54	PB2-4	21-Jun-57	4	7
MN KANDIYOHI	6.09	PB5-5	19-Jul-57	2	7
MN KANDIYOHI	2.45	PB6-7	30-Jul-57	3	7
MN KANDIYOHI	1.20	PB6-10	2-Aug-57	2	7
MN KANDIYOHI	2.58	PB15-3	18-Sep-57	5	7
MN KANDIYOHI	0.10	UE2-7	11-Mar-62	4	7
MN KANDIYOHI	0.11	UE7-5	15-Jul-62	10	7
MN KITTSON	10.86	TP9-3	17-Apr-55	2	4
MN KITTSON	2.43	PB6-9	1-Aug-57	2	4
MN KITTSON	1.18	UE2-4	8-Mar-62	1	4
MN KITTSON	0.22	BN-5	22-Dec-70	1	4
MN KOOCHICHING	2.54	PB2-4	21-Jun-57	4	7
MN KOOCHICHING	6.09	PB5-5	19-Jul-57	2	7
MN KOOCHICHING	1.63	PB6-9	1-Aug-57	3	7
MN KOOCHICHING	6.81	PB15-2	17-Sep-57	4	7
MN KOOCHICHING	1.55	PB15-3	18-Sep-57	10	7
MN KOOCHICHING	0.62	UE2-4	8-Mar-62	6	7
MN KOOCHICHING	0.44	UE2-5	9-Mar-62	1	7
MN LAC QUI PARL	6.09	PB5-5	19-Jul-57	2	6
MN LAC QUI PARL	1.63	PB6-9	1-Aug-57	3	6
MN LAC QUI PARL	1.97	PB8-5	22-Aug-57	9	6
MN LAC QUI PARL	2.58	PB15-3	18-Sep-57	5	6
MN LAC QUI PARL	0.10	UE2-7	11-Mar-62	4	6
MN LAC QUI PARL	0.11	UE7-5	15-Jul-62	10	6

COUNTY	uCi/ sqm	SHOT DAY	DATE	RANK FOR DAY	TIMES IN TOP 10
MN LAKE	2.54	PB2-4	21-Jun-57	4	3
MN LAKE	6.09	PB5-5	19-Jul-57	2	3
MN LAKE	1.55	PB15-3	18-Sep-57	10	3
MN LAKE OF THE	1.50	PB2-4	21-Jun-57	8	4
MN LAKE OF THE	2.01	PB5-5	19-Jul-57	8	4
MN LAKE OF THE	4.11	PB15-2	17-Sep-57	9	4
MN LAKE OF THE	0.62	UE2-4	8-Mar-62	6	4
MN LE SUEUR	2.54	PB2-4	21-Jun-57	4	11
MN LE SUEUR	0.86	PB2-6	23-Jun-57	8	11
MN LE SUEUR	1.61	PB4-3	7-Jul-57	1	11
MN LE SUEUR	10.11	PB5-5	19-Jul-57	1	11
MN LE SUEUR	1.20	PB6-10	2-Aug-57	2	11
MN LE SUEUR	1.97	PB8-5	22-Aug-57	9	11
MN LE SUEUR	1.55	PB15-3	18-Sep-57	10	11
MN LE SUEUR	0.10	UE2-7	11-Mar-62	4	11
MN LE SUEUR	325.19	S-3	8-Jul-62	10	11
MN LE SUEUR	0.12	UE7-5	15-Jul-62	9	11
MN LE SUEUR	9.32	UE12-4	17-Apr-65	6	11
MN LINCOLN	3.97	UpK2-2	25-Mar-53	10	8
MN LINCOLN	2.54	PB2-4	21-Jun-57	4	8
MN LINCOLN	6.09	PB5-5	19-Jul-57	2	8
MN LINCOLN	1.97	PB8-5	22-Aug-57	9	8
MN LINCOLN	2.58	PB15-3	18-Sep-57	5	8
MN LINCOLN	0.10	UE2-7	11-Mar-62	4	8
MN LINCOLN	0.11	UE7-5	15-Jul-62	10	8
MN LINCOLN	11.43	UE12-4	17-Apr-65	3	8
MN LYON	3.97	UpK2-2	25-Mar-53	10	8
MN LYON	2.54	PB2-4	21-Jun-57	4	8
MN LYON	10.11	PB5-5	19-Jul-57	1	8
MN LYON	1.97	PB8-5	22-Aug-57	9	8
MN LYON	2.58	PB15-3	18-Sep-57	5	8
MN LYON	0.10	UE2-7	11-Mar-62	4	8
MN LYON	0.11	UE7-5	15-Jul-62	10	8
MN LYON	9.84	UE12-4	17-Apr-65	4	8
MN MAHNOMEN	2.54	PB2-4	21-Jun-57	4	4

COUNTY	uCi/ sqm	SHOT DAY	DATE	RANK FOR DAY	TIMES IN TOP 10
MN MAHNOMEN	4.02	PB5-5	19-Jul-57	3	4
MN MAHNOMEN	2.58	PB15-3	18-Sep-57	5	4
MN MAHNOMEN	0.93	UE2-4	8-Mar-62	3	4
MN MARSHALL	1.63	PB6-9	1-Aug-57	3	4
MN MARSHALL	4.11	PB15-2	17-Sep-57	9	4
MN MARSHALL	1.18	UE2-4	8-Mar-62	1	4
MN MARSHALL	0.11	BN-5	22-Dec-70	5	4
MN MARTIN	1.50	PB2-4	21-Jun-57	8	9
MN MARTIN	0.85	PB4-3	7-Jul-57	10	9
MN MARTIN	10.11	PB5-5	19-Jul-57	1	9
MN MARTIN	2.45	PB6-7	30-Jul-57	3	9
MN MARTIN	1.55	PB15-3	18-Sep-57	10	9
MN MARTIN	0.10	UE2-7	11-Mar-62	3	9
MN MARTIN	327.38	S-3	8-Jul-62	9	9
MN MARTIN	0.13	UE7-5	15-Jul-62	8	9
MN MARTIN	9.84	UE12-4	17-Apr-65	4	9
MN MCLEOD	2.54	PB2-4	21-Jun-57	4	4
MN MCLEOD	1.20	PB6-10	2-Aug-57	2	4
MN MCLEOD	1.55	PB15-3	18-Sep-57	10	4
MN MCLEOD	0.10	UE2-7	11-Mar-62	4	4
MN MEEKER	2.54	PB2-4	21-Jun-57	4	9
MN MEEKER	6.09	PB5-5	19-Jul-57	2	9
MN MEEKER	2.45	PB6-7	30-Jul-57	3	9
MN MEEKER	1.20	PB6-10	2-Aug-57	2	9
MN MEEKER	1.97	PB8-5	22-Aug-57	9	9
MN MEEKER	1.11	PB9-8	30-Aug-57	1	9
MN MEEKER	1.55	PB15-3	18-Sep-57	10	9
MN MEEKER	0.10	UE2-7	11-Mar-62	3	9
MN MEEKER	0.11	UE7-5	15-Jul-62	10	9
MN MILLE LACS	2.54	PB2-4	21-Jun-57	4	5
MN MILLE LACS	6.09	PB5-5	19-Jul-57	2	5
MN MILLE LACS	1.11	PB9-8	30-Aug-57	1	5
MN MILLE LACS	2.58	PB15-3	18-Sep-57	5	5
MN MILLE LACS	0.10	UE2-7	11-Mar-62	4	5
MN MORRISON	2.54	PB2-4	21-Jun-57	4	5

COUNTY	uCi/ sqm	SHOT DAY	DATE	RANK FOR DAY	TIMES IN TOP 10
MN MORRISON	10.11	PB5-5	19-Jul-57	1	5
MN MORRISON	2.58	PB15-3	18-Sep-57	5	5
MN MORRISON	0.49	UE2-4	8-Mar-62	8	5
MN MORRISON	0.12	UE2-5	9-Mar-62	8	5
MN MOWER	5.92	TS3-8	29-Apr-52	2	10
MN MOWER	5.92	TS3-8	29-Apr-52	2	10
MN MOWER	2.54	PB2-4	21-Jun-57	4	10
MN MOWER	0.93	PB4-3	7-Jul-57	9	10
MN MOWER	1.18	PB5-6	20-Jul-57	7	10
MN MOWER	1.16	PB15-5	20-Sep-57	7	10
MN MOWER	0.17	UE2-7	11-Mar-62	2	10
MN MOWER	584.61	S-3	8-Jul-62	4	10
MN MOWER	0.13	UE7-5	15-Jul-62	8	10
MN MOWER	9.84	UE12-4	17-Apr-65	4	10
MN MURRAY	3.97	UpK2-2	25-Mar-53	10	9
MN MURRAY	2.54	PB2-4	21-Jun-57	4	9
MN MURRAY	4.02	PB5-5	19-Jul-57	3	9
MN MURRAY	1.97	PB8-5	22-Aug-57	9	9
MN MURRAY	1.55	PB15-3	18-Sep-57	10	9
MN MURRAY	0.10	UE2-7	11-Mar-62	4	9
MN MURRAY	327.38	S-3	8-Jul-62	9	9
MN MURRAY	0.22	UE7-5	15-Jul-62	6	9
MN MURRAY	9.84	UE12-4	17-Apr-65	4	9
MN NICOLLET	2.54	PB2-4	21-Jun-57	4	10
MN NICOLLET	0.86	PB2-6	23-Jun-57	8	10
MN NICOLLET	0.85	PB4-3	7-Jul-57	10	10
MN NICOLLET	6.09	PB5-5	19-Jul-57	2	10
MN NICOLLET	1.97	PB8-5	22-Aug-57	9	10
MN NICOLLET	1.16	PB15-5	20-Sep-57	7	10
MN NICOLLET	0.17	UE2-7	11-Mar-62	2	10
MN NICOLLET	325.19	S-3	8-Jul-62	10	10
MN NICOLLET	0.22	UE7-5	15-Jul-62	6	10
MN NICOLLET	9.32	UE12-4	17-Apr-65	6	10
MN NOBLES	0.93	TS8-10	14-Jun-52	1	10
MN NOBLES	2.87	TS8-11	15-Jun-52	1	10

COUNTY	uCi/ sqm	SHOT DAY	DATE	RANK FOR DAY	TIMES IN TOP 10
MN NOBLES	3.97	UpK2-2	25-Mar-53	10	10
MN NOBLES	2.54	PB2-4	21-Jun-57	4	10
MN NOBLES	4.02	PB5-5	19-Jul-57	3	10
MN NOBLES	1.97	PB8-5	22-Aug-57	9	10
MN NOBLES	1.55	PB15-3	18-Sep-57	10	10
MN NOBLES	0.17	UE2-7	11-Mar-62	2	10
MN NOBLES	0.13	UE7-5	15-Jul-62	8	10
MN NOBLES	11.43	UE12-4	17-Apr-65	3	10
MN NORMAN	2.54	PB2-4	21-Jun-57	4	6
MN NORMAN	1.55	PB15-3	18-Sep-57	10	6
MN NORMAN	0.98	UE2-4	8-Mar-62	2	6
MN NORMAN	0.10	UE2-7	11-Mar-62	3	6
MN NORMAN	0.13	BN-4	21-Dec-70	10	6
MN NORMAN	0.09	BN-5	22-Dec-70	7	6
MN OLMSTED	2.54	PB2-4	21-Jun-57	4	7
MN OLMSTED	6.09	PB5-5	19-Jul-57	2	7
MN OLMSTED	1.20	PB6-10	2-Aug-57	2	7
MN OLMSTED	1.55	PB15-3	18-Sep-57	10	7
MN OLMSTED	0.17	UE2-7	11-Mar-62	2	7
MN OLMSTED	0.24	UE7-5	15-Jul-62	5	7
MN OLMSTED	9.32	UE12-4	17-Apr-65	6	7
MN OTTER TAIL	2.54	PB2-4	21-Jun-57	4	5
MN OTTER TAIL	2.01	PB5-5	19-Jul-57	8	5
MN OTTER TAIL	4.11	PB15-2	17-Sep-57	9	5
MN OTTER TAIL	0.98	UE2-4	8-Mar-62	2	5
MN OTTER TAIL	0.10	UE2-7	11-Mar-62	4	5
MN PENNINGTON	1.50	PB2-4	21-Jun-57	8	7
MN PENNINGTON	1.63	PB6-9	1-Aug-57	3	7
MN PENNINGTON	4.11	PB15-2	17-Sep-57	9	7
MN PENNINGTON	2.54	PB17-2	29-Sep-57	2	7
MN PENNINGTON	0.93	UE2-4	8-Mar-62	3	7
MN PENNINGTON	0.10	UE2-7	11-Mar-62	3	7
MN PENNINGTON	0.24	BN-4	21-Dec-70	7	7
MN PINE	2.54	PB2-4	21-Jun-57	4	4
MN PINE	10.11	PB5-5	19-Jul-57	1	4

COUNTY	uCi/ sqm	SHOT DAY	DATE	RANK FOR DAY	TIMES IN TOP 10
MN PINE	1.97	PB8-5	22-Aug-57	9	4
MN PINE	2.58	PB15-3	18-Sep-57	5	4
MN PIPESTONE	3.97	UpK2-2	25-Mar-53	10	9
MN PIPESTONE	1.50	PB2-4	21-Jun-57	8	9
MN PIPESTONE	10.11	PB5-5	19-Jul-57	1	9
MN PIPESTONE	1.97	PB8-5	22-Aug-57	9	9
MN PIPESTONE	1.55	PB15-3	18-Sep-57	10	9
MN PIPESTONE	0.10	UE2-7	11-Mar-62	4	9
MN PIPESTONE	327.38	S-3	8-Jul-62	9	9
MN PIPESTONE	0.22	UE7-5	15-Jul-62	6	9
MN PIPESTONE	11.43	UE12-4	17-Apr-65	3	9
MN POLK	2.54	PB2-4	21-Jun-57	4	6
MN POLK	2.01	PB5-5	19-Jul-57	8	6
MN POLK	4.11	PB15-2	17-Sep-57	9	6
MN POLK	0.98	UE2-4	8-Mar-62	2	6
MN POLK	0.10	UE2-7	11-Mar-62	3	6
MN POLK	0.26	BN-4	21-Dec-70	6	6
MN POPE	2.54	PB2-4	21-Jun-57	4	5
MN POPE	2.43	PB6-9	1-Aug-57	2	5
MN POPE	1.20	PB6-10	2-Aug-57	2	5
MN POPE	2.58	PB15-3	18-Sep-57	5	5
MN POPE	0.10	UE2-7	11-Mar-62	4	5
MN RAMSEY	0.99	TS3-8	29-Apr-52	8	8
MN RAMSEY	0.99	TS3-8	29-Apr-52	8	8
MN RAMSEY	2.54	PB2-4	21-Jun-57	4	8
MN RAMSEY	1.27	PB4-3	7-Jul-57	5	8
MN RAMSEY	10.11	PB5-5	19-Jul-57	1	8
MN RAMSEY	1.97	PB8-5	22-Aug-57	9	8
MN RAMSEY	0.10	UE2-7	11-Mar-62	4	8
MN RAMSEY	0.22	UE7-5	15-Jul-62	6	8
MN RED LAKE	2.54	PB2-4	21-Jun-57	4	6
MN RED LAKE	2.01	PB5-5	19-Jul-57	8	6
MN RED LAKE	1.11	PB9-8	30-Aug-57	1	6
MN RED LAKE	4.11	PB15-2	17-Sep-57	9	6
MN RED LAKE	0.93	UE2-4	8-Mar-62	3	6

COUNTY	uCi/ sqm	SHOT DAY	DATE	RANK FOR DAY	TIMES IN TOP 10
MN RED LAKE	0.23	BN-4	21-Dec-70	8	6
MN REDWOOD	7.20	UpK7-5	29-Apr-53	5	8
MN REDWOOD	2.54	PB2-4	21-Jun-57	4	8
MN REDWOOD	4.02	PB5-5	19-Jul-57	3	8
MN REDWOOD	1.63	PB6-9	1-Aug-57	3	8
MN REDWOOD	1.55	PB15-3	18-Sep-57	10	8
MN REDWOOD	0.10	UE2-7	11-Mar-62	4	8
MN REDWOOD	0.12	UE7-5	15-Jul-62	9	8
MN REDWOOD	9.84	UE12-4	17-Apr-65	4	8
MN RENVILLE	1.50	PB2-4	21-Jun-57	8	7
MN RENVILLE	6.09	PB5-5	19-Jul-57	2	7
MN RENVILLE	1.20	PB6-10	2-Aug-57	2	7
MN RENVILLE	1.97	PB8-5	22-Aug-57	9	7
MN RENVILLE	1.55	PB15-3	18-Sep-57	10	7
MN RENVILLE	0.10	UE2-7	11-Mar-62	4	7
MN RENVILLE	0.11	UE7-5	15-Jul-62	10	7
MN RICE	0.99	TS3-8	29-Apr-52	8	9
MN RICE	0.99	TS3-8	29-Apr-52	8	9
MN RICE	2.54	PB2-4	21-Jun-57	4	9
MN RICE	1.19	PB4-3	7-Jul-57	6	9
MN RICE	10.11	PB5-5	19-Jul-57	1	9
MN RICE	0.10	UE2-7	11-Mar-62	4	9
MN RICE	325.19	S-3	8-Jul-62	10	9
MN RICE	0.12	UE7-5	15-Jul-62	9	9
MN RICE	22.99	UE12-4	17-Apr-65	1	9
MN ROCK	0.93	TS8-10	14-Jun-52	1	10
MN ROCK	1.72	TS8-11	15-Jun-52	2	10
MN ROCK	3.97	UpK2-2	25-Mar-53	10	10
MN ROCK	1.50	PB2-4	21-Jun-57	8	10
MN ROCK	10.11	PB5-5	19-Jul-57	1	10
MN ROCK	1.97	PB8-5	22-Aug-57	9	10
MN ROCK	1.55	PB15-3	18-Sep-57	10	10
MN ROCK	0.10	UE2-7	11-Mar-62	4	10
MN ROCK	527.44	S-3	8-Jul-62	5	10
MN ROCK	0.13	UE7-5	15-Jul-62	8	10

COUNTY	uCi/ sqm	SHOT DAY	DATE	RANK FOR DAY	TIMES IN TOP 10
MN ROSEAU	8.39	TP9-3	17-Apr-55	4	7
MN ROSEAU	2.01	PB5-5	19-Jul-57	8	7
MN ROSEAU	4.11	PB15-2	17-Sep-57	9	7
MN ROSEAU	0.93	UE2-4	8-Mar-62	3	7
MN ROSEAU	0.44	UE2-5	9-Mar-62	1	7
MN ROSEAU	0.16	BN-4	21-Dec-70	9	7
MN ROSEAU	0.11	BN-5	22-Dec-70	5	7
MN SCOTT	2.54	PB2-4	21-Jun-57	4	8
MN SCOTT	0.86	PB2-6	23-Jun-57	8	8
MN SCOTT	10.11	PB5-5	19-Jul-57	1	8
MN SCOTT	2.45	PB6-7	30-Jul-57	3	8
MN SCOTT	1.20	PB6-10	2-Aug-57	2	8
MN SCOTT	1.97	PB8-5	22-Aug-57	9	8
MN SCOTT	0.17	UE2-7	11-Mar-62	2	8
MN SCOTT	0.22	UE7-5	15-Jul-62	6	8
MN SHERBURNE	2.54	PB2-4	21-Jun-57	4	4
MN SHERBURNE	10.11	PB5-5	19-Jul-57	1	4
MN SHERBURNE	1.97	PB8-5	22-Aug-57	9	4
MN SHERBURNE	1.55	PB15-3	18-Sep-57	10	4
MN SIBLEY	2.54	PB2-4	21-Jun-57	4	4
MN SIBLEY	10.11	PB5-5	19-Jul-57	1	4
MN SIBLEY	0.17	UE2-7	11-Mar-62	2	4
MN SIBLEY	0.22	UE7-5	15-Jul-62	6	4
MN ST LOUIS	1.50	PB2-4	21-Jun-57	8	5
MN ST LOUIS	10.11	PB5-5	19-Jul-57	1	5
MN ST LOUIS	2.58	PB15-3	18-Sep-57	5	5
MN ST LOUIS	0.59	UE2-4	8-Mar-62	7	5
MN ST LOUIS	0.44	UE2-5	9-Mar-62	1	5
MN STEARNS	1.50	PB2-4	21-Jun-57	8	6
MN STEARNS	2.01	PB5-5	19-Jul-57	8	6
MN STEARNS	1.63	PB6-9	1-Aug-57	3	6
MN STEARNS	4.11	PB15-2	17-Sep-57	9	6
MN STEARNS	2.58	PB15-3	18-Sep-57	5	6
MN STEARNS	0.17	UE2-7	11-Mar-62	2	6
MN STEELE	0.99	TS3-8	29-Apr-52	8	9

COUNTY	uCi/ sqm	SHOT DAY	DATE	RANK FOR DAY	TIMES IN TOP 10
MN STEELE	0.99	TS3-8	29-Apr-52	8	9
MN STEELE	2.54	PB2-4	21-Jun-57	4	9
MN STEELE	1.19	PB4-3	7-Jul-57	6	9
MN STEELE	10.11	PB5-5	19-Jul-57	1	9
MN STEELE	1.20	PB6-10	2-Aug-57	2	9
MN STEELE	0.10	UE2-7	11-Mar-62	3	9
MN STEELE	0.12	UE7-5	15-Jul-62	9	9
MN STEELE	22.99	UE12-4	17-Apr-65	1	9
MN STEVENS	1.50	PB2-4	21-Jun-57	8	5
MN STEVENS	2.01	PB5-5	19-Jul-57	8	5
MN STEVENS	1.47	PB6-7	30-Jul-57	10	5
MN STEVENS	1.11	PB9-8	30-Aug-57	1	5
MN STEVENS	2.58	PB15-3	18-Sep-57	5	5
MN SWIFT	1.50	PB2-4	21-Jun-57	8	5
MN SWIFT	2.01	PB5-5	19-Jul-57	8	5
MN SWIFT	1.20	PB6-10	2-Aug-57	2	5
MN SWIFT	2.58	PB15-3	18-Sep-57	5	5
MN SWIFT	0.10	UE2-7	11-Mar-62	4	5
MN TODD	2.88	UpK1-4	20-Mar-53	6	5
MN TODD	2.54	PB2-4	21-Jun-57	4	5
MN TODD	10.11	PB5-5	19-Jul-57	1	5
MN TODD	2.58	PB15-3	18-Sep-57	5	5
MN TODD	0.22	UE2-5	9-Mar-62	7	5
MN TRAVERSE	1.50	PB2-4	21-Jun-57	8	4
MN TRAVERSE	1.55	PB15-3	18-Sep-57	10	4
MN TRAVERSE	0.98	UE2-4	8-Mar-62	2	4
MN TRAVERSE	0.17	UE2-7	11-Mar-62	2	4
MN WABASHA	0.99	TS3-8	29-Apr-52	8	11
MN WABASHA	0.99	TS3-8	29-Apr-52	8	11
MN WABASHA	2.54	PB2-4	21-Jun-57	4	11
MN WABASHA	1.19	PB4-3	7-Jul-57	6	11
MN WABASHA	4.02	PB5-5	19-Jul-57	3	11
MN WABASHA	1.47	PB6-7	30-Jul-57	10	11
MN WABASHA	1.16	PB15-5	20-Sep-57	7	11
MN WABASHA	0.17	UE2-7	11-Mar-62	2	11

COUNTY	uCi/ sqm	SHOT DAY	DATE	RANK FOR DAY	TIMES IN TOP 10
MN WABASHA	325.19	S-3	8-Jul-62	10	11
MN WABASHA	0.12	UE7-5	15-Jul-62	9	11
MN WABASHA	14.56	UE12-4	17-Apr-65	2	11
MN WADENA	2.54	PB2-4	21-Jun-57	4	5
MN WADENA	10.11	PB5-5	19-Jul-57	1	5
MN WADENA	2.58	PB15-3	18-Sep-57	5	5
MN WADENA	0.93	UE2-4	8-Mar-62	3	5
MN WADENA	0.44	UE2-5	9-Mar-62	1	5
MN WASECA	0.99	TS3-8	29-Apr-52	8	8
MN WASECA	0.99	TS3-8	29-Apr-52	8	8
MN WASECA	2.54	PB2-4	21-Jun-57	4	8
MN WASECA	1.61	PB4-3	7-Jul-57	1	8
MN WASECA	10.11	PB5-5	19-Jul-57	1	8
MN WASECA	0.10	UE2-7	11-Mar-62	4	8
MN WASECA	325.19	S-3	8-Jul-62	10	8
MN WASECA	9.32	UE12-4	17-Apr-65	6	8
MN WASHINGTON	0.99	TS3-8	29-Apr-52	8	8
MN WASHINGTON	0.99	TS3-8	29-Apr-52	8	8
MN WASHINGTON	2.54	PB2-4	21-Jun-57	4	8
MN WASHINGTON	1.19	PB4-3	7-Jul-57	6	8
MN WASHINGTON	10.11	PB5-5	19-Jul-57	1	8
MN WASHINGTON	1.97	PB8-5	22-Aug-57	9	8
MN WASHINGTON	0.10	UE2-7	11-Mar-62	4	8
MN WASHINGTON	0.12	UE7-5	15-Jul-62	9	8
MN WATONWAN	6.93	UpK7-5	29-Apr-53	10	11
MN WATONWAN	2.88	UpK10-4	28-May-53	5	11
MN WATONWAN	2.54	PB2-4	21-Jun-57	4	11
MN WATONWAN	1.10	PB4-3	7-Jul-57	7	11
MN WATONWAN	10.11	PB5-5	19-Jul-57	1	11
MN WATONWAN	1.97	PB8-5	22-Aug-57	9	11
MN WATONWAN	1.55	PB15-3	18-Sep-57	10	11
MN WATONWAN	0.17	UE2-7	11-Mar-62	2	11
MN WATONWAN	584.61	S-3	8-Jul-62	4	11
MN WATONWAN	0.12	UE7-5	15-Jul-62	9	11
MN WATONWAN	9.84	UE12-4	17-Apr-65	4	11

COUNTY	uCi/ sqm	SHOT DAY	DATE	RANK FOR DAY	TIMES IN TOP 10
MN WILKIN	2.54	PB2-4	21-Jun-57	4	4
MN WILKIN	4.02	PB5-5	19-Jul-57	3	4
MN WILKIN	1.55	PB15-3	18-Sep-57	10	4
MN WILKIN	0.10	UE2-7	11-Mar-62	4	4
MN WINONA	0.99	TS3-8	29-Apr-52	8	10
MN WINONA	0.99	TS3-8	29-Apr-52	8	10
MN WINONA	2.54	PB2-4	21-Jun-57	4	10
MN WINONA	0.85	PB4-3	7-Jul-57	10	10
MN WINONA	1.18	PB5-6	20-Jul-57	7	10
MN WINONA	1.47	PB6-7	30-Jul-57	10	10
MN WINONA	1.16	PB15-5	20-Sep-57	7	10
MN WINONA	0.10	UE2-7	11-Mar-62	4	10
MN WINONA	0.12	UE7-5	15-Jul-62	9	10
MN WINONA	22.99	UE12-4	17-Apr-65	1	10
MN WRIGHT	2.54	PB2-4	21-Jun-57	4	8
MN WRIGHT	0.86	PB2-6	23-Jun-57	8	8
MN WRIGHT	10.11	PB5-5	19-Jul-57	1	8
MN WRIGHT	1.47	PB6-7	30-Jul-57	10	8
MN WRIGHT	1.20	PB6-10	2-Aug-57	2	8
MN WRIGHT	1.97	PB8-5	22-Aug-57	9	8
MN WRIGHT	2.58	PB15-3	18-Sep-57	5	8
MN WRIGHT	0.10	UE2-7	11-Mar-62	4	8
MN YELLOW MEDIC	6.09	PB5-5	19-Jul-57	2	5
MN YELLOW MEDIC	2.58	PB15-3	18-Sep-57	5	5
MN YELLOW MEDIC	0.10	UE2-7	11-Mar-62	4	5
MN YELLOW MEDIC	0.11	UE7-5	15-Jul-62	10	5
MN YELLOW MEDIC	9.32	UE12-4	17-Apr-65	6	5

COUNTY	uCi/ sqm	SHOT DAY	DATE	RANK FOR DAY	TIMES IN TOP 10
MO ADAIR	3.09	PB4-10	14-Jul-57	5	8
MO ADAIR	2.48	PB5-7	21-Jul-57	6	8
MO ADAIR	7.99	PB8-3	20-Aug-57	9	8
MO ADAIR	16.67	PB9-4	26-Aug-57	4	8
MO ADAIR	1.21	PB15-5	20-Sep-57	6	8
MO ADAIR	8.80	PB18-8	14-Oct-57	1	8
MO ADAIR	0.08	UE2-7	11-Mar-62	5	8
MO ADAIR	0.53	UE7-5	15-Jul-62	1	8
MO ANDREW	0.84	UpK10-7	31-May-53	6	7
MO ANDREW	3.79	UpK11-3	6-Jun-53	6	7
MO ANDREW	2.48	PB5-7	21-Jul-57	6	7
MO ANDREW	1.21	PB15-4	19-Sep-57	5	7
MO ANDREW	0.86	PB18-6	12-Oct-57	7	7
MO ANDREW	0.08	UE2-7	11-Mar-62	5	7
MO ANDREW	0.33	UE7-5	15-Jul-62	3	7
MO ATCHISON	0.98	UpK11-5	8-Jun-53	3	4
MO ATCHISON	0.99	TP8-5	13-Apr-55	3	4
MO ATCHISON	1.21	PB15-4	19-Sep-57	5	4
MO ATCHISON	0.86	PB18-6	12-Oct-57	7	4
MO AUDRAIN	39.50	R1-2	29-Jan-51	4	10
MO AUDRAIN	131.72	TS7-3	3-Jun-52	9	10
MO AUDRAIN	0.87	UpK6-4	21-Apr-53	6	10
MO AUDRAIN	1.86	PB4-10	14-Jul-57	8	10
MO AUDRAIN	15.90	PB8-3	20-Aug-57	7	10
MO AUDRAIN	13.72	PB9-3	25-Aug-57	6	10
MO AUDRAIN	8.37	PB9-4	26-Aug-57	7	10
MO AUDRAIN	1.21	PB15-5	20-Sep-57	6	10
MO AUDRAIN	5.26	PB18-8	14-Oct-57	3	10
MO AUDRAIN	0.13	UE7-5	15-Jul-62	8	10
MO BARRY	8.76	BJ2-2	31-Oct-51	3	8
MO BARRY	2.23	UpK2-10	2-Apr-53	3	8
MO BARRY	2.10	UpK3-3	2-Apr-53	4	8
MO BARRY	4.10	UpK3-5	4-Apr-53	5	8
MO BARRY	2.45	PB3-8	1-Jul-57	1	8
MO BARRY	3.55	PB13-4	9-Sep-57	5	8

COUNTY	uCi/ sqm	SHOT DAY	DATE	RANK FOR DAY	TIMES IN TOP 10
MO BARRY	1.21	PB15-5	20-Sep-57	6	8
MO BARRY	5.26	PB18-8	14-Oct-57	3	8
MO BARTON	8.76	BJ2-2	31-Oct-51	3	5
MO BARTON	1.48	PB3-8	1-Jul-57	2	5
MO BARTON	2.42	PB14-7	20-Sep-57	2	5
MO BARTON	1.21	PB15-5	20-Sep-57	6	5
MO BARTON	3.53	PB18-8	14-Oct-57	5	5
MO BATES	1.48	PB3-8	1-Jul-57	2	7
MO BATES	17.72	PB4-5	9-Jul-57	3	7
MO BATES	7.99	PB8-3	20-Aug-57	9	7
MO BATES	2.42	PB14-7	20-Sep-57	2	7
MO BATES	1.21	PB15-5	20-Sep-57	6	7
MO BATES	5.26	PB18-8	14-Oct-57	3	7
MO BATES	0.22	UE7-5	15-Jul-62	6	7
MO BENTON	29.96	R1-2	29-Jan-51	9	7
MO BENTON	10.65	PB4-5	9-Jul-57	10	7
MO BENTON	23.89	PB8-3	20-Aug-57	5	7
MO BENTON	8.37	PB9-4	26-Aug-57	7	7
MO BENTON	2.02	PB15-5	20-Sep-57	3	7
MO BENTON	5.26	PB18-8	14-Oct-57	3	7
MO BENTON	0.22	UE7-5	15-Jul-62	6	7
MO BOLLINGER	0.87	UpK6-4	21-Apr-53	6	5
MO BOLLINGER	1.06	TP7-8	5-Apr-55	5	5
MO BOLLINGER	1.27	PB3-9	2-Jul-57	2	5
MO BOLLINGER	5.46	PB9-2	24-Aug-57	4	5
MO BOLLINGER	1.43	PB15-7	22-Sep-57	3	5
MO BOONE	0.87	UpK6-4	21-Apr-53	6	9
MO BOONE	2.48	PB5-7	21-Jul-57	6	9
MO BOONE	7.72	PB6-3	26-Jul-57	8	9
MO BOONE	15.90	PB8-3	20-Aug-57	7	9
MO BOONE	13.72	PB9-3	25-Aug-57	6	9
MO BOONE	8.37	PB9-4	26-Aug-57	7	9
MO BOONE	1.21	PB15-5	20-Sep-57	6	9
MO BOONE	5.26	PB18-8	14-Oct-57	3	9
MO BOONE	0.12	UE7-5	15-Jul-62	9	9

COUNTY	uCi/ sqm	SHOT DAY	DATE	RANK FOR DAY	TIMES IN TOP 10
MO BUCHANAN	1.48	TS4-5	5-May-52	6	7
MO BUCHANAN	1.48	TS4-5	5-May-52	6	7
MO BUCHANAN	1.17	UpK7-9	3-May-53	2	7
MO BUCHANAN	5.43	UpK11-3	6-Jun-53	1	7
MO BUCHANAN	2.48	PB5-7	21-Jul-57	6	7
MO BUCHANAN	1.21	PB15-4	19-Sep-57	5	7
MO BUCHANAN	0.13	UE7-5	15-Jul-62	8	7
MO BUTLER	6.07	UpK2-7	30-Mar-53	7	7
MO BUTLER	2.17	UpK3-6	5-Apr-53	2	7
MO BUTLER	3.68	UpK8-3	10-May-53	4	7
MO BUTLER	2.78	PB2-5	22-Jun-57	4	7
MO BUTLER	2.10	PB3-9	2-Jul-57	1	7
MO BUTLER	8.19	PB9-2	24-Aug-57	3	7
MO BUTLER	13.72	PB9-3	25-Aug-57	6	7
MO CALDWELL	1.96	UpK7-9	3-May-53	1	8
MO CALDWELL	4.15	PB5-7	21-Jul-57	2	8
MO CALDWELL	39.79	PB8-3	20-Aug-57	3	8
MO CALDWELL	8.37	PB9-4	26-Aug-57	7	8
MO CALDWELL	1.21	PB15-4	19-Sep-57	5	8
MO CALDWELL	1.33	PB15-5	20-Sep-57	5	8
MO CALDWELL	0.86	PB18-6	12-Oct-57	7	8
MO CALDWELL	0.33	UE7-5	15-Jul-62	3	8
MO CALLAWAY	27.53	R1-2	29-Jan-51	10	10
MO CALLAWAY	0.87	UpK6-4	21-Apr-53	6	10
MO CALLAWAY	3.31	PB3-6	29-Jun-57	4	10
MO CALLAWAY	15.25	PB4-4	8-Jul-57	4	10
MO CALLAWAY	15.90	PB8-3	20-Aug-57	7	10
MO CALLAWAY	13.72	PB9-3	25-Aug-57	6	10
MO CALLAWAY	8.37	PB9-4	26-Aug-57	7	10
MO CALLAWAY	1.21	PB15-5	20-Sep-57	6	10
MO CALLAWAY	5.26	PB18-8	14-Oct-57	3	10
MO CALLAWAY	0.12	UE7-5	15-Jul-62	9	10
MO CAMDEN	59.65	TP10-3	7-May-55	1	7
MO CAMDEN	0.97	PB3-8	1-Jul-57	4	7
MO CAMDEN	10.65	PB4-5	9-Jul-57	10	7

COUNTY	uCi/ sqm	SHOT DAY	DATE	RANK FOR DAY	TIMES IN TOP 10
MO CAMDEN	0.99	PB5-6	20-Jul-57	9	7
MO CAMDEN	23.89	PB8-3	20-Aug-57	5	7
MO CAMDEN	8.37	PB9-4	26-Aug-57	7	7
MO CAMDEN	2.02	PB15-5	20-Sep-57	3	7
MO CAPE GIRARDE	0.87	UpK6-4	21-Apr-53	6	5
MO CAPE GIRARDE	1.06	TP7-8	5-Apr-55	5	5
MO CAPE GIRARDE	0.83	PB3-9	2-Jul-57	4	5
MO CAPE GIRARDE	2.73	PB9-2	24-Aug-57	6	5
MO CAPE GIRARDE	0.89	PB13-7	12-Sep-57	5	5
MO CARROLL	0.98	UpK11-5	8-Jun-53	3	7
MO CARROLL	4.15	PB5-7	21-Jul-57	2	7
MO CARROLL	23.89	PB8-3	20-Aug-57	5	7
MO CARROLL	8.37	PB9-4	26-Aug-57	7	7
MO CARROLL	2.02	PB15-5	20-Sep-57	3	7
MO CARROLL	8.80	PB18-8	14-Oct-57	1	7
MO CARROLL	0.53	UE7-5	15-Jul-62	1	7
MO CARTER	1.22	BJ5-6	24-Nov-51	6	8
MO CARTER	1.22	BJ5-6	24-Nov-51	6	8
MO CARTER	0.87	UpK6-4	21-Apr-53	6	8
MO CARTER	1.06	TP7-8	5-Apr-55	5	8
MO CARTER	7.27	PB5-3	17-Jul-57	5	8
MO CARTER	2.73	PB9-2	24-Aug-57	6	8
MO CARTER	13.72	PB9-3	25-Aug-57	6	8
MO CARTER	2.25	PB18-7	13-Oct-57	2	8
MO CASS	10.65	PB4-5	9-Jul-57	10	5
MO CASS	15.90	PB8-3	20-Aug-57	7	5
MO CASS	2.02	PB15-5	20-Sep-57	3	5
MO CASS	3.53	PB18-8	14-Oct-57	5	5
MO CASS	0.13	UE7-5	15-Jul-62	8	5
MO CEDAR	1.48	PB3-8	1-Jul-57	2	6
MO CEDAR	10.65	PB4-5	9-Jul-57	10	6
MO CEDAR	7.99	PB8-3	20-Aug-57	9	6
MO CEDAR	20.95	PB12-4	5-Sep-57	5	6
MO CEDAR	1.21	PB15-5	20-Sep-57	6	6
MO CEDAR	3.53	PB18-8	14-Oct-57	5	6

COUNTY	uCi/ sqm	SHOT DAY	DATE	RANK FOR DAY	TIMES IN TOP 10
MO CHARITON	2.59	TP3-3	3-Mar-55	5	7
MO CHARITON	2.48	PB5-7	21-Jul-57	6	7
MO CHARITON	23.89	PB8-3	20-Aug-57	5	7
MO CHARITON	8.37	PB9-4	26-Aug-57	7	7
MO CHARITON	2.02	PB15-5	20-Sep-57	3	7
MO CHARITON	5.26	PB18-8	14-Oct-57	3	7
MO CHARITON	0.53	UE7-5	15-Jul-62	1	7
MO CHRISTIAN	5.84	BJ2-2	31-Oct-51	7	7
MO CHRISTIAN	1.47	UpK2-10	2-Apr-53	5	7
MO CHRISTIAN	1.40	UpK3-3	2-Apr-53	6	7
MO CHRISTIAN	0.97	PB3-8	1-Jul-57	4	7
MO CHRISTIAN	23.89	PB8-3	20-Aug-57	5	7
MO CHRISTIAN	1.21	PB15-5	20-Sep-57	6	7
MO CHRISTIAN	3.53	PB18-8	14-Oct-57	5	7
MO CLARK	1.86	PB4-10	14-Jul-57	8	7
MO CLARK	2.48	PB5-7	21-Jul-57	6	7
MO CLARK	54.86	PB9-3	25-Aug-57	1	7
MO CLARK	8.37	PB9-4	26-Aug-57	7	7
MO CLARK	6.63	PB18-8	14-Oct-57	2	7
MO CLARK	358.08	S-3	8-Jul-62	7	7
MO CLARK	0.53	UE7-5	15-Jul-62	1	7
MO CLAY	1.17	UpK7-9	3-May-53	2	4
MO CLAY	0.97	PB3-8	1-Jul-57	4	4
MO CLAY	10.65	PB4-5	9-Jul-57	10	4
MO CLAY	3.53	PB18-8	14-Oct-57	5	4
MO CLINTON	4.15	PB5-7	21-Jul-57	2	4
MO CLINTON	2.01	PB15-4	19-Sep-57	3	4
MO CLINTON	0.81	PB15-5	20-Sep-57	9	4
MO CLINTON	0.86	PB18-6	12-Oct-57	7	4
MO COLE	1.42	UpK5-4	14-Apr-53	6	8
MO COLE	0.87	UpK6-4	21-Apr-53	6	8
MO COLE	28.11	PB4-4	8-Jul-57	2	8
MO COLE	2.38	PB4-6	10-Jul-57	3	8
MO COLE	6.60	PB8-3	20-Aug-57	10	8
MO COLE	13.72	PB9-3	25-Aug-57	6	8

COUNTY	uCi/ sqm	SHOT DAY	DATE	RANK FOR DAY	TIMES IN TOP 10
MO COLE	8.37	PB9-4	26-Aug-57	7	8
MO COLE	3.65	PB18-8	14-Oct-57	5	8
MO COOPER	11.00	PB4-4	8-Jul-57	9	8
MO COOPER	2.80	PB4-6	10-Jul-57	1	8
MO COOPER	2.48	PB5-7	21-Jul-57	6	8
MO COOPER	15.90	PB8-3	20-Aug-57	7	8
MO COOPER	8.37	PB9-4	26-Aug-57	7	8
MO COOPER	2.02	PB15-5	20-Sep-57	3	8
MO COOPER	5.26	PB18-8	14-Oct-57	3	8
MO COOPER	0.12	UE7-5	15-Jul-62	9	8
MO CRAWFORD	0.87	UpK6-4	21-Apr-53	6	5
MO CRAWFORD	7.99	PB8-3	20-Aug-57	9	5
MO CRAWFORD	13.72	PB9-3	25-Aug-57	6	5
MO CRAWFORD	1.21	PB15-5	20-Sep-57	6	5
MO CRAWFORD	5.26	PB18-8	14-Oct-57	3	5
MO DADE	8.76	BJ2-2	31-Oct-51	3	3
MO DADE	1.21	PB15-5	20-Sep-57	6	3
MO DADE	3.53	PB18-8	14-Oct-57	5	3
MO DALLAS	2.89	PB3-6	29-Jun-57	7	7
MO DALLAS	0.97	PB3-8	1-Jul-57	4	7
MO DALLAS	17.72	PB4-5	9-Jul-57	3	7
MO DALLAS	2.18	PB4-6	10-Jul-57	4	7
MO DALLAS	7.99	PB8-3	20-Aug-57	9	7
MO DALLAS	1.21	PB15-5	20-Sep-57	6	7
MO DALLAS	3.53	PB18-8	14-Oct-57	5	7
MO DAVIESS	0.98	UpK11-5	8-Jun-53	3	9
MO DAVIESS	4.15	PB5-7	21-Jul-57	2	9
MO DAVIESS	39.79	PB8-3	20-Aug-57	3	9
MO DAVIESS	8.37	PB9-4	26-Aug-57	7	9
MO DAVIESS	2.01	PB15-4	19-Sep-57	3	9
MO DAVIESS	0.81	PB15-5	20-Sep-57	9	9
MO DAVIESS	0.86	PB18-6	12-Oct-57	7	9
MO DAVIESS	0.08	UE2-7	11-Mar-62	5	9
MO DAVIESS	0.53	UE7-5	15-Jul-62	1	9
MO DE KALB	1.63	UpK11-5	8-Jun-53	1	8

COUNTY	uCi/ sqm	SHOT DAY	DATE	RANK FOR DAY	TIMES IN TOP 10
MO DE KALB	2.48	PB5-7	21-Jul-57	6	8
MO DE KALB	39.79	PB8-3	20-Aug-57	3	8
MO DE KALB	8.37	PB9-4	26-Aug-57	7	8
MO DE KALB	2.04	PB9-5	27-Aug-57	8	8
MO DE KALB	1.21	PB15-4	19-Sep-57	5	8
MO DE KALB	0.86	PB18-6	12-Oct-57	7	8
MO DE KALB	0.24	UE7-5	15-Jul-62	5	8
MO DENT	0.87	UpK6-4	21-Apr-53	6	7
MO DENT	2.45	PB3-8	1-Jul-57	1	7
MO DENT	7.99	PB8-3	20-Aug-57	9	7
MO DENT	8.19	PB9-2	24-Aug-57	3	7
MO DENT	13.72	PB9-3	25-Aug-57	6	7
MO DENT	1.21	PB15-5	20-Sep-57	6	7
MO DENT	5.26	PB18-8	14-Oct-57	3	7
MO DOUGLAS	54.22	TP10-3	7-May-55	9	4
MO DOUGLAS	7.99	PB8-3	20-Aug-57	9	4
MO DOUGLAS	1.21	PB15-5	20-Sep-57	6	4
MO DOUGLAS	5.26	PB18-8	14-Oct-57	3	4
MO DUNKLIN	1.22	BJ5-6	24-Nov-51	6	9
MO DUNKLIN	1.22	BJ5-6	24-Nov-51	6	9
MO DUNKLIN	5.67	UpK1-4	20-Mar-53	1	9
MO DUNKLIN	12.19	UpK2-7	30-Mar-53	4	9
MO DUNKLIN	1.49	UpK3-6	5-Apr-53	5	9
MO DUNKLIN	3.68	UpK8-3	10-May-53	4	9
MO DUNKLIN	1.44	TP10-7	11-May-55	3	9
MO DUNKLIN	2.78	PB2-5	22-Jun-57	4	9
MO DUNKLIN	2.10	PB3-9	2-Jul-57	1	9
MO FRANKLIN	0.87	UpK6-4	21-Apr-53	6	7
MO FRANKLIN	1.94	PB3-3	26-Jun-57	9	7
MO FRANKLIN	4.17	PB3-6	29-Jun-57	1	7
MO FRANKLIN	7.99	PB8-3	20-Aug-57	9	7
MO FRANKLIN	13.72	PB9-3	25-Aug-57	6	7
MO FRANKLIN	8.37	PB9-4	26-Aug-57	7	7
MO FRANKLIN	1.13	PB18-7	13-Oct-57	6	7
MO GASCONADE	32.70	R1-2	29-Jan-51	7	8

COUNTY	uCi/ sqm	SHOT DAY	DATE	RANK FOR DAY	TIMES IN TOP 10
MO GASCONADE	0.87	UpK6-4	21-Apr-53	6	8
MO GASCONADE	3.91	PB3-5	28-Jun-57	1	8
MO GASCONADE	11.53	PB4-4	8-Jul-57	8	8
MO GASCONADE	13.72	PB9-3	25-Aug-57	6	8
MO GASCONADE	8.37	PB9-4	26-Aug-57	7	8
MO GASCONADE	1.21	PB15-5	20-Sep-57	6	8
MO GASCONADE	5.26	PB18-8	14-Oct-57	3	8
MO GENTRY	1.63	UpK11-5	8-Jun-53	1	8
MO GENTRY	1.67	PB5-7	21-Jul-57	7	8
MO GENTRY	39.79	PB8-3	20-Aug-57	3	8
MO GENTRY	16.67	PB9-4	26-Aug-57	4	8
MO GENTRY	2.04	PB9-5	27-Aug-57	8	8
MO GENTRY	2.01	PB15-4	19-Sep-57	3	8
MO GENTRY	0.81	PB15-5	20-Sep-57	9	8
MO GENTRY	0.12	UE7-5	15-Jul-62	9	8
MO GREENE	5.84	BJ2-2	31-Oct-51	7	6
MO GREENE	1.38	UpK10-7	31-May-53	5	6
MO GREENE	2.45	PB3-8	1-Jul-57	1	6
MO GREENE	23.89	PB8-3	20-Aug-57	5	6
MO GREENE	1.21	PB15-5	20-Sep-57	6	6
MO GREENE	3.53	PB18-8	14-Oct-57	5	6
MO GRUNDY	3.09	PB4-10	14-Jul-57	5	8
MO GRUNDY	0.99	PB5-6	20-Jul-57	9	8
MO GRUNDY	2.48	PB5-7	21-Jul-57	6	8
MO GRUNDY	39.79	PB8-3	20-Aug-57	3	8
MO GRUNDY	16.67	PB9-4	26-Aug-57	4	8
MO GRUNDY	1.21	PB15-4	19-Sep-57	5	8
MO GRUNDY	0.81	PB15-5	20-Sep-57	9	8
MO GRUNDY	0.53	UE7-5	15-Jul-62	1	8
MO HARRISON	39.79	PB8-3	20-Aug-57	3	6
MO HARRISON	16.67	PB9-4	26-Aug-57	4	6
MO HARRISON	1.21	PB15-4	19-Sep-57	5	6
MO HARRISON	0.81	PB15-5	20-Sep-57	9	6
MO HARRISON	1.44	PB18-6	12-Oct-57	5	6
MO HARRISON	0.53	UE7-5	15-Jul-62	1	6

COUNTY	uCi/ sqm	SHOT DAY	DATE	RANK FOR DAY	TIMES IN TOP 10
MO HENRY	0.84	UpK10-7	31-May-53	6	7
MO HENRY	52.53	TP10-3	7-May-55	10	7
MO HENRY	10.65	PB4-5	9-Jul-57	10	7
MO HENRY	39.79	PB8-3	20-Aug-57	3	7
MO HENRY	2.02	PB15-5	20-Sep-57	3	7
MO HENRY	5.26	PB18-8	14-Oct-57	3	7
MO HENRY	0.12	UE7-5	15-Jul-62	9	7
MO HICKORY	5.84	BJ2-2	31-Oct-51	7	6
MO HICKORY	10.65	PB4-5	9-Jul-57	10	6
MO HICKORY	2.08	PB4-6	10-Jul-57	6	6
MO HICKORY	23.89	PB8-3	20-Aug-57	5	6
MO HICKORY	2.02	PB15-5	20-Sep-57	3	6
MO HICKORY	5.26	PB18-8	14-Oct-57	3	6
MO HOLT	0.84	UpK10-7	31-May-53	6	5
MO HOLT	1.67	PB5-7	21-Jul-57	7	5
MO HOLT	1.21	PB15-4	19-Sep-57	5	5
MO HOLT	0.86	PB18-6	12-Oct-57	7	5
MO HOLT	0.08	UE2-7	11-Mar-62	5	5
MO HOWARD	2.48	PB5-7	21-Jul-57	6	7
MO HOWARD	7.72	PB6-3	26-Jul-57	8	7
MO HOWARD	23.89	PB8-3	20-Aug-57	5	7
MO HOWARD	8.37	PB9-4	26-Aug-57	7	7
MO HOWARD	2.02	PB15-5	20-Sep-57	3	7
MO HOWARD	5.26	PB18-8	14-Oct-57	3	7
MO HOWARD	0.12	UE7-5	15-Jul-62	9	7
MO HOWELL	12.19	UpK2-7	30-Mar-53	4	7
MO HOWELL	2.17	UpK3-6	5-Apr-53	2	7
MO HOWELL	2.50	UpK8-3	10-May-53	6	7
MO HOWELL	0.97	PB3-8	1-Jul-57	4	7
MO HOWELL	7.99	PB8-3	20-Aug-57	9	7
MO HOWELL	13.72	PB9-3	25-Aug-57	6	7
MO HOWELL	5.26	PB18-8	14-Oct-57	3	7
MO IRON	0.87	UpK6-4	21-Apr-53	6	6
MO IRON	1.06	TP7-8	5-Apr-55	5	6
MO IRON	3.16	PB3-6	29-Jun-57	5	6

COUNTY	uCi/ sqm	SHOT DAY	DATE	RANK FOR DAY	TIMES IN TOP 10
MO IRON	5.46	PB9-2	24-Aug-57	4	6
MO IRON	13.72	PB9-3	25-Aug-57	6	6
MO IRON	1.43	PB15-7	22-Sep-57	3	6
MO JACKSON	8.76	BJ2-2	31-Oct-51	3	3
MO JACKSON	10.65	PB4-5	9-Jul-57	10	3
MO JACKSON	3.53	PB18-8	14-Oct-57	5	3
MO JASPER	8.76	BJ2-2	31-Oct-51	3	5
MO JASPER	1.48	PB3-8	1-Jul-57	2	5
MO JASPER	2.42	PB14-7	20-Sep-57	2	5
MO JASPER	1.21	PB15-5	20-Sep-57	6	5
MO JASPER	1.77	PB18-8	14-Oct-57	10	5
MO JEFFERSON	0.87	UpK6-4	21-Apr-53	6	4
MO JEFFERSON	1.49	PB2-5	22-Jun-57	7	4
MO JEFFERSON	5.46	PB9-2	24-Aug-57	4	4
MO JEFFERSON	13.72	PB9-3	25-Aug-57	6	4
MO JOHNSON	1.38	UpK10-7	31-May-53	5	8
MO JOHNSON	5.17	TP3-3	3-Mar-55	2	8
MO JOHNSON	10.65	PB4-5	9-Jul-57	10	8
MO JOHNSON	1.67	PB5-7	21-Jul-57	7	8
MO JOHNSON	39.79	PB8-3	20-Aug-57	3	8
MO JOHNSON	2.02	PB15-5	20-Sep-57	3	8
MO JOHNSON	5.26	PB18-8	14-Oct-57	3	8
MO JOHNSON	0.22	UE7-5	15-Jul-62	6	8
MO KNOX	225.94	TS7-3	3-Jun-52	1	10
MO KNOX	1.86	PB4-10	14-Jul-57	8	10
MO KNOX	4.15	PB5-7	21-Jul-57	2	10
MO KNOX	7.72	PB6-3	26-Jul-57	8	10
MO KNOX	13.72	PB9-3	25-Aug-57	6	10
MO KNOX	8.37	PB9-4	26-Aug-57	7	10
MO KNOX	2.02	PB15-5	20-Sep-57	3	10
MO KNOX	5.26	PB18-8	14-Oct-57	3	10
MO KNOX	358.08	S-3	8-Jul-62	7	10
MO KNOX	0.53	UE7-5	15-Jul-62	1	10
MO LACLEDE	58.33	TP10-3	7-May-55	3	6
MO LACLEDE	10.65	PB4-5	9-Jul-57	10	6

COUNTY	uCi/ sqm	SHOT DAY	DATE	RANK FOR DAY	TIMES IN TOP 10
MO LACLEDE	2.71	PB4-6	10-Jul-57	2	6
MO LACLEDE	15.90	PB8-3	20-Aug-57	7	6
MO LACLEDE	2.02	PB15-5	20-Sep-57	3	6
MO LACLEDE	3.53	PB18-8	14-Oct-57	5	6
MO LAFAYETTE	0.97	PB3-8	1-Jul-57	4	8
MO LAFAYETTE	1.86	PB4-10	14-Jul-57	8	8
MO LAFAYETTE	1.67	PB5-7	21-Jul-57	7	8
MO LAFAYETTE	39.79	PB8-3	20-Aug-57	3	8
MO LAFAYETTE	8.37	PB9-4	26-Aug-57	7	8
MO LAFAYETTE	2.02	PB15-5	20-Sep-57	3	8
MO LAFAYETTE	5.26	PB18-8	14-Oct-57	3	8
MO LAFAYETTE	0.22	UE7-5	15-Jul-62	6	8
MO LAWRENCE	8.76	BJ2-2	31-Oct-51	3	8
MO LAWRENCE	1.47	UpK2-10	2-Apr-53	5	8
MO LAWRENCE	1.40	UpK3-3	2-Apr-53	6	8
MO LAWRENCE	4.10	UpK3-5	4-Apr-53	5	8
MO LAWRENCE	2.45	PB3-8	1-Jul-57	1	8
MO LAWRENCE	4.85	PB5-3	17-Jul-57	9	8
MO LAWRENCE	0.81	PB15-5	20-Sep-57	9	8
MO LAWRENCE	3.53	PB18-8	14-Oct-57	5	8
MO LEWIS	210.74	TS7-3	3-Jun-52	2	11
MO LEWIS	1.42	UpK5-4	14-Apr-53	6	11
MO LEWIS	0.87	UpK6-4	21-Apr-53	6	11
MO LEWIS	3.09	PB4-10	14-Jul-57	5	11
MO LEWIS	2.48	PB5-7	21-Jul-57	6	11
MO LEWIS	7.99	PB8-3	20-Aug-57	9	11
MO LEWIS	13.72	PB9-3	25-Aug-57	6	11
MO LEWIS	16.67	PB9-4	26-Aug-57	4	11
MO LEWIS	3.98	PB18-8	14-Oct-57	4	11
MO LEWIS	1.25	PB18-9	15-Oct-57	3	11
MO LEWIS	0.33	UE7-5	15-Jul-62	3	11
MO LINCOLN	0.87	UpK6-4	21-Apr-53	6	3
MO LINCOLN	13.72	PB9-3	25-Aug-57	6	3
MO LINCOLN	8.37	PB9-4	26-Aug-57	7	3
MO LINN	1.86	PB4-10	14-Jul-57	8	7

COUNTY	uCi/ sqm	SHOT DAY	DATE	RANK FOR DAY	TIMES IN TOP 10
MO LINN	4.15	PB5-7	21-Jul-57	2	7
MO LINN	23.89	PB8-3	20-Aug-57	5	7
MO LINN	16.67	PB9-4	26-Aug-57	4	7
MO LINN	2.02	PB15-5	20-Sep-57	3	7
MO LINN	5.26	PB18-8	14-Oct-57	3	7
MO LINN	0.53	UE7-5	15-Jul-62	1	7
MO LIVINGSTON	1.86	PB4-10	14-Jul-57	8	7
MO LIVINGSTON	4.15	PB5-7	21-Jul-57	2	7
MO LIVINGSTON	39.79	PB8-3	20-Aug-57	3	7
MO LIVINGSTON	8.37	PB9-4	26-Aug-57	7	7
MO LIVINGSTON	2.01	PB15-4	19-Sep-57	3	7
MO LIVINGSTON	1.33	PB15-5	20-Sep-57	5	7
MO LIVINGSTON	0.53	UE7-5	15-Jul-62	1	7
MO MACON	126.79	TS7-3	3-Jun-52	10	10
MO MACON	2.51	UpK1-4	20-Mar-53	10	10
MO MACON	1.86	PB4-10	14-Jul-57	8	10
MO MACON	4.15	PB5-7	21-Jul-57	2	10
MO MACON	7.72	PB6-3	26-Jul-57	8	10
MO MACON	23.89	PB8-3	20-Aug-57	5	10
MO MACON	8.37	PB9-4	26-Aug-57	7	10
MO MACON	2.02	PB15-5	20-Sep-57	3	10
MO MACON	5.26	PB18-8	14-Oct-57	3	10
MO MACON	0.53	UE7-5	15-Jul-62	1	10
MO MADISON	0.87	UpK6-4	21-Apr-53	6	4
MO MADISON	1.06	TP7-8	5-Apr-55	5	4
MO MADISON	2.73	PB9-2	24-Aug-57	6	4
MO MADISON	13.72	PB9-3	25-Aug-57	6	4
MO MARIES	3.03	UpK1-4	20-Mar-53	5	10
MO MARIES	0.87	UpK6-4	21-Apr-53	6	10
MO MARIES	1.48	PB3-8	1-Jul-57	2	10
MO MARIES	13.83	PB4-4	8-Jul-57	6	10
MO MARIES	7.72	PB6-3	26-Jul-57	8	10
MO MARIES	15.90	PB8-3	20-Aug-57	7	10
MO MARIES	13.72	PB9-3	25-Aug-57	6	10
MO MARIES	8.37	PB9-4	26-Aug-57	7	10

COUNTY	uCi/ sqm	SHOT DAY	DATE	RANK FOR DAY	TIMES IN TOP 10
MO MARIES	2.02	PB15-5	20-Sep-57	3	10
MO MARIES	5.26	PB18-8	14-Oct-57	3	10
MO MARION	1.42	UpK5-4	14-Apr-53	6	8
MO MARION	0.87	UpK6-4	21-Apr-53	6	8
MO MARION	1.86	PB4-10	14-Jul-57	8	8
MO MARION	13.72	PB9-3	25-Aug-57	6	8
MO MARION	16.67	PB9-4	26-Aug-57	4	8
MO MARION	2.02	PB15-5	20-Sep-57	3	8
MO MARION	8.80	PB18-8	14-Oct-57	1	8
MO MARION	0.33	UE7-5	15-Jul-62	3	8
MO MCDONALD	8.76	BJ2-2	31-Oct-51	3	5
MO MCDONALD	2.23	UpK2-10	2-Apr-53	3	5
MO MCDONALD	2.10	UpK3-3	2-Apr-53	4	5
MO MCDONALD	4.10	UpK3-5	4-Apr-53	5	5
MO MCDONALD	1.48	PB3-8	1-Jul-57	2	5
MO MERCER	3.09	PB4-10	14-Jul-57	5	8
MO MERCER	0.99	PB5-6	20-Jul-57	9	8
MO MERCER	1.67	PB5-7	21-Jul-57	7	8
MO MERCER	39.79	PB8-3	20-Aug-57	3	8
MO MERCER	8.37	PB9-4	26-Aug-57	7	8
MO MERCER	1.21	PB15-4	19-Sep-57	5	8
MO MERCER	0.86	PB18-6	12-Oct-57	7	8
MO MERCER	0.33	UE7-5	15-Jul-62	3	8
MO MILLER	1.42	UpK5-4	14-Apr-53	6	10
MO MILLER	0.87	UpK6-4	21-Apr-53	6	10
MO MILLER	0.97	PB3-8	1-Jul-57	4	10
MO MILLER	12.43	PB4-4	8-Jul-57	7	10
MO MILLER	10.65	PB4-5	9-Jul-57	10	10
MO MILLER	2.14	PB4-6	10-Jul-57	5	10
MO MILLER	23.89	PB8-3	20-Aug-57	5	10
MO MILLER	8.37	PB9-4	26-Aug-57	7	10
MO MILLER	2.02	PB15-5	20-Sep-57	3	10
MO MILLER	5.26	PB18-8	14-Oct-57	3	10
MO MISSISSIPPI	1.22	BJ5-6	24-Nov-51	6	5
MO MISSISSIPPI	1.22	BJ5-6	24-Nov-51	6	5

COUNTY	uCi/ sqm	SHOT DAY	DATE	RANK FOR DAY	TIMES IN TOP 10
MO MISSISSIPPI	12.19	UpK2-7	30-Mar-53	4	5
MO MISSISSIPPI	2.50	UpK8-3	10-May-53	6	5
MO MISSISSIPPI	1.27	PB3-9	2-Jul-57	2	5
MO MONITEAU	0.84	UpK10-7	31-May-53	6	9
MO MONITEAU	1.48	PB3-8	1-Jul-57	2	9
MO MONITEAU	24.30	PB4-4	8-Jul-57	3	9
MO MONITEAU	7.72	PB6-3	26-Jul-57	8	9
MO MONITEAU	15.90	PB8-3	20-Aug-57	7	9
MO MONITEAU	8.37	PB9-4	26-Aug-57	7	9
MO MONITEAU	1.21	PB15-5	20-Sep-57	6	9
MO MONITEAU	5.26	PB18-8	14-Oct-57	3	9
MO MONITEAU	0.13	UE7-5	15-Jul-62	8	9
MO MONROE	0.87	UpK6-4	21-Apr-53	6	9
MO MONROE	1.86	PB4-10	14-Jul-57	8	9
MO MONROE	7.72	PB6-3	26-Jul-57	8	9
MO MONROE	15.90	PB8-3	20-Aug-57	7	9
MO MONROE	13.72	PB9-3	25-Aug-57	6	9
MO MONROE	8.37	PB9-4	26-Aug-57	7	9
MO MONROE	1.21	PB15-5	20-Sep-57	6	9
MO MONROE	8.80	PB18-8	14-Oct-57	1	9
MO MONROE	0.39	UE7-5	15-Jul-62	2	9
MO MONTGOMERY	0.87	UpK6-4	21-Apr-53	6	9
MO MONTGOMERY	1.86	PB4-10	14-Jul-57	8	9
MO MONTGOMERY	7.72	PB6-3	26-Jul-57	8	9
MO MONTGOMERY	7.99	PB8-3	20-Aug-57	9	9
MO MONTGOMERY	13.72	PB9-3	25-Aug-57	6	9
MO MONTGOMERY	8.37	PB9-4	26-Aug-57	7	9
MO MONTGOMERY	1.21	PB15-5	20-Sep-57	6	9
MO MONTGOMERY	5.26	PB18-8	14-Oct-57	3	9
MO MONTGOMERY	0.12	UE7-5	15-Jul-62	9	9
MO MORGAN	59.50	TP10-3	7-May-55	2	10
MO MORGAN	2.79	PB3-6	29-Jun-57	8	10
MO MORGAN	9.95	PB4-4	8-Jul-57	10	10
MO MORGAN	10.65	PB4-5	9-Jul-57	10	10
MO MORGAN	1.32	PB4-6	10-Jul-57	9	10

COUNTY	uCi/ sqm	SHOT DAY	DATE	RANK FOR DAY	TIMES IN TOP 10
MO MORGAN	23.89	PB8-3	20-Aug-57	5	10
MO MORGAN	8.37	PB9-4	26-Aug-57	7	10
MO MORGAN	2.02	PB15-5	20-Sep-57	3	10
MO MORGAN	5.26	PB18-8	14-Oct-57	3	10
MO MORGAN	0.12	UE7-5	15-Jul-62	9	10
MO NEW MADRID	12.19	UpK2-7	30-Mar-53	4	5
MO NEW MADRID	3.68	UpK8-3	10-May-53	4	5
MO NEW MADRID	2.78	PB2-5	22-Jun-57	4	5
MO NEW MADRID	1.27	PB3-9	2-Jul-57	2	5
MO NEW MADRID	2.73	PB9-2	24-Aug-57	6	5
MO NEWTON	8.76	BJ2-2	31-Oct-51	3	8
MO NEWTON	2.23	UpK2-10	2-Apr-53	3	8
MO NEWTON	2.10	UpK3-3	2-Apr-53	4	8
MO NEWTON	6.83	UpK3-5	4-Apr-53	3	8
MO NEWTON	2.45	PB3-8	1-Jul-57	1	8
MO NEWTON	0.99	PB5-6	20-Jul-57	9	8
MO NEWTON	1.21	PB15-5	20-Sep-57	6	8
MO NEWTON	1.77	PB18-8	14-Oct-57	10	8
MO NODAWAY	0.98	UpK11-5	8-Jun-53	3	8
MO NODAWAY	23.89	PB8-3	20-Aug-57	5	8
MO NODAWAY	33.34	PB9-4	26-Aug-57	3	8
MO NODAWAY	2.04	PB9-5	27-Aug-57	8	8
MO NODAWAY	1.21	PB15-4	19-Sep-57	5	8
MO NODAWAY	0.86	PB18-6	12-Oct-57	7	8
MO NODAWAY	0.08	UE2-7	11-Mar-62	5	8
MO NODAWAY	0.33	UE7-5	15-Jul-62	3	8
MO OREGON	1.22	BJ5-6	24-Nov-51	6	10
MO OREGON	1.22	BJ5-6	24-Nov-51	6	10
MO OREGON	12.19	UpK2-7	30-Mar-53	4	10
MO OREGON	2.17	UpK3-6	5-Apr-53	2	10
MO OREGON	3.68	UpK8-3	10-May-53	4	10
MO OREGON	0.89	PB1-9	5-Jun-57	7	10
MO OREGON	0.97	PB3-8	1-Jul-57	4	10
MO OREGON	2.73	PB9-2	24-Aug-57	6	10
MO OREGON	13.72	PB9-3	25-Aug-57	6	10

COUNTY	uCi/ sqm	SHOT DAY	DATE	RANK FOR DAY	TIMES IN TOP 10
MO OREGON	3.53	PB18-8	14-Oct-57	5	10
MO OSAGE	0.87	UpK6-4	21-Apr-53	6	12
MO OSAGE	58.18	TP10-3	7-May-55	4	12
MO OSAGE	3.73	PB3-6	29-Jun-57	2	12
MO OSAGE	0.97	PB3-8	1-Jul-57	4	12
MO OSAGE	28.17	PB4-4	8-Jul-57	1	12
MO OSAGE	10.65	PB4-5	9-Jul-57	10	12
MO OSAGE	1.37	PB4-6	10-Jul-57	7	12
MO OSAGE	15.90	PB8-3	20-Aug-57	7	12
MO OSAGE	13.72	PB9-3	25-Aug-57	6	12
MO OSAGE	8.37	PB9-4	26-Aug-57	7	12
MO OSAGE	1.21	PB15-5	20-Sep-57	6	12
MO OSAGE	5.26	PB18-8	14-Oct-57	3	12
MO OZARK	2.60	UpK1-4	20-Mar-53	8	7
MO OZARK	12.19	UpK2-7	30-Mar-53	4	7
MO OZARK	2.73	UpK3-5	4-Apr-53	8	7
MO OZARK	7.99	PB8-3	20-Aug-57	9	7
MO OZARK	2.73	PB9-2	24-Aug-57	6	7
MO OZARK	0.81	PB15-5	20-Sep-57	9	7
MO OZARK	5.26	PB18-8	14-Oct-57	3	7
MO PEMISCOT	12.19	UpK2-7	30-Mar-53	4	8
MO PEMISCOT	21.07	UpK2-8	31-Mar-53	3	8
MO PEMISCOT	1.49	UpK3-6	5-Apr-53	5	8
MO PEMISCOT	3.68	UpK8-3	10-May-53	4	8
MO PEMISCOT	1.21	PB1-5	1-Jun-57	10	8
MO PEMISCOT	1.86	PB2-5	22-Jun-57	5	8
MO PEMISCOT	2.10	PB3-9	2-Jul-57	1	8
MO PEMISCOT	2.73	PB9-2	24-Aug-57	6	8
MO PERRY	0.87	UpK6-4	21-Apr-53	6	4
MO PERRY	1.06	TP7-8	5-Apr-55	5	4
MO PERRY	1.49	PB2-5	22-Jun-57	7	4
MO PERRY	8.19	PB9-2	24-Aug-57	3	4
MO PETTIS	5.17	TP3-3	3-Mar-55	2	7
MO PETTIS	57.16	TP10-3	7-May-55	6	7
MO PETTIS	39.79	PB8-3	20-Aug-57	3	7

COUNTY	uCi/ sqm	SHOT DAY	DATE	RANK FOR DAY	TIMES IN TOP 10
MO PETTIS	8.37	PB9-4	26-Aug-57	7	7
MO PETTIS	2.02	PB15-5	20-Sep-57	3	7
MO PETTIS	5.26	PB18-8	14-Oct-57	3	7
MO PETTIS	0.12	UE7-5	15-Jul-62	9	7
MO PHELPS	0.87	UpK6-4	21-Apr-53	6	8
MO PHELPS	3.08	PB3-6	29-Jun-57	6	8
MO PHELPS	0.97	PB3-8	1-Jul-57	4	8
MO PHELPS	0.99	PB5-6	20-Jul-57	9	8
MO PHELPS	7.99	PB8-3	20-Aug-57	9	8
MO PHELPS	13.72	PB9-3	25-Aug-57	6	8
MO PHELPS	2.02	PB15-5	20-Sep-57	3	8
MO PHELPS	5.26	PB18-8	14-Oct-57	3	8
MO PIKE	39.50	R1-2	29-Jan-51	4	10
MO PIKE	126.79	TS7-3	3-Jun-52	10	10
MO PIKE	0.87	UpK6-4	21-Apr-53	6	10
MO PIKE	1.86	PB4-10	14-Jul-57	8	10
MO PIKE	1.67	PB5-7	21-Jul-57	7	10
MO PIKE	7.99	PB8-3	20-Aug-57	9	10
MO PIKE	13.72	PB9-3	25-Aug-57	6	10
MO PIKE	8.37	PB9-4	26-Aug-57	7	10
MO PIKE	1.13	PB18-7	13-Oct-57	6	10
MO PIKE	0.13	UE7-5	15-Jul-62	8	10
MO PLATTE	1.17	UpK7-9	3-May-53	2	6
MO PLATTE	4.15	PB5-7	21-Jul-57	2	6
MO PLATTE	23.89	PB8-3	20-Aug-57	5	6
MO PLATTE	2.01	PB15-4	19-Sep-57	3	6
MO PLATTE	0.81	PB15-5	20-Sep-57	9	6
MO PLATTE	0.86	PB18-6	12-Oct-57	7	6
MO POLK	0.84	UpK10-7	31-May-53	6	6
MO POLK	1.48	PB3-8	1-Jul-57	2	6
MO POLK	17.72	PB4-5	9-Jul-57	3	6
MO POLK	39.79	PB8-3	20-Aug-57	3	6
MO POLK	1.21	PB15-5	20-Sep-57	6	6
MO POLK	3.53	PB18-8	14-Oct-57	5	6
MO PULASKI	0.87	UpK6-4	21-Apr-53	6	6

COUNTY	uCi/ sqm	SHOT DAY	DATE	RANK FOR DAY	TIMES IN TOP 10
MO PULASKI	10.65	PB4-5	9-Jul-57	10	6
MO PULASKI	23.89	PB8-3	20-Aug-57	5	6
MO PULASKI	13.72	PB9-3	25-Aug-57	6	6
MO PULASKI	2.02	PB15-5	20-Sep-57	3	6
MO PULASKI	3.53	PB18-8	14-Oct-57	5	6
MO PUTNAM	1.86	PB4-10	14-Jul-57	8	6
MO PUTNAM	23.89	PB8-3	20-Aug-57	5	6
MO PUTNAM	8.37	PB9-4	26-Aug-57	7	6
MO PUTNAM	1.21	PB15-4	19-Sep-57	5	6
MO PUTNAM	0.81	PB15-5	20-Sep-57	9	6
MO PUTNAM	0.33	UE7-5	15-Jul-62	3	6
MO RALLS	132.14	TS7-3	3-Jun-52	8	9
MO RALLS	0.87	UpK6-4	21-Apr-53	6	9
MO RALLS	6.53	PB6-4	27-Jul-57	5	9
MO RALLS	7.99	PB8-3	20-Aug-57	9	9
MO RALLS	13.72	PB9-3	25-Aug-57	6	9
MO RALLS	8.37	PB9-4	26-Aug-57	7	9
MO RALLS	1.21	PB15-5	20-Sep-57	6	9
MO RALLS	5.26	PB18-8	14-Oct-57	3	9
MO RALLS	0.22	UE7-5	15-Jul-62	6	9
MO RANDOLPH	1.86	PB4-10	14-Jul-57	8	7
MO RANDOLPH	7.72	PB6-3	26-Jul-57	8	7
MO RANDOLPH	23.89	PB8-3	20-Aug-57	5	7
MO RANDOLPH	8.37	PB9-4	26-Aug-57	7	7
MO RANDOLPH	2.02	PB15-5	20-Sep-57	3	7
MO RANDOLPH	5.26	PB18-8	14-Oct-57	3	7
MO RANDOLPH	0.53	UE7-5	15-Jul-62	1	7
MO RAY	2.59	TP3-3	3-Mar-55	5	7
MO RAY	4.15	PB5-7	21-Jul-57	2	7
MO RAY	23.89	PB8-3	20-Aug-57	5	7
MO RAY	16.67	PB9-4	26-Aug-57	4	7
MO RAY	2.02	PB15-5	20-Sep-57	3	7
MO RAY	5.26	PB18-8	14-Oct-57	3	7
MO RAY	0.33	UE7-5	15-Jul-62	3	7
MO REYNOLDS	0.87	UpK6-4	21-Apr-53	6	7

COUNTY	uCi/ sqm	SHOT DAY	DATE	RANK FOR DAY	TIMES IN TOP 10
MO REYNOLDS	1.06	TP7-8	5-Apr-55	5	7
MO REYNOLDS	0.89	PB1-9	5-Jun-57	7	7
MO REYNOLDS	1.49	PB2-5	22-Jun-57	7	7
MO REYNOLDS	2.73	PB9-2	24-Aug-57	6	7
MO REYNOLDS	13.72	PB9-3	25-Aug-57	6	7
MO REYNOLDS	0.87	PB18-7	13-Oct-57	8	7
MO RIPLEY	1.22	BJ5-6	24-Nov-51	6	8
MO RIPLEY	1.22	BJ5-6	24-Nov-51	6	8
MO RIPLEY	18.26	UpK2-7	30-Mar-53	2	8
MO RIPLEY	2.50	UpK8-3	10-May-53	6	8
MO RIPLEY	2.78	PB2-5	22-Jun-57	4	8
MO RIPLEY	2.10	PB3-9	2-Jul-57	1	8
MO RIPLEY	8.19	PB9-2	24-Aug-57	3	8
MO RIPLEY	13.72	PB9-3	25-Aug-57	6	8
MO SALINE	0.84	UpK10-7	31-May-53	6	8
MO SALINE	2.59	TP3-3	3-Mar-55	5	8
MO SALINE	2.48	PB5-7	21-Jul-57	6	8
MO SALINE	39.79	PB8-3	20-Aug-57	3	8
MO SALINE	8.37	PB9-4	26-Aug-57	7	8
MO SALINE	2.02	PB15-5	20-Sep-57	3	8
MO SALINE	5.26	PB18-8	14-Oct-57	3	8
MO SALINE	0.39	UE7-5	15-Jul-62	2	8
MO SCHUYLER	1.67	PB5-7	21-Jul-57	7	6
MO SCHUYLER	23.89	PB8-3	20-Aug-57	5	6
MO SCHUYLER	8.37	PB9-4	26-Aug-57	7	6
MO SCHUYLER	1.21	PB15-4	19-Sep-57	5	6
MO SCHUYLER	0.81	PB15-5	20-Sep-57	9	6
MO SCHUYLER	0.33	UE7-5	15-Jul-62	3	6
MO SCOTLAND	181.55	TS7-3	3-Jun-52	5	7
MO SCOTLAND	2.48	PB5-7	21-Jul-57	6	7
MO SCOTLAND	13.72	PB9-3	25-Aug-57	6	7
MO SCOTLAND	8.37	PB9-4	26-Aug-57	7	7
MO SCOTLAND	3.98	PB18-8	14-Oct-57	4	7
MO SCOTLAND	358.08	S-3	8-Jul-62	7	7
MO SCOTLAND	0.33	UE7-5	15-Jul-62	3	7

COUNTY	uCi/ sqm	SHOT DAY	DATE	RANK FOR DAY	TIMES IN TOP 10
MO SCOTT	0.87	UpK6-4	21-Apr-53	6	5
MO SCOTT	1.24	TP10-7	11-May-55	8	5
MO SCOTT	1.27	PB3-9	2-Jul-57	2	5
MO SCOTT	2.73	PB9-2	24-Aug-57	6	5
MO SCOTT	0.89	PB13-7	12-Sep-57	5	5
MO SHANNON	0.87	UpK6-4	21-Apr-53	6	6
MO SHANNON	1.61	TP7-8	5-Apr-55	3	6
MO SHANNON	7.99	PB8-3	20-Aug-57	9	6
MO SHANNON	13.72	PB9-3	25-Aug-57	6	6
MO SHANNON	0.81	PB15-5	20-Sep-57	9	6
MO SHANNON	3.53	PB18-8	14-Oct-57	5	6
MO SHELBY	2.13	UpK5-4	14-Apr-53	3	9
MO SHELBY	0.87	UpK6-4	21-Apr-53	6	9
MO SHELBY	3.09	PB4-10	14-Jul-57	5	9
MO SHELBY	15.90	PB8-3	20-Aug-57	7	9
MO SHELBY	13.72	PB9-3	25-Aug-57	6	9
MO SHELBY	8.37	PB9-4	26-Aug-57	7	9
MO SHELBY	2.02	PB15-5	20-Sep-57	3	9
MO SHELBY	8.80	PB18-8	14-Oct-57	1	9
MO SHELBY	0.33	UE7-5	15-Jul-62	3	9
MO ST CHARLES	0.87	UpK6-4	21-Apr-53	6	5
MO ST CHARLES	1.06	TP7-8	5-Apr-55	5	5
MO ST CHARLES	1.94	PB3-3	26-Jun-57	9	5
MO ST CHARLES	13.72	PB9-3	25-Aug-57	6	5
MO ST CHARLES	8.37	PB9-4	26-Aug-57	7	5
MO ST CLAIR	1.42	UpK5-4	14-Apr-53	6	7
MO ST CLAIR	0.87	UpK6-4	21-Apr-53	6	7
MO ST CLAIR	2.48	PB5-7	21-Jul-57	6	7
MO ST CLAIR	13.72	PB9-3	25-Aug-57	6	7
MO ST CLAIR	8.37	PB9-4	26-Aug-57	7	7
MO ST CLAIR	0.98	PB15-5	20-Sep-57	8	7
MO ST CLAIR	1.13	PB18-7	13-Oct-57	6	7
MO ST FRANCOIS	0.87	UpK6-4	21-Apr-53	6	5
MO ST FRANCOIS	2.92	PB3-5	28-Jun-57	8	5
MO ST FRANCOIS	2.78	PB3-6	29-Jun-57	9	5

COUNTY	uCi/ sqm	SHOT DAY	DATE	RANK FOR DAY	TIMES IN TOP 10
MO ST FRANCOIS	13.72	PB9-3	25-Aug-57	6	5
MO ST FRANCOIS	1.13	PB18-7	13-Oct-57	6	5
MO ST LOUIS	13.72	PB9-3	25-Aug-57	6	2
MO ST LOUIS	8.37	PB9-4	26-Aug-57	7	2
MO STE GENEVIEV	0.87	UpK6-4	21-Apr-53	6	2
MO STE GENEVIEV	13.72	PB9-3	25-Aug-57	6	2
MO STODDARD	6.07	UpK2-7	30-Mar-53	7	6
MO STODDARD	14.03	UpK2-8	31-Mar-53	8	6
MO STODDARD	3.68	UpK8-3	10-May-53	4	6
MO STODDARD	1.86	PB2-5	22-Jun-57	5	6
MO STODDARD	1.27	PB3-9	2-Jul-57	2	6
MO STODDARD	5.46	PB9-2	24-Aug-57	4	6
MO STONE	8.76	BJ2-2	31-Oct-51	3	8
MO STONE	1.47	UpK2-10	2-Apr-53	5	8
MO STONE	1.40	UpK3-3	2-Apr-53	6	8
MO STONE	2.73	UpK3-5	4-Apr-53	8	8
MO STONE	2.45	PB3-8	1-Jul-57	1	8
MO STONE	7.99	PB8-3	20-Aug-57	9	8
MO STONE	0.81	PB15-5	20-Sep-57	9	8
MO STONE	5.26	PB18-8	14-Oct-57	3	8
MO SULLIVAN	3.09	PB4-10	14-Jul-57	5	8
MO SULLIVAN	1.67	PB5-7	21-Jul-57	7	8
MO SULLIVAN	23.89	PB8-3	20-Aug-57	5	8
MO SULLIVAN	8.37	PB9-4	26-Aug-57	7	8
MO SULLIVAN	2.01	PB15-4	19-Sep-57	3	8
MO SULLIVAN	0.81	PB15-5	20-Sep-57	9	8
MO SULLIVAN	0.86	PB18-6	12-Oct-57	7	8
MO SULLIVAN	0.33	UE7-5	15-Jul-62	3	8
MO TANEY	3.61	UpK1-4	20-Mar-53	4	5
MO TANEY	4.10	UpK3-5	4-Apr-53	5	5
MO TANEY	1.48	PB3-8	1-Jul-57	2	5
MO TANEY	1.21	PB15-5	20-Sep-57	6	5
MO TANEY	5.26	PB18-8	14-Oct-57	3	5
MO TEXAS	0.87	UpK6-4	21-Apr-53	6	7
MO TEXAS	1.31	PB4-6	10-Jul-57	10	7

COUNTY	uCi/ sqm	SHOT DAY	DATE	RANK FOR DAY	TIMES IN TOP 10
MO TEXAS	15.90	PB8-3	20-Aug-57	7	7
MO TEXAS	8.19	PB9-2	24-Aug-57	3	7
MO TEXAS	13.72	PB9-3	25-Aug-57	6	7
MO TEXAS	1.21	PB15-5	20-Sep-57	6	7
MO TEXAS	5.26	PB18-8	14-Oct-57	3	7
MO VERNON	5.84	BJ2-2	31-Oct-51	7	8
MO VERNON	1.48	PB3-8	1-Jul-57	2	8
MO VERNON	10.65	PB4-5	9-Jul-57	10	8
MO VERNON	21.83	PB12-4	5-Sep-57	3	8
MO VERNON	4.04	PB14-7	20-Sep-57	1	8
MO VERNON	2.02	PB15-5	20-Sep-57	3	8
MO VERNON	5.26	PB18-8	14-Oct-57	3	8
MO VERNON	0.13	UE7-5	15-Jul-62	8	8
MO WARREN	0.87	UpK6-4	21-Apr-53	6	8
MO WARREN	2.59	PB3-6	29-Jun-57	10	8
MO WARREN	3.09	PB4-10	14-Jul-57	5	8
MO WARREN	7.99	PB8-3	20-Aug-57	9	8
MO WARREN	13.72	PB9-3	25-Aug-57	6	8
MO WARREN	8.37	PB9-4	26-Aug-57	7	8
MO WARREN	1.13	PB18-7	13-Oct-57	6	8
MO WARREN	1.93	PB18-8	14-Oct-57	9	8
MO WASHINGTON	0.87	UpK6-4	21-Apr-53	6	6
MO WASHINGTON	1.06	TP7-8	5-Apr-55	5	6
MO WASHINGTON	3.63	PB3-6	29-Jun-57	3	6
MO WASHINGTON	5.46	PB9-2	24-Aug-57	4	6
MO WASHINGTON	13.72	PB9-3	25-Aug-57	6	6
MO WASHINGTON	1.43	PB15-7	22-Sep-57	3	6
MO WAYNE	0.87	UpK6-4	21-Apr-53	6	9
MO WAYNE	1.06	TP7-8	5-Apr-55	5	9
MO WAYNE	0.89	PB1-9	5-Jun-57	7	9
MO WAYNE	1.49	PB2-5	22-Jun-57	7	9
MO WAYNE	1.27	PB3-9	2-Jul-57	2	9
MO WAYNE	8.19	PB9-2	24-Aug-57	3	9
MO WAYNE	13.72	PB9-3	25-Aug-57	6	9
MO WAYNE	1.43	PB15-7	22-Sep-57	3	9

COUNTY	uCi/ sqm	SHOT DAY	DATE	RANK FOR DAY	TIMES IN TOP 10
MO WAYNE	2.25	PB18-7	13-Oct-57	2	9
MO WEBSTER	0.84	UpK10-7	31-May-53	6	7
MO WEBSTER	57.96	TP10-3	7-May-55	5	7
MO WEBSTER	1.48	PB3-8	1-Jul-57	2	7
MO WEBSTER	10.65	PB4-5	9-Jul-57	10	7
MO WEBSTER	23.89	PB8-3	20-Aug-57	5	7
MO WEBSTER	1.21	PB15-5	20-Sep-57	6	7
MO WEBSTER	1.77	PB18-8	14-Oct-57	10	7
MO WORTH	0.99	TP8-5	13-Apr-55	3	6
MO WORTH	2.48	PB5-7	21-Jul-57	6	6
MO WORTH	39.79	PB8-3	20-Aug-57	3	6
MO WORTH	33.34	PB9-4	26-Aug-57	3	6
MO WORTH	1.21	PB15-4	19-Sep-57	5	6
MO WORTH	0.33	UE7-5	15-Jul-62	3	6
MO WRIGHT	0.87	UpK6-4	21-Apr-53	6	9
MO WRIGHT	54.66	TP10-3	7-May-55	8	9
MO WRIGHT	1.48	PB3-8	1-Jul-57	2	9
MO WRIGHT	1.35	PB4-6	10-Jul-57	8	9
MO WRIGHT	7.99	PB8-3	20-Aug-57	9	9
MO WRIGHT	2.73	PB9-2	24-Aug-57	6	9
MO WRIGHT	13.72	PB9-3	25-Aug-57	6	9
MO WRIGHT	0.81	PB15-5	20-Sep-57	9	9
MO WRIGHT	5.26	PB18-8	14-Oct-57	3	9

COUNTY	uCi/ sqm	SHOT DAY	DATE	RANK FOR DAY	TIMES IN TOP 10
MS ADAMS	49.22	UpK7-4	28-Apr-53	1	5
MS ADAMS	1.56	UpK8-9	16-May-53	1	5
MS ADAMS	1.47	TP6-2	23-Mar-55	9	5
MS ADAMS	4.11	TP6-3	24-Mar-55	4	5
MS ADAMS	114.67	PB3-3	26-Jun-57	1	5
MS ALCORN	6.07	UpK2-7	30-Mar-53	7	5
MS ALCORN	14.03	UpK2-8	31-Mar-53	8	5
MS ALCORN	3.67	UpK3-6	5-Apr-53	1	5
MS ALCORN	1.18	UpK8-3	10-May-53	10	5
MS ALCORN	1.86	PB2-5	22-Jun-57	5	5
MS AMITE	1.16	UpK8-5	12-May-53	5	5
MS AMITE	1.47	TP6-2	23-Mar-55	9	5
MS AMITE	4.11	TP6-3	24-Mar-55	4	5
MS AMITE	1.09	TP8-4	12-Apr-55	9	5
MS AMITE	22.97	PB3-3	26-Jun-57	4	5
MS ATTALA	0.91	BJ1-1	28-Oct-51	4	9
MS ATTALA	1.56	UpK8-9	16-May-53	1	9
MS ATTALA	1.47	TP6-2	23-Mar-55	9	9
MS ATTALA	4.11	TP6-3	24-Mar-55	4	9
MS ATTALA	2.87	TP6-4	25-Mar-55	4	9
MS ATTALA	1.21	PB1-5	1-Jun-57	10	9
MS ATTALA	4.64	PB2-5	22-Jun-57	2	9
MS ATTALA	0.83	PB3-9	2-Jul-57	4	9
MS ATTALA	1.00	PB6-11	3-Aug-57	3	9
MS BENTON	3.06	UpK2-7	30-Mar-53	9	8
MS BENTON	21.07	UpK2-8	31-Mar-53	3	8
MS BENTON	3.67	UpK3-6	5-Apr-53	1	8
MS BENTON	3.68	UpK8-3	10-May-53	4	8
MS BENTON	1.52	UpK8-4	11-May-53	6	8
MS BENTON	2.78	PB2-5	22-Jun-57	4	8
MS BENTON	1.25	PB3-8	1-Jul-57	3	8
MS BENTON	1.27	PB3-9	2-Jul-57	2	8
MS BOLIVAR	6.07	UpK2-7	30-Mar-53	7	7
MS BOLIVAR	2.17	UpK3-6	5-Apr-53	2	7
MS BOLIVAR	3.68	UpK8-3	10-May-53	4	7

COUNTY	uCi/ sqm	SHOT DAY	DATE	RANK FOR DAY	TIMES IN TOP 10
MS BOLIVAR	0.89	PB1-9	5-Jun-57	7	7
MS BOLIVAR	2.78	PB2-5	22-Jun-57	4	7
MS BOLIVAR	1.27	PB3-9	2-Jul-57	2	7
MS BOLIVAR	2.73	PB9-2	24-Aug-57	6	7
MS CALHOUN	6.07	UpK2-7	30-Mar-53	7	6
MS CALHOUN	2.17	UpK3-6	5-Apr-53	2	6
MS CALHOUN	3.68	UpK8-3	10-May-53	4	6
MS CALHOUN	4.64	PB2-5	22-Jun-57	2	6
MS CALHOUN	1.25	PB3-8	1-Jul-57	3	6
MS CALHOUN	1.27	PB3-9	2-Jul-57	2	6
MS CARROLL	0.91	BJ1-1	28-Oct-51	4	8
MS CARROLL	0.93	UpK8-9	16-May-53	2	8
MS CARROLL	1.47	TP6-2	23-Mar-55	9	8
MS CARROLL	8.23	TP6-3	24-Mar-55	2	8
MS CARROLL	0.89	PB1-9	5-Jun-57	7	8
MS CARROLL	4.64	PB2-5	22-Jun-57	2	8
MS CARROLL	1.27	PB3-9	2-Jul-57	2	8
MS CARROLL	5.46	PB9-2	24-Aug-57	4	8
MS CHICKASAW	6.07	UpK2-7	30-Mar-53	7	9
MS CHICKASAW	14.03	UpK2-8	31-Mar-53	8	9
MS CHICKASAW	2.17	UpK3-6	5-Apr-53	2	9
MS CHICKASAW	2.50	UpK8-3	10-May-53	6	9
MS CHICKASAW	4.64	PB2-5	22-Jun-57	2	9
MS CHICKASAW	1.25	PB3-8	1-Jul-57	3	9
MS CHICKASAW	1.27	PB3-9	2-Jul-57	2	9
MS CHICKASAW	5.46	PB9-2	24-Aug-57	4	9
MS CHICKASAW	14.72	PB17-5	2-Oct-57	2	9
MS CHOCTAW	1.32	TS5-5	11-May-52	2	8
MS CHOCTAW	1.56	UpK8-9	16-May-53	1	8
MS CHOCTAW	1.47	TP6-2	23-Mar-55	9	8
MS CHOCTAW	4.11	TP6-3	24-Mar-55	4	8
MS CHOCTAW	0.89	PB1-9	5-Jun-57	7	8
MS CHOCTAW	4.64	PB2-5	22-Jun-57	2	8
MS CHOCTAW	0.83	PB3-9	2-Jul-57	4	8
MS CHOCTAW	5.46	PB9-2	24-Aug-57	4	8

COUNTY	uCi/ sqm	SHOT DAY	DATE	RANK FOR DAY	TIMES IN TOP 10
MS CLAIBORNE	47.76	UpK7-4	28-Apr-53	3	7
MS CLAIBORNE	1.16	UpK8-5	12-May-53	5	7
MS CLAIBORNE	1.56	UpK8-9	16-May-53	1	7
MS CLAIBORNE	1.47	TP6-2	23-Mar-55	9	7
MS CLAIBORNE	4.11	TP6-3	24-Mar-55	4	7
MS CLAIBORNE	1.12	TP8-4	12-Apr-55	8	7
MS CLAIBORNE	45.85	PB3-3	26-Jun-57	3	7
MS CLARKE	1.47	TP6-2	23-Mar-55	9	4
MS CLARKE	4.11	TP6-3	24-Mar-55	4	4
MS CLARKE	3.14	TP6-5	26-Mar-55	2	4
MS CLARKE	45.85	PB3-3	26-Jun-57	3	4
MS CLAY	0.93	UpK8-9	16-May-53	2	6
MS CLAY	1.47	TP6-2	23-Mar-55	9	6
MS CLAY	4.11	TP6-3	24-Mar-55	4	6
MS CLAY	2.87	TP6-4	25-Mar-55	4	6
MS CLAY	4.64	PB2-5	22-Jun-57	2	6
MS CLAY	13.49	PB12-6	7-Sep-57	5	6
MS COAHOMA	6.07	UpK2-7	30-Mar-53	7	10
MS COAHOMA	3.67	UpK3-6	5-Apr-53	1	10
MS COAHOMA	6.18	UpK8-3	10-May-53	2	10
MS COAHOMA	1.52	UpK8-4	11-May-53	6	10
MS COAHOMA	3.38	TP7-4	1-Apr-55	2	10
MS COAHOMA	0.89	PB1-9	5-Jun-57	7	10
MS COAHOMA	2.78	PB2-5	22-Jun-57	4	10
MS COAHOMA	1.25	PB3-8	1-Jul-57	3	10
MS COAHOMA	1.27	PB3-9	2-Jul-57	2	10
MS COAHOMA	8.19	PB9-2	24-Aug-57	3	10
MS COPIAH	1.56	UpK8-9	16-May-53	1	4
MS COPIAH	1.47	TP6-2	23-Mar-55	9	4
MS COPIAH	4.11	TP6-3	24-Mar-55	4	4
MS COPIAH	45.85	PB3-3	26-Jun-57	3	4
MS COVINGTON	1.47	TP6-2	23-Mar-55	9	6
MS COVINGTON	4.11	TP6-3	24-Mar-55	4	6
MS COVINGTON	11.44	PB3-3	26-Jun-57	6	6
MS COVINGTON	1.40	PB6-8	31-Jul-57	8	6

COUNTY	uCi/ sqm	SHOT DAY	DATE	RANK FOR DAY	TIMES IN TOP 10
MS COVINGTON	1.06	PB6-11	3-Aug-57	2	6
MS COVINGTON	21.82	PB12-6	7-Sep-57	3	6
MS DE SOTO	18.26	UpK2-7	30-Mar-53	2	10
MS DE SOTO	3.67	UpK3-6	5-Apr-53	1	10
MS DE SOTO	6.18	UpK8-3	10-May-53	2	10
MS DE SOTO	1.52	UpK8-4	11-May-53	6	10
MS DE SOTO	0.89	UpK8-5	12-May-53	6	10
MS DE SOTO	3.16	TP7-4	1-Apr-55	4	10
MS DE SOTO	4.64	PB2-5	22-Jun-57	2	10
MS DE SOTO	1.25	PB3-8	1-Jul-57	3	10
MS DE SOTO	0.83	PB3-9	2-Jul-57	4	10
MS DE SOTO	2.73	PB9-2	24-Aug-57	6	10
MS FORREST	1.10	TS6-5	29-May-52	4	6
MS FORREST	0.81	UpK3-6	5-Apr-53	9	6
MS FORREST	38.63	UpK6-2	19-Apr-53	2	6
MS FORREST	3.14	TP6-5	26-Mar-55	2	6
MS FORREST	45.85	PB3-3	26-Jun-57	3	6
MS FORREST	1.40	PB6-8	31-Jul-57	8	6
MS FRANKLIN	1.97	UpK4-1	6-Apr-53	2	7
MS FRANKLIN	1.16	UpK8-5	12-May-53	5	7
MS FRANKLIN	1.56	UpK8-9	16-May-53	1	7
MS FRANKLIN	1.47	TP6-2	23-Mar-55	9	7
MS FRANKLIN	4.11	TP6-3	24-Mar-55	4	7
MS FRANKLIN	1.09	TP8-4	12-Apr-55	9	7
MS FRANKLIN	114.67	PB3-3	26-Jun-57	1	7
MS GEORGE	38.63	UpK6-2	19-Apr-53	2	6
MS GEORGE	3.14	TP6-5	26-Mar-55	2	6
MS GEORGE	1.03	TP8-5	13-Apr-55	2	6
MS GEORGE	68.82	PB3-3	26-Jun-57	2	6
MS GEORGE	1.99	PB6-7	30-Jul-57	6	6
MS GEORGE	2.34	PB6-8	31-Jul-57	3	6
MS GREENE	38.63	UpK6-2	19-Apr-53	2	4
MS GREENE	3.14	TP6-5	26-Mar-55	2	4
MS GREENE	1.03	TP8-5	13-Apr-55	2	4
MS GREENE	45.85	PB3-3	26-Jun-57	3	4

COUNTY	uCi/ sqm	SHOT DAY	DATE	RANK FOR DAY	TIMES IN TOP 10
MS GRENADA	3.06	UpK2-7	30-Mar-53	9	7
MS GRENADA	2.17	UpK3-6	5-Apr-53	2	7
MS GRENADA	3.68	UpK8-3	10-May-53	4	7
MS GRENADA	4.64	PB2-5	22-Jun-57	2	7
MS GRENADA	1.25	PB3-8	1-Jul-57	3	7
MS GRENADA	8.19	PB9-2	24-Aug-57	3	7
MS GRENADA	9.38	PB17-5	2-Oct-57	5	7
MS HANCOCK	38.63	UpK6-2	19-Apr-53	2	4
MS HANCOCK	1.28	TP10-7	11-May-55	7	4
MS HANCOCK	68.82	PB3-3	26-Jun-57	2	4
MS HANCOCK	1.40	PB6-8	31-Jul-57	8	4
MS HARRISON	38.63	UpK6-2	19-Apr-53	2	6
MS HARRISON	3.14	TP6-5	26-Mar-55	2	6
MS HARRISON	45.85	PB3-3	26-Jun-57	3	6
MS HARRISON	1.40	PB6-8	31-Jul-57	8	6
MS HARRISON	1.06	PB6-11	3-Aug-57	2	6
MS HARRISON	1.07	PB9-8	30-Aug-57	2	6
MS HINDS	1.71	TP6-3	24-Mar-55	8	3
MS HINDS	3.33	TP10-4	8-May-55	1	3
MS HINDS	68.82	PB3-3	26-Jun-57	2	3
MS HOLMES	0.91	BJ1-1	28-Oct-51	4	6
MS HOLMES	0.97	TS5-5	11-May-52	8	6
MS HOLMES	1.47	TP6-2	23-Mar-55	9	6
MS HOLMES	4.11	TP6-3	24-Mar-55	4	6
MS HOLMES	2.87	TP6-4	25-Mar-55	4	6
MS HOLMES	4.64	PB2-5	22-Jun-57	2	6
MS HUMPHREYS	0.91	BJ1-1	28-Oct-51	4	5
MS HUMPHREYS	0.93	UpK8-9	16-May-53	2	5
MS HUMPHREYS	1.47	TP6-2	23-Mar-55	9	5
MS HUMPHREYS	8.23	TP6-3	24-Mar-55	2	5
MS HUMPHREYS	1.86	PB2-5	22-Jun-57	5	5
MS ISSAQUENA	44.84	UpK7-4	28-Apr-53	5	6
MS ISSAQUENA	1.16	UpK8-5	12-May-53	5	6
MS ISSAQUENA	1.56	UpK8-9	16-May-53	1	6
MS ISSAQUENA	1.47	TP6-2	23-Mar-55	9	6

COUNTY	uCi/ sqm	SHOT DAY	DATE	RANK FOR DAY	TIMES IN TOP 10
MS ISSAQUENA	4.11	TP6-3	24-Mar-55	4	6
MS ISSAQUENA	3.21	PB18-8	14-Oct-57	6	6
MS ITAWAMBA	3.06	UpK2-7	30-Mar-53	9	7
MS ITAWAMBA	2.17	UpK3-6	5-Apr-53	2	7
MS ITAWAMBA	2.50	UpK8-3	10-May-53	6	7
MS ITAWAMBA	0.86	TP8-5	13-Apr-55	4	7
MS ITAWAMBA	4.64	PB2-5	22-Jun-57	2	7
MS ITAWAMBA	1.25	PB3-8	1-Jul-57	3	7
MS ITAWAMBA	1.27	PB3-9	2-Jul-57	2	7
MS JACKSON	0.81	UpK3-6	5-Apr-53	9	6
MS JACKSON	38.63	UpK6-2	19-Apr-53	2	6
MS JACKSON	3.14	TP6-5	26-Mar-55	2	6
MS JACKSON	45.85	PB3-3	26-Jun-57	3	6
MS JACKSON	1.99	PB6-7	30-Jul-57	6	6
MS JACKSON	2.34	PB6-8	31-Jul-57	3	6
MS JASPER	0.99	UpK5-5	15-Apr-53	2	6
MS JASPER	1.47	TP6-2	23-Mar-55	9	6
MS JASPER	4.11	TP6-3	24-Mar-55	4	6
MS JASPER	22.97	PB3-3	26-Jun-57	4	6
MS JASPER	1.40	PB6-8	31-Jul-57	8	6
MS JASPER	8.38	PB17-5	2-Oct-57	9	6
MS JEFFERSON	45.67	UpK7-4	28-Apr-53	4	7
MS JEFFERSON	1.16	UpK8-5	12-May-53	5	7
MS JEFFERSON	1.56	UpK8-9	16-May-53	1	7
MS JEFFERSON	1.47	TP6-2	23-Mar-55	9	7
MS JEFFERSON	4.11	TP6-3	24-Mar-55	4	7
MS JEFFERSON	1.09	TP8-4	12-Apr-55	9	7
MS JEFFERSON	45.85	PB3-3	26-Jun-57	3	7
MS JEFFERSON DA	1.36	TS5-4	10-May-52	10	5
MS JEFFERSON DA	1.56	UpK8-9	16-May-53	1	5
MS JEFFERSON DA	1.47	TP6-2	23-Mar-55	9	5
MS JEFFERSON DA	4.11	TP6-3	24-Mar-55	4	5
MS JEFFERSON DA	11.44	PB3-3	26-Jun-57	6	5
MS JONES	1.03	TS5-5	11-May-52	6	6
MS JONES	0.99	UpK5-5	15-Apr-53	2	6

COUNTY	uCi/ sqm	SHOT DAY	DATE	RANK FOR DAY	TIMES IN TOP 10
MS JONES	1.47	TP6-2	23-Mar-55	9	6
MS JONES	4.11	TP6-3	24-Mar-55	4	6
MS JONES	11.44	PB3-3	26-Jun-57	6	6
MS JONES	10.51	PB12-6	7-Sep-57	9	6
MS KEMPER	0.93	UpK8-9	16-May-53	2	5
MS KEMPER	1.47	TP6-2	23-Mar-55	9	5
MS KEMPER	4.11	TP6-3	24-Mar-55	4	5
MS KEMPER	2.87	TP6-4	25-Mar-55	4	5
MS KEMPER	2.78	PB2-5	22-Jun-57	4	5
MS LAFAYETTE	6.07	UpK2-7	30-Mar-53	7	7
MS LAFAYETTE	14.03	UpK2-8	31-Mar-53	8	7
MS LAFAYETTE	3.67	UpK3-6	5-Apr-53	1	7
MS LAFAYETTE	3.68	UpK8-3	10-May-53	4	7
MS LAFAYETTE	4.64	PB2-5	22-Jun-57	2	7
MS LAFAYETTE	1.25	PB3-8	1-Jul-57	3	7
MS LAFAYETTE	0.83	PB3-9	2-Jul-57	4	7
MS LAMAR	1.48	UpK5-5	15-Apr-53	1	5
MS LAMAR	1.47	TP6-2	23-Mar-55	9	5
MS LAMAR	4.11	TP6-3	24-Mar-55	4	5
MS LAMAR	45.85	PB3-3	26-Jun-57	3	5
MS LAMAR	10.80	PB12-6	7-Sep-57	7	5
MS LAUDERDALE	0.93	UpK8-9	16-May-53	2	5
MS LAUDERDALE	1.47	TP6-2	23-Mar-55	9	5
MS LAUDERDALE	4.11	TP6-3	24-Mar-55	4	5
MS LAUDERDALE	68.82	PB3-3	26-Jun-57	2	5
MS LAUDERDALE	1.40	PB6-8	31-Jul-57	8	5
MS LAWRENCE	1.48	UpK5-5	15-Apr-53	1	6
MS LAWRENCE	0.93	UpK8-9	16-May-53	2	6
MS LAWRENCE	1.47	TP6-2	23-Mar-55	9	6
MS LAWRENCE	4.11	TP6-3	24-Mar-55	4	6
MS LAWRENCE	1.09	TP8-4	12-Apr-55	9	6
MS LAWRENCE	11.44	PB3-3	26-Jun-57	6	6
MS LEAKE	0.99	UpK5-5	15-Apr-53	2	7
MS LEAKE	1.56	UpK8-9	16-May-53	1	7
MS LEAKE	1.47	TP6-2	23-Mar-55	9	7

COUNTY	uCi/ sqm	SHOT DAY	DATE	RANK FOR DAY	TIMES IN TOP 10
MS LEAKE	8.23	TP6-3	24-Mar-55	2	7
MS LEAKE	2.87	TP6-4	25-Mar-55	4	7
MS LEAKE	2.78	PB2-5	22-Jun-57	4	7
MS LEAKE	0.83	PB3-9	2-Jul-57	4	7
MS LEE	6.07	UpK2-7	30-Mar-53	7	6
MS LEE	3.67	UpK3-6	5-Apr-53	1	6
MS LEE	1.18	UpK8-3	10-May-53	10	6
MS LEE	4.64	PB2-5	22-Jun-57	2	6
MS LEE	1.25	PB3-8	1-Jul-57	3	6
MS LEE	0.83	PB3-9	2-Jul-57	4	6
MS LEFLORE	0.93	UpK8-9	16-May-53	2	9
MS LEFLORE	1.47	TP6-2	23-Mar-55	9	9
MS LEFLORE	4.11	TP6-3	24-Mar-55	4	9
MS LEFLORE	2.94	TP7-4	1-Apr-55	9	9
MS LEFLORE	0.89	PB1-9	5-Jun-57	7	9
MS LEFLORE	4.64	PB2-5	22-Jun-57	2	9
MS LEFLORE	2.10	PB3-9	2-Jul-57	1	9
MS LEFLORE	1.00	PB6-11	3-Aug-57	3	9
MS LEFLORE	2.73	PB9-2	24-Aug-57	6	9
MS LINCOLN	1.97	UpK4-1	6-Apr-53	2	8
MS LINCOLN	1.16	UpK8-5	12-May-53	5	8
MS LINCOLN	1.56	UpK8-9	16-May-53	1	8
MS LINCOLN	1.47	TP6-2	23-Mar-55	9	8
MS LINCOLN	4.11	TP6-3	24-Mar-55	4	8
MS LINCOLN	1.15	TP8-4	12-Apr-55	6	8
MS LINCOLN	11.44	PB3-3	26-Jun-57	6	8
MS LINCOLN	1.06	PB6-11	3-Aug-57	2	8
MS LOWNDES	0.93	UpK8-9	16-May-53	2	5
MS LOWNDES	1.47	TP6-2	23-Mar-55	9	5
MS LOWNDES	4.11	TP6-3	24-Mar-55	4	5
MS LOWNDES	4.64	PB2-5	22-Jun-57	2	5
MS LOWNDES	0.83	PB3-9	2-Jul-57	4	5
MS MADISON	38.28	UpK7-4	28-Apr-53	9	4
MS MADISON	1.56	UpK8-9	16-May-53	1	4
MS MADISON	1.47	TP6-2	23-Mar-55	9	4

COUNTY	uCi/ sqm	SHOT DAY	DATE	RANK FOR DAY	TIMES IN TOP 10
MS MADISON	4.11	TP6-3	24-Mar-55	4	4
MS MARION	1.47	TP6-2	23-Mar-55	9	4
MS MARION	4.11	TP6-3	24-Mar-55	4	4
MS MARION	68.82	PB3-3	26-Jun-57	2	4
MS MARION	11.68	PB12-6	7-Sep-57	6	4
MS MARSHALL	18.26	UpK2-7	30-Mar-53	2	8
MS MARSHALL	14.03	UpK2-8	31-Mar-53	8	8
MS MARSHALL	3.67	UpK3-6	5-Apr-53	1	8
MS MARSHALL	3.68	UpK8-3	10-May-53	4	8
MS MARSHALL	1.52	UpK8-4	11-May-53	6	8
MS MARSHALL	2.78	PB2-5	22-Jun-57	4	8
MS MARSHALL	1.25	PB3-8	1-Jul-57	3	8
MS MARSHALL	1.27	PB3-9	2-Jul-57	2	8
MS MONROE	1.60	BJ1-1	28-Oct-51	3	8
MS MONROE	3.41	UpK2-4	27-Mar-53	10	8
MS MONROE	3.06	UpK2-7	30-Mar-53	9	8
MS MONROE	2.17	UpK3-6	5-Apr-53	2	8
MS MONROE	1.18	UpK8-3	10-May-53	10	8
MS MONROE	2.78	PB2-5	22-Jun-57	4	8
MS MONROE	1.25	PB3-8	1-Jul-57	3	8
MS MONROE	1.27	PB3-9	2-Jul-57	2	8
MS MONTGOMERY	0.93	UpK8-9	16-May-53	2	6
MS MONTGOMERY	1.47	TP6-2	23-Mar-55	9	6
MS MONTGOMERY	16.45	TP6-3	24-Mar-55	1	6
MS MONTGOMERY	2.78	PB2-5	22-Jun-57	4	6
MS MONTGOMERY	1.27	PB3-9	2-Jul-57	2	6
MS MONTGOMERY	8.19	PB9-2	24-Aug-57	3	6
MS NESHOBA	0.93	UpK8-9	16-May-53	2	7
MS NESHOBA	1.47	TP6-2	23-Mar-55	9	7
MS NESHOBA	4.11	TP6-3	24-Mar-55	4	7
MS NESHOBA	2.87	TP6-4	25-Mar-55	4	7
MS NESHOBA	4.72	TP10-5	9-May-55	3	7
MS NESHOBA	1.86	PB2-5	22-Jun-57	5	7
MS NESHOBA	0.83	PB3-9	2-Jul-57	4	7
MS NEWTON	0.93	UpK8-9	16-May-53	2	5

COUNTY	uCi/ sqm	SHOT DAY	DATE	RANK FOR DAY	TIMES IN TOP 10
MS NEWTON	1.47	TP6-2	23-Mar-55	9	5
MS NEWTON	4.11	TP6-3	24-Mar-55	4	5
MS NEWTON	22.97	PB3-3	26-Jun-57	4	5
MS NEWTON	8.48	PB17-5	2-Oct-57	8	5
MS NOXUBEE	0.93	UpK8-9	16-May-53	2	7
MS NOXUBEE	1.47	TP6-2	23-Mar-55	9	7
MS NOXUBEE	4.11	TP6-3	24-Mar-55	4	7
MS NOXUBEE	0.89	PB1-9	5-Jun-57	7	7
MS NOXUBEE	2.78	PB2-5	22-Jun-57	4	7
MS NOXUBEE	0.83	PB3-9	2-Jul-57	4	7
MS NOXUBEE	2.73	PB9-2	24-Aug-57	6	7
MS OKTIBBEHA	0.93	UpK8-9	16-May-53	2	5
MS OKTIBBEHA	1.47	TP6-2	23-Mar-55	9	5
MS OKTIBBEHA	4.11	TP6-3	24-Mar-55	4	5
MS OKTIBBEHA	4.64	PB2-5	22-Jun-57	2	5
MS OKTIBBEHA	0.83	PB3-9	2-Jul-57	4	5
MS PANOLA	12.19	UpK2-7	30-Mar-53	4	7
MS PANOLA	14.03	UpK2-8	31-Mar-53	8	7
MS PANOLA	3.67	UpK3-6	5-Apr-53	1	7
MS PANOLA	6.18	UpK8-3	10-May-53	2	7
MS PANOLA	1.52	UpK8-4	11-May-53	6	7
MS PANOLA	4.64	PB2-5	22-Jun-57	2	7
MS PANOLA	1.27	PB3-9	2-Jul-57	2	7
MS PEARL RIVER	38.63	UpK6-2	19-Apr-53	2	3
MS PEARL RIVER	68.82	PB3-3	26-Jun-57	2	3
MS PEARL RIVER	1.40	PB6-8	31-Jul-57	8	3
MS PERRY	0.81	UpK3-6	5-Apr-53	9	4
MS PERRY	38.63	UpK6-2	19-Apr-53	2	4
MS PERRY	3.14	TP6-5	26-Mar-55	2	4
MS PERRY	22.97	PB3-3	26-Jun-57	4	4
MS PIKE	0.93	UpK8-9	16-May-53	2	5
MS PIKE	1.47	TP6-2	23-Mar-55	9	5
MS PIKE	4.11	TP6-3	24-Mar-55	4	5
MS PIKE	2.22	TP8-3	11-Apr-55	9	5
MS PIKE	68.82	PB3-3	26-Jun-57	2	5

COUNTY	uCi/ sqm	SHOT DAY	DATE	RANK FOR DAY	TIMES IN TOP 10
MS PONTOTOC	12.19	UpK2-7	30-Mar-53	4	7
MS PONTOTOC	14.03	UpK2-8	31-Mar-53	8	7
MS PONTOTOC	3.67	UpK3-6	5-Apr-53	1	7
MS PONTOTOC	2.50	UpK8-3	10-May-53	6	7
MS PONTOTOC	2.78	PB2-5	22-Jun-57	4	7
MS PONTOTOC	1.25	PB3-8	1-Jul-57	3	7
MS PONTOTOC	0.83	PB3-9	2-Jul-57	4	7
MS PRENTISS	6.07	UpK2-7	30-Mar-53	7	7
MS PRENTISS	21.07	UpK2-8	31-Mar-53	3	7
MS PRENTISS	3.67	UpK3-6	5-Apr-53	1	7
MS PRENTISS	1.18	UpK8-3	10-May-53	10	7
MS PRENTISS	0.89	PB1-9	5-Jun-57	7	7
MS PRENTISS	4.64	PB2-5	22-Jun-57	2	7
MS PRENTISS	1.27	PB3-9	2-Jul-57	2	7
MS QUITMAN	6.07	UpK2-7	30-Mar-53	7	8
MS QUITMAN	21.07	UpK2-8	31-Mar-53	3	8
MS QUITMAN	3.67	UpK3-6	5-Apr-53	1	8
MS QUITMAN	6.18	UpK8-3	10-May-53	2	8
MS QUITMAN	5.18	TP7-4	1-Apr-55	1	8
MS QUITMAN	1.86	PB2-5	22-Jun-57	5	8
MS QUITMAN	0.83	PB3-9	2-Jul-57	4	8
MS QUITMAN	5.46	PB9-2	24-Aug-57	4	8
MS RANKIN	0.93	UpK8-9	16-May-53	2	4
MS RANKIN	1.47	TP6-2	23-Mar-55	9	4
MS RANKIN	4.11	TP6-3	24-Mar-55	4	4
MS RANKIN	11.44	PB3-3	26-Jun-57	6	4
MS SCOTT	1.05	TS5-5	11-May-52	5	7
MS SCOTT	0.93	UpK8-9	16-May-53	2	7
MS SCOTT	1.47	TP6-2	23-Mar-55	9	7
MS SCOTT	4.11	TP6-3	24-Mar-55	4	7
MS SCOTT	2.87	TP6-4	25-Mar-55	4	7
MS SCOTT	22.97	PB3-3	26-Jun-57	4	7
MS SCOTT	8.30	PB17-5	2-Oct-57	10	7
MS SHARKEY	0.91	BJ1-1	28-Oct-51	4	6
MS SHARKEY	1.56	UpK8-9	16-May-53	1	6

COUNTY	uCi/ sqm	SHOT DAY	DATE	RANK FOR DAY	TIMES IN TOP 10
MS SHARKEY	1.47	TP6-2	23-Mar-55	9	6
MS SHARKEY	4.11	TP6-3	24-Mar-55	4	6
MS SHARKEY	4.26	PB11-2	1-Sep-57	10	6
MS SHARKEY	1.25	PB18-9	15-Oct-57	3	6
MS SIMPSON	1.47	TP6-2	23-Mar-55	9	3
MS SIMPSON	4.11	TP6-3	24-Mar-55	4	3
MS SIMPSON	11.44	PB3-3	26-Jun-57	6	3
MS SMITH	0.95	TS5-5	11-May-52	9	5
MS SMITH	0.99	UpK5-5	15-Apr-53	2	5
MS SMITH	1.47	TP6-2	23-Mar-55	9	5
MS SMITH	4.11	TP6-3	24-Mar-55	4	5
MS SMITH	22.97	PB3-3	26-Jun-57	4	5
MS STONE	38.63	UpK6-2	19-Apr-53	2	6
MS STONE	3.14	TP6-5	26-Mar-55	2	6
MS STONE	1.03	TP8-5	13-Apr-55	2	6
MS STONE	45.85	PB3-3	26-Jun-57	3	6
MS STONE	1.99	PB6-7	30-Jul-57	6	6
MS STONE	2.34	PB6-8	31-Jul-57	3	6
MS SUNFLOWER	0.93	UpK8-9	16-May-53	2	6
MS SUNFLOWER	1.47	TP6-2	23-Mar-55	9	6
MS SUNFLOWER	4.11	TP6-3	24-Mar-55	4	6
MS SUNFLOWER	3.06	TP7-4	1-Apr-55	6	6
MS SUNFLOWER	4.64	PB2-5	22-Jun-57	2	6
MS SUNFLOWER	2.73	PB9-2	24-Aug-57	6	6
MS TALLAHATCHIE	12.19	UpK2-7	30-Mar-53	4	8
MS TALLAHATCHIE	3.67	UpK3-6	5-Apr-53	1	8
MS TALLAHATCHIE	3.68	UpK8-3	10-May-53	4	8
MS TALLAHATCHIE	3.14	TP7-4	1-Apr-55	5	8
MS TALLAHATCHIE	1.86	PB2-5	22-Jun-57	5	8
MS TALLAHATCHIE	1.25	PB3-8	1-Jul-57	3	8
MS TALLAHATCHIE	0.83	PB3-9	2-Jul-57	4	8
MS TALLAHATCHIE	8.19	PB9-2	24-Aug-57	3	8
MS TATE	0.91	BJ1-1	28-Oct-51	4	10
MS TATE	12.19	UpK2-7	30-Mar-53	4	10
MS TATE	14.03	UpK2-8	31-Mar-53	8	10

COUNTY	uCi/ sqm	SHOT DAY	DATE	RANK FOR DAY	TIMES IN TOP 10
MS TATE	3.67	UpK3-6	5-Apr-53	1	10
MS TATE	6.18	UpK8-3	10-May-53	2	10
MS TATE	1.52	UpK8-4	11-May-53	6	10
MS TATE	2.99	TP7-4	1-Apr-55	8	10
MS TATE	2.78	PB2-5	22-Jun-57	4	10
MS TATE	1.27	PB3-9	2-Jul-57	2	10
MS TATE	2.73	PB9-2	24-Aug-57	6	10
MS TIPPAH	12.19	UpK2-7	30-Mar-53	4	7
MS TIPPAH	21.07	UpK2-8	31-Mar-53	3	7
MS TIPPAH	3.67	UpK3-6	5-Apr-53	1	7
MS TIPPAH	3.68	UpK8-3	10-May-53	4	7
MS TIPPAH	2.78	PB2-5	22-Jun-57	4	7
MS TIPPAH	1.25	PB3-8	1-Jul-57	3	7
MS TIPPAH	1.27	PB3-9	2-Jul-57	2	7
MS TISHOMINGO	3.06	UpK2-7	30-Mar-53	9	7
MS TISHOMINGO	21.07	UpK2-8	31-Mar-53	3	7
MS TISHOMINGO	3.67	UpK3-6	5-Apr-53	1	7
MS TISHOMINGO	0.89	PB1-9	5-Jun-57	7	7
MS TISHOMINGO	2.78	PB2-5	22-Jun-57	4	7
MS TISHOMINGO	0.83	PB3-9	2-Jul-57	4	7
MS TISHOMINGO	1.89	PB6-7	30-Jul-57	7	7
MS TUNICA	12.19	UpK2-7	30-Mar-53	4	9
MS TUNICA	3.67	UpK3-6	5-Apr-53	1	9
MS TUNICA	3.68	UpK8-3	10-May-53	4	9
MS TUNICA	1.52	UpK8-4	11-May-53	6	9
MS TUNICA	3.02	TP7-4	1-Apr-55	7	9
MS TUNICA	1.21	PB1-5	1-Jun-57	10	9
MS TUNICA	2.78	PB2-5	22-Jun-57	4	9
MS TUNICA	1.27	PB3-9	2-Jul-57	2	9
MS TUNICA	2.73	PB9-2	24-Aug-57	6	9
MS UNION	12.19	UpK2-7	30-Mar-53	4	8
MS UNION	14.03	UpK2-8	31-Mar-53	8	8
MS UNION	3.67	UpK3-6	5-Apr-53	1	8
MS UNION	2.50	UpK8-3	10-May-53	6	8
MS UNION	0.89	PB1-9	5-Jun-57	7	8

COUNTY	uCi/ sqm	SHOT DAY	DATE	RANK FOR DAY	TIMES IN TOP 10
MS UNION	4.64	PB2-5	22-Jun-57	2	8
MS UNION	1.25	PB3-8	1-Jul-57	3	8
MS UNION	2.10	PB3-9	2-Jul-57	1	8
MS WALTHALL	38.63	UpK6-2	19-Apr-53	2	4
MS WALTHALL	0.80	UpK8-5	12-May-53	7	4
MS WALTHALL	1.09	TP8-4	12-Apr-55	9	4
MS WALTHALL	22.97	PB3-3	26-Jun-57	4	4
MS WARREN	48.58	UpK7-4	28-Apr-53	2	6
MS WARREN	1.16	UpK8-5	12-May-53	5	6
MS WARREN	1.56	UpK8-9	16-May-53	1	6
MS WARREN	1.47	TP6-2	23-Mar-55	9	6
MS WARREN	4.11	TP6-3	24-Mar-55	4	6
MS WARREN	114.67	PB3-3	26-Jun-57	1	6
MS WASHINGTON	0.93	UpK8-9	16-May-53	2	5
MS WASHINGTON	1.47	TP6-2	23-Mar-55	9	5
MS WASHINGTON	4.11	TP6-3	24-Mar-55	4	5
MS WASHINGTON	0.89	PB1-9	5-Jun-57	7	5
MS WASHINGTON	1.25	PB18-9	15-Oct-57	3	5
MS WAYNE	0.81	UpK3-6	5-Apr-53	9	6
MS WAYNE	1.47	TP6-2	23-Mar-55	9	6
MS WAYNE	4.11	TP6-3	24-Mar-55	4	6
MS WAYNE	3.14	TP6-5	26-Mar-55	2	6
MS WAYNE	45.85	PB3-3	26-Jun-57	3	6
MS WAYNE	1.40	PB6-8	31-Jul-57	8	6
MS WEBSTER	0.93	UpK8-9	16-May-53	2	7
MS WEBSTER	1.47	TP6-2	23-Mar-55	9	7
MS WEBSTER	4.11	TP6-3	24-Mar-55	4	7
MS WEBSTER	2.87	TP6-4	25-Mar-55	4	7
MS WEBSTER	4.64	PB2-5	22-Jun-57	2	7
MS WEBSTER	1.27	PB3-9	2-Jul-57	2	7
MS WEBSTER	16.10	PB12-6	7-Sep-57	4	7
MS WILKINSON	1.16	UpK8-5	12-May-53	5	6
MS WILKINSON	1.47	TP6-2	23-Mar-55	9	6
MS WILKINSON	4.11	TP6-3	24-Mar-55	4	6
MS WILKINSON	114.67	PB3-3	26-Jun-57	1	6

COUNTY	uCi/ sqm	SHOT DAY	DATE	RANK FOR DAY	TIMES IN TOP 10
MS WILKINSON	1.99	PB6-7	30-Jul-57	6	6
MS WILKINSON	2.34	PB6-8	31-Jul-57	3	6
MS WINSTON	0.91	BJ1-1	28-Oct-51	4	6
MS WINSTON	0.93	UpK8-9	16-May-53	2	6
MS WINSTON	1.47	TP6-2	23-Mar-55	9	6
MS WINSTON	4.11	TP6-3	24-Mar-55	4	6
MS WINSTON	2.78	PB2-5	22-Jun-57	4	6
MS WINSTON	0.83	PB3-9	2-Jul-57	4	6
MS YALOBUSHA	12.19	UpK2-7	30-Mar-53	4	8
MS YALOBUSHA	14.03	UpK2-8	31-Mar-53	8	8
MS YALOBUSHA	3.67	UpK3-6	5-Apr-53	1	8
MS YALOBUSHA	3.68	UpK8-3	10-May-53	4	8
MS YALOBUSHA	4.64	PB2-5	22-Jun-57	2	8
MS YALOBUSHA	1.25	PB3-8	1-Jul-57	3	8
MS YALOBUSHA	1.27	PB3-9	2-Jul-57	2	8
MS YALOBUSHA	2.73	PB9-2	24-Aug-57	6	8
MS YAZOO	1.03	TS5-5	11-May-52	6	5
MS YAZOO	1.56	UpK8-9	16-May-53	1	5
MS YAZOO	1.47	TP6-2	23-Mar-55	9	5
MS YAZOO	8.23	TP6-3	24-Mar-55	2	5
MS YAZOO	2.78	PB2-5	22-Jun-57	4	5

COUNTY	uCi/ sqm	SHOT DAY	DATE	RANK FOR DAY	TIMES IN TOP 10
MT BEAVERHEAD	2.30	TS5-8	14-May-52	1	14
MT BEAVERHEAD	2.62	TS5-9	15-May-52	2	14
MT BEAVERHEAD	22.18	TS8-2	6-Jun-52	9	14
MT BEAVERHEAD	5.46	UpK8-2	9-May-53	10	14
MT BEAVERHEAD	21.00	PB6-2	25-Jul-57	1	14
MT BEAVERHEAD	24.56	PB6-3	26-Jul-57	2	14
MT BEAVERHEAD	5.34	PB15-3	18-Sep-57	1	14
MT BEAVERHEAD	7.58	PB16-3	25-Sep-57	1	14
MT BEAVERHEAD	1.29	PB16-4	26-Sep-57	9	14
MT BEAVERHEAD	0.85	PB18-4	10-Oct-57	7	14
MT BEAVERHEAD	38.04	UE12-3	16-Apr-65	5	14
MT BEAVERHEAD	0.61	BN-3	20-Dec-70	10	14
MT BEAVERHEAD	0.46	BN-4	21-Dec-70	2	14
MT BEAVERHEAD	0.20	BN-5	22-Dec-70	3	14
MT BIG HORN	5.17	UpK2-2	25-Mar-53	9	17
MT BIG HORN	32.75	UpK8-2	9-May-53	2	17
MT BIG HORN	7.89	UpK8-3	10-May-53	1	17
MT BIG HORN	1.22	UpK8-4	11-May-53	8	17
MT BIG HORN	2.07	UpK10-7	31-May-53	3	17
MT BIG HORN	5.09	PB5-3	17-Jul-57	8	17
MT BIG HORN	1.65	PB8-2	19-Aug-57	6	17
MT BIG HORN	3.07	PB8-4	21-Aug-57	10	17
MT BIG HORN	3.07	PB8-4	21-Aug-57	10	17
MT BIG HORN	1.86	PB9-2	24-Aug-57	9	17
MT BIG HORN	3.68	PB9-4	26-Aug-57	9	17
MT BIG HORN	5.14	PB15-2	17-Sep-57	7	17
MT BIG HORN	0.37	UE2-5	9-Mar-62	2	17
MT BIG HORN	0.23	UE2-6	10-Mar-62	1	17
MT BIG HORN	1.75	UE7-3	13-Jul-62	7	17
MT BIG HORN	29.29	UE12-3	16-Apr-65	10	17
MT BIG HORN	0.21	BN-5	22-Dec-70	2	17
MT BLAINE	32.75	UpK8-2	9-May-53	2	9
MT BLAINE	2.63	UpK8-3	10-May-53	5	9
MT BLAINE	4.13	UpK10-7	31-May-53	2	9
MT BLAINE	1.91	PB2-2	19-Jun-57	8	9

COUNTY	uCi/ sqm	SHOT DAY	DATE	RANK FOR DAY	TIMES IN TOP 10
MT BLAINE	29.83	PB8-2	19-Aug-57	2	9
MT BLAINE	1.86	PB9-2	24-Aug-57	9	9
MT BLAINE	5.14	PB15-2	17-Sep-57	7	9
MT BLAINE	0.11	UE2-6	10-Mar-62	6	9
MT BLAINE	0.38	BN-4	21-Dec-70	3	9
MT BROADWATER	1.32	PB2-2	19-Jun-57	9	4
MT BROADWATER	24.56	PB6-3	26-Jul-57	2	4
MT BROADWATER	1.86	PB9-2	24-Aug-57	9	4
MT BROADWATER	1.72	PB15-3	18-Sep-57	8	4
MT CARBON	16.38	UpK8-2	9-May-53	3	13
MT CARBON	5.26	UpK8-3	10-May-53	3	13
MT CARBON	2.43	UpK8-4	11-May-53	3	13
MT CARBON	2.07	UpK10-7	31-May-53	3	13
MT CARBON	1.97	PB4-2	6-Jul-57	8	13
MT CARBON	1.30	PB6-8	31-Jul-57	9	13
MT CARBON	29.83	PB8-2	19-Aug-57	2	13
MT CARBON	1.86	PB9-2	24-Aug-57	9	13
MT CARBON	1.72	PB15-3	18-Sep-57	8	13
MT CARBON	0.24	UE2-5	9-Mar-62	5	13
MT CARBON	0.23	UE2-6	10-Mar-62	1	13
MT CARBON	47.38	UE12-3	16-Apr-65	4	13
MT CARBON	0.22	BN-5	22-Dec-70	1	13
MT CARTER	45.45	UpK2-2	25-Mar-53	4	13
MT CARTER	4.73	UpK8-1	8-May-53	6	13
MT CARTER	5.09	PB5-3	17-Jul-57	8	13
MT CARTER	7.20	PB6-5	28-Jul-57	2	13
MT CARTER	1.30	PB8-2	19-Aug-57	7	13
MT CARTER	39.46	PB8-3	20-Aug-57	4	13
MT CARTER	8.05	PB8-4	21-Aug-57	4	13
MT CARTER	8.05	PB8-4	21-Aug-57	4	13
MT CARTER	5.22	PB9-4	26-Aug-57	8	13
MT CARTER	6.81	PB11-2	1-Sep-57	6	13
MT CARTER	0.22	UE2-6	10-Mar-62	2	13
MT CARTER	1.85	UE7-3	13-Jul-62	6	13
MT CARTER	0.14	BN-5	22-Dec-70	4	13

COUNTY	uCi/ sqm	SHOT DAY	DATE	RANK FOR DAY	TIMES IN TOP 10
MT CASCADE	1.94	PB3-3	26-Jun-57	9	10
MT CASCADE	24.56	PB6-3	26-Jul-57	2	10
MT CASCADE	1.30	PB6-8	31-Jul-57	9	10
MT CASCADE	1.86	PB9-2	24-Aug-57	9	10
MT CASCADE	5.14	PB15-2	17-Sep-57	7	10
MT CASCADE	0.12	UE2-6	10-Mar-62	5	10
MT CASCADE	1.74	UE7-2	12-Jul-62	10	10
MT CASCADE	94.25	UE12-3	16-Apr-65	1	10
MT CASCADE	0.38	BN-4	21-Dec-70	3	10
MT CASCADE	0.20	BN-5	22-Dec-70	3	10
MT CHOUTEAU	323.70	TS8-1	5-Jun-52	9	14
MT CHOUTEAU	5.34	TS8-3	7-Jun-52	6	14
MT CHOUTEAU	0.86	TS8-11	15-Jun-52	4	14
MT CHOUTEAU	1.32	PB2-2	19-Jun-57	9	14
MT CHOUTEAU	18.45	PB6-3	26-Jul-57	4	14
MT CHOUTEAU	1.30	PB6-8	31-Jul-57	9	14
MT CHOUTEAU	1.86	PB9-2	24-Aug-57	9	14
MT CHOUTEAU	6.02	PB12-2	3-Sep-57	10	14
MT CHOUTEAU	5.14	PB15-2	17-Sep-57	7	14
MT CHOUTEAU	4.76	PB17-6	3-Oct-57	9	14
MT CHOUTEAU	0.12	UE2-6	10-Mar-62	5	14
MT CHOUTEAU	3.68	UE7-2	12-Jul-62	1	14
MT CHOUTEAU	0.38	BN-4	21-Dec-70	3	14
MT CHOUTEAU	0.20	BN-5	22-Dec-70	3	14
MT CUSTER	122.17	UpK2-2	25-Mar-53	1	17
MT CUSTER	89.57	UpK8-1	8-May-53	1	17
MT CUSTER	5.46	UpK8-2	9-May-53	10	17
MT CUSTER	1.56	PB2-4	21-Jun-57	7	17
MT CUSTER	2.63	PB4-2	6-Jul-57	5	17
MT CUSTER	1.30	PB6-8	31-Jul-57	9	17
MT CUSTER	0.83	PB8-2	19-Aug-57	8	17
MT CUSTER	3.07	PB8-4	21-Aug-57	10	17
MT CUSTER	3.07	PB8-4	21-Aug-57	10	17
MT CUSTER	1.86	PB9-2	24-Aug-57	9	17
MT CUSTER	7.36	PB9-4	26-Aug-57	7	17

COUNTY	uCi/ sqm	SHOT DAY	DATE	RANK FOR DAY	TIMES IN TOP 10
MT CUSTER	6.02	PB12-2	3-Sep-57	10	17
MT CUSTER	3.47	PB15-2	17-Sep-57	10	17
MT CUSTER	0.44	UE2-5	9-Mar-62	1	17
MT CUSTER	0.23	UE2-6	10-Mar-62	1	17
MT CUSTER	4.20	UE7-3	13-Jul-62	1	17
MT CUSTER	0.20	BN-5	22-Dec-70	3	17
MT DANIELS	61.09	UpK2-2	25-Mar-53	3	11
MT DANIELS	17.76	UpK8-1	8-May-53	4	11
MT DANIELS	5.46	UpK8-2	9-May-53	10	11
MT DANIELS	1.15	UpK10-9	2-Jun-53	1	11
MT DANIELS	2.28	PB5-5	19-Jul-57	7	11
MT DANIELS	0.83	PB8-2	19-Aug-57	8	11
MT DANIELS	19.73	PB8-3	20-Aug-57	6	11
MT DANIELS	1.92	PB8-5	22-Aug-57	10	11
MT DANIELS	1.86	PB9-2	24-Aug-57	9	11
MT DANIELS	0.22	UE2-6	10-Mar-62	2	11
MT DANIELS	0.38	BN-4	21-Dec-70	3	11
MT DAWSON	122.17	UpK2-2	25-Mar-53	1	15
MT DAWSON	53.66	UpK8-1	8-May-53	1	15
MT DAWSON	5.46	UpK8-2	9-May-53	10	15
MT DAWSON	2.09	UpK10-3	27-May-53	8	15
MT DAWSON	1.86	PB4-10	14-Jul-57	8	15
MT DAWSON	19.36	PB5-2	16-Jul-57	3	15
MT DAWSON	3.30	PB8-2	19-Aug-57	5	15
MT DAWSON	19.73	PB8-3	20-Aug-57	6	15
MT DAWSON	1.92	PB8-5	22-Aug-57	10	15
MT DAWSON	3.48	PB9-4	26-Aug-57	10	15
MT DAWSON	15.54	PB11-4	3-Sep-57	3	15
MT DAWSON	9.03	PB12-2	3-Sep-57	7	15
MT DAWSON	5.14	PB15-2	17-Sep-57	7	15
MT DAWSON	0.22	UE2-6	10-Mar-62	2	15
MT DAWSON	0.22	BN-5	22-Dec-70	1	15
MT DEER LODGE	493.49	TS8-1	5-Jun-52	5	13
MT DEER LODGE	5.54	TS8-6	10-Jun-52	1	13
MT DEER LODGE	1.85	PB1-7	3-Jun-57	1	13

COUNTY	uCi/ sqm	SHOT DAY	DATE	RANK FOR DAY	TIMES IN TOP 10
MT DEER LODGE	21.00	PB6-2	25-Jul-57	1	13
MT DEER LODGE	12.28	PB6-3	26-Jul-57	6	13
MT DEER LODGE	2.67	PB15-3	18-Sep-57	4	13
MT DEER LODGE	1.16	PB15-5	20-Sep-57	7	13
MT DEER LODGE	7.58	PB16-3	25-Sep-57	1	13
MT DEER LODGE	1.29	PB16-4	26-Sep-57	9	13
MT DEER LODGE	9.12	PB17-5	2-Oct-57	6	13
MT DEER LODGE	0.85	PB18-4	10-Oct-57	7	13
MT DEER LODGE	33.29	UE12-3	16-Apr-65	8	13
MT DEER LODGE	0.38	BN-4	21-Dec-70	3	13
MT FALLON	61.09	UpK2-2	25-Mar-53	3	13
MT FALLON	53.66	UpK8-1	8-May-53	1	13
MT FALLON	5.46	UpK8-2	9-May-53	10	13
MT FALLON	9.68	PB5-2	16-Jul-57	6	13
MT FALLON	10.10	PB5-3	17-Jul-57	4	13
MT FALLON	0.83	PB8-2	19-Aug-57	8	13
MT FALLON	39.46	PB8-3	20-Aug-57	4	13
MT FALLON	3.78	PB8-5	22-Aug-57	4	13
MT FALLON	3.48	PB9-4	26-Aug-57	10	13
MT FALLON	3.47	PB15-2	17-Sep-57	10	13
MT FALLON	0.23	UE2-6	10-Mar-62	1	13
MT FALLON	1.85	UE7-3	13-Jul-62	6	13
MT FALLON	0.20	BN-5	22-Dec-70	3	13
MT FERGUS	25.47	TS8-2	6-Jun-52	3	15
MT FERGUS	32.75	UpK8-2	9-May-53	2	15
MT FERGUS	5.26	UpK8-3	10-May-53	3	15
MT FERGUS	4.13	UpK10-7	31-May-53	2	15
MT FERGUS	1.32	PB2-2	19-Jun-57	9	15
MT FERGUS	1.56	PB2-4	21-Jun-57	7	15
MT FERGUS	2.63	PB4-2	6-Jul-57	5	15
MT FERGUS	15.87	PB6-2	25-Jul-57	3	15
MT FERGUS	6.10	PB6-4	27-Jul-57	9	15
MT FERGUS	29.83	PB8-2	19-Aug-57	2	15
MT FERGUS	3.60	PB9-2	24-Aug-57	5	15
MT FERGUS	5.14	PB15-2	17-Sep-57	7	15

COUNTY	uCi/ sqm	SHOT DAY	DATE	RANK FOR DAY	TIMES IN TOP 10
MT FERGUS	0.23	UE2-6	10-Mar-62	1	15
MT FERGUS	0.46	BN-4	21-Dec-70	2	15
MT FERGUS	0.22	BN-5	22-Dec-70	1	15
MT FLATHEAD	2.34	TS8-6	10-Jun-52	5	10
MT FLATHEAD	1.91	TS8-7	11-Jun-52	7	10
MT FLATHEAD	0.93	TS8-10	14-Jun-52	1	10
MT FLATHEAD	1.72	TS8-11	15-Jun-52	2	10
MT FLATHEAD	12.28	PB6-3	26-Jul-57	6	10
MT FLATHEAD	7.58	PB16-3	25-Sep-57	1	10
MT FLATHEAD	1.29	PB16-4	26-Sep-57	9	10
MT FLATHEAD	0.85	PB18-4	10-Oct-57	7	10
MT FLATHEAD	0.36	BN-4	21-Dec-70	4	10
MT FLATHEAD	0.11	BN-5	22-Dec-70	5	10
MT GALLATIN	1.76	TS8-6	10-Jun-52	7	17
MT GALLATIN	0.86	TS8-11	15-Jun-52	4	17
MT GALLATIN	1.81	PB1-6	2-Jun-57	7	17
MT GALLATIN	1.91	PB2-2	19-Jun-57	8	17
MT GALLATIN	1.97	PB4-2	6-Jul-57	8	17
MT GALLATIN	15.87	PB6-2	25-Jul-57	3	17
MT GALLATIN	1.30	PB6-8	31-Jul-57	9	17
MT GALLATIN	29.83	PB8-2	19-Aug-57	2	17
MT GALLATIN	1.86	PB9-2	24-Aug-57	9	17
MT GALLATIN	0.11	UE2-6	10-Mar-62	6	17
MT GALLATIN	2.90	UE7-2	12-Jul-62	4	17
MT GALLATIN	28.39	UE12-2	15-Apr-65	1	17
MT GALLATIN	36.05	UE12-3	16-Apr-65	6	17
MT GALLATIN	9.04	UE12-4	17-Apr-65	7	17
MT GALLATIN	0.88	BN-3	20-Dec-70	5	17
MT GALLATIN	0.38	BN-4	21-Dec-70	3	17
MT GALLATIN	0.22	BN-5	22-Dec-70	1	17
MT GARFIELD	5.17	UpK2-2	25-Mar-53	9	17
MT GARFIELD	49.13	UpK8-2	9-May-53	1	17
MT GARFIELD	2.63	UpK8-3	10-May-53	5	17
MT GARFIELD	1.22	UpK8-4	11-May-53	8	17
MT GARFIELD	2.23	UpK10-3	27-May-53	5	17

COUNTY	uCi/ sqm	SHOT DAY	DATE	RANK FOR DAY	TIMES IN TOP 10
MT GARFIELD	8.26	UpK10-7	31-May-53	1	17
MT GARFIELD	13.03	TP11-2	16-May-55	10	17
MT GARFIELD	1.91	PB2-2	19-Jun-57	8	17
MT GARFIELD	1.56	PB2-4	21-Jun-57	7	17
MT GARFIELD	1.30	PB6-8	31-Jul-57	9	17
MT GARFIELD	1.65	PB8-2	19-Aug-57	6	17
MT GARFIELD	1.86	PB9-2	24-Aug-57	9	17
MT GARFIELD	3.68	PB9-4	26-Aug-57	9	17
MT GARFIELD	5.14	PB15-2	17-Sep-57	7	17
MT GARFIELD	0.22	UE2-6	10-Mar-62	2	17
MT GARFIELD	2.88	UE7-3	13-Jul-62	2	17
MT GARFIELD	0.22	BN-5	22-Dec-70	1	17
MT GLACIER	4.95	TS8-6	10-Jun-52	2	10
MT GLACIER	1.84	TS8-7	11-Jun-52	9	10
MT GLACIER	0.93	TS8-10	14-Jun-52	1	10
MT GLACIER	0.86	TS8-11	15-Jun-52	4	10
MT GLACIER	1.19	PB1-9	5-Jun-57	4	10
MT GLACIER	12.28	PB6-3	26-Jul-57	6	10
MT GLACIER	2.73	PB9-2	24-Aug-57	6	10
MT GLACIER	5.14	PB15-2	17-Sep-57	7	10
MT GLACIER	0.38	BN-4	21-Dec-70	3	10
MT GLACIER	0.20	BN-5	22-Dec-70	3	10
MT GOLDEN VALLE	16.38	UpK8-2	9-May-53	3	15
MT GOLDEN VALLE	5.26	UpK8-3	10-May-53	3	15
MT GOLDEN VALLE	1.22	UpK8-4	11-May-53	8	15
MT GOLDEN VALLE	2.07	UpK10-7	31-May-53	3	15
MT GOLDEN VALLE	2.28	PB5-5	19-Jul-57	7	15
MT GOLDEN VALLE	15.87	PB6-2	25-Jul-57	3	15
MT GOLDEN VALLE	6.18	PB6-4	27-Jul-57	8	15
MT GOLDEN VALLE	1.30	PB6-8	31-Jul-57	9	15
MT GOLDEN VALLE	29.83	PB8-2	19-Aug-57	2	15
MT GOLDEN VALLE	1.86	PB9-2	24-Aug-57	9	15
MT GOLDEN VALLE	3.47	PB15-2	17-Sep-57	10	15
MT GOLDEN VALLE	1.72	PB15-3	18-Sep-57	8	15
MT GOLDEN VALLE	5.92	PB17-6	3-Oct-57	4	15

COUNTY	uCi/ sqm	SHOT DAY	DATE	RANK FOR DAY	TIMES IN TOP 10
MT GOLDEN VALLE	0.23	UE2-6	10-Mar-62	1	15
MT GOLDEN VALLE	0.22	BN-5	22-Dec-70	1	15
MT GRANITE	21.00	PB6-2	25-Jul-57	1	8
MT GRANITE	24.56	PB6-3	26-Jul-57	2	8
MT GRANITE	5.34	PB15-3	18-Sep-57	1	8
MT GRANITE	7.58	PB16-3	25-Sep-57	1	8
MT GRANITE	1.29	PB16-4	26-Sep-57	9	8
MT GRANITE	0.85	PB18-4	10-Oct-57	7	8
MT GRANITE	33.29	UE12-3	16-Apr-65	8	8
MT GRANITE	0.20	BN-5	22-Dec-70	3	8
MT HILL	0.86	TS8-11	15-Jun-52	4	8
MT HILL	12.28	PB6-3	26-Jul-57	6	8
MT HILL	1.30	PB6-8	31-Jul-57	9	8
MT HILL	1.86	PB9-2	24-Aug-57	9	8
MT HILL	5.14	PB15-2	17-Sep-57	7	8
MT HILL	0.12	UE2-6	10-Mar-62	5	8
MT HILL	0.38	BN-4	21-Dec-70	3	8
MT HILL	0.22	BN-5	22-Dec-70	1	8
MT JEFFERSON	3.87	PB1-6	2-Jun-57	1	7
MT JEFFERSON	15.87	PB6-2	25-Jul-57	3	7
MT JEFFERSON	12.28	PB6-3	26-Jul-57	6	7
MT JEFFERSON	1.86	PB9-2	24-Aug-57	9	7
MT JEFFERSON	5.14	PB15-2	17-Sep-57	7	7
MT JEFFERSON	0.12	UE2-6	10-Mar-62	5	7
MT JEFFERSON	0.22	BN-5	22-Dec-70	1	7
MT JUDITH BASIN	26.71	TS8-2	6-Jun-52	1	10
MT JUDITH BASIN	1.91	PB2-2	19-Jun-57	8	10
MT JUDITH BASIN	15.87	PB6-2	25-Jul-57	3	10
MT JUDITH BASIN	1.30	PB6-8	31-Jul-57	9	10
MT JUDITH BASIN	44.80	PB8-2	19-Aug-57	1	10
MT JUDITH BASIN	1.86	PB9-2	24-Aug-57	9	10
MT JUDITH BASIN	3.47	PB15-2	17-Sep-57	10	10
MT JUDITH BASIN	0.21	UE2-6	10-Mar-62	3	10
MT JUDITH BASIN	58.91	UE12-3	16-Apr-65	2	10
MT JUDITH BASIN	0.22	BN-5	22-Dec-70	1	10

COUNTY	uCi/ sqm	SHOT DAY	DATE	RANK FOR DAY	TIMES IN TOP 10
MT LAKE	2.85	TS8-5	9-Jun-52	7	15
MT LAKE	2.93	TS8-6	10-Jun-52	3	15
MT LAKE	2.28	TS8-7	11-Jun-52	2	15
MT LAKE	0.93	TS8-10	14-Jun-52	1	15
MT LAKE	1.72	TS8-11	15-Jun-52	2	15
MT LAKE	1.94	PB1-5	1-Jun-57	7	15
MT LAKE	0.83	PB1-9	5-Jun-57	8	15
MT LAKE	15.78	PB6-2	25-Jul-57	4	15
MT LAKE	12.28	PB6-3	26-Jul-57	6	15
MT LAKE	7.58	PB16-3	25-Sep-57	1	15
MT LAKE	1.29	PB16-4	26-Sep-57	9	15
MT LAKE	0.85	PB18-4	10-Oct-57	7	15
MT LAKE	19.79	UE12-2	15-Apr-65	2	15
MT LAKE	0.36	BN-4	21-Dec-70	4	15
MT LAKE	0.22	BN-5	22-Dec-70	1	15
MT LEWIS AND CL	5.01	TS8-3	7-Jun-52	10	11
MT LEWIS AND CL	1.79	PB1-6	2-Jun-57	8	11
MT LEWIS AND CL	15.87	PB6-2	25-Jul-57	3	11
MT LEWIS AND CL	18.45	PB6-3	26-Jul-57	4	11
MT LEWIS AND CL	1.86	PB9-2	24-Aug-57	9	11
MT LEWIS AND CL	5.14	PB15-2	17-Sep-57	7	11
MT LEWIS AND CL	1.74	UE7-2	12-Jul-62	10	11
MT LEWIS AND CL	28.39	UE12-2	15-Apr-65	1	11
MT LEWIS AND CL	94.25	UE12-3	16-Apr-65	1	11
MT LEWIS AND CL	0.46	BN-4	21-Dec-70	2	11
MT LEWIS AND CL	0.22	BN-5	22-Dec-70	1	11
MT LIBERTY	5.23	TS8-3	7-Jun-52	7	7
MT LIBERTY	12.28	PB6-3	26-Jul-57	6	7
MT LIBERTY	1.86	PB9-2	24-Aug-57	9	7
MT LIBERTY	6.02	PB12-2	3-Sep-57	10	7
MT LIBERTY	5.14	PB15-2	17-Sep-57	7	7
MT LIBERTY	0.38	BN-4	21-Dec-70	3	7
MT LIBERTY	0.20	BN-5	22-Dec-70	3	7
MT LINCOLN	0.83	PB1-9	5-Jun-57	8	6
MT LINCOLN	12.28	PB6-3	26-Jul-57	6	6

COUNTY	uCi/ sqm	SHOT DAY	DATE	RANK FOR DAY	TIMES IN TOP 10
MT LINCOLN	7.58	PB16-3	25-Sep-57	1	6
MT LINCOLN	1.29	PB16-4	26-Sep-57	9	6
MT LINCOLN	0.85	PB18-4	10-Oct-57	7	6
MT LINCOLN	0.36	BN-4	21-Dec-70	4	6
MT MADISON	26.29	TS8-2	6-Jun-52	2	13
MT MADISON	1.86	PB1-6	2-Jun-57	6	13
MT MADISON	1.91	PB2-2	19-Jun-57	8	13
MT MADISON	2.63	PB4-2	6-Jul-57	5	13
MT MADISON	15.87	PB6-2	25-Jul-57	3	13
MT MADISON	29.83	PB8-2	19-Aug-57	2	13
MT MADISON	1.86	PB9-2	24-Aug-57	9	13
MT MADISON	1.74	UE7-2	12-Jul-62	10	13
MT MADISON	19.79	UE12-2	15-Apr-65	2	13
MT MADISON	38.04	UE12-3	16-Apr-65	5	13
MT MADISON	0.61	BN-3	20-Dec-70	10	13
MT MADISON	0.32	BN-4	21-Dec-70	5	13
MT MADISON	0.20	BN-5	22-Dec-70	3	13
MT MCCONE	61.09	UpK2-2	25-Mar-53	3	15
MT MCCONE	53.66	UpK8-1	8-May-53	1	15
MT MCCONE	9.10	UpK8-2	9-May-53	5	15
MT MCCONE	1.32	PB2-2	19-Jun-57	9	15
MT MCCONE	2.06	PB4-10	14-Jul-57	7	15
MT MCCONE	15.87	PB6-2	25-Jul-57	3	15
MT MCCONE	1.30	PB6-8	31-Jul-57	9	15
MT MCCONE	3.30	PB8-2	19-Aug-57	5	15
MT MCCONE	19.73	PB8-3	20-Aug-57	6	15
MT MCCONE	1.86	PB9-2	24-Aug-57	9	15
MT MCCONE	6.02	PB12-2	3-Sep-57	10	15
MT MCCONE	5.14	PB15-2	17-Sep-57	7	15
MT MCCONE	0.22	UE2-5	9-Mar-62	7	15
MT MCCONE	0.22	UE2-6	10-Mar-62	2	15
MT MCCONE	0.22	BN-5	22-Dec-70	1	15
MT MEAGHER	15.87	PB6-2	25-Jul-57	3	8
MT MEAGHER	1.30	PB6-8	31-Jul-57	9	8
MT MEAGHER	29.83	PB8-2	19-Aug-57	2	8

COUNTY	uCi/ sqm	SHOT DAY	DATE	RANK FOR DAY	TIMES IN TOP 10
MT MEAGHER	1.86	PB9-2	24-Aug-57	9	8
MT MEAGHER	3.47	PB15-2	17-Sep-57	10	8
MT MEAGHER	2.75	UE7-2	12-Jul-62	5	8
MT MEAGHER	38.04	UE12-3	16-Apr-65	5	8
MT MEAGHER	0.38	BN-4	21-Dec-70	3	8
MT MINERAL	9.86	BJ5-2	20-Nov-51	3	16
MT MINERAL	9.82	BJ5-3	21-Nov-51	4	16
MT MINERAL	2.98	BJ5-4	22-Nov-51	1	16
MT MINERAL	2.98	BJ5-4	22-Nov-51	1	16
MT MINERAL	2.14	BJ5-5	23-Nov-51	1	16
MT MINERAL	2.14	BJ5-5	23-Nov-51	1	16
MT MINERAL	2.25	BJ5-8	26-Nov-51	5	16
MT MINERAL	2.25	BJ5-8	26-Nov-51	5	16
MT MINERAL	1.60	TS8-6	10-Jun-52	10	16
MT MINERAL	1.94	PB1-5	1-Jun-57	7	16
MT MINERAL	1.40	PB1-9	5-Jun-57	3	16
MT MINERAL	12.28	PB6-3	26-Jul-57	6	16
MT MINERAL	7.58	PB16-3	25-Sep-57	1	16
MT MINERAL	1.29	PB16-4	26-Sep-57	9	16
MT MINERAL	0.85	PB18-4	10-Oct-57	7	16
MT MINERAL	0.38	BN-4	21-Dec-70	3	16
MT MISSOULA	1.74	TS8-7	11-Jun-52	10	12
MT MISSOULA	0.93	TS8-10	14-Jun-52	1	12
MT MISSOULA	0.86	TS8-11	15-Jun-52	4	12
MT MISSOULA	15.78	PB6-2	25-Jul-57	4	12
MT MISSOULA	12.28	PB6-3	26-Jul-57	6	12
MT MISSOULA	2.67	PB15-3	18-Sep-57	4	12
MT MISSOULA	7.58	PB16-3	25-Sep-57	1	12
MT MISSOULA	1.29	PB16-4	26-Sep-57	9	12
MT MISSOULA	0.85	PB18-4	10-Oct-57	7	12
MT MISSOULA	33.38	UE12-3	16-Apr-65	7	12
MT MISSOULA	0.36	BN-4	21-Dec-70	4	12
MT MISSOULA	0.21	BN-5	22-Dec-70	2	12
MT MUSSELSHELL	5.17	UpK2-2	25-Mar-53	9	20
MT MUSSELSHELL	32.75	UpK8-2	9-May-53	2	20

COUNTY	uCi/ sqm	SHOT DAY	DATE	RANK FOR DAY	TIMES IN TOP 10
MT MUSSELSHELL	5.26	UpK8-3	10-May-53	3	20
MT MUSSELSHELL	2.26	UpK10-3	27-May-53	4	20
MT MUSSELSHELL	2.07	UpK10-7	31-May-53	3	20
MT MUSSELSHELL	1.32	PB2-2	19-Jun-57	9	20
MT MUSSELSHELL	1.56	PB2-4	21-Jun-57	7	20
MT MUSSELSHELL	7.32	PB6-4	27-Jul-57	3	20
MT MUSSELSHELL	1.30	PB6-8	31-Jul-57	9	20
MT MUSSELSHELL	0.83	PB8-2	19-Aug-57	8	20
MT MUSSELSHELL	3.07	PB8-4	21-Aug-57	10	20
MT MUSSELSHELL	3.07	PB8-4	21-Aug-57	10	20
MT MUSSELSHELL	1.86	PB9-2	24-Aug-57	9	20
MT MUSSELSHELL	3.68	PB9-4	26-Aug-57	9	20
MT MUSSELSHELL	3.47	PB15-2	17-Sep-57	10	20
MT MUSSELSHELL	0.22	UE2-6	10-Mar-62	2	20
MT MUSSELSHELL	3.15	UE7-2	12-Jul-62	3	20
MT MUSSELSHELL	58.91	UE12-3	16-Apr-65	2	20
MT MUSSELSHELL	0.46	BN-4	21-Dec-70	2	20
MT MUSSELSHELL	0.20	BN-5	22-Dec-70	3	20
MT PARK	1.32	PB2-2	19-Jun-57	9	12
MT PARK	5.26	PB4-2	6-Jul-57	1	12
MT PARK	2.28	PB5-5	19-Jul-57	7	12
MT PARK	15.87	PB6-2	25-Jul-57	3	12
MT PARK	1.30	PB6-8	31-Jul-57	9	12
MT PARK	29.83	PB8-2	19-Aug-57	2	12
MT PARK	1.86	PB9-2	24-Aug-57	9	12
MT PARK	0.15	UE2-6	10-Mar-62	4	12
MT PARK	2.75	UE7-2	12-Jul-62	5	12
MT PARK	38.04	UE12-3	16-Apr-65	5	12
MT PARK	0.32	BN-4	21-Dec-70	5	12
MT PARK	0.20	BN-5	22-Dec-70	3	12
MT PETROLEUM	32.75	UpK8-2	9-May-53	2	15
MT PETROLEUM	5.26	UpK8-3	10-May-53	3	15
MT PETROLEUM	1.22	UpK8-4	11-May-53	8	15
MT PETROLEUM	8.26	UpK10-7	31-May-53	1	15
MT PETROLEUM	1.91	PB2-2	19-Jun-57	8	15

COUNTY	uCi/ sqm	SHOT DAY	DATE	RANK FOR DAY	TIMES IN TOP 10
MT PETROLEUM	1.56	PB2-4	21-Jun-57	7	15
MT PETROLEUM	7.11	PB6-4	27-Jul-57	4	15
MT PETROLEUM	0.83	PB8-2	19-Aug-57	8	15
MT PETROLEUM	1.86	PB9-2	24-Aug-57	9	15
MT PETROLEUM	7.36	PB9-4	26-Aug-57	7	15
MT PETROLEUM	6.02	PB12-2	3-Sep-57	10	15
MT PETROLEUM	5.14	PB15-2	17-Sep-57	7	15
MT PETROLEUM	0.22	UE2-6	10-Mar-62	2	15
MT PETROLEUM	2.20	UE7-2	12-Jul-62	8	15
MT PETROLEUM	0.22	BN-5	22-Dec-70	1	15
MT PHILLIPS	1.73	TS8-6	10-Jun-52	8	12
MT PHILLIPS	32.75	UpK8-2	9-May-53	2	12
MT PHILLIPS	2.63	UpK8-3	10-May-53	5	12
MT PHILLIPS	1.22	UpK8-4	11-May-53	8	12
MT PHILLIPS	4.13	UpK10-7	31-May-53	2	12
MT PHILLIPS	1.65	PB8-2	19-Aug-57	6	12
MT PHILLIPS	1.86	PB9-2	24-Aug-57	9	12
MT PHILLIPS	6.02	PB12-2	3-Sep-57	10	12
MT PHILLIPS	5.14	PB15-2	17-Sep-57	7	12
MT PHILLIPS	0.22	UE2-6	10-Mar-62	2	12
MT PHILLIPS	0.46	BN-4	21-Dec-70	2	12
MT PHILLIPS	0.20	BN-5	22-Dec-70	3	12
MT PONDERA	5.34	TS8-3	7-Jun-52	6	9
MT PONDERA	2.53	TS8-6	10-Jun-52	4	9
MT PONDERA	0.93	TS8-10	14-Jun-52	1	9
MT PONDERA	0.86	TS8-11	15-Jun-52	4	9
MT PONDERA	12.28	PB6-3	26-Jul-57	6	9
MT PONDERA	1.86	PB9-2	24-Aug-57	9	9
MT PONDERA	5.14	PB15-2	17-Sep-57	7	9
MT PONDERA	0.46	BN-4	21-Dec-70	2	9
MT PONDERA	0.20	BN-5	22-Dec-70	3	9
MT POWDER RIVER	23.57	UpK2-2	25-Mar-53	5	14
MT POWDER RIVER	4.73	UpK8-1	8-May-53	6	14
MT POWDER RIVER	1.89	TP3-2	2-Mar-55	9	14
MT POWDER RIVER	20.60	TP11-2	16-May-55	5	14

COUNTY	uCi/ sqm	SHOT DAY	DATE	RANK FOR DAY	TIMES IN TOP 10
MT POWDER RIVER	6.44	PB6-5	28-Jul-57	5	14
MT POWDER RIVER	0.83	PB8-2	19-Aug-57	8	14
MT POWDER RIVER	4.60	PB8-4	21-Aug-57	8	14
MT POWDER RIVER	4.60	PB8-4	21-Aug-57	8	14
MT POWDER RIVER	7.36	PB9-4	26-Aug-57	7	14
MT POWDER RIVER	3.47	PB15-2	17-Sep-57	10	14
MT POWDER RIVER	0.44	UE2-5	9-Mar-62	1	14
MT POWDER RIVER	0.23	UE2-6	10-Mar-62	1	14
MT POWDER RIVER	4.20	UE7-3	13-Jul-62	1	14
MT POWDER RIVER	0.14	BN-5	22-Dec-70	4	14
MT POWELL	3.43	TS8-5	9-Jun-52	4	9
MT POWELL	7.28	UpK8-2	9-May-53	7	9
MT POWELL	1.81	PB1-6	2-Jun-57	7	9
MT POWELL	2.63	PB4-2	6-Jul-57	5	9
MT POWELL	15.87	PB6-2	25-Jul-57	3	9
MT POWELL	24.56	PB6-3	26-Jul-57	2	9
MT POWELL	1.86	PB9-2	24-Aug-57	9	9
MT POWELL	5.14	PB15-2	17-Sep-57	7	9
MT POWELL	0.21	BN-5	22-Dec-70	2	9
MT PRAIRIE	61.09	UpK2-2	25-Mar-53	3	11
MT PRAIRIE	53.66	UpK8-1	8-May-53	1	11
MT PRAIRIE	5.46	UpK8-2	9-May-53	10	11
MT PRAIRIE	1.56	PB2-4	21-Jun-57	7	11
MT PRAIRIE	0.83	PB8-2	19-Aug-57	8	11
MT PRAIRIE	19.73	PB8-3	20-Aug-57	6	11
MT PRAIRIE	1.86	PB9-2	24-Aug-57	9	11
MT PRAIRIE	3.48	PB9-4	26-Aug-57	10	11
MT PRAIRIE	5.14	PB15-2	17-Sep-57	7	11
MT PRAIRIE	0.23	UE2-6	10-Mar-62	1	11
MT PRAIRIE	0.20	BN-5	22-Dec-70	3	11
MT RAVALLI	15.78	PB6-2	25-Jul-57	4	7
MT RAVALLI	12.28	PB6-3	26-Jul-57	6	7
MT RAVALLI	2.67	PB15-3	18-Sep-57	4	7
MT RAVALLI	7.58	PB16-3	25-Sep-57	1	7
MT RAVALLI	1.29	PB16-4	26-Sep-57	9	7

COUNTY	uCi/ sqm	SHOT DAY	DATE	RANK FOR DAY	TIMES IN TOP 10
MT RAVALLI	0.85	PB18-4	10-Oct-57	7	7
MT RAVALLI	29.29	UE12-3	16-Apr-65	10	7
MT RICHLAND	61.09	UpK2-2	25-Mar-53	3	14
MT RICHLAND	89.57	UpK8-1	8-May-53	1	14
MT RICHLAND	5.46	UpK8-2	9-May-53	10	14
MT RICHLAND	1.15	UpK10-9	2-Jun-53	1	14
MT RICHLAND	19.36	PB5-2	16-Jul-57	3	14
MT RICHLAND	1.65	PB8-2	19-Aug-57	6	14
MT RICHLAND	19.73	PB8-3	20-Aug-57	6	14
MT RICHLAND	1.92	PB8-5	22-Aug-57	10	14
MT RICHLAND	5.22	PB9-4	26-Aug-57	8	14
MT RICHLAND	15.54	PB11-4	3-Sep-57	3	14
MT RICHLAND	9.03	PB12-2	3-Sep-57	7	14
MT RICHLAND	5.14	PB15-2	17-Sep-57	7	14
MT RICHLAND	0.23	UE2-6	10-Mar-62	1	14
MT RICHLAND	0.22	BN-5	22-Dec-70	1	14
MT ROOSEVELT	61.09	UpK2-2	25-Mar-53	3	14
MT ROOSEVELT	53.66	UpK8-1	8-May-53	1	14
MT ROOSEVELT	5.46	UpK8-2	9-May-53	10	14
MT ROOSEVELT	1.15	UpK10-9	2-Jun-53	1	14
MT ROOSEVELT	1.32	PB2-2	19-Jun-57	9	14
MT ROOSEVELT	7.00	PB5-2	16-Jul-57	9	14
MT ROOSEVELT	3.30	PB8-2	19-Aug-57	5	14
MT ROOSEVELT	19.73	PB8-3	20-Aug-57	6	14
MT ROOSEVELT	1.92	PB8-5	22-Aug-57	10	14
MT ROOSEVELT	1.86	PB9-2	24-Aug-57	9	14
MT ROOSEVELT	3.47	PB15-2	17-Sep-57	10	14
MT ROOSEVELT	0.22	UE2-6	10-Mar-62	2	14
MT ROOSEVELT	0.32	BN-4	21-Dec-70	5	14
MT ROOSEVELT	0.22	BN-5	22-Dec-70	1	14
MT ROSEBUD	5.17	UpK2-2	25-Mar-53	9	19
MT ROSEBUD	32.75	UpK8-2	9-May-53	2	19
MT ROSEBUD	5.26	UpK8-3	10-May-53	3	19
MT ROSEBUD	2.02	UpK10-3	27-May-53	10	19
MT ROSEBUD	2.07	UpK10-7	31-May-53	3	19

COUNTY	uCi/ sqm	SHOT DAY	DATE	RANK FOR DAY	TIMES IN TOP 10
MT ROSEBUD	13.03	TP11-2	16-May-55	10	19
MT ROSEBUD	1.56	PB2-4	21-Jun-57	7	19
MT ROSEBUD	1.30	PB6-8	31-Jul-57	9	19
MT ROSEBUD	0.83	PB8-2	19-Aug-57	8	19
MT ROSEBUD	3.07	PB8-4	21-Aug-57	10	19
MT ROSEBUD	3.07	PB8-4	21-Aug-57	10	19
MT ROSEBUD	1.86	PB9-2	24-Aug-57	9	19
MT ROSEBUD	3.68	PB9-4	26-Aug-57	9	19
MT ROSEBUD	5.14	PB15-2	17-Sep-57	7	19
MT ROSEBUD	0.44	UE2-5	9-Mar-62	1	19
MT ROSEBUD	0.23	UE2-6	10-Mar-62	1	19
MT ROSEBUD	1.85	UE7-3	13-Jul-62	6	19
MT ROSEBUD	0.46	BN-4	21-Dec-70	2	19
MT ROSEBUD	0.20	BN-5	22-Dec-70	3	19
MT SANDERS	1.40	PB1-9	5-Jun-57	3	6
MT SANDERS	12.28	PB6-3	26-Jul-57	6	6
MT SANDERS	7.58	PB16-3	25-Sep-57	1	6
MT SANDERS	1.29	PB16-4	26-Sep-57	9	6
MT SANDERS	0.85	PB18-4	10-Oct-57	7	6
MT SANDERS	0.36	BN-4	21-Dec-70	4	6
MT SHERIDAN	61.09	UpK2-2	25-Mar-53	3	14
MT SHERIDAN	53.66	UpK8-1	8-May-53	1	14
MT SHERIDAN	5.46	UpK8-2	9-May-53	10	14
MT SHERIDAN	1.15	UpK10-9	2-Jun-53	1	14
MT SHERIDAN	9.68	PB5-2	16-Jul-57	6	14
MT SHERIDAN	0.85	PB5-6	20-Jul-57	10	14
MT SHERIDAN	4.64	PB6-5	28-Jul-57	8	14
MT SHERIDAN	1.65	PB8-2	19-Aug-57	6	14
MT SHERIDAN	19.73	PB8-3	20-Aug-57	6	14
MT SHERIDAN	1.92	PB8-5	22-Aug-57	10	14
MT SHERIDAN	0.22	UE2-5	9-Mar-62	7	14
MT SHERIDAN	0.22	UE2-6	10-Mar-62	2	14
MT SHERIDAN	0.46	BN-4	21-Dec-70	2	14
MT SHERIDAN	0.20	BN-5	22-Dec-70	3	14
MT SILVER BOW	0.86	TS8-11	15-Jun-52	4	10

COUNTY	uCi/ sqm	SHOT DAY	DATE	RANK FOR DAY	TIMES IN TOP 10
MT SILVER BOW	1.91	PB2-2	19-Jun-57	8	10
MT SILVER BOW	2.63	PB4-2	6-Jul-57	5	10
MT SILVER BOW	15.87	PB6-2	25-Jul-57	3	10
MT SILVER BOW	12.28	PB6-3	26-Jul-57	6	10
MT SILVER BOW	1.86	PB9-2	24-Aug-57	9	10
MT SILVER BOW	5.14	PB15-2	17-Sep-57	7	10
MT SILVER BOW	8.62	PB17-5	2-Oct-57	7	10
MT SILVER BOW	0.21	UE2-6	10-Mar-62	3	10
MT SILVER BOW	36.05	UE12-3	16-Apr-65	6	10
MT STILLWATER	5.17	UpK2-2	25-Mar-53	9	16
MT STILLWATER	16.38	UpK8-2	9-May-53	3	16
MT STILLWATER	5.26	UpK8-3	10-May-53	3	16
MT STILLWATER	2.07	UpK10-7	31-May-53	3	16
MT STILLWATER	1.97	PB4-2	6-Jul-57	8	16
MT STILLWATER	15.87	PB6-2	25-Jul-57	3	16
MT STILLWATER	1.30	PB6-8	31-Jul-57	9	16
MT STILLWATER	29.83	PB8-2	19-Aug-57	2	16
MT STILLWATER	2.73	PB9-2	24-Aug-57	6	16
MT STILLWATER	3.47	PB15-2	17-Sep-57	10	16
MT STILLWATER	1.72	PB15-3	18-Sep-57	8	16
MT STILLWATER	0.22	UE2-6	10-Mar-62	2	16
MT STILLWATER	2.86	UE7-3	13-Jul-62	3	16
MT STILLWATER	47.38	UE12-3	16-Apr-65	4	16
MT STILLWATER	0.38	BN-4	21-Dec-70	3	16
MT STILLWATER	0.22	BN-5	22-Dec-70	1	16
MT SWEET GRASS	5.17	UpK2-2	25-Mar-53	9	15
MT SWEET GRASS	8.29	UpK8-2	9-May-53	6	15
MT SWEET GRASS	2.63	UpK8-3	10-May-53	5	15
MT SWEET GRASS	1.22	UpK8-4	11-May-53	8	15
MT SWEET GRASS	2.07	UpK10-7	31-May-53	3	15
MT SWEET GRASS	2.28	PB5-5	19-Jul-57	7	15
MT SWEET GRASS	15.87	PB6-2	25-Jul-57	3	15
MT SWEET GRASS	1.30	PB6-8	31-Jul-57	9	15
MT SWEET GRASS	29.83	PB8-2	19-Aug-57	2	15
MT SWEET GRASS	1.86	PB9-2	24-Aug-57	9	15

COUNTY	uCi/ sqm	SHOT DAY	DATE	RANK FOR DAY	TIMES IN TOP 10
MT SWEET GRASS	1.72	PB15-3	18-Sep-57	8	15
MT SWEET GRASS	5.15	PB17-6	3-Oct-57	7	15
MT SWEET GRASS	0.23	UE2-6	10-Mar-62	1	15
MT SWEET GRASS	36.05	UE12-3	16-Apr-65	6	15
MT SWEET GRASS	0.21	BN-5	22-Dec-70	2	15
MT TETON	5.62	TS8-3	7-Jun-52	3	10
MT TETON	0.86	TS8-11	15-Jun-52	4	10
MT TETON	12.28	PB6-3	26-Jul-57	6	10
MT TETON	1.30	PB6-8	31-Jul-57	9	10
MT TETON	1.86	PB9-2	24-Aug-57	9	10
MT TETON	6.02	PB12-2	3-Sep-57	10	10
MT TETON	5.14	PB15-2	17-Sep-57	7	10
MT TETON	58.91	UE12-3	16-Apr-65	3	10
MT TETON	0.46	BN-4	21-Dec-70	2	10
MT TETON	0.20	BN-5	22-Dec-70	3	10
MT TOOLE	5.18	TS8-3	7-Jun-52	8	10
MT TOOLE	0.86	TS8-11	15-Jun-52	4	10
MT TOOLE	1.32	PB2-2	19-Jun-57	9	10
MT TOOLE	15.87	PB6-2	25-Jul-57	3	10
MT TOOLE	12.28	PB6-3	26-Jul-57	6	10
MT TOOLE	3.60	PB9-2	24-Aug-57	5	10
MT TOOLE	6.02	PB12-2	3-Sep-57	10	10
MT TOOLE	5.14	PB15-2	17-Sep-57	7	10
MT TOOLE	0.38	BN-4	21-Dec-70	3	10
MT TOOLE	0.20	BN-5	22-Dec-70	3	10
MT TREASURE	5.17	UpK2-2	25-Mar-53	9	17
MT TREASURE	16.38	UpK8-2	9-May-53	3	17
MT TREASURE	7.89	UpK8-3	10-May-53	1	17
MT TREASURE	1.22	UpK8-4	11-May-53	8	17
MT TREASURE	2.07	UpK10-7	31-May-53	3	17
MT TREASURE	13.03	TP11-2	16-May-55	10	17
MT TREASURE	2.60	PB2-4	21-Jun-57	3	17
MT TREASURE	6.23	PB6-4	27-Jul-57	7	17
MT TREASURE	0.83	PB8-2	19-Aug-57	8	17
MT TREASURE	4.60	PB8-4	21-Aug-57	8	17

COUNTY	uCi/ sqm	SHOT DAY	DATE	RANK FOR DAY	TIMES IN TOP 10
MT TREASURE	4.60	PB8-4	21-Aug-57	8	17
MT TREASURE	1.86	PB9-2	24-Aug-57	9	17
MT TREASURE	5.14	PB15-2	17-Sep-57	7	17
MT TREASURE	0.35	UE2-5	9-Mar-62	3	17
MT TREASURE	0.23	UE2-6	10-Mar-62	1	17
MT TREASURE	4.20	UE7-3	13-Jul-62	1	17
MT TREASURE	0.22	BN-5	22-Dec-70	1	17
MT VALLEY	61.09	UpK2-2	25-Mar-53	3	11
MT VALLEY	17.76	UpK8-1	8-May-53	4	11
MT VALLEY	1.30	PB6-8	31-Jul-57	9	11
MT VALLEY	1.30	PB8-2	19-Aug-57	7	11
MT VALLEY	3.60	PB9-2	24-Aug-57	5	11
MT VALLEY	3.68	PB9-4	26-Aug-57	9	11
MT VALLEY	9.03	PB12-2	3-Sep-57	7	11
MT VALLEY	3.47	PB15-2	17-Sep-57	10	11
MT VALLEY	0.22	UE2-6	10-Mar-62	2	11
MT VALLEY	0.46	BN-4	21-Dec-70	2	11
MT VALLEY	0.22	BN-5	22-Dec-70	1	11
MT WHEATLAND	16.38	UpK8-2	9-May-53	3	13
MT WHEATLAND	2.63	UpK8-3	10-May-53	5	13
MT WHEATLAND	1.22	UpK8-4	11-May-53	8	13
MT WHEATLAND	8.26	UpK10-7	31-May-53	1	13
MT WHEATLAND	2.63	PB4-2	6-Jul-57	5	13
MT WHEATLAND	1.30	PB6-8	31-Jul-57	9	13
MT WHEATLAND	29.83	PB8-2	19-Aug-57	2	13
MT WHEATLAND	1.86	PB9-2	24-Aug-57	9	13
MT WHEATLAND	6.02	PB12-2	3-Sep-57	10	13
MT WHEATLAND	3.47	PB15-2	17-Sep-57	10	13
MT WHEATLAND	0.15	UE2-6	10-Mar-62	4	13
MT WHEATLAND	3.35	UE7-2	12-Jul-62	2	13
MT WHEATLAND	0.20	BN-5	22-Dec-70	3	13
MT WIBAUX	61.09	UpK2-2	25-Mar-53	3	13
MT WIBAUX	53.66	UpK8-1	8-May-53	1	13
MT WIBAUX	5.46	UpK8-2	9-May-53	10	13
MT WIBAUX	29.04	PB5-2	16-Jul-57	1	13

COUNTY	uCi/ sqm	SHOT DAY	DATE	RANK FOR DAY	TIMES IN TOP 10
MT WIBAUX	3.30	PB8-2	19-Aug-57	5	13
MT WIBAUX	19.73	PB8-3	20-Aug-57	6	13
MT WIBAUX	5.36	PB8-4	21-Aug-57	6	13
MT WIBAUX	5.36	PB8-4	21-Aug-57	6	13
MT WIBAUX	1.92	PB8-5	22-Aug-57	10	13
MT WIBAUX	25.90	PB11-4	3-Sep-57	1	13
MT WIBAUX	15.05	PB12-2	3-Sep-57	4	13
MT WIBAUX	5.14	PB15-2	17-Sep-57	7	13
MT WIBAUX	1.53	UE7-3	13-Jul-62	10	13
MT YELLOWSTONE	6.80	UpK8-2	9-May-53	8	11
MT YELLOWSTONE	1.52	UpK8-4	11-May-53	6	11
MT YELLOWSTONE	0.86	UpK10-7	31-May-53	6	11
MT YELLOWSTONE	0.83	PB8-2	19-Aug-57	8	11
MT YELLOWSTONE	4.60	PB8-4	21-Aug-57	8	11
MT YELLOWSTONE	4.60	PB8-4	21-Aug-57	8	11
MT YELLOWSTONE	3.68	PB9-4	26-Aug-57	9	11
MT YELLOWSTONE	0.12	UE2-5	9-Mar-62	9	11
MT YELLOWSTONE	0.22	UE2-6	10-Mar-62	2	11
MT YELLOWSTONE	4.20	UE7-3	13-Jul-62	1	11
MT YELLOWSTONE	0.21	BN-5	22-Dec-70	2	11

COUNTY	uCi/ sqm	SHOT DAY	DATE	RANK FOR DAY	TIMES IN TOP 10
NC ALAMANCE	6.42	BJ2-2	31-Oct-51	6	5
NC ALAMANCE	8.96	PB5-8	22-Jul-57	2	5
NC ALAMANCE	4.72	PB5-9	23-Jul-57	1	5
NC ALAMANCE	2.54	PB8-7	24-Aug-57	7	5
NC ALAMANCE	0.16	UE8-8	21-Jul-62	7	5
NC ALEXANDER	6.96	PB2-5	22-Jun-57	1	4
NC ALEXANDER	2.20	PB2-8	25-Jun-57	6	4
NC ALEXANDER	5.97	PB5-8	22-Jul-57	3	4
NC ALEXANDER	1.89	PB5-9	23-Jul-57	3	4
NC ALLEGHANY	3.89	BJ2-2	31-Oct-51	10	4
NC ALLEGHANY	1.51	PB5-8	22-Jul-57	7	4
NC ALLEGHANY	2.83	PB5-9	23-Jul-57	2	4
NC ALLEGHANY	0.93	PB18-5	11-Oct-57	3	4
NC ANSON	0.88	TP7-8	5-Apr-55	6	5
NC ANSON	1.62	PB2-9	26-Jun-57	10	5
NC ANSON	1.51	PB5-8	22-Jul-57	7	5
NC ANSON	0.94	PB5-9	23-Jul-57	4	5
NC ANSON	2.54	PB8-7	24-Aug-57	7	5
NC ASHE	1.04	UpK5-2	12-Apr-53	4	9
NC ASHE	2.03	TP6-4	25-Mar-55	8	9
NC ASHE	0.97	TP6-5	26-Mar-55	8	9
NC ASHE	4.18	PB2-5	22-Jun-57	3	9
NC ASHE	1.62	PB2-9	26-Jun-57	10	9
NC ASHE	5.97	PB5-8	22-Jul-57	3	9
NC ASHE	2.83	PB5-9	23-Jul-57	2	9
NC ASHE	0.93	PB18-5	11-Oct-57	3	9
NC ASHE	1.74	UE8-5	18-Jul-62	10	9
NC AVERY	3.89	BJ2-2	31-Oct-51	10	8
NC AVERY	1.55	UpK5-2	12-Apr-53	2	8
NC AVERY	3.11	TP6-4	25-Mar-55	3	8
NC AVERY	1.45	TP6-5	26-Mar-55	5	8
NC AVERY	2.78	PB2-5	22-Jun-57	4	8
NC AVERY	1.62	PB2-9	26-Jun-57	10	8
NC AVERY	5.97	PB5-8	22-Jul-57	3	8
NC AVERY	0.93	PB18-5	11-Oct-57	3	8

COUNTY	uCi/ sqm	SHOT DAY	DATE	RANK FOR DAY	TIMES IN TOP 10
NC BEAUFORT	11.58	BJ3-6	6-Nov-51	3	9
NC BEAUFORT	3.43	BJ4-7	11-Nov-51	1	9
NC BEAUFORT	14.27	PB4-5	9-Jul-57	6	9
NC BEAUFORT	1.51	PB5-8	22-Jul-57	7	9
NC BEAUFORT	2.83	PB5-9	23-Jul-57	2	9
NC BEAUFORT	0.92	PB6-9	1-Aug-57	6	9
NC BEAUFORT	1.89	PB16-7	29-Sep-57	10	9
NC BEAUFORT	38.35	S-5	10-Jul-62	4	9
NC BEAUFORT	2.58	UE8-6	19-Jul-62	4	9
NC BERTIE	11.58	BJ3-6	6-Nov-51	3	7
NC BERTIE	3.43	BJ4-7	11-Nov-51	1	7
NC BERTIE	1.28	PB2-8	25-Jun-57	10	7
NC BERTIE	1.51	PB5-8	22-Jul-57	7	7
NC BERTIE	4.72	PB5-9	23-Jul-57	1	7
NC BERTIE	2.58	UE8-6	19-Jul-62	4	7
NC BERTIE	0.16	UE8-8	21-Jul-62	6	7
NC BLADEN	1.31	TP7-8	5-Apr-55	4	5
NC BLADEN	1.51	PB5-8	22-Jul-57	7	5
NC BLADEN	4.72	PB5-9	23-Jul-57	1	5
NC BLADEN	1.51	PB8-7	24-Aug-57	10	5
NC BLADEN	2.58	UE8-6	19-Jul-62	4	5
NC BRUNSWICK	2.20	PB2-8	25-Jun-57	6	6
NC BRUNSWICK	1.62	PB2-9	26-Jun-57	10	6
NC BRUNSWICK	1.51	PB5-8	22-Jul-57	7	6
NC BRUNSWICK	2.83	PB5-9	23-Jul-57	2	6
NC BRUNSWICK	1.51	PB8-7	24-Aug-57	10	6
NC BRUNSWICK	0.16	UE8-8	21-Jul-62	7	6
NC BUNCOMBE	3.89	BJ2-2	31-Oct-51	10	7
NC BUNCOMBE	1.55	UpK5-2	12-Apr-53	2	7
NC BUNCOMBE	3.11	TP6-4	25-Mar-55	3	7
NC BUNCOMBE	0.97	TP6-5	26-Mar-55	8	7
NC BUNCOMBE	4.18	PB2-5	22-Jun-57	3	7
NC BUNCOMBE	2.99	PB5-8	22-Jul-57	4	7
NC BUNCOMBE	0.93	PB18-5	11-Oct-57	3	7
NC BURKE	2.03	TP6-4	25-Mar-55	8	6

COUNTY	uCi/ sqm	SHOT DAY	DATE	RANK FOR DAY	TIMES IN TOP 10
NC BURKE	0.97	TP6-5	26-Mar-55	8	6
NC BURKE	4.18	PB2-5	22-Jun-57	3	6
NC BURKE	1.32	PB2-8	25-Jun-57	9	6
NC BURKE	5.97	PB5-8	22-Jul-57	3	6
NC BURKE	1.89	PB5-9	23-Jul-57	3	6
NC CABARRUS	4.07	UpK9-5	23-May-53	10	3
NC CABARRUS	1.51	PB5-8	22-Jul-57	7	3
NC CABARRUS	1.89	PB5-9	23-Jul-57	3	3
NC CALDWELL	3.89	BJ2-2	31-Oct-51	10	6
NC CALDWELL	0.97	TP6-5	26-Mar-55	8	6
NC CALDWELL	4.18	PB2-5	22-Jun-57	3	6
NC CALDWELL	1.62	PB2-9	26-Jun-57	10	6
NC CALDWELL	2.99	PB5-8	22-Jul-57	4	6
NC CALDWELL	2.83	PB5-9	23-Jul-57	2	6
NC CAMDEN	11.58	BJ3-6	6-Nov-51	3	7
NC CAMDEN	3.43	BJ4-7	11-Nov-51	1	7
NC CAMDEN	8.96	PB5-8	22-Jul-57	2	7
NC CAMDEN	2.83	PB5-9	23-Jul-57	2	7
NC CAMDEN	1.51	PB8-7	24-Aug-57	10	7
NC CAMDEN	94.45	S-4	9-Jul-62	7	7
NC CAMDEN	96.55	S-5	10-Jul-62	1	7
NC CARTERET	11.58	BJ3-6	6-Nov-51	3	9
NC CARTERET	3.43	BJ4-7	11-Nov-51	1	9
NC CARTERET	1.23	UpK3-7	6-Apr-53	10	9
NC CARTERET	1.28	PB2-8	25-Jun-57	10	9
NC CARTERET	1.51	PB5-8	22-Jul-57	7	9
NC CARTERET	1.89	PB5-9	23-Jul-57	3	9
NC CARTERET	1.51	PB8-7	24-Aug-57	10	9
NC CARTERET	2.40	PB16-7	29-Sep-57	3	9
NC CARTERET	0.16	UE8-8	21-Jul-62	7	9
NC CASWELL	3.89	BJ2-2	31-Oct-51	10	4
NC CASWELL	2.99	PB5-8	22-Jul-57	4	4
NC CASWELL	2.83	PB5-9	23-Jul-57	2	4
NC CASWELL	2.54	PB8-7	24-Aug-57	7	4
NC CATAWBA	4.18	PB2-5	22-Jun-57	3	6

COUNTY	uCi/ sqm	SHOT DAY	DATE	RANK FOR DAY	TIMES IN TOP 10
NC CATAWBA	1.32	PB2-8	25-Jun-57	9	6
NC CATAWBA	1.62	PB2-9	26-Jun-57	10	6
NC CATAWBA	2.99	PB5-8	22-Jul-57	4	6
NC CATAWBA	0.93	PB18-5	11-Oct-57	3	6
NC CATAWBA	1.32	UE8-6	19-Jul-62	9	6
NC CHATHAM	3.89	BJ2-2	31-Oct-51	10	6
NC CHATHAM	1.32	PB2-8	25-Jun-57	9	6
NC CHATHAM	1.51	PB5-8	22-Jul-57	7	6
NC CHATHAM	4.72	PB5-9	23-Jul-57	1	6
NC CHATHAM	2.54	PB8-7	24-Aug-57	7	6
NC CHATHAM	1.32	UE8-6	19-Jul-62	9	6
NC CHEROKEE	8.10	BJ3-6	6-Nov-51	5	6
NC CHEROKEE	1.55	UpK5-2	12-Apr-53	2	6
NC CHEROKEE	3.11	TP6-4	25-Mar-55	3	6
NC CHEROKEE	0.97	TP6-5	26-Mar-55	8	6
NC CHEROKEE	2.78	PB2-5	22-Jun-57	4	6
NC CHEROKEE	14.98	PB17-3	30-Sep-57	2	6
NC CHOWAN	11.58	BJ3-6	6-Nov-51	3	4
NC CHOWAN	3.43	BJ4-7	11-Nov-51	1	4
NC CHOWAN	14.93	PB5-8	22-Jul-57	1	4
NC CHOWAN	38.35	S-5	10-Jul-62	4	4
NC CLAY	2.59	UpK5-2	12-Apr-53	1	6
NC CLAY	3.11	TP6-4	25-Mar-55	3	6
NC CLAY	1.93	TP6-5	26-Mar-55	4	6
NC CLAY	4.10	TP10-8	12-May-55	9	6
NC CLAY	4.18	PB2-5	22-Jun-57	3	6
NC CLAY	22.47	PB17-3	30-Sep-57	1	6
NC CLEVELAND	1.04	UpK5-2	12-Apr-53	4	7
NC CLEVELAND	2.03	TP6-4	25-Mar-55	8	7
NC CLEVELAND	0.97	TP6-5	26-Mar-55	8	7
NC CLEVELAND	1.39	PB2-5	22-Jun-57	8	7
NC CLEVELAND	1.51	PB5-8	22-Jul-57	7	7
NC CLEVELAND	0.94	PB5-9	23-Jul-57	4	7
NC CLEVELAND	2.58	UE8-6	19-Jul-62	4	7
NC COLUMBUS	1.51	PB5-8	22-Jul-57	7	4

COUNTY	uCi/ sqm	SHOT DAY	DATE	RANK FOR DAY	TIMES IN TOP 10
NC COLUMBUS	4.72	PB5-9	23-Jul-57	1	4
NC COLUMBUS	2.58	UE8-6	19-Jul-62	4	4
NC COLUMBUS	1.37	UE8-7	20-Jul-62	4	4
NC CRAVEN	1.92	PB2-8	25-Jun-57	7	5
NC CRAVEN	1.51	PB5-8	22-Jul-57	7	5
NC CRAVEN	2.83	PB5-9	23-Jul-57	2	5
NC CRAVEN	1.92	PB16-7	29-Sep-57	8	5
NC CRAVEN	1.32	UE8-6	19-Jul-62	9	5
NC CUMBERLAND	1.32	PB2-8	25-Jun-57	9	6
NC CUMBERLAND	1.51	PB5-8	22-Jul-57	7	6
NC CUMBERLAND	4.72	PB5-9	23-Jul-57	1	6
NC CUMBERLAND	2.54	PB8-7	24-Aug-57	7	6
NC CUMBERLAND	1.32	UE8-6	19-Jul-62	9	6
NC CUMBERLAND	0.30	UE8-8	21-Jul-62	3	6
NC CURRITUCK	11.58	BJ3-6	6-Nov-51	3	7
NC CURRITUCK	3.43	BJ4-7	11-Nov-51	1	7
NC CURRITUCK	8.96	PB5-8	22-Jul-57	2	7
NC CURRITUCK	2.83	PB5-9	23-Jul-57	2	7
NC CURRITUCK	1.51	PB8-7	24-Aug-57	10	7
NC CURRITUCK	94.45	S-4	9-Jul-62	7	7
NC CURRITUCK	96.55	S-5	10-Jul-62	1	7
NC DARE	11.58	BJ3-6	6-Nov-51	3	8
NC DARE	3.43	BJ4-7	11-Nov-51	1	8
NC DARE	1.67	UpK3-7	6-Apr-53	5	8
NC DARE	14.84	PB4-5	9-Jul-57	5	8
NC DARE	1.51	PB5-8	22-Jul-57	7	8
NC DARE	0.94	PB5-9	23-Jul-57	4	8
NC DARE	1.51	PB8-7	24-Aug-57	10	8
NC DARE	2.03	PB16-7	29-Sep-57	7	8
NC DAVIDSON	6.42	BJ2-2	31-Oct-51	6	6
NC DAVIDSON	15.91	TS5-3	9-May-52	1	6
NC DAVIDSON	1.51	PB5-8	22-Jul-57	7	6
NC DAVIDSON	1.89	PB5-9	23-Jul-57	3	6
NC DAVIDSON	1.51	PB8-7	24-Aug-57	10	6
NC DAVIDSON	5.84	UE8-5	18-Jul-62	1	6

COUNTY	uCi/ sqm	SHOT DAY	DATE	RANK FOR DAY	TIMES IN TOP 10
NC DAVIE	3.89	BJ2-2	31-Oct-51	10	7
NC DAVIE	1.60	TS5-4	10-May-52	5	7
NC DAVIE	1.39	PB2-5	22-Jun-57	8	7
NC DAVIE	2.20	PB2-8	25-Jun-57	6	7
NC DAVIE	1.51	PB5-8	22-Jul-57	7	7
NC DAVIE	2.83	PB5-9	23-Jul-57	2	7
NC DAVIE	0.93	PB18-5	11-Oct-57	3	7
NC DUPLIN	1.51	PB5-8	22-Jul-57	7	3
NC DUPLIN	2.83	PB5-9	23-Jul-57	2	3
NC DUPLIN	0.16	UE8-8	21-Jul-62	7	3
NC DURHAM	3.89	BJ2-2	31-Oct-51	10	5
NC DURHAM	1.28	PB2-8	25-Jun-57	10	5
NC DURHAM	8.96	PB5-8	22-Jul-57	2	5
NC DURHAM	2.83	PB5-9	23-Jul-57	2	5
NC DURHAM	2.54	PB8-7	24-Aug-57	7	5
NC EDGECOMBE	0.85	UpK11-5	8-Jun-53	4	5
NC EDGECOMBE	1.92	PB2-8	25-Jun-57	7	5
NC EDGECOMBE	1.51	PB5-8	22-Jul-57	7	5
NC EDGECOMBE	2.83	PB5-9	23-Jul-57	2	5
NC EDGECOMBE	2.58	UE8-6	19-Jul-62	4	5
NC FORSYTH	6.42	BJ2-2	31-Oct-51	6	6
NC FORSYTH	1.32	PB2-8	25-Jun-57	9	6
NC FORSYTH	1.54	PB5-7	21-Jul-57	9	6
NC FORSYTH	1.51	PB5-8	22-Jul-57	7	6
NC FORSYTH	1.89	PB5-9	23-Jul-57	3	6
NC FORSYTH	0.93	PB18-5	11-Oct-57	3	6
NC FRANKLIN	1.62	PB2-9	26-Jun-57	10	4
NC FRANKLIN	1.51	PB5-8	22-Jul-57	7	4
NC FRANKLIN	4.72	PB5-9	23-Jul-57	1	4
NC FRANKLIN	1.51	PB8-7	24-Aug-57	10	4
NC GASTON	1.31	TP7-8	5-Apr-55	4	5
NC GASTON	5.97	PB5-8	22-Jul-57	3	5
NC GASTON	1.89	PB5-9	23-Jul-57	3	5
NC GASTON	1.32	UE8-6	19-Jul-62	9	5
NC GASTON	0.30	UE8-8	21-Jul-62	3	5

COUNTY	uCi/ sqm	SHOT DAY	DATE	RANK FOR DAY	TIMES IN TOP 10
NC GATES	11.58	BJ3-6	6-Nov-51	3	6
NC GATES	3.43	BJ4-7	11-Nov-51	1	6
NC GATES	1.55	TS5-4	10-May-52	6	6
NC GATES	1.28	PB2-8	25-Jun-57	10	6
NC GATES	1.51	PB5-8	22-Jul-57	7	6
NC GATES	4.72	PB5-9	23-Jul-57	1	6
NC GRAHAM	8.10	BJ3-6	6-Nov-51	5	6
NC GRAHAM	2.59	UpK5-2	12-Apr-53	1	6
NC GRAHAM	3.11	TP6-4	25-Mar-55	3	6
NC GRAHAM	0.97	TP6-5	26-Mar-55	8	6
NC GRAHAM	1.39	PB2-5	22-Jun-57	8	6
NC GRAHAM	7.49	PB17-3	30-Sep-57	5	6
NC GRANVILLE	3.89	BJ2-2	31-Oct-51	10	5
NC GRANVILLE	1.92	PB2-8	25-Jun-57	7	5
NC GRANVILLE	8.96	PB5-8	22-Jul-57	2	5
NC GRANVILLE	2.83	PB5-9	23-Jul-57	2	5
NC GRANVILLE	2.54	PB8-7	24-Aug-57	7	5
NC GREENE	8.96	PB5-8	22-Jul-57	2	3
NC GREENE	2.83	PB5-9	23-Jul-57	2	3
NC GREENE	1.51	PB8-7	24-Aug-57	10	3
NC GUILFORD	3.89	BJ2-2	31-Oct-51	10	6
NC GUILFORD	1.32	PB2-8	25-Jun-57	9	6
NC GUILFORD	2.99	PB5-8	22-Jul-57	4	6
NC GUILFORD	1.89	PB5-9	23-Jul-57	3	6
NC GUILFORD	1.51	PB8-7	24-Aug-57	10	6
NC GUILFORD	0.21	UE8-8	21-Jul-62	5	6
NC HALIFAX	1.28	PB2-8	25-Jun-57	10	4
NC HALIFAX	1.51	PB5-8	22-Jul-57	7	4
NC HALIFAX	4.72	PB5-9	23-Jul-57	1	4
NC HALIFAX	2.58	UE8-6	19-Jul-62	4	4
NC HARNETT	3.89	BJ2-2	31-Oct-51	10	6
NC HARNETT	1.51	PB5-8	22-Jul-57	7	6
NC HARNETT	4.72	PB5-9	23-Jul-57	1	6
NC HARNETT	2.54	PB8-7	24-Aug-57	7	6
NC HARNETT	1.32	UE8-6	19-Jul-62	9	6

COUNTY	uCi/ sqm	SHOT DAY	DATE	RANK FOR DAY	TIMES IN TOP 10
NC HARNETT	0.30	UE8-8	21-Jul-62	3	6
NC HAYWOOD	0.86	BJ1-2	29-Oct-51	5	9
NC HAYWOOD	1.55	UpK5-2	12-Apr-53	2	9
NC HAYWOOD	1.83	TP6-3	24-Mar-55	7	9
NC HAYWOOD	3.11	TP6-4	25-Mar-55	3	9
NC HAYWOOD	1.45	TP6-5	26-Mar-55	5	9
NC HAYWOOD	2.78	PB2-5	22-Jun-57	4	9
NC HAYWOOD	2.99	PB5-8	22-Jul-57	4	9
NC HAYWOOD	14.98	PB17-3	30-Sep-57	2	9
NC HAYWOOD	1.74	UE8-5	18-Jul-62	10	9
NC HENDERSON	3.89	BJ2-2	31-Oct-51	10	9
NC HENDERSON	2.12	TS6-6	30-May-52	9	9
NC HENDERSON	1.04	UpK5-2	12-Apr-53	4	9
NC HENDERSON	3.11	TP6-4	25-Mar-55	3	9
NC HENDERSON	0.97	TP6-5	26-Mar-55	8	9
NC HENDERSON	2.78	PB2-5	22-Jun-57	4	9
NC HENDERSON	1.32	PB2-8	25-Jun-57	9	9
NC HENDERSON	2.99	PB5-8	22-Jul-57	4	9
NC HENDERSON	0.93	PB18-5	11-Oct-57	3	9
NC HERTFORD	11.58	BJ3-6	6-Nov-51	3	6
NC HERTFORD	3.43	BJ4-7	11-Nov-51	1	6
NC HERTFORD	1.60	TS5-4	10-May-52	5	6
NC HERTFORD	1.28	PB2-8	25-Jun-57	10	6
NC HERTFORD	1.51	PB5-8	22-Jul-57	7	6
NC HERTFORD	4.72	PB5-9	23-Jul-57	1	6
NC HOKE	3.89	BJ2-2	31-Oct-51	10	7
NC HOKE	1.31	TP7-8	5-Apr-55	4	7
NC HOKE	1.51	PB5-8	22-Jul-57	7	7
NC HOKE	4.72	PB5-9	23-Jul-57	1	7
NC HOKE	2.54	PB8-7	24-Aug-57	7	7
NC HOKE	1.32	UE8-6	19-Jul-62	9	7
NC HOKE	0.30	UE8-8	21-Jul-62	3	7
NC HYDE	11.58	BJ3-6	6-Nov-51	3	9
NC HYDE	3.43	BJ4-7	11-Nov-51	1	9
NC HYDE	1.23	UpK3-7	6-Apr-53	10	9

COUNTY	uCi/ sqm	SHOT DAY	DATE	RANK FOR DAY	TIMES IN TOP 10
NC HYDE	21.46	PB4-5	9-Jul-57	2	9
NC HYDE	1.51	PB5-8	22-Jul-57	7	9
NC HYDE	0.94	PB5-9	23-Jul-57	4	9
NC HYDE	1.51	PB8-7	24-Aug-57	10	9
NC HYDE	2.05	PB16-7	29-Sep-57	6	9
NC HYDE	2.58	UE8-6	19-Jul-62	4	9
NC IREDELL	3.89	BJ2-2	31-Oct-51	10	8
NC IREDELL	9.21	TS5-3	9-May-52	7	8
NC IREDELL	4.18	PB2-5	22-Jun-57	3	8
NC IREDELL	1.32	PB2-8	25-Jun-57	9	8
NC IREDELL	5.97	PB5-8	22-Jul-57	3	8
NC IREDELL	0.94	PB5-9	23-Jul-57	4	8
NC IREDELL	0.93	PB18-5	11-Oct-57	3	8
NC IREDELL	3.35	UE8-5	18-Jul-62	4	8
NC JACKSON	1.55	UpK5-2	12-Apr-53	2	8
NC JACKSON	16.93	UpK9-4	22-May-53	8	8
NC JACKSON	3.11	TP6-4	25-Mar-55	3	8
NC JACKSON	0.97	TP6-5	26-Mar-55	8	8
NC JACKSON	4.18	PB2-5	22-Jun-57	3	8
NC JACKSON	1.51	PB5-8	22-Jul-57	7	8
NC JACKSON	14.98	PB17-3	30-Sep-57	2	8
NC JACKSON	2.58	UE8-6	19-Jul-62	4	8
NC JOHNSTON	1.51	PB5-8	22-Jul-57	7	5
NC JOHNSTON	4.72	PB5-9	23-Jul-57	1	5
NC JOHNSTON	1.51	PB8-7	24-Aug-57	10	5
NC JOHNSTON	2.58	UE8-6	19-Jul-62	4	5
NC JOHNSTON	0.30	UE8-8	21-Jul-62	3	5
NC JONES	1.51	PB5-8	22-Jul-57	7	5
NC JONES	2.83	PB5-9	23-Jul-57	2	5
NC JONES	1.51	PB8-7	24-Aug-57	10	5
NC JONES	2.11	PB16-7	29-Sep-57	5	5
NC JONES	27.88	S-5	10-Jul-62	6	5
NC LEE	3.89	BJ2-2	31-Oct-51	10	4
NC LEE	1.51	PB5-8	22-Jul-57	7	4
NC LEE	4.72	PB5-9	23-Jul-57	1	4

COUNTY	uCi/ sqm	SHOT DAY	DATE	RANK FOR DAY	TIMES IN TOP 10
NC LEE	2.54	PB8-7	24-Aug-57	7	4
NC LENOIR	1.55	TS5-4	10-May-52	6	4
NC LENOIR	1.51	PB5-8	22-Jul-57	7	4
NC LENOIR	2.83	PB5-9	23-Jul-57	2	4
NC LENOIR	2.22	PB16-7	29-Sep-57	4	4
NC LINCOLN	1.04	UpK5-2	12-Apr-53	4	6
NC LINCOLN	3.11	TP6-4	25-Mar-55	3	6
NC LINCOLN	0.97	TP6-5	26-Mar-55	8	6
NC LINCOLN	1.51	PB5-8	22-Jul-57	7	6
NC LINCOLN	3.35	UE8-5	18-Jul-62	4	6
NC LINCOLN	2.58	UE8-6	19-Jul-62	4	6
NC MACON	1.29	BJ1-2	29-Oct-51	3	8
NC MACON	1.55	UpK5-2	12-Apr-53	2	8
NC MACON	3.11	TP6-4	25-Mar-55	3	8
NC MACON	0.97	TP6-5	26-Mar-55	8	8
NC MACON	4.21	TP10-8	12-May-55	7	8
NC MACON	2.78	PB2-5	22-Jun-57	4	8
NC MACON	2.99	PB5-8	22-Jul-57	4	8
NC MACON	14.98	PB17-3	30-Sep-57	2	8
NC MADISON	3.89	BJ2-2	31-Oct-51	10	7
NC MADISON	1.55	UpK5-2	12-Apr-53	2	7
NC MADISON	3.11	TP6-4	25-Mar-55	3	7
NC MADISON	0.97	TP6-5	26-Mar-55	8	7
NC MADISON	4.17	TP10-8	12-May-55	8	7
NC MADISON	1.51	PB5-8	22-Jul-57	7	7
NC MADISON	7.49	PB17-3	30-Sep-57	5	7
NC MARTIN	11.58	BJ3-6	6-Nov-51	3	5
NC MARTIN	3.43	BJ4-7	11-Nov-51	1	5
NC MARTIN	1.51	PB5-8	22-Jul-57	7	5
NC MARTIN	2.83	PB5-9	23-Jul-57	2	5
NC MARTIN	38.35	S-5	10-Jul-62	4	5
NC MCDOWELL	3.89	BJ2-2	31-Oct-51	10	8
NC MCDOWELL	1.45	TS5-4	10-May-52	8	8
NC MCDOWELL	0.97	TP6-5	26-Mar-55	8	8
NC MCDOWELL	4.18	PB2-5	22-Jun-57	3	8

COUNTY	uCi/ sqm	SHOT DAY	DATE	RANK FOR DAY	TIMES IN TOP 10
NC MCDOWELL	1.32	PB2-8	25-Jun-57	9	8
NC MCDOWELL	2.99	PB5-8	22-Jul-57	4	8
NC MCDOWELL	0.93	PB18-5	11-Oct-57	3	8
NC MCDOWELL	5.84	UE8-5	18-Jul-62	1	8
NC MECKLENBURG	1.31	TP7-8	5-Apr-55	4	3
NC MECKLENBURG	8.96	PB5-8	22-Jul-57	2	3
NC MECKLENBURG	1.89	PB5-9	23-Jul-57	3	3
NC MITCHELL	3.89	BJ2-2	31-Oct-51	10	10
NC MITCHELL	1.55	UpK5-2	12-Apr-53	2	10
NC MITCHELL	3.11	TP6-4	25-Mar-55	3	10
NC MITCHELL	0.97	TP6-5	26-Mar-55	8	10
NC MITCHELL	2.78	PB2-5	22-Jun-57	4	10
NC MITCHELL	2.72	PB2-9	26-Jun-57	7	10
NC MITCHELL	1.51	PB5-8	22-Jul-57	7	10
NC MITCHELL	7.85	PB6-3	26-Jul-57	7	10
NC MITCHELL	0.93	PB18-5	11-Oct-57	3	10
NC MITCHELL	0.08	UE8-8	21-Jul-62	10	10
NC MONTGOMERY	3.89	BJ2-2	31-Oct-51	10	6
NC MONTGOMERY	1.32	PB2-8	25-Jun-57	9	6
NC MONTGOMERY	2.99	PB5-8	22-Jul-57	4	6
NC MONTGOMERY	2.83	PB5-9	23-Jul-57	2	6
NC MONTGOMERY	2.54	PB8-7	24-Aug-57	7	6
NC MONTGOMERY	0.48	UE8-8	21-Jul-62	1	6
NC MOORE	13.93	TS5-3	9-May-52	2	5
NC MOORE	1.32	PB2-8	25-Jun-57	9	5
NC MOORE	1.51	PB5-8	22-Jul-57	7	5
NC MOORE	4.72	PB5-9	23-Jul-57	1	5
NC MOORE	2.54	PB8-7	24-Aug-57	7	5
NC NASH	1.28	PB2-8	25-Jun-57	10	3
NC NASH	2.99	PB5-8	22-Jul-57	4	3
NC NASH	2.83	PB5-9	23-Jul-57	2	3
NC NEW HANOVER	1.85	UpK3-7	6-Apr-53	4	8
NC NEW HANOVER	1.92	PB2-8	25-Jun-57	7	8
NC NEW HANOVER	14.27	PB4-5	9-Jul-57	6	8
NC NEW HANOVER	1.51	PB5-8	22-Jul-57	7	8

COUNTY	uCi/ sqm	SHOT DAY	DATE	RANK FOR DAY	TIMES IN TOP 10
NC NEW HANOVER	4.72	PB5-9	23-Jul-57	1	8
NC NEW HANOVER	1.51	PB8-7	24-Aug-57	10	8
NC NEW HANOVER	1.32	UE8-6	19-Jul-62	9	8
NC NEW HANOVER	0.30	UE8-8	21-Jul-62	3	8
NC NORTHAMPTON	11.58	BJ3-6	6-Nov-51	3	6
NC NORTHAMPTON	3.43	BJ4-7	11-Nov-51	1	6
NC NORTHAMPTON	1.28	PB2-8	25-Jun-57	10	6
NC NORTHAMPTON	8.96	PB5-8	22-Jul-57	2	6
NC NORTHAMPTON	4.72	PB5-9	23-Jul-57	1	6
NC NORTHAMPTON	1.51	PB8-7	24-Aug-57	10	6
NC ONSLOW	14.27	PB4-5	9-Jul-57	6	5
NC ONSLOW	1.51	PB5-8	22-Jul-57	7	5
NC ONSLOW	2.83	PB5-9	23-Jul-57	2	5
NC ONSLOW	1.51	PB8-7	24-Aug-57	10	5
NC ONSLOW	2.11	PB16-7	29-Sep-57	5	5
NC ORANGE	3.89	BJ2-2	31-Oct-51	10	5
NC ORANGE	1.32	PB2-8	25-Jun-57	9	5
NC ORANGE	1.51	PB5-8	22-Jul-57	7	5
NC ORANGE	4.72	PB5-9	23-Jul-57	1	5
NC ORANGE	2.54	PB8-7	24-Aug-57	7	5
NC PAMLICO	3.83	BJ3-5	5-Nov-51	9	8
NC PAMLICO	6.95	BJ3-6	6-Nov-51	6	8
NC PAMLICO	3.43	BJ4-7	11-Nov-51	1	8
NC PAMLICO	1.28	PB2-8	25-Jun-57	10	8
NC PAMLICO	14.27	PB4-5	9-Jul-57	6	8
NC PAMLICO	1.51	PB5-8	22-Jul-57	7	8
NC PAMLICO	1.89	PB5-9	23-Jul-57	3	8
NC PAMLICO	1.32	UE8-6	19-Jul-62	9	8
NC PASQUOTANK	11.58	BJ3-6	6-Nov-51	3	7
NC PASQUOTANK	3.43	BJ4-7	11-Nov-51	1	7
NC PASQUOTANK	8.96	PB5-8	22-Jul-57	2	7
NC PASQUOTANK	2.83	PB5-9	23-Jul-57	2	7
NC PASQUOTANK	1.51	PB8-7	24-Aug-57	10	7
NC PASQUOTANK	94.45	S-4	9-Jul-62	7	7
NC PASQUOTANK	96.55	S-5	10-Jul-62	1	7

COUNTY	uCi/ sqm	SHOT DAY	DATE	RANK FOR DAY	TIMES IN TOP 10
NC PENDER	1.51	PB5-8	22-Jul-57	7	4
NC PENDER	2.83	PB5-9	23-Jul-57	2	4
NC PENDER	1.51	PB8-7	24-Aug-57	10	4
NC PENDER	0.48	UE8-8	21-Jul-62	1	4
NC PERQUIMANS	11.58	BJ3-6	6-Nov-51	3	7
NC PERQUIMANS	3.43	BJ4-7	11-Nov-51	1	7
NC PERQUIMANS	8.96	PB5-8	22-Jul-57	2	7
NC PERQUIMANS	2.83	PB5-9	23-Jul-57	2	7
NC PERQUIMANS	1.51	PB8-7	24-Aug-57	10	7
NC PERQUIMANS	94.45	S-4	9-Jul-62	7	7
NC PERQUIMANS	96.55	S-5	10-Jul-62	1	7
NC PERSON	3.89	BJ2-2	31-Oct-51	10	5
NC PERSON	1.28	PB2-8	25-Jun-57	10	5
NC PERSON	1.51	PB5-8	22-Jul-57	7	5
NC PERSON	4.72	PB5-9	23-Jul-57	1	5
NC PERSON	2.54	PB8-7	24-Aug-57	7	5
NC PITT	1.92	PB2-8	25-Jun-57	7	3
NC PITT	1.51	PB5-8	22-Jul-57	7	3
NC PITT	2.83	PB5-9	23-Jul-57	2	3
NC POLK	3.89	BJ2-2	31-Oct-51	10	7
NC POLK	3.47	TS6-6	30-May-52	3	7
NC POLK	3.11	TP6-4	25-Mar-55	3	7
NC POLK	0.97	TP6-5	26-Mar-55	8	7
NC POLK	6.96	PB2-5	22-Jun-57	1	7
NC POLK	1.51	PB5-8	22-Jul-57	7	7
NC POLK	0.93	PB18-5	11-Oct-57	3	7
NC RANDOLPH	3.89	BJ2-2	31-Oct-51	10	6
NC RANDOLPH	1.32	PB2-8	25-Jun-57	9	6
NC RANDOLPH	1.51	PB5-8	22-Jul-57	7	6
NC RANDOLPH	4.72	PB5-9	23-Jul-57	1	6
NC RANDOLPH	2.54	PB8-7	24-Aug-57	7	6
NC RANDOLPH	0.30	UE8-8	21-Jul-62	3	6
NC RICHMOND	1.31	TP7-8	5-Apr-55	4	3
NC RICHMOND	8.96	PB5-8	22-Jul-57	2	3
NC RICHMOND	2.54	PB8-7	24-Aug-57	7	3

COUNTY	uCi/ sqm	SHOT DAY	DATE	RANK FOR DAY	TIMES IN TOP 10
NC ROBESON	0.88	TP7-8	5-Apr-55	6	4
NC ROBESON	1.51	PB5-8	22-Jul-57	7	4
NC ROBESON	2.83	PB5-9	23-Jul-57	2	4
NC ROBESON	2.54	PB8-7	24-Aug-57	7	4
NC ROCKINGHAM	3.89	BJ2-2	31-Oct-51	10	4
NC ROCKINGHAM	1.32	PB2-8	25-Jun-57	9	4
NC ROCKINGHAM	1.51	PB5-8	22-Jul-57	7	4
NC ROCKINGHAM	2.83	PB5-9	23-Jul-57	2	4
NC ROWAN	3.89	BJ2-2	31-Oct-51	10	4
NC ROWAN	1.51	PB5-8	22-Jul-57	7	4
NC ROWAN	4.72	PB5-9	23-Jul-57	1	4
NC ROWAN	1.51	PB8-7	24-Aug-57	10	4
NC RUTHERFORD	2.03	TP6-4	25-Mar-55	8	6
NC RUTHERFORD	0.97	TP6-5	26-Mar-55	8	6
NC RUTHERFORD	2.78	PB2-5	22-Jun-57	4	6
NC RUTHERFORD	1.51	PB5-8	22-Jul-57	7	6
NC RUTHERFORD	0.93	PB18-5	11-Oct-57	3	6
NC RUTHERFORD	1.32	UE8-6	19-Jul-62	9	6
NC SAMPSON	1.51	PB5-8	22-Jul-57	7	4
NC SAMPSON	2.83	PB5-9	23-Jul-57	2	4
NC SAMPSON	2.11	UE8-5	18-Jul-62	7	4
NC SAMPSON	2.58	UE8-6	19-Jul-62	4	4
NC SCOTLAND	3.89	BJ2-2	31-Oct-51	10	6
NC SCOTLAND	1.31	TP7-8	5-Apr-55	4	6
NC SCOTLAND	1.51	PB5-8	22-Jul-57	7	6
NC SCOTLAND	4.72	PB5-9	23-Jul-57	1	6
NC SCOTLAND	2.54	PB8-7	24-Aug-57	7	6
NC SCOTLAND	0.16	UE8-8	21-Jul-62	7	6
NC STANLY	8.96	PB5-8	22-Jul-57	2	2
NC STANLY	2.54	PB8-7	24-Aug-57	7	2
NC STOKES	3.89	BJ2-2	31-Oct-51	10	4
NC STOKES	5.97	PB5-8	22-Jul-57	3	4
NC STOKES	2.83	PB5-9	23-Jul-57	2	4
NC STOKES	0.93	PB18-5	11-Oct-57	3	4
NC SURRY	6.42	BJ2-2	31-Oct-51	6	4

COUNTY	uCi/ sqm	SHOT DAY	DATE	RANK FOR DAY	TIMES IN TOP 10
NC SURRY	1.39	PB2-5	22-Jun-57	8	4
NC SURRY	2.99	PB5-8	22-Jul-57	4	4
NC SURRY	1.89	PB5-9	23-Jul-57	3	4
NC SWAIN	8.10	BJ3-6	6-Nov-51	5	7
NC SWAIN	1.55	UpK5-2	12-Apr-53	2	7
NC SWAIN	3.11	TP6-4	25-Mar-55	3	7
NC SWAIN	0.97	TP6-5	26-Mar-55	8	7
NC SWAIN	1.39	PB2-5	22-Jun-57	8	7
NC SWAIN	1.51	PB5-8	22-Jul-57	7	7
NC SWAIN	14.98	PB17-3	30-Sep-57	2	7
NC TRANSYLVANIA	1.55	UpK5-2	12-Apr-53	2	6
NC TRANSYLVANIA	3.11	TP6-4	25-Mar-55	3	6
NC TRANSYLVANIA	0.97	TP6-5	26-Mar-55	8	6
NC TRANSYLVANIA	4.18	PB2-5	22-Jun-57	3	6
NC TRANSYLVANIA	1.51	PB5-8	22-Jul-57	7	6
NC TRANSYLVANIA	14.98	PB17-3	30-Sep-57	2	6
NC TYRRELL	3.43	BJ4-7	11-Nov-51	1	4
NC TYRRELL	1.51	PB5-8	22-Jul-57	7	4
NC TYRRELL	1.89	PB5-9	23-Jul-57	3	4
NC TYRRELL	1.51	PB8-7	24-Aug-57	10	4
NC UNION	1.31	TP7-8	5-Apr-55	4	5
NC UNION	1.51	PB5-8	22-Jul-57	7	5
NC UNION	2.83	PB5-9	23-Jul-57	2	5
NC UNION	2.06	PB6-7	30-Jul-57	5	5
NC UNION	1.51	PB8-7	24-Aug-57	10	5
NC VANCE	6.42	BJ2-2	31-Oct-51	6	6
NC VANCE	1.92	PB2-8	25-Jun-57	7	6
NC VANCE	1.51	PB5-8	22-Jul-57	7	6
NC VANCE	2.83	PB5-9	23-Jul-57	2	6
NC VANCE	2.54	PB8-7	24-Aug-57	7	6
NC VANCE	2.58	UE8-6	19-Jul-62	4	6
NC WAKE	1.28	PB2-8	25-Jun-57	10	5
NC WAKE	5.97	PB5-8	22-Jul-57	3	5
NC WAKE	4.72	PB5-9	23-Jul-57	1	5
NC WAKE	1.51	PB8-7	24-Aug-57	10	5

COUNTY	uCi/ sqm	SHOT DAY	DATE	RANK FOR DAY	TIMES IN TOP 10
NC WAKE	0.16	UE8-8	21-Jul-62	7	5
NC WARREN	6.42	BJ2-2	31-Oct-51	6	5
NC WARREN	1.92	PB2-8	25-Jun-57	7	5
NC WARREN	8.96	PB5-8	22-Jul-57	2	5
NC WARREN	2.83	PB5-9	23-Jul-57	2	5
NC WARREN	2.54	PB8-7	24-Aug-57	7	5
NC WASHINGTON	11.58	BJ3-6	6-Nov-51	3	5
NC WASHINGTON	3.43	BJ4-7	11-Nov-51	1	5
NC WASHINGTON	1.51	PB5-8	22-Jul-57	7	5
NC WASHINGTON	2.83	PB5-9	23-Jul-57	2	5
NC WASHINGTON	94.45	S-4	9-Jul-62	7	5
NC WATAUGA	3.89	BJ2-2	31-Oct-51	10	10
NC WATAUGA	1.55	UpK5-2	12-Apr-53	2	10
NC WATAUGA	1.15	UpK6-5	22-Apr-53	5	10
NC WATAUGA	3.11	TP6-4	25-Mar-55	3	10
NC WATAUGA	0.97	TP6-5	26-Mar-55	8	10
NC WATAUGA	2.78	PB2-5	22-Jun-57	4	10
NC WATAUGA	1.62	PB2-9	26-Jun-57	10	10
NC WATAUGA	1.51	PB5-8	22-Jul-57	7	10
NC WATAUGA	2.83	PB5-9	23-Jul-57	2	10
NC WATAUGA	0.93	PB18-5	11-Oct-57	3	10
NC WAYNE	1.60	TS5-4	10-May-52	5	4
NC WAYNE	1.51	PB5-8	22-Jul-57	7	4
NC WAYNE	4.72	PB5-9	23-Jul-57	1	4
NC WAYNE	1.32	UE8-6	19-Jul-62	9	4
NC WILKES	4.18	PB2-5	22-Jun-57	3	5
NC WILKES	1.62	PB2-9	26-Jun-57	10	5
NC WILKES	8.96	PB5-8	22-Jul-57	2	5
NC WILKES	2.83	PB5-9	23-Jul-57	2	5
NC WILKES	0.93	PB18-5	11-Oct-57	3	5
NC WILSON	8.96	PB5-8	22-Jul-57	2	4
NC WILSON	2.83	PB5-9	23-Jul-57	2	4
NC WILSON	1.51	PB8-7	24-Aug-57	10	4
NC WILSON	0.30	UE8-8	21-Jul-62	3	4
NC YADKIN	3.89	BJ2-2	31-Oct-51	10	7

COUNTY	uCi/ sqm	SHOT DAY	DATE	RANK FOR DAY	TIMES IN TOP 10
NC YADKIN	9.21	TS5-3	9-May-52	7	7
NC YADKIN	4.18	PB2-5	22-Jun-57	3	7
NC YADKIN	1.32	PB2-8	25-Jun-57	9	7
NC YADKIN	2.99	PB5-8	22-Jul-57	4	7
NC YADKIN	2.83	PB5-9	23-Jul-57	2	7
NC YADKIN	3.35	UE8-5	18-Jul-62	4	7
NC YANCEY	3.89	BJ2-2	31-Oct-51	10	9
NC YANCEY	1.55	UpK5-2	12-Apr-53	2	9
NC YANCEY	5.14	TP6-4	25-Mar-55	1	9
NC YANCEY	1.93	TP6-5	26-Mar-55	4	9
NC YANCEY	4.18	PB2-5	22-Jun-57	3	9
NC YANCEY	1.62	PB2-9	26-Jun-57	10	9
NC YANCEY	1.51	PB5-8	22-Jul-57	7	9
NC YANCEY	0.93	PB18-5	11-Oct-57	3	9
NC YANCEY	1.32	UE8-6	19-Jul-62	9	9

COUNTY	uCi/ sqm	SHOT DAY	DATE	RANK FOR DAY	TIMES IN TOP 10
ND ADAMS	45.45	UpK2-2	25-Mar-53	4	15
ND ADAMS	4.73	UpK8-1	8-May-53	6	15
ND ADAMS	2.43	UpK10-4	28-May-53	9	15
ND ADAMS	19.36	PB5-2	16-Jul-57	3	15
ND ADAMS	10.10	PB5-3	17-Jul-57	4	15
ND ADAMS	7.56	PB6-5	28-Jul-57	1	15
ND ADAMS	0.83	PB8-2	19-Aug-57	8	15
ND ADAMS	19.73	PB8-3	20-Aug-57	6	15
ND ADAMS	5.36	PB8-4	21-Aug-57	6	15
ND ADAMS	5.36	PB8-4	21-Aug-57	6	15
ND ADAMS	1.92	PB8-5	22-Aug-57	10	15
ND ADAMS	18.31	PB11-2	1-Sep-57	3	15
ND ADAMS	20.04	PB15-2	17-Sep-57	2	15
ND ADAMS	1.44	PB18-4	10-Oct-57	6	15
ND ADAMS	0.22	UE2-6	10-Mar-62	2	15
ND BARNES	15.08	PB2-3	20-Jun-57	5	13
ND BARNES	8.16	PB5-2	16-Jul-57	8	13
ND BARNES	8.86	PB5-4	18-Jul-57	3	13
ND BARNES	19.73	PB8-3	20-Aug-57	6	13
ND BARNES	3.78	PB8-5	22-Aug-57	4	13
ND BARNES	5.22	PB9-4	26-Aug-57	8	13
ND BARNES	45.77	PB11-2	1-Sep-57	1	13
ND BARNES	20.04	PB15-2	17-Sep-57	2	13
ND BARNES	1.64	PB15-3	18-Sep-57	9	13
ND BARNES	1.48	PB15-4	19-Sep-57	4	13
ND BARNES	0.21	UE2-6	10-Mar-62	3	13
ND BARNES	0.10	UE2-7	11-Mar-62	3	13
ND BARNES	0.23	BN-4	21-Dec-70	8	13
ND BENSON	39.46	PB8-3	20-Aug-57	4	9
ND BENSON	1.92	PB8-5	22-Aug-57	10	9
ND BENSON	5.22	PB9-4	26-Aug-57	8	9
ND BENSON	27.46	PB11-2	1-Sep-57	2	9
ND BENSON	20.04	PB15-2	17-Sep-57	2	9
ND BENSON	0.87	PB17-7	4-Oct-57	2	9
ND BENSON	0.23	UE2-6	10-Mar-62	1	9

COUNTY	uCi/ sqm	SHOT DAY	DATE	RANK FOR DAY	TIMES IN TOP 10
ND BENSON	0.26	BN-4	21-Dec-70	6	9
ND BENSON	0.22	BN-5	22-Dec-70	1	9
ND BILLINGS	61.09	UpK2-2	25-Mar-53	3	13
ND BILLINGS	53.66	UpK8-1	8-May-53	1	13
ND BILLINGS	5.46	UpK8-2	9-May-53	10	13
ND BILLINGS	19.36	PB5-2	16-Jul-57	3	13
ND BILLINGS	5.09	PB5-3	17-Jul-57	8	13
ND BILLINGS	3.30	PB8-2	19-Aug-57	5	13
ND BILLINGS	118.39	PB8-3	20-Aug-57	1	13
ND BILLINGS	1.92	PB8-5	22-Aug-57	10	13
ND BILLINGS	3.48	PB9-4	26-Aug-57	10	13
ND BILLINGS	18.31	PB11-2	1-Sep-57	3	13
ND BILLINGS	20.04	PB15-2	17-Sep-57	2	13
ND BILLINGS	0.23	UE2-6	10-Mar-62	1	13
ND BILLINGS	1.62	UE7-3	13-Jul-62	9	13
ND BOTTINEAU	122.17	UpK2-2	25-Mar-53	1	10
ND BOTTINEAU	35.91	UpK8-1	8-May-53	2	10
ND BOTTINEAU	15.08	PB2-3	20-Jun-57	5	10
ND BOTTINEAU	19.73	PB8-3	20-Aug-57	6	10
ND BOTTINEAU	3.78	PB8-5	22-Aug-57	4	10
ND BOTTINEAU	3.48	PB9-4	26-Aug-57	10	10
ND BOTTINEAU	9.15	PB11-2	1-Sep-57	5	10
ND BOTTINEAU	6.68	PB15-2	17-Sep-57	5	10
ND BOTTINEAU	0.22	UE2-6	10-Mar-62	2	10
ND BOTTINEAU	0.20	BN-5	22-Dec-70	3	10
ND BOWMAN	45.45	UpK2-2	25-Mar-53	4	15
ND BOWMAN	3.16	UpK8-1	8-May-53	7	15
ND BOWMAN	2.09	UpK10-3	27-May-53	7	15
ND BOWMAN	9.68	PB5-2	16-Jul-57	6	15
ND BOWMAN	10.10	PB5-3	17-Jul-57	4	15
ND BOWMAN	4.14	PB6-5	28-Jul-57	9	15
ND BOWMAN	0.83	PB8-2	19-Aug-57	8	15
ND BOWMAN	19.73	PB8-3	20-Aug-57	6	15
ND BOWMAN	8.05	PB8-4	21-Aug-57	4	15
ND BOWMAN	8.05	PB8-4	21-Aug-57	4	15

COUNTY	uCi/ sqm	SHOT DAY	DATE	RANK FOR DAY	TIMES IN TOP 10
ND BOWMAN	1.92	PB8-5	22-Aug-57	10	15
ND BOWMAN	18.31	PB11-2	1-Sep-57	3	15
ND BOWMAN	20.04	PB15-2	17-Sep-57	2	15
ND BOWMAN	1.85	UE7-3	13-Jul-62	6	15
ND BOWMAN	0.20	BN-5	22-Dec-70	3	15
ND BURKE	91.63	UpK2-2	25-Mar-53	2	13
ND BURKE	53.66	UpK8-1	8-May-53	1	13
ND BURKE	5.46	UpK8-2	9-May-53	10	13
ND BURKE	1.15	UpK10-9	2-Jun-53	1	13
ND BURKE	9.68	PB5-2	16-Jul-57	6	13
ND BURKE	1.65	PB8-2	19-Aug-57	6	13
ND BURKE	39.46	PB8-3	20-Aug-57	4	13
ND BURKE	1.92	PB8-5	22-Aug-57	10	13
ND BURKE	5.22	PB9-4	26-Aug-57	8	13
ND BURKE	9.15	PB11-2	1-Sep-57	5	13
ND BURKE	6.68	PB15-2	17-Sep-57	5	13
ND BURKE	0.21	UE2-6	10-Mar-62	3	13
ND BURKE	0.20	BN-5	22-Dec-70	3	13
ND BURLEIGH	61.09	UpK2-2	25-Mar-53	3	13
ND BURLEIGH	9.08	UpK8-1	8-May-53	5	13
ND BURLEIGH	30.08	PB2-3	20-Jun-57	3	13
ND BURLEIGH	1.65	PB4-10	14-Jul-57	10	13
ND BURLEIGH	4.95	PB8-2	19-Aug-57	4	13
ND BURLEIGH	19.73	PB8-3	20-Aug-57	6	13
ND BURLEIGH	3.78	PB8-5	22-Aug-57	4	13
ND BURLEIGH	9.15	PB11-2	1-Sep-57	5	13
ND BURLEIGH	20.04	PB15-2	17-Sep-57	2	13
ND BURLEIGH	1.44	PB18-4	10-Oct-57	6	13
ND BURLEIGH	0.23	UE2-6	10-Mar-62	1	13
ND BURLEIGH	0.23	BN-4	21-Dec-70	8	13
ND BURLEIGH	0.11	BN-5	22-Dec-70	5	13
ND CASS	45.16	PB2-3	20-Jun-57	2	10
ND CASS	5.89	PB5-4	18-Jul-57	5	10
ND CASS	19.73	PB8-3	20-Aug-57	6	10
ND CASS	2.85	PB8-5	22-Aug-57	7	10

COUNTY	uCi/ sqm	SHOT DAY	DATE	RANK FOR DAY	TIMES IN TOP 10
ND CASS	5.22	PB9-4	26-Aug-57	8	10
ND CASS	45.77	PB11-2	1-Sep-57	1	10
ND CASS	20.04	PB15-2	17-Sep-57	2	10
ND CASS	1.64	PB15-3	18-Sep-57	9	10
ND CASS	0.21	UE2-6	10-Mar-62	3	10
ND CASS	0.10	UE2-7	11-Mar-62	3	10
ND CAVALIER	8.08	TP9-3	17-Apr-55	6	10
ND CAVALIER	3.88	PB5-3	17-Jul-57	10	10
ND CAVALIER	19.73	PB8-3	20-Aug-57	6	10
ND CAVALIER	3.78	PB8-5	22-Aug-57	4	10
ND CAVALIER	5.22	PB9-4	26-Aug-57	8	10
ND CAVALIER	27.46	PB11-2	1-Sep-57	2	10
ND CAVALIER	6.68	PB15-2	17-Sep-57	5	10
ND CAVALIER	0.21	UE2-6	10-Mar-62	3	10
ND CAVALIER	0.23	BN-4	21-Dec-70	8	10
ND CAVALIER	0.20	BN-5	22-Dec-70	3	10
ND DICKEY	3.97	UpK2-2	25-Mar-53	10	12
ND DICKEY	15.08	PB2-3	20-Jun-57	5	12
ND DICKEY	8.86	PB5-4	18-Jul-57	3	12
ND DICKEY	39.46	PB8-3	20-Aug-57	4	12
ND DICKEY	7.56	PB8-5	22-Aug-57	3	12
ND DICKEY	3.48	PB9-4	26-Aug-57	10	12
ND DICKEY	45.77	PB11-2	1-Sep-57	1	12
ND DICKEY	20.04	PB15-2	17-Sep-57	2	12
ND DICKEY	1.64	PB15-3	18-Sep-57	9	12
ND DICKEY	0.98	UE2-4	8-Mar-62	2	12
ND DICKEY	0.21	UE2-6	10-Mar-62	3	12
ND DICKEY	0.10	UE2-7	11-Mar-62	4	12
ND DIVIDE	61.09	UpK2-2	25-Mar-53	3	13
ND DIVIDE	53.66	UpK8-1	8-May-53	1	13
ND DIVIDE	5.46	UpK8-2	9-May-53	10	13
ND DIVIDE	1.15	UpK10-9	2-Jun-53	1	13
ND DIVIDE	9.68	PB5-2	16-Jul-57	6	13
ND DIVIDE	0.83	PB8-2	19-Aug-57	8	13
ND DIVIDE	78.92	PB8-3	20-Aug-57	2	13

COUNTY	uCi/ sqm	SHOT DAY	DATE	RANK FOR DAY	TIMES IN TOP 10
ND DIVIDE	1.92	PB8-5	22-Aug-57	10	13
ND DIVIDE	3.48	PB9-4	26-Aug-57	10	13
ND DIVIDE	4.58	PB11-2	1-Sep-57	9	13
ND DIVIDE	6.68	PB15-2	17-Sep-57	5	13
ND DIVIDE	0.21	UE2-6	10-Mar-62	3	13
ND DIVIDE	0.20	BN-5	22-Dec-70	3	13
ND DUNN	91.63	UpK2-2	25-Mar-53	2	15
ND DUNN	35.91	UpK8-1	8-May-53	2	15
ND DUNN	1.15	UpK10-9	2-Jun-53	1	15
ND DUNN	29.04	PB5-2	16-Jul-57	1	15
ND DUNN	5.09	PB5-3	17-Jul-57	8	15
ND DUNN	4.95	PB8-2	19-Aug-57	4	15
ND DUNN	19.73	PB8-3	20-Aug-57	6	15
ND DUNN	5.36	PB8-4	21-Aug-57	6	15
ND DUNN	5.36	PB8-4	21-Aug-57	6	15
ND DUNN	3.78	PB8-5	22-Aug-57	4	15
ND DUNN	27.46	PB11-2	1-Sep-57	2	15
ND DUNN	20.04	PB15-2	17-Sep-57	2	15
ND DUNN	0.23	UE2-6	10-Mar-62	1	15
ND DUNN	0.46	BN-4	21-Dec-70	2	15
ND DUNN	0.14	BN-5	22-Dec-70	4	15
ND EDDY	1.49	TS4-6	6-May-52	8	11
ND EDDY	1.49	TS4-6	6-May-52	8	11
ND EDDY	8.16	PB5-2	16-Jul-57	8	11
ND EDDY	19.73	PB8-3	20-Aug-57	6	11
ND EDDY	3.78	PB8-5	22-Aug-57	4	11
ND EDDY	3.48	PB9-4	26-Aug-57	10	11
ND EDDY	1.24	PB9-7	29-Aug-57	3	11
ND EDDY	45.77	PB11-2	1-Sep-57	1	11
ND EDDY	20.04	PB15-2	17-Sep-57	2	11
ND EDDY	0.91	PB17-7	4-Oct-57	1	11
ND EDDY	0.21	UE2-6	10-Mar-62	3	11
ND EMMONS	3.97	UpK2-2	25-Mar-53	10	14
ND EMMONS	45.16	PB2-3	20-Jun-57	2	14
ND EMMONS	8.86	PB5-4	18-Jul-57	3	14

COUNTY	uCi/ sqm	SHOT DAY	DATE	RANK FOR DAY	TIMES IN TOP 10
ND EMMONS	4.95	PB8-2	19-Aug-57	4	14
ND EMMONS	39.46	PB8-3	20-Aug-57	4	14
ND EMMONS	5.36	PB8-4	21-Aug-57	6	14
ND EMMONS	5.36	PB8-4	21-Aug-57	6	14
ND EMMONS	3.78	PB8-5	22-Aug-57	4	14
ND EMMONS	3.48	PB9-4	26-Aug-57	10	14
ND EMMONS	27.46	PB11-2	1-Sep-57	2	14
ND EMMONS	20.04	PB15-2	17-Sep-57	2	14
ND EMMONS	1.64	PB15-3	18-Sep-57	9	14
ND EMMONS	1.44	PB18-4	10-Oct-57	6	14
ND EMMONS	0.10	UE2-7	11-Mar-62	3	14
ND FOSTER	1.49	TS4-6	6-May-52	8	12
ND FOSTER	1.49	TS4-6	6-May-52	8	12
ND FOSTER	45.16	PB2-3	20-Jun-57	2	12
ND FOSTER	8.16	PB5-2	16-Jul-57	8	12
ND FOSTER	39.46	PB8-3	20-Aug-57	4	12
ND FOSTER	1.92	PB8-5	22-Aug-57	10	12
ND FOSTER	1.24	PB9-7	29-Aug-57	3	12
ND FOSTER	27.46	PB11-2	1-Sep-57	2	12
ND FOSTER	13.36	PB15-2	17-Sep-57	3	12
ND FOSTER	2.50	PB15-3	18-Sep-57	6	12
ND FOSTER	0.21	UE2-6	10-Mar-62	3	12
ND FOSTER	0.11	BN-5	22-Dec-70	5	12
ND GOLDEN VALLE	61.09	UpK2-2	25-Mar-53	3	15
ND GOLDEN VALLE	53.66	UpK8-1	8-May-53	1	15
ND GOLDEN VALLE	5.46	UpK8-2	9-May-53	10	15
ND GOLDEN VALLE	29.04	PB5-2	16-Jul-57	1	15
ND GOLDEN VALLE	6.65	PB6-5	28-Jul-57	4	15
ND GOLDEN VALLE	1.65	PB8-2	19-Aug-57	6	15
ND GOLDEN VALLE	19.73	PB8-3	20-Aug-57	6	15
ND GOLDEN VALLE	3.78	PB8-5	22-Aug-57	4	15
ND GOLDEN VALLE	3.48	PB9-4	26-Aug-57	10	15
ND GOLDEN VALLE	18.31	PB11-2	1-Sep-57	3	15
ND GOLDEN VALLE	10.36	PB11-4	3-Sep-57	6	15
ND GOLDEN VALLE	20.04	PB15-2	17-Sep-57	2	15

COUNTY	uCi/ sqm	SHOT DAY	DATE	RANK FOR DAY	TIMES IN TOP 10
ND GOLDEN VALLE	0.21	UE2-6	10-Mar-62	3	15
ND GOLDEN VALLE	1.67	UE7-3	13-Jul-62	8	15
ND GOLDEN VALLE	0.20	BN-5	22-Dec-70	3	15
ND GRAND FORKS	8.13	TP9-3	17-Apr-55	5	12
ND GRAND FORKS	15.08	PB2-3	20-Jun-57	5	12
ND GRAND FORKS	8.16	PB5-2	16-Jul-57	8	12
ND GRAND FORKS	19.73	PB8-3	20-Aug-57	6	12
ND GRAND FORKS	1.92	PB8-5	22-Aug-57	10	12
ND GRAND FORKS	3.48	PB9-4	26-Aug-57	10	12
ND GRAND FORKS	45.77	PB11-2	1-Sep-57	1	12
ND GRAND FORKS	20.04	PB15-2	17-Sep-57	2	12
ND GRAND FORKS	0.23	UE2-6	10-Mar-62	1	12
ND GRAND FORKS	0.10	UE2-7	11-Mar-62	3	12
ND GRAND FORKS	0.26	BN-4	21-Dec-70	6	12
ND GRAND FORKS	0.11	BN-5	22-Dec-70	6	12
ND GRANT	61.09	UpK2-2	25-Mar-53	3	12
ND GRANT	17.76	UpK8-1	8-May-53	4	12
ND GRANT	15.08	PB2-3	20-Jun-57	5	12
ND GRANT	2.10	PB4-2	6-Jul-57	7	12
ND GRANT	1.65	PB8-2	19-Aug-57	6	12
ND GRANT	19.73	PB8-3	20-Aug-57	6	12
ND GRANT	3.78	PB8-5	22-Aug-57	4	12
ND GRANT	9.15	PB11-2	1-Sep-57	5	12
ND GRANT	20.04	PB15-2	17-Sep-57	2	12
ND GRANT	1.44	PB18-4	10-Oct-57	6	12
ND GRANT	0.21	UE2-6	10-Mar-62	3	12
ND GRANT	0.16	BN-4	21-Dec-70	9	12
ND GRIGGS	8.16	PB5-2	16-Jul-57	8	13
ND GRIGGS	5.89	PB5-4	18-Jul-57	5	13
ND GRIGGS	19.73	PB8-3	20-Aug-57	6	13
ND GRIGGS	1.92	PB8-5	22-Aug-57	10	13
ND GRIGGS	1.24	PB9-7	29-Aug-57	3	13
ND GRIGGS	45.77	PB11-2	1-Sep-57	1	13
ND GRIGGS	10.36	PB11-4	3-Sep-57	6	13
ND GRIGGS	20.04	PB15-2	17-Sep-57	2	13

COUNTY	uCi/ sqm	SHOT DAY	DATE	RANK FOR DAY	TIMES IN TOP 10
ND GRIGGS	1.64	PB15-3	18-Sep-57	9	13
ND GRIGGS	1.18	UE2-4	8-Mar-62	1	13
ND GRIGGS	0.23	UE2-6	10-Mar-62	1	13
ND GRIGGS	0.10	UE2-7	11-Mar-62	3	13
ND GRIGGS	0.23	BN-4	21-Dec-70	8	13
ND HETTINGER	61.09	UpK2-2	25-Mar-53	3	14
ND HETTINGER	17.76	UpK8-1	8-May-53	4	14
ND HETTINGER	9.68	PB5-2	16-Jul-57	6	14
ND HETTINGER	10.10	PB5-3	17-Jul-57	4	14
ND HETTINGER	0.83	PB8-2	19-Aug-57	8	14
ND HETTINGER	19.73	PB8-3	20-Aug-57	6	14
ND HETTINGER	8.05	PB8-4	21-Aug-57	4	14
ND HETTINGER	8.05	PB8-4	21-Aug-57	4	14
ND HETTINGER	1.92	PB8-5	22-Aug-57	10	14
ND HETTINGER	18.31	PB11-2	1-Sep-57	3	14
ND HETTINGER	10.36	PB11-4	3-Sep-57	6	14
ND HETTINGER	20.04	PB15-2	17-Sep-57	2	14
ND HETTINGER	1.44	PB18-4	10-Oct-57	6	14
ND HETTINGER	0.20	BN-5	22-Dec-70	3	14
ND KIDDER	1.49	TS4-6	6-May-52	8	17
ND KIDDER	1.49	TS4-6	6-May-52	8	17
ND KIDDER	15.08	PB2-3	20-Jun-57	5	17
ND KIDDER	2.10	PB4-2	6-Jul-57	7	17
ND KIDDER	8.16	PB5-2	16-Jul-57	8	17
ND KIDDER	19.73	PB8-3	20-Aug-57	6	17
ND KIDDER	1.92	PB8-5	22-Aug-57	10	17
ND KIDDER	3.48	PB9-4	26-Aug-57	10	17
ND KIDDER	1.24	PB9-7	29-Aug-57	3	17
ND KIDDER	27.46	PB11-2	1-Sep-57	2	17
ND KIDDER	20.04	PB15-2	17-Sep-57	2	17
ND KIDDER	1.64	PB15-3	18-Sep-57	9	17
ND KIDDER	0.21	UE2-6	10-Mar-62	3	17
ND KIDDER	0.10	UE2-7	11-Mar-62	4	17
ND KIDDER	2.48	UE7-3	13-Jul-62	4	17
ND KIDDER	0.26	BN-4	21-Dec-70	6	17

COUNTY	uCi/ sqm	SHOT DAY	DATE	RANK FOR DAY	TIMES IN TOP 10
ND KIDDER	0.11	BN-5	22-Dec-70	5	17
ND LA MOURE	15.08	PB2-3	20-Jun-57	5	9
ND LA MOURE	8.86	PB5-4	18-Jul-57	3	9
ND LA MOURE	19.73	PB8-3	20-Aug-57	6	9
ND LA MOURE	3.78	PB8-5	22-Aug-57	4	9
ND LA MOURE	3.48	PB9-4	26-Aug-57	10	9
ND LA MOURE	45.77	PB11-2	1-Sep-57	1	9
ND LA MOURE	20.04	PB15-2	17-Sep-57	2	9
ND LA MOURE	2.50	PB15-3	18-Sep-57	6	9
ND LA MOURE	0.17	UE2-7	11-Mar-62	2	9
ND LOGAN	3.97	UpK2-2	25-Mar-53	10	12
ND LOGAN	30.08	PB2-3	20-Jun-57	3	12
ND LOGAN	8.16	PB5-2	16-Jul-57	8	12
ND LOGAN	8.86	PB5-4	18-Jul-57	3	12
ND LOGAN	19.73	PB8-3	20-Aug-57	6	12
ND LOGAN	1.92	PB8-5	22-Aug-57	10	12
ND LOGAN	5.22	PB9-4	26-Aug-57	8	12
ND LOGAN	27.46	PB11-2	1-Sep-57	2	12
ND LOGAN	20.04	PB15-2	17-Sep-57	2	12
ND LOGAN	2.50	PB15-3	18-Sep-57	6	12
ND LOGAN	0.10	UE2-7	11-Mar-62	3	12
ND LOGAN	0.23	BN-4	21-Dec-70	8	12
ND MCHENRY	122.17	UpK2-2	25-Mar-53	1	11
ND MCHENRY	17.76	UpK8-1	8-May-53	4	11
ND MCHENRY	2.16	UpK10-3	27-May-53	6	11
ND MCHENRY	30.08	PB2-3	20-Jun-57	3	11
ND MCHENRY	19.73	PB8-3	20-Aug-57	6	11
ND MCHENRY	1.92	PB8-5	22-Aug-57	10	11
ND MCHENRY	3.48	PB9-4	26-Aug-57	10	11
ND MCHENRY	9.15	PB11-2	1-Sep-57	5	11
ND MCHENRY	6.68	PB15-2	17-Sep-57	5	11
ND MCHENRY	0.21	UE2-6	10-Mar-62	3	11
ND MCHENRY	0.22	BN-5	22-Dec-70	1	11
ND MCINTOSH	3.97	UpK2-2	25-Mar-53	10	10
ND MCINTOSH	30.08	PB2-3	20-Jun-57	3	10

COUNTY	uCi/ sqm	SHOT DAY	DATE	RANK FOR DAY	TIMES IN TOP 10
ND MCINTOSH	8.86	PB5-4	18-Jul-57	3	10
ND MCINTOSH	19.73	PB8-3	20-Aug-57	6	10
ND MCINTOSH	3.78	PB8-5	22-Aug-57	4	10
ND MCINTOSH	5.22	PB9-4	26-Aug-57	8	10
ND MCINTOSH	27.46	PB11-2	1-Sep-57	2	10
ND MCINTOSH	13.36	PB15-2	17-Sep-57	3	10
ND MCINTOSH	1.64	PB15-3	18-Sep-57	9	10
ND MCINTOSH	0.10	UE2-7	11-Mar-62	3	10
ND MCKENZIE	91.63	UpK2-2	25-Mar-53	2	14
ND MCKENZIE	53.66	UpK8-1	8-May-53	1	14
ND MCKENZIE	5.46	UpK8-2	9-May-53	10	14
ND MCKENZIE	9.68	PB5-2	16-Jul-57	6	14
ND MCKENZIE	3.30	PB8-2	19-Aug-57	5	14
ND MCKENZIE	39.46	PB8-3	20-Aug-57	4	14
ND MCKENZIE	5.36	PB8-4	21-Aug-57	6	14
ND MCKENZIE	5.36	PB8-4	21-Aug-57	6	14
ND MCKENZIE	1.92	PB8-5	22-Aug-57	10	14
ND MCKENZIE	3.48	PB9-4	26-Aug-57	10	14
ND MCKENZIE	4.58	PB11-2	1-Sep-57	9	14
ND MCKENZIE	20.04	PB15-2	17-Sep-57	2	14
ND MCKENZIE	0.21	UE2-6	10-Mar-62	3	14
ND MCKENZIE	0.20	BN-5	22-Dec-70	3	14
ND MCLEAN	61.09	UpK2-2	25-Mar-53	3	14
ND MCLEAN	17.76	UpK8-1	8-May-53	4	14
ND MCLEAN	30.08	PB2-3	20-Jun-57	3	14
ND MCLEAN	3.88	PB5-3	17-Jul-57	10	14
ND MCLEAN	1.65	PB8-2	19-Aug-57	6	14
ND MCLEAN	19.73	PB8-3	20-Aug-57	6	14
ND MCLEAN	3.78	PB8-5	22-Aug-57	4	14
ND MCLEAN	5.22	PB9-4	26-Aug-57	8	14
ND MCLEAN	18.31	PB11-2	1-Sep-57	3	14
ND MCLEAN	20.04	PB15-2	17-Sep-57	2	14
ND MCLEAN	1.44	PB18-4	10-Oct-57	6	14
ND MCLEAN	0.23	UE2-6	10-Mar-62	1	14
ND MCLEAN	2.86	UE7-3	13-Jul-62	3	14

COUNTY	uCi/ sqm	SHOT DAY	DATE	RANK FOR DAY	TIMES IN TOP 10
ND MCLEAN	0.22	BN-5	22-Dec-70	1	14
ND MERCER	61.09	UpK2-2	25-Mar-53	3	11
ND MERCER	53.66	UpK8-1	8-May-53	1	11
ND MERCER	45.16	PB2-3	20-Jun-57	2	11
ND MERCER	1.65	PB4-10	14-Jul-57	10	11
ND MERCER	3.30	PB8-2	19-Aug-57	5	11
ND MERCER	19.73	PB8-3	20-Aug-57	6	11
ND MERCER	3.78	PB8-5	22-Aug-57	4	11
ND MERCER	9.15	PB11-2	1-Sep-57	5	11
ND MERCER	20.04	PB15-2	17-Sep-57	2	11
ND MERCER	1.64	PB15-3	18-Sep-57	9	11
ND MERCER	0.23	UE2-6	10-Mar-62	1	11
ND MORTON	61.09	UpK2-2	25-Mar-53	3	14
ND MORTON	17.76	UpK8-1	8-May-53	4	14
ND MORTON	30.08	PB2-3	20-Jun-57	3	14
ND MORTON	2.10	PB4-2	6-Jul-57	7	14
ND MORTON	3.88	PB5-3	17-Jul-57	10	14
ND MORTON	3.30	PB8-2	19-Aug-57	5	14
ND MORTON	19.73	PB8-3	20-Aug-57	6	14
ND MORTON	3.78	PB8-5	22-Aug-57	4	14
ND MORTON	9.15	PB11-2	1-Sep-57	5	14
ND MORTON	20.04	PB15-2	17-Sep-57	2	14
ND MORTON	1.64	PB15-3	18-Sep-57	9	14
ND MORTON	1.44	PB18-4	10-Oct-57	6	14
ND MORTON	0.21	UE2-6	10-Mar-62	3	14
ND MORTON	0.20	BN-5	22-Dec-70	3	14
ND MOUNTRAIL	61.09	UpK2-2	25-Mar-53	3	12
ND MOUNTRAIL	53.66	UpK8-1	8-May-53	1	12
ND MOUNTRAIL	1.15	UpK10-9	2-Jun-53	1	12
ND MOUNTRAIL	29.04	PB5-2	16-Jul-57	1	12
ND MOUNTRAIL	1.65	PB8-2	19-Aug-57	6	12
ND MOUNTRAIL	19.73	PB8-3	20-Aug-57	6	12
ND MOUNTRAIL	1.92	PB8-5	22-Aug-57	10	12
ND MOUNTRAIL	3.48	PB9-4	26-Aug-57	10	12
ND MOUNTRAIL	18.31	PB11-2	1-Sep-57	3	12

COUNTY	uCi/ sqm	SHOT DAY	DATE	RANK FOR DAY	TIMES IN TOP 10
ND MOUNTRAIL	13.36	PB15-2	17-Sep-57	3	12
ND MOUNTRAIL	0.23	UE2-6	10-Mar-62	1	12
ND MOUNTRAIL	0.22	BN-5	22-Dec-70	1	12
ND NELSON	8.16	PB5-2	16-Jul-57	8	10
ND NELSON	19.73	PB8-3	20-Aug-57	6	10
ND NELSON	1.92	PB8-5	22-Aug-57	10	10
ND NELSON	3.48	PB9-4	26-Aug-57	10	10
ND NELSON	45.77	PB11-2	1-Sep-57	1	10
ND NELSON	20.04	PB15-2	17-Sep-57	2	10
ND NELSON	0.21	UE2-6	10-Mar-62	3	10
ND NELSON	0.10	UE2-7	11-Mar-62	3	10
ND NELSON	0.26	BN-4	21-Dec-70	6	10
ND NELSON	0.11	BN-5	22-Dec-70	6	10
ND OLIVER	1.49	TS4-6	6-May-52	8	15
ND OLIVER	1.49	TS4-6	6-May-52	8	15
ND OLIVER	61.09	UpK2-2	25-Mar-53	3	15
ND OLIVER	53.66	UpK8-1	8-May-53	1	15
ND OLIVER	45.16	PB2-3	20-Jun-57	2	15
ND OLIVER	2.10	PB4-2	6-Jul-57	7	15
ND OLIVER	1.65	PB4-10	14-Jul-57	10	15
ND OLIVER	3.30	PB8-2	19-Aug-57	5	15
ND OLIVER	19.73	PB8-3	20-Aug-57	6	15
ND OLIVER	1.92	PB8-5	22-Aug-57	10	15
ND OLIVER	9.15	PB11-2	1-Sep-57	5	15
ND OLIVER	20.04	PB15-2	17-Sep-57	2	15
ND OLIVER	1.44	PB18-4	10-Oct-57	6	15
ND OLIVER	0.21	UE2-6	10-Mar-62	3	15
ND OLIVER	0.22	BN-5	22-Dec-70	1	15
ND PEMBINA	1.12	TS3-9	30-Apr-52	1	10
ND PEMBINA	1.12	TS3-9	30-Apr-52	1	10
ND PEMBINA	11.63	TP9-3	17-Apr-55	1	10
ND PEMBINA	15.08	PB2-3	20-Jun-57	5	10
ND PEMBINA	4.21	PB4-2	6-Jul-57	3	10
ND PEMBINA	19.73	PB8-3	20-Aug-57	6	10
ND PEMBINA	1.92	PB8-5	22-Aug-57	10	10

COUNTY	uCi/ sqm	SHOT DAY	DATE	RANK FOR DAY	TIMES IN TOP 10
ND PEMBINA	3.48	PB9-4	26-Aug-57	10	10
ND PEMBINA	45.77	PB11-2	1-Sep-57	1	10
ND PEMBINA	0.22	BN-5	22-Dec-70	1	10
ND PIERCE	3.88	PB5-3	17-Jul-57	10	10
ND PIERCE	19.73	PB8-3	20-Aug-57	6	10
ND PIERCE	3.78	PB8-5	22-Aug-57	4	10
ND PIERCE	3.48	PB9-4	26-Aug-57	10	10
ND PIERCE	18.31	PB11-2	1-Sep-57	3	10
ND PIERCE	6.68	PB15-2	17-Sep-57	5	10
ND PIERCE	0.87	PB17-7	4-Oct-57	2	10
ND PIERCE	0.21	UE2-6	10-Mar-62	3	10
ND PIERCE	0.23	BN-4	21-Dec-70	8	10
ND PIERCE	0.22	BN-5	22-Dec-70	1	10
ND RAMSEY	7.70	TP9-3	17-Apr-55	7	10
ND RAMSEY	19.73	PB8-3	20-Aug-57	6	10
ND RAMSEY	1.92	PB8-5	22-Aug-57	10	10
ND RAMSEY	3.48	PB9-4	26-Aug-57	10	10
ND RAMSEY	45.77	PB11-2	1-Sep-57	1	10
ND RAMSEY	13.36	PB15-2	17-Sep-57	3	10
ND RAMSEY	0.87	PB17-7	4-Oct-57	2	10
ND RAMSEY	0.23	UE2-6	10-Mar-62	1	10
ND RAMSEY	0.23	BN-4	21-Dec-70	8	10
ND RAMSEY	0.22	BN-5	22-Dec-70	1	10
ND RANSOM	5.89	PB5-4	18-Jul-57	5	9
ND RANSOM	78.92	PB8-3	20-Aug-57	2	9
ND RANSOM	7.56	PB8-5	22-Aug-57	3	9
ND RANSOM	3.48	PB9-4	26-Aug-57	10	9
ND RANSOM	45.77	PB11-2	1-Sep-57	1	9
ND RANSOM	6.68	PB15-2	17-Sep-57	5	9
ND RANSOM	2.50	PB15-3	18-Sep-57	6	9
ND RANSOM	0.21	UE2-6	10-Mar-62	3	9
ND RANSOM	0.10	UE2-7	11-Mar-62	4	9
ND RENVILLE	61.09	UpK2-2	25-Mar-53	3	12
ND RENVILLE	53.66	UpK8-1	8-May-53	1	12
ND RENVILLE	5.46	UpK8-2	9-May-53	10	12

COUNTY	uCi/ sqm	SHOT DAY	DATE	RANK FOR DAY	TIMES IN TOP 10
ND RENVILLE	1.15	UpK10-9	2-Jun-53	1	12
ND RENVILLE	0.83	PB8-2	19-Aug-57	8	12
ND RENVILLE	39.46	PB8-3	20-Aug-57	4	12
ND RENVILLE	1.92	PB8-5	22-Aug-57	10	12
ND RENVILLE	5.22	PB9-4	26-Aug-57	8	12
ND RENVILLE	18.31	PB11-2	1-Sep-57	3	12
ND RENVILLE	6.68	PB15-2	17-Sep-57	5	12
ND RENVILLE	0.21	UE2-6	10-Mar-62	3	12
ND RENVILLE	0.20	BN-5	22-Dec-70	3	12
ND RICHLAND	75.24	PB2-3	20-Jun-57	1	9
ND RICHLAND	8.86	PB5-4	18-Jul-57	3	9
ND RICHLAND	39.46	PB8-3	20-Aug-57	4	9
ND RICHLAND	7.56	PB8-5	22-Aug-57	3	9
ND RICHLAND	45.77	PB11-2	1-Sep-57	1	9
ND RICHLAND	20.04	PB15-2	17-Sep-57	2	9
ND RICHLAND	2.50	PB15-3	18-Sep-57	6	9
ND RICHLAND	2.54	PB17-2	29-Sep-57	2	9
ND RICHLAND	0.31	UE2-5	9-Mar-62	4	9
ND ROLETTE	2.30	UpK10-3	27-May-53	3	8
ND ROLETTE	19.73	PB8-3	20-Aug-57	6	8
ND ROLETTE	1.92	PB8-5	22-Aug-57	10	8
ND ROLETTE	5.22	PB9-4	26-Aug-57	8	8
ND ROLETTE	27.46	PB11-2	1-Sep-57	2	8
ND ROLETTE	0.21	UE2-6	10-Mar-62	3	8
ND ROLETTE	0.26	BN-4	21-Dec-70	6	8
ND ROLETTE	0.22	BN-5	22-Dec-70	1	8
ND SARGENT	45.16	PB2-3	20-Jun-57	2	10
ND SARGENT	8.86	PB5-4	18-Jul-57	3	10
ND SARGENT	19.73	PB8-3	20-Aug-57	6	10
ND SARGENT	7.56	PB8-5	22-Aug-57	3	10
ND SARGENT	3.48	PB9-4	26-Aug-57	10	10
ND SARGENT	45.77	PB11-2	1-Sep-57	1	10
ND SARGENT	20.04	PB15-2	17-Sep-57	2	10
ND SARGENT	1.64	PB15-3	18-Sep-57	9	10
ND SARGENT	0.21	UE2-6	10-Mar-62	3	10

COUNTY	uCi/ sqm	SHOT DAY	DATE	RANK FOR DAY	TIMES IN TOP 10
ND SARGENT	0.10	UE2-7	11-Mar-62	4	10
ND SHERIDAN	61.09	UpK2-2	25-Mar-53	3	11
ND SHERIDAN	9.08	UpK8-1	8-May-53	5	11
ND SHERIDAN	15.08	PB2-3	20-Jun-57	5	11
ND SHERIDAN	19.73	PB8-3	20-Aug-57	6	11
ND SHERIDAN	1.92	PB8-5	22-Aug-57	10	11
ND SHERIDAN	5.22	PB9-4	26-Aug-57	8	11
ND SHERIDAN	18.31	PB11-2	1-Sep-57	3	11
ND SHERIDAN	20.04	PB15-2	17-Sep-57	2	11
ND SHERIDAN	0.23	UE2-6	10-Mar-62	1	11
ND SHERIDAN	0.23	BN-4	21-Dec-70	8	11
ND SHERIDAN	0.20	BN-5	22-Dec-70	3	11
ND SIOUX	45.45	UpK2-2	25-Mar-53	4	14
ND SIOUX	0.83	PB8-2	19-Aug-57	8	14
ND SIOUX	19.73	PB8-3	20-Aug-57	6	14
ND SIOUX	5.36	PB8-4	21-Aug-57	6	14
ND SIOUX	5.36	PB8-4	21-Aug-57	6	14
ND SIOUX	1.92	PB8-5	22-Aug-57	10	14
ND SIOUX	3.48	PB9-4	26-Aug-57	10	14
ND SIOUX	4.58	PB11-2	1-Sep-57	9	14
ND SIOUX	10.36	PB11-4	3-Sep-57	6	14
ND SIOUX	6.68	PB15-2	17-Sep-57	5	14
ND SIOUX	1.64	PB15-3	18-Sep-57	9	14
ND SIOUX	1.44	PB18-4	10-Oct-57	6	14
ND SIOUX	0.21	UE2-6	10-Mar-62	3	14
ND SIOUX	0.10	UE2-7	11-Mar-62	3	14
ND SLOPE	61.09	UpK2-2	25-Mar-53	3	14
ND SLOPE	35.91	UpK8-1	8-May-53	2	14
ND SLOPE	2.16	UpK10-3	27-May-53	6	14
ND SLOPE	9.68	PB5-2	16-Jul-57	6	14
ND SLOPE	5.09	PB5-3	17-Jul-57	8	14
ND SLOPE	1.65	PB8-2	19-Aug-57	6	14
ND SLOPE	39.46	PB8-3	20-Aug-57	4	14
ND SLOPE	5.36	PB8-4	21-Aug-57	6	14
ND SLOPE	5.36	PB8-4	21-Aug-57	6	14

COUNTY	uCi/ sqm	SHOT DAY	DATE	RANK FOR DAY	TIMES IN TOP 10
ND SLOPE	3.78	PB8-5	22-Aug-57	4	14
ND SLOPE	3.48	PB9-4	26-Aug-57	10	14
ND SLOPE	18.31	PB11-2	1-Sep-57	3	14
ND SLOPE	13.36	PB15-2	17-Sep-57	3	14
ND SLOPE	0.20	BN-5	22-Dec-70	3	14
ND STARK	61.09	UpK2-2	25-Mar-53	3	12
ND STARK	53.66	UpK8-1	8-May-53	1	12
ND STARK	29.04	PB5-2	16-Jul-57	1	12
ND STARK	15.19	PB5-3	17-Jul-57	1	12
ND STARK	4.95	PB8-2	19-Aug-57	4	12
ND STARK	39.46	PB8-3	20-Aug-57	4	12
ND STARK	3.78	PB8-5	22-Aug-57	4	12
ND STARK	18.31	PB11-2	1-Sep-57	3	12
ND STARK	10.36	PB11-4	3-Sep-57	6	12
ND STARK	20.04	PB15-2	17-Sep-57	2	12
ND STARK	0.22	UE2-6	10-Mar-62	2	12
ND STARK	0.14	BN-5	22-Dec-70	4	12
ND STEELE	8.16	PB5-2	16-Jul-57	8	11
ND STEELE	19.73	PB8-3	20-Aug-57	6	11
ND STEELE	3.78	PB8-5	22-Aug-57	4	11
ND STEELE	3.48	PB9-4	26-Aug-57	10	11
ND STEELE	45.77	PB11-2	1-Sep-57	1	11
ND STEELE	20.04	PB15-2	17-Sep-57	2	11
ND STEELE	1.64	PB15-3	18-Sep-57	9	11
ND STEELE	1.18	UE2-4	8-Mar-62	1	11
ND STEELE	0.23	UE2-6	10-Mar-62	1	11
ND STEELE	0.10	UE2-7	11-Mar-62	3	11
ND STEELE	0.23	BN-4	21-Dec-70	8	11
ND STUTSMAN	15.08	PB2-3	20-Jun-57	5	12
ND STUTSMAN	5.89	PB5-4	18-Jul-57	5	12
ND STUTSMAN	39.46	PB8-3	20-Aug-57	4	12
ND STUTSMAN	3.78	PB8-5	22-Aug-57	4	12
ND STUTSMAN	3.48	PB9-4	26-Aug-57	10	12
ND STUTSMAN	1.24	PB9-7	29-Aug-57	3	12
ND STUTSMAN	27.46	PB11-2	1-Sep-57	2	12

COUNTY	uCi/ sqm	SHOT DAY	DATE	RANK FOR DAY	TIMES IN TOP 10
ND STUTSMAN	20.04	PB15-2	17-Sep-57	2	12
ND STUTSMAN	1.64	PB15-3	18-Sep-57	9	12
ND STUTSMAN	0.23	UE2-6	10-Mar-62	1	12
ND STUTSMAN	0.10	UE2-7	11-Mar-62	3	12
ND STUTSMAN	0.23	BN-4	21-Dec-70	8	12
ND TOWNER	3.88	PB5-3	17-Jul-57	10	9
ND TOWNER	19.73	PB8-3	20-Aug-57	6	9
ND TOWNER	1.92	PB8-5	22-Aug-57	10	9
ND TOWNER	5.22	PB9-4	26-Aug-57	8	9
ND TOWNER	27.46	PB11-2	1-Sep-57	2	9
ND TOWNER	6.68	PB15-2	17-Sep-57	5	9
ND TOWNER	0.21	UE2-6	10-Mar-62	3	9
ND TOWNER	0.26	BN-4	21-Dec-70	6	9
ND TOWNER	0.22	BN-5	22-Dec-70	1	9
ND TRAILL	30.08	PB2-3	20-Jun-57	3	11
ND TRAILL	19.73	PB8-3	20-Aug-57	6	11
ND TRAILL	1.92	PB8-5	22-Aug-57	10	11
ND TRAILL	45.77	PB11-2	1-Sep-57	1	11
ND TRAILL	20.04	PB15-2	17-Sep-57	2	11
ND TRAILL	1.64	PB15-3	18-Sep-57	9	11
ND TRAILL	1.48	PB15-4	19-Sep-57	4	11
ND TRAILL	1.18	UE2-4	8-Mar-62	1	11
ND TRAILL	0.21	UE2-6	10-Mar-62	3	11
ND TRAILL	0.10	UE2-7	11-Mar-62	4	11
ND TRAILL	0.13	BN-4	21-Dec-70	10	11
ND WALSH	7.55	TP9-3	17-Apr-55	8	9
ND WALSH	30.08	PB2-3	20-Jun-57	3	9
ND WALSH	19.73	PB8-3	20-Aug-57	6	9
ND WALSH	1.92	PB8-5	22-Aug-57	10	9
ND WALSH	3.48	PB9-4	26-Aug-57	10	9
ND WALSH	45.77	PB11-2	1-Sep-57	1	9
ND WALSH	13.36	PB15-2	17-Sep-57	3	9
ND WALSH	0.10	UE2-7	11-Mar-62	4	9
ND WALSH	0.22	BN-5	22-Dec-70	1	9
ND WARD	91.63	UpK2-2	25-Mar-53	2	14

COUNTY	uCi/ sqm	SHOT DAY	DATE	RANK FOR DAY	TIMES IN TOP 10
ND WARD	53.66	UpK8-1	8-May-53	1	14
ND WARD	5.46	UpK8-2	9-May-53	10	14
ND WARD	2.09	UpK10-3	27-May-53	7	14
ND WARD	30.08	PB2-3	20-Jun-57	3	14
ND WARD	0.83	PB8-2	19-Aug-57	8	14
ND WARD	19.73	PB8-3	20-Aug-57	6	14
ND WARD	1.92	PB8-5	22-Aug-57	10	14
ND WARD	5.22	PB9-4	26-Aug-57	8	14
ND WARD	9.15	PB11-2	1-Sep-57	5	14
ND WARD	13.36	PB15-2	17-Sep-57	3	14
ND WARD	0.23	UE2-6	10-Mar-62	1	14
ND WARD	0.16	BN-4	21-Dec-70	9	14
ND WARD	0.22	BN-5	22-Dec-70	1	14
ND WELLS	1.49	TS4-6	6-May-52	8	12
ND WELLS	1.49	TS4-6	6-May-52	8	12
ND WELLS	15.08	PB2-3	20-Jun-57	5	12
ND WELLS	3.88	PB5-3	17-Jul-57	10	12
ND WELLS	19.73	PB8-3	20-Aug-57	6	12
ND WELLS	1.92	PB8-5	22-Aug-57	10	12
ND WELLS	5.22	PB9-4	26-Aug-57	8	12
ND WELLS	18.31	PB11-2	1-Sep-57	3	12
ND WELLS	13.36	PB15-2	17-Sep-57	3	12
ND WELLS	0.23	UE2-6	10-Mar-62	1	12
ND WELLS	0.23	BN-4	21-Dec-70	8	12
ND WELLS	0.22	BN-5	22-Dec-70	1	12
ND WILLIAMS	25.37	UpK2-2	25-Mar-53	5	13
ND WILLIAMS	22.29	UpK8-1	8-May-53	3	13
ND WILLIAMS	19.36	PB5-2	16-Jul-57	3	13
ND WILLIAMS	3.30	PB8-2	19-Aug-57	5	13
ND WILLIAMS	19.73	PB8-3	20-Aug-57	6	13
ND WILLIAMS	1.92	PB8-5	22-Aug-57	10	13
ND WILLIAMS	5.22	PB9-4	26-Aug-57	8	13
ND WILLIAMS	6.92	PB11-2	1-Sep-57	6	13
ND WILLIAMS	10.36	PB11-4	3-Sep-57	6	13
ND WILLIAMS	13.36	PB15-2	17-Sep-57	3	13

COUNTY	uCi/ sqm	SHOT DAY	DATE	RANK FOR DAY	TIMES IN TOP 10
ND WILLIAMS	0.21	UE2-6	10-Mar-62	3	13
ND WILLIAMS	0.32	BN-4	21-Dec-70	5	13
ND WILLIAMS	0.22	BN-5	22-Dec-70	1	13

COUNTY	uCi/ sqm	SHOT DAY	DATE	RANK FOR DAY	TIMES IN TOP 10
NE ADAMS	12.24	TP4-2	8-Mar-55	9	12
NE ADAMS	1.54	PB3-2	25-Jun-57	8	12
NE ADAMS	2.33	PB3-3	26-Jun-57	8	12
NE ADAMS	5.89	PB5-4	18-Jul-57	5	12
NE ADAMS	1.65	PB8-2	19-Aug-57	6	12
NE ADAMS	19.73	PB8-3	20-Aug-57	6	12
NE ADAMS	3.78	PB8-5	22-Aug-57	4	12
NE ADAMS	12.75	PB13-3	8-Sep-57	4	12
NE ADAMS	3.55	PB13-4	9-Sep-57	5	12
NE ADAMS	0.81	PB14-7	20-Sep-57	9	12
NE ADAMS	0.17	UE2-7	11-Mar-62	2	12
NE ADAMS	8.02	UE8-4	17-Jul-62	3	12
NE ANTELOPE	1.37	TS4-6	6-May-52	9	11
NE ANTELOPE	1.37	TS4-6	6-May-52	9	11
NE ANTELOPE	3.97	UpK2-2	25-Mar-53	10	11
NE ANTELOPE	19.73	PB8-3	20-Aug-57	6	11
NE ANTELOPE	3.78	PB8-5	22-Aug-57	4	11
NE ANTELOPE	3.48	PB9-4	26-Aug-57	10	11
NE ANTELOPE	27.46	PB11-2	1-Sep-57	2	11
NE ANTELOPE	6.68	PB15-2	17-Sep-57	5	11
NE ANTELOPE	4.13	PB15-3	18-Sep-57	2	11
NE ANTELOPE	0.17	UE2-7	11-Mar-62	2	11
NE ANTELOPE	4.12	UE8-4	17-Jul-62	9	11
NE ARTHUR	1.06	TS5-8	14-May-52	4	16
NE ARTHUR	9.92	UpK2-7	30-Mar-53	5	16
NE ARTHUR	4.00	TP3-2	2-Mar-55	5	16
NE ARTHUR	14.24	TP11-2	16-May-55	7	16
NE ARTHUR	0.97	PB1-10	6-Jun-57	3	16
NE ARTHUR	7.28	PB5-4	18-Jul-57	4	16
NE ARTHUR	1.96	PB5-5	19-Jul-57	9	16
NE ARTHUR	0.85	PB5-6	20-Jul-57	10	16
NE ARTHUR	0.83	PB8-2	19-Aug-57	8	16
NE ARTHUR	19.73	PB8-3	20-Aug-57	6	16
NE ARTHUR	6.72	PB13-3	8-Sep-57	8	16
NE ARTHUR	0.81	PB15-4	19-Sep-57	7	16

COUNTY	uCi/ sqm	SHOT DAY	DATE	RANK FOR DAY	TIMES IN TOP 10
NE ARTHUR	0.97	PB18-2	8-Oct-57	7	16
NE ARTHUR	4.73	PB18-3	9-Oct-57	1	16
NE ARTHUR	1.44	PB18-4	10-Oct-57	6	16
NE ARTHUR	18.17	UE8-3	16-Jul-62	3	16
NE BANNER	0.82	TS5-9	15-May-52	6	16
NE BANNER	2.49	UpK2-7	30-Mar-53	10	16
NE BANNER	4.00	TP3-2	2-Mar-55	5	16
NE BANNER	1.71	PB4-2	6-Jul-57	10	16
NE BANNER	12.15	PB5-4	18-Jul-57	2	16
NE BANNER	0.83	PB8-2	19-Aug-57	8	16
NE BANNER	6.85	PB8-3	20-Aug-57	10	16
NE BANNER	3.07	PB8-4	21-Aug-57	10	16
NE BANNER	3.07	PB8-4	21-Aug-57	10	16
NE BANNER	3.68	PB9-4	26-Aug-57	9	16
NE BANNER	1.62	PB9-6	28-Aug-57	2	16
NE BANNER	2.37	PB18-3	9-Oct-57	3	16
NE BANNER	1.44	PB18-4	10-Oct-57	6	16
NE BANNER	0.59	UE2-4	8-Mar-62	7	16
NE BANNER	0.12	UE2-6	10-Mar-62	5	16
NE BANNER	13.28	UE8-3	16-Jul-62	5	16
NE BLAINE	1.44	BJ5-3	21-Nov-51	6	14
NE BLAINE	3.97	UpK2-2	25-Mar-53	10	14
NE BLAINE	1.31	UpK6-3	20-Apr-53	9	14
NE BLAINE	45.16	PB2-3	20-Jun-57	2	14
NE BLAINE	3.30	PB8-2	19-Aug-57	5	14
NE BLAINE	19.73	PB8-3	20-Aug-57	6	14
NE BLAINE	11.34	PB8-5	22-Aug-57	2	14
NE BLAINE	4.58	PB11-2	1-Sep-57	9	14
NE BLAINE	2.50	PB15-3	18-Sep-57	6	14
NE BLAINE	2.37	PB18-3	9-Oct-57	3	14
NE BLAINE	1.44	PB18-4	10-Oct-57	6	14
NE BLAINE	0.10	UE2-6	10-Mar-62	7	14
NE BLAINE	7.94	UE8-4	17-Jul-62	4	14
NE BLAINE	0.68	BN-2	19-Dec-70	9	14
NE BOONE	1.32	TS3-8	29-Apr-52	7	13

COUNTY	uCi/ sqm	SHOT DAY	DATE	RANK FOR DAY	TIMES IN TOP 10
NE BOONE	1.32	TS3-8	29-Apr-52	7	13
NE BOONE	1.37	TS4-6	6-May-52	9	13
NE BOONE	1.37	TS4-6	6-May-52	9	13
NE BOONE	12.24	TP4-2	8-Mar-55	9	13
NE BOONE	1.02	PB4-3	7-Jul-57	8	13
NE BOONE	19.73	PB8-3	20-Aug-57	6	13
NE BOONE	3.78	PB8-5	22-Aug-57	4	13
NE BOONE	5.22	PB9-4	26-Aug-57	8	13
NE BOONE	27.46	PB11-2	1-Sep-57	2	13
NE BOONE	1.64	PB15-3	18-Sep-57	9	13
NE BOONE	0.17	UE2-7	11-Mar-62	2	13
NE BOONE	0.82	BN-2	19-Dec-70	7	13
NE BOX BUTTE	0.82	TS5-9	15-May-52	6	14
NE BOX BUTTE	2.49	UpK2-7	30-Mar-53	10	14
NE BOX BUTTE	4.00	TP3-2	2-Mar-55	5	14
NE BOX BUTTE	0.95	PB1-9	5-Jun-57	6	14
NE BOX BUTTE	1.30	PB8-2	19-Aug-57	7	14
NE BOX BUTTE	7.76	PB9-3	25-Aug-57	9	14
NE BOX BUTTE	7.36	PB9-4	26-Aug-57	7	14
NE BOX BUTTE	4.73	PB18-3	9-Oct-57	1	14
NE BOX BUTTE	1.44	PB18-4	10-Oct-57	6	14
NE BOX BUTTE	0.41	UE2-4	8-Mar-62	9	14
NE BOX BUTTE	0.44	UE2-5	9-Mar-62	1	14
NE BOX BUTTE	0.12	UE2-6	10-Mar-62	5	14
NE BOX BUTTE	0.10	UE2-7	11-Mar-62	3	14
NE BOX BUTTE	0.82	BN-2	19-Dec-70	7	14
NE BOYD	1.44	BJ5-3	21-Nov-51	6	9
NE BOYD	3.97	UpK2-2	25-Mar-53	10	9
NE BOYD	11.31	PB2-3	20-Jun-57	9	9
NE BOYD	19.73	PB8-3	20-Aug-57	6	9
NE BOYD	1.92	PB8-5	22-Aug-57	10	9
NE BOYD	4.58	PB11-2	1-Sep-57	9	9
NE BOYD	13.36	PB15-2	17-Sep-57	3	9
NE BOYD	2.50	PB15-3	18-Sep-57	6	9
NE BOYD	0.10	UE2-7	11-Mar-62	4	9

COUNTY	uCi/ sqm	SHOT DAY	DATE	RANK FOR DAY	TIMES IN TOP 10
NE BROWN	1.44	BJ5-3	21-Nov-51	6	15
NE BROWN	1.32	TS3-8	29-Apr-52	7	15
NE BROWN	1.32	TS3-8	29-Apr-52	7	15
NE BROWN	3.97	UpK2-2	25-Mar-53	10	15
NE BROWN	1.31	UpK6-3	20-Apr-53	9	15
NE BROWN	45.16	PB2-3	20-Jun-57	2	15
NE BROWN	2.10	PB4-2	6-Jul-57	7	15
NE BROWN	1.65	PB8-2	19-Aug-57	6	15
NE BROWN	19.73	PB8-3	20-Aug-57	6	15
NE BROWN	7.56	PB8-5	22-Aug-57	3	15
NE BROWN	4.58	PB11-2	1-Sep-57	9	15
NE BROWN	2.50	PB15-3	18-Sep-57	6	15
NE BROWN	4.73	PB18-3	9-Oct-57	1	15
NE BROWN	1.44	PB18-4	10-Oct-57	6	15
NE BROWN	0.10	UE2-7	11-Mar-62	3	15
NE BUFFALO	1.68	TS4-6	6-May-52	6	13
NE BUFFALO	1.68	TS4-6	6-May-52	6	13
NE BUFFALO	12.24	TP4-2	8-Mar-55	9	13
NE BUFFALO	1.54	PB3-2	25-Jun-57	8	13
NE BUFFALO	1.02	PB4-3	7-Jul-57	8	13
NE BUFFALO	1.65	PB8-2	19-Aug-57	6	13
NE BUFFALO	19.73	PB8-3	20-Aug-57	6	13
NE BUFFALO	3.78	PB8-5	22-Aug-57	4	13
NE BUFFALO	31.89	PB13-3	8-Sep-57	1	13
NE BUFFALO	5.36	PB13-4	9-Sep-57	3	13
NE BUFFALO	1.34	PB13-7	12-Sep-57	2	13
NE BUFFALO	0.10	UE2-7	11-Mar-62	4	13
NE BUFFALO	14.44	UE8-4	17-Jul-62	1	13
NE BURT	2.30	TS4-6	6-May-52	3	10
NE BURT	2.30	TS4-6	6-May-52	3	10
NE BURT	3.25	UpK11-3	6-Jun-53	8	10
NE BURT	12.24	TP4-2	8-Mar-55	9	10
NE BURT	1.30	PB2-4	21-Jun-57	9	10
NE BURT	0.85	PB4-3	7-Jul-57	10	10
NE BURT	1.86	PB4-10	14-Jul-57	8	10

COUNTY	uCi/ sqm	SHOT DAY	DATE	RANK FOR DAY	TIMES IN TOP 10
NE BURT	6.09	PB5-5	19-Jul-57	2	10
NE BURT	0.80	PB13-7	12-Sep-57	7	10
NE BURT	0.10	UE2-7	11-Mar-62	4	10
NE BUTLER	1.32	TS3-8	29-Apr-52	7	13
NE BUTLER	1.32	TS3-8	29-Apr-52	7	13
NE BUTLER	3.08	UpK11-3	6-Jun-53	10	13
NE BUTLER	12.24	TP4-2	8-Mar-55	9	13
NE BUTLER	1.76	PB2-4	21-Jun-57	6	13
NE BUTLER	0.85	PB4-3	7-Jul-57	10	13
NE BUTLER	14.75	PB5-4	18-Jul-57	1	13
NE BUTLER	1.63	PB5-5	19-Jul-57	10	13
NE BUTLER	19.73	PB8-3	20-Aug-57	6	13
NE BUTLER	8.92	PB13-4	9-Sep-57	2	13
NE BUTLER	0.89	PB13-7	12-Sep-57	5	13
NE BUTLER	0.17	UE2-7	11-Mar-62	2	13
NE BUTLER	4.35	UE8-4	17-Jul-62	8	13
NE CASS	12.24	TP4-2	8-Mar-55	9	4
NE CASS	1.67	PB5-7	21-Jul-57	7	4
NE CASS	0.81	PB15-4	19-Sep-57	7	4
NE CASS	0.86	PB18-6	12-Oct-57	7	4
NE CEDAR	2.30	TS4-6	6-May-52	3	12
NE CEDAR	2.30	TS4-6	6-May-52	3	12
NE CEDAR	3.97	UpK2-2	25-Mar-53	10	12
NE CEDAR	1.44	PB4-3	7-Jul-57	3	12
NE CEDAR	1.65	PB4-10	14-Jul-57	10	12
NE CEDAR	19.73	PB8-3	20-Aug-57	6	12
NE CEDAR	1.92	PB8-5	22-Aug-57	10	12
NE CEDAR	5.22	PB9-4	26-Aug-57	8	12
NE CEDAR	18.31	PB11-2	1-Sep-57	3	12
NE CEDAR	4.13	PB15-3	18-Sep-57	2	12
NE CEDAR	0.17	UE2-7	11-Mar-62	2	12
NE CEDAR	145.87	S-2	7-Jul-62	6	12
NE CHASE	0.87	TS4-3	3-May-52	1	16
NE CHASE	0.87	TS4-3	3-May-52	1	16
NE CHASE	56.70	TS6-3	27-May-52	3	16

COUNTY	uCi/ sqm	SHOT DAY	DATE	RANK FOR DAY	TIMES IN TOP 10
NE CHASE	1.31	UpK6-3	20-Apr-53	9	16
NE CHASE	7.78	TP3-2	2-Mar-55	3	16
NE CHASE	49.44	TP4-2	8-Mar-55	4	16
NE CHASE	32.33	TP10-2	6-May-55	5	16
NE CHASE	4.87	PB5-4	18-Jul-57	8	16
NE CHASE	1.96	PB5-5	19-Jul-57	9	16
NE CHASE	0.83	PB8-2	19-Aug-57	8	16
NE CHASE	7.76	PB9-3	25-Aug-57	9	16
NE CHASE	11.05	PB9-4	26-Aug-57	6	16
NE CHASE	0.97	PB18-2	8-Oct-57	7	16
NE CHASE	2.37	PB18-3	9-Oct-57	3	16
NE CHASE	1.44	PB18-4	10-Oct-57	6	16
NE CHASE	2.54	PB18-6	12-Oct-57	2	16
NE CHERRY	7.52	TS5-3	9-May-52	10	15
NE CHERRY	45.45	UpK2-2	25-Mar-53	4	15
NE CHERRY	15.08	PB2-3	20-Jun-57	5	15
NE CHERRY	8.86	PB5-4	18-Jul-57	3	15
NE CHERRY	3.30	PB8-2	19-Aug-57	5	15
NE CHERRY	78.92	PB8-3	20-Aug-57	2	15
NE CHERRY	5.36	PB8-4	21-Aug-57	6	15
NE CHERRY	5.36	PB8-4	21-Aug-57	6	15
NE CHERRY	2.85	PB8-5	22-Aug-57	7	15
NE CHERRY	18.31	PB11-2	1-Sep-57	3	15
NE CHERRY	13.36	PB15-2	17-Sep-57	3	15
NE CHERRY	1.64	PB15-3	18-Sep-57	9	15
NE CHERRY	3.55	PB18-3	9-Oct-57	2	15
NE CHERRY	1.44	PB18-4	10-Oct-57	6	15
NE CHERRY	0.10	UE2-7	11-Mar-62	3	15
NE CHEYENNE	0.82	TS5-9	15-May-52	6	20
NE CHEYENNE	9.92	UpK2-7	30-Mar-53	5	20
NE CHEYENNE	4.00	TP3-2	2-Mar-55	5	20
NE CHEYENNE	12.15	PB5-4	18-Jul-57	2	20
NE CHEYENNE	3.26	PB5-5	19-Jul-57	5	20
NE CHEYENNE	0.83	PB8-2	19-Aug-57	8	20
NE CHEYENNE	3.07	PB8-4	21-Aug-57	10	20

COUNTY	uCi/ sqm	SHOT DAY	DATE	RANK FOR DAY	TIMES IN TOP 10
NE CHEYENNE	3.07	PB8-4	21-Aug-57	10	20
NE CHEYENNE	15.53	PB9-3	25-Aug-57	5	20
NE CHEYENNE	3.68	PB9-4	26-Aug-57	9	20
NE CHEYENNE	0.83	PB9-6	28-Aug-57	8	20
NE CHEYENNE	6.81	PB11-2	1-Sep-57	6	20
NE CHEYENNE	0.97	PB18-2	8-Oct-57	7	20
NE CHEYENNE	3.55	PB18-3	9-Oct-57	2	20
NE CHEYENNE	2.16	PB18-4	10-Oct-57	4	20
NE CHEYENNE	2.54	PB18-6	12-Oct-57	2	20
NE CHEYENNE	0.22	UE2-5	9-Mar-62	7	20
NE CHEYENNE	0.10	UE2-7	11-Mar-62	3	20
NE CHEYENNE	18.17	UE8-3	16-Jul-62	3	20
NE CHEYENNE	7.97	UE12-4	17-Apr-65	9	20
NE CLAY	12.24	TP4-2	8-Mar-55	9	9
NE CLAY	1.54	PB3-2	25-Jun-57	8	9
NE CLAY	2.33	PB3-3	26-Jun-57	8	9
NE CLAY	5.89	PB5-4	18-Jul-57	5	9
NE CLAY	19.73	PB8-3	20-Aug-57	6	9
NE CLAY	3.78	PB8-5	22-Aug-57	4	9
NE CLAY	3.55	PB13-4	9-Sep-57	5	9
NE CLAY	0.10	UE2-7	11-Mar-62	4	9
NE CLAY	10.54	UE8-4	17-Jul-62	2	9
NE COLFAX	1.37	TS4-6	6-May-52	9	11
NE COLFAX	1.37	TS4-6	6-May-52	9	11
NE COLFAX	5.21	UpK9-5	23-May-53	3	11
NE COLFAX	12.24	TP4-2	8-Mar-55	9	11
NE COLFAX	1.30	PB2-4	21-Jun-57	9	11
NE COLFAX	1.86	PB4-10	14-Jul-57	8	11
NE COLFAX	5.89	PB5-4	18-Jul-57	5	11
NE COLFAX	19.73	PB8-3	20-Aug-57	6	11
NE COLFAX	0.80	PB13-7	12-Sep-57	7	11
NE COLFAX	0.17	UE2-7	11-Mar-62	2	11
NE COLFAX	4.35	UE8-4	17-Jul-62	8	11
NE CUMING	2.30	TS4-6	6-May-52	3	7
NE CUMING	2.30	TS4-6	6-May-52	3	7

COUNTY	uCi/ sqm	SHOT DAY	DATE	RANK FOR DAY	TIMES IN TOP 10
NE CUMING	12.24	TP4-2	8-Mar-55	9	7
NE CUMING	0.85	PB4-3	7-Jul-57	10	7
NE CUMING	19.73	PB8-3	20-Aug-57	6	7
NE CUMING	0.80	PB13-7	12-Sep-57	7	7
NE CUMING	0.17	UE2-7	11-Mar-62	2	7
NE CUSTER	12.24	TP4-2	8-Mar-55	9	12
NE CUSTER	0.93	PB4-3	7-Jul-57	9	12
NE CUSTER	1.65	PB8-2	19-Aug-57	6	12
NE CUSTER	19.73	PB8-3	20-Aug-57	6	12
NE CUSTER	3.78	PB8-5	22-Aug-57	4	12
NE CUSTER	4.58	PB11-2	1-Sep-57	9	12
NE CUSTER	6.68	PB15-2	17-Sep-57	5	12
NE CUSTER	2.50	PB15-3	18-Sep-57	6	12
NE CUSTER	3.55	PB18-3	9-Oct-57	2	12
NE CUSTER	2.88	PB18-4	10-Oct-57	2	12
NE CUSTER	0.10	UE2-7	11-Mar-62	4	12
NE CUSTER	14.44	UE8-4	17-Jul-62	1	12
NE DAKOTA	1.37	TS4-6	6-May-52	9	10
NE DAKOTA	1.37	TS4-6	6-May-52	9	10
NE DAKOTA	3.97	UpK2-2	25-Mar-53	10	10
NE DAKOTA	1.26	UpK11-4	7-Jun-53	9	10
NE DAKOTA	1.30	PB2-4	21-Jun-57	9	10
NE DAKOTA	1.10	PB4-3	7-Jul-57	7	10
NE DAKOTA	2.01	PB5-5	19-Jul-57	8	10
NE DAKOTA	0.80	PB13-7	12-Sep-57	7	10
NE DAKOTA	0.17	UE2-7	11-Mar-62	2	10
NE DAKOTA	527.44	S-3	8-Jul-62	5	10
NE DAWES	25.53	TS5-2	8-May-52	8	12
NE DAWES	11.06	TS5-3	9-May-52	6	12
NE DAWES	2.49	UpK2-7	30-Mar-53	10	12
NE DAWES	4.00	TP3-2	2-Mar-55	5	12
NE DAWES	1.96	PB5-5	19-Jul-57	9	12
NE DAWES	0.83	PB8-2	19-Aug-57	8	12
NE DAWES	3.68	PB9-4	26-Aug-57	9	12
NE DAWES	2.37	PB18-3	9-Oct-57	3	12

COUNTY	uCi/ sqm	SHOT DAY	DATE	RANK FOR DAY	TIMES IN TOP 10
NE DAWES	1.44	PB18-4	10-Oct-57	6	12
NE DAWES	1.18	UE2-4	8-Mar-62	1	12
NE DAWES	0.11	UE2-6	10-Mar-62	6	12
NE DAWES	0.10	UE2-7	11-Mar-62	3	12
NE DAWSON	1.31	UpK6-3	20-Apr-53	9	15
NE DAWSON	7.78	TP3-2	2-Mar-55	3	15
NE DAWSON	49.44	TP4-2	8-Mar-55	4	15
NE DAWSON	32.33	TP10-2	6-May-55	5	15
NE DAWSON	0.93	PB4-3	7-Jul-57	9	15
NE DAWSON	3.30	PB8-2	19-Aug-57	5	15
NE DAWSON	19.73	PB8-3	20-Aug-57	6	15
NE DAWSON	7.56	PB8-5	22-Aug-57	3	15
NE DAWSON	5.22	PB9-4	26-Aug-57	8	15
NE DAWSON	4.58	PB11-2	1-Sep-57	9	15
NE DAWSON	2.50	PB15-3	18-Sep-57	6	15
NE DAWSON	0.97	PB18-2	8-Oct-57	7	15
NE DAWSON	3.55	PB18-3	9-Oct-57	2	15
NE DAWSON	1.44	PB18-4	10-Oct-57	6	15
NE DAWSON	13.69	UE8-3	16-Jul-62	4	15
NE DEUEL	0.82	TS5-9	15-May-52	6	13
NE DEUEL	48.29	TS6-3	27-May-52	6	13
NE DEUEL	14.85	UpK2-7	30-Mar-53	3	13
NE DEUEL	4.00	TP3-2	2-Mar-55	5	13
NE DEUEL	14.24	TP11-2	16-May-55	7	13
NE DEUEL	0.83	PB8-2	19-Aug-57	8	13
NE DEUEL	23.29	PB9-3	25-Aug-57	3	13
NE DEUEL	3.68	PB9-4	26-Aug-57	9	13
NE DEUEL	6.81	PB11-2	1-Sep-57	6	13
NE DEUEL	2.37	PB18-3	9-Oct-57	3	13
NE DEUEL	1.44	PB18-4	10-Oct-57	6	13
NE DEUEL	2.54	PB18-6	12-Oct-57	2	13
NE DEUEL	0.12	UE2-5	9-Mar-62	9	13
NE DIXON	1.08	BJ5-3	21-Nov-51	7	12
NE DIXON	1.37	TS4-6	6-May-52	9	12
NE DIXON	1.37	TS4-6	6-May-52	9	12

COUNTY	uCi/ sqm	SHOT DAY	DATE	RANK FOR DAY	TIMES IN TOP 10
NE DIXON	3.97	UpK2-2	25-Mar-53	10	12
NE DIXON	3.08	UpK11-3	6-Jun-53	10	12
NE DIXON	1.10	PB4-3	7-Jul-57	7	12
NE DIXON	1.86	PB4-10	14-Jul-57	8	12
NE DIXON	19.73	PB8-3	20-Aug-57	6	12
NE DIXON	3.48	PB9-4	26-Aug-57	10	12
NE DIXON	0.80	PB13-7	12-Sep-57	7	12
NE DIXON	0.17	UE2-7	11-Mar-62	2	12
NE DIXON	527.44	S-3	8-Jul-62	5	12
NE DODGE	1.32	TS3-8	29-Apr-52	7	16
NE DODGE	1.32	TS3-8	29-Apr-52	7	16
NE DODGE	1.37	TS4-6	6-May-52	9	16
NE DODGE	1.37	TS4-6	6-May-52	9	16
NE DODGE	4.27	UpK9-5	23-May-53	7	16
NE DODGE	12.24	TP4-2	8-Mar-55	9	16
NE DODGE	1.30	PB2-4	21-Jun-57	9	16
NE DODGE	1.10	PB4-3	7-Jul-57	7	16
NE DODGE	1.63	PB5-5	19-Jul-57	10	16
NE DODGE	19.73	PB8-3	20-Aug-57	6	16
NE DODGE	5.36	PB8-4	21-Aug-57	6	16
NE DODGE	5.36	PB8-4	21-Aug-57	6	16
NE DODGE	0.81	PB15-4	19-Sep-57	7	16
NE DODGE	0.17	UE2-7	11-Mar-62	2	16
NE DODGE	0.24	UE7-5	15-Jul-62	5	16
NE DODGE	0.82	BN-2	19-Dec-70	7	16
NE DOUGLAS	4.37	UpK9-5	23-May-53	7	3
NE DOUGLAS	12.24	TP4-2	8-Mar-55	9	3
NE DOUGLAS	0.10	UE2-7	11-Mar-62	4	3
NE DUNDY	0.87	TS4-3	3-May-52	1	14
NE DUNDY	0.87	TS4-3	3-May-52	1	14
NE DUNDY	1.31	UpK6-3	20-Apr-53	9	14
NE DUNDY	7.78	TP3-2	2-Mar-55	3	14
NE DUNDY	49.44	TP4-2	8-Mar-55	4	14
NE DUNDY	2.59	TP8-3	11-Apr-55	4	14
NE DUNDY	32.33	TP10-2	6-May-55	5	14

COUNTY	uCi/ sqm	SHOT DAY	DATE	RANK FOR DAY	TIMES IN TOP 10
NE DUNDY	1.96	PB5-5	19-Jul-57	9	14
NE DUNDY	0.83	PB8-2	19-Aug-57	8	14
NE DUNDY	7.36	PB9-4	26-Aug-57	7	14
NE DUNDY	0.97	PB18-2	8-Oct-57	7	14
NE DUNDY	4.73	PB18-3	9-Oct-57	1	14
NE DUNDY	1.44	PB18-4	10-Oct-57	6	14
NE DUNDY	0.86	PB18-6	12-Oct-57	7	14
NE FILLMORE	12.24	TP4-2	8-Mar-55	9	6
NE FILLMORE	19.73	PB8-3	20-Aug-57	6	6
NE FILLMORE	3.55	PB13-4	9-Sep-57	5	6
NE FILLMORE	0.81	PB14-7	20-Sep-57	9	6
NE FILLMORE	0.10	UE2-7	11-Mar-62	4	6
NE FILLMORE	4.12	UE8-4	17-Jul-62	9	6
NE FRANKLIN	0.92	TS3-8	29-Apr-52	9	13
NE FRANKLIN	0.92	TS3-8	29-Apr-52	9	13
NE FRANKLIN	12.24	TP4-2	8-Mar-55	9	13
NE FRANKLIN	1.54	PB3-2	25-Jun-57	8	13
NE FRANKLIN	4.95	PB8-2	19-Aug-57	4	13
NE FRANKLIN	19.73	PB8-3	20-Aug-57	6	13
NE FRANKLIN	3.78	PB8-5	22-Aug-57	4	13
NE FRANKLIN	12.75	PB13-3	8-Sep-57	4	13
NE FRANKLIN	5.36	PB13-4	9-Sep-57	3	13
NE FRANKLIN	0.81	PB14-7	20-Sep-57	9	13
NE FRANKLIN	1.29	PB18-6	12-Oct-57	6	13
NE FRANKLIN	0.17	UE2-7	11-Mar-62	2	13
NE FRANKLIN	8.02	UE8-4	17-Jul-62	3	13
NE FRONTIER	0.87	TS4-3	3-May-52	1	18
NE FRONTIER	0.87	TS4-3	3-May-52	1	18
NE FRONTIER	1.31	UpK6-3	20-Apr-53	9	18
NE FRONTIER	7.78	TP3-2	2-Mar-55	3	18
NE FRONTIER	49.44	TP4-2	8-Mar-55	4	18
NE FRONTIER	32.33	TP10-2	6-May-55	5	18
NE FRONTIER	0.85	PB4-3	7-Jul-57	10	18
NE FRONTIER	4.87	PB5-4	18-Jul-57	8	18
NE FRONTIER	1.65	PB8-2	19-Aug-57	6	18

COUNTY	uCi/ sqm	SHOT DAY	DATE	RANK FOR DAY	TIMES IN TOP 10
NE FRONTIER	19.73	PB8-3	20-Aug-57	6	18
NE FRONTIER	6.72	PB13-3	8-Sep-57	8	18
NE FRONTIER	0.97	PB18-2	8-Oct-57	7	18
NE FRONTIER	3.55	PB18-3	9-Oct-57	2	18
NE FRONTIER	1.44	PB18-4	10-Oct-57	6	18
NE FRONTIER	1.68	PB18-6	12-Oct-57	4	18
NE FRONTIER	0.10	UE2-7	11-Mar-62	3	18
NE FRONTIER	18.17	UE8-3	16-Jul-62	3	18
NE FRONTIER	4.55	UE8-4	17-Jul-62	7	18
NE FURNAS	0.87	TS4-3	3-May-52	1	15
NE FURNAS	0.87	TS4-3	3-May-52	1	15
NE FURNAS	1.31	UpK6-3	20-Apr-53	9	15
NE FURNAS	1.32	UpK8-3	10-May-53	9	15
NE FURNAS	7.78	TP3-2	2-Mar-55	3	15
NE FURNAS	49.44	TP4-2	8-Mar-55	4	15
NE FURNAS	32.33	TP10-2	6-May-55	5	15
NE FURNAS	0.93	PB4-3	7-Jul-57	9	15
NE FURNAS	1.65	PB8-2	19-Aug-57	6	15
NE FURNAS	19.73	PB8-3	20-Aug-57	6	15
NE FURNAS	5.36	PB13-4	9-Sep-57	3	15
NE FURNAS	2.37	PB18-3	9-Oct-57	3	15
NE FURNAS	1.44	PB18-4	10-Oct-57	6	15
NE FURNAS	0.86	PB18-6	12-Oct-57	7	15
NE FURNAS	7.94	UE8-4	17-Jul-62	4	15
NE GAGE	12.24	TP4-2	8-Mar-55	9	6
NE GAGE	5.36	PB8-4	21-Aug-57	6	6
NE GAGE	5.36	PB8-4	21-Aug-57	6	6
NE GAGE	3.55	PB13-4	9-Sep-57	5	6
NE GAGE	1.62	PB14-7	20-Sep-57	4	6
NE GAGE	0.24	UE7-5	15-Jul-62	5	6
NE GARDEN	9.92	UpK2-7	30-Mar-53	5	12
NE GARDEN	4.00	TP3-2	2-Mar-55	5	12
NE GARDEN	7.28	PB5-4	18-Jul-57	4	12
NE GARDEN	3.26	PB5-5	19-Jul-57	5	12
NE GARDEN	0.83	PB8-2	19-Aug-57	8	12

COUNTY	uCi/ sqm	SHOT DAY	DATE	RANK FOR DAY	TIMES IN TOP 10
NE GARDEN	23.29	PB9-3	25-Aug-57	3	12
NE GARDEN	3.68	PB9-4	26-Aug-57	9	12
NE GARDEN	0.97	PB18-2	8-Oct-57	7	12
NE GARDEN	4.73	PB18-3	9-Oct-57	1	12
NE GARDEN	1.44	PB18-4	10-Oct-57	6	12
NE GARDEN	2.54	PB18-6	12-Oct-57	2	12
NE GARDEN	0.10	UE2-7	11-Mar-62	3	12
NE GARFIELD	1.44	BJ5-3	21-Nov-51	6	16
NE GARFIELD	1.32	TS3-8	29-Apr-52	7	16
NE GARFIELD	1.32	TS3-8	29-Apr-52	7	16
NE GARFIELD	1.37	TS4-6	6-May-52	9	16
NE GARFIELD	1.37	TS4-6	6-May-52	9	16
NE GARFIELD	3.97	UpK2-2	25-Mar-53	10	16
NE GARFIELD	15.08	PB2-3	20-Jun-57	5	16
NE GARFIELD	5.89	PB5-4	18-Jul-57	5	16
NE GARFIELD	1.65	PB8-2	19-Aug-57	6	16
NE GARFIELD	19.73	PB8-3	20-Aug-57	6	16
NE GARFIELD	1.92	PB8-5	22-Aug-57	10	16
NE GARFIELD	3.48	PB9-4	26-Aug-57	10	16
NE GARFIELD	4.58	PB11-2	1-Sep-57	9	16
NE GARFIELD	33.40	PB15-2	17-Sep-57	1	16
NE GARFIELD	2.50	PB15-3	18-Sep-57	6	16
NE GARFIELD	1.17	PB18-7	13-Oct-57	5	16
NE GOSPER	0.87	TS4-3	3-May-52	1	15
NE GOSPER	0.87	TS4-3	3-May-52	1	15
NE GOSPER	1.31	UpK6-3	20-Apr-53	9	15
NE GOSPER	7.78	TP3-2	2-Mar-55	3	15
NE GOSPER	49.44	TP4-2	8-Mar-55	4	15
NE GOSPER	32.33	TP10-2	6-May-55	5	15
NE GOSPER	2.33	PB3-3	26-Jun-57	8	15
NE GOSPER	0.93	PB4-3	7-Jul-57	9	15
NE GOSPER	0.83	PB8-2	19-Aug-57	8	15
NE GOSPER	19.73	PB8-3	20-Aug-57	6	15
NE GOSPER	3.78	PB8-5	22-Aug-57	4	15
NE GOSPER	5.36	PB13-4	9-Sep-57	3	15

COUNTY	uCi/ sqm	SHOT DAY	DATE	RANK FOR DAY	TIMES IN TOP 10
NE GOSPER	2.37	PB18-3	9-Oct-57	3	15
NE GOSPER	1.44	PB18-4	10-Oct-57	6	15
NE GOSPER	4.55	UE8-4	17-Jul-62	7	15
NE GRANT	2.49	UpK2-7	30-Mar-53	10	11
NE GRANT	4.00	TP3-2	2-Mar-55	5	11
NE GRANT	14.24	TP11-2	16-May-55	7	11
NE GRANT	12.15	PB5-4	18-Jul-57	2	11
NE GRANT	1.65	PB8-2	19-Aug-57	6	11
NE GRANT	19.73	PB8-3	20-Aug-57	6	11
NE GRANT	8.05	PB8-4	21-Aug-57	4	11
NE GRANT	8.05	PB8-4	21-Aug-57	4	11
NE GRANT	2.37	PB18-3	9-Oct-57	3	11
NE GRANT	1.44	PB18-4	10-Oct-57	6	11
NE GRANT	0.10	UE2-7	11-Mar-62	4	11
NE GREELEY	12.24	TP4-2	8-Mar-55	9	7
NE GREELEY	19.73	PB8-3	20-Aug-57	6	7
NE GREELEY	1.92	PB8-5	22-Aug-57	10	7
NE GREELEY	9.15	PB11-2	1-Sep-57	5	7
NE GREELEY	1.64	PB15-3	18-Sep-57	9	7
NE GREELEY	0.10	UE2-7	11-Mar-62	4	7
NE GREELEY	8.02	UE8-4	17-Jul-62	3	7
NE HALL	12.24	TP4-2	8-Mar-55	9	10
NE HALL	1.54	PB3-2	25-Jun-57	8	10
NE HALL	1.02	PB4-3	7-Jul-57	8	10
NE HALL	1.65	PB8-2	19-Aug-57	6	10
NE HALL	19.73	PB8-3	20-Aug-57	6	10
NE HALL	3.78	PB8-5	22-Aug-57	4	10
NE HALL	31.89	PB13-3	8-Sep-57	1	10
NE HALL	5.36	PB13-4	9-Sep-57	3	10
NE HALL	1.34	PB13-7	12-Sep-57	2	10
NE HALL	0.10	UE2-7	11-Mar-62	4	10
NE HAMILTON	12.24	TP4-2	8-Mar-55	9	7
NE HAMILTON	1.63	PB5-5	19-Jul-57	10	7
NE HAMILTON	19.73	PB8-3	20-Aug-57	6	7
NE HAMILTON	3.78	PB8-5	22-Aug-57	4	7

COUNTY	uCi/ sqm	SHOT DAY	DATE	RANK FOR DAY	TIMES IN TOP 10
NE HAMILTON	8.92	PB13-4	9-Sep-57	2	7
NE HAMILTON	0.17	UE2-7	11-Mar-62	2	7
NE HAMILTON	10.54	UE8-4	17-Jul-62	2	7
NE HARLAN	0.92	TS3-8	29-Apr-52	9	14
NE HARLAN	0.92	TS3-8	29-Apr-52	9	14
NE HARLAN	12.24	TP4-2	8-Mar-55	9	14
NE HARLAN	1.65	PB8-2	19-Aug-57	6	14
NE HARLAN	19.73	PB8-3	20-Aug-57	6	14
NE HARLAN	3.78	PB8-5	22-Aug-57	4	14
NE HARLAN	12.75	PB13-3	8-Sep-57	4	14
NE HARLAN	8.92	PB13-4	9-Sep-57	2	14
NE HARLAN	0.81	PB14-7	20-Sep-57	9	14
NE HARLAN	2.37	PB18-3	9-Oct-57	3	14
NE HARLAN	1.44	PB18-4	10-Oct-57	6	14
NE HARLAN	1.29	PB18-6	12-Oct-57	6	14
NE HARLAN	0.10	UE2-7	11-Mar-62	4	14
NE HARLAN	8.02	UE8-4	17-Jul-62	3	14
NE HAYES	0.87	TS4-3	3-May-52	1	13
NE HAYES	0.87	TS4-3	3-May-52	1	13
NE HAYES	40.63	TS6-3	27-May-52	9	13
NE HAYES	1.31	UpK6-3	20-Apr-53	9	13
NE HAYES	7.78	TP3-2	2-Mar-55	3	13
NE HAYES	49.44	TP4-2	8-Mar-55	4	13
NE HAYES	32.33	TP10-2	6-May-55	5	13
NE HAYES	7.28	PB5-4	18-Jul-57	4	13
NE HAYES	1.96	PB5-5	19-Jul-57	9	13
NE HAYES	0.83	PB8-2	19-Aug-57	8	13
NE HAYES	4.73	PB18-3	9-Oct-57	1	13
NE HAYES	1.44	PB18-4	10-Oct-57	6	13
NE HAYES	0.86	PB18-6	12-Oct-57	7	13
NE HITCHCOCK	0.87	TS4-3	3-May-52	1	15
NE HITCHCOCK	0.87	TS4-3	3-May-52	1	15
NE HITCHCOCK	1.31	UpK6-3	20-Apr-53	9	15
NE HITCHCOCK	7.78	TP3-2	2-Mar-55	3	15
NE HITCHCOCK	49.44	TP4-2	8-Mar-55	4	15

COUNTY	uCi/ sqm	SHOT DAY	DATE	RANK FOR DAY	TIMES IN TOP 10
NE HITCHCOCK	32.33	TP10-2	6-May-55	5	15
NE HITCHCOCK	1.10	PB4-3	7-Jul-57	7	15
NE HITCHCOCK	9.10	PB5-2	16-Jul-57	7	15
NE HITCHCOCK	4.87	PB5-4	18-Jul-57	8	15
NE HITCHCOCK	0.83	PB8-2	19-Aug-57	8	15
NE HITCHCOCK	11.05	PB9-4	26-Aug-57	6	15
NE HITCHCOCK	0.83	PB9-6	28-Aug-57	8	15
NE HITCHCOCK	2.37	PB18-3	9-Oct-57	3	15
NE HITCHCOCK	1.44	PB18-4	10-Oct-57	6	15
NE HITCHCOCK	0.86	PB18-6	12-Oct-57	7	15
NE HOLT	1.44	BJ5-3	21-Nov-51	6	14
NE HOLT	1.32	TS3-8	29-Apr-52	7	14
NE HOLT	1.32	TS3-8	29-Apr-52	7	14
NE HOLT	1.37	TS4-6	6-May-52	9	14
NE HOLT	1.37	TS4-6	6-May-52	9	14
NE HOLT	3.97	UpK2-2	25-Mar-53	10	14
NE HOLT	1.02	PB4-3	7-Jul-57	8	14
NE HOLT	19.73	PB8-3	20-Aug-57	6	14
NE HOLT	3.78	PB8-5	22-Aug-57	4	14
NE HOLT	4.58	PB11-2	1-Sep-57	9	14
NE HOLT	6.68	PB15-2	17-Sep-57	5	14
NE HOLT	4.13	PB15-3	18-Sep-57	2	14
NE HOLT	0.10	UE2-7	11-Mar-62	4	14
NE HOLT	145.87	S-2	7-Jul-62	6	14
NE HOOKER	2.49	UpK2-7	30-Mar-53	10	13
NE HOOKER	4.00	TP3-2	2-Mar-55	5	13
NE HOOKER	10.18	PB5-3	17-Jul-57	3	13
NE HOOKER	7.28	PB5-4	18-Jul-57	4	13
NE HOOKER	4.95	PB8-2	19-Aug-57	4	13
NE HOOKER	39.46	PB8-3	20-Aug-57	4	13
NE HOOKER	3.48	PB9-4	26-Aug-57	10	13
NE HOOKER	6.72	PB13-3	8-Sep-57	8	13
NE HOOKER	0.81	PB15-4	19-Sep-57	7	13
NE HOOKER	4.73	PB18-3	9-Oct-57	1	13
NE HOOKER	1.44	PB18-4	10-Oct-57	6	13

COUNTY	uCi/ sqm	SHOT DAY	DATE	RANK FOR DAY	TIMES IN TOP 10
NE HOOKER	0.10	UE2-7	11-Mar-62	4	13
NE HOOKER	31.93	UE8-3	16-Jul-62	1	13
NE HOWARD	1.32	TS3-8	29-Apr-52	7	17
NE HOWARD	1.32	TS3-8	29-Apr-52	7	17
NE HOWARD	1.37	TS4-6	6-May-52	9	17
NE HOWARD	1.37	TS4-6	6-May-52	9	17
NE HOWARD	1.32	UpK8-3	10-May-53	9	17
NE HOWARD	12.24	TP4-2	8-Mar-55	9	17
NE HOWARD	1.02	PB4-3	7-Jul-57	8	17
NE HOWARD	1.65	PB4-10	14-Jul-57	10	17
NE HOWARD	0.83	PB8-2	19-Aug-57	8	17
NE HOWARD	19.73	PB8-3	20-Aug-57	6	17
NE HOWARD	3.78	PB8-5	22-Aug-57	4	17
NE HOWARD	5.22	PB9-4	26-Aug-57	8	17
NE HOWARD	18.31	PB11-2	1-Sep-57	3	17
NE HOWARD	1.64	PB15-3	18-Sep-57	9	17
NE HOWARD	0.10	UE2-7	11-Mar-62	4	17
NE HOWARD	345.56	S-3	8-Jul-62	8	17
NE HOWARD	8.02	UE8-4	17-Jul-62	3	17
NE JEFFERSON	12.24	TP4-2	8-Mar-55	9	6
NE JEFFERSON	2.33	PB3-3	26-Jun-57	8	6
NE JEFFERSON	3.55	PB13-4	9-Sep-57	5	6
NE JEFFERSON	1.62	PB14-7	20-Sep-57	4	6
NE JEFFERSON	0.24	UE7-5	15-Jul-62	5	6
NE JEFFERSON	4.35	UE8-4	17-Jul-62	8	6
NE JOHNSON	12.24	TP4-2	8-Mar-55	9	6
NE JOHNSON	7.99	PB8-3	20-Aug-57	9	6
NE JOHNSON	5.36	PB13-4	9-Sep-57	3	6
NE JOHNSON	0.83	PB13-8	13-Sep-57	9	6
NE JOHNSON	0.81	PB14-7	20-Sep-57	9	6
NE JOHNSON	0.86	PB18-6	12-Oct-57	7	6
NE KEARNEY	1.75	UpK10-5	29-May-53	2	12
NE KEARNEY	12.24	TP4-2	8-Mar-55	9	12
NE KEARNEY	2.24	PB3-2	25-Jun-57	5	12
NE KEARNEY	2.33	PB3-3	26-Jun-57	8	12

COUNTY	uCi/ sqm	SHOT DAY	DATE	RANK FOR DAY	TIMES IN TOP 10
NE KEARNEY	1.02	PB4-3	7-Jul-57	8	12
NE KEARNEY	3.30	PB8-2	19-Aug-57	5	12
NE KEARNEY	19.73	PB8-3	20-Aug-57	6	12
NE KEARNEY	1.92	PB8-5	22-Aug-57	10	12
NE KEARNEY	31.89	PB13-3	8-Sep-57	1	12
NE KEARNEY	5.36	PB13-4	9-Sep-57	3	12
NE KEARNEY	0.10	UE2-7	11-Mar-62	4	12
NE KEARNEY	14.44	UE8-4	17-Jul-62	1	12
NE KEITH	0.87	TS4-3	3-May-52	1	17
NE KEITH	0.87	TS4-3	3-May-52	1	17
NE KEITH	1.31	UpK6-3	20-Apr-53	9	17
NE KEITH	7.78	TP3-2	2-Mar-55	3	17
NE KEITH	49.44	TP4-2	8-Mar-55	4	17
NE KEITH	32.33	TP10-2	6-May-55	5	17
NE KEITH	6.79	PB5-3	17-Jul-57	7	17
NE KEITH	4.87	PB5-4	18-Jul-57	8	17
NE KEITH	1.65	PB8-2	19-Aug-57	6	17
NE KEITH	23.29	PB9-3	25-Aug-57	3	17
NE KEITH	3.68	PB9-4	26-Aug-57	9	17
NE KEITH	1.62	PB9-6	28-Aug-57	2	17
NE KEITH	0.97	PB18-2	8-Oct-57	7	17
NE KEITH	4.73	PB18-3	9-Oct-57	1	17
NE KEITH	1.44	PB18-4	10-Oct-57	6	17
NE KEITH	1.68	PB18-6	12-Oct-57	4	17
NE KEITH	8.97	UE8-3	16-Jul-62	9	17
NE KEYA PAHA	3.97	UpK2-2	25-Mar-53	10	10
NE KEYA PAHA	0.83	PB8-2	19-Aug-57	8	10
NE KEYA PAHA	19.73	PB8-3	20-Aug-57	6	10
NE KEYA PAHA	3.78	PB8-5	22-Aug-57	4	10
NE KEYA PAHA	4.58	PB11-2	1-Sep-57	9	10
NE KEYA PAHA	6.68	PB15-2	17-Sep-57	5	10
NE KEYA PAHA	2.50	PB15-3	18-Sep-57	6	10
NE KEYA PAHA	2.37	PB18-3	9-Oct-57	3	10
NE KEYA PAHA	1.44	PB18-4	10-Oct-57	6	10
NE KEYA PAHA	0.10	UE2-7	11-Mar-62	4	10

COUNTY	uCi/ sqm	SHOT DAY	DATE	RANK FOR DAY	TIMES IN TOP 10
NE KIMBALL	14.85	UpK2-7	30-Mar-53	3	16
NE KIMBALL	4.00	TP3-2	2-Mar-55	5	16
NE KIMBALL	10.18	PB5-3	17-Jul-57	3	16
NE KIMBALL	7.28	PB5-4	18-Jul-57	4	16
NE KIMBALL	0.83	PB8-2	19-Aug-57	8	16
NE KIMBALL	6.85	PB8-3	20-Aug-57	10	16
NE KIMBALL	4.60	PB8-4	21-Aug-57	8	16
NE KIMBALL	4.60	PB8-4	21-Aug-57	8	16
NE KIMBALL	3.68	PB9-4	26-Aug-57	9	16
NE KIMBALL	0.97	PB18-2	8-Oct-57	7	16
NE KIMBALL	4.73	PB18-3	9-Oct-57	1	16
NE KIMBALL	1.44	PB18-4	10-Oct-57	6	16
NE KIMBALL	2.54	PB18-6	12-Oct-57	2	16
NE KIMBALL	0.37	UE2-5	9-Mar-62	2	16
NE KIMBALL	0.10	UE2-7	11-Mar-62	3	16
NE KIMBALL	10.77	UE8-3	16-Jul-62	7	16
NE KNOX	1.44	BJ5-3	21-Nov-51	6	16
NE KNOX	1.32	TS3-8	29-Apr-52	7	16
NE KNOX	1.32	TS3-8	29-Apr-52	7	16
NE KNOX	1.37	TS4-6	6-May-52	9	16
NE KNOX	1.37	TS4-6	6-May-52	9	16
NE KNOX	3.97	UpK2-2	25-Mar-53	10	16
NE KNOX	1.02	PB4-3	7-Jul-57	8	16
NE KNOX	19.73	PB8-3	20-Aug-57	6	16
NE KNOX	3.78	PB8-5	22-Aug-57	4	16
NE KNOX	3.48	PB9-4	26-Aug-57	10	16
NE KNOX	9.15	PB11-2	1-Sep-57	5	16
NE KNOX	10.36	PB11-4	3-Sep-57	6	16
NE KNOX	5.01	PB15-2	17-Sep-57	8	16
NE KNOX	4.13	PB15-3	18-Sep-57	2	16
NE KNOX	0.10	UE2-7	11-Mar-62	4	16
NE KNOX	0.12	UE7-5	15-Jul-62	9	16
NE LANCASTER	1.32	TS3-8	29-Apr-52	7	9
NE LANCASTER	1.32	TS3-8	29-Apr-52	7	9
NE LANCASTER	12.24	TP4-2	8-Mar-55	9	9

COUNTY	uCi/ sqm	SHOT DAY	DATE	RANK FOR DAY	TIMES IN TOP 10
NE LANCASTER	1.63	PB5-5	19-Jul-57	10	9
NE LANCASTER	3.55	PB13-4	9-Sep-57	5	9
NE LANCASTER	0.81	PB14-7	20-Sep-57	9	9
NE LANCASTER	0.86	PB18-6	12-Oct-57	7	9
NE LANCASTER	0.10	UE2-7	11-Mar-62	4	9
NE LANCASTER	5.25	UE8-4	17-Jul-62	6	9
NE LINCOLN	1.31	UpK6-3	20-Apr-53	9	18
NE LINCOLN	7.78	TP3-2	2-Mar-55	3	18
NE LINCOLN	1.65	TP3-3	3-Mar-55	9	18
NE LINCOLN	49.44	TP4-2	8-Mar-55	4	18
NE LINCOLN	32.33	TP10-2	6-May-55	5	18
NE LINCOLN	1.71	PB4-2	6-Jul-57	10	18
NE LINCOLN	7.28	PB5-4	18-Jul-57	4	18
NE LINCOLN	1.96	PB5-5	19-Jul-57	9	18
NE LINCOLN	1.65	PB8-2	19-Aug-57	6	18
NE LINCOLN	39.46	PB8-3	20-Aug-57	4	18
NE LINCOLN	3.48	PB9-4	26-Aug-57	10	18
NE LINCOLN	6.81	PB11-2	1-Sep-57	6	18
NE LINCOLN	3.47	PB15-2	17-Sep-57	10	18
NE LINCOLN	0.97	PB18-2	8-Oct-57	7	18
NE LINCOLN	4.73	PB18-3	9-Oct-57	1	18
NE LINCOLN	2.88	PB18-4	10-Oct-57	2	18
NE LINCOLN	0.10	UE2-7	11-Mar-62	3	18
NE LINCOLN	4.55	UE8-4	17-Jul-62	7	18
NE LOGAN	1.31	UpK6-3	20-Apr-53	9	17
NE LOGAN	7.78	TP3-2	2-Mar-55	3	17
NE LOGAN	49.44	TP4-2	8-Mar-55	4	17
NE LOGAN	32.33	TP10-2	6-May-55	5	17
NE LOGAN	2.10	PB4-2	6-Jul-57	7	17
NE LOGAN	8.86	PB5-4	18-Jul-57	3	17
NE LOGAN	0.83	PB8-2	19-Aug-57	8	17
NE LOGAN	19.73	PB8-3	20-Aug-57	6	17
NE LOGAN	1.92	PB8-5	22-Aug-57	10	17
NE LOGAN	3.48	PB9-4	26-Aug-57	10	17
NE LOGAN	4.58	PB11-2	1-Sep-57	9	17

COUNTY	uCi/ sqm	SHOT DAY	DATE	RANK FOR DAY	TIMES IN TOP 10
NE LOGAN	20.04	PB15-2	17-Sep-57	2	17
NE LOGAN	2.50	PB15-3	18-Sep-57	6	17
NE LOGAN	0.97	PB18-2	8-Oct-57	7	17
NE LOGAN	2.37	PB18-3	9-Oct-57	3	17
NE LOGAN	2.88	PB18-4	10-Oct-57	2	17
NE LOGAN	0.10	UE2-7	11-Mar-62	3	17
NE LOUP	3.97	UpK2-2	25-Mar-53	10	12
NE LOUP	1.65	PB8-2	19-Aug-57	6	12
NE LOUP	19.73	PB8-3	20-Aug-57	6	12
NE LOUP	1.92	PB8-5	22-Aug-57	10	12
NE LOUP	5.22	PB9-4	26-Aug-57	8	12
NE LOUP	4.58	PB11-2	1-Sep-57	9	12
NE LOUP	4.13	PB15-3	18-Sep-57	2	12
NE LOUP	1.44	PB18-4	10-Oct-57	6	12
NE LOUP	0.10	UE2-6	10-Mar-62	7	12
NE LOUP	0.10	UE2-7	11-Mar-62	3	12
NE LOUP	436.50	S-3	8-Jul-62	6	12
NE LOUP	8.02	UE8-4	17-Jul-62	3	12
NE MADISON	12.24	TP4-2	8-Mar-55	9	12
NE MADISON	15.08	PB2-3	20-Jun-57	5	12
NE MADISON	5.89	PB5-4	18-Jul-57	5	12
NE MADISON	39.46	PB8-3	20-Aug-57	4	12
NE MADISON	3.78	PB8-5	22-Aug-57	4	12
NE MADISON	3.48	PB9-4	26-Aug-57	10	12
NE MADISON	27.46	PB11-2	1-Sep-57	2	12
NE MADISON	6.68	PB15-2	17-Sep-57	5	12
NE MADISON	2.50	PB15-3	18-Sep-57	6	12
NE MADISON	0.17	UE2-7	11-Mar-62	2	12
NE MADISON	4.12	UE8-4	17-Jul-62	9	12
NE MADISON	0.82	BN-2	19-Dec-70	7	12
NE MCPHERSON	1.31	UpK6-3	20-Apr-53	9	12
NE MCPHERSON	7.78	TP3-2	2-Mar-55	3	12
NE MCPHERSON	49.44	TP4-2	8-Mar-55	4	12
NE MCPHERSON	32.33	TP10-2	6-May-55	5	12
NE MCPHERSON	4.87	PB5-4	18-Jul-57	8	12

COUNTY	uCi/ sqm	SHOT DAY	DATE	RANK FOR DAY	TIMES IN TOP 10
NE MCPHERSON	1.96	PB5-5	19-Jul-57	9	12
NE MCPHERSON	0.83	PB8-2	19-Aug-57	8	12
NE MCPHERSON	19.73	PB8-3	20-Aug-57	6	12
NE MCPHERSON	0.97	PB18-2	8-Oct-57	7	12
NE MCPHERSON	2.37	PB18-3	9-Oct-57	3	12
NE MCPHERSON	1.44	PB18-4	10-Oct-57	6	12
NE MCPHERSON	18.17	UE8-3	16-Jul-62	3	12
NE MERRICK	1.32	TS3-8	29-Apr-52	7	14
NE MERRICK	1.32	TS3-8	29-Apr-52	7	14
NE MERRICK	3.17	UpK11-3	6-Jun-53	9	14
NE MERRICK	12.24	TP4-2	8-Mar-55	9	14
NE MERRICK	1.02	PB4-3	7-Jul-57	8	14
NE MERRICK	5.89	PB5-4	18-Jul-57	5	14
NE MERRICK	19.73	PB8-3	20-Aug-57	6	14
NE MERRICK	7.56	PB8-5	22-Aug-57	3	14
NE MERRICK	19.13	PB13-3	8-Sep-57	3	14
NE MERRICK	5.36	PB13-4	9-Sep-57	3	14
NE MERRICK	1.34	PB13-7	12-Sep-57	2	14
NE MERRICK	0.81	PB14-7	20-Sep-57	9	14
NE MERRICK	0.10	UE2-7	11-Mar-62	4	14
NE MERRICK	10.54	UE8-4	17-Jul-62	2	14
NE MORRILL	2.49	UpK2-7	30-Mar-53	10	14
NE MORRILL	4.00	TP3-2	2-Mar-55	5	14
NE MORRILL	12.15	PB5-4	18-Jul-57	2	14
NE MORRILL	0.83	PB8-2	19-Aug-57	8	14
NE MORRILL	3.07	PB8-4	21-Aug-57	10	14
NE MORRILL	3.07	PB8-4	21-Aug-57	10	14
NE MORRILL	23.29	PB9-3	25-Aug-57	3	14
NE MORRILL	3.68	PB9-4	26-Aug-57	9	14
NE MORRILL	2.45	PB9-6	28-Aug-57	1	14
NE MORRILL	2.37	PB18-3	9-Oct-57	3	14
NE MORRILL	1.44	PB18-4	10-Oct-57	6	14
NE MORRILL	2.54	PB18-6	12-Oct-57	2	14
NE MORRILL	0.59	UE2-4	8-Mar-62	7	14
NE MORRILL	0.10	UE2-7	11-Mar-62	3	14

COUNTY	uCi/ sqm	SHOT DAY	DATE	RANK FOR DAY	TIMES IN TOP 10
NE NANCE	1.32	TS3-8	29-Apr-52	7	12
NE NANCE	1.32	TS3-8	29-Apr-52	7	12
NE NANCE	12.24	TP4-2	8-Mar-55	9	12
NE NANCE	1.02	PB4-3	7-Jul-57	8	12
NE NANCE	19.73	PB8-3	20-Aug-57	6	12
NE NANCE	7.56	PB8-5	22-Aug-57	3	12
NE NANCE	19.13	PB13-3	8-Sep-57	3	12
NE NANCE	5.36	PB13-4	9-Sep-57	3	12
NE NANCE	1.34	PB13-7	12-Sep-57	2	12
NE NANCE	0.81	PB14-7	20-Sep-57	9	12
NE NANCE	0.17	UE2-7	11-Mar-62	2	12
NE NANCE	4.12	UE8-4	17-Jul-62	9	12
NE NEMAHA	1.17	UpK7-9	3-May-53	2	5
NE NEMAHA	12.24	TP4-2	8-Mar-55	9	5
NE NEMAHA	1.44	PB18-6	12-Oct-57	5	5
NE NEMAHA	0.08	UE2-7	11-Mar-62	5	5
NE NEMAHA	0.13	UE7-5	15-Jul-62	8	5
NE NUCKOLLS	12.24	TP4-2	8-Mar-55	9	9
NE NUCKOLLS	2.24	PB3-2	25-Jun-57	5	9
NE NUCKOLLS	2.33	PB3-3	26-Jun-57	8	9
NE NUCKOLLS	39.46	PB8-3	20-Aug-57	4	9
NE NUCKOLLS	3.55	PB13-4	9-Sep-57	5	9
NE NUCKOLLS	0.83	PB13-8	13-Sep-57	9	9
NE NUCKOLLS	0.81	PB14-7	20-Sep-57	9	9
NE NUCKOLLS	0.10	UE2-7	11-Mar-62	4	9
NE NUCKOLLS	10.54	UE8-4	17-Jul-62	2	9
NE OTOE	12.24	TP4-2	8-Mar-55	9	6
NE OTOE	1.67	PB5-7	21-Jul-57	7	6
NE OTOE	7.99	PB8-3	20-Aug-57	9	6
NE OTOE	0.86	PB18-6	12-Oct-57	7	6
NE OTOE	0.08	UE2-7	11-Mar-62	5	6
NE OTOE	0.24	UE7-5	15-Jul-62	5	6
NE PAWNEE	12.24	TP4-2	8-Mar-55	9	5
NE PAWNEE	7.99	PB8-3	20-Aug-57	9	5
NE PAWNEE	3.55	PB13-4	9-Sep-57	5	5

COUNTY	uCi/ sqm	SHOT DAY	DATE	RANK FOR DAY	TIMES IN TOP 10
NE PAWNEE	0.81	PB14-7	20-Sep-57	9	5
NE PAWNEE	0.13	UE7-5	15-Jul-62	8	5
NE PERKINS	0.87	TS4-3	3-May-52	1	16
NE PERKINS	0.87	TS4-3	3-May-52	1	16
NE PERKINS	47.01	TS6-3	27-May-52	7	16
NE PERKINS	1.31	UpK6-3	20-Apr-53	9	16
NE PERKINS	7.78	TP3-2	2-Mar-55	3	16
NE PERKINS	49.44	TP4-2	8-Mar-55	4	16
NE PERKINS	32.33	TP10-2	6-May-55	5	16
NE PERKINS	4.87	PB5-4	18-Jul-57	8	16
NE PERKINS	1.96	PB5-5	19-Jul-57	9	16
NE PERKINS	1.65	PB8-2	19-Aug-57	6	16
NE PERKINS	11.05	PB9-4	26-Aug-57	6	16
NE PERKINS	0.83	PB9-6	28-Aug-57	8	16
NE PERKINS	2.37	PB18-3	9-Oct-57	3	16
NE PERKINS	1.44	PB18-4	10-Oct-57	6	16
NE PERKINS	2.54	PB18-6	12-Oct-57	2	16
NE PERKINS	18.17	UE8-3	16-Jul-62	3	16
NE PHELPS	38.51	TS6-3	27-May-52	10	16
NE PHELPS	12.24	TP4-2	8-Mar-55	9	16
NE PHELPS	2.24	PB3-2	25-Jun-57	5	16
NE PHELPS	2.33	PB3-3	26-Jun-57	8	16
NE PHELPS	1.27	PB4-3	7-Jul-57	5	16
NE PHELPS	5.89	PB5-4	18-Jul-57	5	16
NE PHELPS	3.30	PB8-2	19-Aug-57	5	16
NE PHELPS	39.46	PB8-3	20-Aug-57	4	16
NE PHELPS	3.78	PB8-5	22-Aug-57	4	16
NE PHELPS	19.13	PB13-3	8-Sep-57	3	16
NE PHELPS	8.92	PB13-4	9-Sep-57	2	16
NE PHELPS	0.89	PB13-7	12-Sep-57	5	16
NE PHELPS	2.37	PB18-3	9-Oct-57	3	16
NE PHELPS	2.88	PB18-4	10-Oct-57	2	16
NE PHELPS	0.10	UE2-7	11-Mar-62	4	16
NE PHELPS	8.02	UE8-4	17-Jul-62	3	16
NE PIERCE	1.44	BJ5-3	21-Nov-51	6	13

COUNTY	uCi/ sqm	SHOT DAY	DATE	RANK FOR DAY	TIMES IN TOP 10
NE PIERCE	1.37	TS4-6	6-May-52	9	13
NE PIERCE	1.37	TS4-6	6-May-52	9	13
NE PIERCE	3.97	UpK2-2	25-Mar-53	10	13
NE PIERCE	3.17	UpK11-3	6-Jun-53	9	13
NE PIERCE	19.73	PB8-3	20-Aug-57	6	13
NE PIERCE	1.92	PB8-5	22-Aug-57	10	13
NE PIERCE	5.22	PB9-4	26-Aug-57	8	13
NE PIERCE	18.31	PB11-2	1-Sep-57	3	13
NE PIERCE	5.01	PB15-2	17-Sep-57	8	13
NE PIERCE	4.13	PB15-3	18-Sep-57	2	13
NE PIERCE	0.10	UE2-7	11-Mar-62	3	13
NE PIERCE	0.13	UE7-5	15-Jul-62	8	13
NE PLATTE	1.32	TS3-8	29-Apr-52	7	9
NE PLATTE	1.32	TS3-8	29-Apr-52	7	9
NE PLATTE	12.24	TP4-2	8-Mar-55	9	9
NE PLATTE	1.76	PB2-4	21-Jun-57	6	9
NE PLATTE	8.86	PB5-4	18-Jul-57	3	9
NE PLATTE	39.46	PB8-3	20-Aug-57	4	9
NE PLATTE	3.55	PB13-4	9-Sep-57	5	9
NE PLATTE	1.34	PB13-7	12-Sep-57	2	9
NE PLATTE	0.17	UE2-7	11-Mar-62	2	9
NE POLK	1.32	TS3-8	29-Apr-52	7	11
NE POLK	1.32	TS3-8	29-Apr-52	7	11
NE POLK	12.24	TP4-2	8-Mar-55	9	11
NE POLK	1.10	PB4-3	7-Jul-57	7	11
NE POLK	19.73	PB8-3	20-Aug-57	6	11
NE POLK	12.75	PB13-3	8-Sep-57	4	11
NE POLK	8.92	PB13-4	9-Sep-57	2	11
NE POLK	0.89	PB13-7	12-Sep-57	5	11
NE POLK	1.62	PB14-7	20-Sep-57	4	11
NE POLK	0.17	UE2-7	11-Mar-62	2	11
NE POLK	4.12	UE8-4	17-Jul-62	9	11
NE RED WILLOW	0.87	TS4-3	3-May-52	1	14
NE RED WILLOW	0.87	TS4-3	3-May-52	1	14
NE RED WILLOW	1.31	UpK6-3	20-Apr-53	9	14

COUNTY	uCi/ sqm	SHOT DAY	DATE	RANK FOR DAY	TIMES IN TOP 10
NE RED WILLOW	7.78	TP3-2	2-Mar-55	3	14
NE RED WILLOW	49.44	TP4-2	8-Mar-55	4	14
NE RED WILLOW	32.33	TP10-2	6-May-55	5	14
NE RED WILLOW	0.85	PB4-3	7-Jul-57	10	14
NE RED WILLOW	1.96	PB5-5	19-Jul-57	9	14
NE RED WILLOW	0.83	PB8-2	19-Aug-57	8	14
NE RED WILLOW	19.73	PB8-3	20-Aug-57	6	14
NE RED WILLOW	6.72	PB13-3	8-Sep-57	8	14
NE RED WILLOW	2.37	PB18-3	9-Oct-57	3	14
NE RED WILLOW	1.44	PB18-4	10-Oct-57	6	14
NE RED WILLOW	4.55	UE8-4	17-Jul-62	7	14
NE RICHARDSON	12.24	TP4-2	8-Mar-55	9	4
NE RICHARDSON	0.83	PB13-8	13-Sep-57	9	4
NE RICHARDSON	1.21	PB15-4	19-Sep-57	5	4
NE RICHARDSON	0.13	UE7-5	15-Jul-62	8	4
NE ROCK	2.88	BJ5-3	21-Nov-51	5	12
NE ROCK	1.32	TS3-8	29-Apr-52	7	12
NE ROCK	1.32	TS3-8	29-Apr-52	7	12
NE ROCK	3.97	UpK2-2	25-Mar-53	10	12
NE ROCK	0.83	PB8-2	19-Aug-57	8	12
NE ROCK	19.73	PB8-3	20-Aug-57	6	12
NE ROCK	1.92	PB8-5	22-Aug-57	10	12
NE ROCK	5.22	PB9-4	26-Aug-57	8	12
NE ROCK	4.58	PB11-2	1-Sep-57	9	12
NE ROCK	2.50	PB15-3	18-Sep-57	6	12
NE ROCK	1.44	PB18-4	10-Oct-57	6	12
NE ROCK	0.10	UE2-7	11-Mar-62	4	12
NE SALINE	0.92	TS3-8	29-Apr-52	9	10
NE SALINE	0.92	TS3-8	29-Apr-52	9	10
NE SALINE	12.24	TP4-2	8-Mar-55	9	10
NE SALINE	1.76	PB2-4	21-Jun-57	6	10
NE SALINE	1.63	PB5-5	19-Jul-57	10	10
NE SALINE	39.46	PB8-3	20-Aug-57	4	10
NE SALINE	3.55	PB13-4	9-Sep-57	5	10
NE SALINE	0.81	PB14-7	20-Sep-57	9	10

COUNTY	uCi/ sqm	SHOT DAY	DATE	RANK FOR DAY	TIMES IN TOP 10
NE SALINE	0.10	UE2-7	11-Mar-62	4	10
NE SALINE	4.35	UE8-4	17-Jul-62	8	10
NE SARPY	1.32	TS3-8	29-Apr-52	7	7
NE SARPY	1.32	TS3-8	29-Apr-52	7	7
NE SARPY	4.47	UpK9-5	23-May-53	5	7
NE SARPY	3.51	UpK11-3	6-Jun-53	7	7
NE SARPY	12.24	TP4-2	8-Mar-55	9	7
NE SARPY	1.21	PB15-4	19-Sep-57	5	7
NE SARPY	0.10	UE2-7	11-Mar-62	4	7
NE SAUNDERS	1.32	TS3-8	29-Apr-52	7	7
NE SAUNDERS	1.32	TS3-8	29-Apr-52	7	7
NE SAUNDERS	12.24	TP4-2	8-Mar-55	9	7
NE SAUNDERS	1.63	PB5-5	19-Jul-57	10	7
NE SAUNDERS	19.73	PB8-3	20-Aug-57	6	7
NE SAUNDERS	1.16	PB16-4	26-Sep-57	10	7
NE SAUNDERS	0.17	UE2-7	11-Mar-62	2	7
NE SCOTTS BLUFF	1.66	TP3-2	2-Mar-55	10	13
NE SCOTTS BLUFF	5.05	PB5-4	18-Jul-57	7	13
NE SCOTTS BLUFF	0.84	PB5-8	22-Jul-57	8	13
NE SCOTTS BLUFF	0.83	PB8-2	19-Aug-57	8	13
NE SCOTTS BLUFF	7.76	PB9-3	25-Aug-57	9	13
NE SCOTTS BLUFF	1.62	PB9-6	28-Aug-57	2	13
NE SCOTTS BLUFF	3.55	PB18-3	9-Oct-57	2	13
NE SCOTTS BLUFF	2.16	PB18-4	10-Oct-57	4	13
NE SCOTTS BLUFF	0.86	PB18-6	12-Oct-57	7	13
NE SCOTTS BLUFF	0.59	UE2-4	8-Mar-62	7	13
NE SCOTTS BLUFF	0.44	UE2-5	9-Mar-62	1	13
NE SCOTTS BLUFF	0.11	UE2-6	10-Mar-62	6	13
NE SCOTTS BLUFF	2.66	S-2	7-Jul-62	10	13
NE SEWARD	1.32	TS3-8	29-Apr-52	7	12
NE SEWARD	1.32	TS3-8	29-Apr-52	7	12
NE SEWARD	12.24	TP4-2	8-Mar-55	9	12
NE SEWARD	1.76	PB2-4	21-Jun-57	6	12
NE SEWARD	1.63	PB5-5	19-Jul-57	10	12
NE SEWARD	19.73	PB8-3	20-Aug-57	6	12

COUNTY	uCi/ sqm	SHOT DAY	DATE	RANK FOR DAY	TIMES IN TOP 10
NE SEWARD	12.75	PB13-3	8-Sep-57	4	12
NE SEWARD	5.36	PB13-4	9-Sep-57	3	12
NE SEWARD	2.42	PB14-7	20-Sep-57	2	12
NE SEWARD	0.86	PB18-6	12-Oct-57	7	12
NE SEWARD	0.10	UE2-7	11-Mar-62	4	12
NE SEWARD	4.35	UE8-4	17-Jul-62	8	12
NE SHERIDAN	11.25	TS5-3	9-May-52	5	12
NE SHERIDAN	0.82	TS5-9	15-May-52	6	12
NE SHERIDAN	2.49	UpK2-7	30-Mar-53	10	12
NE SHERIDAN	4.00	TP3-2	2-Mar-55	5	12
NE SHERIDAN	7.28	PB5-4	18-Jul-57	4	12
NE SHERIDAN	3.30	PB8-2	19-Aug-57	5	12
NE SHERIDAN	15.53	PB9-3	25-Aug-57	5	12
NE SHERIDAN	0.81	PB15-4	19-Sep-57	7	12
NE SHERIDAN	4.73	PB18-3	9-Oct-57	1	12
NE SHERIDAN	1.44	PB18-4	10-Oct-57	6	12
NE SHERIDAN	0.12	UE2-6	10-Mar-62	5	12
NE SHERIDAN	0.10	UE2-7	11-Mar-62	4	12
NE SHERMAN	1.37	TS4-6	6-May-52	9	10
NE SHERMAN	1.37	TS4-6	6-May-52	9	10
NE SHERMAN	12.24	TP4-2	8-Mar-55	9	10
NE SHERMAN	3.30	PB8-2	19-Aug-57	5	10
NE SHERMAN	19.73	PB8-3	20-Aug-57	6	10
NE SHERMAN	3.78	PB8-5	22-Aug-57	4	10
NE SHERMAN	9.15	PB11-2	1-Sep-57	5	10
NE SHERMAN	0.17	UE2-7	11-Mar-62	2	10
NE SHERMAN	345.56	S-3	8-Jul-62	8	10
NE SHERMAN	14.44	UE8-4	17-Jul-62	1	10
NE SIOUX	29.99	TS5-2	8-May-52	5	15
NE SIOUX	11.30	TS5-3	9-May-52	4	15
NE SIOUX	2.49	UpK2-7	30-Mar-53	10	15
NE SIOUX	4.00	TP3-2	2-Mar-55	5	15
NE SIOUX	14.24	TP11-2	16-May-55	7	15
NE SIOUX	14.08	PB2-3	20-Jun-57	6	15
NE SIOUX	1.96	PB5-5	19-Jul-57	9	15

COUNTY	uCi/ sqm	SHOT DAY	DATE	RANK FOR DAY	TIMES IN TOP 10
NE SIOUX	1.65	PB8-2	19-Aug-57	6	15
NE SIOUX	11.05	PB9-4	26-Aug-57	6	15
NE SIOUX	2.37	PB18-3	9-Oct-57	3	15
NE SIOUX	1.44	PB18-4	10-Oct-57	6	15
NE SIOUX	0.11	UE2-6	10-Mar-62	6	15
NE SIOUX	0.10	UE2-7	11-Mar-62	4	15
NE SIOUX	2.66	S-2	7-Jul-62	10	15
NE SIOUX	0.91	BN-2	19-Dec-70	6	15
NE STANTON	2.30	TS4-6	6-May-52	3	10
NE STANTON	2.30	TS4-6	6-May-52	3	10
NE STANTON	12.24	TP4-2	8-Mar-55	9	10
NE STANTON	19.73	PB8-3	20-Aug-57	6	10
NE STANTON	3.48	PB9-4	26-Aug-57	10	10
NE STANTON	0.80	PB13-7	12-Sep-57	7	10
NE STANTON	0.17	UE2-7	11-Mar-62	2	10
NE STANTON	527.44	S-3	8-Jul-62	5	10
NE STANTON	4.35	UE8-4	17-Jul-62	8	10
NE STANTON	0.82	BN-2	19-Dec-70	7	10
NE THAYER	12.24	TP4-2	8-Mar-55	9	5
NE THAYER	19.73	PB8-3	20-Aug-57	6	5
NE THAYER	3.55	PB13-4	9-Sep-57	5	5
NE THAYER	0.81	PB14-7	20-Sep-57	9	5
NE THAYER	10.54	UE8-4	17-Jul-62	2	5
NE THOMAS	1.44	BJ5-3	21-Nov-51	6	18
NE THOMAS	1.31	UpK6-3	20-Apr-53	9	18
NE THOMAS	7.78	TP3-2	2-Mar-55	3	18
NE THOMAS	49.44	TP4-2	8-Mar-55	4	18
NE THOMAS	32.33	TP10-2	6-May-55	5	18
NE THOMAS	15.08	PB2-3	20-Jun-57	5	18
NE THOMAS	5.89	PB5-4	18-Jul-57	5	18
NE THOMAS	1.63	PB5-5	19-Jul-57	10	18
NE THOMAS	8.25	PB8-2	19-Aug-57	3	18
NE THOMAS	19.73	PB8-3	20-Aug-57	6	18
NE THOMAS	3.78	PB8-5	22-Aug-57	4	18
NE THOMAS	5.22	PB9-4	26-Aug-57	8	18

COUNTY	uCi/ sqm	SHOT DAY	DATE	RANK FOR DAY	TIMES IN TOP 10
NE THOMAS	4.58	PB11-2	1-Sep-57	9	18
NE THOMAS	2.50	PB15-3	18-Sep-57	6	18
NE THOMAS	2.37	PB18-3	9-Oct-57	3	18
NE THOMAS	1.44	PB18-4	10-Oct-57	6	18
NE THOMAS	0.10	UE2-7	11-Mar-62	3	18
NE THOMAS	0.64	BN-2	19-Dec-70	10	18
NE THURSTON	2.30	TS4-6	6-May-52	3	11
NE THURSTON	2.30	TS4-6	6-May-52	3	11
NE THURSTON	3.97	UpK2-2	25-Mar-53	10	11
NE THURSTON	4.43	UpK9-5	23-May-53	6	11
NE THURSTON	1.30	PB2-4	21-Jun-57	9	11
NE THURSTON	1.10	PB4-3	7-Jul-57	7	11
NE THURSTON	1.86	PB4-10	14-Jul-57	8	11
NE THURSTON	2.01	PB5-5	19-Jul-57	8	11
NE THURSTON	0.80	PB13-7	12-Sep-57	7	11
NE THURSTON	0.17	UE2-7	11-Mar-62	2	11
NE THURSTON	527.44	S-3	8-Jul-62	5	11
NE VALLEY	1.37	TS4-6	6-May-52	9	10
NE VALLEY	1.37	TS4-6	6-May-52	9	10
NE VALLEY	12.24	TP4-2	8-Mar-55	9	10
NE VALLEY	1.65	PB8-2	19-Aug-57	6	10
NE VALLEY	19.73	PB8-3	20-Aug-57	6	10
NE VALLEY	3.78	PB8-5	22-Aug-57	4	10
NE VALLEY	9.15	PB11-2	1-Sep-57	5	10
NE VALLEY	6.68	PB15-2	17-Sep-57	5	10
NE VALLEY	2.50	PB15-3	18-Sep-57	6	10
NE VALLEY	0.10	UE2-7	11-Mar-62	4	10
NE WASHINGTON	4.22	UpK9-5	23-May-53	8	8
NE WASHINGTON	12.24	TP4-2	8-Mar-55	9	8
NE WASHINGTON	1.30	PB2-4	21-Jun-57	9	8
NE WASHINGTON	1.10	PB4-3	7-Jul-57	7	8
NE WASHINGTON	6.09	PB5-5	19-Jul-57	2	8
NE WASHINGTON	0.17	UE2-7	11-Mar-62	2	8
NE WASHINGTON	345.56	S-3	8-Jul-62	8	8
NE WASHINGTON	0.12	UE7-5	15-Jul-62	9	8

COUNTY	uCi/ sqm	SHOT DAY	DATE	RANK FOR DAY	TIMES IN TOP 10
NE WAYNE	1.32	TS3-8	29-Apr-52	7	11
NE WAYNE	1.32	TS3-8	29-Apr-52	7	11
NE WAYNE	2.30	TS4-6	6-May-52	3	11
NE WAYNE	2.30	TS4-6	6-May-52	3	11
NE WAYNE	3.97	UpK2-2	25-Mar-53	10	11
NE WAYNE	1.30	PB2-4	21-Jun-57	9	11
NE WAYNE	1.10	PB4-3	7-Jul-57	7	11
NE WAYNE	1.86	PB4-10	14-Jul-57	8	11
NE WAYNE	19.73	PB8-3	20-Aug-57	6	11
NE WAYNE	3.48	PB9-4	26-Aug-57	10	11
NE WAYNE	0.10	UE2-7	11-Mar-62	4	11
NE WEBSTER	12.24	TP4-2	8-Mar-55	9	10
NE WEBSTER	2.33	PB3-3	26-Jun-57	8	10
NE WEBSTER	1.65	PB8-2	19-Aug-57	6	10
NE WEBSTER	19.73	PB8-3	20-Aug-57	6	10
NE WEBSTER	1.92	PB8-5	22-Aug-57	10	10
NE WEBSTER	3.55	PB13-4	9-Sep-57	5	10
NE WEBSTER	0.89	PB13-7	12-Sep-57	5	10
NE WEBSTER	1.62	PB14-7	20-Sep-57	4	10
NE WEBSTER	0.10	UE2-7	11-Mar-62	4	10
NE WEBSTER	8.02	UE8-4	17-Jul-62	3	10
NE WHEELER	1.32	TS3-8	29-Apr-52	7	13
NE WHEELER	1.32	TS3-8	29-Apr-52	7	13
NE WHEELER	1.37	TS4-6	6-May-52	9	13
NE WHEELER	1.37	TS4-6	6-May-52	9	13
NE WHEELER	3.97	UpK2-2	25-Mar-53	10	13
NE WHEELER	19.73	PB8-3	20-Aug-57	6	13
NE WHEELER	3.78	PB8-5	22-Aug-57	4	13
NE WHEELER	4.58	PB11-2	1-Sep-57	9	13
NE WHEELER	6.68	PB15-2	17-Sep-57	5	13
NE WHEELER	4.13	PB15-3	18-Sep-57	2	13
NE WHEELER	0.10	UE2-7	11-Mar-62	3	13
NE WHEELER	8.02	UE8-4	17-Jul-62	3	13
NE WHEELER	0.82	BN-2	19-Dec-70	7	13
NE YORK	1.32	TS3-8	29-Apr-52	7	11

COUNTY	uCi/ sqm	SHOT DAY	DATE	RANK FOR DAY	TIMES IN TOP 10
NE YORK	1.32	TS3-8	29-Apr-52	7	11
NE YORK	12.24	TP4-2	8-Mar-55	9	11
NE YORK	1.76	PB2-4	21-Jun-57	6	11
NE YORK	1.63	PB5-5	19-Jul-57	10	11
NE YORK	19.73	PB8-3	20-Aug-57	6	11
NE YORK	12.75	PB13-3	8-Sep-57	4	11
NE YORK	5.36	PB13-4	9-Sep-57	3	11
NE YORK	1.62	PB14-7	20-Sep-57	4	11
NE YORK	0.10	UE2-7	11-Mar-62	4	11
NE YORK	4.12	UE8-4	17-Jul-62	9	11

COUNTY	uCi/ sqm	SHOT DAY	DATE	RANK FOR DAY	TIMES IN TOP 10
NH BELKNAP	0.99	R1-3	30-Jan-51	3	9
NH BELKNAP	7.98	BJ2-2	31-Oct-51	4	9
NH BELKNAP	2.34	UpK4-3	8-Apr-53	6	9
NH BELKNAP	1.99	UpK4-4	9-Apr-53	4	9
NH BELKNAP	1.06	UpK4-5	10-Apr-53	1	9
NH BELKNAP	0.97	TP6-5	26-Mar-55	8	9
NH BELKNAP	1.36	PB5-6	20-Jul-57	6	9
NH BELKNAP	5.97	PB5-8	22-Jul-57	3	9
NH BELKNAP	0.12	UE7-5	15-Jul-62	9	9
NH CARROLL	37.60	R1-2	29-Jan-51	6	9
NH CARROLL	12.07	BJ2-2	31-Oct-51	1	9
NH CARROLL	2.34	UpK4-3	8-Apr-53	6	9
NH CARROLL	3.98	UpK4-4	9-Apr-53	2	9
NH CARROLL	0.99	UpK5-5	15-Apr-53	2	9
NH CARROLL	0.97	TP6-5	26-Mar-55	8	9
NH CARROLL	1.36	PB5-6	20-Jul-57	6	9
NH CARROLL	1.51	PB5-8	22-Jul-57	7	9
NH CARROLL	0.12	UE7-5	15-Jul-62	9	9
NH CHESHIRE	2.34	UpK4-3	8-Apr-53	6	8
NH CHESHIRE	3.98	UpK4-4	9-Apr-53	2	8
NH CHESHIRE	1.06	UpK4-5	10-Apr-53	1	8
NH CHESHIRE	0.97	TP6-5	26-Mar-55	8	8
NH CHESHIRE	1.36	PB5-6	20-Jul-57	6	8
NH CHESHIRE	5.97	PB5-8	22-Jul-57	3	8
NH CHESHIRE	1.27	PB6-8	31-Jul-57	9	8
NH CHESHIRE	0.24	UE7-5	15-Jul-62	5	8
NH COOS	1.17	R1-3	30-Jan-51	2	8
NH COOS	7.98	BJ2-2	31-Oct-51	4	8
NH COOS	2.34	UpK4-3	8-Apr-53	6	8
NH COOS	3.98	UpK4-4	9-Apr-53	2	8
NH COOS	0.97	TP6-5	26-Mar-55	8	8
NH COOS	2.68	PB5-6	20-Jul-57	3	8
NH COOS	8.96	PB5-8	22-Jul-57	2	8
NH COOS	0.33	UE7-5	15-Jul-62	3	8
NH GRAFTON	0.99	R1-3	30-Jan-51	3	11

COUNTY	uCi/ sqm	SHOT DAY	DATE	RANK FOR DAY	TIMES IN TOP 10
NH GRAFTON	7.98	BJ2-2	31-Oct-51	4	11
NH GRAFTON	25.19	TS8-2	6-Jun-52	4	11
NH GRAFTON	2.34	UpK4-3	8-Apr-53	6	11
NH GRAFTON	3.98	UpK4-4	9-Apr-53	2	11
NH GRAFTON	0.99	UpK5-5	15-Apr-53	2	11
NH GRAFTON	0.97	TP6-5	26-Mar-55	8	11
NH GRAFTON	1.14	PB2-6	23-Jun-57	7	11
NH GRAFTON	2.68	PB5-6	20-Jul-57	3	11
NH GRAFTON	5.97	PB5-8	22-Jul-57	3	11
NH GRAFTON	0.33	UE7-5	15-Jul-62	3	11
NH HILLSBOROUGH	2.34	UpK4-3	8-Apr-53	6	8
NH HILLSBOROUGH	3.98	UpK4-4	9-Apr-53	2	8
NH HILLSBOROUGH	1.06	UpK4-5	10-Apr-53	1	8
NH HILLSBOROUGH	0.99	UpK5-5	15-Apr-53	2	8
NH HILLSBOROUGH	0.97	TP6-5	26-Mar-55	8	8
NH HILLSBOROUGH	1.36	PB5-6	20-Jul-57	6	8
NH HILLSBOROUGH	5.97	PB5-8	22-Jul-57	3	8
NH HILLSBOROUGH	0.17	UE7-5	15-Jul-62	7	8
NH MERRIMACK	0.99	R1-3	30-Jan-51	3	7
NH MERRIMACK	2.34	UpK4-3	8-Apr-53	6	7
NH MERRIMACK	3.98	UpK4-4	9-Apr-53	2	7
NH MERRIMACK	1.06	UpK4-5	10-Apr-53	1	7
NH MERRIMACK	0.99	UpK5-5	15-Apr-53	2	7
NH MERRIMACK	2.99	PB5-8	22-Jul-57	4	7
NH MERRIMACK	0.17	UE7-5	15-Jul-62	7	7
NH ROCKINGHAM	37.60	R1-2	29-Jan-51	6	9
NH ROCKINGHAM	2.34	UpK4-3	8-Apr-53	6	9
NH ROCKINGHAM	3.98	UpK4-4	9-Apr-53	2	9
NH ROCKINGHAM	1.06	UpK4-5	10-Apr-53	1	9
NH ROCKINGHAM	0.99	UpK5-5	15-Apr-53	2	9
NH ROCKINGHAM	0.97	TP6-5	26-Mar-55	8	9
NH ROCKINGHAM	1.36	PB5-6	20-Jul-57	6	9
NH ROCKINGHAM	2.99	PB5-8	22-Jul-57	4	9
NH ROCKINGHAM	1.27	PB6-8	31-Jul-57	9	9
NH STRAFFORD	37.60	R1-2	29-Jan-51	6	7

COUNTY	uCi/ sqm	SHOT DAY	DATE	RANK FOR DAY	TIMES IN TOP 10
NH STRAFFORD	2.34	UpK4-3	8-Apr-53	6	7
NH STRAFFORD	1.99	UpK4-4	9-Apr-53	4	7
NH STRAFFORD	1.06	UpK4-5	10-Apr-53	1	7
NH STRAFFORD	0.97	TP6-5	26-Mar-55	8	7
NH STRAFFORD	1.36	PB5-6	20-Jul-57	6	7
NH STRAFFORD	1.51	PB5-8	22-Jul-57	7	7
NH SULLIVAN	1.42	R1-3	30-Jan-51	1	7
NH SULLIVAN	2.34	UpK4-3	8-Apr-53	6	7
NH SULLIVAN	1.99	UpK4-4	9-Apr-53	4	7
NH SULLIVAN	1.06	UpK4-5	10-Apr-53	1	7
NH SULLIVAN	1.45	TP6-5	26-Mar-55	5	7
NH SULLIVAN	1.36	PB5-6	20-Jul-57	6	7
NH SULLIVAN	2.99	PB5-8	22-Jul-57	4	7

COUNTY	uCi/ sqm	SHOT DAY	DATE	RANK FOR DAY	TIMES IN TOP 10
NJ ATLANTIC	1.65	BJ4-6	10-Nov-51	2	5
NJ ATLANTIC	54.77	UpK1-2	18-Mar-53	5	5
NJ ATLANTIC	2.03	TP6-4	25-Mar-55	8	5
NJ ATLANTIC	2.99	PB5-8	22-Jul-57	4	5
NJ ATLANTIC	0.95	PB18-13	19-Oct-57	2	5
NJ BERGEN	0.86	BJ4-9	13-Nov-51	1	4
NJ BERGEN	3.64	TP11-8	22-May-55	4	4
NJ BERGEN	10.79	TP11-9	23-May-55	1	4
NJ BERGEN	2.99	PB5-8	22-Jul-57	4	4
NJ BURLINGTON	3.98	BJ4-6	10-Nov-51	1	6
NJ BURLINGTON	52.08	UpK1-2	18-Mar-53	7	6
NJ BURLINGTON	3.11	TP6-4	25-Mar-55	3	6
NJ BURLINGTON	2.99	PB5-8	22-Jul-57	4	6
NJ BURLINGTON	0.95	PB18-13	19-Oct-57	2	6
NJ BURLINGTON	47.25	S-5	10-Jul-62	2	6
NJ CAMDEN	3.98	BJ4-6	10-Nov-51	1	4
NJ CAMDEN	2.03	TP6-4	25-Mar-55	8	4
NJ CAMDEN	2.99	PB5-8	22-Jul-57	4	4
NJ CAMDEN	1.42	PB18-13	19-Oct-57	1	4
NJ CAPE MAY	3.98	BJ4-6	10-Nov-51	1	4
NJ CAPE MAY	1.42	UpK3-7	6-Apr-53	8	4
NJ CAPE MAY	2.03	TP6-4	25-Mar-55	8	4
NJ CAPE MAY	8.96	PB5-8	22-Jul-57	2	4
NJ CUMBERLAND	3.98	BJ4-6	10-Nov-51	1	4
NJ CUMBERLAND	2.03	TP6-4	25-Mar-55	8	4
NJ CUMBERLAND	2.99	PB5-8	22-Jul-57	4	4
NJ CUMBERLAND	1.42	PB18-13	19-Oct-57	1	4
NJ ESSEX	11.62	BJ3-7	7-Nov-51	3	5
NJ ESSEX	2.43	TP11-8	22-May-55	6	5
NJ ESSEX	7.19	TP11-9	23-May-55	3	5
NJ ESSEX	2.99	PB5-8	22-Jul-57	4	5
NJ ESSEX	1.42	PB18-13	19-Oct-57	1	5
NJ GLOUCESTER	3.98	BJ4-6	10-Nov-51	1	5
NJ GLOUCESTER	1.42	UpK3-7	6-Apr-53	8	5
NJ GLOUCESTER	2.03	TP6-4	25-Mar-55	8	5

COUNTY	uCi/ sqm	SHOT DAY	DATE	RANK FOR DAY	TIMES IN TOP 10
NJ GLOUCESTER	5.97	PB5-8	22-Jul-57	3	5
NJ GLOUCESTER	0.95	PB18-13	19-Oct-57	2	5
NJ HUDSON	0.86	BJ4-9	13-Nov-51	1	5
NJ HUDSON	3.64	TP11-8	22-May-55	4	5
NJ HUDSON	10.79	TP11-9	23-May-55	1	5
NJ HUDSON	1.37	PB5-7	21-Jul-57	10	5
NJ HUDSON	2.99	PB5-8	22-Jul-57	4	5
NJ HUNTERDON	0.86	BJ4-9	13-Nov-51	1	5
NJ HUNTERDON	1.42	UpK3-7	6-Apr-53	8	5
NJ HUNTERDON	3.11	TP6-4	25-Mar-55	3	5
NJ HUNTERDON	5.97	PB5-8	22-Jul-57	3	5
NJ HUNTERDON	0.95	PB18-13	19-Oct-57	2	5
NJ MERCER	3.98	BJ4-6	10-Nov-51	1	4
NJ MERCER	1.42	UpK3-7	6-Apr-53	8	4
NJ MERCER	3.11	TP6-4	25-Mar-55	3	4
NJ MERCER	5.97	PB5-8	22-Jul-57	3	4
NJ MIDDLESEX	3.85	BJ3-7	7-Nov-51	9	5
NJ MIDDLESEX	48.54	UpK1-2	18-Mar-53	9	5
NJ MIDDLESEX	2.15	TP6-4	25-Mar-55	7	5
NJ MIDDLESEX	5.97	PB5-8	22-Jul-57	3	5
NJ MIDDLESEX	1.42	PB18-13	19-Oct-57	1	5
NJ MONMOUTH	0.86	BJ4-9	13-Nov-51	1	8
NJ MONMOUTH	64.09	UpK1-2	18-Mar-53	1	8
NJ MONMOUTH	2.15	TP6-4	25-Mar-55	7	8
NJ MONMOUTH	1.31	TP7-8	5-Apr-55	4	8
NJ MONMOUTH	5.97	PB5-8	22-Jul-57	3	8
NJ MONMOUTH	1.57	PB6-7	30-Jul-57	9	8
NJ MONMOUTH	0.95	PB18-13	19-Oct-57	2	8
NJ MONMOUTH	20.31	S-6	11-Jul-62	3	8
NJ MORRIS	19.31	BJ3-7	7-Nov-51	1	4
NJ MORRIS	2.71	TP11-9	23-May-55	10	4
NJ MORRIS	2.99	PB5-8	22-Jul-57	4	4
NJ MORRIS	1.42	PB18-13	19-Oct-57	1	4
NJ OCEAN	3.98	BJ4-6	10-Nov-51	1	5
NJ OCEAN	62.45	UpK1-2	18-Mar-53	2	5

COUNTY	uCi/ sqm	SHOT DAY	DATE	RANK FOR DAY	TIMES IN TOP 10
NJ OCEAN	2.15	TP6-4	25-Mar-55	7	5
NJ OCEAN	1.97	TP7-8	5-Apr-55	2	5
NJ OCEAN	5.97	PB5-8	22-Jul-57	3	5
NJ PASSAIC	11.62	BJ3-7	7-Nov-51	3	4
NJ PASSAIC	2.71	TP11-9	23-May-55	10	4
NJ PASSAIC	2.99	PB5-8	22-Jul-57	4	4
NJ PASSAIC	1.42	PB18-13	19-Oct-57	1	4
NJ SALEM	3.98	BJ4-6	10-Nov-51	1	5
NJ SALEM	1.42	UpK3-7	6-Apr-53	8	5
NJ SALEM	2.03	TP6-4	25-Mar-55	8	5
NJ SALEM	2.99	PB5-8	22-Jul-57	4	5
NJ SALEM	1.42	PB18-13	19-Oct-57	1	5
NJ SOMERSET	3.85	BJ3-7	7-Nov-51	9	6
NJ SOMERSET	0.86	BJ4-9	13-Nov-51	1	6
NJ SOMERSET	1.42	UpK3-7	6-Apr-53	8	6
NJ SOMERSET	1.21	TP11-8	22-May-55	10	6
NJ SOMERSET	2.99	PB5-8	22-Jul-57	4	6
NJ SOMERSET	0.95	PB18-13	19-Oct-57	2	6
NJ SUSSEX	7.70	BJ3-7	7-Nov-51	6	4
NJ SUSSEX	0.86	BJ4-9	13-Nov-51	1	4
NJ SUSSEX	5.97	PB5-8	22-Jul-57	3	4
NJ SUSSEX	1.42	PB18-13	19-Oct-57	1	4
NJ UNION	3.85	BJ3-7	7-Nov-51	9	6
NJ UNION	0.86	BJ4-9	13-Nov-51	1	6
NJ UNION	2.43	TP11-8	22-May-55	6	6
NJ UNION	7.19	TP11-9	23-May-55	3	6
NJ UNION	2.99	PB5-8	22-Jul-57	4	6
NJ UNION	0.95	PB18-13	19-Oct-57	2	6
NJ WARREN	11.62	BJ3-7	7-Nov-51	3	5
NJ WARREN	1.42	UpK3-7	6-Apr-53	8	5
NJ WARREN	3.11	TP6-4	25-Mar-55	3	5
NJ WARREN	2.99	PB5-8	22-Jul-57	4	5
NJ WARREN	1.42	PB18-13	19-Oct-57	1	5

COUNTY	uCi/ sqm	SHOT DAY	DATE	RANK FOR DAY	TIMES IN TOP 10
NM BERNALILLO	104.09	UpK6-1	18-Apr-53	10	4
NM BERNALILLO	67.97	PB4-1	5-Jul-57	7	4
NM BERNALILLO	67.97	PB4-1	5-Jul-57	7	4
NM BERNALILLO	13.80	PB17-1	28-Sep-57	7	4
NM CHAVES	377.22	UpK7-1	25-Apr-53	5	1
NM COLFAX	130.08	TP11-1	15-May-55	5	3
NM COLFAX	73.63	PB4-1	5-Jul-57	4	3
NM COLFAX	73.63	PB4-1	5-Jul-57	4	3
NM GUADALUPE	54.82	PB4-1	5-Jul-57	9	2
NM GUADALUPE	54.82	PB4-1	5-Jul-57	9	2
NM HARDING	73.63	PB4-1	5-Jul-57	4	2
NM HARDING	73.63	PB4-1	5-Jul-57	4	2
NM LINCOLN	319.99	UpK7-1	25-Apr-53	9	1
NM LOS ALAMOS	54.41	PB4-1	5-Jul-57	10	2
NM LOS ALAMOS	54.41	PB4-1	5-Jul-57	10	2
NM MCKINLEY	4.43	TS4-1	1-May-52	8	7
NM MCKINLEY	4.43	TS4-1	1-May-52	8	7
NM MCKINLEY	202.90	UpK6-1	18-Apr-53	6	7
NM MCKINLEY	419.14	UpK7-1	25-Apr-53	4	7
NM MCKINLEY	33.80	PB3-1	24-Jun-57	6	7
NM MCKINLEY	8.43	PB13-3	8-Sep-57	7	7
NM MCKINLEY	9.82	PB18-1	7-Oct-57	10	7
NM MORA	73.09	PB4-1	5-Jul-57	5	2
NM MORA	73.09	PB4-1	5-Jul-57	5	2
NM QUAY	55.22	PB4-1	5-Jul-57	8	2
NM QUAY	55.22	PB4-1	5-Jul-57	8	2
NM SAN JUAN	4.43	TS4-1	1-May-52	8	5
NM SAN JUAN	4.43	TS4-1	1-May-52	8	5
NM SAN JUAN	27.50	UpK1-1	17-Mar-53	10	5
NM SAN JUAN	23.15	PB3-1	24-Jun-57	10	5
NM SAN JUAN	9.82	PB18-1	7-Oct-57	10	5
NM SAN MIGUEL	73.09	PB4-1	5-Jul-57	5	2
NM SAN MIGUEL	73.09	PB4-1	5-Jul-57	5	2
NM SANDOVAL	150.48	UpK6-1	18-Apr-53	9	4
NM SANDOVAL	324.64	UpK7-1	25-Apr-53	7	4

COUNTY	uCi/ sqm	SHOT DAY	DATE	RANK FOR DAY	TIMES IN TOP 10
NM SANDOVAL	54.41	PB4-1	5-Jul-57	10	4
NM SANDOVAL	54.41	PB4-1	5-Jul-57	10	4
NM SANTA FE	322.26	UpK7-1	25-Apr-53	8	3
NM SANTA FE	73.09	PB4-1	5-Jul-57	5	3
NM SANTA FE	73.09	PB4-1	5-Jul-57	5	3
NM TAOS	55.22	PB4-1	5-Jul-57	8	2
NM TAOS	55.22	PB4-1	5-Jul-57	8	2
NM TORRANCE	72.55	PB4-1	5-Jul-57	6	2
NM TORRANCE	72.55	PB4-1	5-Jul-57	6	2
NM UNION	130.08	TP11-1	15-May-55	6	3
NM UNION	73.63	PB4-1	5-Jul-57	4	3
NM UNION	73.63	PB4-1	5-Jul-57	4	3
NM VALENCIA	4.29	TS4-1	1-May-52	10	5
NM VALENCIA	4.29	TS4-1	1-May-52	10	5
NM VALENCIA	202.90	UpK6-1	18-Apr-53	7	5
NM VALENCIA	251.46	UpK7-1	25-Apr-53	10	5
NM VALENCIA	8.43	PB13-3	8-Sep-57	7	5

COUNTY	uCi/ sqm	SHOT DAY	DATE	RANK FOR DAY	TIMES IN TOP 10
NV CARSON CITY	238.37	PB1-1	28-May-57	7	5
NV CARSON CITY	43.57	PB6-1	24-Jul-57	9	5
NV CARSON CITY	39.80	PB14-1	14-Sep-57	2	5
NV CARSON CITY	45.88	PB16-1	23-Sep-57	8	5
NV CARSON CITY	1.04	BN-1	18-Dec-70	9	5
NV CHURCHILL	279.25	PB1-1	28-May-57	5	5
NV CHURCHILL	121.94	PB6-1	24-Jul-57	4	5
NV CHURCHILL	15.92	PB14-1	14-Sep-57	4	5
NV CHURCHILL	91.75	PB16-1	23-Sep-57	3	5
NV CHURCHILL	0.99	BN-1	18-Dec-70	10	5
NV CLARK1	51.57	UpK1-1	17-Mar-53	6	10
NV CLARK1	311.25	UpK6-1	18-Apr-53	2	10
NV CLARK1	2096.29	UpK7-1	25-Apr-53	1	10
NV CLARK1	87.97	UpK11-1	4-Jun-53	2	10
NV CLARK1	1.99	TP2-1	22-Feb-55	2	10
NV CLARK1	130.07	TP5-1	12-Mar-55	1	10
NV CLARK1	25.17	TP6-2	23-Mar-55	1	10
NV CLARK1	387.19	TP11-1	15-May-55	2	10
NV CLARK1	12.01	PB5-2	16-Jul-57	5	10
NV CLARK1	1.39	UE9	19-Oct-62	4	10
NV CLARK2	245.47	UpK6-1	18-Apr-53	5	7
NV CLARK2	85.77	TP2-1	22-Feb-55	1	7
NV CLARK2	53.66	TP6-1	22-Mar-55	4	7
NV CLARK2	1.39	HT7-1	22-Oct-58	1	7
NV CLARK2	1.41	HT9-1	26-Oct-58	2	7
NV CLARK2	2.98	UE9	19-Oct-62	3	7
NV CLARK2	0.99	UE10-1	13-Mar-64	1	7
NV CLARK3	116.41	TP6-1	22-Mar-55	1	1
NV DOUGLAS	238.37	PB1-1	28-May-57	8	5
NV DOUGLAS	55.22	PB6-1	24-Jul-57	7	5
NV DOUGLAS	39.80	PB14-1	14-Sep-57	2	5
NV DOUGLAS	45.88	PB16-1	23-Sep-57	8	5
NV DOUGLAS	1.04	BN-1	18-Dec-70	9	5
NV ELKO	105.23	BJ5-1	19-Nov-51	3	8
NV ELKO	8.56	BJ6-1	29-Nov-51	6	8

COUNTY	uCi/ sqm	SHOT DAY	DATE	RANK FOR DAY	TIMES IN TOP 10
NV ELKO	8.56	BJ6-1	29-Nov-51	6	8
NV ELKO	184.19	TS7-1	1-Jun-52	2	8
NV ELKO	15.18	PB6-2	25-Jul-57	6	8
NV ELKO	55.23	PB8-1	18-Aug-57	7	8
NV ELKO	20.12	PB17-1	28-Sep-57	3	8
NV ELKO	0.99	BN-1	18-Dec-70	10	8
NV ESMERALDA1	256.26	PB6-1	24-Jul-57	2	2
NV ESMERALDA1	107.32	PB16-1	23-Sep-57	2	2
NV ESMERALDA2	328.70	PB6-1	24-Jul-57	1	4
NV ESMERALDA2	14.05	PB13-1	6-Sep-57	1	4
NV ESMERALDA2	274.35	PB16-1	23-Sep-57	1	4
NV ESMERALDA2	6.70	HT9-1	26-Oct-58	1	4
NV EUREKA	106.82	BJ5-1	19-Nov-51	2	8
NV EUREKA	4.65	BJ6-1	29-Nov-51	8	8
NV EUREKA	4.65	BJ6-1	29-Nov-51	8	8
NV EUREKA	231.27	TS7-1	1-Jun-52	1	8
NV EUREKA	15.08	PB6-2	25-Jul-57	7	8
NV EUREKA	482.83	PB8-1	18-Aug-57	1	8
NV EUREKA	3.54	S-1	6-Jul-62	10	8
NV EUREKA	4.39	BN-1	18-Dec-70	2	8
NV HUMBOLDT	10.12	PB17-1	28-Sep-57	10	2
NV HUMBOLDT	3.88	BN-1	18-Dec-70	3	2
NV LANDER1	56.41	BJ5-1	19-Nov-51	6	3
NV LANDER1	36.74	PB16-1	23-Sep-57	10	3
NV LANDER1	0.99	BN-1	18-Dec-70	10	3
NV LANDER2	283.24	PB1-1	28-May-57	3	3
NV LANDER2	78.71	PB16-1	23-Sep-57	5	3
NV LANDER2	4.39	BN-1	18-Dec-70	2	3
NV LINCOLN1	302.36	TS6-1	25-May-52	3	12
NV LINCOLN1	119.55	UpK1-1	17-Mar-53	4	12
NV LINCOLN1	1716.20	UpK9-1	19-May-53	7	12
NV LINCOLN1	49.78	TP3-1	1-Mar-55	6	12
NV LINCOLN1	31.35	TP5-1	12-Mar-55	8	12
NV LINCOLN1	7.16	TP6-2	23-Mar-55	3	12
NV LINCOLN1	74.08	TP7-1	29-Mar-55	5	12

COUNTY	uCi/ sqm	SHOT DAY	DATE	RANK FOR DAY	TIMES IN TOP 10
NV LINCOLN1	449.00	TP9-1	15-Apr-55	1	12
NV LINCOLN1	56.77	PB3-1	24-Jun-57	4	12
NV LINCOLN1	393.35	PB11-1	31-Aug-57	5	12
NV LINCOLN1	0.26	HT1-1	19-Sep-58	3	12
NV LINCOLN1	0.37	HT02-1	5-Oct-58	2	12
NV LINCOLN2	85.19	BJ6-1	29-Nov-51	2	15
NV LINCOLN2	85.19	BJ6-1	29-Nov-51	2	15
NV LINCOLN2	521.62	TS6-1	25-May-52	1	15
NV LINCOLN2	57.19	TP3-1	1-Mar-55	5	15
NV LINCOLN2	344.83	TP7-1	29-Mar-55	1	15
NV LINCOLN2	216.79	TP9-1	15-Apr-55	7	15
NV LINCOLN2	21.88	PB2-1	18-Jun-57	5	15
NV LINCOLN2	209.95	PB5-1	15-Jul-57	3	15
NV LINCOLN2	0.58	HT1-1	19-Sep-58	1	15
NV LINCOLN2	2.09	HT02-1	5-Oct-58	1	15
NV LINCOLN2	1.45	HT6-1	18-Oct-58	2	15
NV LINCOLN2	0.12	HT8-1	22-Oct-58	2	15
NV LINCOLN2	25.53	UE8-1	14-Jul-62	3	15
NV LINCOLN2	3.28	UE9	19-Oct-62	2	15
NV LINCOLN2	31.85	UE13-1	25-Apr-66	5	15
NV LYON	233.11	PB1-1	28-May-57	10	5
NV LYON	69.69	PB6-1	24-Jul-57	6	5
NV LYON	39.80	PB14-1	14-Sep-57	2	5
NV LYON	45.88	PB16-1	23-Sep-57	8	5
NV LYON	1.04	BN-1	18-Dec-70	9	5
NV MINERAL	282.27	PB1-1	28-May-57	4	6
NV MINERAL	184.78	PB6-1	24-Jul-57	3	6
NV MINERAL	9.07	PB13-1	6-Sep-57	10	6
NV MINERAL	57.61	PB14-1	14-Sep-57	1	6
NV MINERAL	86.53	PB16-1	23-Sep-57	4	6
NV MINERAL	2.16	BN-1	18-Dec-70	5	6
NV NYE1	519.96	PB1-1	28-May-57	2	5
NV NYE1	7.89	PB14-1	14-Sep-57	6	5
NV NYE1	39.08	PB16-1	23-Sep-57	9	5
NV NYE1	4.81	UE12-1	14-Apr-65	2	5

COUNTY	uCi/ sqm	SHOT DAY	DATE	RANK FOR DAY	TIMES IN TOP 10
NV NYE1	12.40	UE16-1	8-Dec-68	8	5
NV NYE2	152.95	BJ5-1	19-Nov-51	1	32
NV NYE2	79.78	BJ6-1	29-Nov-51	3	32
NV NYE2	79.78	BJ6-1	29-Nov-51	3	32
NV NYE2	138.58	TS7-1	1-Jun-52	3	32
NV NYE2	26.47	UpK10-1	25-May-53	10	32
NV NYE2	334.71	TP4-1	7-Mar-55	2	32
NV NYE2	326.92	TP10-1	5-May-55	2	32
NV NYE2	766.45	PB1-1	28-May-57	1	32
NV NYE2	544.75	PB5-1	15-Jul-57	1	32
NV NYE2	17.91	PB6-2	25-Jul-57	2	32
NV NYE2	254.90	PB8-1	18-Aug-57	4	32
NV NYE2	100.99	PB12-1	2-Sep-57	1	32
NV NYE2	12.09	PB14-1	14-Sep-57	5	32
NV NYE2	0.27	HT1-1	19-Sep-58	2	32
NV NYE2	6.35	HT4-1	13-Oct-58	1	32
NV NYE2	0.08	HT5-1	17-Oct-58	1	32
NV NYE2	5.12	HT6-1	18-Oct-58	1	32
NV NYE2	0.12	HT8-1	22-Oct-58	2	32
NV NYE2	2.51	UE-1	15-Sep-61	1	32
NV NYE2	0.55	UE-2-1	5-Mar-62	1	32
NV NYE2	55.90	UE3	14-Apr-62	1	32
NV NYE2	6.63	UE4	19-May-62	1	32
NV NYE2	104.73	UE5	13-Jun-62	1	32
NV NYE2	120.28	S-1	6-Jul-62	2	32
NV NYE2	9.59	UE7-1	11-Jul-62	4	32
NV NYE2	0.46	UE9	19-Oct-62	5	32
NV NYE2	0.17	UE11-1	18-Dec-64	1	32
NV NYE2	5.62	UE12-1	14-Apr-65	1	32
NV NYE2	3.66	UE14-1	26-Jan-68	1	32
NV NYE2	0.98	UE15-1	12-Mar-68	1	32
NV NYE2	71.49	UE16-1	8-Dec-68	1	32
NV NYE2	11.08	BN-1	18-Dec-70	1	32
NV NYE3	20.33	PB2-1	18-Jun-57	8	3
NV NYE3	7.03	HT3-1	10-Oct-58	1	3

COUNTY	uCi/ sqm	SHOT DAY	DATE	RANK FOR DAY	TIMES IN TOP 10
NV NYE3	46.78	UE9	19-Oct-62	1	3
NV PERSHING	27.80	UpK10-1	25-May-53	9	4
NV PERSHING	37.35	PB6-1	24-Jul-57	10	4
NV PERSHING	54.49	PB16-1	23-Sep-57	6	4
NV PERSHING	0.99	BN-1	18-Dec-70	10	4
NV STOREY	237.50	PB1-1	28-May-57	9	4
NV STOREY	39.80	PB14-1	14-Sep-57	2	4
NV STOREY	45.88	PB16-1	23-Sep-57	8	4
NV STOREY	1.04	BN-1	18-Dec-70	9	4
NV WASHOE	253.86	PB1-1	28-May-57	6	5
NV WASHOE	37.35	PB6-1	24-Jul-57	10	5
NV WASHOE	39.80	PB14-1	14-Sep-57	2	5
NV WASHOE	45.88	PB16-1	23-Sep-57	8	5
NV WASHOE	0.99	BN-1	18-Dec-70	10	5
NV WHITE PINE1	40.00	BJ5-1	19-Nov-51	10	13
NV WHITE PINE1	46.98	BJ6-1	29-Nov-51	4	13
NV WHITE PINE1	46.98	BJ6-1	29-Nov-51	4	13
NV WHITE PINE1	365.94	TS5-1	7-May-52	4	13
NV WHITE PINE1	246.10	TP10-1	5-May-55	4	13
NV WHITE PINE1	105.36	PB5-1	15-Jul-57	5	13
NV WHITE PINE1	12.22	PB6-2	25-Jul-57	10	13
NV WHITE PINE1	249.74	PB8-1	18-Aug-57	5	13
NV WHITE PINE1	52.61	PB12-1	2-Sep-57	2	13
NV WHITE PINE1	10.37	PB17-1	28-Sep-57	9	13
NV WHITE PINE1	77.06	S-1	6-Jul-62	4	13
NV WHITE PINE1	12.72	UE16-1	8-Dec-68	7	13
NV WHITE PINE1	3.09	BN-1	18-Dec-70	4	13
NV WHITE PINE2	40.22	BJ5-1	19-Nov-51	9	16
NV WHITE PINE2	88.82	BJ6-1	29-Nov-51	1	16
NV WHITE PINE2	88.82	BJ6-1	29-Nov-51	1	16
NV WHITE PINE2	335.33	TS5-1	7-May-52	6	16
NV WHITE PINE2	213.36	TP4-1	7-Mar-55	3	16
NV WHITE PINE2	29.26	TP8-1	9-Apr-55	8	16
NV WHITE PINE2	444.49	TP10-1	5-May-55	1	16
NV WHITE PINE2	283.42	PB5-1	15-Jul-57	2	16

COUNTY	uCi/ sqm	SHOT DAY	DATE	RANK FOR DAY	TIMES IN TOP 10
NV WHITE PINE2	21.87	PB6-2	25-Jul-57	1	16
NV WHITE PINE2	335.38	PB8-1	18-Aug-57	2	16
NV WHITE PINE2	79.88	PB9-1	23-Aug-57	2	16
NV WHITE PINE2	52.61	PB12-1	2-Sep-57	2	16
NV WHITE PINE2	10.37	PB17-1	28-Sep-57	9	16
NV WHITE PINE2	143.84	S-1	6-Jul-62	1	16
NV WHITE PINE2	63.61	UE16-1	8-Dec-68	2	16
NV WHITE PINE2	2.04	BN-1	18-Dec-70	6	16
NV WHITE PINE3	28.14	BJ6-1	29-Nov-51	5	16
NV WHITE PINE3	28.14	BJ6-1	29-Nov-51	5	16
NV WHITE PINE3	337.36	TS5-1	7-May-52	5	16
NV WHITE PINE3	492.92	UpK2-1	24-Mar-53	10	16
NV WHITE PINE3	35.62	UpK10-1	25-May-53	1	16
NV WHITE PINE3	143.76	TP4-1	7-Mar-55	5	16
NV WHITE PINE3	228.59	TP10-1	5-May-55	9	16
NV WHITE PINE3	206.23	PB5-1	15-Jul-57	4	16
NV WHITE PINE3	15.26	PB6-2	25-Jul-57	5	16
NV WHITE PINE3	263.86	PB8-1	18-Aug-57	3	16
NV WHITE PINE3	52.61	PB12-1	2-Sep-57	2	16
NV WHITE PINE3	10.13	PB15-1	16-Sep-57	5	16
NV WHITE PINE3	10.37	PB17-1	28-Sep-57	9	16
NV WHITE PINE3	117.11	S-1	6-Jul-62	3	16
NV WHITE PINE3	26.98	UE16-1	8-Dec-68	3	16
NV WHITE PINE3	2.04	BN-1	18-Dec-70	6	16

COUNTY	uCi/ sqm	SHOT DAY	DATE	RANK FOR DAY	TIMES IN TOP 10
NY ALBANY	181.55	UpK7-2	26-Apr-53	8	4
NY ALBANY	1.33	UpK7-6	30-Apr-53	8	4
NY ALBANY	2.74	PB5-7	21-Jul-57	4	4
NY ALBANY	5.97	PB5-8	22-Jul-57	3	4
NY ALLEGANY	0.99	R1-3	30-Jan-51	3	15
NY ALLEGANY	6.74	BJ2-6	4-Nov-51	3	15
NY ALLEGANY	16.67	BJ3-6	6-Nov-51	2	15
NY ALLEGANY	8.40	BJ3-7	7-Nov-51	4	15
NY ALLEGANY	2.27	BJ3-8	8-Nov-51	2	15
NY ALLEGANY	1.41	BJ3-9	9-Nov-51	1	15
NY ALLEGANY	0.82	BJ3-10	10-Nov-51	4	15
NY ALLEGANY	2.30	TP3-3	3-Mar-55	6	15
NY ALLEGANY	0.90	TP3-4	4-Mar-55	3	15
NY ALLEGANY	2.57	TP11-8	22-May-55	5	15
NY ALLEGANY	1.14	PB2-6	23-Jun-57	7	15
NY ALLEGANY	3.08	PB3-5	28-Jun-57	5	15
NY ALLEGANY	1.54	PB5-7	21-Jul-57	9	15
NY ALLEGANY	14.93	PB5-8	22-Jul-57	1	15
NY ALLEGANY	1.26	PB15-7	22-Sep-57	4	15
NY BRONX	11.62	BJ3-7	7-Nov-51	3	2
NY BRONX	1.51	PB5-8	22-Jul-57	7	2
NY BROOME	0.99	R1-3	30-Jan-51	3	4
NY BROOME	2.10	BJ2-6	4-Nov-51	7	4
NY BROOME	4.11	PB5-7	21-Jul-57	3	4
NY BROOME	5.97	PB5-8	22-Jul-57	3	4
NY CATTARAUGUS	1.17	R1-3	30-Jan-51	2	13
NY CATTARAUGUS	1.46	BJ2-6	4-Nov-51	10	13
NY CATTARAUGUS	16.67	BJ3-6	6-Nov-51	2	13
NY CATTARAUGUS	4.20	BJ3-7	7-Nov-51	8	13
NY CATTARAUGUS	2.27	BJ3-8	8-Nov-51	2	13
NY CATTARAUGUS	1.41	BJ3-9	9-Nov-51	1	13
NY CATTARAUGUS	0.82	BJ3-10	10-Nov-51	4	13
NY CATTARAUGUS	1.01	UpK8-4	11-May-53	10	13
NY CATTARAUGUS	2.28	UpK10-5	29-May-53	1	13
NY CATTARAUGUS	2.29	PB2-6	23-Jun-57	3	13

COUNTY	uCi/ sqm	SHOT DAY	DATE	RANK FOR DAY	TIMES IN TOP 10
NY CATTARAUGUS	3.21	PB3-5	28-Jun-57	3	13
NY CATTARAUGUS	8.96	PB5-8	22-Jul-57	2	13
NY CATTARAUGUS	1.26	PB15-7	22-Sep-57	4	13
NY CAYUGA	1.42	R1-3	30-Jan-51	1	6
NY CAYUGA	3.37	BJ2-6	4-Nov-51	5	6
NY CAYUGA	1.10	BJ3-8	8-Nov-51	4	6
NY CAYUGA	1.51	TP3-3	3-Mar-55	10	6
NY CAYUGA	4.11	PB5-7	21-Jul-57	3	6
NY CAYUGA	2.99	PB5-8	22-Jul-57	4	6
NY CHAUTAUQUA	1.42	R1-3	30-Jan-51	1	10
NY CHAUTAUQUA	1.76	BJ2-6	4-Nov-51	8	10
NY CHAUTAUQUA	27.78	BJ3-6	6-Nov-51	1	10
NY CHAUTAUQUA	8.40	BJ3-7	7-Nov-51	4	10
NY CHAUTAUQUA	2.27	BJ3-8	8-Nov-51	2	10
NY CHAUTAUQUA	1.41	BJ3-9	9-Nov-51	1	10
NY CHAUTAUQUA	0.82	BJ3-10	10-Nov-51	4	10
NY CHAUTAUQUA	5.97	PB5-8	22-Jul-57	3	10
NY CHAUTAUQUA	0.24	UE7-5	15-Jul-62	5	10
NY CHAUTAUQUA	7.97	UE12-4	17-Apr-65	8	10
NY CHEMUNG	1.42	R1-3	30-Jan-51	1	7
NY CHEMUNG	34.20	BJ2-4	2-Nov-51	8	7
NY CHEMUNG	3.37	BJ2-6	4-Nov-51	5	7
NY CHEMUNG	4.13	BJ3-7	7-Nov-51	8	7
NY CHEMUNG	2.74	PB5-7	21-Jul-57	4	7
NY CHEMUNG	8.96	PB5-8	22-Jul-57	2	7
NY CHEMUNG	0.94	PB15-7	22-Sep-57	6	7
NY CHENANGO	1.42	R1-3	30-Jan-51	1	6
NY CHENANGO	6.74	BJ2-6	4-Nov-51	3	6
NY CHENANGO	1.10	BJ3-8	8-Nov-51	4	6
NY CHENANGO	1.37	PB5-7	21-Jul-57	10	6
NY CHENANGO	8.96	PB5-8	22-Jul-57	2	6
NY CHENANGO	0.94	PB15-7	22-Sep-57	6	6
NY CLINTON	0.99	R1-3	30-Jan-51	3	9
NY CLINTON	7.98	BJ2-2	31-Oct-51	4	9
NY CLINTON	1.37	TS4-6	6-May-52	9	9

COUNTY	uCi/ sqm	SHOT DAY	DATE	RANK FOR DAY	TIMES IN TOP 10
NY CLINTON	1.37	TS4-6	6-May-52	9	9
NY CLINTON	1.32	UpK8-3	10-May-53	9	9
NY CLINTON	1.32	UpK8-4	11-May-53	7	9
NY CLINTON	5.41	PB5-6	20-Jul-57	2	9
NY CLINTON	2.23	PB5-8	22-Jul-57	6	9
NY CLINTON	0.12	UE7-5	15-Jul-62	9	9
NY COLUMBIA	172.79	UpK7-2	26-Apr-53	9	6
NY COLUMBIA	1.28	UpK7-6	30-Apr-53	9	6
NY COLUMBIA	1.32	UpK8-3	10-May-53	9	6
NY COLUMBIA	1.26	TP7-6	3-Apr-55	4	6
NY COLUMBIA	5.97	PB5-8	22-Jul-57	3	6
NY COLUMBIA	1.57	PB6-7	30-Jul-57	9	6
NY CORTLAND	1.42	R1-3	30-Jan-51	1	7
NY CORTLAND	3.37	BJ2-6	4-Nov-51	5	7
NY CORTLAND	1.10	BJ3-8	8-Nov-51	4	7
NY CORTLAND	0.96	TP3-4	4-Mar-55	2	7
NY CORTLAND	4.11	PB5-7	21-Jul-57	3	7
NY CORTLAND	14.93	PB5-8	22-Jul-57	1	7
NY CORTLAND	0.94	PB15-7	22-Sep-57	6	7
NY DELAWARE	0.99	R1-3	30-Jan-51	3	7
NY DELAWARE	32.39	BJ2-4	2-Nov-51	9	7
NY DELAWARE	3.37	BJ2-6	4-Nov-51	5	7
NY DELAWARE	4.13	BJ3-7	7-Nov-51	8	7
NY DELAWARE	0.84	BJ3-8	8-Nov-51	5	7
NY DELAWARE	1.37	PB5-7	21-Jul-57	10	7
NY DELAWARE	8.96	PB5-8	22-Jul-57	2	7
NY DUTCHESS	1.32	UpK8-3	10-May-53	9	4
NY DUTCHESS	1.72	UpK8-4	11-May-53	5	4
NY DUTCHESS	1.37	PB5-7	21-Jul-57	10	4
NY DUTCHESS	5.97	PB5-8	22-Jul-57	3	4
NY ERIE	1.17	R1-3	30-Jan-51	2	15
NY ERIE	20.85	BJ2-3	1-Nov-51	4	15
NY ERIE	6.08	BJ2-5	3-Nov-51	4	15
NY ERIE	20.21	BJ2-6	4-Nov-51	1	15
NY ERIE	27.78	BJ3-6	6-Nov-51	1	15

COUNTY	uCi/ sqm	SHOT DAY	DATE	RANK FOR DAY	TIMES IN TOP 10
NY ERIE	4.20	BJ3-7	7-Nov-51	8	15
NY ERIE	2.27	BJ3-8	8-Nov-51	2	15
NY ERIE	1.41	BJ3-9	9-Nov-51	1	15
NY ERIE	0.82	BJ3-10	10-Nov-51	4	15
NY ERIE	1.65	UpK10-5	29-May-53	4	15
NY ERIE	4.58	PB2-6	23-Jun-57	2	15
NY ERIE	3.34	PB3-5	28-Jun-57	2	15
NY ERIE	1.54	PB5-7	21-Jul-57	9	15
NY ERIE	8.96	PB5-8	22-Jul-57	2	15
NY ERIE	1.26	PB15-7	22-Sep-57	4	15
NY ESSEX	1.42	R1-3	30-Jan-51	1	7
NY ESSEX	7.98	BJ2-2	31-Oct-51	4	7
NY ESSEX	1.97	UpK8-3	10-May-53	7	7
NY ESSEX	1.72	UpK8-4	11-May-53	5	7
NY ESSEX	2.02	PB5-6	20-Jul-57	4	7
NY ESSEX	5.97	PB5-8	22-Jul-57	3	7
NY ESSEX	0.24	UE7-5	15-Jul-62	5	7
NY FRANKLIN	1.42	R1-3	30-Jan-51	1	7
NY FRANKLIN	7.98	BJ2-2	31-Oct-51	4	7
NY FRANKLIN	1.30	UpK6-5	22-Apr-53	4	7
NY FRANKLIN	2.68	PB5-6	20-Jul-57	3	7
NY FRANKLIN	1.51	PB5-8	22-Jul-57	7	7
NY FRANKLIN	1.57	PB6-7	30-Jul-57	9	7
NY FRANKLIN	0.13	UE7-5	15-Jul-62	8	7
NY FULTON	0.99	R1-3	30-Jan-51	3	9
NY FULTON	6.74	BJ2-6	4-Nov-51	3	9
NY FULTON	4.13	BJ3-7	7-Nov-51	8	9
NY FULTON	1.10	BJ3-8	8-Nov-51	4	9
NY FULTON	288.95	UpK7-2	26-Apr-53	5	9
NY FULTON	1.32	UpK8-3	10-May-53	9	9
NY FULTON	1.51	TP3-3	3-Mar-55	10	9
NY FULTON	2.99	PB5-8	22-Jul-57	4	9
NY FULTON	1.57	PB6-7	30-Jul-57	9	9
NY GENESEE	1.17	R1-3	30-Jan-51	2	15
NY GENESEE	23.09	BJ2-3	1-Nov-51	2	15

COUNTY	uCi/ sqm	SHOT DAY	DATE	RANK FOR DAY	TIMES IN TOP 10
NY GENESEE	8.75	BJ2-5	3-Nov-51	1	15
NY GENESEE	6.74	BJ2-6	4-Nov-51	3	15
NY GENESEE	27.78	BJ3-6	6-Nov-51	1	15
NY GENESEE	1.10	BJ3-8	8-Nov-51	4	15
NY GENESEE	1.41	BJ3-9	9-Nov-51	1	15
NY GENESEE	0.82	BJ3-10	10-Nov-51	4	15
NY GENESEE	2.30	TP3-3	3-Mar-55	6	15
NY GENESEE	2.57	TP11-8	22-May-55	5	15
NY GENESEE	2.29	PB2-6	23-Jun-57	3	15
NY GENESEE	3.03	PB3-5	28-Jun-57	6	15
NY GENESEE	1.54	PB5-7	21-Jul-57	9	15
NY GENESEE	14.93	PB5-8	22-Jul-57	1	15
NY GENESEE	1.26	PB15-7	22-Sep-57	4	15
NY GREENE	30.19	R1-2	29-Jan-51	8	6
NY GREENE	3.71	BJ3-7	7-Nov-51	10	6
NY GREENE	1.97	UpK8-3	10-May-53	7	6
NY GREENE	1.32	UpK8-4	11-May-53	7	6
NY GREENE	1.37	PB5-7	21-Jul-57	10	6
NY GREENE	5.97	PB5-8	22-Jul-57	3	6
NY HAMILTON	0.99	R1-3	30-Jan-51	3	6
NY HAMILTON	7.98	BJ2-2	31-Oct-51	4	6
NY HAMILTON	1.32	UpK8-3	10-May-53	9	6
NY HAMILTON	1.72	UpK8-4	11-May-53	5	6
NY HAMILTON	2.99	PB5-8	22-Jul-57	4	6
NY HAMILTON	0.17	UE7-5	15-Jul-62	7	6
NY HERKIMER	1.42	R1-3	30-Jan-51	1	8
NY HERKIMER	6.74	BJ2-6	4-Nov-51	3	8
NY HERKIMER	1.10	BJ3-8	8-Nov-51	4	8
NY HERKIMER	0.87	UpK6-5	22-Apr-53	8	8
NY HERKIMER	2.74	PB5-7	21-Jul-57	4	8
NY HERKIMER	5.97	PB5-8	22-Jul-57	3	8
NY HERKIMER	1.57	PB6-7	30-Jul-57	9	8
NY HERKIMER	0.24	UE7-5	15-Jul-62	5	8
NY JEFFERSON	1.17	R1-3	30-Jan-51	2	12
NY JEFFERSON	13.47	BJ2-6	4-Nov-51	2	12

COUNTY	uCi/ sqm	SHOT DAY	DATE	RANK FOR DAY	TIMES IN TOP 10
NY JEFFERSON	16.67	BJ3-6	6-Nov-51	2	12
NY JEFFERSON	8.40	BJ3-7	7-Nov-51	4	12
NY JEFFERSON	2.27	BJ3-8	8-Nov-51	2	12
NY JEFFERSON	1.41	BJ3-9	9-Nov-51	1	12
NY JEFFERSON	1.63	BJ3-10	10-Nov-51	3	12
NY JEFFERSON	2.29	PB2-6	23-Jun-57	3	12
NY JEFFERSON	2.74	PB5-7	21-Jul-57	4	12
NY JEFFERSON	2.99	PB5-8	22-Jul-57	4	12
NY JEFFERSON	1.26	PB15-7	22-Sep-57	4	12
NY JEFFERSON	0.13	UE7-5	15-Jul-62	8	12
NY KINGS	0.86	BJ4-9	13-Nov-51	1	6
NY KINGS	60.02	UpK1-2	18-Mar-53	3	6
NY KINGS	2.15	TP6-4	25-Mar-55	7	6
NY KINGS	1.21	TP11-8	22-May-55	10	6
NY KINGS	3.60	TP11-9	23-May-55	7	6
NY KINGS	2.99	PB5-8	22-Jul-57	4	6
NY LEWIS	1.42	R1-3	30-Jan-51	1	5
NY LEWIS	6.74	BJ2-6	4-Nov-51	3	5
NY LEWIS	1.10	BJ3-8	8-Nov-51	4	5
NY LEWIS	1.37	PB5-7	21-Jul-57	10	5
NY LEWIS	5.97	PB5-8	22-Jul-57	3	5
NY LIVINGSTON	1.17	R1-3	30-Jan-51	2	13
NY LIVINGSTON	20.85	BJ2-3	1-Nov-51	5	13
NY LIVINGSTON	6.74	BJ2-6	4-Nov-51	3	13
NY LIVINGSTON	16.67	BJ3-6	6-Nov-51	2	13
NY LIVINGSTON	1.10	BJ3-8	8-Nov-51	4	13
NY LIVINGSTON	1.41	BJ3-9	9-Nov-51	1	13
NY LIVINGSTON	0.82	BJ3-10	10-Nov-51	4	13
NY LIVINGSTON	2.30	TP3-3	3-Mar-55	6	13
NY LIVINGSTON	4.58	PB2-6	23-Jun-57	2	13
NY LIVINGSTON	2.78	PB3-5	28-Jun-57	10	13
NY LIVINGSTON	2.74	PB5-7	21-Jul-57	4	13
NY LIVINGSTON	8.96	PB5-8	22-Jul-57	2	13
NY LIVINGSTON	1.26	PB15-7	22-Sep-57	4	13
NY MADISON	1.17	R1-3	30-Jan-51	2	5

COUNTY	uCi/ sqm	SHOT DAY	DATE	RANK FOR DAY	TIMES IN TOP 10
NY MADISON	3.37	BJ2-6	4-Nov-51	5	5
NY MADISON	1.10	BJ3-8	8-Nov-51	4	5
NY MADISON	5.97	PB5-8	22-Jul-57	3	5
NY MADISON	0.94	PB15-7	22-Sep-57	6	5
NY MONROE	1.42	R1-3	30-Jan-51	1	7
NY MONROE	23.77	BJ2-3	1-Nov-51	1	7
NY MONROE	3.37	BJ2-6	4-Nov-51	5	7
NY MONROE	4.69	BJ3-8	8-Nov-51	1	7
NY MONROE	1.37	PB5-7	21-Jul-57	10	7
NY MONROE	1.51	PB5-8	22-Jul-57	7	7
NY MONROE	7.97	UE12-4	17-Apr-65	8	7
NY MONTGOMERY	3.37	BJ2-6	4-Nov-51	5	5
NY MONTGOMERY	4.13	BJ3-7	7-Nov-51	8	5
NY MONTGOMERY	1.32	UpK8-3	10-May-53	9	5
NY MONTGOMERY	5.97	PB5-8	22-Jul-57	3	5
NY MONTGOMERY	1.57	PB6-7	30-Jul-57	9	5
NY NASSAU	11.62	BJ3-7	7-Nov-51	3	6
NY NASSAU	52.47	UpK1-2	18-Mar-53	6	6
NY NASSAU	0.91	UpK11-5	8-Jun-53	3	6
NY NASSAU	2.15	TP6-4	25-Mar-55	7	6
NY NASSAU	2.71	TP11-9	23-May-55	10	6
NY NASSAU	2.99	PB5-8	22-Jul-57	4	6
NY NEW YORK	8.02	BJ3-7	7-Nov-51	5	4
NY NEW YORK	50.60	UpK1-2	18-Mar-53	8	4
NY NEW YORK	6.07	TP11-8	22-May-55	2	4
NY NEW YORK	2.99	PB5-8	22-Jul-57	4	4
NY NIAGARA	1.17	R1-3	30-Jan-51	2	16
NY NIAGARA	17.81	BJ2-3	1-Nov-51	10	16
NY NIAGARA	6.77	BJ2-5	3-Nov-51	2	16
NY NIAGARA	6.74	BJ2-6	4-Nov-51	3	16
NY NIAGARA	27.78	BJ3-6	6-Nov-51	1	16
NY NIAGARA	12.60	BJ3-7	7-Nov-51	2	16
NY NIAGARA	1.10	BJ3-8	8-Nov-51	4	16
NY NIAGARA	1.41	BJ3-9	9-Nov-51	1	16
NY NIAGARA	0.82	BJ3-10	10-Nov-51	4	16

COUNTY	uCi/ sqm	SHOT DAY	DATE	RANK FOR DAY	TIMES IN TOP 10
NY NIAGARA	1.01	UpK8-4	11-May-53	10	16
NY NIAGARA	1.65	UpK10-5	29-May-53	4	16
NY NIAGARA	1.14	PB2-6	23-Jun-57	7	16
NY NIAGARA	3.17	PB3-5	28-Jun-57	4	16
NY NIAGARA	2.23	PB5-8	22-Jul-57	6	16
NY NIAGARA	1.26	PB15-7	22-Sep-57	4	16
NY NIAGARA	7.97	UE12-4	17-Apr-65	8	16
NY ONEIDA	1.42	R1-3	30-Jan-51	1	5
NY ONEIDA	5.05	BJ2-6	4-Nov-51	4	5
NY ONEIDA	1.10	BJ3-8	8-Nov-51	4	5
NY ONEIDA	2.74	PB5-7	21-Jul-57	4	5
NY ONEIDA	5.97	PB5-8	22-Jul-57	3	5
NY ONONDAGA	1.17	R1-3	30-Jan-51	2	4
NY ONONDAGA	3.37	BJ2-6	4-Nov-51	5	4
NY ONONDAGA	0.84	BJ3-8	8-Nov-51	5	4
NY ONONDAGA	2.99	PB5-8	22-Jul-57	4	4
NY ONTARIO	1.17	R1-3	30-Jan-51	2	14
NY ONTARIO	3.37	BJ2-6	4-Nov-51	5	14
NY ONTARIO	16.67	BJ3-6	6-Nov-51	2	14
NY ONTARIO	8.40	BJ3-7	7-Nov-51	4	14
NY ONTARIO	2.27	BJ3-8	8-Nov-51	2	14
NY ONTARIO	1.41	BJ3-9	9-Nov-51	1	14
NY ONTARIO	0.82	BJ3-10	10-Nov-51	4	14
NY ONTARIO	1.51	TP3-3	3-Mar-55	10	14
NY ONTARIO	0.90	TP3-4	4-Mar-55	3	14
NY ONTARIO	1.29	TP11-8	22-May-55	9	14
NY ONTARIO	4.58	PB2-6	23-Jun-57	2	14
NY ONTARIO	4.11	PB5-7	21-Jul-57	3	14
NY ONTARIO	8.96	PB5-8	22-Jul-57	2	14
NY ONTARIO	1.26	PB15-7	22-Sep-57	4	14
NY ORANGE	3.85	BJ3-7	7-Nov-51	9	8
NY ORANGE	0.86	BJ4-9	13-Nov-51	1	8
NY ORANGE	1.32	UpK8-3	10-May-53	9	8
NY ORANGE	1.72	UpK8-4	11-May-53	5	8
NY ORANGE	1.21	TP11-8	22-May-55	10	8

COUNTY	uCi/ sqm	SHOT DAY	DATE	RANK FOR DAY	TIMES IN TOP 10
NY ORANGE	3.60	TP11-9	23-May-55	7	8
NY ORANGE	1.37	PB5-7	21-Jul-57	10	8
NY ORANGE	5.97	PB5-8	22-Jul-57	3	8
NY ORLEANS	17.87	BJ2-3	1-Nov-51	9	15
NY ORLEANS	4.54	BJ2-5	3-Nov-51	6	15
NY ORLEANS	3.37	BJ2-6	4-Nov-51	5	15
NY ORLEANS	27.78	BJ3-6	6-Nov-51	1	15
NY ORLEANS	4.20	BJ3-7	7-Nov-51	8	15
NY ORLEANS	1.10	BJ3-8	8-Nov-51	4	15
NY ORLEANS	1.41	BJ3-9	9-Nov-51	1	15
NY ORLEANS	0.82	BJ3-10	10-Nov-51	4	15
NY ORLEANS	1.52	UpK10-5	29-May-53	8	15
NY ORLEANS	1.51	TP3-3	3-Mar-55	10	15
NY ORLEANS	1.14	PB2-6	23-Jun-57	7	15
NY ORLEANS	1.37	PB5-7	21-Jul-57	10	15
NY ORLEANS	1.51	PB5-8	22-Jul-57	7	15
NY ORLEANS	0.85	PB15-7	22-Sep-57	8	15
NY ORLEANS	7.97	UE12-4	17-Apr-65	8	15
NY OSWEGO	1.17	R1-3	30-Jan-51	2	7
NY OSWEGO	55.93	BJ2-4	2-Nov-51	2	7
NY OSWEGO	3.37	BJ2-6	4-Nov-51	5	7
NY OSWEGO	1.10	BJ3-8	8-Nov-51	4	7
NY OSWEGO	1.37	PB5-7	21-Jul-57	10	7
NY OSWEGO	2.99	PB5-8	22-Jul-57	4	7
NY OSWEGO	0.13	UE7-5	15-Jul-62	8	7
NY OTSEGO	1.42	R1-3	30-Jan-51	1	6
NY OTSEGO	37.23	BJ2-4	2-Nov-51	7	6
NY OTSEGO	3.37	BJ2-6	4-Nov-51	5	6
NY OTSEGO	0.84	BJ3-8	8-Nov-51	5	6
NY OTSEGO	4.11	PB5-7	21-Jul-57	3	6
NY OTSEGO	8.96	PB5-8	22-Jul-57	2	6
NY PUTNAM	19.31	BJ3-7	7-Nov-51	1	2
NY PUTNAM	2.99	PB5-8	22-Jul-57	4	2
NY QUEENS	0.86	BJ4-9	13-Nov-51	1	8
NY QUEENS	55.86	UpK1-2	18-Mar-53	4	8

COUNTY	uCi/ sqm	SHOT DAY	DATE	RANK FOR DAY	TIMES IN TOP 10
NY QUEENS	0.91	UpK11-5	8-Jun-53	3	8
NY QUEENS	2.15	TP6-4	25-Mar-55	7	8
NY QUEENS	1.31	TP7-8	5-Apr-55	4	8
NY QUEENS	1.51	TP11-8	22-May-55	8	8
NY QUEENS	10.79	TP11-9	23-May-55	1	8
NY QUEENS	2.99	PB5-8	22-Jul-57	4	8
NY RENSSELAER	30.19	R1-2	29-Jan-51	8	8
NY RENSSELAER	0.99	R1-3	30-Jan-51	3	8
NY RENSSELAER	481.92	UpK7-2	26-Apr-53	3	8
NY RENSSELAER	1.68	UpK7-6	30-Apr-53	2	8
NY RENSSELAER	1.32	UpK8-3	10-May-53	9	8
NY RENSSELAER	1.24	TP7-6	3-Apr-55	5	8
NY RENSSELAER	1.13	PB5-6	20-Jul-57	8	8
NY RENSSELAER	5.97	PB5-8	22-Jul-57	3	8
NY RICHMOND	0.86	BJ4-9	13-Nov-51	1	8
NY RICHMOND	47.94	UpK1-2	18-Mar-53	10	8
NY RICHMOND	2.15	TP6-4	25-Mar-55	7	8
NY RICHMOND	2.43	TP11-8	22-May-55	6	8
NY RICHMOND	7.19	TP11-9	23-May-55	3	8
NY RICHMOND	2.99	PB5-8	22-Jul-57	4	8
NY RICHMOND	1.57	PB6-7	30-Jul-57	9	8
NY RICHMOND	0.95	PB18-13	19-Oct-57	2	8
NY ROCKLAND	0.86	BJ4-9	13-Nov-51	1	5
NY ROCKLAND	1.21	TP11-8	22-May-55	10	5
NY ROCKLAND	3.60	TP11-9	23-May-55	7	5
NY ROCKLAND	1.37	PB5-7	21-Jul-57	10	5
NY ROCKLAND	2.99	PB5-8	22-Jul-57	4	5
NY SARATOGA	0.99	R1-3	30-Jan-51	3	8
NY SARATOGA	12.07	BJ2-2	31-Oct-51	1	8
NY SARATOGA	295.34	UpK7-2	26-Apr-53	4	8
NY SARATOGA	1.79	UpK7-6	30-Apr-53	1	8
NY SARATOGA	2.63	UpK8-3	10-May-53	5	8
NY SARATOGA	1.26	TP7-6	3-Apr-55	4	8
NY SARATOGA	1.36	PB5-6	20-Jul-57	6	8
NY SARATOGA	2.99	PB5-8	22-Jul-57	4	8

COUNTY	uCi/ sqm	SHOT DAY	DATE	RANK FOR DAY	TIMES IN TOP 10
NY SCHENECTADY	225.40	UpK7-2	26-Apr-53	7	4
NY SCHENECTADY	1.54	UpK7-6	30-Apr-53	3	4
NY SCHENECTADY	1.32	UpK8-3	10-May-53	9	4
NY SCHENECTADY	2.99	PB5-8	22-Jul-57	4	4
NY SCHOHARIE	3.37	BJ2-6	4-Nov-51	5	5
NY SCHOHARIE	4.13	BJ3-7	7-Nov-51	8	5
NY SCHOHARIE	1.32	UpK8-3	10-May-53	9	5
NY SCHOHARIE	8.96	PB5-8	22-Jul-57	2	5
NY SCHOHARIE	1.57	PB6-7	30-Jul-57	9	5
NY SCHUYLER	3.37	BJ2-6	4-Nov-51	5	5
NY SCHUYLER	1.51	TP3-3	3-Mar-55	10	5
NY SCHUYLER	6.84	PB5-7	21-Jul-57	1	5
NY SCHUYLER	1.51	PB5-8	22-Jul-57	7	5
NY SCHUYLER	7.97	UE12-4	17-Apr-65	8	5
NY SENECA	1.42	R1-3	30-Jan-51	1	11
NY SENECA	3.37	BJ2-6	4-Nov-51	5	11
NY SENECA	11.11	BJ3-6	6-Nov-51	4	11
NY SENECA	8.40	BJ3-7	7-Nov-51	4	11
NY SENECA	1.10	BJ3-8	8-Nov-51	4	11
NY SENECA	1.41	BJ3-9	9-Nov-51	1	11
NY SENECA	0.82	BJ3-10	10-Nov-51	4	11
NY SENECA	1.51	TP3-3	3-Mar-55	10	11
NY SENECA	6.87	PB2-6	23-Jun-57	1	11
NY SENECA	8.96	PB5-8	22-Jul-57	2	11
NY SENECA	1.26	PB15-7	22-Sep-57	4	11
NY ST LAWRENCE	1.42	R1-3	30-Jan-51	1	8
NY ST LAWRENCE	4.09	BJ2-2	31-Oct-51	9	8
NY ST LAWRENCE	68.31	TS7-4	4-Jun-52	2	8
NY ST LAWRENCE	0.87	UpK6-5	22-Apr-53	8	8
NY ST LAWRENCE	2.68	PB5-6	20-Jul-57	3	8
NY ST LAWRENCE	4.11	PB5-7	21-Jul-57	3	8
NY ST LAWRENCE	2.99	PB5-8	22-Jul-57	4	8
NY ST LAWRENCE	0.12	UE7-5	15-Jul-62	9	8
NY STEUBEN	1.42	R1-3	30-Jan-51	1	14
NY STEUBEN	5.05	BJ2-6	4-Nov-51	4	14

COUNTY	uCi/ sqm	SHOT DAY	DATE	RANK FOR DAY	TIMES IN TOP 10
NY STEUBEN	16.67	BJ3-6	6-Nov-51	2	14
NY STEUBEN	8.40	BJ3-7	7-Nov-51	4	14
NY STEUBEN	1.69	BJ3-8	8-Nov-51	3	14
NY STEUBEN	1.41	BJ3-9	9-Nov-51	1	14
NY STEUBEN	0.82	BJ3-10	10-Nov-51	4	14
NY STEUBEN	2.30	TP3-3	3-Mar-55	6	14
NY STEUBEN	0.90	TP3-4	4-Mar-55	3	14
NY STEUBEN	1.29	TP11-8	22-May-55	9	14
NY STEUBEN	1.72	PB2-6	23-Jun-57	5	14
NY STEUBEN	4.11	PB5-7	21-Jul-57	3	14
NY STEUBEN	14.93	PB5-8	22-Jul-57	1	14
NY STEUBEN	1.26	PB15-7	22-Sep-57	4	14
NY SUFFOLK	1.04	UpK4-4	9-Apr-53	5	3
NY SUFFOLK	2.71	TP11-9	23-May-55	10	3
NY SUFFOLK	2.99	PB5-8	22-Jul-57	4	3
NY SULLIVAN	30.19	R1-2	29-Jan-51	8	6
NY SULLIVAN	3.37	BJ2-6	4-Nov-51	5	6
NY SULLIVAN	6.86	BJ3-7	7-Nov-51	7	6
NY SULLIVAN	1.21	TP11-8	22-May-55	10	6
NY SULLIVAN	8.96	PB5-8	22-Jul-57	2	6
NY SULLIVAN	0.94	PB15-7	22-Sep-57	6	6
NY TIOGA	1.42	R1-3	30-Jan-51	1	4
NY TIOGA	3.37	BJ2-6	4-Nov-51	5	4
NY TIOGA	8.96	PB5-8	22-Jul-57	2	4
NY TIOGA	0.94	PB15-7	22-Sep-57	6	4
NY TOMPKINS	1.42	R1-3	30-Jan-51	1	7
NY TOMPKINS	48.54	BJ2-4	2-Nov-51	3	7
NY TOMPKINS	3.37	BJ2-6	4-Nov-51	5	7
NY TOMPKINS	0.84	BJ3-8	8-Nov-51	5	7
NY TOMPKINS	1.37	PB5-7	21-Jul-57	10	7
NY TOMPKINS	14.93	PB5-8	22-Jul-57	1	7
NY TOMPKINS	0.94	PB15-7	22-Sep-57	6	7
NY ULSTER	30.19	R1-2	29-Jan-51	8	6
NY ULSTER	19.31	BJ3-7	7-Nov-51	1	6
NY ULSTER	2.63	UpK8-3	10-May-53	5	6

COUNTY	uCi/ sqm	SHOT DAY	DATE	RANK FOR DAY	TIMES IN TOP 10
NY ULSTER	1.72	UpK8-4	11-May-53	5	6
NY ULSTER	1.37	PB5-7	21-Jul-57	10	6
NY ULSTER	5.97	PB5-8	22-Jul-57	3	6
NY WARREN	7.98	BJ2-2	31-Oct-51	4	6
NY WARREN	501.27	UpK7-2	26-Apr-53	2	6
NY WARREN	1.32	UpK8-3	10-May-53	9	6
NY WARREN	1.36	PB5-6	20-Jul-57	6	6
NY WARREN	1.51	PB5-8	22-Jul-57	7	6
NY WARREN	2.62	PB6-7	30-Jul-57	2	6
NY WASHINGTON	4.09	BJ2-2	31-Oct-51	9	6
NY WASHINGTON	527.06	UpK7-2	26-Apr-53	1	6
NY WASHINGTON	1.32	UpK8-3	10-May-53	9	6
NY WASHINGTON	2.02	PB5-6	20-Jul-57	4	6
NY WASHINGTON	5.97	PB5-8	22-Jul-57	3	6
NY WASHINGTON	1.57	PB6-7	30-Jul-57	9	6
NY WAYNE	1.17	R1-3	30-Jan-51	2	16
NY WAYNE	29.53	BJ2-4	2-Nov-51	10	16
NY WAYNE	6.22	BJ2-5	3-Nov-51	3	16
NY WAYNE	3.37	BJ2-6	4-Nov-51	5	16
NY WAYNE	16.67	BJ3-6	6-Nov-51	2	16
NY WAYNE	8.40	BJ3-7	7-Nov-51	4	16
NY WAYNE	1.10	BJ3-8	8-Nov-51	4	16
NY WAYNE	1.41	BJ3-9	9-Nov-51	1	16
NY WAYNE	0.82	BJ3-10	10-Nov-51	4	16
NY WAYNE	1.51	TP3-3	3-Mar-55	10	16
NY WAYNE	1.29	TP11-8	22-May-55	9	16
NY WAYNE	2.29	PB2-6	23-Jun-57	3	16
NY WAYNE	1.37	PB5-7	21-Jul-57	10	16
NY WAYNE	2.99	PB5-8	22-Jul-57	4	16
NY WAYNE	1.26	PB15-7	22-Sep-57	4	16
NY WAYNE	0.17	UE7-5	15-Jul-62	7	16
NY WESTCHESTER	11.62	BJ3-7	7-Nov-51	3	4
NY WESTCHESTER	1.21	TP11-8	22-May-55	10	4
NY WESTCHESTER	3.60	TP11-9	23-May-55	7	4
NY WESTCHESTER	2.99	PB5-8	22-Jul-57	4	4

COUNTY	uCi/ sqm	SHOT DAY	DATE	RANK FOR DAY	TIMES IN TOP 10
NY WYOMING	13.47	BJ2-6	4-Nov-51	2	11
NY WYOMING	16.67	BJ3-6	6-Nov-51	2	11
NY WYOMING	8.40	BJ3-7	7-Nov-51	4	11
NY WYOMING	1.10	BJ3-8	8-Nov-51	4	11
NY WYOMING	1.41	BJ3-9	9-Nov-51	1	11
NY WYOMING	0.82	BJ3-10	10-Nov-51	4	11
NY WYOMING	4.58	PB2-6	23-Jun-57	2	11
NY WYOMING	2.95	PB3-5	28-Jun-57	7	11
NY WYOMING	1.54	PB5-7	21-Jul-57	9	11
NY WYOMING	8.96	PB5-8	22-Jul-57	2	11
NY WYOMING	1.26	PB15-7	22-Sep-57	4	11
NY YATES	3.37	BJ2-6	4-Nov-51	5	9
NY YATES	16.67	BJ3-6	6-Nov-51	2	9
NY YATES	4.20	BJ3-7	7-Nov-51	8	9
NY YATES	1.10	BJ3-8	8-Nov-51	4	9
NY YATES	1.41	BJ3-9	9-Nov-51	1	9
NY YATES	0.82	BJ3-10	10-Nov-51	4	9
NY YATES	2.30	TP3-3	3-Mar-55	6	9
NY YATES	1.14	PB2-6	23-Jun-57	7	9
NY YATES	1.51	PB5-8	22-Jul-57	7	9

COUNTY	uCi/ sqm	SHOT DAY	DATE	RANK FOR DAY	TIMES IN TOP 10
OH ADAMS	1.02	UpK3-6	5-Apr-53	7	5
OH ADAMS	15.70	PB6-3	26-Jul-57	5	5
OH ADAMS	0.89	PB13-7	12-Sep-57	5	5
OH ADAMS	1.08	PB15-7	22-Sep-57	5	5
OH ADAMS	208.78	S-4	9-Jul-62	3	5
OH ALLEN	1.17	R1-3	30-Jan-51	2	4
OH ALLEN	1.08	PB15-7	22-Sep-57	5	4
OH ALLEN	208.78	S-4	9-Jul-62	3	4
OH ALLEN	0.12	UE7-5	15-Jul-62	9	4
OH ASHLAND	0.99	R1-3	30-Jan-51	3	4
OH ASHLAND	7.20	BJ2-2	31-Oct-51	5	4
OH ASHLAND	0.90	PB15-7	22-Sep-57	7	4
OH ASHLAND	103.59	S-4	9-Jul-62	6	4
OH ASHTABULA	7.20	BJ2-2	31-Oct-51	5	5
OH ASHTABULA	0.95	BJ4-8	12-Nov-51	2	5
OH ASHTABULA	12.03	TP4-4	10-Mar-55	4	5
OH ASHTABULA	8.96	PB5-8	22-Jul-57	2	5
OH ASHTABULA	0.17	UE7-5	15-Jul-62	7	5
OH ATHENS	38.66	R1-2	29-Jan-51	5	5
OH ATHENS	7.20	BJ2-2	31-Oct-51	5	5
OH ATHENS	14.93	PB5-8	22-Jul-57	1	5
OH ATHENS	0.90	PB15-7	22-Sep-57	7	5
OH ATHENS	85.56	S-4	9-Jul-62	9	5
OH AUGLAIZE	0.89	PB13-7	12-Sep-57	5	3
OH AUGLAIZE	85.56	S-4	9-Jul-62	9	3
OH AUGLAIZE	0.22	UE7-5	15-Jul-62	6	3
OH BELMONT	48.19	R1-2	29-Jan-51	1	6
OH BELMONT	1.72	TP3-3	3-Mar-55	8	6
OH BELMONT	2.57	TP11-8	22-May-55	5	6
OH BELMONT	3.12	TP11-9	23-May-55	8	6
OH BELMONT	8.96	PB5-8	22-Jul-57	2	6
OH BELMONT	90.29	S-4	9-Jul-62	8	6
OH BROWN	1.02	UpK3-6	5-Apr-53	7	4
OH BROWN	15.70	PB6-3	26-Jul-57	5	4
OH BROWN	1.08	PB15-7	22-Sep-57	5	4

COUNTY	uCi/ sqm	SHOT DAY	DATE	RANK FOR DAY	TIMES IN TOP 10
OH BROWN	90.29	S-4	9-Jul-62	8	4
OH BUTLER	1.02	UpK3-6	5-Apr-53	7	5
OH BUTLER	15.70	PB6-3	26-Jul-57	5	5
OH BUTLER	1.34	PB13-7	12-Sep-57	2	5
OH BUTLER	208.78	S-4	9-Jul-62	3	5
OH BUTLER	0.13	UE7-5	15-Jul-62	8	5
OH CARROLL	7.20	BJ2-2	31-Oct-51	5	3
OH CARROLL	8.96	PB5-8	22-Jul-57	2	3
OH CARROLL	0.90	PB15-7	22-Sep-57	7	3
OH CHAMPAIGN	1.14	TP9-8	22-Apr-55	1	5
OH CHAMPAIGN	2.23	PB13-7	12-Sep-57	1	5
OH CHAMPAIGN	1.08	PB15-7	22-Sep-57	5	5
OH CHAMPAIGN	208.78	S-4	9-Jul-62	3	5
OH CHAMPAIGN	0.11	UE7-5	15-Jul-62	10	5
OH CLARK	1.34	PB13-7	12-Sep-57	2	3
OH CLARK	208.78	S-4	9-Jul-62	3	3
OH CLARK	0.13	UE7-5	15-Jul-62	8	3
OH CLERMONT	1.02	UpK3-6	5-Apr-53	7	5
OH CLERMONT	23.55	PB6-3	26-Jul-57	3	5
OH CLERMONT	1.34	PB13-7	12-Sep-57	2	5
OH CLERMONT	1.08	PB15-7	22-Sep-57	5	5
OH CLERMONT	85.56	S-4	9-Jul-62	9	5
OH CLINTON	1.02	UpK3-6	5-Apr-53	7	3
OH CLINTON	1.34	PB13-7	12-Sep-57	2	3
OH CLINTON	208.78	S-4	9-Jul-62	3	3
OH COLUMBIANA	7.20	BJ2-2	31-Oct-51	5	7
OH COLUMBIANA	1.72	TP3-3	3-Mar-55	8	7
OH COLUMBIANA	2.57	TP11-8	22-May-55	5	7
OH COLUMBIANA	3.12	TP11-9	23-May-55	8	7
OH COLUMBIANA	8.96	PB5-8	22-Jul-57	2	7
OH COLUMBIANA	2.38	PB6-7	30-Jul-57	4	7
OH COLUMBIANA	85.56	S-4	9-Jul-62	9	7
OH COSHOCTON	7.20	BJ2-2	31-Oct-51	5	5
OH COSHOCTON	14.93	PB5-8	22-Jul-57	1	5
OH COSHOCTON	0.89	PB13-7	12-Sep-57	5	5

COUNTY	uCi/ sqm	SHOT DAY	DATE	RANK FOR DAY	TIMES IN TOP 10
OH COSHOCTON	0.90	PB15-7	22-Sep-57	7	5
OH COSHOCTON	85.56	S-4	9-Jul-62	9	5
OH CRAWFORD	7.20	BJ2-2	31-Oct-51	5	4
OH CRAWFORD	0.90	PB15-7	22-Sep-57	7	4
OH CRAWFORD	90.29	S-4	9-Jul-62	8	4
OH CRAWFORD	0.24	UE7-5	15-Jul-62	5	4
OH CUYAHOGA	0.99	R1-3	30-Jan-51	3	6
OH CUYAHOGA	0.85	BJ4-7	11-Nov-51	4	6
OH CUYAHOGA	11.70	TP4-4	10-Mar-55	5	6
OH CUYAHOGA	72.24	TP9-2	16-Apr-55	3	6
OH CUYAHOGA	5.97	PB5-8	22-Jul-57	3	6
OH CUYAHOGA	0.24	UE7-5	15-Jul-62	5	6
OH DARKE	2.00	TS4-5	5-May-52	4	6
OH DARKE	2.00	TS4-5	5-May-52	4	6
OH DARKE	1.02	UpK3-6	5-Apr-53	7	6
OH DARKE	1.51	PB9-5	27-Aug-57	9	6
OH DARKE	85.56	S-4	9-Jul-62	9	6
OH DARKE	0.26	UE7-5	15-Jul-62	4	6
OH DEFIANCE	0.85	PB13-7	12-Sep-57	6	2
OH DEFIANCE	0.12	UE7-5	15-Jul-62	9	2
OH DELAWARE	7.20	BJ2-2	31-Oct-51	5	5
OH DELAWARE	0.95	BJ4-8	12-Nov-51	2	5
OH DELAWARE	0.90	PB15-7	22-Sep-57	7	5
OH DELAWARE	90.29	S-4	9-Jul-62	8	5
OH DELAWARE	0.11	UE7-5	15-Jul-62	10	5
OH ERIE	10.53	BJ3-5	5-Nov-51	1	5
OH ERIE	1.43	BJ4-8	12-Nov-51	1	5
OH ERIE	45.07	TP9-2	16-Apr-55	6	5
OH ERIE	1.51	PB9-5	27-Aug-57	9	5
OH ERIE	0.24	UE7-5	15-Jul-62	5	5
OH FAIRFIELD	5.90	PB6-3	26-Jul-57	10	3
OH FAIRFIELD	1.34	PB13-7	12-Sep-57	2	3
OH FAIRFIELD	90.29	S-4	9-Jul-62	8	3
OH FAYETTE	1.02	UpK3-6	5-Apr-53	7	3
OH FAYETTE	1.34	PB13-7	12-Sep-57	2	3

COUNTY	uCi/ sqm	SHOT DAY	DATE	RANK FOR DAY	TIMES IN TOP 10
OH FAYETTE	90.29	S-4	9-Jul-62	8	3
OH FRANKLIN	2.00	TS4-5	5-May-52	4	4
OH FRANKLIN	2.00	TS4-5	5-May-52	4	4
OH FRANKLIN	2.23	PB13-7	12-Sep-57	1	4
OH FRANKLIN	0.13	UE7-5	15-Jul-62	8	4
OH FULTON	4.52	PB9-5	27-Aug-57	2	4
OH FULTON	0.85	PB13-7	12-Sep-57	6	4
OH FULTON	208.78	S-4	9-Jul-62	3	4
OH FULTON	0.13	UE7-5	15-Jul-62	8	4
OH GALLIA	8.96	PB5-8	22-Jul-57	2	4
OH GALLIA	23.55	PB6-3	26-Jul-57	3	4
OH GALLIA	1.34	PB13-7	12-Sep-57	2	4
OH GALLIA	85.56	S-4	9-Jul-62	9	4
OH GEAUGA	4.21	BJ3-5	5-Nov-51	7	8
OH GEAUGA	0.95	BJ4-8	12-Nov-51	2	8
OH GEAUGA	11.65	UpK9-3	21-May-53	3	8
OH GEAUGA	11.49	TP4-4	10-Mar-55	6	8
OH GEAUGA	40.00	TP9-2	16-Apr-55	8	8
OH GEAUGA	0.85	TP9-5	19-Apr-55	1	8
OH GEAUGA	5.97	PB5-8	22-Jul-57	3	8
OH GEAUGA	7.97	UE12-4	17-Apr-65	8	8
OH GREENE	1.02	UpK3-6	5-Apr-53	7	3
OH GREENE	2.23	PB13-7	12-Sep-57	1	3
OH GREENE	0.13	UE7-5	15-Jul-62	8	3
OH GUERNSEY	48.19	R1-2	29-Jan-51	1	6
OH GUERNSEY	7.20	BJ2-2	31-Oct-51	5	6
OH GUERNSEY	0.91	TP9-8	22-Apr-55	2	6
OH GUERNSEY	8.96	PB5-8	22-Jul-57	2	6
OH GUERNSEY	0.90	PB15-7	22-Sep-57	7	6
OH GUERNSEY	103.59	S-4	9-Jul-62	6	6
OH HAMILTON	1.56	UpK3-6	5-Apr-53	3	2
OH HAMILTON	6.52	PB6-3	26-Jul-57	9	2
OH HANCOCK	0.12	UE7-5	15-Jul-62	9	1
OH HARDIN	0.12	UE7-5	15-Jul-62	9	1
OH HARRISON	7.20	BJ2-2	31-Oct-51	5	8

COUNTY	uCi/ sqm	SHOT DAY	DATE	RANK FOR DAY	TIMES IN TOP 10
OH HARRISON	1.72	TP3-3	3-Mar-55	8	8
OH HARRISON	1.29	TP11-8	22-May-55	9	8
OH HARRISON	8.96	PB5-8	22-Jul-57	2	8
OH HARRISON	2.38	PB6-7	30-Jul-57	4	8
OH HARRISON	0.81	PB15-7	22-Sep-57	9	8
OH HARRISON	85.56	S-4	9-Jul-62	9	8
OH HARRISON	2.38	UE8-5	18-Jul-62	5	8
OH HENRY	4.52	PB9-5	27-Aug-57	2	3
OH HENRY	0.85	PB13-7	12-Sep-57	6	3
OH HENRY	0.12	UE7-5	15-Jul-62	9	3
OH HIGHLAND	1.02	UpK3-6	5-Apr-53	7	4
OH HIGHLAND	23.55	PB6-3	26-Jul-57	3	4
OH HIGHLAND	1.08	PB15-7	22-Sep-57	5	4
OH HIGHLAND	208.78	S-4	9-Jul-62	3	4
OH HOCKING	14.93	PB5-8	22-Jul-57	1	5
OH HOCKING	0.89	PB13-7	12-Sep-57	5	5
OH HOCKING	1.08	PB15-7	22-Sep-57	5	5
OH HOCKING	85.56	S-4	9-Jul-62	9	5
OH HOCKING	4.76	UE8-5	18-Jul-62	2	5
OH HOLMES	7.20	BJ2-2	31-Oct-51	5	4
OH HOLMES	10.71	TP4-4	10-Mar-55	8	4
OH HOLMES	8.96	PB5-8	22-Jul-57	2	4
OH HOLMES	0.90	PB15-7	22-Sep-57	7	4
OH HURON	7.20	BJ2-2	31-Oct-51	5	3
OH HURON	6.32	BJ3-5	5-Nov-51	4	3
OH HURON	1.43	BJ4-8	12-Nov-51	1	3
OH JACKSON	8.96	PB5-8	22-Jul-57	2	4
OH JACKSON	2.23	PB13-7	12-Sep-57	1	4
OH JACKSON	1.08	PB15-7	22-Sep-57	5	4
OH JACKSON	208.78	S-4	9-Jul-62	3	4
OH JEFFERSON	10.90	BJ2-2	31-Oct-51	2	6
OH JEFFERSON	10.33	UpK9-3	21-May-53	10	6
OH JEFFERSON	1.72	TP3-3	3-Mar-55	8	6
OH JEFFERSON	2.57	TP11-8	22-May-55	5	6
OH JEFFERSON	3.12	TP11-9	23-May-55	8	6

COUNTY	uCi/ sqm	SHOT DAY	DATE	RANK FOR DAY	TIMES IN TOP 10
OH JEFFERSON	8.96	PB5-8	22-Jul-57	2	6
OH KNOX	7.20	BJ2-2	31-Oct-51	5	5
OH KNOX	0.89	PB13-7	12-Sep-57	5	5
OH KNOX	0.90	PB15-7	22-Sep-57	7	5
OH KNOX	208.78	S-4	9-Jul-62	3	5
OH KNOX	0.11	UE7-5	15-Jul-62	10	5
OH LAKE	10.90	BJ2-2	31-Oct-51	2	7
OH LAKE	6.32	BJ3-5	5-Nov-51	4	7
OH LAKE	1.43	BJ4-8	12-Nov-51	1	7
OH LAKE	11.21	UpK9-3	21-May-53	7	7
OH LAKE	12.55	TP4-4	10-Mar-55	3	7
OH LAKE	0.85	TP9-5	19-Apr-55	1	7
OH LAKE	5.97	PB5-8	22-Jul-57	3	7
OH LAWRENCE	0.97	TP6-5	26-Mar-55	8	6
OH LAWRENCE	5.97	PB5-8	22-Jul-57	3	6
OH LAWRENCE	23.55	PB6-3	26-Jul-57	3	6
OH LAWRENCE	1.34	PB13-7	12-Sep-57	2	6
OH LAWRENCE	1.08	PB15-7	22-Sep-57	5	6
OH LAWRENCE	85.56	S-4	9-Jul-62	9	6
OH LICKING	7.20	BJ2-2	31-Oct-51	5	4
OH LICKING	4.21	BJ3-5	5-Nov-51	7	4
OH LICKING	0.90	PB15-7	22-Sep-57	7	4
OH LICKING	208.78	S-4	9-Jul-62	3	4
OH LOGAN	1.42	R1-3	30-Jan-51	1	5
OH LOGAN	0.89	PB13-7	12-Sep-57	5	5
OH LOGAN	1.08	PB15-7	22-Sep-57	5	5
OH LOGAN	85.56	S-4	9-Jul-62	9	5
OH LOGAN	0.11	UE7-5	15-Jul-62	10	5
OH LORAIN	7.20	BJ2-2	31-Oct-51	5	4
OH LORAIN	4.21	BJ3-5	5-Nov-51	7	4
OH LORAIN	1.43	BJ4-8	12-Nov-51	1	4
OH LORAIN	10.20	TP4-4	10-Mar-55	10	4
OH LUCAS	1.51	PB9-5	27-Aug-57	9	3
OH LUCAS	0.85	PB13-7	12-Sep-57	6	3
OH LUCAS	85.56	S-4	9-Jul-62	9	3

COUNTY	uCi/ sqm	SHOT DAY	DATE	RANK FOR DAY	TIMES IN TOP 10
OH MADISON	2.00	TS4-5	5-May-52	4	5
OH MADISON	2.00	TS4-5	5-May-52	4	5
OH MADISON	2.23	PB13-7	12-Sep-57	1	5
OH MADISON	103.59	S-4	9-Jul-62	6	5
OH MADISON	0.13	UE7-5	15-Jul-62	8	5
OH MAHONING	13.53	UpK9-3	21-May-53	1	6
OH MAHONING	3.86	TP11-8	22-May-55	3	6
OH MAHONING	4.68	TP11-9	23-May-55	5	6
OH MAHONING	5.97	PB5-8	22-Jul-57	3	6
OH MAHONING	1.57	PB6-7	30-Jul-57	9	6
OH MAHONING	103.59	S-4	9-Jul-62	6	6
OH MARION	7.20	BJ2-2	31-Oct-51	5	5
OH MARION	0.95	BJ4-8	12-Nov-51	2	5
OH MARION	0.90	PB15-7	22-Sep-57	7	5
OH MARION	85.56	S-4	9-Jul-62	9	5
OH MARION	0.13	UE7-5	15-Jul-62	8	5
OH MEDINA	7.20	BJ2-2	31-Oct-51	5	8
OH MEDINA	0.95	BJ4-8	12-Nov-51	2	8
OH MEDINA	16.73	TP4-4	10-Mar-55	1	8
OH MEDINA	77.85	TP9-2	16-Apr-55	2	8
OH MEDINA	8.96	PB5-8	22-Jul-57	2	8
OH MEDINA	0.90	PB15-7	22-Sep-57	7	8
OH MEDINA	208.78	S-4	9-Jul-62	3	8
OH MEDINA	0.12	UE7-5	15-Jul-62	9	8
OH MEIGS	2.57	TP11-8	22-May-55	5	6
OH MEIGS	3.12	TP11-9	23-May-55	8	6
OH MEIGS	8.96	PB5-8	22-Jul-57	2	6
OH MEIGS	1.57	PB6-7	30-Jul-57	9	6
OH MEIGS	90.29	S-4	9-Jul-62	8	6
OH MEIGS	3.35	UE8-5	18-Jul-62	4	6
OH MERCER	0.33	UE7-5	15-Jul-62	3	1
OH MIAMI	1.02	UpK3-6	5-Apr-53	7	6
OH MIAMI	3.01	PB9-5	27-Aug-57	6	6
OH MIAMI	1.34	PB13-7	12-Sep-57	2	6
OH MIAMI	1.08	PB15-7	22-Sep-57	5	6

COUNTY	uCi/ sqm	SHOT DAY	DATE	RANK FOR DAY	TIMES IN TOP 10
OH MIAMI	208.78	S-4	9-Jul-62	3	6
OH MIAMI	0.13	UE7-5	15-Jul-62	8	6
OH MONROE	2.57	TP11-8	22-May-55	5	5
OH MONROE	3.12	TP11-9	23-May-55	8	5
OH MONROE	8.96	PB5-8	22-Jul-57	2	5
OH MONROE	103.59	S-4	9-Jul-62	6	5
OH MONROE	3.35	UE8-5	18-Jul-62	4	5
OH MONTGOMERY	1.23	TS4-5	5-May-52	8	8
OH MONTGOMERY	1.23	TS4-5	5-May-52	8	8
OH MONTGOMERY	1.02	UpK3-6	5-Apr-53	7	8
OH MONTGOMERY	1.51	PB9-5	27-Aug-57	9	8
OH MONTGOMERY	1.34	PB13-7	12-Sep-57	2	8
OH MONTGOMERY	1.08	PB15-7	22-Sep-57	5	8
OH MONTGOMERY	208.78	S-4	9-Jul-62	3	8
OH MONTGOMERY	0.13	UE7-5	15-Jul-62	8	8
OH MORGAN	7.20	BJ2-2	31-Oct-51	5	4
OH MORGAN	8.96	PB5-8	22-Jul-57	2	4
OH MORGAN	0.90	PB15-7	22-Sep-57	7	4
OH MORGAN	208.78	S-4	9-Jul-62	3	4
OH MORROW	7.20	BJ2-2	31-Oct-51	5	6
OH MORROW	0.95	BJ4-8	12-Nov-51	2	6
OH MORROW	1.39	PB2-5	22-Jun-57	8	6
OH MORROW	0.99	PB15-6	21-Sep-57	1	6
OH MORROW	0.90	PB15-7	22-Sep-57	7	6
OH MORROW	85.56	S-4	9-Jul-62	9	6
OH MUSKINGUM	7.20	BJ2-2	31-Oct-51	5	4
OH MUSKINGUM	4.21	BJ3-5	5-Nov-51	7	4
OH MUSKINGUM	8.96	PB5-8	22-Jul-57	2	4
OH MUSKINGUM	90.29	S-4	9-Jul-62	8	4
OH NOBLE	7.20	BJ2-2	31-Oct-51	5	5
OH NOBLE	2.57	TP11-8	22-May-55	5	5
OH NOBLE	3.12	TP11-9	23-May-55	8	5
OH NOBLE	14.93	PB5-8	22-Jul-57	1	5
OH NOBLE	85.56	S-4	9-Jul-62	9	5
OH OTTAWA	1.43	BJ4-8	12-Nov-51	1	4

COUNTY	uCi/ sqm	SHOT DAY	DATE	RANK FOR DAY	TIMES IN TOP 10
OH OTTAWA	0.85	PB4-3	7-Jul-57	10	4
OH OTTAWA	3.01	PB9-5	27-Aug-57	6	4
OH OTTAWA	1.25	PB13-7	12-Sep-57	3	4
OH PAULDING	1.51	PB9-5	27-Aug-57	9	4
OH PAULDING	1.34	PB13-7	12-Sep-57	2	4
OH PAULDING	85.56	S-4	9-Jul-62	9	4
OH PAULDING	0.13	UE7-5	15-Jul-62	8	4
OH PERRY	7.20	BJ2-2	31-Oct-51	5	7
OH PERRY	6.32	BJ3-5	5-Nov-51	4	7
OH PERRY	0.95	BJ4-8	12-Nov-51	2	7
OH PERRY	8.96	PB5-8	22-Jul-57	2	7
OH PERRY	0.90	PB15-7	22-Sep-57	7	7
OH PERRY	208.78	S-4	9-Jul-62	3	7
OH PERRY	2.03	UE8-5	18-Jul-62	8	7
OH PICKAWAY	0.89	PB13-7	12-Sep-57	5	3
OH PICKAWAY	1.08	PB15-7	22-Sep-57	5	3
OH PICKAWAY	2.32	UE8-5	18-Jul-62	6	3
OH PIKE	7.85	PB6-3	26-Jul-57	7	4
OH PIKE	1.34	PB13-7	12-Sep-57	2	4
OH PIKE	1.08	PB15-7	22-Sep-57	5	4
OH PIKE	208.78	S-4	9-Jul-62	3	4
OH PORTAGE	7.20	BJ2-2	31-Oct-51	5	6
OH PORTAGE	15.54	TP4-4	10-Mar-55	2	6
OH PORTAGE	0.82	TP9-5	19-Apr-55	2	6
OH PORTAGE	5.97	PB5-8	22-Jul-57	3	6
OH PORTAGE	85.56	S-4	9-Jul-62	9	6
OH PORTAGE	0.24	UE7-5	15-Jul-62	5	6
OH PREBLE	1.02	UpK3-6	5-Apr-53	7	6
OH PREBLE	1.19	PB2-5	22-Jun-57	9	6
OH PREBLE	3.01	PB9-5	27-Aug-57	6	6
OH PREBLE	1.34	PB13-7	12-Sep-57	2	6
OH PREBLE	208.78	S-4	9-Jul-62	3	6
OH PREBLE	0.13	UE7-5	15-Jul-62	8	6
OH PUTNAM	1.51	PB9-5	27-Aug-57	9	2
OH PUTNAM	0.13	UE7-5	15-Jul-62	8	2

COUNTY	uCi/ sqm	SHOT DAY	DATE	RANK FOR DAY	TIMES IN TOP 10
OH RICHLAND	6.32	BJ3-5	5-Nov-51	4	4
OH RICHLAND	1.43	BJ4-8	12-Nov-51	1	4
OH RICHLAND	103.59	S-4	9-Jul-62	6	4
OH RICHLAND	0.13	UE7-5	15-Jul-62	8	4
OH ROSS	7.85	PB6-3	26-Jul-57	7	5
OH ROSS	0.89	PB13-7	12-Sep-57	5	5
OH ROSS	1.08	PB15-7	22-Sep-57	5	5
OH ROSS	90.29	S-4	9-Jul-62	8	5
OH ROSS	2.32	UE8-5	18-Jul-62	6	5
OH SANDUSKY	1.43	BJ4-8	12-Nov-51	1	5
OH SANDUSKY	37.47	TP9-2	16-Apr-55	9	5
OH SANDUSKY	0.85	PB13-7	12-Sep-57	6	5
OH SANDUSKY	1.08	PB15-7	22-Sep-57	5	5
OH SANDUSKY	90.29	S-4	9-Jul-62	8	5
OH SCIOTO	1.02	UpK3-6	5-Apr-53	7	4
OH SCIOTO	1.47	PB6-7	30-Jul-57	10	4
OH SCIOTO	1.34	PB13-7	12-Sep-57	2	4
OH SCIOTO	359.52	S-4	9-Jul-62	1	4
OH SENECA	0.89	UpK8-5	12-May-53	6	4
OH SENECA	66.19	TP9-2	16-Apr-55	4	4
OH SENECA	1.08	PB15-7	22-Sep-57	5	4
OH SENECA	208.78	S-4	9-Jul-62	3	4
OH SHELBY	1.42	R1-3	30-Jan-51	1	7
OH SHELBY	2.00	TS4-5	5-May-52	4	7
OH SHELBY	2.00	TS4-5	5-May-52	4	7
OH SHELBY	3.01	PB9-5	27-Aug-57	6	7
OH SHELBY	0.89	PB13-7	12-Sep-57	5	7
OH SHELBY	1.08	PB15-7	22-Sep-57	5	7
OH SHELBY	0.11	UE7-5	15-Jul-62	10	7
OH STARK	8.96	PB5-8	22-Jul-57	2	3
OH STARK	0.90	PB15-7	22-Sep-57	7	3
OH STARK	90.29	S-4	9-Jul-62	8	3
OH SUMMIT	7.20	BJ2-2	31-Oct-51	5	6
OH SUMMIT	6.32	BJ3-5	5-Nov-51	4	6
OH SUMMIT	0.95	BJ4-8	12-Nov-51	2	6

COUNTY	uCi/ sqm	SHOT DAY	DATE	RANK FOR DAY	TIMES IN TOP 10
OH SUMMIT	11.58	UpK9-3	21-May-53	4	6
OH SUMMIT	8.96	PB5-8	22-Jul-57	2	6
OH SUMMIT	0.24	UE7-5	15-Jul-62	5	6
OH TRUMBULL	4.21	BJ3-5	5-Nov-51	7	4
OH TRUMBULL	5.97	PB5-8	22-Jul-57	3	4
OH TRUMBULL	90.29	S-4	9-Jul-62	8	4
OH TRUMBULL	0.24	UE7-5	15-Jul-62	5	4
OH TUSCARAWAS	7.20	BJ2-2	31-Oct-51	5	4
OH TUSCARAWAS	8.96	PB5-8	22-Jul-57	2	4
OH TUSCARAWAS	0.90	PB15-7	22-Sep-57	7	4
OH TUSCARAWAS	103.59	S-4	9-Jul-62	6	4
OH UNION	2.00	TS4-5	5-May-52	4	6
OH UNION	2.00	TS4-5	5-May-52	4	6
OH UNION	1.34	PB13-7	12-Sep-57	2	6
OH UNION	1.08	PB15-7	22-Sep-57	5	6
OH UNION	85.56	S-4	9-Jul-62	9	6
OH UNION	0.11	UE7-5	15-Jul-62	10	6
OH VAN WERT	1.51	PB9-5	27-Aug-57	9	2
OH VAN WERT	0.13	UE7-5	15-Jul-62	8	2
OH VINTON	8.96	PB5-8	22-Jul-57	2	6
OH VINTON	7.85	PB6-3	26-Jul-57	7	6
OH VINTON	0.89	PB13-7	12-Sep-57	5	6
OH VINTON	1.08	PB15-7	22-Sep-57	5	6
OH VINTON	85.56	S-4	9-Jul-62	9	6
OH VINTON	4.76	UE8-5	18-Jul-62	2	6
OH WARREN	7.85	PB6-3	26-Jul-57	7	3
OH WARREN	1.34	PB13-7	12-Sep-57	2	3
OH WARREN	208.78	S-4	9-Jul-62	3	3
OH WASHINGTON	2.57	TP11-8	22-May-55	5	4
OH WASHINGTON	3.12	TP11-9	23-May-55	8	4
OH WASHINGTON	14.93	PB5-8	22-Jul-57	1	4
OH WASHINGTON	90.29	S-4	9-Jul-62	8	4
OH WAYNE	14.93	PB5-8	22-Jul-57	1	1
OH WILLIAMS	1.25	PB13-7	12-Sep-57	3	3
OH WILLIAMS	0.89	PB15-6	21-Sep-57	2	3

COUNTY	uCi/ sqm	SHOT DAY	DATE	RANK FOR DAY	TIMES IN TOP 10
OH WILLIAMS	0.12	UE7-5	15-Jul-62	9	3
OH WOOD	1.25	PB13-7	12-Sep-57	3	3
OH WOOD	85.56	S-4	9-Jul-62	9	3
OH WOOD	0.24	UE7-5	15-Jul-62	5	3
OH WYANDOT	7.20	BJ2-2	31-Oct-51	5	4
OH WYANDOT	6.32	BJ3-5	5-Nov-51	4	4
OH WYANDOT	1.43	BJ4-8	12-Nov-51	1	4
OH WYANDOT	208.78	S-4	9-Jul-62	3	4

COUNTY	uCi/ sqm	SHOT DAY	DATE	RANK FOR DAY	TIMES IN TOP 10
OK ADAIR	2.23	UpK2-10	2-Apr-53	3	6
OK ADAIR	2.10	UpK3-3	2-Apr-53	4	6
OK ADAIR	4.10	UpK3-5	4-Apr-53	5	6
OK ADAIR	1.83	UpK8-4	11-May-53	4	6
OK ADAIR	6.53	PB13-6	11-Sep-57	9	6
OK ADAIR	4.15	PB13-9	14-Sep-57	7	6
OK ALFALFA	3.70	BJ5-6	24-Nov-51	2	7
OK ALFALFA	3.70	BJ5-6	24-Nov-51	2	7
OK ALFALFA	3.78	PB3-2	25-Jun-57	3	7
OK ALFALFA	1.63	PB5-7	21-Jul-57	8	7
OK ALFALFA	2.11	PB13-8	13-Sep-57	4	7
OK ALFALFA	4.15	PB13-9	14-Sep-57	7	7
OK ALFALFA	2.42	PB14-7	20-Sep-57	2	7
OK ATOKA	2.48	BJ5-6	24-Nov-51	4	9
OK ATOKA	2.48	BJ5-6	24-Nov-51	4	9
OK ATOKA	3.04	UpK8-4	11-May-53	2	9
OK ATOKA	1.16	UpK8-5	12-May-53	5	9
OK ATOKA	12.96	TP6-2	23-Mar-55	2	9
OK ATOKA	13.82	TP11-5	19-May-55	8	9
OK ATOKA	1.78	PB1-5	1-Jun-57	8	9
OK ATOKA	4.15	PB13-9	14-Sep-57	7	9
OK ATOKA	2.60	PB18-7	13-Oct-57	1	9
OK BEAVER	1.22	BJ5-6	24-Nov-51	6	13
OK BEAVER	1.22	BJ5-6	24-Nov-51	6	13
OK BEAVER	1.02	TS4-4	4-May-52	1	13
OK BEAVER	1.02	TS4-4	4-May-52	1	13
OK BEAVER	1.74	TS4-6	6-May-52	5	13
OK BEAVER	1.74	TS4-6	6-May-52	5	13
OK BEAVER	8.19	UpK3-5	4-Apr-53	2	13
OK BEAVER	4.23	UpK6-3	20-Apr-53	2	13
OK BEAVER	25.62	TP7-2	30-Mar-55	4	13
OK BEAVER	31.48	TP10-2	6-May-55	6	13
OK BEAVER	0.93	PB3-8	1-Jul-57	4	13
OK BEAVER	2.11	PB13-8	13-Sep-57	4	13
OK BEAVER	1.92	PB18-6	12-Oct-57	3	13

COUNTY	uCi/ sqm	SHOT DAY	DATE	RANK FOR DAY	TIMES IN TOP 10
OK BECKHAM	0.95	BJ5-6	24-Nov-51	9	9
OK BECKHAM	0.95	BJ5-6	24-Nov-51	9	9
OK BECKHAM	13.66	UpK3-5	4-Apr-53	1	9
OK BECKHAM	4.23	UpK6-3	20-Apr-53	2	9
OK BECKHAM	6.07	UpK8-2	9-May-53	9	9
OK BECKHAM	25.62	TP7-2	30-Mar-55	4	9
OK BECKHAM	31.48	TP10-2	6-May-55	6	9
OK BECKHAM	2.42	PB14-7	20-Sep-57	2	9
OK BECKHAM	3.21	PB18-6	12-Oct-57	1	9
OK BLAINE	3.70	BJ5-6	24-Nov-51	2	9
OK BLAINE	3.70	BJ5-6	24-Nov-51	2	9
OK BLAINE	1.71	UpK3-5	4-Apr-53	10	9
OK BLAINE	1.54	PB3-2	25-Jun-57	8	9
OK BLAINE	2.69	PB5-7	21-Jul-57	5	9
OK BLAINE	2.11	PB13-8	13-Sep-57	4	9
OK BLAINE	4.15	PB13-9	14-Sep-57	7	9
OK BLAINE	2.42	PB14-7	20-Sep-57	2	9
OK BLAINE	1.92	PB18-6	12-Oct-57	3	9
OK BRYAN	3.70	BJ5-6	24-Nov-51	2	10
OK BRYAN	3.70	BJ5-6	24-Nov-51	2	10
OK BRYAN	41.05	UpK7-3	27-Apr-53	10	10
OK BRYAN	1.22	UpK8-4	11-May-53	8	10
OK BRYAN	1.16	UpK8-5	12-May-53	5	10
OK BRYAN	12.96	TP6-2	23-Mar-55	2	10
OK BRYAN	14.37	TP11-5	19-May-55	3	10
OK BRYAN	1.78	PB1-5	1-Jun-57	8	10
OK BRYAN	0.83	PB13-8	13-Sep-57	9	10
OK BRYAN	2.60	PB18-7	13-Oct-57	1	10
OK CADDO	2.48	BJ5-6	24-Nov-51	4	7
OK CADDO	2.48	BJ5-6	24-Nov-51	4	7
OK CADDO	1.71	UpK3-5	4-Apr-53	10	7
OK CADDO	0.89	TP9-13	27-Apr-55	5	7
OK CADDO	1.63	PB5-7	21-Jul-57	8	7
OK CADDO	4.04	PB14-7	20-Sep-57	1	7
OK CADDO	3.21	PB18-6	12-Oct-57	1	7

COUNTY	uCi/ sqm	SHOT DAY	DATE	RANK FOR DAY	TIMES IN TOP 10
OK CANADIAN	3.70	BJ5-6	24-Nov-51	2	9
OK CANADIAN	3.70	BJ5-6	24-Nov-51	2	9
OK CANADIAN	1.71	UpK3-5	4-Apr-53	10	9
OK CANADIAN	0.89	TP9-13	27-Apr-55	5	9
OK CANADIAN	1.54	PB3-2	25-Jun-57	8	9
OK CANADIAN	1.63	PB5-7	21-Jul-57	8	9
OK CANADIAN	4.15	PB13-9	14-Sep-57	7	9
OK CANADIAN	4.04	PB14-7	20-Sep-57	1	9
OK CANADIAN	3.21	PB18-6	12-Oct-57	1	9
OK CARTER	2.48	BJ5-6	24-Nov-51	4	7
OK CARTER	2.48	BJ5-6	24-Nov-51	4	7
OK CARTER	1.83	UpK8-4	11-May-53	4	7
OK CARTER	12.96	TP6-2	23-Mar-55	2	7
OK CARTER	2.96	PB1-5	1-Jun-57	5	7
OK CARTER	2.50	PB13-9	14-Sep-57	8	7
OK CARTER	1.92	PB18-6	12-Oct-57	3	7
OK CHEROKEE	3.70	BJ5-6	24-Nov-51	2	10
OK CHEROKEE	3.70	BJ5-6	24-Nov-51	2	10
OK CHEROKEE	2.23	UpK2-10	2-Apr-53	3	10
OK CHEROKEE	2.10	UpK3-3	2-Apr-53	4	10
OK CHEROKEE	4.10	UpK3-5	4-Apr-53	5	10
OK CHEROKEE	1.83	UpK8-4	11-May-53	4	10
OK CHEROKEE	0.92	TP9-13	27-Apr-55	4	10
OK CHEROKEE	2.11	PB13-8	13-Sep-57	4	10
OK CHEROKEE	4.15	PB13-9	14-Sep-57	7	10
OK CHEROKEE	4.04	PB14-7	20-Sep-57	1	10
OK CHOCTAW	1.22	UpK3-6	5-Apr-53	6	10
OK CHOCTAW	3.91	UpK5-4	14-Apr-53	1	10
OK CHOCTAW	1.32	UpK8-3	10-May-53	9	10
OK CHOCTAW	1.83	UpK8-4	11-May-53	4	10
OK CHOCTAW	12.96	TP6-2	23-Mar-55	2	10
OK CHOCTAW	0.92	TP9-13	27-Apr-55	4	10
OK CHOCTAW	13.79	TP11-5	19-May-55	9	10
OK CHOCTAW	1.78	PB1-5	1-Jun-57	8	10
OK CHOCTAW	4.15	PB13-9	14-Sep-57	7	10

COUNTY	uCi/ sqm	SHOT DAY	DATE	RANK FOR DAY	TIMES IN TOP 10
OK CHOCTAW	2.60	PB18-7	13-Oct-57	1	10
OK CIMARRON	1.02	TS4-4	4-May-52	1	13
OK CIMARRON	1.02	TS4-4	4-May-52	1	13
OK CIMARRON	3.48	TS4-6	6-May-52	1	13
OK CIMARRON	3.48	TS4-6	6-May-52	1	13
OK CIMARRON	2.73	UpK3-5	4-Apr-53	8	13
OK CIMARRON	4.23	UpK6-3	20-Apr-53	2	13
OK CIMARRON	6.07	UpK8-2	9-May-53	9	13
OK CIMARRON	1.22	UpK8-4	11-May-53	8	13
OK CIMARRON	51.23	TP7-2	30-Mar-55	3	13
OK CIMARRON	31.48	TP10-2	6-May-55	6	13
OK CIMARRON	0.89	PB1-11	7-Jun-57	4	13
OK CIMARRON	0.98	PB1-12	8-Jun-57	1	13
OK CIMARRON	1.35	PB18-5	11-Oct-57	1	13
OK CLEVELAND	2.48	BJ5-6	24-Nov-51	4	9
OK CLEVELAND	2.48	BJ5-6	24-Nov-51	4	9
OK CLEVELAND	1.83	UpK8-4	11-May-53	4	9
OK CLEVELAND	12.96	TP6-2	23-Mar-55	2	9
OK CLEVELAND	3.82	TP10-5	9-May-55	9	9
OK CLEVELAND	1.78	PB1-5	1-Jun-57	8	9
OK CLEVELAND	1.28	PB13-8	13-Sep-57	6	9
OK CLEVELAND	4.15	PB13-9	14-Sep-57	7	9
OK CLEVELAND	1.92	PB18-6	12-Oct-57	3	9
OK COAL	1.83	UpK8-4	11-May-53	4	7
OK COAL	1.16	UpK8-5	12-May-53	5	7
OK COAL	12.96	TP6-2	23-Mar-55	2	7
OK COAL	2.96	PB1-5	1-Jun-57	5	7
OK COAL	1.63	PB5-7	21-Jul-57	8	7
OK COAL	4.15	PB13-9	14-Sep-57	7	7
OK COAL	1.92	PB18-6	12-Oct-57	3	7
OK COMANCHE	1.22	BJ5-6	24-Nov-51	6	7
OK COMANCHE	1.22	BJ5-6	24-Nov-51	6	7
OK COMANCHE	1.83	UpK8-4	11-May-53	4	7
OK COMANCHE	12.96	TP6-2	23-Mar-55	2	7
OK COMANCHE	1.78	PB1-5	1-Jun-57	8	7

COUNTY	uCi/ sqm	SHOT DAY	DATE	RANK FOR DAY	TIMES IN TOP 10
OK COMANCHE	1.63	PB5-7	21-Jul-57	8	7
OK COMANCHE	3.21	PB18-6	12-Oct-57	1	7
OK COTTON	1.22	BJ5-6	24-Nov-51	6	7
OK COTTON	1.22	BJ5-6	24-Nov-51	6	7
OK COTTON	12.96	TP6-2	23-Mar-55	2	7
OK COTTON	1.63	PB5-7	21-Jul-57	8	7
OK COTTON	1.28	PB13-8	13-Sep-57	6	7
OK COTTON	2.50	PB13-9	14-Sep-57	8	7
OK COTTON	3.21	PB18-6	12-Oct-57	1	7
OK CRAIG	8.76	BJ2-2	31-Oct-51	3	9
OK CRAIG	1.47	UpK2-10	2-Apr-53	5	9
OK CRAIG	1.40	UpK3-3	2-Apr-53	6	9
OK CRAIG	6.83	UpK3-5	4-Apr-53	3	9
OK CRAIG	0.92	TP9-13	27-Apr-55	4	9
OK CRAIG	1.76	PB2-4	21-Jun-57	6	9
OK CRAIG	1.28	PB13-8	13-Sep-57	6	9
OK CRAIG	2.50	PB13-9	14-Sep-57	8	9
OK CRAIG	2.42	PB14-7	20-Sep-57	2	9
OK CREEK	3.70	BJ5-6	24-Nov-51	2	10
OK CREEK	3.70	BJ5-6	24-Nov-51	2	10
OK CREEK	2.23	UpK2-10	2-Apr-53	3	10
OK CREEK	2.10	UpK3-3	2-Apr-53	4	10
OK CREEK	4.10	UpK3-5	4-Apr-53	5	10
OK CREEK	48.33	UpK7-3	27-Apr-53	5	10
OK CREEK	0.92	TP9-13	27-Apr-55	4	10
OK CREEK	1.28	PB13-8	13-Sep-57	6	10
OK CREEK	4.15	PB13-9	14-Sep-57	7	10
OK CREEK	2.42	PB14-7	20-Sep-57	2	10
OK CUSTER	2.48	BJ5-6	24-Nov-51	4	9
OK CUSTER	2.48	BJ5-6	24-Nov-51	4	9
OK CUSTER	1.71	UpK3-5	4-Apr-53	10	9
OK CUSTER	0.87	UpK6-5	22-Apr-53	8	9
OK CUSTER	1.63	PB5-7	21-Jul-57	8	9
OK CUSTER	2.11	PB13-8	13-Sep-57	4	9
OK CUSTER	1.64	PB13-9	14-Sep-57	10	9

COUNTY	uCi/ sqm	SHOT DAY	DATE	RANK FOR DAY	TIMES IN TOP 10
OK CUSTER	2.42	PB14-7	20-Sep-57	2	9
OK CUSTER	1.92	PB18-6	12-Oct-57	3	9
OK DELAWARE	8.76	BJ2-2	31-Oct-51	3	8
OK DELAWARE	2.23	UpK2-10	2-Apr-53	3	8
OK DELAWARE	2.10	UpK3-3	2-Apr-53	4	8
OK DELAWARE	6.83	UpK3-5	4-Apr-53	3	8
OK DELAWARE	5.09	TP10-5	9-May-55	1	8
OK DELAWARE	0.83	PB13-8	13-Sep-57	9	8
OK DELAWARE	4.15	PB13-9	14-Sep-57	7	8
OK DELAWARE	2.42	PB14-7	20-Sep-57	2	8
OK DEWEY	3.70	BJ5-6	24-Nov-51	2	8
OK DEWEY	3.70	BJ5-6	24-Nov-51	2	8
OK DEWEY	1.71	UpK3-5	4-Apr-53	10	8
OK DEWEY	1.63	PB5-7	21-Jul-57	8	8
OK DEWEY	1.28	PB13-8	13-Sep-57	6	8
OK DEWEY	1.64	PB13-9	14-Sep-57	10	8
OK DEWEY	0.81	PB14-7	20-Sep-57	9	8
OK DEWEY	1.92	PB18-6	12-Oct-57	3	8
OK ELLIS	2.48	BJ5-6	24-Nov-51	4	13
OK ELLIS	2.48	BJ5-6	24-Nov-51	4	13
OK ELLIS	1.02	TS4-4	4-May-52	1	13
OK ELLIS	1.02	TS4-4	4-May-52	1	13
OK ELLIS	1.30	TS4-6	6-May-52	10	13
OK ELLIS	1.30	TS4-6	6-May-52	10	13
OK ELLIS	8.19	UpK3-5	4-Apr-53	2	13
OK ELLIS	4.23	UpK6-3	20-Apr-53	2	13
OK ELLIS	1.22	UpK8-4	11-May-53	8	13
OK ELLIS	25.62	TP7-2	30-Mar-55	4	13
OK ELLIS	31.48	TP10-2	6-May-55	6	13
OK ELLIS	2.11	PB13-8	13-Sep-57	4	13
OK ELLIS	1.92	PB18-6	12-Oct-57	3	13
OK GARFIELD	6.18	BJ5-6	24-Nov-51	1	9
OK GARFIELD	6.18	BJ5-6	24-Nov-51	1	9
OK GARFIELD	1.71	UpK3-5	4-Apr-53	10	9
OK GARFIELD	3.78	PB3-2	25-Jun-57	3	9

COUNTY	uCi/ sqm	SHOT DAY	DATE	RANK FOR DAY	TIMES IN TOP 10
OK GARFIELD	0.93	PB3-8	1-Jul-57	4	9
OK GARFIELD	2.69	PB5-7	21-Jul-57	5	9
OK GARFIELD	2.11	PB13-8	13-Sep-57	4	9
OK GARFIELD	4.15	PB13-9	14-Sep-57	7	9
OK GARFIELD	1.62	PB14-7	20-Sep-57	4	9
OK GARVIN	1.22	BJ5-6	24-Nov-51	6	9
OK GARVIN	1.22	BJ5-6	24-Nov-51	6	9
OK GARVIN	2.30	TS6-4	28-May-52	10	9
OK GARVIN	1.83	UpK8-4	11-May-53	4	9
OK GARVIN	12.96	TP6-2	23-Mar-55	2	9
OK GARVIN	3.94	TP10-5	9-May-55	8	9
OK GARVIN	1.78	PB1-5	1-Jun-57	8	9
OK GARVIN	2.50	PB13-9	14-Sep-57	8	9
OK GARVIN	1.92	PB18-6	12-Oct-57	3	9
OK GRADY	3.70	BJ5-6	24-Nov-51	2	8
OK GRADY	3.70	BJ5-6	24-Nov-51	2	8
OK GRADY	1.22	UpK8-4	11-May-53	8	8
OK GRADY	12.96	TP6-2	23-Mar-55	2	8
OK GRADY	1.78	PB1-5	1-Jun-57	8	8
OK GRADY	1.63	PB5-7	21-Jul-57	8	8
OK GRADY	2.50	PB13-9	14-Sep-57	8	8
OK GRADY	3.21	PB18-6	12-Oct-57	1	8
OK GRANT	3.70	BJ5-6	24-Nov-51	2	8
OK GRANT	3.70	BJ5-6	24-Nov-51	2	8
OK GRANT	3.78	PB3-2	25-Jun-57	3	8
OK GRANT	0.93	PB3-8	1-Jul-57	4	8
OK GRANT	2.69	PB5-7	21-Jul-57	5	8
OK GRANT	2.11	PB13-8	13-Sep-57	4	8
OK GRANT	2.50	PB13-9	14-Sep-57	8	8
OK GRANT	4.04	PB14-7	20-Sep-57	1	8
OK GREER	13.66	UpK3-5	4-Apr-53	1	7
OK GREER	4.23	UpK6-3	20-Apr-53	2	7
OK GREER	6.07	UpK8-2	9-May-53	9	7
OK GREER	25.62	TP7-2	30-Mar-55	4	7
OK GREER	31.48	TP10-2	6-May-55	6	7

COUNTY	uCi/ sqm	SHOT DAY	DATE	RANK FOR DAY	TIMES IN TOP 10
OK GREER	1.78	PB1-5	1-Jun-57	8	7
OK GREER	3.21	PB18-6	12-Oct-57	1	7
OK HARMON	13.66	UpK3-5	4-Apr-53	1	7
OK HARMON	4.23	UpK6-3	20-Apr-53	2	7
OK HARMON	6.07	UpK8-2	9-May-53	9	7
OK HARMON	25.62	TP7-2	30-Mar-55	4	7
OK HARMON	31.48	TP10-2	6-May-55	6	7
OK HARMON	2.96	PB1-5	1-Jun-57	5	7
OK HARMON	1.92	PB18-6	12-Oct-57	3	7
OK HARPER	1.22	BJ5-6	24-Nov-51	6	9
OK HARPER	1.22	BJ5-6	24-Nov-51	6	9
OK HARPER	1.99	TS4-6	6-May-52	4	9
OK HARPER	1.99	TS4-6	6-May-52	4	9
OK HARPER	1.76	PB2-4	21-Jun-57	6	9
OK HARPER	0.93	PB3-8	1-Jul-57	4	9
OK HARPER	2.11	PB13-8	13-Sep-57	4	9
OK HARPER	1.64	PB13-9	14-Sep-57	10	9
OK HARPER	1.92	PB18-6	12-Oct-57	3	9
OK HASKELL	2.23	UpK2-10	2-Apr-53	3	8
OK HASKELL	2.10	UpK3-3	2-Apr-53	4	8
OK HASKELL	1.07	UpK5-4	14-Apr-53	8	8
OK HASKELL	1.83	UpK8-4	11-May-53	4	8
OK HASKELL	1.16	UpK8-5	12-May-53	5	8
OK HASKELL	12.96	TP6-2	23-Mar-55	2	8
OK HASKELL	0.92	TP9-13	27-Apr-55	4	8
OK HASKELL	4.15	PB13-9	14-Sep-57	7	8
OK HUGHES	2.48	BJ5-6	24-Nov-51	4	13
OK HUGHES	2.48	BJ5-6	24-Nov-51	4	13
OK HUGHES	2.23	UpK2-10	2-Apr-53	3	13
OK HUGHES	2.10	UpK3-3	2-Apr-53	4	13
OK HUGHES	4.10	UpK3-5	4-Apr-53	5	13
OK HUGHES	1.07	UpK5-4	14-Apr-53	8	13
OK HUGHES	1.83	UpK8-4	11-May-53	4	13
OK HUGHES	12.96	TP6-2	23-Mar-55	2	13
OK HUGHES	0.92	TP9-13	27-Apr-55	4	13

COUNTY	uCi/ sqm	SHOT DAY	DATE	RANK FOR DAY	TIMES IN TOP 10
OK HUGHES	2.96	PB1-5	1-Jun-57	5	13
OK HUGHES	1.28	PB13-8	13-Sep-57	6	13
OK HUGHES	4.15	PB13-9	14-Sep-57	7	13
OK HUGHES	1.92	PB18-6	12-Oct-57	3	13
OK JACKSON	2.48	BJ5-6	24-Nov-51	4	8
OK JACKSON	2.48	BJ5-6	24-Nov-51	4	8
OK JACKSON	5.43	UpK6-2	19-Apr-53	9	8
OK JACKSON	1.58	UpK6-4	21-Apr-53	3	8
OK JACKSON	1.30	UpK6-5	22-Apr-53	4	8
OK JACKSON	3.18	TP6-2	23-Mar-55	6	8
OK JACKSON	2.96	PB1-5	1-Jun-57	5	8
OK JACKSON	3.21	PB18-6	12-Oct-57	1	8
OK JEFFERSON	1.22	BJ5-6	24-Nov-51	6	7
OK JEFFERSON	1.22	BJ5-6	24-Nov-51	6	7
OK JEFFERSON	1.83	UpK8-4	11-May-53	4	7
OK JEFFERSON	12.96	TP6-2	23-Mar-55	2	7
OK JEFFERSON	2.96	PB1-5	1-Jun-57	5	7
OK JEFFERSON	2.50	PB13-9	14-Sep-57	8	7
OK JEFFERSON	3.21	PB18-6	12-Oct-57	1	7
OK JOHNSTON	2.48	BJ5-6	24-Nov-51	4	7
OK JOHNSTON	2.48	BJ5-6	24-Nov-51	4	7
OK JOHNSTON	12.96	TP6-2	23-Mar-55	2	7
OK JOHNSTON	2.96	PB1-5	1-Jun-57	5	7
OK JOHNSTON	1.63	PB5-7	21-Jul-57	8	7
OK JOHNSTON	4.15	PB13-9	14-Sep-57	7	7
OK JOHNSTON	1.92	PB18-6	12-Oct-57	3	7
OK KAY	3.70	BJ5-6	24-Nov-51	2	10
OK KAY	3.70	BJ5-6	24-Nov-51	2	10
OK KAY	0.87	UpK6-5	22-Apr-53	8	10
OK KAY	1.32	TP9-13	27-Apr-55	3	10
OK KAY	1.76	PB2-4	21-Jun-57	6	10
OK KAY	3.78	PB3-2	25-Jun-57	3	10
OK KAY	1.63	PB5-7	21-Jul-57	8	10
OK KAY	2.11	PB13-8	13-Sep-57	4	10
OK KAY	2.50	PB13-9	14-Sep-57	8	10

COUNTY	uCi/ sqm	SHOT DAY	DATE	RANK FOR DAY	TIMES IN TOP 10
OK KAY	2.42	PB14-7	20-Sep-57	2	10
OK KINGFISHER	2.48	BJ5-6	24-Nov-51	4	10
OK KINGFISHER	2.48	BJ5-6	24-Nov-51	4	10
OK KINGFISHER	1.71	UpK3-5	4-Apr-53	10	10
OK KINGFISHER	1.76	PB2-4	21-Jun-57	6	10
OK KINGFISHER	1.54	PB3-2	25-Jun-57	8	10
OK KINGFISHER	2.69	PB5-7	21-Jul-57	5	10
OK KINGFISHER	2.11	PB13-8	13-Sep-57	4	10
OK KINGFISHER	4.15	PB13-9	14-Sep-57	7	10
OK KINGFISHER	2.42	PB14-7	20-Sep-57	2	10
OK KINGFISHER	3.21	PB18-6	12-Oct-57	1	10
OK KIOWA	2.48	BJ5-6	24-Nov-51	4	7
OK KIOWA	2.48	BJ5-6	24-Nov-51	4	7
OK KIOWA	5.43	UpK6-2	19-Apr-53	9	7
OK KIOWA	1.30	UpK6-5	22-Apr-53	4	7
OK KIOWA	3.18	TP6-2	23-Mar-55	6	7
OK KIOWA	2.96	PB1-5	1-Jun-57	5	7
OK KIOWA	1.29	PB18-6	12-Oct-57	6	7
OK LATIMER	2.23	UpK2-10	2-Apr-53	3	9
OK LATIMER	2.10	UpK3-3	2-Apr-53	4	9
OK LATIMER	0.81	UpK3-6	5-Apr-53	9	9
OK LATIMER	1.78	UpK5-4	14-Apr-53	4	9
OK LATIMER	1.12	UpK8-4	11-May-53	9	9
OK LATIMER	1.16	UpK8-5	12-May-53	5	9
OK LATIMER	12.96	TP6-2	23-Mar-55	2	9
OK LATIMER	0.92	TP9-13	27-Apr-55	4	9
OK LATIMER	4.15	PB13-9	14-Sep-57	7	9
OK LE FLORE	2.23	UpK2-10	2-Apr-53	3	8
OK LE FLORE	2.10	UpK3-3	2-Apr-53	4	8
OK LE FLORE	0.81	UpK3-6	5-Apr-53	9	8
OK LE FLORE	1.07	UpK5-4	14-Apr-53	8	8
OK LE FLORE	1.83	UpK8-4	11-May-53	4	8
OK LE FLORE	1.16	UpK8-5	12-May-53	5	8
OK LE FLORE	12.96	TP6-2	23-Mar-55	2	8
OK LE FLORE	4.15	PB13-9	14-Sep-57	7	8

COUNTY	uCi/ sqm	SHOT DAY	DATE	RANK FOR DAY	TIMES IN TOP 10
OK LINCOLN	3.70	BJ5-6	24-Nov-51	2	9
OK LINCOLN	3.70	BJ5-6	24-Nov-51	2	9
OK LINCOLN	1.71	UpK3-5	4-Apr-53	10	9
OK LINCOLN	0.87	UpK6-5	22-Apr-53	8	9
OK LINCOLN	0.89	TP9-13	27-Apr-55	5	9
OK LINCOLN	4.15	TP10-5	9-May-55	6	9
OK LINCOLN	2.11	PB13-8	13-Sep-57	4	9
OK LINCOLN	4.15	PB13-9	14-Sep-57	7	9
OK LINCOLN	4.04	PB14-7	20-Sep-57	1	9
OK LOGAN	3.70	BJ5-6	24-Nov-51	2	9
OK LOGAN	3.70	BJ5-6	24-Nov-51	2	9
OK LOGAN	1.71	UpK3-5	4-Apr-53	10	9
OK LOGAN	3.78	PB3-2	25-Jun-57	3	9
OK LOGAN	2.69	PB5-7	21-Jul-57	5	9
OK LOGAN	6.37	PB13-6	11-Sep-57	10	9
OK LOGAN	2.11	PB13-8	13-Sep-57	4	9
OK LOGAN	4.15	PB13-9	14-Sep-57	7	9
OK LOGAN	1.62	PB14-7	20-Sep-57	4	9
OK LOVE	2.48	BJ5-6	24-Nov-51	4	9
OK LOVE	2.48	BJ5-6	24-Nov-51	4	9
OK LOVE	5.74	UpK1-9	25-Mar-53	8	9
OK LOVE	1.22	UpK8-4	11-May-53	8	9
OK LOVE	12.96	TP6-2	23-Mar-55	2	9
OK LOVE	13.50	TP11-5	19-May-55	10	9
OK LOVE	1.78	PB1-5	1-Jun-57	8	9
OK LOVE	2.50	PB13-9	14-Sep-57	8	9
OK LOVE	2.60	PB18-7	13-Oct-57	1	9
OK MAJOR	6.18	BJ5-6	24-Nov-51	1	8
OK MAJOR	6.18	BJ5-6	24-Nov-51	1	8
OK MAJOR	1.71	UpK3-5	4-Apr-53	10	8
OK MAJOR	1.32	UpK8-3	10-May-53	9	8
OK MAJOR	2.24	PB3-2	25-Jun-57	5	8
OK MAJOR	2.11	PB13-8	13-Sep-57	4	8
OK MAJOR	4.15	PB13-9	14-Sep-57	7	8
OK MAJOR	0.81	PB14-7	20-Sep-57	9	8

COUNTY	uCi/ sqm	SHOT DAY	DATE	RANK FOR DAY	TIMES IN TOP 10
OK MARSHALL	2.48	BJ5-6	24-Nov-51	4	7
OK MARSHALL	2.48	BJ5-6	24-Nov-51	4	7
OK MARSHALL	1.83	UpK8-4	11-May-53	4	7
OK MARSHALL	12.96	TP6-2	23-Mar-55	2	7
OK MARSHALL	13.85	TP11-5	19-May-55	7	7
OK MARSHALL	4.15	PB13-9	14-Sep-57	7	7
OK MARSHALL	1.56	PB18-7	13-Oct-57	4	7
OK MAYES	3.70	BJ5-6	24-Nov-51	2	8
OK MAYES	3.70	BJ5-6	24-Nov-51	2	8
OK MAYES	2.23	UpK2-10	2-Apr-53	3	8
OK MAYES	2.10	UpK3-3	2-Apr-53	4	8
OK MAYES	6.83	UpK3-5	4-Apr-53	3	8
OK MAYES	1.28	PB13-8	13-Sep-57	6	8
OK MAYES	4.15	PB13-9	14-Sep-57	7	8
OK MAYES	2.42	PB14-7	20-Sep-57	2	8
OK MCCLAIN	2.48	BJ5-6	24-Nov-51	4	11
OK MCCLAIN	2.48	BJ5-6	24-Nov-51	4	11
OK MCCLAIN	2.53	TS6-4	28-May-52	8	11
OK MCCLAIN	1.83	UpK8-4	11-May-53	4	11
OK MCCLAIN	12.96	TP6-2	23-Mar-55	2	11
OK MCCLAIN	3.74	TP10-5	9-May-55	10	11
OK MCCLAIN	1.78	PB1-5	1-Jun-57	8	11
OK MCCLAIN	1.63	PB5-7	21-Jul-57	8	11
OK MCCLAIN	1.28	PB13-8	13-Sep-57	6	11
OK MCCLAIN	4.15	PB13-9	14-Sep-57	7	11
OK MCCLAIN	3.21	PB18-6	12-Oct-57	1	11
OK MCCURTAIN	1.22	UpK3-6	5-Apr-53	6	10
OK MCCURTAIN	3.91	UpK5-4	14-Apr-53	1	10
OK MCCURTAIN	1.32	UpK8-3	10-May-53	9	10
OK MCCURTAIN	1.83	UpK8-4	11-May-53	4	10
OK MCCURTAIN	12.96	TP6-2	23-Mar-55	2	10
OK MCCURTAIN	0.92	TP9-13	27-Apr-55	4	10
OK MCCURTAIN	3.80	PB5-4	18-Jul-57	10	10
OK MCCURTAIN	7.99	PB8-3	20-Aug-57	9	10
OK MCCURTAIN	2.50	PB13-9	14-Sep-57	8	10

COUNTY	uCi/ sqm	SHOT DAY	DATE	RANK FOR DAY	TIMES IN TOP 10
OK MCCURTAIN	1.56	PB18-7	13-Oct-57	4	10
OK MCINTOSH	2.48	BJ5-6	24-Nov-51	4	14
OK MCINTOSH	2.48	BJ5-6	24-Nov-51	4	14
OK MCINTOSH	3.70	UpK2-10	2-Apr-53	1	14
OK MCINTOSH	3.51	UpK3-3	2-Apr-53	2	14
OK MCINTOSH	2.73	UpK3-5	4-Apr-53	8	14
OK MCINTOSH	1.07	UpK5-4	14-Apr-53	8	14
OK MCINTOSH	1.83	UpK8-4	11-May-53	4	14
OK MCINTOSH	12.96	TP6-2	23-Mar-55	2	14
OK MCINTOSH	0.92	TP9-13	27-Apr-55	4	14
OK MCINTOSH	4.00	TP10-5	9-May-55	7	14
OK MCINTOSH	2.96	PB1-5	1-Jun-57	5	14
OK MCINTOSH	7.31	PB13-6	11-Sep-57	4	14
OK MCINTOSH	0.83	PB13-8	13-Sep-57	9	14
OK MCINTOSH	4.15	PB13-9	14-Sep-57	7	14
OK MURRAY	2.48	BJ5-6	24-Nov-51	4	8
OK MURRAY	2.48	BJ5-6	24-Nov-51	4	8
OK MURRAY	3.04	UpK8-4	11-May-53	2	8
OK MURRAY	12.96	TP6-2	23-Mar-55	2	8
OK MURRAY	1.78	PB1-5	1-Jun-57	8	8
OK MURRAY	1.63	PB5-7	21-Jul-57	8	8
OK MURRAY	4.15	PB13-9	14-Sep-57	7	8
OK MURRAY	1.92	PB18-6	12-Oct-57	3	8
OK MUSKOGEE	3.70	BJ5-6	24-Nov-51	2	11
OK MUSKOGEE	3.70	BJ5-6	24-Nov-51	2	11
OK MUSKOGEE	2.23	UpK2-10	2-Apr-53	3	11
OK MUSKOGEE	2.10	UpK3-3	2-Apr-53	4	11
OK MUSKOGEE	4.10	UpK3-5	4-Apr-53	5	11
OK MUSKOGEE	1.83	UpK8-4	11-May-53	4	11
OK MUSKOGEE	0.92	TP9-13	27-Apr-55	4	11
OK MUSKOGEE	4.31	TP10-5	9-May-55	5	11
OK MUSKOGEE	7.23	PB13-6	11-Sep-57	5	11
OK MUSKOGEE	4.15	PB13-9	14-Sep-57	7	11
OK MUSKOGEE	4.04	PB14-7	20-Sep-57	1	11
OK NOBLE	3.70	BJ5-6	24-Nov-51	2	9

COUNTY	uCi/ sqm	SHOT DAY	DATE	RANK FOR DAY	TIMES IN TOP 10
OK NOBLE	3.70	BJ5-6	24-Nov-51	2	9
OK NOBLE	1.71	UpK3-5	4-Apr-53	10	9
OK NOBLE	2.24	PB3-2	25-Jun-57	5	9
OK NOBLE	0.93	PB3-8	1-Jul-57	4	9
OK NOBLE	1.63	PB5-7	21-Jul-57	8	9
OK NOBLE	2.11	PB13-8	13-Sep-57	4	9
OK NOBLE	4.15	PB13-9	14-Sep-57	7	9
OK NOBLE	0.81	PB14-7	20-Sep-57	9	9
OK NOWATA	8.76	BJ2-2	31-Oct-51	3	8
OK NOWATA	2.23	UpK2-10	2-Apr-53	3	8
OK NOWATA	2.10	UpK3-3	2-Apr-53	4	8
OK NOWATA	6.83	UpK3-5	4-Apr-53	3	8
OK NOWATA	6.77	PB13-6	11-Sep-57	7	8
OK NOWATA	1.28	PB13-8	13-Sep-57	6	8
OK NOWATA	2.50	PB13-9	14-Sep-57	8	8
OK NOWATA	2.42	PB14-7	20-Sep-57	2	8
OK OKFUSKEE	2.48	BJ5-6	24-Nov-51	4	11
OK OKFUSKEE	2.48	BJ5-6	24-Nov-51	4	11
OK OKFUSKEE	2.23	UpK2-10	2-Apr-53	3	11
OK OKFUSKEE	2.10	UpK3-3	2-Apr-53	4	11
OK OKFUSKEE	4.10	UpK3-5	4-Apr-53	5	11
OK OKFUSKEE	1.07	UpK5-4	14-Apr-53	8	11
OK OKFUSKEE	1.38	TP9-13	27-Apr-55	2	11
OK OKFUSKEE	0.89	PB13-7	12-Sep-57	5	11
OK OKFUSKEE	1.28	PB13-8	13-Sep-57	6	11
OK OKFUSKEE	4.15	PB13-9	14-Sep-57	7	11
OK OKFUSKEE	4.04	PB14-7	20-Sep-57	1	11
OK OKLAHOMA	1.54	BJ5-6	24-Nov-51	5	8
OK OKLAHOMA	1.54	BJ5-6	24-Nov-51	5	8
OK OKLAHOMA	1.71	UpK3-5	4-Apr-53	10	8
OK OKLAHOMA	0.87	UpK6-5	22-Apr-53	8	8
OK OKLAHOMA	1.76	PB2-4	21-Jun-57	6	8
OK OKLAHOMA	0.88	PB13-8	13-Sep-57	8	8
OK OKLAHOMA	1.72	PB13-9	14-Sep-57	9	8
OK OKLAHOMA	4.04	PB14-7	20-Sep-57	1	8

COUNTY	uCi/ sqm	SHOT DAY	DATE	RANK FOR DAY	TIMES IN TOP 10
OK OKMULGEE	3.70	BJ5-6	24-Nov-51	2	12
OK OKMULGEE	3.70	BJ5-6	24-Nov-51	2	12
OK OKMULGEE	2.23	UpK2-10	2-Apr-53	3	12
OK OKMULGEE	2.10	UpK3-3	2-Apr-53	4	12
OK OKMULGEE	4.10	UpK3-5	4-Apr-53	5	12
OK OKMULGEE	1.07	UpK5-4	14-Apr-53	8	12
OK OKMULGEE	4.99	TP10-5	9-May-55	2	12
OK OKMULGEE	0.93	PB3-8	1-Jul-57	4	12
OK OKMULGEE	7.89	PB13-6	11-Sep-57	1	12
OK OKMULGEE	0.83	PB13-8	13-Sep-57	9	12
OK OKMULGEE	4.15	PB13-9	14-Sep-57	7	12
OK OKMULGEE	2.42	PB14-7	20-Sep-57	2	12
OK OSAGE	3.70	BJ5-6	24-Nov-51	2	8
OK OSAGE	3.70	BJ5-6	24-Nov-51	2	8
OK OSAGE	0.87	UpK6-5	22-Apr-53	8	8
OK OSAGE	1.32	TP9-13	27-Apr-55	3	8
OK OSAGE	1.76	PB2-4	21-Jun-57	6	8
OK OSAGE	2.11	PB13-8	13-Sep-57	4	8
OK OSAGE	2.50	PB13-9	14-Sep-57	8	8
OK OSAGE	0.81	PB14-7	20-Sep-57	9	8
OK OTTAWA	5.84	BJ2-2	31-Oct-51	7	6
OK OTTAWA	2.23	UpK2-10	2-Apr-53	3	6
OK OTTAWA	2.10	UpK3-3	2-Apr-53	4	6
OK OTTAWA	4.10	UpK3-5	4-Apr-53	5	6
OK OTTAWA	4.15	PB13-9	14-Sep-57	7	6
OK OTTAWA	2.42	PB14-7	20-Sep-57	2	6
OK PAWNEE	3.70	BJ5-6	24-Nov-51	2	8
OK PAWNEE	3.70	BJ5-6	24-Nov-51	2	8
OK PAWNEE	1.71	UpK3-5	4-Apr-53	10	8
OK PAWNEE	2.24	PB3-2	25-Jun-57	5	8
OK PAWNEE	6.62	PB13-6	11-Sep-57	8	8
OK PAWNEE	2.11	PB13-8	13-Sep-57	4	8
OK PAWNEE	4.15	PB13-9	14-Sep-57	7	8
OK PAWNEE	1.62	PB14-7	20-Sep-57	4	8
OK PAYNE	3.70	BJ5-6	24-Nov-51	2	9

COUNTY	uCi/ sqm	SHOT DAY	DATE	RANK FOR DAY	TIMES IN TOP 10
OK PAYNE	3.70	BJ5-6	24-Nov-51	2	9
OK PAYNE	1.71	UpK3-5	4-Apr-53	10	9
OK PAYNE	1.54	PB3-2	25-Jun-57	8	9
OK PAYNE	0.93	PB3-8	1-Jul-57	4	9
OK PAYNE	7.34	PB13-6	11-Sep-57	3	9
OK PAYNE	2.11	PB13-8	13-Sep-57	4	9
OK PAYNE	4.15	PB13-9	14-Sep-57	7	9
OK PAYNE	2.42	PB14-7	20-Sep-57	2	9
OK PITTSBURG	1.22	BJ5-6	24-Nov-51	6	12
OK PITTSBURG	1.22	BJ5-6	24-Nov-51	6	12
OK PITTSBURG	2.23	UpK2-10	2-Apr-53	3	12
OK PITTSBURG	2.10	UpK3-3	2-Apr-53	4	12
OK PITTSBURG	4.10	UpK3-5	4-Apr-53	5	12
OK PITTSBURG	1.83	UpK8-4	11-May-53	4	12
OK PITTSBURG	12.96	TP6-2	23-Mar-55	2	12
OK PITTSBURG	0.92	TP9-13	27-Apr-55	4	12
OK PITTSBURG	1.78	PB1-5	1-Jun-57	8	12
OK PITTSBURG	3.80	PB5-4	18-Jul-57	10	12
OK PITTSBURG	1.28	PB13-8	13-Sep-57	6	12
OK PITTSBURG	4.15	PB13-9	14-Sep-57	7	12
OK PONTOTOC	2.48	BJ5-6	24-Nov-51	4	8
OK PONTOTOC	2.48	BJ5-6	24-Nov-51	4	8
OK PONTOTOC	3.04	UpK8-4	11-May-53	2	8
OK PONTOTOC	12.96	TP6-2	23-Mar-55	2	8
OK PONTOTOC	25.20	TP11-5	19-May-55	1	8
OK PONTOTOC	1.78	PB1-5	1-Jun-57	8	8
OK PONTOTOC	4.15	PB13-9	14-Sep-57	7	8
OK PONTOTOC	1.92	PB18-6	12-Oct-57	3	8
OK POTTAWATOMIE	2.48	BJ5-6	24-Nov-51	4	10
OK POTTAWATOMIE	2.48	BJ5-6	24-Nov-51	4	10
OK POTTAWATOMIE	1.83	UpK8-4	11-May-53	4	10
OK POTTAWATOMIE	12.96	TP6-2	23-Mar-55	2	10
OK POTTAWATOMIE	4.43	TP10-5	9-May-55	4	10
OK POTTAWATOMIE	1.78	PB1-5	1-Jun-57	8	10
OK POTTAWATOMIE	1.63	PB5-7	21-Jul-57	8	10

COUNTY	uCi/ sqm	SHOT DAY	DATE	RANK FOR DAY	TIMES IN TOP 10
OK POTTAWATOMIE	2.11	PB13-8	13-Sep-57	4	10
OK POTTAWATOMIE	4.15	PB13-9	14-Sep-57	7	10
OK POTTAWATOMIE	1.92	PB18-6	12-Oct-57	3	10
OK PUSHMATAHA	0.81	UpK3-6	5-Apr-53	9	8
OK PUSHMATAHA	1.07	UpK5-4	14-Apr-53	8	8
OK PUSHMATAHA	1.83	UpK8-4	11-May-53	4	8
OK PUSHMATAHA	1.16	UpK8-5	12-May-53	5	8
OK PUSHMATAHA	12.96	TP6-2	23-Mar-55	2	8
OK PUSHMATAHA	3.80	PB5-4	18-Jul-57	10	8
OK PUSHMATAHA	4.15	PB13-9	14-Sep-57	7	8
OK PUSHMATAHA	2.60	PB18-7	13-Oct-57	1	8
OK ROGER MILLS	3.70	BJ5-6	24-Nov-51	2	11
OK ROGER MILLS	3.70	BJ5-6	24-Nov-51	2	11
OK ROGER MILLS	4.27	TS6-7	31-May-52	3	11
OK ROGER MILLS	13.66	UpK3-5	4-Apr-53	1	11
OK ROGER MILLS	4.23	UpK6-3	20-Apr-53	2	11
OK ROGER MILLS	6.07	UpK8-2	9-May-53	9	11
OK ROGER MILLS	1.83	UpK8-4	11-May-53	4	11
OK ROGER MILLS	20.41	UpK9-4	22-May-53	5	11
OK ROGER MILLS	25.62	TP7-2	30-Mar-55	4	11
OK ROGER MILLS	31.48	TP10-2	6-May-55	6	11
OK ROGER MILLS	1.92	PB18-6	12-Oct-57	3	11
OK ROGERS	3.70	BJ5-6	24-Nov-51	2	10
OK ROGERS	3.70	BJ5-6	24-Nov-51	2	10
OK ROGERS	2.23	UpK2-10	2-Apr-53	3	10
OK ROGERS	2.10	UpK3-3	2-Apr-53	4	10
OK ROGERS	6.83	UpK3-5	4-Apr-53	3	10
OK ROGERS	1.07	UpK5-4	14-Apr-53	8	10
OK ROGERS	21.62	PB12-4	5-Sep-57	4	10
OK ROGERS	3.55	PB13-4	9-Sep-57	5	10
OK ROGERS	1.28	PB13-8	13-Sep-57	6	10
OK ROGERS	2.50	PB13-9	14-Sep-57	8	10
OK SEMINOLE	2.48	BJ5-6	24-Nov-51	4	12
OK SEMINOLE	2.48	BJ5-6	24-Nov-51	4	12
OK SEMINOLE	2.23	UpK2-10	2-Apr-53	3	12

COUNTY	uCi/ sqm	SHOT DAY	DATE	RANK FOR DAY	TIMES IN TOP 10
OK SEMINOLE	2.10	UpK3-3	2-Apr-53	4	12
OK SEMINOLE	4.10	UpK3-5	4-Apr-53	5	12
OK SEMINOLE	1.07	UpK5-4	14-Apr-53	8	12
OK SEMINOLE	1.83	UpK8-4	11-May-53	4	12
OK SEMINOLE	12.96	TP6-2	23-Mar-55	2	12
OK SEMINOLE	1.78	PB1-5	1-Jun-57	8	12
OK SEMINOLE	2.11	PB13-8	13-Sep-57	4	12
OK SEMINOLE	4.15	PB13-9	14-Sep-57	7	12
OK SEMINOLE	1.92	PB18-6	12-Oct-57	3	12
OK SEQUOYAH	2.23	UpK2-10	2-Apr-53	3	9
OK SEQUOYAH	2.10	UpK3-3	2-Apr-53	4	9
OK SEQUOYAH	2.73	UpK3-5	4-Apr-53	8	9
OK SEQUOYAH	1.07	UpK5-4	14-Apr-53	8	9
OK SEQUOYAH	1.83	UpK8-4	11-May-53	4	9
OK SEQUOYAH	1.16	UpK8-5	12-May-53	5	9
OK SEQUOYAH	1.79	PB1-6	2-Jun-57	9	9
OK SEQUOYAH	0.93	PB3-8	1-Jul-57	4	9
OK SEQUOYAH	4.15	PB13-9	14-Sep-57	7	9
OK STEPHENS	1.22	BJ5-6	24-Nov-51	6	8
OK STEPHENS	1.22	BJ5-6	24-Nov-51	6	8
OK STEPHENS	2.72	TS6-4	28-May-52	6	8
OK STEPHENS	1.83	UpK8-4	11-May-53	4	8
OK STEPHENS	12.96	TP6-2	23-Mar-55	2	8
OK STEPHENS	1.78	PB1-5	1-Jun-57	8	8
OK STEPHENS	2.50	PB13-9	14-Sep-57	8	8
OK STEPHENS	3.21	PB18-6	12-Oct-57	1	8
OK TEXAS	1.02	TS4-4	4-May-52	1	12
OK TEXAS	1.02	TS4-4	4-May-52	1	12
OK TEXAS	1.74	TS4-6	6-May-52	5	12
OK TEXAS	1.74	TS4-6	6-May-52	5	12
OK TEXAS	8.19	UpK3-5	4-Apr-53	2	12
OK TEXAS	4.23	UpK6-3	20-Apr-53	2	12
OK TEXAS	1.22	UpK8-4	11-May-53	8	12
OK TEXAS	153.69	TP7-2	30-Mar-55	1	12
OK TEXAS	31.48	TP10-2	6-May-55	6	12

COUNTY	uCi/ sqm	SHOT DAY	DATE	RANK FOR DAY	TIMES IN TOP 10
OK TEXAS	0.89	PB13-7	12-Sep-57	5	12
OK TEXAS	1.28	PB13-8	13-Sep-57	6	12
OK TEXAS	1.92	PB18-6	12-Oct-57	3	12
OK TILLMAN	1.22	BJ5-6	24-Nov-51	6	10
OK TILLMAN	1.22	BJ5-6	24-Nov-51	6	10
OK TILLMAN	3.37	TS6-4	28-May-52	2	10
OK TILLMAN	5.43	UpK6-2	19-Apr-53	9	10
OK TILLMAN	1.30	UpK6-5	22-Apr-53	4	10
OK TILLMAN	1.22	UpK8-4	11-May-53	8	10
OK TILLMAN	3.18	TP6-2	23-Mar-55	6	10
OK TILLMAN	2.96	PB1-5	1-Jun-57	5	10
OK TILLMAN	2.50	PB13-9	14-Sep-57	8	10
OK TILLMAN	1.29	PB18-6	12-Oct-57	6	10
OK TULSA	2.48	BJ5-6	24-Nov-51	4	11
OK TULSA	2.48	BJ5-6	24-Nov-51	4	11
OK TULSA	2.23	UpK2-10	2-Apr-53	3	11
OK TULSA	2.10	UpK3-3	2-Apr-53	4	11
OK TULSA	6.83	UpK3-5	4-Apr-53	3	11
OK TULSA	0.92	TP9-13	27-Apr-55	4	11
OK TULSA	3.55	PB13-4	9-Sep-57	5	11
OK TULSA	0.89	PB13-7	12-Sep-57	5	11
OK TULSA	2.11	PB13-8	13-Sep-57	4	11
OK TULSA	2.50	PB13-9	14-Sep-57	8	11
OK TULSA	2.42	PB14-7	20-Sep-57	2	11
OK WAGONER	3.70	BJ5-6	24-Nov-51	2	12
OK WAGONER	3.70	BJ5-6	24-Nov-51	2	12
OK WAGONER	2.23	UpK2-10	2-Apr-53	3	12
OK WAGONER	2.10	UpK3-3	2-Apr-53	4	12
OK WAGONER	4.10	UpK3-5	4-Apr-53	5	12
OK WAGONER	0.92	TP9-13	27-Apr-55	4	12
OK WAGONER	20.68	PB12-4	5-Sep-57	7	12
OK WAGONER	7.48	PB13-6	11-Sep-57	2	12
OK WAGONER	0.89	PB13-7	12-Sep-57	5	12
OK WAGONER	0.83	PB13-8	13-Sep-57	9	12
OK WAGONER	4.15	PB13-9	14-Sep-57	7	12

COUNTY	uCi/ sqm	SHOT DAY	DATE	RANK FOR DAY	TIMES IN TOP 10
OK WAGONER	2.42	PB14-7	20-Sep-57	2	12
OK WASHINGTON	3.70	BJ5-6	24-Nov-51	2	8
OK WASHINGTON	3.70	BJ5-6	24-Nov-51	2	8
OK WASHINGTON	0.87	UpK6-5	22-Apr-53	8	8
OK WASHINGTON	1.76	PB2-4	21-Jun-57	6	8
OK WASHINGTON	6.98	PB13-6	11-Sep-57	6	8
OK WASHINGTON	2.11	PB13-8	13-Sep-57	4	8
OK WASHINGTON	2.50	PB13-9	14-Sep-57	8	8
OK WASHINGTON	0.81	PB14-7	20-Sep-57	9	8
OK WASHITA	13.66	UpK3-5	4-Apr-53	1	9
OK WASHITA	4.23	UpK6-3	20-Apr-53	2	9
OK WASHITA	6.07	UpK8-2	9-May-53	9	9
OK WASHITA	1.22	UpK8-4	11-May-53	8	9
OK WASHITA	25.62	TP7-2	30-Mar-55	4	9
OK WASHITA	31.48	TP10-2	6-May-55	6	9
OK WASHITA	1.63	PB5-7	21-Jul-57	8	9
OK WASHITA	4.04	PB14-7	20-Sep-57	1	9
OK WASHITA	1.29	PB18-6	12-Oct-57	6	9
OK WOODS	2.48	BJ5-6	24-Nov-51	4	6
OK WOODS	2.48	BJ5-6	24-Nov-51	4	6
OK WOODS	2.24	PB3-2	25-Jun-57	5	6
OK WOODS	1.63	PB5-7	21-Jul-57	8	6
OK WOODS	2.11	PB13-8	13-Sep-57	4	6
OK WOODS	4.15	PB13-9	14-Sep-57	7	6
OK WOODWARD	3.70	BJ5-6	24-Nov-51	2	7
OK WOODWARD	3.70	BJ5-6	24-Nov-51	2	7
OK WOODWARD	1.76	PB2-4	21-Jun-57	6	7
OK WOODWARD	2.11	PB13-8	13-Sep-57	4	7
OK WOODWARD	1.64	PB13-9	14-Sep-57	10	7
OK WOODWARD	0.81	PB14-7	20-Sep-57	9	7
OK WOODWARD	1.92	PB18-6	12-Oct-57	3	7

COUNTY	uCi/ sqm	SHOT DAY	DATE	RANK FOR DAY	TIMES IN TOP 10
OR BAKER	0.93	TS8-8	12-Jun-52	3	13
OR BAKER	1.94	PB1-5	1-Jun-57	7	13
OR BAKER	21.00	PB6-2	25-Jul-57	1	13
OR BAKER	12.28	PB6-3	26-Jul-57	6	13
OR BAKER	1.98	PB15-3	18-Sep-57	7	13
OR BAKER	7.58	PB16-3	25-Sep-57	1	13
OR BAKER	1.94	PB16-4	26-Sep-57	8	13
OR BAKER	9.52	PB17-3	30-Sep-57	4	13
OR BAKER	0.89	PB17-4	1-Oct-57	3	13
OR BAKER	0.85	PB18-4	10-Oct-57	7	13
OR BAKER	0.65	BN-3	20-Dec-70	8	13
OR BAKER	0.38	BN-4	21-Dec-70	3	13
OR BAKER	0.20	BN-5	22-Dec-70	3	13
OR CLACKAMAS	2.80	UpK2-12	4-Apr-53	7	3
OR CLACKAMAS	1.47	TP6-2	23-Mar-55	9	3
OR CLACKAMAS	1.83	TP6-3	24-Mar-55	7	3
OR CLATSOP	1.83	TP6-3	24-Mar-55	7	1
OR COLUMBIA	1.83	TP6-3	24-Mar-55	7	1
OR COOS	1.42	TP6-6	27-Mar-55	2	2
OR COOS	2.47	PB16-3	25-Sep-57	2	2
OR CROOK	0.23	BN-4	21-Dec-70	8	1
OR CURRY	1.42	TP6-6	27-Mar-55	2	2
OR CURRY	2.47	PB16-3	25-Sep-57	2	2
OR DESCHUTES	0.99	PB16-3	25-Sep-57	4	1
OR DOUGLAS	1.42	TP6-6	27-Mar-55	2	2
OR DOUGLAS	1.48	PB16-3	25-Sep-57	3	2
OR GILLIAM	0.26	BN-4	21-Dec-70	6	1
OR GRANT	1.94	PB1-5	1-Jun-57	7	10
OR GRANT	21.00	PB6-2	25-Jul-57	1	10
OR GRANT	12.28	PB6-3	26-Jul-57	6	10
OR GRANT	2.67	PB15-3	18-Sep-57	4	10
OR GRANT	7.58	PB16-3	25-Sep-57	1	10
OR GRANT	2.58	PB16-4	26-Sep-57	7	10
OR GRANT	9.52	PB17-3	30-Sep-57	4	10
OR GRANT	0.85	PB18-4	10-Oct-57	7	10

COUNTY	uCi/ sqm	SHOT DAY	DATE	RANK FOR DAY	TIMES IN TOP 10
OR GRANT	0.38	BN-4	21-Dec-70	3	10
OR GRANT	0.20	BN-5	22-Dec-70	3	10
OR HOOD RIVER	1.83	TP6-3	24-Mar-55	7	1
OR JOSEPHINE	1.42	TP6-6	27-Mar-55	2	2
OR JOSEPHINE	2.47	PB16-3	25-Sep-57	2	2
OR KLAMATH	1.42	TP6-6	27-Mar-55	2	3
OR KLAMATH	0.99	PB16-3	25-Sep-57	4	3
OR KLAMATH	1.29	PB16-4	26-Sep-57	9	3
OR LAKE	0.98	TP6-6	27-Mar-55	5	1
OR LANE	1.42	TP6-6	27-Mar-55	2	2
OR LANE	0.99	PB16-3	25-Sep-57	4	2
OR LINCOLN	1.47	TP6-2	23-Mar-55	9	2
OR LINCOLN	1.48	PB16-3	25-Sep-57	3	2
OR LINN	1.47	TP6-2	23-Mar-55	9	2
OR LINN	1.83	TP6-3	24-Mar-55	7	2
OR MARION	1.47	TP6-2	23-Mar-55	9	3
OR MARION	1.83	TP6-3	24-Mar-55	7	3
OR MARION	0.99	PB16-3	25-Sep-57	4	3
OR MORROW	0.26	BN-4	21-Dec-70	6	1
OR POLK	1.47	TP6-2	23-Mar-55	9	2
OR POLK	0.99	PB16-3	25-Sep-57	4	2
OR SHERMAN	0.26	BN-4	21-Dec-70	6	1
OR TILLAMOOK	1.83	TP6-3	24-Mar-55	7	1
OR UMATILLA	1.94	PB1-5	1-Jun-57	7	9
OR UMATILLA	0.83	PB1-9	5-Jun-57	8	9
OR UMATILLA	12.28	PB6-3	26-Jul-57	6	9
OR UMATILLA	2.67	PB15-3	18-Sep-57	4	9
OR UMATILLA	7.58	PB16-3	25-Sep-57	1	9
OR UMATILLA	5.16	PB16-4	26-Sep-57	6	9
OR UMATILLA	9.52	PB17-3	30-Sep-57	4	9
OR UMATILLA	0.85	PB18-4	10-Oct-57	7	9
OR UMATILLA	0.38	BN-4	21-Dec-70	3	9
OR UNION	1.94	PB1-5	1-Jun-57	7	9
OR UNION	21.00	PB6-2	25-Jul-57	1	9
OR UNION	12.28	PB6-3	26-Jul-57	6	9

COUNTY	uCi/ sqm	SHOT DAY	DATE	RANK FOR DAY	TIMES IN TOP 10
OR UNION	2.67	PB15-3	18-Sep-57	4	9
OR UNION	7.58	PB16-3	25-Sep-57	1	9
OR UNION	1.94	PB16-4	26-Sep-57	8	9
OR UNION	9.52	PB17-3	30-Sep-57	4	9
OR UNION	0.85	PB18-4	10-Oct-57	7	9
OR UNION	0.46	BN-4	21-Dec-70	2	9
OR WALLOWA	2.70	TS8-5	9-Jun-52	10	8
OR WALLOWA	1.94	PB1-5	1-Jun-57	7	8
OR WALLOWA	12.28	PB6-3	26-Jul-57	6	8
OR WALLOWA	7.58	PB16-3	25-Sep-57	1	8
OR WALLOWA	2.58	PB16-4	26-Sep-57	7	8
OR WALLOWA	6.34	PB17-3	30-Sep-57	6	8
OR WALLOWA	0.85	PB18-4	10-Oct-57	7	8
OR WALLOWA	0.14	BN-5	22-Dec-70	4	8
OR WASHINGTON	1.83	TP6-3	24-Mar-55	7	1
OR WHEELER	1.83	TP6-3	24-Mar-55	7	2
OR WHEELER	0.23	BN-4	21-Dec-70	8	2

COUNTY	uCi/ sqm	SHOT DAY	DATE	RANK FOR DAY	TIMES IN TOP 10
PA ADAMS	4.63	BJ3-6	6-Nov-51	9	4
PA ADAMS	1.67	TP6-4	25-Mar-55	10	4
PA ADAMS	2.99	PB5-8	22-Jul-57	4	4
PA ADAMS	2.83	PB5-9	23-Jul-57	2	4
PA ALLEGHENY	3.45	BJ3-5	5-Nov-51	10	3
PA ALLEGHENY	1.60	TP11-8	22-May-55	7	3
PA ALLEGHENY	5.97	PB5-8	22-Jul-57	3	3
PA ARMSTRONG	1.72	TP3-3	3-Mar-55	8	7
PA ARMSTRONG	3.86	TP11-8	22-May-55	3	7
PA ARMSTRONG	4.68	TP11-9	23-May-55	5	7
PA ARMSTRONG	8.96	PB5-8	22-Jul-57	2	7
PA ARMSTRONG	0.81	PB15-7	22-Sep-57	9	7
PA ARMSTRONG	90.29	S-4	9-Jul-62	8	7
PA ARMSTRONG	0.11	UE7-5	15-Jul-62	10	7
PA BEAVER	7.20	BJ2-2	31-Oct-51	5	6
PA BEAVER	2.57	TP11-8	22-May-55	5	6
PA BEAVER	3.12	TP11-9	23-May-55	8	6
PA BEAVER	8.96	PB5-8	22-Jul-57	2	6
PA BEAVER	0.81	PB15-7	22-Sep-57	9	6
PA BEAVER	103.59	S-4	9-Jul-62	6	6
PA BEDFORD	1.72	TP3-3	3-Mar-55	8	5
PA BEDFORD	1.29	TP11-8	22-May-55	9	5
PA BEDFORD	8.96	PB5-8	22-Jul-57	2	5
PA BEDFORD	1.89	PB5-9	23-Jul-57	3	5
PA BEDFORD	2.11	UE8-5	18-Jul-62	7	5
PA BERKS	4.63	BJ3-6	6-Nov-51	9	6
PA BERKS	1.42	UpK3-7	6-Apr-53	8	6
PA BERKS	2.03	TP6-4	25-Mar-55	8	6
PA BERKS	1.09	TP6-5	26-Mar-55	7	6
PA BERKS	5.70	TP11-9	23-May-55	4	6
PA BERKS	5.97	PB5-8	22-Jul-57	3	6
PA BLAIR	1.01	UpK6-5	22-Apr-53	7	4
PA BLAIR	1.29	TP11-8	22-May-55	9	4
PA BLAIR	8.96	PB5-8	22-Jul-57	2	4
PA BLAIR	0.94	PB5-9	23-Jul-57	4	4

COUNTY	uCi/ sqm	SHOT DAY	DATE	RANK FOR DAY	TIMES IN TOP 10
PA BRADFORD	3.37	BJ2-6	4-Nov-51	5	4
PA BRADFORD	2.74	PB5-7	21-Jul-57	4	4
PA BRADFORD	8.96	PB5-8	22-Jul-57	2	4
PA BRADFORD	0.94	PB15-7	22-Sep-57	6	4
PA BUCKS	3.98	BJ4-6	10-Nov-51	1	4
PA BUCKS	1.42	UpK3-7	6-Apr-53	8	4
PA BUCKS	3.11	TP6-4	25-Mar-55	3	4
PA BUCKS	5.97	PB5-8	22-Jul-57	3	4
PA BUTLER	7.20	BJ2-2	31-Oct-51	5	6
PA BUTLER	3.86	TP11-8	22-May-55	3	6
PA BUTLER	4.68	TP11-9	23-May-55	5	6
PA BUTLER	8.96	PB5-8	22-Jul-57	2	6
PA BUTLER	1.57	PB6-7	30-Jul-57	9	6
PA BUTLER	85.56	S-4	9-Jul-62	9	6
PA CAMBRIA	1.46	BJ2-6	4-Nov-51	10	6
PA CAMBRIA	1.01	UpK6-5	22-Apr-53	7	6
PA CAMBRIA	1.72	TP3-3	3-Mar-55	8	6
PA CAMBRIA	5.97	PB5-8	22-Jul-57	3	6
PA CAMBRIA	1.89	PB5-9	23-Jul-57	3	6
PA CAMBRIA	90.29	S-4	9-Jul-62	8	6
PA CAMERON	1.14	PB2-6	23-Jun-57	7	3
PA CAMERON	8.96	PB5-8	22-Jul-57	2	3
PA CAMERON	2.11	PB15-7	22-Sep-57	1	3
PA CARBON	3.37	BJ2-6	4-Nov-51	5	6
PA CARBON	4.13	BJ3-7	7-Nov-51	8	6
PA CARBON	1.42	UpK3-7	6-Apr-53	8	6
PA CARBON	2.03	TP6-4	25-Mar-55	8	6
PA CARBON	5.70	TP11-9	23-May-55	4	6
PA CARBON	5.97	PB5-8	22-Jul-57	3	6
PA CENTRE	1.67	TP6-4	25-Mar-55	10	3
PA CENTRE	8.96	PB5-8	22-Jul-57	2	3
PA CENTRE	0.94	PB5-9	23-Jul-57	4	3
PA CHESTER	4.63	BJ3-6	6-Nov-51	9	5
PA CHESTER	1.42	UpK3-7	6-Apr-53	8	5
PA CHESTER	3.11	TP6-4	25-Mar-55	3	5

COUNTY	uCi/ sqm	SHOT DAY	DATE	RANK FOR DAY	TIMES IN TOP 10
PA CHESTER	3.80	TP11-9	23-May-55	6	5
PA CHESTER	5.97	PB5-8	22-Jul-57	3	5
PA CLARION	7.20	BJ2-2	31-Oct-51	5	10
PA CLARION	6.12	UpK10-6	30-May-53	2	10
PA CLARION	1.72	TP3-3	3-Mar-55	8	10
PA CLARION	3.86	TP11-8	22-May-55	3	10
PA CLARION	4.68	TP11-9	23-May-55	5	10
PA CLARION	1.54	PB5-7	21-Jul-57	9	10
PA CLARION	8.96	PB5-8	22-Jul-57	2	10
PA CLARION	2.38	PB6-7	30-Jul-57	4	10
PA CLARION	0.81	PB15-7	22-Sep-57	9	10
PA CLARION	85.56	S-4	9-Jul-62	9	10
PA CLEARFIELD	1.01	UpK6-5	22-Apr-53	7	6
PA CLEARFIELD	2.57	TP11-8	22-May-55	5	6
PA CLEARFIELD	3.12	TP11-9	23-May-55	8	6
PA CLEARFIELD	8.96	PB5-8	22-Jul-57	2	6
PA CLEARFIELD	0.94	PB5-9	23-Jul-57	4	6
PA CLEARFIELD	103.59	S-4	9-Jul-62	6	6
PA CLINTON	1.46	BJ2-6	4-Nov-51	10	3
PA CLINTON	8.96	PB5-8	22-Jul-57	2	3
PA CLINTON	0.94	PB5-9	23-Jul-57	4	3
PA COLUMBIA	5.97	PB5-8	22-Jul-57	3	1
PA CRAWFORD	0.99	R1-3	30-Jan-51	3	7
PA CRAWFORD	3.86	TP11-8	22-May-55	3	7
PA CRAWFORD	4.68	TP11-9	23-May-55	5	7
PA CRAWFORD	5.97	PB5-8	22-Jul-57	3	7
PA CRAWFORD	1.57	PB6-7	30-Jul-57	9	7
PA CRAWFORD	85.56	S-4	9-Jul-62	9	7
PA CRAWFORD	0.24	UE7-5	15-Jul-62	5	7
PA CUMBERLAND	1.67	TP6-4	25-Mar-55	10	4
PA CUMBERLAND	1.19	PB2-6	23-Jun-57	6	4
PA CUMBERLAND	8.96	PB5-8	22-Jul-57	2	4
PA CUMBERLAND	1.89	PB5-9	23-Jul-57	3	4
PA DAUPHIN	4.63	BJ3-6	6-Nov-51	9	4
PA DAUPHIN	1.67	TP6-4	25-Mar-55	10	4

COUNTY	uCi/ sqm	SHOT DAY	DATE	RANK FOR DAY	TIMES IN TOP 10
PA DAUPHIN	8.96	PB5-8	22-Jul-57	2	4
PA DAUPHIN	1.89	PB5-9	23-Jul-57	3	4
PA DELAWARE	3.98	BJ4-6	10-Nov-51	1	4
PA DELAWARE	1.42	UpK3-7	6-Apr-53	8	4
PA DELAWARE	3.11	TP6-4	25-Mar-55	3	4
PA DELAWARE	5.97	PB5-8	22-Jul-57	3	4
PA ELK	2.93	BJ2-6	4-Nov-51	6	11
PA ELK	16.67	BJ3-6	6-Nov-51	2	11
PA ELK	4.20	BJ3-7	7-Nov-51	8	11
PA ELK	2.27	BJ3-8	8-Nov-51	2	11
PA ELK	1.41	BJ3-9	9-Nov-51	1	11
PA ELK	0.82	BJ3-10	10-Nov-51	4	11
PA ELK	8.96	PB5-8	22-Jul-57	2	11
PA ELK	1.57	PB6-7	30-Jul-57	9	11
PA ELK	0.81	PB15-7	22-Sep-57	9	11
PA ELK	85.56	S-4	9-Jul-62	9	11
PA ELK	47.25	S-5	10-Jul-62	2	11
PA ERIE	0.99	R1-3	30-Jan-51	3	7
PA ERIE	6.32	BJ3-5	5-Nov-51	4	7
PA ERIE	1.02	TS4-4	4-May-52	1	7
PA ERIE	1.02	TS4-4	4-May-52	1	7
PA ERIE	3.57	UpK10-6	30-May-53	7	7
PA ERIE	10.36	TP4-4	10-Mar-55	9	7
PA ERIE	5.97	PB5-8	22-Jul-57	3	7
PA FAYETTE	1.72	TP3-3	3-Mar-55	8	8
PA FAYETTE	1.29	TP11-8	22-May-55	9	8
PA FAYETTE	8.96	PB5-8	22-Jul-57	2	8
PA FAYETTE	2.83	PB5-9	23-Jul-57	2	8
PA FAYETTE	1.57	PB6-7	30-Jul-57	9	8
PA FAYETTE	85.56	S-4	9-Jul-62	9	8
PA FAYETTE	32.93	S-5	10-Jul-62	5	8
PA FAYETTE	3.35	UE8-5	18-Jul-62	4	8
PA FOREST	16.67	BJ3-6	6-Nov-51	2	11
PA FOREST	4.20	BJ3-7	7-Nov-51	8	11
PA FOREST	2.27	BJ3-8	8-Nov-51	2	11

COUNTY	uCi/ sqm	SHOT DAY	DATE	RANK FOR DAY	TIMES IN TOP 10
PA FOREST	1.41	BJ3-9	9-Nov-51	1	11
PA FOREST	0.82	BJ3-10	10-Nov-51	4	11
PA FOREST	1.72	TP3-3	3-Mar-55	8	11
PA FOREST	6.43	TP11-8	22-May-55	1	11
PA FOREST	7.81	TP11-9	23-May-55	2	11
PA FOREST	5.97	PB5-8	22-Jul-57	3	11
PA FOREST	85.56	S-4	9-Jul-62	9	11
PA FOREST	0.13	UE7-5	15-Jul-62	8	11
PA FRANKLIN	4.63	BJ3-6	6-Nov-51	9	5
PA FRANKLIN	1.04	UpK4-4	9-Apr-53	5	5
PA FRANKLIN	5.97	PB5-8	22-Jul-57	3	5
PA FRANKLIN	2.83	PB5-9	23-Jul-57	2	5
PA FRANKLIN	2.11	UE8-5	18-Jul-62	7	5
PA FULTON	1.04	UpK4-4	9-Apr-53	5	5
PA FULTON	8.96	PB5-8	22-Jul-57	2	5
PA FULTON	1.89	PB5-9	23-Jul-57	3	5
PA FULTON	149.13	S-4	9-Jul-62	5	5
PA FULTON	2.11	UE8-5	18-Jul-62	7	5
PA GREENE	1.72	TP3-3	3-Mar-55	8	7
PA GREENE	2.57	TP11-8	22-May-55	5	7
PA GREENE	3.12	TP11-9	23-May-55	8	7
PA GREENE	8.96	PB5-8	22-Jul-57	2	7
PA GREENE	4.72	PB5-9	23-Jul-57	1	7
PA GREENE	85.56	S-4	9-Jul-62	9	7
PA GREENE	5.84	UE8-5	18-Jul-62	1	7
PA HUNTINGDON	6.02	UpK10-6	30-May-53	3	4
PA HUNTINGDON	1.54	PB5-7	21-Jul-57	9	4
PA HUNTINGDON	5.97	PB5-8	22-Jul-57	3	4
PA HUNTINGDON	1.89	PB5-9	23-Jul-57	3	4
PA INDIANA	4.59	BJ3-5	5-Nov-51	6	7
PA INDIANA	1.72	TP3-3	3-Mar-55	8	7
PA INDIANA	1.29	TP11-8	22-May-55	9	7
PA INDIANA	1.54	PB5-7	21-Jul-57	9	7
PA INDIANA	8.96	PB5-8	22-Jul-57	2	7
PA INDIANA	0.94	PB5-9	23-Jul-57	4	7

COUNTY	uCi/ sqm	SHOT DAY	DATE	RANK FOR DAY	TIMES IN TOP 10
PA INDIANA	90.29	S-4	9-Jul-62	8	7
PA JEFFERSON	1.02	TS4-4	4-May-52	1	7
PA JEFFERSON	1.02	TS4-4	4-May-52	1	7
PA JEFFERSON	3.86	TP11-8	22-May-55	3	7
PA JEFFERSON	4.68	TP11-9	23-May-55	5	7
PA JEFFERSON	1.54	PB5-7	21-Jul-57	9	7
PA JEFFERSON	8.96	PB5-8	22-Jul-57	2	7
PA JEFFERSON	90.29	S-4	9-Jul-62	8	7
PA JUNIATA	1.67	TP6-4	25-Mar-55	10	4
PA JUNIATA	1.19	PB2-6	23-Jun-57	6	4
PA JUNIATA	2.99	PB5-8	22-Jul-57	4	4
PA JUNIATA	1.89	PB5-9	23-Jul-57	3	4
PA LACKAWANNA	3.37	BJ2-6	4-Nov-51	5	5
PA LACKAWANNA	3.80	TP11-9	23-May-55	6	5
PA LACKAWANNA	5.97	PB5-8	22-Jul-57	3	5
PA LACKAWANNA	0.94	PB15-7	22-Sep-57	6	5
PA LACKAWANNA	40.85	S-5	10-Jul-62	3	5
PA LANCASTER	4.63	BJ3-6	6-Nov-51	9	3
PA LANCASTER	1.67	TP6-4	25-Mar-55	10	3
PA LANCASTER	8.96	PB5-8	22-Jul-57	2	3
PA LAWRENCE	10.33	UpK9-3	21-May-53	10	7
PA LAWRENCE	2.57	TP11-8	22-May-55	5	7
PA LAWRENCE	3.12	TP11-9	23-May-55	8	7
PA LAWRENCE	5.97	PB5-8	22-Jul-57	3	7
PA LAWRENCE	2.38	PB6-7	30-Jul-57	4	7
PA LAWRENCE	103.59	S-4	9-Jul-62	6	7
PA LAWRENCE	0.11	UE7-5	15-Jul-62	10	7
PA LEBANON	4.63	BJ3-6	6-Nov-51	9	3
PA LEBANON	1.67	TP6-4	25-Mar-55	10	3
PA LEBANON	2.99	PB5-8	22-Jul-57	4	3
PA LEHIGH	4.63	BJ3-6	6-Nov-51	9	5
PA LEHIGH	1.42	UpK3-7	6-Apr-53	8	5
PA LEHIGH	3.11	TP6-4	25-Mar-55	3	5
PA LEHIGH	5.70	TP11-9	23-May-55	4	5
PA LEHIGH	5.97	PB5-8	22-Jul-57	3	5

COUNTY	uCi/ sqm	SHOT DAY	DATE	RANK FOR DAY	TIMES IN TOP 10
PA LUZERNE	3.37	BJ2-6	4-Nov-51	5	4
PA LUZERNE	4.13	BJ3-7	7-Nov-51	8	4
PA LUZERNE	3.80	TP11-9	23-May-55	6	4
PA LUZERNE	8.96	PB5-8	22-Jul-57	2	4
PA LYCOMING	5.97	PB5-8	22-Jul-57	3	1
PA MCKEAN	0.99	R1-3	30-Jan-51	3	13
PA MCKEAN	3.57	BJ2-5	3-Nov-51	10	13
PA MCKEAN	16.67	BJ3-6	6-Nov-51	2	13
PA MCKEAN	4.20	BJ3-7	7-Nov-51	8	13
PA MCKEAN	2.27	BJ3-8	8-Nov-51	2	13
PA MCKEAN	1.41	BJ3-9	9-Nov-51	1	13
PA MCKEAN	0.82	BJ3-10	10-Nov-51	4	13
PA MCKEAN	1.52	UpK10-5	29-May-53	8	13
PA MCKEAN	1.72	PB2-6	23-Jun-57	5	13
PA MCKEAN	2.82	PB3-5	28-Jun-57	9	13
PA MCKEAN	5.97	PB5-8	22-Jul-57	3	13
PA MCKEAN	1.26	PB15-7	22-Sep-57	4	13
PA MCKEAN	0.24	UE7-5	15-Jul-62	5	13
PA MERCER	1.17	R1-3	30-Jan-51	2	7
PA MERCER	1.72	TP3-3	3-Mar-55	8	7
PA MERCER	3.86	TP11-8	22-May-55	3	7
PA MERCER	4.68	TP11-9	23-May-55	5	7
PA MERCER	5.97	PB5-8	22-Jul-57	3	7
PA MERCER	85.56	S-4	9-Jul-62	9	7
PA MERCER	0.13	UE7-5	15-Jul-62	8	7
PA MIFFLIN	1.54	PB5-7	21-Jul-57	9	4
PA MIFFLIN	5.97	PB5-8	22-Jul-57	3	4
PA MIFFLIN	1.89	PB5-9	23-Jul-57	3	4
PA MIFFLIN	0.11	UE7-5	15-Jul-62	10	4
PA MONROE	3.37	BJ2-6	4-Nov-51	5	6
PA MONROE	4.13	BJ3-7	7-Nov-51	8	6
PA MONROE	1.42	UpK3-7	6-Apr-53	8	6
PA MONROE	2.03	TP6-4	25-Mar-55	8	6
PA MONROE	1.37	PB5-7	21-Jul-57	10	6
PA MONROE	5.97	PB5-8	22-Jul-57	3	6

COUNTY	uCi/ sqm	SHOT DAY	DATE	RANK FOR DAY	TIMES IN TOP 10
PA MONTGOMERY	3.98	BJ4-6	10-Nov-51	1	4
PA MONTGOMERY	1.42	UpK3-7	6-Apr-53	8	4
PA MONTGOMERY	3.11	TP6-4	25-Mar-55	3	4
PA MONTGOMERY	5.97	PB5-8	22-Jul-57	3	4
PA MONTOUR	8.96	PB5-8	22-Jul-57	2	1
PA NORTHAMPTON	3.37	BJ2-6	4-Nov-51	5	5
PA NORTHAMPTON	4.13	BJ3-7	7-Nov-51	8	5
PA NORTHAMPTON	1.42	UpK3-7	6-Apr-53	8	5
PA NORTHAMPTON	2.03	TP6-4	25-Mar-55	8	5
PA NORTHAMPTON	5.97	PB5-8	22-Jul-57	3	5
PA NORTHUMBERLA	3.45	BJ3-5	5-Nov-51	10	2
PA NORTHUMBERLA	8.96	PB5-8	22-Jul-57	2	2
PA PERRY	1.67	TP6-4	25-Mar-55	10	4
PA PERRY	1.19	PB2-6	23-Jun-57	6	4
PA PERRY	2.99	PB5-8	22-Jul-57	4	4
PA PERRY	1.89	PB5-9	23-Jul-57	3	4
PA PHILADELPHIA	3.98	BJ4-6	10-Nov-51	1	2
PA PHILADELPHIA	2.99	PB5-8	22-Jul-57	4	2
PA PIKE	3.37	BJ2-6	4-Nov-51	5	4
PA PIKE	6.86	BJ3-7	7-Nov-51	7	4
PA PIKE	1.21	TP11-8	22-May-55	10	4
PA PIKE	5.97	PB5-8	22-Jul-57	3	4
PA POTTER	1.17	R1-3	30-Jan-51	2	15
PA POTTER	1.46	BJ2-6	4-Nov-51	10	15
PA POTTER	16.67	BJ3-6	6-Nov-51	2	15
PA POTTER	8.40	BJ3-7	7-Nov-51	4	15
PA POTTER	2.27	BJ3-8	8-Nov-51	2	15
PA POTTER	1.41	BJ3-9	9-Nov-51	1	15
PA POTTER	0.82	BJ3-10	10-Nov-51	4	15
PA POTTER	3.70	UpK10-6	30-May-53	6	15
PA POTTER	2.30	TP3-3	3-Mar-55	6	15
PA POTTER	1.35	TP3-4	4-Mar-55	1	15
PA POTTER	2.57	TP11-8	22-May-55	5	15
PA POTTER	2.29	PB2-6	23-Jun-57	3	15
PA POTTER	5.97	PB5-8	22-Jul-57	3	15

COUNTY	uCi/ sqm	SHOT DAY	DATE	RANK FOR DAY	TIMES IN TOP 10
PA POTTER	1.26	PB15-7	22-Sep-57	4	15
PA POTTER	7.38	S-5	10-Jul-62	9	15
PA SCHUYLKILL	4.63	BJ3-6	6-Nov-51	9	5
PA SCHUYLKILL	1.42	UpK3-7	6-Apr-53	8	5
PA SCHUYLKILL	2.03	TP6-4	25-Mar-55	8	5
PA SCHUYLKILL	3.80	TP11-9	23-May-55	6	5
PA SCHUYLKILL	8.96	PB5-8	22-Jul-57	2	5
PA SNYDER	1.67	TP6-4	25-Mar-55	10	3
PA SNYDER	8.96	PB5-8	22-Jul-57	2	3
PA SNYDER	149.13	S-4	9-Jul-62	5	3
PA SOMERSET	1.72	TP3-3	3-Mar-55	8	6
PA SOMERSET	1.29	TP11-8	22-May-55	9	6
PA SOMERSET	8.96	PB5-8	22-Jul-57	2	6
PA SOMERSET	2.83	PB5-9	23-Jul-57	2	6
PA SOMERSET	85.56	S-4	9-Jul-62	9	6
PA SOMERSET	2.11	UE8-5	18-Jul-62	7	6
PA SULLIVAN	3.37	BJ2-6	4-Nov-51	5	4
PA SULLIVAN	4.13	BJ3-7	7-Nov-51	8	4
PA SULLIVAN	5.97	PB5-8	22-Jul-57	3	4
PA SULLIVAN	0.94	PB15-7	22-Sep-57	6	4
PA SUSQUEHANNA	3.37	BJ2-6	4-Nov-51	5	4
PA SUSQUEHANNA	2.74	PB5-7	21-Jul-57	4	4
PA SUSQUEHANNA	8.96	PB5-8	22-Jul-57	2	4
PA SUSQUEHANNA	0.94	PB15-7	22-Sep-57	6	4
PA TIOGA	3.37	BJ2-6	4-Nov-51	5	5
PA TIOGA	2.74	UpK10-6	30-May-53	8	5
PA TIOGA	1.37	PB5-7	21-Jul-57	10	5
PA TIOGA	8.96	PB5-8	22-Jul-57	2	5
PA TIOGA	0.94	PB15-7	22-Sep-57	6	5
PA UNION	8.96	PB5-8	22-Jul-57	2	1
PA VENANGO	1.72	TP3-3	3-Mar-55	8	8
PA VENANGO	11.29	TP4-4	10-Mar-55	7	8
PA VENANGO	3.86	TP11-8	22-May-55	3	8
PA VENANGO	4.68	TP11-9	23-May-55	5	8
PA VENANGO	8.96	PB5-8	22-Jul-57	2	8

COUNTY	uCi/ sqm	SHOT DAY	DATE	RANK FOR DAY	TIMES IN TOP 10
PA VENANGO	0.81	PB15-7	22-Sep-57	9	8
PA VENANGO	208.78	S-4	9-Jul-62	3	8
PA VENANGO	0.13	UE7-5	15-Jul-62	8	8
PA WARREN	16.67	BJ3-6	6-Nov-51	2	9
PA WARREN	8.40	BJ3-7	7-Nov-51	4	9
PA WARREN	2.27	BJ3-8	8-Nov-51	2	9
PA WARREN	1.41	BJ3-9	9-Nov-51	1	9
PA WARREN	0.82	BJ3-10	10-Nov-51	4	9
PA WARREN	5.07	UpK10-6	30-May-53	5	9
PA WARREN	8.96	PB5-8	22-Jul-57	2	9
PA WARREN	2.38	PB6-7	30-Jul-57	4	9
PA WARREN	0.13	UE7-5	15-Jul-62	8	9
PA WASHINGTON	1.72	TP3-3	3-Mar-55	8	7
PA WASHINGTON	2.57	TP11-8	22-May-55	5	7
PA WASHINGTON	3.12	TP11-9	23-May-55	8	7
PA WASHINGTON	8.96	PB5-8	22-Jul-57	2	7
PA WASHINGTON	2.38	PB6-7	30-Jul-57	4	7
PA WASHINGTON	85.56	S-4	9-Jul-62	9	7
PA WASHINGTON	2.38	UE8-5	18-Jul-62	5	7
PA WAYNE	32.31	R1-2	29-Jan-51	7	6
PA WAYNE	3.37	BJ2-6	4-Nov-51	5	6
PA WAYNE	4.13	BJ3-7	7-Nov-51	8	6
PA WAYNE	5.97	PB5-8	22-Jul-57	3	6
PA WAYNE	0.94	PB15-7	22-Sep-57	6	6
PA WAYNE	8.99	S-5	10-Jul-62	8	6
PA WESTMORELAND	3.45	BJ3-5	5-Nov-51	10	7
PA WESTMORELAND	1.72	TP3-3	3-Mar-55	8	7
PA WESTMORELAND	2.57	TP11-8	22-May-55	5	7
PA WESTMORELAND	3.12	TP11-9	23-May-55	8	7
PA WESTMORELAND	8.96	PB5-8	22-Jul-57	2	7
PA WESTMORELAND	0.94	PB5-9	23-Jul-57	4	7
PA WESTMORELAND	90.29	S-4	9-Jul-62	8	7
PA WYOMING	3.37	BJ2-6	4-Nov-51	5	6
PA WYOMING	4.13	BJ3-7	7-Nov-51	8	6
PA WYOMING	4.11	PB5-7	21-Jul-57	3	6

COUNTY	uCi/ sqm	SHOT DAY	DATE	RANK FOR DAY	TIMES IN TOP 10
PA WYOMING	8.96	PB5-8	22-Jul-57	2	6
PA WYOMING	0.94	PB15-7	22-Sep-57	6	6
PA WYOMING	16.19	S-5	10-Jul-62	7	6
PA YORK	2.51	TP6-4	25-Mar-55	5	3
PA YORK	8.96	PB5-8	22-Jul-57	2	3
PA YORK	2.83	PB5-9	23-Jul-57	2	3

COUNTY	uCi/ sqm	SHOT DAY	DATE	RANK FOR DAY	TIMES IN TOP 10
RI BRISTOL	99.38	UpK4-2	7-Apr-53	6	6
RI BRISTOL	1.00	UpK4-3	8-Apr-53	10	6
RI BRISTOL	0.99	UpK5-5	15-Apr-53	2	6
RI BRISTOL	1.28	UpK7-6	30-Apr-53	10	6
RI BRISTOL	1.83	UpK8-4	11-May-53	4	6
RI BRISTOL	2.99	PB5-8	22-Jul-57	4	6
RI KENT	3.71	BJ3-7	7-Nov-51	10	5
RI KENT	1.00	UpK4-3	8-Apr-53	10	5
RI KENT	1.47	UpK7-6	30-Apr-53	4	5
RI KENT	1.83	UpK8-4	11-May-53	4	5
RI KENT	5.97	PB5-8	22-Jul-57	3	5
RI NEWPORT	1.00	UpK4-3	8-Apr-53	10	4
RI NEWPORT	0.99	UpK5-5	15-Apr-53	2	4
RI NEWPORT	1.83	UpK8-4	11-May-53	4	4
RI NEWPORT	2.99	PB5-8	22-Jul-57	4	4
RI PROVIDENCE	1.47	UpK7-6	30-Apr-53	5	2
RI PROVIDENCE	2.99	PB5-8	22-Jul-57	4	2
RI WASHINGTON	3.71	BJ3-7	7-Nov-51	10	5
RI WASHINGTON	99.16	UpK4-2	7-Apr-53	7	5
RI WASHINGTON	1.00	UpK4-3	8-Apr-53	10	5
RI WASHINGTON	1.83	UpK8-4	11-May-53	4	5
RI WASHINGTON	5.97	PB5-8	22-Jul-57	3	5

COUNTY	uCi/ sqm	SHOT DAY	DATE	RANK FOR DAY	TIMES IN TOP 10
SC ABBEVILLE	2.22	TS6-6	30-May-52	8	3
SC ABBEVILLE	1.32	PB2-8	25-Jun-57	9	3
SC ABBEVILLE	1.51	PB5-8	22-Jul-57	7	3
SC AIKEN	1.51	PB5-8	22-Jul-57	7	2
SC AIKEN	0.94	PB5-9	23-Jul-57	4	2
SC ALLENDALE	2.63	TS3-7	28-Apr-52	4	4
SC ALLENDALE	2.63	TS3-7	28-Apr-52	4	4
SC ALLENDALE	1.31	TP7-8	5-Apr-55	4	4
SC ALLENDALE	1.51	PB5-8	22-Jul-57	7	4
SC ANDERSON	1.91	TP6-4	25-Mar-55	9	5
SC ANDERSON	2.73	PB2-4	21-Jun-57	2	5
SC ANDERSON	1.51	PB5-8	22-Jul-57	7	5
SC ANDERSON	0.93	PB18-5	11-Oct-57	3	5
SC ANDERSON	2.58	UE8-6	19-Jul-62	4	5
SC BAMBERG	1.35	TS3-7	28-Apr-52	5	6
SC BAMBERG	1.35	TS3-7	28-Apr-52	5	6
SC BAMBERG	2.26	TS6-6	30-May-52	7	6
SC BAMBERG	1.32	PB2-8	25-Jun-57	9	6
SC BAMBERG	1.51	PB5-8	22-Jul-57	7	6
SC BAMBERG	2.83	PB5-9	23-Jul-57	2	6
SC BARNWELL	2.63	TS3-7	28-Apr-52	4	4
SC BARNWELL	2.63	TS3-7	28-Apr-52	4	4
SC BARNWELL	1.31	TP7-8	5-Apr-55	4	4
SC BARNWELL	1.51	PB5-8	22-Jul-57	7	4
SC BEAUFORT	2.63	TS3-7	28-Apr-52	4	5
SC BEAUFORT	2.63	TS3-7	28-Apr-52	4	5
SC BEAUFORT	1.51	PB5-8	22-Jul-57	7	5
SC BEAUFORT	2.83	PB5-9	23-Jul-57	2	5
SC BEAUFORT	0.48	UE8-8	21-Jul-62	1	5
SC BERKELEY	2.63	TS3-7	28-Apr-52	4	6
SC BERKELEY	2.63	TS3-7	28-Apr-52	4	6
SC BERKELEY	1.51	PB5-8	22-Jul-57	7	6
SC BERKELEY	7.77	PB11-4	3-Sep-57	8	6
SC BERKELEY	2.58	UE8-6	19-Jul-62	4	6
SC BERKELEY	0.16	UE8-8	21-Jul-62	6	6

COUNTY	uCi/ sqm	SHOT DAY	DATE	RANK FOR DAY	TIMES IN TOP 10
SC CALHOUN	2.63	TS3-7	28-Apr-52	4	4
SC CALHOUN	2.63	TS3-7	28-Apr-52	4	4
SC CALHOUN	1.62	PB2-9	26-Jun-57	10	4
SC CALHOUN	1.51	PB5-8	22-Jul-57	7	4
SC CHARLESTON	5.26	TS3-7	28-Apr-52	3	6
SC CHARLESTON	5.26	TS3-7	28-Apr-52	3	6
SC CHARLESTON	2.20	PB2-8	25-Jun-57	6	6
SC CHARLESTON	1.51	PB5-8	22-Jul-57	7	6
SC CHARLESTON	4.72	PB5-9	23-Jul-57	1	6
SC CHARLESTON	0.16	UE8-8	21-Jul-62	7	6
SC CHEROKEE	2.03	TP6-4	25-Mar-55	8	4
SC CHEROKEE	0.97	TP6-5	26-Mar-55	8	4
SC CHEROKEE	1.51	PB5-8	22-Jul-57	7	4
SC CHEROKEE	2.83	PB5-9	23-Jul-57	2	4
SC CHESTER	2.19	TP7-8	5-Apr-55	1	3
SC CHESTER	2.99	PB5-8	22-Jul-57	4	3
SC CHESTER	2.83	PB5-9	23-Jul-57	2	3
SC CHESTERFIELD	1.51	PB5-8	22-Jul-57	7	6
SC CHESTERFIELD	2.83	PB5-9	23-Jul-57	2	6
SC CHESTERFIELD	2.54	PB8-7	24-Aug-57	7	6
SC CHESTERFIELD	4.62	UE8-6	19-Jul-62	1	6
SC CHESTERFIELD	1.37	UE8-7	20-Jul-62	4	6
SC CHESTERFIELD	0.16	UE8-8	21-Jul-62	6	6
SC CLARENDON	7.89	TS3-7	28-Apr-52	1	4
SC CLARENDON	7.89	TS3-7	28-Apr-52	1	4
SC CLARENDON	1.32	PB2-8	25-Jun-57	9	4
SC CLARENDON	1.51	PB5-8	22-Jul-57	7	4
SC COLLETON	1.35	TS3-7	28-Apr-52	5	6
SC COLLETON	1.35	TS3-7	28-Apr-52	5	6
SC COLLETON	2.20	PB2-8	25-Jun-57	6	6
SC COLLETON	1.51	PB5-8	22-Jul-57	7	6
SC COLLETON	2.83	PB5-9	23-Jul-57	2	6
SC COLLETON	0.16	UE8-8	21-Jul-62	7	6
SC DARLINGTON	0.84	BJ4-5	9-Nov-51	3	6
SC DARLINGTON	0.88	TP7-8	5-Apr-55	6	6

COUNTY	uCi/ sqm	SHOT DAY	DATE	RANK FOR DAY	TIMES IN TOP 10
SC DARLINGTON	1.51	PB5-8	22-Jul-57	7	6
SC DARLINGTON	2.83	PB5-9	23-Jul-57	2	6
SC DARLINGTON	1.51	PB8-7	24-Aug-57	10	6
SC DARLINGTON	1.37	UE8-7	20-Jul-62	4	6
SC DILLON	2.63	TS3-7	28-Apr-52	4	5
SC DILLON	2.63	TS3-7	28-Apr-52	4	5
SC DILLON	1.51	PB5-8	22-Jul-57	7	5
SC DILLON	2.83	PB5-9	23-Jul-57	2	5
SC DILLON	2.54	PB8-7	24-Aug-57	7	5
SC DORCHESTER	1.35	TS3-7	28-Apr-52	5	3
SC DORCHESTER	1.35	TS3-7	28-Apr-52	5	3
SC DORCHESTER	1.51	PB5-8	22-Jul-57	7	3
SC EDGEFIELD	1.62	PB2-9	26-Jun-57	10	3
SC EDGEFIELD	1.51	PB5-8	22-Jul-57	7	3
SC EDGEFIELD	4.62	UE8-6	19-Jul-62	1	3
SC FAIRFIELD	0.88	TP7-8	5-Apr-55	6	5
SC FAIRFIELD	1.51	PB5-8	22-Jul-57	7	5
SC FAIRFIELD	0.94	PB5-9	23-Jul-57	4	5
SC FAIRFIELD	1.32	UE8-6	19-Jul-62	9	5
SC FAIRFIELD	0.30	UE8-8	21-Jul-62	3	5
SC FLORENCE	5.26	TS3-7	28-Apr-52	3	5
SC FLORENCE	5.26	TS3-7	28-Apr-52	3	5
SC FLORENCE	1.51	PB5-8	22-Jul-57	7	5
SC FLORENCE	1.89	PB5-9	23-Jul-57	3	5
SC FLORENCE	19.43	PB11-4	3-Sep-57	2	5
SC GEORGETOWN	2.63	TS3-7	28-Apr-52	4	5
SC GEORGETOWN	2.63	TS3-7	28-Apr-52	4	5
SC GEORGETOWN	1.62	PB2-9	26-Jun-57	10	5
SC GEORGETOWN	1.51	PB5-8	22-Jul-57	7	5
SC GEORGETOWN	1.32	UE8-6	19-Jul-62	9	5
SC GREENVILLE	3.89	BJ2-2	31-Oct-51	10	8
SC GREENVILLE	4.26	TS6-6	30-May-52	1	8
SC GREENVILLE	2.03	TP6-4	25-Mar-55	8	8
SC GREENVILLE	0.97	TP6-5	26-Mar-55	8	8
SC GREENVILLE	4.18	PB2-5	22-Jun-57	3	8

COUNTY	uCi/ sqm	SHOT DAY	DATE	RANK FOR DAY	TIMES IN TOP 10
SC GREENVILLE	1.32	PB2-8	25-Jun-57	9	8
SC GREENVILLE	2.99	PB5-8	22-Jul-57	4	8
SC GREENVILLE	1.56	PB18-5	11-Oct-57	1	8
SC GREENWOOD	4.20	TS6-6	30-May-52	2	5
SC GREENWOOD	1.32	PB2-8	25-Jun-57	9	5
SC GREENWOOD	1.51	PB5-8	22-Jul-57	7	5
SC GREENWOOD	2.58	UE8-6	19-Jul-62	4	5
SC GREENWOOD	0.10	UE8-8	21-Jul-62	8	5
SC HAMPTON	2.63	TS3-7	28-Apr-52	4	4
SC HAMPTON	2.63	TS3-7	28-Apr-52	4	4
SC HAMPTON	1.51	PB5-8	22-Jul-57	7	4
SC HAMPTON	2.83	PB5-9	23-Jul-57	2	4
SC HORRY	2.63	TS3-7	28-Apr-52	4	8
SC HORRY	2.63	TS3-7	28-Apr-52	4	8
SC HORRY	1.62	PB2-9	26-Jun-57	10	8
SC HORRY	1.51	PB5-8	22-Jul-57	7	8
SC HORRY	0.94	PB5-9	23-Jul-57	4	8
SC HORRY	1.51	PB8-7	24-Aug-57	10	8
SC HORRY	0.93	PB18-5	11-Oct-57	3	8
SC HORRY	0.30	UE8-8	21-Jul-62	3	8
SC JASPER	1.35	TS3-7	28-Apr-52	5	7
SC JASPER	1.35	TS3-7	28-Apr-52	5	7
SC JASPER	5.05	PB4-14	18-Jul-57	6	7
SC JASPER	1.51	PB5-8	22-Jul-57	7	7
SC JASPER	1.89	PB5-9	23-Jul-57	3	7
SC JASPER	11.66	PB11-4	3-Sep-57	5	7
SC JASPER	0.16	UE8-8	21-Jul-62	7	7
SC KERSHAW	1.51	PB5-8	22-Jul-57	7	6
SC KERSHAW	1.89	PB5-9	23-Jul-57	3	6
SC KERSHAW	7.67	S-6	11-Jul-62	6	6
SC KERSHAW	1.32	UE8-6	19-Jul-62	9	6
SC KERSHAW	2.43	UE8-7	20-Jul-62	1	6
SC KERSHAW	0.30	UE8-8	21-Jul-62	3	6
SC LANCASTER	1.31	TP7-8	5-Apr-55	4	4
SC LANCASTER	1.51	PB5-8	22-Jul-57	7	4

COUNTY	uCi/ sqm	SHOT DAY	DATE	RANK FOR DAY	TIMES IN TOP 10
SC LANCASTER	1.51	PB8-7	24-Aug-57	10	4
SC LANCASTER	0.16	UE8-8	21-Jul-62	6	4
SC LAURENS	0.97	TP6-5	26-Mar-55	8	4
SC LAURENS	2.20	PB2-8	25-Jun-57	6	4
SC LAURENS	1.62	PB2-9	26-Jun-57	10	4
SC LAURENS	1.51	PB5-8	22-Jul-57	7	4
SC LEE	2.63	TS3-7	28-Apr-52	4	5
SC LEE	2.63	TS3-7	28-Apr-52	4	5
SC LEE	1.65	PB4-10	14-Jul-57	10	5
SC LEE	1.51	PB5-8	22-Jul-57	7	5
SC LEE	1.37	UE8-7	20-Jul-62	4	5
SC LEXINGTON	0.88	TP7-8	5-Apr-55	6	6
SC LEXINGTON	1.32	PB2-8	25-Jun-57	9	6
SC LEXINGTON	1.62	PB2-9	26-Jun-57	10	6
SC LEXINGTON	1.51	PB5-8	22-Jul-57	7	6
SC LEXINGTON	1.37	UE8-7	20-Jul-62	4	6
SC LEXINGTON	0.16	UE8-8	21-Jul-62	7	6
SC MARION	2.63	TS3-7	28-Apr-52	4	5
SC MARION	2.63	TS3-7	28-Apr-52	4	5
SC MARION	1.51	PB5-8	22-Jul-57	7	5
SC MARION	2.83	PB5-9	23-Jul-57	2	5
SC MARION	2.54	PB8-7	24-Aug-57	7	5
SC MARLBORO	9.17	TS5-3	9-May-52	8	7
SC MARLBORO	1.31	TP7-8	5-Apr-55	4	7
SC MARLBORO	1.51	PB5-8	22-Jul-57	7	7
SC MARLBORO	2.83	PB5-9	23-Jul-57	2	7
SC MARLBORO	2.54	PB8-7	24-Aug-57	7	7
SC MARLBORO	1.32	UE8-6	19-Jul-62	9	7
SC MARLBORO	0.48	UE8-8	21-Jul-62	1	7
SC MCCORMICK	1.00	TS5-5	11-May-52	7	3
SC MCCORMICK	2.42	TS6-6	30-May-52	6	3
SC MCCORMICK	1.51	PB5-8	22-Jul-57	7	3
SC NEWBERRY	0.88	TP7-8	5-Apr-55	6	4
SC NEWBERRY	1.51	PB5-8	22-Jul-57	7	4
SC NEWBERRY	2.83	PB5-9	23-Jul-57	2	4

COUNTY	uCi/ sqm	SHOT DAY	DATE	RANK FOR DAY	TIMES IN TOP 10
SC NEWBERRY	0.10	UE8-8	21-Jul-62	8	4
SC OCONEE	0.86	BJ1-2	29-Oct-51	5	8
SC OCONEE	1.04	UpK5-2	12-Apr-53	4	8
SC OCONEE	2.03	TP6-4	25-Mar-55	8	8
SC OCONEE	0.97	TP6-5	26-Mar-55	8	8
SC OCONEE	4.18	PB2-5	22-Jun-57	3	8
SC OCONEE	2.23	PB5-8	22-Jul-57	6	8
SC OCONEE	22.47	PB17-3	30-Sep-57	1	8
SC OCONEE	2.58	UE8-6	19-Jul-62	4	8
SC ORANGEBURG	2.63	TS3-7	28-Apr-52	4	5
SC ORANGEBURG	2.63	TS3-7	28-Apr-52	4	5
SC ORANGEBURG	1.62	PB2-9	26-Jun-57	10	5
SC ORANGEBURG	1.51	PB5-8	22-Jul-57	7	5
SC ORANGEBURG	1.37	UE8-7	20-Jul-62	4	5
SC PICKENS	3.89	BJ2-2	31-Oct-51	10	8
SC PICKENS	3.17	TS6-6	30-May-52	4	8
SC PICKENS	1.04	UpK5-2	12-Apr-53	4	8
SC PICKENS	3.11	TP6-4	25-Mar-55	3	8
SC PICKENS	0.97	TP6-5	26-Mar-55	8	8
SC PICKENS	4.18	PB2-5	22-Jun-57	3	8
SC PICKENS	2.99	PB5-8	22-Jul-57	4	8
SC PICKENS	1.56	PB18-5	11-Oct-57	1	8
SC RICHLAND	1.51	PB5-8	22-Jul-57	7	2
SC RICHLAND	0.94	PB5-9	23-Jul-57	4	2
SC SALUDA	0.88	TP7-8	5-Apr-55	6	4
SC SALUDA	1.51	PB5-8	22-Jul-57	7	4
SC SALUDA	2.58	UE8-6	19-Jul-62	4	4
SC SALUDA	0.27	UE8-8	21-Jul-62	4	4
SC SPARTANBURG	3.89	BJ2-2	31-Oct-51	10	7
SC SPARTANBURG	3.13	TS6-6	30-May-52	5	7
SC SPARTANBURG	2.03	TP6-4	25-Mar-55	8	7
SC SPARTANBURG	0.97	TP6-5	26-Mar-55	8	7
SC SPARTANBURG	1.39	PB2-5	22-Jun-57	8	7
SC SPARTANBURG	1.51	PB5-8	22-Jul-57	7	7
SC SPARTANBURG	0.93	PB18-5	11-Oct-57	3	7

COUNTY	uCi/ sqm	SHOT DAY	DATE	RANK FOR DAY	TIMES IN TOP 10
SC SUMTER	5.26	TS3-7	28-Apr-52	3	6
SC SUMTER	5.26	TS3-7	28-Apr-52	3	6
SC SUMTER	1.51	PB5-8	22-Jul-57	7	6
SC SUMTER	4.72	PB5-9	23-Jul-57	1	6
SC SUMTER	1.32	UE8-6	19-Jul-62	9	6
SC SUMTER	2.43	UE8-7	20-Jul-62	1	6
SC UNION	1.31	TP7-8	5-Apr-55	4	3
SC UNION	1.51	PB5-8	22-Jul-57	7	3
SC UNION	0.94	PB5-9	23-Jul-57	4	3
SC WILLIAMSBURG	2.63	TS3-7	28-Apr-52	4	3
SC WILLIAMSBURG	2.63	TS3-7	28-Apr-52	4	3
SC WILLIAMSBURG	1.51	PB5-8	22-Jul-57	7	3
SC YORK	1.31	TP7-8	5-Apr-55	4	5
SC YORK	1.51	PB5-8	22-Jul-57	7	5
SC YORK	2.83	PB5-9	23-Jul-57	2	5
SC YORK	1.32	UE8-6	19-Jul-62	9	5
SC YORK	0.21	UE8-8	21-Jul-62	5	5

COUNTY	uCi/ sqm	SHOT DAY	DATE	RANK FOR DAY	TIMES IN TOP 10
SD AURORA	3.97	UpK2-2	25-Mar-53	10	11
SD AURORA	45.16	PB2-3	20-Jun-57	2	11
SD AURORA	0.93	PB3-8	1-Jul-57	4	11
SD AURORA	5.89	PB5-4	18-Jul-57	5	11
SD AURORA	19.73	PB8-3	20-Aug-57	6	11
SD AURORA	11.34	PB8-5	22-Aug-57	2	11
SD AURORA	27.46	PB11-2	1-Sep-57	2	11
SD AURORA	20.04	PB15-2	17-Sep-57	2	11
SD AURORA	1.64	PB15-3	18-Sep-57	9	11
SD AURORA	0.10	UE2-7	11-Mar-62	4	11
SD AURORA	9.84	UE12-4	17-Apr-65	4	11
SD BEADLE	2.53	TS7-2	2-Jun-52	3	13
SD BEADLE	45.16	PB2-3	20-Jun-57	2	13
SD BEADLE	5.89	PB5-4	18-Jul-57	5	13
SD BEADLE	19.73	PB8-3	20-Aug-57	6	13
SD BEADLE	7.56	PB8-5	22-Aug-57	3	13
SD BEADLE	5.22	PB9-4	26-Aug-57	8	13
SD BEADLE	9.15	PB11-2	1-Sep-57	5	13
SD BEADLE	20.04	PB15-2	17-Sep-57	2	13
SD BEADLE	2.50	PB15-3	18-Sep-57	6	13
SD BEADLE	0.59	UE2-4	8-Mar-62	7	13
SD BEADLE	0.12	UE2-6	10-Mar-62	5	13
SD BEADLE	0.10	UE2-7	11-Mar-62	4	13
SD BEADLE	9.84	UE12-4	17-Apr-65	4	13
SD BENNETT	9.21	TS5-3	9-May-52	7	15
SD BENNETT	45.45	UpK2-2	25-Mar-53	4	15
SD BENNETT	30.08	PB2-3	20-Jun-57	3	15
SD BENNETT	8.86	PB5-4	18-Jul-57	3	15
SD BENNETT	0.83	PB8-2	19-Aug-57	8	15
SD BENNETT	19.73	PB8-3	20-Aug-57	6	15
SD BENNETT	13.41	PB8-4	21-Aug-57	2	15
SD BENNETT	13.41	PB8-4	21-Aug-57	2	15
SD BENNETT	3.78	PB8-5	22-Aug-57	4	15
SD BENNETT	4.58	PB11-2	1-Sep-57	9	15
SD BENNETT	13.36	PB15-2	17-Sep-57	3	15

COUNTY	uCi/ sqm	SHOT DAY	DATE	RANK FOR DAY	TIMES IN TOP 10
SD BENNETT	2.37	PB18-3	9-Oct-57	3	15
SD BENNETT	1.44	PB18-4	10-Oct-57	6	15
SD BENNETT	0.11	UE2-6	10-Mar-62	6	15
SD BENNETT	0.10	UE2-7	11-Mar-62	4	15
SD BON HOMME	1.44	BJ5-3	21-Nov-51	6	12
SD BON HOMME	1.37	TS4-6	6-May-52	9	12
SD BON HOMME	1.37	TS4-6	6-May-52	9	12
SD BON HOMME	1.72	TS8-11	15-Jun-52	2	12
SD BON HOMME	3.97	UpK2-2	25-Mar-53	10	12
SD BON HOMME	1.65	PB4-10	14-Jul-57	10	12
SD BON HOMME	19.73	PB8-3	20-Aug-57	6	12
SD BON HOMME	1.92	PB8-5	22-Aug-57	10	12
SD BON HOMME	4.58	PB11-2	1-Sep-57	9	12
SD BON HOMME	4.13	PB15-3	18-Sep-57	2	12
SD BON HOMME	0.10	UE2-7	11-Mar-62	4	12
SD BON HOMME	11.43	UE12-4	17-Apr-65	3	12
SD BROOKINGS	3.97	UpK2-2	25-Mar-53	10	8
SD BROOKINGS	15.08	PB2-3	20-Jun-57	5	8
SD BROOKINGS	78.92	PB8-3	20-Aug-57	2	8
SD BROOKINGS	11.34	PB8-5	22-Aug-57	2	8
SD BROOKINGS	4.58	PB11-2	1-Sep-57	9	8
SD BROOKINGS	2.50	PB15-3	18-Sep-57	6	8
SD BROOKINGS	0.10	UE2-7	11-Mar-62	3	8
SD BROOKINGS	0.11	UE7-5	15-Jul-62	10	8
SD BROWN	3.97	UpK2-2	25-Mar-53	10	9
SD BROWN	30.08	PB2-3	20-Jun-57	3	9
SD BROWN	19.73	PB8-3	20-Aug-57	6	9
SD BROWN	3.78	PB8-5	22-Aug-57	4	9
SD BROWN	3.48	PB9-4	26-Aug-57	10	9
SD BROWN	27.46	PB11-2	1-Sep-57	2	9
SD BROWN	13.36	PB15-2	17-Sep-57	3	9
SD BROWN	1.18	UE2-4	8-Mar-62	1	9
SD BROWN	0.21	UE2-6	10-Mar-62	3	9
SD BRULE	3.97	UpK2-2	25-Mar-53	10	10
SD BRULE	45.16	PB2-3	20-Jun-57	2	10

COUNTY	uCi/ sqm	SHOT DAY	DATE	RANK FOR DAY	TIMES IN TOP 10
SD BRULE	19.73	PB8-3	20-Aug-57	6	10
SD BRULE	7.56	PB8-5	22-Aug-57	3	10
SD BRULE	27.46	PB11-2	1-Sep-57	2	10
SD BRULE	13.36	PB15-2	17-Sep-57	3	10
SD BRULE	2.50	PB15-3	18-Sep-57	6	10
SD BRULE	0.10	UE2-7	11-Mar-62	4	10
SD BRULE	0.11	UE7-5	15-Jul-62	10	10
SD BRULE	9.84	UE12-4	17-Apr-65	4	10
SD BUFFALO	3.97	UpK2-2	25-Mar-53	10	12
SD BUFFALO	45.16	PB2-3	20-Jun-57	2	12
SD BUFFALO	19.73	PB8-3	20-Aug-57	6	12
SD BUFFALO	3.78	PB8-5	22-Aug-57	4	12
SD BUFFALO	3.48	PB9-4	26-Aug-57	10	12
SD BUFFALO	9.15	PB11-2	1-Sep-57	5	12
SD BUFFALO	13.36	PB15-2	17-Sep-57	3	12
SD BUFFALO	2.50	PB15-3	18-Sep-57	6	12
SD BUFFALO	1.00	PB17-9	6-Oct-57	9	12
SD BUFFALO	0.12	UE2-6	10-Mar-62	5	12
SD BUFFALO	0.10	UE2-7	11-Mar-62	4	12
SD BUFFALO	9.84	UE12-4	17-Apr-65	4	12
SD BUTTE	45.45	UpK2-2	25-Mar-53	4	14
SD BUTTE	1.58	UpK8-1	8-May-53	8	14
SD BUTTE	9.68	PB5-2	16-Jul-57	6	14
SD BUTTE	15.19	PB5-3	17-Jul-57	1	14
SD BUTTE	1.30	PB8-2	19-Aug-57	7	14
SD BUTTE	19.73	PB8-3	20-Aug-57	6	14
SD BUTTE	8.05	PB8-4	21-Aug-57	4	14
SD BUTTE	8.05	PB8-4	21-Aug-57	4	14
SD BUTTE	5.22	PB9-4	26-Aug-57	8	14
SD BUTTE	6.81	PB11-2	1-Sep-57	6	14
SD BUTTE	2.16	PB18-4	10-Oct-57	4	14
SD BUTTE	0.22	UE2-6	10-Mar-62	2	14
SD BUTTE	2.48	UE7-3	13-Jul-62	4	14
SD BUTTE	9.84	UE12-4	17-Apr-65	4	14
SD CAMPBELL	3.97	UpK2-2	25-Mar-53	10	13

COUNTY	uCi/ sqm	SHOT DAY	DATE	RANK FOR DAY	TIMES IN TOP 10
SD CAMPBELL	45.16	PB2-3	20-Jun-57	2	13
SD CAMPBELL	8.86	PB5-4	18-Jul-57	3	13
SD CAMPBELL	3.30	PB8-2	19-Aug-57	5	13
SD CAMPBELL	19.73	PB8-3	20-Aug-57	6	13
SD CAMPBELL	5.36	PB8-4	21-Aug-57	6	13
SD CAMPBELL	5.36	PB8-4	21-Aug-57	6	13
SD CAMPBELL	1.92	PB8-5	22-Aug-57	10	13
SD CAMPBELL	3.48	PB9-4	26-Aug-57	10	13
SD CAMPBELL	18.31	PB11-2	1-Sep-57	3	13
SD CAMPBELL	20.04	PB15-2	17-Sep-57	2	13
SD CAMPBELL	1.44	PB18-4	10-Oct-57	6	13
SD CAMPBELL	0.10	UE2-7	11-Mar-62	4	13
SD CHARLES MIX	3.97	UpK2-2	25-Mar-53	10	12
SD CHARLES MIX	15.08	PB2-3	20-Jun-57	5	12
SD CHARLES MIX	19.73	PB8-3	20-Aug-57	6	12
SD CHARLES MIX	5.36	PB8-4	21-Aug-57	6	12
SD CHARLES MIX	5.36	PB8-4	21-Aug-57	6	12
SD CHARLES MIX	3.78	PB8-5	22-Aug-57	4	12
SD CHARLES MIX	9.15	PB11-2	1-Sep-57	5	12
SD CHARLES MIX	13.36	PB15-2	17-Sep-57	3	12
SD CHARLES MIX	2.50	PB15-3	18-Sep-57	6	12
SD CHARLES MIX	0.17	UE2-7	11-Mar-62	2	12
SD CHARLES MIX	0.12	UE7-5	15-Jul-62	9	12
SD CHARLES MIX	11.43	UE12-4	17-Apr-65	3	12
SD CLARK	3.97	UpK2-2	25-Mar-53	10	13
SD CLARK	30.08	PB2-3	20-Jun-57	3	13
SD CLARK	5.89	PB5-4	18-Jul-57	5	13
SD CLARK	19.73	PB8-3	20-Aug-57	6	13
SD CLARK	7.56	PB8-5	22-Aug-57	3	13
SD CLARK	5.22	PB9-4	26-Aug-57	8	13
SD CLARK	1.24	PB9-7	29-Aug-57	3	13
SD CLARK	18.31	PB11-2	1-Sep-57	3	13
SD CLARK	20.04	PB15-2	17-Sep-57	2	13
SD CLARK	2.50	PB15-3	18-Sep-57	6	13
SD CLARK	0.12	UE2-6	10-Mar-62	5	13

COUNTY	uCi/ sqm	SHOT DAY	DATE	RANK FOR DAY	TIMES IN TOP 10
SD CLARK	0.10	UE2-7	11-Mar-62	3	13
SD CLARK	9.84	UE12-4	17-Apr-65	4	13
SD CLAY	1.37	TS4-6	6-May-52	9	8
SD CLAY	1.37	TS4-6	6-May-52	9	8
SD CLAY	3.97	UpK2-2	25-Mar-53	10	8
SD CLAY	3.06	UpK11-3	6-Jun-53	10	8
SD CLAY	1.44	PB4-3	7-Jul-57	3	8
SD CLAY	19.73	PB8-3	20-Aug-57	6	8
SD CLAY	5.22	PB9-4	26-Aug-57	8	8
SD CLAY	0.10	UE2-7	11-Mar-62	4	8
SD CODINGTON	3.97	UpK2-2	25-Mar-53	10	14
SD CODINGTON	45.16	PB2-3	20-Jun-57	2	14
SD CODINGTON	19.73	PB8-3	20-Aug-57	6	14
SD CODINGTON	5.36	PB8-4	21-Aug-57	6	14
SD CODINGTON	5.36	PB8-4	21-Aug-57	6	14
SD CODINGTON	7.56	PB8-5	22-Aug-57	3	14
SD CODINGTON	5.22	PB9-4	26-Aug-57	8	14
SD CODINGTON	4.58	PB11-2	1-Sep-57	9	14
SD CODINGTON	33.40	PB15-2	17-Sep-57	1	14
SD CODINGTON	1.64	PB15-3	18-Sep-57	9	14
SD CODINGTON	1.18	UE2-4	8-Mar-62	1	14
SD CODINGTON	0.12	UE2-6	10-Mar-62	5	14
SD CODINGTON	0.10	UE2-7	11-Mar-62	4	14
SD CODINGTON	9.84	UE12-4	17-Apr-65	4	14
SD CORSON	45.45	UpK2-2	25-Mar-53	4	16
SD CORSON	1.58	UpK8-1	8-May-53	8	16
SD CORSON	30.08	PB2-3	20-Jun-57	3	16
SD CORSON	0.83	PB8-2	19-Aug-57	8	16
SD CORSON	19.73	PB8-3	20-Aug-57	6	16
SD CORSON	5.36	PB8-4	21-Aug-57	6	16
SD CORSON	5.36	PB8-4	21-Aug-57	6	16
SD CORSON	7.56	PB8-5	22-Aug-57	3	16
SD CORSON	3.48	PB9-4	26-Aug-57	10	16
SD CORSON	9.15	PB11-2	1-Sep-57	5	16
SD CORSON	20.04	PB15-2	17-Sep-57	2	16

COUNTY	uCi/ sqm	SHOT DAY	DATE	RANK FOR DAY	TIMES IN TOP 10
SD CORSON	2.37	PB18-3	9-Oct-57	3	16
SD CORSON	1.44	PB18-4	10-Oct-57	6	16
SD CORSON	0.21	UE2-6	10-Mar-62	3	16
SD CORSON	0.10	UE2-7	11-Mar-62	4	16
SD CORSON	2.48	UE7-3	13-Jul-62	4	16
SD CUSTER	154.39	TS7-3	3-Jun-52	7	13
SD CUSTER	45.45	UpK2-2	25-Mar-53	4	13
SD CUSTER	1.58	UpK8-1	8-May-53	8	13
SD CUSTER	0.83	PB8-2	19-Aug-57	8	13
SD CUSTER	7.76	PB9-3	25-Aug-57	9	13
SD CUSTER	3.68	PB9-4	26-Aug-57	9	13
SD CUSTER	6.81	PB11-2	1-Sep-57	6	13
SD CUSTER	2.37	PB18-3	9-Oct-57	3	13
SD CUSTER	2.88	PB18-4	10-Oct-57	2	13
SD CUSTER	0.22	UE2-6	10-Mar-62	2	13
SD CUSTER	0.10	UE2-7	11-Mar-62	3	13
SD CUSTER	434.17	S-2	7-Jul-62	2	13
SD CUSTER	7.97	UE12-4	17-Apr-65	9	13
SD DAVISON	3.97	UpK2-2	25-Mar-53	10	13
SD DAVISON	0.81	TP7-5	2-Apr-55	2	13
SD DAVISON	9.91	TP9-3	17-Apr-55	3	13
SD DAVISON	15.08	PB2-3	20-Jun-57	5	13
SD DAVISON	19.73	PB8-3	20-Aug-57	6	13
SD DAVISON	18.89	PB8-5	22-Aug-57	1	13
SD DAVISON	3.48	PB9-4	26-Aug-57	10	13
SD DAVISON	9.15	PB11-2	1-Sep-57	5	13
SD DAVISON	6.68	PB15-2	17-Sep-57	5	13
SD DAVISON	2.50	PB15-3	18-Sep-57	6	13
SD DAVISON	0.10	UE2-7	11-Mar-62	4	13
SD DAVISON	0.11	UE7-5	15-Jul-62	10	13
SD DAVISON	11.43	UE12-4	17-Apr-65	3	13
SD DAY	3.97	UpK2-2	25-Mar-53	10	11
SD DAY	45.16	PB2-3	20-Jun-57	2	11
SD DAY	5.89	PB5-4	18-Jul-57	5	11
SD DAY	19.73	PB8-3	20-Aug-57	6	11

COUNTY	uCi/ sqm	SHOT DAY	DATE	RANK FOR DAY	TIMES IN TOP 10
SD DAY	2.85	PB8-5	22-Aug-57	7	11
SD DAY	27.46	PB11-2	1-Sep-57	2	11
SD DAY	20.04	PB15-2	17-Sep-57	2	11
SD DAY	1.64	PB15-3	18-Sep-57	9	11
SD DAY	1.51	UE2-3	7-Mar-62	4	11
SD DAY	0.98	UE2-4	8-Mar-62	2	11
SD DAY	0.10	UE2-7	11-Mar-62	4	11
SD DEUEL	3.97	UpK2-2	25-Mar-53	10	4
SD DEUEL	2.58	PB15-3	18-Sep-57	5	4
SD DEUEL	0.10	UE2-7	11-Mar-62	4	4
SD DEUEL	9.84	UE12-4	17-Apr-65	4	4
SD DEWEY	45.45	UpK2-2	25-Mar-53	4	15
SD DEWEY	1.58	UpK8-1	8-May-53	8	15
SD DEWEY	45.16	PB2-3	20-Jun-57	2	15
SD DEWEY	8.86	PB5-4	18-Jul-57	3	15
SD DEWEY	0.83	PB8-2	19-Aug-57	8	15
SD DEWEY	19.73	PB8-3	20-Aug-57	6	15
SD DEWEY	7.56	PB8-5	22-Aug-57	3	15
SD DEWEY	3.48	PB9-4	26-Aug-57	10	15
SD DEWEY	9.15	PB11-2	1-Sep-57	5	15
SD DEWEY	20.04	PB15-2	17-Sep-57	2	15
SD DEWEY	2.37	PB18-3	9-Oct-57	3	15
SD DEWEY	1.44	PB18-4	10-Oct-57	6	15
SD DEWEY	0.21	UE2-6	10-Mar-62	3	15
SD DEWEY	0.10	UE2-7	11-Mar-62	4	15
SD DEWEY	9.32	UE12-4	17-Apr-65	6	15
SD DOUGLAS	3.97	UpK2-2	25-Mar-53	10	10
SD DOUGLAS	15.08	PB2-3	20-Jun-57	5	10
SD DOUGLAS	1.65	PB4-10	14-Jul-57	10	10
SD DOUGLAS	5.89	PB5-4	18-Jul-57	5	10
SD DOUGLAS	19.73	PB8-3	20-Aug-57	6	10
SD DOUGLAS	3.78	PB8-5	22-Aug-57	4	10
SD DOUGLAS	4.58	PB11-2	1-Sep-57	9	10
SD DOUGLAS	2.50	PB15-3	18-Sep-57	6	10
SD DOUGLAS	0.10	UE2-7	11-Mar-62	4	10

COUNTY	uCi/ sqm	SHOT DAY	DATE	RANK FOR DAY	TIMES IN TOP 10
SD DOUGLAS	11.43	UE12-4	17-Apr-65	3	10
SD EDMUNDS	3.97	UpK2-2	25-Mar-53	10	12
SD EDMUNDS	15.08	PB2-3	20-Jun-57	5	12
SD EDMUNDS	5.89	PB5-4	18-Jul-57	5	12
SD EDMUNDS	19.73	PB8-3	20-Aug-57	6	12
SD EDMUNDS	7.56	PB8-5	22-Aug-57	3	12
SD EDMUNDS	3.48	PB9-4	26-Aug-57	10	12
SD EDMUNDS	1.24	PB9-7	29-Aug-57	3	12
SD EDMUNDS	27.46	PB11-2	1-Sep-57	2	12
SD EDMUNDS	20.04	PB15-2	17-Sep-57	2	12
SD EDMUNDS	1.48	PB15-4	19-Sep-57	4	12
SD EDMUNDS	0.21	UE2-6	10-Mar-62	3	12
SD EDMUNDS	0.10	UE2-7	11-Mar-62	3	12
SD FALL RIVER	30.40	TS5-2	8-May-52	4	11
SD FALL RIVER	45.45	UpK2-2	25-Mar-53	4	11
SD FALL RIVER	1.58	UpK8-1	8-May-53	8	11
SD FALL RIVER	9.68	PB5-2	16-Jul-57	6	11
SD FALL RIVER	0.83	PB8-2	19-Aug-57	8	11
SD FALL RIVER	7.76	PB9-3	25-Aug-57	9	11
SD FALL RIVER	2.37	PB18-3	9-Oct-57	3	11
SD FALL RIVER	1.44	PB18-4	10-Oct-57	6	11
SD FALL RIVER	0.22	UE2-6	10-Mar-62	2	11
SD FALL RIVER	0.10	UE2-7	11-Mar-62	4	11
SD FALL RIVER	345.60	S-2	7-Jul-62	4	11
SD FAULK	3.97	UpK2-2	25-Mar-53	10	11
SD FAULK	45.16	PB2-3	20-Jun-57	2	11
SD FAULK	39.46	PB8-3	20-Aug-57	4	11
SD FAULK	3.78	PB8-5	22-Aug-57	4	11
SD FAULK	5.22	PB9-4	26-Aug-57	8	11
SD FAULK	27.46	PB11-2	1-Sep-57	2	11
SD FAULK	20.04	PB15-2	17-Sep-57	2	11
SD FAULK	1.64	PB15-3	18-Sep-57	9	11
SD FAULK	0.23	UE2-6	10-Mar-62	1	11
SD FAULK	0.10	UE2-7	11-Mar-62	4	11
SD FAULK	9.32	UE12-4	17-Apr-65	6	11

COUNTY	uCi/ sqm	SHOT DAY	DATE	RANK FOR DAY	TIMES IN TOP 10
SD GRANT	3.97	UpK2-2	25-Mar-53	10	5
SD GRANT	1.50	PB2-4	21-Jun-57	8	5
SD GRANT	2.45	PB6-7	30-Jul-57	3	5
SD GRANT	2.58	PB15-3	18-Sep-57	5	5
SD GRANT	0.10	UE2-7	11-Mar-62	4	5
SD GREGORY	3.97	UpK2-2	25-Mar-53	10	11
SD GREGORY	15.08	PB2-3	20-Jun-57	5	11
SD GREGORY	0.93	PB3-8	1-Jul-57	4	11
SD GREGORY	0.83	PB8-2	19-Aug-57	8	11
SD GREGORY	19.73	PB8-3	20-Aug-57	6	11
SD GREGORY	7.56	PB8-5	22-Aug-57	3	11
SD GREGORY	4.58	PB11-2	1-Sep-57	9	11
SD GREGORY	13.36	PB15-2	17-Sep-57	3	11
SD GREGORY	2.50	PB15-3	18-Sep-57	6	11
SD GREGORY	1.48	PB15-4	19-Sep-57	4	11
SD GREGORY	0.10	UE2-7	11-Mar-62	4	11
SD HAAKON	45.45	UpK2-2	25-Mar-53	4	17
SD HAAKON	30.08	PB2-3	20-Jun-57	3	17
SD HAAKON	8.86	PB5-4	18-Jul-57	3	17
SD HAAKON	2.60	PB6-8	31-Jul-57	2	17
SD HAAKON	1.30	PB8-2	19-Aug-57	7	17
SD HAAKON	19.73	PB8-3	20-Aug-57	6	17
SD HAAKON	8.05	PB8-4	21-Aug-57	4	17
SD HAAKON	8.05	PB8-4	21-Aug-57	4	17
SD HAAKON	7.56	PB8-5	22-Aug-57	3	17
SD HAAKON	3.48	PB9-4	26-Aug-57	10	17
SD HAAKON	9.15	PB11-2	1-Sep-57	5	17
SD HAAKON	20.04	PB15-2	17-Sep-57	2	17
SD HAAKON	2.37	PB18-3	9-Oct-57	3	17
SD HAAKON	1.44	PB18-4	10-Oct-57	6	17
SD HAAKON	0.23	UE2-6	10-Mar-62	1	17
SD HAAKON	0.10	UE2-7	11-Mar-62	4	17
SD HAAKON	9.32	UE12-4	17-Apr-65	6	17
SD HAMLIN	3.97	UpK2-2	25-Mar-53	10	12
SD HAMLIN	30.08	PB2-3	20-Jun-57	3	12

COUNTY	uCi/ sqm	SHOT DAY	DATE	RANK FOR DAY	TIMES IN TOP 10
SD HAMLIN	5.89	PB5-4	18-Jul-57	5	12
SD HAMLIN	19.73	PB8-3	20-Aug-57	6	12
SD HAMLIN	11.34	PB8-5	22-Aug-57	2	12
SD HAMLIN	3.48	PB9-4	26-Aug-57	10	12
SD HAMLIN	9.15	PB11-2	1-Sep-57	5	12
SD HAMLIN	20.04	PB15-2	17-Sep-57	2	12
SD HAMLIN	4.13	PB15-3	18-Sep-57	2	12
SD HAMLIN	0.10	UE2-7	11-Mar-62	4	12
SD HAMLIN	361.42	S-2	7-Jul-62	3	12
SD HAMLIN	9.84	UE12-4	17-Apr-65	4	12
SD HAND	1.70	TS7-2	2-Jun-52	7	11
SD HAND	3.97	UpK2-2	25-Mar-53	10	11
SD HAND	30.08	PB2-3	20-Jun-57	3	11
SD HAND	19.73	PB8-3	20-Aug-57	6	11
SD HAND	1.92	PB8-5	22-Aug-57	10	11
SD HAND	5.22	PB9-4	26-Aug-57	8	11
SD HAND	9.15	PB11-2	1-Sep-57	5	11
SD HAND	20.04	PB15-2	17-Sep-57	2	11
SD HAND	0.11	UE2-6	10-Mar-62	6	11
SD HAND	0.10	UE2-7	11-Mar-62	4	11
SD HAND	9.84	UE12-4	17-Apr-65	4	11
SD HANSON	3.97	UpK2-2	25-Mar-53	10	11
SD HANSON	6.79	TP9-3	17-Apr-55	10	11
SD HANSON	15.08	PB2-3	20-Jun-57	5	11
SD HANSON	1.65	PB4-10	14-Jul-57	10	11
SD HANSON	19.73	PB8-3	20-Aug-57	6	11
SD HANSON	18.89	PB8-5	22-Aug-57	1	11
SD HANSON	5.22	PB9-4	26-Aug-57	8	11
SD HANSON	4.58	PB11-2	1-Sep-57	9	11
SD HANSON	2.50	PB15-3	18-Sep-57	6	11
SD HANSON	0.17	UE2-7	11-Mar-62	2	11
SD HANSON	9.84	UE12-4	17-Apr-65	4	11
SD HARDING	45.45	UpK2-2	25-Mar-53	4	15
SD HARDING	3.16	UpK8-1	8-May-53	7	15
SD HARDING	9.68	PB5-2	16-Jul-57	6	15

COUNTY	uCi/ sqm	SHOT DAY	DATE	RANK FOR DAY	TIMES IN TOP 10
SD HARDING	10.10	PB5-3	17-Jul-57	4	15
SD HARDING	5.87	PB6-5	28-Jul-57	6	15
SD HARDING	0.83	PB8-2	19-Aug-57	8	15
SD HARDING	39.46	PB8-3	20-Aug-57	4	15
SD HARDING	5.36	PB8-4	21-Aug-57	6	15
SD HARDING	5.36	PB8-4	21-Aug-57	6	15
SD HARDING	3.78	PB8-5	22-Aug-57	4	15
SD HARDING	3.48	PB9-4	26-Aug-57	10	15
SD HARDING	18.31	PB11-2	1-Sep-57	3	15
SD HARDING	20.04	PB15-2	17-Sep-57	2	15
SD HARDING	1.44	PB18-4	10-Oct-57	6	15
SD HARDING	0.22	UE2-6	10-Mar-62	2	15
SD HUGHES	3.97	UpK2-2	25-Mar-53	10	11
SD HUGHES	12.49	PB2-3	20-Jun-57	8	11
SD HUGHES	4.80	PB6-5	28-Jul-57	7	11
SD HUGHES	0.83	PB8-2	19-Aug-57	8	11
SD HUGHES	8.19	PB8-3	20-Aug-57	8	11
SD HUGHES	3.14	PB8-5	22-Aug-57	6	11
SD HUGHES	5.55	PB15-2	17-Sep-57	6	11
SD HUGHES	1.99	PB17-9	6-Oct-57	1	11
SD HUGHES	1.44	PB18-4	10-Oct-57	6	11
SD HUGHES	0.17	UE2-7	11-Mar-62	2	11
SD HUGHES	9.32	UE12-4	17-Apr-65	6	11
SD HUTCHINSON	1.44	BJ5-3	21-Nov-51	6	14
SD HUTCHINSON	57.45	TS7-4	4-Jun-52	7	14
SD HUTCHINSON	3.97	UpK2-2	25-Mar-53	10	14
SD HUTCHINSON	15.08	PB2-3	20-Jun-57	5	14
SD HUTCHINSON	1.65	PB4-10	14-Jul-57	10	14
SD HUTCHINSON	19.73	PB8-3	20-Aug-57	6	14
SD HUTCHINSON	3.78	PB8-5	22-Aug-57	4	14
SD HUTCHINSON	3.48	PB9-4	26-Aug-57	10	14
SD HUTCHINSON	4.58	PB11-2	1-Sep-57	9	14
SD HUTCHINSON	20.04	PB15-2	17-Sep-57	2	14
SD HUTCHINSON	2.50	PB15-3	18-Sep-57	6	14
SD HUTCHINSON	0.10	UE2-7	11-Mar-62	4	14

COUNTY	uCi/ sqm	SHOT DAY	DATE	RANK FOR DAY	TIMES IN TOP 10
SD HUTCHINSON	138.24	S-2	7-Jul-62	7	14
SD HUTCHINSON	7.97	UE12-4	17-Apr-65	9	14
SD HYDE	3.97	UpK2-2	25-Mar-53	10	12
SD HYDE	45.16	PB2-3	20-Jun-57	2	12
SD HYDE	19.73	PB8-3	20-Aug-57	6	12
SD HYDE	1.92	PB8-5	22-Aug-57	10	12
SD HYDE	5.22	PB9-4	26-Aug-57	8	12
SD HYDE	9.15	PB11-2	1-Sep-57	5	12
SD HYDE	20.04	PB15-2	17-Sep-57	2	12
SD HYDE	1.48	PB15-4	19-Sep-57	4	12
SD HYDE	1.13	PB17-9	6-Oct-57	6	12
SD HYDE	0.12	UE2-6	10-Mar-62	5	12
SD HYDE	0.10	UE2-7	11-Mar-62	4	12
SD HYDE	9.84	UE12-4	17-Apr-65	4	12
SD JACKSON	45.45	UpK2-2	25-Mar-53	4	15
SD JACKSON	30.08	PB2-3	20-Jun-57	3	15
SD JACKSON	8.86	PB5-4	18-Jul-57	3	15
SD JACKSON	0.83	PB8-2	19-Aug-57	8	15
SD JACKSON	19.73	PB8-3	20-Aug-57	6	15
SD JACKSON	13.41	PB8-4	21-Aug-57	2	15
SD JACKSON	13.41	PB8-4	21-Aug-57	2	15
SD JACKSON	3.78	PB8-5	22-Aug-57	4	15
SD JACKSON	4.58	PB11-2	1-Sep-57	9	15
SD JACKSON	20.04	PB15-2	17-Sep-57	2	15
SD JACKSON	2.37	PB18-3	9-Oct-57	3	15
SD JACKSON	1.44	PB18-4	10-Oct-57	6	15
SD JACKSON	0.22	UE2-6	10-Mar-62	2	15
SD JACKSON	0.10	UE2-7	11-Mar-62	4	15
SD JACKSON	9.32	UE12-4	17-Apr-65	6	15
SD JERAULD	2.12	TS7-2	2-Jun-52	5	16
SD JERAULD	3.97	UpK2-2	25-Mar-53	10	16
SD JERAULD	7.22	TP9-3	17-Apr-55	9	16
SD JERAULD	30.08	PB2-3	20-Jun-57	3	16
SD JERAULD	0.93	PB3-8	1-Jul-57	4	16
SD JERAULD	5.89	PB5-4	18-Jul-57	5	16

COUNTY	uCi/ sqm	SHOT DAY	DATE	RANK FOR DAY	TIMES IN TOP 10
SD JERAULD	19.73	PB8-3	20-Aug-57	6	16
SD JERAULD	7.56	PB8-5	22-Aug-57	3	16
SD JERAULD	3.48	PB9-4	26-Aug-57	10	16
SD JERAULD	4.58	PB11-2	1-Sep-57	9	16
SD JERAULD	13.36	PB15-2	17-Sep-57	3	16
SD JERAULD	2.50	PB15-3	18-Sep-57	6	16
SD JERAULD	0.59	UE2-4	8-Mar-62	7	16
SD JERAULD	0.10	UE2-7	11-Mar-62	4	16
SD JERAULD	145.87	S-2	7-Jul-62	6	16
SD JERAULD	11.43	UE12-4	17-Apr-65	3	16
SD JONES	3.97	UpK2-2	25-Mar-53	10	14
SD JONES	30.08	PB2-3	20-Jun-57	3	14
SD JONES	3.30	PB8-2	19-Aug-57	5	14
SD JONES	19.73	PB8-3	20-Aug-57	6	14
SD JONES	7.56	PB8-5	22-Aug-57	3	14
SD JONES	9.75	PB9-3	25-Aug-57	8	14
SD JONES	4.58	PB11-2	1-Sep-57	9	14
SD JONES	13.36	PB15-2	17-Sep-57	3	14
SD JONES	1.52	PB17-9	6-Oct-57	4	14
SD JONES	2.37	PB18-3	9-Oct-57	3	14
SD JONES	1.44	PB18-4	10-Oct-57	6	14
SD JONES	0.10	UE2-7	11-Mar-62	4	14
SD JONES	2.42	UE7-3	13-Jul-62	5	14
SD JONES	0.11	UE7-5	15-Jul-62	10	14
SD KINGSBURY	3.97	UpK2-2	25-Mar-53	10	10
SD KINGSBURY	30.08	PB2-3	20-Jun-57	3	10
SD KINGSBURY	39.46	PB8-3	20-Aug-57	4	10
SD KINGSBURY	18.89	PB8-5	22-Aug-57	1	10
SD KINGSBURY	9.15	PB11-2	1-Sep-57	5	10
SD KINGSBURY	20.04	PB15-2	17-Sep-57	2	10
SD KINGSBURY	4.13	PB15-3	18-Sep-57	2	10
SD KINGSBURY	0.10	UE2-7	11-Mar-62	4	10
SD KINGSBURY	361.42	S-2	7-Jul-62	3	10
SD KINGSBURY	11.43	UE12-4	17-Apr-65	3	10
SD LAKE	3.97	UpK2-2	25-Mar-53	10	10

COUNTY	uCi/ sqm	SHOT DAY	DATE	RANK FOR DAY	TIMES IN TOP 10
SD LAKE	15.08	PB2-3	20-Jun-57	5	10
SD LAKE	19.73	PB8-3	20-Aug-57	6	10
SD LAKE	18.89	PB8-5	22-Aug-57	1	10
SD LAKE	3.48	PB9-4	26-Aug-57	10	10
SD LAKE	18.31	PB11-2	1-Sep-57	3	10
SD LAKE	20.04	PB15-2	17-Sep-57	2	10
SD LAKE	2.50	PB15-3	18-Sep-57	6	10
SD LAKE	0.10	UE2-7	11-Mar-62	4	10
SD LAKE	327.38	S-3	8-Jul-62	9	10
SD LAWRENCE	45.45	UpK2-2	25-Mar-53	4	16
SD LAWRENCE	1.58	UpK8-1	8-May-53	8	16
SD LAWRENCE	1.67	PB4-9	13-Jul-57	8	16
SD LAWRENCE	15.19	PB5-3	17-Jul-57	1	16
SD LAWRENCE	6.05	PB6-4	27-Jul-57	10	16
SD LAWRENCE	1.65	PB8-2	19-Aug-57	6	16
SD LAWRENCE	4.60	PB8-4	21-Aug-57	8	16
SD LAWRENCE	4.60	PB8-4	21-Aug-57	8	16
SD LAWRENCE	7.76	PB9-3	25-Aug-57	9	16
SD LAWRENCE	7.36	PB9-4	26-Aug-57	7	16
SD LAWRENCE	2.37	PB18-3	9-Oct-57	3	16
SD LAWRENCE	1.44	PB18-4	10-Oct-57	6	16
SD LAWRENCE	0.22	UE2-6	10-Mar-62	2	16
SD LAWRENCE	0.10	UE2-7	11-Mar-62	4	16
SD LAWRENCE	2.48	UE7-3	13-Jul-62	4	16
SD LAWRENCE	9.84	UE12-4	17-Apr-65	4	16
SD LINCOLN	1.44	BJ5-3	21-Nov-51	6	10
SD LINCOLN	1.37	TS4-6	6-May-52	9	10
SD LINCOLN	1.37	TS4-6	6-May-52	9	10
SD LINCOLN	1.72	TS8-11	15-Jun-52	2	10
SD LINCOLN	3.97	UpK2-2	25-Mar-53	10	10
SD LINCOLN	1.09	TP7-6	3-Apr-55	10	10
SD LINCOLN	19.73	PB8-3	20-Aug-57	6	10
SD LINCOLN	5.22	PB9-4	26-Aug-57	8	10
SD LINCOLN	0.17	UE2-7	11-Mar-62	2	10
SD LINCOLN	9.32	UE12-4	17-Apr-65	6	10

COUNTY	uCi/ sqm	SHOT DAY	DATE	RANK FOR DAY	TIMES IN TOP 10
SD LYMAN	3.97	UpK2-2	25-Mar-53	10	14
SD LYMAN	45.16	PB2-3	20-Jun-57	2	14
SD LYMAN	1.65	PB8-2	19-Aug-57	6	14
SD LYMAN	19.73	PB8-3	20-Aug-57	6	14
SD LYMAN	11.34	PB8-5	22-Aug-57	2	14
SD LYMAN	5.22	PB9-4	26-Aug-57	8	14
SD LYMAN	18.31	PB11-2	1-Sep-57	3	14
SD LYMAN	13.36	PB15-2	17-Sep-57	3	14
SD LYMAN	1.64	PB15-3	18-Sep-57	9	14
SD LYMAN	2.37	PB18-3	9-Oct-57	3	14
SD LYMAN	1.44	PB18-4	10-Oct-57	6	14
SD LYMAN	0.12	UE2-6	10-Mar-62	5	14
SD LYMAN	0.10	UE2-7	11-Mar-62	4	14
SD LYMAN	9.84	UE12-4	17-Apr-65	4	14
SD MARSHALL	30.08	PB2-3	20-Jun-57	3	9
SD MARSHALL	5.89	PB5-4	18-Jul-57	5	9
SD MARSHALL	19.73	PB8-3	20-Aug-57	6	9
SD MARSHALL	1.92	PB8-5	22-Aug-57	10	9
SD MARSHALL	3.48	PB9-4	26-Aug-57	10	9
SD MARSHALL	45.77	PB11-2	1-Sep-57	1	9
SD MARSHALL	20.04	PB15-2	17-Sep-57	2	9
SD MARSHALL	1.18	UE2-4	8-Mar-62	1	9
SD MARSHALL	0.10	UE2-7	11-Mar-62	3	9
SD MCCOOK	3.97	UpK2-2	25-Mar-53	10	12
SD MCCOOK	15.08	PB2-3	20-Jun-57	5	12
SD MCCOOK	19.73	PB8-3	20-Aug-57	6	12
SD MCCOOK	5.36	PB8-4	21-Aug-57	6	12
SD MCCOOK	5.36	PB8-4	21-Aug-57	6	12
SD MCCOOK	18.89	PB8-5	22-Aug-57	1	12
SD MCCOOK	4.58	PB11-2	1-Sep-57	9	12
SD MCCOOK	13.36	PB15-2	17-Sep-57	3	12
SD MCCOOK	2.50	PB15-3	18-Sep-57	6	12
SD MCCOOK	0.10	UE2-7	11-Mar-62	4	12
SD MCCOOK	138.24	S-2	7-Jul-62	7	12
SD MCCOOK	9.84	UE12-4	17-Apr-65	4	12

COUNTY	uCi/ sqm	SHOT DAY	DATE	RANK FOR DAY	TIMES IN TOP 10
SD MCPHERSON	3.97	UpK2-2	25-Mar-53	10	10
SD MCPHERSON	15.08	PB2-3	20-Jun-57	5	10
SD MCPHERSON	5.89	PB5-4	18-Jul-57	5	10
SD MCPHERSON	19.73	PB8-3	20-Aug-57	6	10
SD MCPHERSON	7.56	PB8-5	22-Aug-57	3	10
SD MCPHERSON	5.22	PB9-4	26-Aug-57	8	10
SD MCPHERSON	27.46	PB11-2	1-Sep-57	2	10
SD MCPHERSON	20.04	PB15-2	17-Sep-57	2	10
SD MCPHERSON	1.48	PB15-4	19-Sep-57	4	10
SD MCPHERSON	0.10	UE2-7	11-Mar-62	4	10
SD MEADE	45.45	UpK2-2	25-Mar-53	4	15
SD MEADE	1.58	UpK8-1	8-May-53	8	15
SD MEADE	5.09	PB5-3	17-Jul-57	8	15
SD MEADE	0.83	PB8-2	19-Aug-57	8	15
SD MEADE	19.73	PB8-3	20-Aug-57	6	15
SD MEADE	8.05	PB8-4	21-Aug-57	4	15
SD MEADE	8.05	PB8-4	21-Aug-57	4	15
SD MEADE	1.92	PB8-5	22-Aug-57	10	15
SD MEADE	6.92	PB11-2	1-Sep-57	6	15
SD MEADE	20.04	PB15-2	17-Sep-57	2	15
SD MEADE	2.37	PB18-3	9-Oct-57	3	15
SD MEADE	1.44	PB18-4	10-Oct-57	6	15
SD MEADE	0.23	UE2-6	10-Mar-62	1	15
SD MEADE	0.10	UE2-7	11-Mar-62	4	15
SD MEADE	9.84	UE12-4	17-Apr-65	4	15
SD MELLETTE	12.12	TS5-3	9-May-52	3	16
SD MELLETTE	45.45	UpK2-2	25-Mar-53	4	16
SD MELLETTE	45.16	PB2-3	20-Jun-57	2	16
SD MELLETTE	14.75	PB5-4	18-Jul-57	1	16
SD MELLETTE	0.83	PB8-2	19-Aug-57	8	16
SD MELLETTE	19.73	PB8-3	20-Aug-57	6	16
SD MELLETTE	11.34	PB8-5	22-Aug-57	2	16
SD MELLETTE	9.15	PB11-2	1-Sep-57	5	16
SD MELLETTE	6.68	PB15-2	17-Sep-57	5	16
SD MELLETTE	1.04	PB17-9	6-Oct-57	7	16

COUNTY	uCi/ sqm	SHOT DAY	DATE	RANK FOR DAY	TIMES IN TOP 10
SD MELLETTE	2.37	PB18-3	9-Oct-57	3	16
SD MELLETTE	1.44	PB18-4	10-Oct-57	6	16
SD MELLETTE	0.12	UE2-6	10-Mar-62	5	16
SD MELLETTE	0.10	UE2-7	11-Mar-62	4	16
SD MELLETTE	0.11	UE7-5	15-Jul-62	10	16
SD MELLETTE	9.84	UE12-4	17-Apr-65	4	16
SD MINER	3.97	UpK2-2	25-Mar-53	10	9
SD MINER	30.08	PB2-3	20-Jun-57	3	9
SD MINER	19.73	PB8-3	20-Aug-57	6	9
SD MINER	18.89	PB8-5	22-Aug-57	1	9
SD MINER	9.15	PB11-2	1-Sep-57	5	9
SD MINER	20.04	PB15-2	17-Sep-57	2	9
SD MINER	4.13	PB15-3	18-Sep-57	2	9
SD MINER	0.10	UE2-7	11-Mar-62	4	9
SD MINER	177.44	S-2	7-Jul-62	5	9
SD MINNEHAHA	3.97	UpK2-2	25-Mar-53	10	13
SD MINNEHAHA	15.08	PB2-3	20-Jun-57	5	13
SD MINNEHAHA	19.73	PB8-3	20-Aug-57	6	13
SD MINNEHAHA	13.41	PB8-4	21-Aug-57	2	13
SD MINNEHAHA	13.41	PB8-4	21-Aug-57	2	13
SD MINNEHAHA	7.56	PB8-5	22-Aug-57	3	13
SD MINNEHAHA	5.22	PB9-4	26-Aug-57	8	13
SD MINNEHAHA	4.58	PB11-2	1-Sep-57	9	13
SD MINNEHAHA	20.04	PB15-2	17-Sep-57	2	13
SD MINNEHAHA	2.50	PB15-3	18-Sep-57	6	13
SD MINNEHAHA	0.17	UE2-7	11-Mar-62	2	13
SD MINNEHAHA	138.24	S-2	7-Jul-62	7	13
SD MINNEHAHA	9.84	UE12-4	17-Apr-65	4	13
SD MOODY	3.97	UpK2-2	25-Mar-53	10	7
SD MOODY	2.01	PB5-5	19-Jul-57	8	7
SD MOODY	1.97	PB8-5	22-Aug-57	9	7
SD MOODY	4.11	PB15-2	17-Sep-57	9	7
SD MOODY	1.55	PB15-3	18-Sep-57	10	7
SD MOODY	0.17	UE2-7	11-Mar-62	2	7
SD MOODY	361.42	S-2	7-Jul-62	3	7

COUNTY	uCi/ sqm	SHOT DAY	DATE	RANK FOR DAY	TIMES IN TOP 10
SD PENNINGTON	18.88	UpK2-2	25-Mar-53	6	13
SD PENNINGTON	1.31	UpK8-1	8-May-53	9	13
SD PENNINGTON	1.30	PB8-2	19-Aug-57	7	13
SD PENNINGTON	29.60	PB8-3	20-Aug-57	5	13
SD PENNINGTON	8.05	PB8-4	21-Aug-57	4	13
SD PENNINGTON	8.05	PB8-4	21-Aug-57	4	13
SD PENNINGTON	5.22	PB9-4	26-Aug-57	8	13
SD PENNINGTON	5.11	PB11-2	1-Sep-57	7	13
SD PENNINGTON	2.37	PB18-3	9-Oct-57	3	13
SD PENNINGTON	1.44	PB18-4	10-Oct-57	6	13
SD PENNINGTON	0.22	UE2-6	10-Mar-62	2	13
SD PENNINGTON	0.10	UE2-7	11-Mar-62	3	13
SD PENNINGTON	9.84	UE12-4	17-Apr-65	4	13
SD PERKINS	45.45	UpK2-2	25-Mar-53	4	15
SD PERKINS	3.16	UpK8-1	8-May-53	7	15
SD PERKINS	2.43	UpK10-4	28-May-53	9	15
SD PERKINS	10.10	PB5-3	17-Jul-57	4	15
SD PERKINS	6.66	PB6-5	28-Jul-57	3	15
SD PERKINS	0.83	PB8-2	19-Aug-57	8	15
SD PERKINS	19.73	PB8-3	20-Aug-57	6	15
SD PERKINS	8.05	PB8-4	21-Aug-57	4	15
SD PERKINS	8.05	PB8-4	21-Aug-57	4	15
SD PERKINS	1.92	PB8-5	22-Aug-57	10	15
SD PERKINS	9.15	PB11-2	1-Sep-57	5	15
SD PERKINS	20.04	PB15-2	17-Sep-57	2	15
SD PERKINS	2.16	PB18-4	10-Oct-57	4	15
SD PERKINS	0.44	UE2-5	9-Mar-62	1	15
SD PERKINS	0.22	UE2-6	10-Mar-62	2	15
SD POTTER	3.97	UpK2-2	25-Mar-53	10	14
SD POTTER	45.16	PB2-3	20-Jun-57	2	14
SD POTTER	5.89	PB5-4	18-Jul-57	5	14
SD POTTER	4.08	PB6-5	28-Jul-57	10	14
SD POTTER	1.65	PB8-2	19-Aug-57	6	14
SD POTTER	19.73	PB8-3	20-Aug-57	6	14
SD POTTER	3.78	PB8-5	22-Aug-57	4	14

COUNTY	uCi/ sqm	SHOT DAY	DATE	RANK FOR DAY	TIMES IN TOP 10
SD POTTER	5.22	PB9-4	26-Aug-57	8	14
SD POTTER	27.46	PB11-2	1-Sep-57	2	14
SD POTTER	20.04	PB15-2	17-Sep-57	2	14
SD POTTER	1.44	PB18-4	10-Oct-57	6	14
SD POTTER	0.23	UE2-6	10-Mar-62	1	14
SD POTTER	0.10	UE2-7	11-Mar-62	4	14
SD POTTER	9.32	UE12-4	17-Apr-65	6	14
SD ROBERTS	75.24	PB2-3	20-Jun-57	1	12
SD ROBERTS	1.65	PB4-10	14-Jul-57	10	12
SD ROBERTS	5.89	PB5-4	18-Jul-57	5	12
SD ROBERTS	0.81	PB6-10	2-Aug-57	6	12
SD ROBERTS	19.73	PB8-3	20-Aug-57	6	12
SD ROBERTS	2.85	PB8-5	22-Aug-57	7	12
SD ROBERTS	3.48	PB9-4	26-Aug-57	10	12
SD ROBERTS	27.46	PB11-2	1-Sep-57	2	12
SD ROBERTS	20.04	PB15-2	17-Sep-57	2	12
SD ROBERTS	2.50	PB15-3	18-Sep-57	6	12
SD ROBERTS	1.18	UE2-4	8-Mar-62	1	12
SD ROBERTS	0.10	UE2-7	11-Mar-62	4	12
SD SANBORN	2.34	TS7-2	2-Jun-52	4	10
SD SANBORN	3.97	UpK2-2	25-Mar-53	10	10
SD SANBORN	45.16	PB2-3	20-Jun-57	2	10
SD SANBORN	19.73	PB8-3	20-Aug-57	6	10
SD SANBORN	18.89	PB8-5	22-Aug-57	1	10
SD SANBORN	3.48	PB9-4	26-Aug-57	10	10
SD SANBORN	4.58	PB11-2	1-Sep-57	9	10
SD SANBORN	13.36	PB15-2	17-Sep-57	3	10
SD SANBORN	2.50	PB15-3	18-Sep-57	6	10
SD SANBORN	0.17	UE2-7	11-Mar-62	2	10
SD SHANNON	26.09	TS5-2	8-May-52	7	10
SD SHANNON	45.45	UpK2-2	25-Mar-53	4	10
SD SHANNON	0.85	PB5-6	20-Jul-57	10	10
SD SHANNON	0.83	PB8-2	19-Aug-57	8	10
SD SHANNON	19.73	PB8-3	20-Aug-57	6	10
SD SHANNON	2.37	PB18-3	9-Oct-57	3	10

COUNTY	uCi/ sqm	SHOT DAY	DATE	RANK FOR DAY	TIMES IN TOP 10
SD SHANNON	1.44	PB18-4	10-Oct-57	6	10
SD SHANNON	0.23	UE2-6	10-Mar-62	1	10
SD SHANNON	0.17	UE2-7	11-Mar-62	2	10
SD SHANNON	345.60	S-2	7-Jul-62	4	10
SD SPINK	1.91	TS7-2	2-Jun-52	6	12
SD SPINK	3.97	UpK2-2	25-Mar-53	10	12
SD SPINK	45.16	PB2-3	20-Jun-57	2	12
SD SPINK	19.73	PB8-3	20-Aug-57	6	12
SD SPINK	3.78	PB8-5	22-Aug-57	4	12
SD SPINK	5.22	PB9-4	26-Aug-57	8	12
SD SPINK	27.46	PB11-2	1-Sep-57	2	12
SD SPINK	20.04	PB15-2	17-Sep-57	2	12
SD SPINK	1.64	PB15-3	18-Sep-57	9	12
SD SPINK	0.83	UE2-4	8-Mar-62	4	12
SD SPINK	0.10	UE2-7	11-Mar-62	4	12
SD SPINK	9.32	UE12-4	17-Apr-65	6	12
SD STANLEY	45.45	UpK2-2	25-Mar-53	4	15
SD STANLEY	30.08	PB2-3	20-Jun-57	3	15
SD STANLEY	8.86	PB5-4	18-Jul-57	3	15
SD STANLEY	1.65	PB8-2	19-Aug-57	6	15
SD STANLEY	19.73	PB8-3	20-Aug-57	6	15
SD STANLEY	11.34	PB8-5	22-Aug-57	2	15
SD STANLEY	3.48	PB9-4	26-Aug-57	10	15
SD STANLEY	18.31	PB11-2	1-Sep-57	3	15
SD STANLEY	20.04	PB15-2	17-Sep-57	2	15
SD STANLEY	1.78	PB17-9	6-Oct-57	2	15
SD STANLEY	2.37	PB18-3	9-Oct-57	3	15
SD STANLEY	1.44	PB18-4	10-Oct-57	6	15
SD STANLEY	0.10	UE2-7	11-Mar-62	4	15
SD STANLEY	345.60	S-2	7-Jul-62	4	15
SD STANLEY	9.32	UE12-4	17-Apr-65	6	15
SD SULLY	3.97	UpK2-2	25-Mar-53	10	11
SD SULLY	45.16	PB2-3	20-Jun-57	2	11
SD SULLY	0.83	PB8-2	19-Aug-57	8	11
SD SULLY	19.73	PB8-3	20-Aug-57	6	11

COUNTY	uCi/ sqm	SHOT DAY	DATE	RANK FOR DAY	TIMES IN TOP 10
SD SULLY	7.56	PB8-5	22-Aug-57	3	11
SD SULLY	5.22	PB9-4	26-Aug-57	8	11
SD SULLY	9.15	PB11-2	1-Sep-57	5	11
SD SULLY	20.04	PB15-2	17-Sep-57	2	11
SD SULLY	1.15	PB17-9	6-Oct-57	5	11
SD SULLY	1.44	PB18-4	10-Oct-57	6	11
SD SULLY	9.32	UE12-4	17-Apr-65	6	11
SD TODD	45.45	UpK2-2	25-Mar-53	4	13
SD TODD	45.16	PB2-3	20-Jun-57	2	13
SD TODD	0.83	PB8-2	19-Aug-57	8	13
SD TODD	19.73	PB8-3	20-Aug-57	6	13
SD TODD	5.36	PB8-4	21-Aug-57	6	13
SD TODD	5.36	PB8-4	21-Aug-57	6	13
SD TODD	11.34	PB8-5	22-Aug-57	2	13
SD TODD	4.58	PB11-2	1-Sep-57	9	13
SD TODD	2.50	PB15-3	18-Sep-57	6	13
SD TODD	2.37	PB18-3	9-Oct-57	3	13
SD TODD	1.44	PB18-4	10-Oct-57	6	13
SD TODD	0.12	UE2-6	10-Mar-62	5	13
SD TODD	0.10	UE2-7	11-Mar-62	3	13
SD TRIPP	1.44	BJ5-3	21-Nov-51	6	14
SD TRIPP	3.97	UpK2-2	25-Mar-53	10	14
SD TRIPP	30.08	PB2-3	20-Jun-57	3	14
SD TRIPP	0.83	PB8-2	19-Aug-57	8	14
SD TRIPP	19.73	PB8-3	20-Aug-57	6	14
SD TRIPP	11.34	PB8-5	22-Aug-57	2	14
SD TRIPP	4.58	PB11-2	1-Sep-57	9	14
SD TRIPP	13.36	PB15-2	17-Sep-57	3	14
SD TRIPP	2.50	PB15-3	18-Sep-57	6	14
SD TRIPP	1.03	PB17-9	6-Oct-57	8	14
SD TRIPP	2.37	PB18-3	9-Oct-57	3	14
SD TRIPP	1.44	PB18-4	10-Oct-57	6	14
SD TRIPP	0.17	UE2-7	11-Mar-62	2	14
SD TRIPP	0.11	UE7-5	15-Jul-62	10	14
SD TURNER	1.44	BJ5-3	21-Nov-51	6	12

COUNTY	uCi/ sqm	SHOT DAY	DATE	RANK FOR DAY	TIMES IN TOP 10
SD TURNER	59.83	TS7-4	4-Jun-52	4	12
SD TURNER	3.97	UpK2-2	25-Mar-53	10	12
SD TURNER	19.73	PB8-3	20-Aug-57	6	12
SD TURNER	3.78	PB8-5	22-Aug-57	4	12
SD TURNER	3.48	PB9-4	26-Aug-57	10	12
SD TURNER	4.58	PB11-2	1-Sep-57	9	12
SD TURNER	6.68	PB15-2	17-Sep-57	5	12
SD TURNER	2.50	PB15-3	18-Sep-57	6	12
SD TURNER	0.17	UE2-7	11-Mar-62	2	12
SD TURNER	527.44	S-3	8-Jul-62	5	12
SD TURNER	11.43	UE12-4	17-Apr-65	3	12
SD UNION	1.44	BJ5-3	21-Nov-51	6	8
SD UNION	1.37	TS4-6	6-May-52	9	8
SD UNION	1.37	TS4-6	6-May-52	9	8
SD UNION	3.97	UpK2-2	25-Mar-53	10	8
SD UNION	1.15	TP7-6	3-Apr-55	7	8
SD UNION	19.73	PB8-3	20-Aug-57	6	8
SD UNION	5.22	PB9-4	26-Aug-57	8	8
SD UNION	0.10	UE2-7	11-Mar-62	4	8
SD WALWORTH	3.97	UpK2-2	25-Mar-53	10	13
SD WALWORTH	45.16	PB2-3	20-Jun-57	2	13
SD WALWORTH	8.86	PB5-4	18-Jul-57	3	13
SD WALWORTH	0.83	PB8-2	19-Aug-57	8	13
SD WALWORTH	19.73	PB8-3	20-Aug-57	6	13
SD WALWORTH	3.78	PB8-5	22-Aug-57	4	13
SD WALWORTH	27.46	PB11-2	1-Sep-57	2	13
SD WALWORTH	20.04	PB15-2	17-Sep-57	2	13
SD WALWORTH	1.64	PB15-3	18-Sep-57	9	13
SD WALWORTH	1.48	PB15-4	19-Sep-57	4	13
SD WALWORTH	1.44	PB18-4	10-Oct-57	6	13
SD WALWORTH	0.21	UE2-6	10-Mar-62	3	13
SD WALWORTH	0.10	UE2-7	11-Mar-62	3	13
SD WASHABAUGH	8.69	TS5-3	9-May-52	9	16
SD WASHABAUGH	45.45	UpK2-2	25-Mar-53	4	16
SD WASHABAUGH	15.08	PB2-3	20-Jun-57	5	16

COUNTY	uCi/ sqm	SHOT DAY	DATE	RANK FOR DAY	TIMES IN TOP 10
SD WASHABAUGH	2.76	PB4-10	14-Jul-57	6	16
SD WASHABAUGH	8.86	PB5-4	18-Jul-57	3	16
SD WASHABAUGH	0.83	PB8-2	19-Aug-57	8	16
SD WASHABAUGH	19.73	PB8-3	20-Aug-57	6	16
SD WASHABAUGH	8.05	PB8-4	21-Aug-57	4	16
SD WASHABAUGH	8.05	PB8-4	21-Aug-57	4	16
SD WASHABAUGH	1.92	PB8-5	22-Aug-57	10	16
SD WASHABAUGH	4.58	PB11-2	1-Sep-57	9	16
SD WASHABAUGH	2.37	PB18-3	9-Oct-57	3	16
SD WASHABAUGH	1.44	PB18-4	10-Oct-57	6	16
SD WASHABAUGH	0.12	UE2-6	10-Mar-62	5	16
SD WASHABAUGH	0.17	UE2-7	11-Mar-62	2	16
SD WASHABAUGH	860.52	S-2	7-Jul-62	1	16
SD YANKTON	3.97	UpK2-2	25-Mar-53	10	9
SD YANKTON	1.02	PB4-3	7-Jul-57	8	9
SD YANKTON	1.65	PB4-10	14-Jul-57	10	9
SD YANKTON	19.73	PB8-3	20-Aug-57	6	9
SD YANKTON	7.56	PB8-5	22-Aug-57	3	9
SD YANKTON	3.48	PB9-4	26-Aug-57	10	9
SD YANKTON	9.15	PB11-2	1-Sep-57	5	9
SD YANKTON	2.50	PB15-3	18-Sep-57	6	9
SD YANKTON	0.10	UE2-7	11-Mar-62	3	9
SD ZIEBACH	45.45	UpK2-2	25-Mar-53	4	16
SD ZIEBACH	1.58	UpK8-1	8-May-53	8	16
SD ZIEBACH	45.16	PB2-3	20-Jun-57	2	16
SD ZIEBACH	5.89	PB5-4	18-Jul-57	5	16
SD ZIEBACH	0.83	PB8-2	19-Aug-57	8	16
SD ZIEBACH	19.73	PB8-3	20-Aug-57	6	16
SD ZIEBACH	5.36	PB8-4	21-Aug-57	6	16
SD ZIEBACH	5.36	PB8-4	21-Aug-57	6	16
SD ZIEBACH	3.78	PB8-5	22-Aug-57	4	16
SD ZIEBACH	9.15	PB11-2	1-Sep-57	5	16
SD ZIEBACH	20.04	PB15-2	17-Sep-57	2	16
SD ZIEBACH	2.37	PB18-3	9-Oct-57	3	16
SD ZIEBACH	1.44	PB18-4	10-Oct-57	6	16

COUNTY	uCi/ sqm	SHOT DAY	DATE	RANK FOR DAY	TIMES IN TOP 10
SD ZIEBACH	0.21	UE2-6	10-Mar-62	3	16
SD ZIEBACH	0.10	UE2-7	11-Mar-62	4	16
SD ZIEBACH	9.32	UE12-4	17-Apr-65	6	16

COUNTY	uCi/ sqm	SHOT DAY	DATE	RANK FOR DAY	TIMES IN TOP 10
TN ANDERSON	4.59	BJ3-5	5-Nov-51	6	8
TN ANDERSON	8.10	BJ3-6	6-Nov-51	5	8
TN ANDERSON	1.04	UpK5-2	12-Apr-53	4	8
TN ANDERSON	3.11	TP6-4	25-Mar-55	3	8
TN ANDERSON	0.97	TP6-5	26-Mar-55	8	8
TN ANDERSON	4.83	TP10-8	12-May-55	2	8
TN ANDERSON	2.78	PB2-5	22-Jun-57	4	8
TN ANDERSON	7.49	PB17-3	30-Sep-57	5	8
TN BEDFORD	1.83	TP6-3	24-Mar-55	7	4
TN BEDFORD	2.03	TP6-4	25-Mar-55	8	4
TN BEDFORD	0.97	TP6-5	26-Mar-55	8	4
TN BEDFORD	14.98	PB17-3	30-Sep-57	2	4
TN BENTON	9.94	UpK1-3	19-Mar-53	4	5
TN BENTON	2.39	TP6-4	25-Mar-55	6	5
TN BENTON	1.49	PB1-9	5-Jun-57	2	5
TN BENTON	2.78	PB2-5	22-Jun-57	4	5
TN BENTON	1.27	PB3-9	2-Jul-57	2	5
TN BLEDSOE	8.10	BJ3-6	6-Nov-51	5	6
TN BLEDSOE	1.83	TP6-3	24-Mar-55	7	6
TN BLEDSOE	3.11	TP6-4	25-Mar-55	3	6
TN BLEDSOE	0.97	TP6-5	26-Mar-55	8	6
TN BLEDSOE	1.39	PB2-5	22-Jun-57	8	6
TN BLEDSOE	14.98	PB17-3	30-Sep-57	2	6
TN BLOUNT	4.86	BJ3-6	6-Nov-51	7	7
TN BLOUNT	1.55	UpK5-2	12-Apr-53	2	7
TN BLOUNT	1.83	TP6-3	24-Mar-55	7	7
TN BLOUNT	3.11	TP6-4	25-Mar-55	3	7
TN BLOUNT	1.45	TP6-5	26-Mar-55	5	7
TN BLOUNT	1.39	PB2-5	22-Jun-57	8	7
TN BLOUNT	14.98	PB17-3	30-Sep-57	2	7
TN BRADLEY	8.10	BJ3-6	6-Nov-51	5	6
TN BRADLEY	1.55	UpK5-2	12-Apr-53	2	6
TN BRADLEY	5.14	TP6-4	25-Mar-55	1	6
TN BRADLEY	0.97	TP6-5	26-Mar-55	8	6
TN BRADLEY	2.78	PB2-5	22-Jun-57	4	6

COUNTY	uCi/ sqm	SHOT DAY	DATE	RANK FOR DAY	TIMES IN TOP 10
TN BRADLEY	14.98	PB17-3	30-Sep-57	2	6
TN CAMPBELL	1.04	UpK5-2	12-Apr-53	4	7
TN CAMPBELL	1.83	TP6-3	24-Mar-55	7	7
TN CAMPBELL	3.11	TP6-4	25-Mar-55	3	7
TN CAMPBELL	1.93	TP6-5	26-Mar-55	4	7
TN CAMPBELL	4.24	TP10-8	12-May-55	6	7
TN CAMPBELL	1.39	PB2-5	22-Jun-57	8	7
TN CAMPBELL	3.79	PB17-3	30-Sep-57	10	7
TN CANNON	2.03	TP6-4	25-Mar-55	8	5
TN CANNON	0.97	TP6-5	26-Mar-55	8	5
TN CANNON	4.18	PB2-5	22-Jun-57	3	5
TN CANNON	0.99	PB6-10	2-Aug-57	3	5
TN CANNON	3.79	PB17-3	30-Sep-57	10	5
TN CARROLL	1.86	PB2-5	22-Jun-57	5	3
TN CARROLL	1.27	PB3-9	2-Jul-57	2	3
TN CARROLL	2.11	PB9-2	24-Aug-57	8	3
TN CARTER	3.89	BJ2-2	31-Oct-51	10	8
TN CARTER	1.04	BJ4-5	9-Nov-51	2	8
TN CARTER	1.55	UpK5-2	12-Apr-53	2	8
TN CARTER	3.11	TP6-4	25-Mar-55	3	8
TN CARTER	1.93	TP6-5	26-Mar-55	4	8
TN CARTER	2.78	PB2-5	22-Jun-57	4	8
TN CARTER	1.51	PB5-8	22-Jul-57	7	8
TN CARTER	0.93	PB18-5	11-Oct-57	3	8
TN CHEATHAM	1.51	TS6-5	29-May-52	2	5
TN CHEATHAM	9.23	UpK1-3	19-Mar-53	10	5
TN CHEATHAM	1.57	UpK1-6	22-Mar-53	2	5
TN CHEATHAM	11.16	PB4-5	9-Jul-57	9	5
TN CHEATHAM	2.73	PB9-2	24-Aug-57	6	5
TN CHESTER	3.06	UpK2-7	30-Mar-53	9	6
TN CHESTER	21.07	UpK2-8	31-Mar-53	3	6
TN CHESTER	3.67	UpK3-6	5-Apr-53	1	6
TN CHESTER	1.83	TP6-3	24-Mar-55	7	6
TN CHESTER	2.78	PB2-5	22-Jun-57	4	6
TN CHESTER	1.89	PB6-7	30-Jul-57	7	6

COUNTY	uCi/ sqm	SHOT DAY	DATE	RANK FOR DAY	TIMES IN TOP 10
TN CLAIBORNE	1.55	UpK5-2	12-Apr-53	2	7
TN CLAIBORNE	1.83	TP6-3	24-Mar-55	7	7
TN CLAIBORNE	3.11	TP6-4	25-Mar-55	3	7
TN CLAIBORNE	0.97	TP6-5	26-Mar-55	8	7
TN CLAIBORNE	2.78	PB2-5	22-Jun-57	4	7
TN CLAIBORNE	1.51	PB5-8	22-Jul-57	7	7
TN CLAIBORNE	3.79	PB17-3	30-Sep-57	10	7
TN CLAY	1.83	TP6-3	24-Mar-55	7	5
TN CLAY	2.03	TP6-4	25-Mar-55	8	5
TN CLAY	0.97	TP6-5	26-Mar-55	8	5
TN CLAY	4.18	PB2-5	22-Jun-57	3	5
TN CLAY	7.49	PB17-3	30-Sep-57	5	5
TN COCKE	4.86	BJ3-6	6-Nov-51	7	7
TN COCKE	2.59	UpK5-2	12-Apr-53	1	7
TN COCKE	3.11	TP6-4	25-Mar-55	3	7
TN COCKE	1.93	TP6-5	26-Mar-55	4	7
TN COCKE	4.18	PB2-5	22-Jun-57	3	7
TN COCKE	1.51	PB5-8	22-Jul-57	7	7
TN COCKE	7.49	PB17-3	30-Sep-57	5	7
TN COFFEE	4.86	BJ3-6	6-Nov-51	7	6
TN COFFEE	1.83	TP6-3	24-Mar-55	7	6
TN COFFEE	2.03	TP6-4	25-Mar-55	8	6
TN COFFEE	0.97	TP6-5	26-Mar-55	8	6
TN COFFEE	4.18	PB2-5	22-Jun-57	3	6
TN COFFEE	22.47	PB17-3	30-Sep-57	1	6
TN CROCKETT	6.07	UpK2-7	30-Mar-53	7	3
TN CROCKETT	35.10	UpK2-8	31-Mar-53	1	3
TN CROCKETT	3.68	UpK8-3	10-May-53	4	3
TN CUMBERLAND	8.10	BJ3-6	6-Nov-51	5	6
TN CUMBERLAND	1.04	UpK5-2	12-Apr-53	4	6
TN CUMBERLAND	3.11	TP6-4	25-Mar-55	3	6
TN CUMBERLAND	0.97	TP6-5	26-Mar-55	8	6
TN CUMBERLAND	2.78	PB2-5	22-Jun-57	4	6
TN CUMBERLAND	14.98	PB17-3	30-Sep-57	2	6
TN DAVIDSON	1.64	TS6-5	29-May-52	1	5

COUNTY	uCi/ sqm	SHOT DAY	DATE	RANK FOR DAY	TIMES IN TOP 10
TN DAVIDSON	1.64	UpK1-6	22-Mar-53	1	5
TN DAVIDSON	11.16	PB4-5	9-Jul-57	9	5
TN DAVIDSON	2.73	PB9-2	24-Aug-57	6	5
TN DAVIDSON	2.54	PB17-2	29-Sep-57	2	5
TN DE KALB	1.83	TP6-3	24-Mar-55	7	5
TN DE KALB	2.03	TP6-4	25-Mar-55	8	5
TN DE KALB	0.97	TP6-5	26-Mar-55	8	5
TN DE KALB	1.39	PB2-5	22-Jun-57	8	5
TN DE KALB	7.49	PB17-3	30-Sep-57	5	5
TN DECATUR	0.89	PB1-9	5-Jun-57	7	3
TN DECATUR	1.86	PB2-5	22-Jun-57	5	3
TN DECATUR	0.83	PB3-9	2-Jul-57	4	3
TN DICKSON	10.72	UpK1-3	19-Mar-53	2	4
TN DICKSON	1.01	UpK1-6	22-Mar-53	6	4
TN DICKSON	1.17	TP10-7	11-May-55	9	4
TN DICKSON	1.17	PB2-4	21-Jun-57	10	4
TN DYER	30.45	UpK2-7	30-Mar-53	1	6
TN DYER	2.17	UpK3-6	5-Apr-53	2	6
TN DYER	6.18	UpK8-3	10-May-53	2	6
TN DYER	1.21	PB1-5	1-Jun-57	10	6
TN DYER	1.86	PB2-5	22-Jun-57	5	6
TN DYER	2.10	PB3-9	2-Jul-57	1	6
TN FAYETTE	18.26	UpK2-7	30-Mar-53	2	4
TN FAYETTE	2.17	UpK3-6	5-Apr-53	2	4
TN FAYETTE	6.18	UpK8-3	10-May-53	2	4
TN FAYETTE	1.25	PB3-8	1-Jul-57	3	4
TN FENTRESS	4.86	BJ3-6	6-Nov-51	7	7
TN FENTRESS	1.04	UpK5-2	12-Apr-53	4	7
TN FENTRESS	1.83	TP6-3	24-Mar-55	7	7
TN FENTRESS	3.11	TP6-4	25-Mar-55	3	7
TN FENTRESS	0.97	TP6-5	26-Mar-55	8	7
TN FENTRESS	4.18	PB2-5	22-Jun-57	3	7
TN FENTRESS	7.49	PB17-3	30-Sep-57	5	7
TN FRANKLIN	4.86	BJ3-6	6-Nov-51	7	3
TN FRANKLIN	2.87	TP6-4	25-Mar-55	4	3

COUNTY	uCi/ sqm	SHOT DAY	DATE	RANK FOR DAY	TIMES IN TOP 10
TN FRANKLIN	5.99	PB17-3	30-Sep-57	7	3
TN GIBSON	30.45	UpK2-7	30-Mar-53	1	8
TN GIBSON	2.17	UpK3-6	5-Apr-53	2	8
TN GIBSON	3.68	UpK8-3	10-May-53	4	8
TN GIBSON	1.83	TP6-3	24-Mar-55	7	8
TN GIBSON	1.58	TP10-7	11-May-55	2	8
TN GIBSON	1.21	PB1-5	1-Jun-57	10	8
TN GIBSON	1.86	PB2-5	22-Jun-57	5	8
TN GIBSON	2.10	PB3-9	2-Jul-57	1	8
TN GILES	2.39	TP6-4	25-Mar-55	6	2
TN GILES	1.82	PB2-4	21-Jun-57	5	2
TN GRAINGER	1.04	UpK5-2	12-Apr-53	4	7
TN GRAINGER	3.11	TP6-4	25-Mar-55	3	7
TN GRAINGER	0.97	TP6-5	26-Mar-55	8	7
TN GRAINGER	4.53	TP10-8	12-May-55	4	7
TN GRAINGER	2.78	PB2-5	22-Jun-57	4	7
TN GRAINGER	1.51	PB5-8	22-Jul-57	7	7
TN GRAINGER	3.79	PB17-3	30-Sep-57	10	7
TN GREENE	1.55	UpK5-2	12-Apr-53	2	6
TN GREENE	3.11	TP6-4	25-Mar-55	3	6
TN GREENE	1.93	TP6-5	26-Mar-55	4	6
TN GREENE	1.39	PB2-5	22-Jun-57	8	6
TN GREENE	2.99	PB5-8	22-Jul-57	4	6
TN GREENE	7.49	PB17-3	30-Sep-57	5	6
TN GRUNDY	4.59	BJ3-5	5-Nov-51	6	7
TN GRUNDY	4.86	BJ3-6	6-Nov-51	7	7
TN GRUNDY	1.83	TP6-3	24-Mar-55	7	7
TN GRUNDY	2.03	TP6-4	25-Mar-55	8	7
TN GRUNDY	0.97	TP6-5	26-Mar-55	8	7
TN GRUNDY	4.18	PB2-5	22-Jun-57	3	7
TN GRUNDY	14.98	PB17-3	30-Sep-57	2	7
TN HAMBLEN	1.55	UpK5-2	12-Apr-53	2	6
TN HAMBLEN	3.11	TP6-4	25-Mar-55	3	6
TN HAMBLEN	1.93	TP6-5	26-Mar-55	4	6
TN HAMBLEN	1.39	PB2-5	22-Jun-57	8	6

COUNTY	uCi/ sqm	SHOT DAY	DATE	RANK FOR DAY	TIMES IN TOP 10
TN HAMBLEN	1.51	PB5-8	22-Jul-57	7	6
TN HAMBLEN	7.49	PB17-3	30-Sep-57	5	6
TN HAMILTON	1.04	UpK5-2	12-Apr-53	4	5
TN HAMILTON	3.11	TP6-4	25-Mar-55	3	5
TN HAMILTON	0.97	TP6-5	26-Mar-55	8	5
TN HAMILTON	4.18	PB2-5	22-Jun-57	3	5
TN HAMILTON	14.98	PB17-3	30-Sep-57	2	5
TN HANCOCK	1.55	UpK5-2	12-Apr-53	2	5
TN HANCOCK	3.11	TP6-4	25-Mar-55	3	5
TN HANCOCK	0.97	TP6-5	26-Mar-55	8	5
TN HANCOCK	1.51	PB5-8	22-Jul-57	7	5
TN HANCOCK	7.49	PB17-3	30-Sep-57	5	5
TN HARDEMAN	3.06	UpK2-7	30-Mar-53	9	5
TN HARDEMAN	3.67	UpK3-6	5-Apr-53	1	5
TN HARDEMAN	6.18	UpK8-3	10-May-53	2	5
TN HARDEMAN	1.86	PB2-5	22-Jun-57	5	5
TN HARDEMAN	2.10	PB3-9	2-Jul-57	1	5
TN HARDIN	0.88	UpK3-6	5-Apr-53	8	3
TN HARDIN	1.86	PB2-5	22-Jun-57	5	3
TN HARDIN	0.83	PB3-9	2-Jul-57	4	3
TN HAWKINS	1.55	UpK5-2	12-Apr-53	2	6
TN HAWKINS	3.11	TP6-4	25-Mar-55	3	6
TN HAWKINS	0.97	TP6-5	26-Mar-55	8	6
TN HAWKINS	1.39	PB2-5	22-Jun-57	8	6
TN HAWKINS	1.51	PB5-8	22-Jul-57	7	6
TN HAWKINS	3.79	PB17-3	30-Sep-57	10	6
TN HAYWOOD	6.07	UpK2-7	30-Mar-53	7	6
TN HAYWOOD	35.10	UpK2-8	31-Mar-53	1	6
TN HAYWOOD	2.17	UpK3-6	5-Apr-53	2	6
TN HAYWOOD	3.68	UpK8-3	10-May-53	4	6
TN HAYWOOD	0.89	UpK8-5	12-May-53	6	6
TN HAYWOOD	2.10	PB3-9	2-Jul-57	1	6
TN HENDERSON	0.89	PB1-9	5-Jun-57	7	3
TN HENDERSON	1.86	PB2-5	22-Jun-57	5	3
TN HENDERSON	1.27	PB3-9	2-Jul-57	2	3

COUNTY	uCi/ sqm	SHOT DAY	DATE	RANK FOR DAY	TIMES IN TOP 10
TN HENRY	1.09	TP4-5	11-Mar-55	1	4
TN HENRY	2.78	PB2-5	22-Jun-57	4	4
TN HENRY	1.27	PB3-9	2-Jul-57	2	4
TN HENRY	2.73	PB9-2	24-Aug-57	6	4
TN HICKMAN	0.95	TS6-5	29-May-52	10	6
TN HICKMAN	9.85	UpK1-3	19-Mar-53	5	6
TN HICKMAN	0.85	UpK1-6	22-Mar-53	9	6
TN HICKMAN	1.83	TP6-3	24-Mar-55	7	6
TN HICKMAN	2.78	PB2-5	22-Jun-57	4	6
TN HICKMAN	0.83	PB3-9	2-Jul-57	4	6
TN HOUSTON	10.39	UpK1-3	19-Mar-53	3	2
TN HOUSTON	1.86	PB2-5	22-Jun-57	5	2
TN HUMPHREYS	0.97	TS6-5	29-May-52	9	6
TN HUMPHREYS	11.05	UpK1-3	19-Mar-53	1	6
TN HUMPHREYS	1.83	TP6-3	24-Mar-55	7	6
TN HUMPHREYS	2.78	PB2-5	22-Jun-57	4	6
TN HUMPHREYS	0.83	PB3-9	2-Jul-57	4	6
TN HUMPHREYS	2.11	PB9-2	24-Aug-57	8	6
TN JACKSON	1.83	TP6-3	24-Mar-55	7	5
TN JACKSON	2.03	TP6-4	25-Mar-55	8	5
TN JACKSON	0.97	TP6-5	26-Mar-55	8	5
TN JACKSON	4.18	PB2-5	22-Jun-57	3	5
TN JACKSON	7.49	PB17-3	30-Sep-57	5	5
TN JEFFERSON	4.86	BJ3-6	6-Nov-51	7	7
TN JEFFERSON	1.55	UpK5-2	12-Apr-53	2	7
TN JEFFERSON	3.11	TP6-4	25-Mar-55	3	7
TN JEFFERSON	0.97	TP6-5	26-Mar-55	8	7
TN JEFFERSON	1.39	PB2-5	22-Jun-57	8	7
TN JEFFERSON	1.51	PB5-8	22-Jul-57	7	7
TN JEFFERSON	7.49	PB17-3	30-Sep-57	5	7
TN JOHNSON	1.55	UpK5-2	12-Apr-53	2	5
TN JOHNSON	3.11	TP6-4	25-Mar-55	3	5
TN JOHNSON	0.97	TP6-5	26-Mar-55	8	5
TN JOHNSON	1.51	PB5-8	22-Jul-57	7	5
TN JOHNSON	0.93	PB18-5	11-Oct-57	3	5

COUNTY	uCi/ sqm	SHOT DAY	DATE	RANK FOR DAY	TIMES IN TOP 10
TN KNOX	8.10	BJ3-6	6-Nov-51	5	3
TN KNOX	1.07	UpK5-2	12-Apr-53	3	3
TN KNOX	6.07	TP10-8	12-May-55	1	3
TN LAKE	12.19	UpK2-7	30-Mar-53	4	4
TN LAKE	3.68	UpK8-3	10-May-53	4	4
TN LAKE	2.78	PB2-5	22-Jun-57	4	4
TN LAKE	2.73	PB9-2	24-Aug-57	6	4
TN LAUDERDALE	12.19	UpK2-7	30-Mar-53	4	5
TN LAUDERDALE	21.07	UpK2-8	31-Mar-53	3	5
TN LAUDERDALE	6.18	UpK8-3	10-May-53	2	5
TN LAUDERDALE	1.86	PB2-5	22-Jun-57	5	5
TN LAUDERDALE	2.73	PB9-2	24-Aug-57	6	5
TN LAWRENCE	0.88	UpK3-6	5-Apr-53	8	2
TN LAWRENCE	1.86	PB2-5	22-Jun-57	5	2
TN LEWIS	1.27	TS6-5	29-May-52	3	5
TN LEWIS	9.61	UpK1-3	19-Mar-53	6	5
TN LEWIS	1.83	TP6-3	24-Mar-55	7	5
TN LEWIS	0.89	PB1-9	5-Jun-57	7	5
TN LEWIS	2.78	PB2-5	22-Jun-57	4	5
TN LINCOLN	2.87	TP6-4	25-Mar-55	4	2
TN LINCOLN	3.96	PB17-3	30-Sep-57	9	2
TN LOUDON	8.10	BJ3-6	6-Nov-51	5	7
TN LOUDON	1.55	UpK5-2	12-Apr-53	2	7
TN LOUDON	3.11	TP6-4	25-Mar-55	3	7
TN LOUDON	0.97	TP6-5	26-Mar-55	8	7
TN LOUDON	4.29	TP10-8	12-May-55	5	7
TN LOUDON	1.39	PB2-5	22-Jun-57	8	7
TN LOUDON	7.49	PB17-3	30-Sep-57	5	7
TN MADISON	18.26	UpK2-7	30-Mar-53	2	6
TN MADISON	21.07	UpK2-8	31-Mar-53	3	6
TN MADISON	2.17	UpK3-6	5-Apr-53	2	6
TN MADISON	2.50	UpK8-3	10-May-53	6	6
TN MADISON	2.10	PB3-9	2-Jul-57	1	6
TN MADISON	2.11	PB9-2	24-Aug-57	8	6
TN MARION	8.10	BJ3-6	6-Nov-51	5	4

COUNTY	uCi/ sqm	SHOT DAY	DATE	RANK FOR DAY	TIMES IN TOP 10
TN MARION	2.87	TP6-4	25-Mar-55	4	4
TN MARION	1.16	PB1-9	5-Jun-57	5	4
TN MARION	3.96	PB17-3	30-Sep-57	9	4
TN MARSHALL	3.11	TP6-4	25-Mar-55	3	3
TN MARSHALL	0.97	TP6-5	26-Mar-55	8	3
TN MARSHALL	14.98	PB17-3	30-Sep-57	2	3
TN MAURY	0.98	TS6-5	29-May-52	8	4
TN MAURY	9.25	UpK1-3	19-Mar-53	9	4
TN MAURY	1.07	UpK1-6	22-Mar-53	5	4
TN MAURY	2.54	PB17-2	29-Sep-57	2	4
TN MCMINN	8.10	BJ3-6	6-Nov-51	5	6
TN MCMINN	1.55	UpK5-2	12-Apr-53	2	6
TN MCMINN	5.14	TP6-4	25-Mar-55	1	6
TN MCMINN	0.97	TP6-5	26-Mar-55	8	6
TN MCMINN	2.78	PB2-5	22-Jun-57	4	6
TN MCMINN	14.98	PB17-3	30-Sep-57	2	6
TN MCNAIRY	12.19	UpK2-7	30-Mar-53	4	5
TN MCNAIRY	14.03	UpK2-8	31-Mar-53	8	5
TN MCNAIRY	3.67	UpK3-6	5-Apr-53	1	5
TN MCNAIRY	4.64	PB2-5	22-Jun-57	2	5
TN MCNAIRY	1.89	PB6-7	30-Jul-57	7	5
TN MEIGS	8.10	BJ3-6	6-Nov-51	5	6
TN MEIGS	1.55	UpK5-2	12-Apr-53	2	6
TN MEIGS	5.14	TP6-4	25-Mar-55	1	6
TN MEIGS	0.97	TP6-5	26-Mar-55	8	6
TN MEIGS	1.39	PB2-5	22-Jun-57	8	6
TN MEIGS	7.49	PB17-3	30-Sep-57	5	6
TN MONROE	4.86	BJ3-6	6-Nov-51	7	6
TN MONROE	1.55	UpK5-2	12-Apr-53	2	6
TN MONROE	5.14	TP6-4	25-Mar-55	1	6
TN MONROE	0.97	TP6-5	26-Mar-55	8	6
TN MONROE	2.78	PB2-5	22-Jun-57	4	6
TN MONROE	7.49	PB17-3	30-Sep-57	5	6
TN MONTGOMERY	1.00	UpK1-6	22-Mar-53	7	3
TN MONTGOMERY	1.86	PB2-5	22-Jun-57	5	3

COUNTY	uCi/ sqm	SHOT DAY	DATE	RANK FOR DAY	TIMES IN TOP 10
TN MONTGOMERY	16.75	PB4-5	9-Jul-57	4	3
TN MOORE	2.87	TP6-4	25-Mar-55	4	2
TN MOORE	5.99	PB17-3	30-Sep-57	7	2
TN MORGAN	8.10	BJ3-6	6-Nov-51	5	6
TN MORGAN	1.55	UpK5-2	12-Apr-53	2	6
TN MORGAN	3.11	TP6-4	25-Mar-55	3	6
TN MORGAN	0.97	TP6-5	26-Mar-55	8	6
TN MORGAN	4.04	TP10-8	12-May-55	10	6
TN MORGAN	7.49	PB17-3	30-Sep-57	5	6
TN OBION	12.19	UpK2-7	30-Mar-53	4	10
TN OBION	21.07	UpK2-8	31-Mar-53	3	10
TN OBION	1.49	UpK3-6	5-Apr-53	5	10
TN OBION	3.68	UpK8-3	10-May-53	4	10
TN OBION	1.09	TP4-5	11-Mar-55	1	10
TN OBION	1.21	PB1-5	1-Jun-57	10	10
TN OBION	2.78	PB2-5	22-Jun-57	4	10
TN OBION	1.27	PB3-9	2-Jul-57	2	10
TN OBION	1.89	PB6-7	30-Jul-57	7	10
TN OBION	2.73	PB9-2	24-Aug-57	6	10
TN OVERTON	8.10	BJ3-6	6-Nov-51	5	6
TN OVERTON	1.83	TP6-3	24-Mar-55	7	6
TN OVERTON	2.03	TP6-4	25-Mar-55	8	6
TN OVERTON	0.97	TP6-5	26-Mar-55	8	6
TN OVERTON	4.18	PB2-5	22-Jun-57	3	6
TN OVERTON	3.79	PB17-3	30-Sep-57	10	6
TN PERRY	1.10	TS6-5	29-May-52	4	5
TN PERRY	9.41	UpK1-3	19-Mar-53	8	5
TN PERRY	2.78	PB2-5	22-Jun-57	4	5
TN PERRY	0.83	PB3-9	2-Jul-57	4	5
TN PERRY	2.11	PB9-2	24-Aug-57	8	5
TN PICKETT	4.59	BJ3-5	5-Nov-51	6	6
TN PICKETT	4.86	BJ3-6	6-Nov-51	7	6
TN PICKETT	3.11	TP6-4	25-Mar-55	3	6
TN PICKETT	0.97	TP6-5	26-Mar-55	8	6
TN PICKETT	4.18	PB2-5	22-Jun-57	3	6

COUNTY	uCi/ sqm	SHOT DAY	DATE	RANK FOR DAY	TIMES IN TOP 10
TN PICKETT	3.79	PB17-3	30-Sep-57	10	6
TN POLK	8.10	BJ3-6	6-Nov-51	5	6
TN POLK	1.55	UpK5-2	12-Apr-53	2	6
TN POLK	3.11	TP6-4	25-Mar-55	3	6
TN POLK	1.45	TP6-5	26-Mar-55	5	6
TN POLK	1.39	PB2-5	22-Jun-57	8	6
TN POLK	14.98	PB17-3	30-Sep-57	2	6
TN PUTNAM	8.10	BJ3-6	6-Nov-51	5	6
TN PUTNAM	1.83	TP6-3	24-Mar-55	7	6
TN PUTNAM	3.11	TP6-4	25-Mar-55	3	6
TN PUTNAM	0.97	TP6-5	26-Mar-55	8	6
TN PUTNAM	6.96	PB2-5	22-Jun-57	1	6
TN PUTNAM	7.49	PB17-3	30-Sep-57	5	6
TN RHEA	8.10	BJ3-6	6-Nov-51	5	6
TN RHEA	1.55	UpK5-2	12-Apr-53	2	6
TN RHEA	3.11	TP6-4	25-Mar-55	3	6
TN RHEA	0.97	TP6-5	26-Mar-55	8	6
TN RHEA	4.18	PB2-5	22-Jun-57	3	6
TN RHEA	14.98	PB17-3	30-Sep-57	2	6
TN ROANE	4.59	BJ3-5	5-Nov-51	6	7
TN ROANE	4.86	BJ3-6	6-Nov-51	7	7
TN ROANE	1.55	UpK5-2	12-Apr-53	2	7
TN ROANE	3.11	TP6-4	25-Mar-55	3	7
TN ROANE	0.97	TP6-5	26-Mar-55	8	7
TN ROANE	1.39	PB2-5	22-Jun-57	8	7
TN ROANE	14.98	PB17-3	30-Sep-57	2	7
TN ROBERTSON	1.14	UpK1-6	22-Mar-53	3	2
TN ROBERTSON	16.75	PB4-5	9-Jul-57	4	2
TN RUTHERFORD	0.83	UpK1-6	22-Mar-53	10	2
TN RUTHERFORD	5.07	PB17-2	29-Sep-57	1	2
TN SCOTT	3.11	TP6-4	25-Mar-55	3	3
TN SCOTT	0.97	TP6-5	26-Mar-55	8	3
TN SCOTT	3.79	PB17-3	30-Sep-57	10	3
TN SEQUATCHIE	4.59	BJ3-5	5-Nov-51	6	7
TN SEQUATCHIE	8.10	BJ3-6	6-Nov-51	5	7

COUNTY	uCi/ sqm	SHOT DAY	DATE	RANK FOR DAY	TIMES IN TOP 10
TN SEQUATCHIE	1.83	TP6-3	24-Mar-55	7	7
TN SEQUATCHIE	3.11	TP6-4	25-Mar-55	3	7
TN SEQUATCHIE	0.97	TP6-5	26-Mar-55	8	7
TN SEQUATCHIE	2.78	PB2-5	22-Jun-57	4	7
TN SEQUATCHIE	14.98	PB17-3	30-Sep-57	2	7
TN SEVIER	4.86	BJ3-6	6-Nov-51	7	8
TN SEVIER	2.59	UpK5-2	12-Apr-53	1	8
TN SEVIER	3.11	TP6-4	25-Mar-55	3	8
TN SEVIER	0.97	TP6-5	26-Mar-55	8	8
TN SEVIER	4.74	TP10-8	12-May-55	3	8
TN SEVIER	2.78	PB2-5	22-Jun-57	4	8
TN SEVIER	2.99	PB5-8	22-Jul-57	4	8
TN SEVIER	14.98	PB17-3	30-Sep-57	2	8
TN SHELBY	7.58	UpK2-7	30-Mar-53	6	4
TN SHELBY	1.52	UpK3-6	5-Apr-53	4	4
TN SHELBY	1.53	UpK8-3	10-May-53	8	4
TN SHELBY	1.16	PB2-5	22-Jun-57	10	4
TN SMITH	16.75	PB4-5	9-Jul-57	4	2
TN SMITH	5.07	PB17-2	29-Sep-57	1	2
TN STEWART	1.49	PB1-9	5-Jun-57	2	3
TN STEWART	1.86	PB2-5	22-Jun-57	5	3
TN STEWART	0.83	PB3-9	2-Jul-57	4	3
TN SULLIVAN	1.55	UpK5-2	12-Apr-53	2	7
TN SULLIVAN	27.51	UpK9-4	22-May-53	3	7
TN SULLIVAN	1.83	TP6-3	24-Mar-55	7	7
TN SULLIVAN	3.11	TP6-4	25-Mar-55	3	7
TN SULLIVAN	1.45	TP6-5	26-Mar-55	5	7
TN SULLIVAN	4.18	PB2-5	22-Jun-57	3	7
TN SULLIVAN	2.99	PB5-8	22-Jul-57	4	7
TN SUMNER	1.11	UpK1-6	22-Mar-53	4	3
TN SUMNER	16.75	PB4-5	9-Jul-57	4	3
TN SUMNER	2.54	PB17-2	29-Sep-57	2	3
TN TIPTON	12.19	UpK2-7	30-Mar-53	4	10
TN TIPTON	21.07	UpK2-8	31-Mar-53	3	10
TN TIPTON	2.17	UpK3-6	5-Apr-53	2	10

COUNTY	uCi/ sqm	SHOT DAY	DATE	RANK FOR DAY	TIMES IN TOP 10
TN TIPTON	6.18	UpK8-3	10-May-53	2	10
TN TIPTON	0.89	UpK8-5	12-May-53	6	10
TN TIPTON	2.92	TP7-4	1-Apr-55	10	10
TN TIPTON	1.21	PB1-5	1-Jun-57	10	10
TN TIPTON	1.86	PB2-5	22-Jun-57	5	10
TN TIPTON	0.83	PB3-9	2-Jul-57	4	10
TN TIPTON	2.73	PB9-2	24-Aug-57	6	10
TN UNICOI	3.89	BJ2-2	31-Oct-51	10	6
TN UNICOI	1.55	UpK5-2	12-Apr-53	2	6
TN UNICOI	3.11	TP6-4	25-Mar-55	3	6
TN UNICOI	5.79	TP6-5	26-Mar-55	1	6
TN UNICOI	1.51	PB5-8	22-Jul-57	7	6
TN UNICOI	7.49	PB17-3	30-Sep-57	5	6
TN UNION	1.04	UpK5-2	12-Apr-53	4	5
TN UNION	3.11	TP6-4	25-Mar-55	3	5
TN UNION	0.97	TP6-5	26-Mar-55	8	5
TN UNION	1.39	PB2-5	22-Jun-57	8	5
TN UNION	3.79	PB17-3	30-Sep-57	10	5
TN VAN BUREN	8.10	BJ3-6	6-Nov-51	5	6
TN VAN BUREN	1.83	TP6-3	24-Mar-55	7	6
TN VAN BUREN	3.11	TP6-4	25-Mar-55	3	6
TN VAN BUREN	0.97	TP6-5	26-Mar-55	8	6
TN VAN BUREN	4.18	PB2-5	22-Jun-57	3	6
TN VAN BUREN	14.98	PB17-3	30-Sep-57	2	6
TN WARREN	4.86	BJ3-6	6-Nov-51	7	5
TN WARREN	2.03	TP6-4	25-Mar-55	8	5
TN WARREN	0.97	TP6-5	26-Mar-55	8	5
TN WARREN	6.96	PB2-5	22-Jun-57	1	5
TN WARREN	14.98	PB17-3	30-Sep-57	2	5
TN WASHINGTON	1.55	UpK5-2	12-Apr-53	2	7
TN WASHINGTON	3.11	TP6-4	25-Mar-55	3	7
TN WASHINGTON	1.45	TP6-5	26-Mar-55	5	7
TN WASHINGTON	2.78	PB2-5	22-Jun-57	4	7
TN WASHINGTON	1.51	PB5-8	22-Jul-57	7	7
TN WASHINGTON	7.49	PB17-3	30-Sep-57	5	7

COUNTY	uCi/ sqm	SHOT DAY	DATE	RANK FOR DAY	TIMES IN TOP 10
TN WASHINGTON	77.90	S-4	9-Jul-62	10	7
TN WAYNE	0.88	UpK3-6	5-Apr-53	8	2
TN WAYNE	2.78	PB2-5	22-Jun-57	4	2
TN WEAKLEY	12.19	UpK2-7	30-Mar-53	4	8
TN WEAKLEY	21.07	UpK2-8	31-Mar-53	3	8
TN WEAKLEY	1.49	UpK3-6	5-Apr-53	5	8
TN WEAKLEY	3.68	UpK8-3	10-May-53	4	8
TN WEAKLEY	1.09	TP4-5	11-Mar-55	1	8
TN WEAKLEY	2.78	PB2-5	22-Jun-57	4	8
TN WEAKLEY	1.27	PB3-9	2-Jul-57	2	8
TN WEAKLEY	2.73	PB9-2	24-Aug-57	6	8
TN WHITE	4.86	BJ3-6	6-Nov-51	7	5
TN WHITE	3.11	TP6-4	25-Mar-55	3	5
TN WHITE	0.97	TP6-5	26-Mar-55	8	5
TN WHITE	6.96	PB2-5	22-Jun-57	1	5
TN WHITE	7.49	PB17-3	30-Sep-57	5	5
TN WILLIAMSON	1.05	TS6-5	29-May-52	6	4
TN WILLIAMSON	9.59	UpK1-3	19-Mar-53	7	4
TN WILLIAMSON	1.57	UpK1-6	22-Mar-53	2	4
TN WILLIAMSON	5.07	PB17-2	29-Sep-57	1	4
TN WILSON	1.05	TS6-5	29-May-52	6	4
TN WILSON	0.86	UpK1-6	22-Mar-53	8	4
TN WILSON	11.16	PB4-5	9-Jul-57	9	4
TN WILSON	5.07	PB17-2	29-Sep-57	1	4

COUNTY	uCi/ sqm	SHOT DAY	DATE	RANK FOR DAY	TIMES IN TOP 10
TX ANDERSON	3.04	UpK8-4	11-May-53	2	5
TX ANDERSON	1.16	UpK8-5	12-May-53	5	5
TX ANDERSON	12.96	TP6-2	23-Mar-55	2	5
TX ANDERSON	2.96	PB1-5	1-Jun-57	5	5
TX ANDERSON	2.60	PB18-7	13-Oct-57	1	5
TX ANDREWS	3.46	TS5-11	17-May-52	1	6
TX ANDREWS	16.75	UpK6-2	19-Apr-53	6	6
TX ANDREWS	1.51	UpK6-3	20-Apr-53	8	6
TX ANDREWS	1.25	UpK8-5	12-May-53	4	6
TX ANDREWS	5.38	TP7-2	30-Mar-55	10	6
TX ANDREWS	11.19	PB3-2	25-Jun-57	2	6
TX ANGELINA	79.98	UpK6-2	19-Apr-53	1	6
TX ANGELINA	2.72	UpK6-3	20-Apr-53	4	6
TX ANGELINA	1.01	UpK6-5	22-Apr-53	7	6
TX ANGELINA	6.40	TP6-3	24-Mar-55	3	6
TX ANGELINA	1.78	PB1-5	1-Jun-57	8	6
TX ANGELINA	2.60	PB18-7	13-Oct-57	1	6
TX ARANSAS	4.68	UpK6-2	19-Apr-53	10	2
TX ARANSAS	4.94	PB1-5	1-Jun-57	2	2
TX ARCHER	5.43	UpK6-2	19-Apr-53	9	5
TX ARCHER	1.30	UpK6-5	22-Apr-53	4	5
TX ARCHER	1.83	UpK8-4	11-May-53	4	5
TX ARCHER	3.18	TP6-2	23-Mar-55	6	5
TX ARCHER	1.64	PB13-9	14-Sep-57	10	5
TX ARMSTRONG	1.02	TS4-4	4-May-52	1	11
TX ARMSTRONG	1.02	TS4-4	4-May-52	1	11
TX ARMSTRONG	2.45	TS6-4	28-May-52	9	11
TX ARMSTRONG	13.66	UpK3-5	4-Apr-53	1	11
TX ARMSTRONG	4.23	UpK6-3	20-Apr-53	2	11
TX ARMSTRONG	25.62	TP7-2	30-Mar-55	4	11
TX ARMSTRONG	31.48	TP10-2	6-May-55	6	11
TX ARMSTRONG	12.43	TP11-4	18-May-55	2	11
TX ARMSTRONG	2.69	PB5-7	21-Jul-57	5	11
TX ARMSTRONG	1.23	PB9-6	28-Aug-57	4	11
TX ARMSTRONG	3.21	PB18-6	12-Oct-57	1	11

COUNTY	uCi/ sqm	SHOT DAY	DATE	RANK FOR DAY	TIMES IN TOP 10
TX ATASCOSA	4.68	UpK6-2	19-Apr-53	10	1
TX AUSTIN	79.98	UpK6-2	19-Apr-53	1	5
TX AUSTIN	2.72	UpK6-3	20-Apr-53	4	5
TX AUSTIN	1.34	UpK8-5	12-May-53	3	5
TX AUSTIN	6.40	TP6-3	24-Mar-55	3	5
TX AUSTIN	1.66	PB1-5	1-Jun-57	9	5
TX BAILEY	1.02	TS4-4	4-May-52	1	9
TX BAILEY	1.02	TS4-4	4-May-52	1	9
TX BAILEY	4.23	UpK6-3	20-Apr-53	2	9
TX BAILEY	25.62	TP7-2	30-Mar-55	4	9
TX BAILEY	31.48	TP10-2	6-May-55	6	9
TX BAILEY	9.55	TP11-4	18-May-55	10	9
TX BAILEY	11.19	PB3-2	25-Jun-57	2	9
TX BAILEY	1.23	PB9-6	28-Aug-57	4	9
TX BAILEY	9.99	PB13-4	9-Sep-57	1	9
TX BANDERA	6.11	TP6-2	23-Mar-55	4	3
TX BANDERA	4.94	PB1-5	1-Jun-57	2	3
TX BANDERA	1.64	PB9-7	29-Aug-57	2	3
TX BASTROP	4.68	UpK6-2	19-Apr-53	10	2
TX BASTROP	3.32	PB1-5	1-Jun-57	4	2
TX BAYLOR	5.43	UpK6-2	19-Apr-53	9	6
TX BAYLOR	1.30	UpK6-5	22-Apr-53	4	6
TX BAYLOR	1.83	UpK8-4	11-May-53	4	6
TX BAYLOR	3.18	TP6-2	23-Mar-55	6	6
TX BAYLOR	1.78	PB1-5	1-Jun-57	8	6
TX BAYLOR	3.21	PB18-6	12-Oct-57	1	6
TX BEE	4.68	UpK6-2	19-Apr-53	10	1
TX BELL	3.04	UpK8-4	11-May-53	2	5
TX BELL	1.16	UpK8-5	12-May-53	5	5
TX BELL	12.96	TP6-2	23-Mar-55	2	5
TX BELL	1.78	PB1-5	1-Jun-57	8	5
TX BELL	2.60	PB18-7	13-Oct-57	1	5
TX BEXAR	4.68	UpK6-2	19-Apr-53	10	2
TX BEXAR	3.32	PB1-5	1-Jun-57	4	2
TX BLANCO	5.43	UpK6-2	19-Apr-53	9	5

COUNTY	uCi/ sqm	SHOT DAY	DATE	RANK FOR DAY	TIMES IN TOP 10
TX BLANCO	1.30	UpK6-5	22-Apr-53	4	5
TX BLANCO	0.80	UpK8-5	12-May-53	7	5
TX BLANCO	3.18	TP6-2	23-Mar-55	6	5
TX BLANCO	4.94	PB1-5	1-Jun-57	2	5
TX BORDEN	5.43	UpK6-2	19-Apr-53	9	4
TX BORDEN	1.30	UpK6-5	22-Apr-53	4	4
TX BORDEN	3.18	TP6-2	23-Mar-55	6	4
TX BORDEN	1.28	PB13-8	13-Sep-57	6	4
TX BOSQUE	3.04	UpK8-4	11-May-53	2	4
TX BOSQUE	12.96	TP6-2	23-Mar-55	2	4
TX BOSQUE	1.78	PB1-5	1-Jun-57	8	4
TX BOSQUE	2.60	PB18-7	13-Oct-57	1	4
TX BOWIE	4.02	BJ3-5	5-Nov-51	8	9
TX BOWIE	1.22	UpK3-6	5-Apr-53	6	9
TX BOWIE	3.91	UpK5-4	14-Apr-53	1	9
TX BOWIE	1.32	UpK8-3	10-May-53	9	9
TX BOWIE	1.83	UpK8-4	11-May-53	4	9
TX BOWIE	1.16	UpK8-5	12-May-53	5	9
TX BOWIE	12.96	TP6-2	23-Mar-55	2	9
TX BOWIE	1.78	PB1-5	1-Jun-57	8	9
TX BOWIE	2.60	PB18-7	13-Oct-57	1	9
TX BRAZORIA	79.98	UpK6-2	19-Apr-53	1	5
TX BRAZORIA	2.72	UpK6-3	20-Apr-53	4	5
TX BRAZORIA	1.34	UpK8-5	12-May-53	3	5
TX BRAZORIA	6.40	TP6-3	24-Mar-55	3	5
TX BRAZORIA	3.32	PB1-5	1-Jun-57	4	5
TX BRAZOS	3.04	UpK8-4	11-May-53	2	4
TX BRAZOS	12.96	TP6-2	23-Mar-55	2	4
TX BRAZOS	2.69	PB5-7	21-Jul-57	5	4
TX BRAZOS	2.60	PB18-7	13-Oct-57	1	4
TX BREWSTER	6.11	TP6-2	23-Mar-55	4	2
TX BREWSTER	1.66	PB1-5	1-Jun-57	9	2
TX BRISCOE	1.02	TS4-4	4-May-52	1	9
TX BRISCOE	1.02	TS4-4	4-May-52	1	9
TX BRISCOE	8.19	UpK3-5	4-Apr-53	2	9

COUNTY	uCi/ sqm	SHOT DAY	DATE	RANK FOR DAY	TIMES IN TOP 10
TX BRISCOE	4.23	UpK6-3	20-Apr-53	2	9
TX BRISCOE	6.07	UpK8-2	9-May-53	9	9
TX BRISCOE	23.30	UpK9-4	22-May-53	4	9
TX BRISCOE	51.23	TP7-2	30-Mar-55	3	9
TX BRISCOE	31.48	TP10-2	6-May-55	6	9
TX BRISCOE	3.21	PB18-6	12-Oct-57	1	9
TX BROOKS	4.68	UpK6-2	19-Apr-53	10	2
TX BROOKS	8.26	PB1-5	1-Jun-57	1	2
TX BROWN	5.43	UpK6-2	19-Apr-53	9	7
TX BROWN	1.30	UpK6-5	22-Apr-53	4	7
TX BROWN	1.83	UpK8-4	11-May-53	4	7
TX BROWN	0.80	UpK8-5	12-May-53	7	7
TX BROWN	3.18	TP6-2	23-Mar-55	6	7
TX BROWN	2.96	PB1-5	1-Jun-57	5	7
TX BROWN	2.60	PB18-7	13-Oct-57	1	7
TX BURLESON	1.83	UpK8-4	11-May-53	4	4
TX BURLESON	1.16	UpK8-5	12-May-53	5	4
TX BURLESON	12.96	TP6-2	23-Mar-55	2	4
TX BURLESON	2.60	PB18-7	13-Oct-57	1	4
TX BURNET	5.43	UpK6-2	19-Apr-53	9	8
TX BURNET	2.16	UpK6-5	22-Apr-53	1	8
TX BURNET	3.04	UpK8-4	11-May-53	2	8
TX BURNET	0.80	UpK8-5	12-May-53	7	8
TX BURNET	3.18	TP6-2	23-Mar-55	6	8
TX BURNET	1.78	PB1-5	1-Jun-57	8	8
TX BURNET	1.63	PB5-7	21-Jul-57	8	8
TX BURNET	2.60	PB18-7	13-Oct-57	1	8
TX CALDWELL	4.68	UpK6-2	19-Apr-53	10	2
TX CALDWELL	4.94	PB1-5	1-Jun-57	2	2
TX CALHOUN	4.68	UpK6-2	19-Apr-53	10	2
TX CALHOUN	8.26	PB1-5	1-Jun-57	1	2
TX CALLAHAN	21.73	UpK6-2	19-Apr-53	4	6
TX CALLAHAN	1.58	UpK6-4	21-Apr-53	3	6
TX CALLAHAN	1.83	UpK8-4	11-May-53	4	6
TX CALLAHAN	3.18	TP6-2	23-Mar-55	6	6

COUNTY	uCi/ sqm	SHOT DAY	DATE	RANK FOR DAY	TIMES IN TOP 10
TX CALLAHAN	1.78	PB1-5	1-Jun-57	8	6
TX CALLAHAN	2.60	PB18-7	13-Oct-57	1	6
TX CAMERON	4.68	UpK6-2	19-Apr-53	10	2
TX CAMERON	8.26	PB1-5	1-Jun-57	1	2
TX CAMP	4.02	BJ3-5	5-Nov-51	8	9
TX CAMP	1.22	UpK3-6	5-Apr-53	6	9
TX CAMP	2.58	UpK5-4	14-Apr-53	2	9
TX CAMP	1.32	UpK8-3	10-May-53	9	9
TX CAMP	1.83	UpK8-4	11-May-53	4	9
TX CAMP	1.16	UpK8-5	12-May-53	5	9
TX CAMP	12.96	TP6-2	23-Mar-55	2	9
TX CAMP	2.96	PB1-5	1-Jun-57	5	9
TX CAMP	2.60	PB18-7	13-Oct-57	1	9
TX CARSON	1.02	TS4-4	4-May-52	1	11
TX CARSON	1.02	TS4-4	4-May-52	1	11
TX CARSON	3.10	TS6-4	28-May-52	4	11
TX CARSON	13.66	UpK3-5	4-Apr-53	1	11
TX CARSON	4.23	UpK6-3	20-Apr-53	2	11
TX CARSON	25.62	TP7-2	30-Mar-55	4	11
TX CARSON	31.48	TP10-2	6-May-55	6	11
TX CARSON	11.12	TP11-4	18-May-55	5	11
TX CARSON	2.69	PB5-7	21-Jul-57	5	11
TX CARSON	0.83	PB13-8	13-Sep-57	9	11
TX CARSON	3.21	PB18-6	12-Oct-57	1	11
TX CASS	1.22	UpK3-6	5-Apr-53	6	6
TX CASS	2.58	UpK5-4	14-Apr-53	2	6
TX CASS	1.83	UpK8-4	11-May-53	4	6
TX CASS	12.96	TP6-2	23-Mar-55	2	6
TX CASS	1.34	PB1-8	4-Jun-57	9	6
TX CASS	1.56	PB18-7	13-Oct-57	4	6
TX CASTRO	1.02	TS4-4	4-May-52	1	10
TX CASTRO	1.02	TS4-4	4-May-52	1	10
TX CASTRO	8.19	UpK3-5	4-Apr-53	2	10
TX CASTRO	4.23	UpK6-3	20-Apr-53	2	10
TX CASTRO	25.62	TP7-2	30-Mar-55	4	10

COUNTY	uCi/ sqm	SHOT DAY	DATE	RANK FOR DAY	TIMES IN TOP 10
TX CASTRO	31.48	TP10-2	6-May-55	6	10
TX CASTRO	9.55	TP11-4	18-May-55	10	10
TX CASTRO	11.19	PB3-2	25-Jun-57	2	10
TX CASTRO	9.99	PB13-4	9-Sep-57	1	10
TX CASTRO	1.35	PB18-5	11-Oct-57	1	10
TX CHAMBERS	79.98	UpK6-2	19-Apr-53	1	6
TX CHAMBERS	2.72	UpK6-3	20-Apr-53	4	6
TX CHAMBERS	1.34	UpK8-5	12-May-53	3	6
TX CHAMBERS	6.40	TP6-3	24-Mar-55	3	6
TX CHAMBERS	1.66	PB1-5	1-Jun-57	9	6
TX CHAMBERS	0.98	PB9-7	29-Aug-57	5	6
TX CHEROKEE	4.02	BJ3-5	5-Nov-51	8	5
TX CHEROKEE	3.04	UpK8-4	11-May-53	2	5
TX CHEROKEE	12.96	TP6-2	23-Mar-55	2	5
TX CHEROKEE	2.96	PB1-5	1-Jun-57	5	5
TX CHEROKEE	2.60	PB18-7	13-Oct-57	1	5
TX CHILDRESS	1.02	TS4-4	4-May-52	1	10
TX CHILDRESS	1.02	TS4-4	4-May-52	1	10
TX CHILDRESS	8.19	UpK3-5	4-Apr-53	2	10
TX CHILDRESS	4.23	UpK6-3	20-Apr-53	2	10
TX CHILDRESS	6.07	UpK8-2	9-May-53	9	10
TX CHILDRESS	1.22	UpK8-4	11-May-53	8	10
TX CHILDRESS	25.62	TP7-2	30-Mar-55	4	10
TX CHILDRESS	31.48	TP10-2	6-May-55	6	10
TX CHILDRESS	2.96	PB1-5	1-Jun-57	5	10
TX CHILDRESS	1.92	PB18-6	12-Oct-57	3	10
TX CLAY	1.22	BJ5-6	24-Nov-51	6	7
TX CLAY	1.22	BJ5-6	24-Nov-51	6	7
TX CLAY	1.83	UpK8-4	11-May-53	4	7
TX CLAY	12.96	TP6-2	23-Mar-55	2	7
TX CLAY	1.63	PB5-7	21-Jul-57	8	7
TX CLAY	2.50	PB13-9	14-Sep-57	8	7
TX CLAY	1.56	PB18-7	13-Oct-57	4	7
TX COCHRAN	16.75	UpK6-2	19-Apr-53	6	7
TX COCHRAN	1.51	UpK6-3	20-Apr-53	8	7

COUNTY	uCi/ sqm	SHOT DAY	DATE	RANK FOR DAY	TIMES IN TOP 10
TX COCHRAN	1.25	UpK8-5	12-May-53	4	7
TX COCHRAN	1.48	UpK10-7	31-May-53	4	7
TX COCHRAN	5.38	TP7-2	30-Mar-55	10	7
TX COCHRAN	11.19	PB3-2	25-Jun-57	2	7
TX COCHRAN	9.99	PB13-4	9-Sep-57	1	7
TX COKE	5.43	UpK6-2	19-Apr-53	9	5
TX COKE	1.30	UpK6-5	22-Apr-53	4	5
TX COKE	3.18	TP6-2	23-Mar-55	6	5
TX COKE	1.78	PB1-5	1-Jun-57	8	5
TX COKE	2.60	PB18-7	13-Oct-57	1	5
TX COLEMAN	5.43	UpK6-2	19-Apr-53	9	6
TX COLEMAN	1.58	UpK6-4	21-Apr-53	3	6
TX COLEMAN	1.30	UpK6-5	22-Apr-53	4	6
TX COLEMAN	3.18	TP6-2	23-Mar-55	6	6
TX COLEMAN	2.96	PB1-5	1-Jun-57	5	6
TX COLEMAN	1.56	PB18-7	13-Oct-57	4	6
TX COLLIN	2.48	BJ5-6	24-Nov-51	4	6
TX COLLIN	2.48	BJ5-6	24-Nov-51	4	6
TX COLLIN	1.83	UpK8-4	11-May-53	4	6
TX COLLIN	12.96	TP6-2	23-Mar-55	2	6
TX COLLIN	2.11	TP10-6	10-May-55	9	6
TX COLLIN	2.60	PB18-7	13-Oct-57	1	6
TX COLLINGSWORT	1.02	TS4-4	4-May-52	1	9
TX COLLINGSWORT	1.02	TS4-4	4-May-52	1	9
TX COLLINGSWORT	13.66	UpK3-5	4-Apr-53	1	9
TX COLLINGSWORT	4.23	UpK6-3	20-Apr-53	2	9
TX COLLINGSWORT	6.07	UpK8-2	9-May-53	9	9
TX COLLINGSWORT	15.39	UpK9-4	22-May-53	10	9
TX COLLINGSWORT	25.62	TP7-2	30-Mar-55	4	9
TX COLLINGSWORT	31.48	TP10-2	6-May-55	6	9
TX COLLINGSWORT	3.21	PB18-6	12-Oct-57	1	9
TX COLORADO	4.68	UpK6-2	19-Apr-53	10	2
TX COLORADO	8.26	PB1-5	1-Jun-57	1	2
TX COMAL	4.68	UpK6-2	19-Apr-53	10	3
TX COMAL	4.94	PB1-5	1-Jun-57	2	3

COUNTY	uCi/ sqm	SHOT DAY	DATE	RANK FOR DAY	TIMES IN TOP 10
TX COMAL	1.63	PB5-7	21-Jul-57	8	3
TX COMANCHE	5.43	UpK6-2	19-Apr-53	9	6
TX COMANCHE	0.87	UpK6-5	22-Apr-53	8	6
TX COMANCHE	3.04	UpK8-4	11-May-53	2	6
TX COMANCHE	0.80	UpK8-5	12-May-53	7	6
TX COMANCHE	3.18	TP6-2	23-Mar-55	6	6
TX COMANCHE	2.60	PB18-7	13-Oct-57	1	6
TX CONCHO	5.43	UpK6-2	19-Apr-53	9	7
TX CONCHO	2.37	UpK6-4	21-Apr-53	1	7
TX CONCHO	1.30	UpK6-5	22-Apr-53	4	7
TX CONCHO	0.80	UpK8-5	12-May-53	7	7
TX CONCHO	3.18	TP6-2	23-Mar-55	6	7
TX CONCHO	1.78	PB1-5	1-Jun-57	8	7
TX CONCHO	2.60	PB18-7	13-Oct-57	1	7
TX COOKE	2.48	BJ5-6	24-Nov-51	4	8
TX COOKE	2.48	BJ5-6	24-Nov-51	4	8
TX COOKE	1.83	UpK8-4	11-May-53	4	8
TX COOKE	12.96	TP6-2	23-Mar-55	2	8
TX COOKE	14.69	TP11-5	19-May-55	2	8
TX COOKE	1.78	PB1-5	1-Jun-57	8	8
TX COOKE	4.15	PB13-9	14-Sep-57	7	8
TX COOKE	2.60	PB18-7	13-Oct-57	1	8
TX CORYELL	3.04	UpK8-4	11-May-53	2	4
TX CORYELL	12.96	TP6-2	23-Mar-55	2	4
TX CORYELL	2.96	PB1-5	1-Jun-57	5	4
TX CORYELL	1.56	PB18-7	13-Oct-57	4	4
TX COTTLE	5.43	UpK6-2	19-Apr-53	9	4
TX COTTLE	1.30	UpK6-5	22-Apr-53	4	4
TX COTTLE	3.18	TP6-2	23-Mar-55	6	4
TX COTTLE	1.92	PB18-6	12-Oct-57	3	4
TX CRANE	6.11	TP6-2	23-Mar-55	4	1
TX CROCKETT	6.11	TP6-2	23-Mar-55	4	1
TX CROSBY	1.02	TS4-4	4-May-52	1	7
TX CROSBY	1.02	TS4-4	4-May-52	1	7
TX CROSBY	4.23	UpK6-3	20-Apr-53	2	7

COUNTY	uCi/ sqm	SHOT DAY	DATE	RANK FOR DAY	TIMES IN TOP 10
TX CROSBY	25.62	TP7-2	30-Mar-55	4	7
TX CROSBY	31.48	TP10-2	6-May-55	6	7
TX CROSBY	1.78	PB1-5	1-Jun-57	8	7
TX CROSBY	3.21	PB18-6	12-Oct-57	1	7
TX CULBERSON	16.75	UpK6-2	19-Apr-53	6	6
TX CULBERSON	1.51	UpK6-3	20-Apr-53	8	6
TX CULBERSON	1.25	UpK8-5	12-May-53	4	6
TX CULBERSON	5.38	TP7-2	30-Mar-55	10	6
TX CULBERSON	11.19	PB3-2	25-Jun-57	2	6
TX CULBERSON	1.64	PB9-7	29-Aug-57	2	6
TX DALLAM	46.74	R-1-1	28-Jan-51	2	11
TX DALLAM	1.02	TS4-4	4-May-52	1	11
TX DALLAM	1.02	TS4-4	4-May-52	1	11
TX DALLAM	2.73	UpK3-5	4-Apr-53	8	11
TX DALLAM	4.23	UpK6-3	20-Apr-53	2	11
TX DALLAM	102.46	TP7-2	30-Mar-55	2	11
TX DALLAM	1.87	TP9-14	28-Apr-55	2	11
TX DALLAM	31.48	TP10-2	6-May-55	6	11
TX DALLAM	11.19	PB3-2	25-Jun-57	2	11
TX DALLAM	9.99	PB13-4	9-Sep-57	1	11
TX DALLAM	1.35	PB18-5	11-Oct-57	1	11
TX DALLAS	1.22	BJ5-6	24-Nov-51	6	8
TX DALLAS	1.22	BJ5-6	24-Nov-51	6	8
TX DALLAS	46.24	UpK7-3	27-Apr-53	6	8
TX DALLAS	1.83	UpK8-4	11-May-53	4	8
TX DALLAS	12.96	TP6-2	23-Mar-55	2	8
TX DALLAS	2.70	TP10-6	10-May-55	3	8
TX DALLAS	9.96	PB12-5	6-Sep-57	6	8
TX DALLAS	2.60	PB18-7	13-Oct-57	1	8
TX DAWSON	5.43	UpK6-2	19-Apr-53	9	5
TX DAWSON	3.18	TP6-2	23-Mar-55	6	5
TX DAWSON	1.78	PB1-5	1-Jun-57	8	5
TX DAWSON	3.59	PB5-5	19-Jul-57	4	5
TX DAWSON	1.63	PB5-7	21-Jul-57	8	5
TX DE WITT	4.68	UpK6-2	19-Apr-53	10	2

COUNTY	uCi/ sqm	SHOT DAY	DATE	RANK FOR DAY	TIMES IN TOP 10
TX DE WITT	8.26	PB1-5	1-Jun-57	1	2
TX DEAF SMITH	1.02	TS4-4	4-May-52	1	12
TX DEAF SMITH	1.02	TS4-4	4-May-52	1	12
TX DEAF SMITH	1.74	TS4-6	6-May-52	5	12
TX DEAF SMITH	1.74	TS4-6	6-May-52	5	12
TX DEAF SMITH	1.15	TS5-11	17-May-52	3	12
TX DEAF SMITH	2.68	TS6-4	28-May-52	7	12
TX DEAF SMITH	8.19	UpK3-5	4-Apr-53	2	12
TX DEAF SMITH	4.23	UpK6-3	20-Apr-53	2	12
TX DEAF SMITH	25.62	TP7-2	30-Mar-55	4	12
TX DEAF SMITH	31.48	TP10-2	6-May-55	6	12
TX DEAF SMITH	11.19	PB3-2	25-Jun-57	2	12
TX DEAF SMITH	9.99	PB13-4	9-Sep-57	1	12
TX DELTA	12.96	TP6-2	23-Mar-55	2	2
TX DELTA	2.60	PB18-7	13-Oct-57	1	2
TX DENTON	2.48	BJ5-6	24-Nov-51	4	7
TX DENTON	2.48	BJ5-6	24-Nov-51	4	7
TX DENTON	1.83	UpK8-4	11-May-53	4	7
TX DENTON	12.96	TP6-2	23-Mar-55	2	7
TX DENTON	1.78	PB1-5	1-Jun-57	8	7
TX DENTON	1.64	PB13-9	14-Sep-57	10	7
TX DENTON	2.60	PB18-7	13-Oct-57	1	7
TX DICKENS	5.43	UpK6-2	19-Apr-53	9	6
TX DICKENS	1.30	UpK6-5	22-Apr-53	4	6
TX DICKENS	3.18	TP6-2	23-Mar-55	6	6
TX DICKENS	1.78	PB1-5	1-Jun-57	8	6
TX DICKENS	1.63	PB5-7	21-Jul-57	8	6
TX DICKENS	3.21	PB18-6	12-Oct-57	1	6
TX DIMMIT	6.11	TP6-2	23-Mar-55	4	2
TX DIMMIT	1.66	PB1-5	1-Jun-57	9	2
TX DONLEY	1.02	TS4-4	4-May-52	1	10
TX DONLEY	1.02	TS4-4	4-May-52	1	10
TX DONLEY	13.66	UpK3-5	4-Apr-53	1	10
TX DONLEY	4.23	UpK6-3	20-Apr-53	2	10
TX DONLEY	27.68	UpK9-4	22-May-53	2	10

COUNTY	uCi/ sqm	SHOT DAY	DATE	RANK FOR DAY	TIMES IN TOP 10
TX DONLEY	25.62	TP7-2	30-Mar-55	4	10
TX DONLEY	31.48	TP10-2	6-May-55	6	10
TX DONLEY	10.02	TP11-4	18-May-55	7	10
TX DONLEY	1.63	PB5-7	21-Jul-57	8	10
TX DONLEY	1.92	PB18-6	12-Oct-57	3	10
TX DUVAL	4.68	UpK6-2	19-Apr-53	10	2
TX DUVAL	4.94	PB1-5	1-Jun-57	2	2
TX EASTLAND	5.43	UpK6-2	19-Apr-53	9	6
TX EASTLAND	0.87	UpK6-5	22-Apr-53	8	6
TX EASTLAND	3.04	UpK8-4	11-May-53	2	6
TX EASTLAND	3.18	TP6-2	23-Mar-55	6	6
TX EASTLAND	1.78	PB1-5	1-Jun-57	8	6
TX EASTLAND	1.56	PB18-7	13-Oct-57	4	6
TX ECTOR	16.75	UpK6-2	19-Apr-53	6	5
TX ECTOR	1.51	UpK6-3	20-Apr-53	8	5
TX ECTOR	1.03	UpK6-4	21-Apr-53	4	5
TX ECTOR	1.25	UpK8-5	12-May-53	4	5
TX ECTOR	5.38	TP7-2	30-Mar-55	10	5
TX EDWARDS	6.11	TP6-2	23-Mar-55	4	1
TX EL PASO	1.15	TS5-11	17-May-52	3	8
TX EL PASO	16.75	UpK6-2	19-Apr-53	6	8
TX EL PASO	6.04	UpK6-3	20-Apr-53	1	8
TX EL PASO	1.03	UpK6-4	21-Apr-53	4	8
TX EL PASO	1.25	UpK8-5	12-May-53	4	8
TX EL PASO	5.38	TP7-2	30-Mar-55	10	8
TX EL PASO	11.19	PB3-2	25-Jun-57	2	8
TX EL PASO	1.01	PB9-7	29-Aug-57	4	8
TX ELLIS	1.83	UpK8-4	11-May-53	4	7
TX ELLIS	1.16	UpK8-5	12-May-53	5	7
TX ELLIS	12.96	TP6-2	23-Mar-55	2	7
TX ELLIS	2.48	TP8-3	11-Apr-55	6	7
TX ELLIS	2.06	TP10-6	10-May-55	10	7
TX ELLIS	1.63	PB5-7	21-Jul-57	8	7
TX ELLIS	2.60	PB18-7	13-Oct-57	1	7
TX ERATH	1.83	UpK8-4	11-May-53	4	3

COUNTY	uCi/ sqm	SHOT DAY	DATE	RANK FOR DAY	TIMES IN TOP 10
TX ERATH	12.96	TP6-2	23-Mar-55	2	3
TX ERATH	2.60	PB18-7	13-Oct-57	1	3
TX FALLS	3.04	UpK8-4	11-May-53	2	3
TX FALLS	12.96	TP6-2	23-Mar-55	2	3
TX FALLS	2.60	PB18-7	13-Oct-57	1	3
TX FANNIN	5.74	UpK1-9	25-Mar-53	8	8
TX FANNIN	55.85	UpK7-3	27-Apr-53	3	8
TX FANNIN	1.83	UpK8-4	11-May-53	4	8
TX FANNIN	12.96	TP6-2	23-Mar-55	2	8
TX FANNIN	14.27	TP11-5	19-May-55	5	8
TX FANNIN	1.78	PB1-5	1-Jun-57	8	8
TX FANNIN	2.50	PB13-9	14-Sep-57	8	8
TX FANNIN	2.60	PB18-7	13-Oct-57	1	8
TX FAYETTE	4.68	UpK6-2	19-Apr-53	10	2
TX FAYETTE	4.94	PB1-5	1-Jun-57	2	2
TX FISHER	5.43	UpK6-2	19-Apr-53	9	8
TX FISHER	1.30	UpK6-5	22-Apr-53	4	8
TX FISHER	1.34	UpK8-5	12-May-53	3	8
TX FISHER	3.18	TP6-2	23-Mar-55	6	8
TX FISHER	3.34	TP7-3	31-Mar-55	6	8
TX FISHER	2.96	PB1-5	1-Jun-57	5	8
TX FISHER	1.28	PB13-8	13-Sep-57	6	8
TX FISHER	1.04	PB18-7	13-Oct-57	7	8
TX FLOYD	1.02	TS4-4	4-May-52	1	9
TX FLOYD	1.02	TS4-4	4-May-52	1	9
TX FLOYD	2.73	UpK3-5	4-Apr-53	8	9
TX FLOYD	4.23	UpK6-3	20-Apr-53	2	9
TX FLOYD	25.62	TP7-2	30-Mar-55	4	9
TX FLOYD	31.48	TP10-2	6-May-55	6	9
TX FLOYD	1.78	PB1-5	1-Jun-57	8	9
TX FLOYD	1.63	PB5-7	21-Jul-57	8	9
TX FLOYD	3.21	PB18-6	12-Oct-57	1	9
TX FOARD	5.43	UpK6-2	19-Apr-53	9	5
TX FOARD	1.30	UpK6-5	22-Apr-53	4	5
TX FOARD	3.18	TP6-2	23-Mar-55	6	5

COUNTY	uCi/ sqm	SHOT DAY	DATE	RANK FOR DAY	TIMES IN TOP 10
TX FOARD	14.33	TP11-5	19-May-55	4	5
TX FOARD	1.92	PB18-6	12-Oct-57	3	5
TX FORT BEND	79.98	UpK6-2	19-Apr-53	1	5
TX FORT BEND	2.72	UpK6-3	20-Apr-53	4	5
TX FORT BEND	1.34	UpK8-5	12-May-53	3	5
TX FORT BEND	6.40	TP6-3	24-Mar-55	3	5
TX FORT BEND	4.94	PB1-5	1-Jun-57	2	5
TX FRANKLIN	2.58	UpK5-4	14-Apr-53	2	6
TX FRANKLIN	1.32	UpK8-3	10-May-53	9	6
TX FRANKLIN	1.12	UpK8-4	11-May-53	9	6
TX FRANKLIN	12.96	TP6-2	23-Mar-55	2	6
TX FRANKLIN	2.96	PB1-5	1-Jun-57	5	6
TX FRANKLIN	1.56	PB18-7	13-Oct-57	4	6
TX FREESTONE	3.04	UpK8-4	11-May-53	2	3
TX FREESTONE	12.96	TP6-2	23-Mar-55	2	3
TX FREESTONE	2.60	PB18-7	13-Oct-57	1	3
TX FRIO	6.11	TP6-2	23-Mar-55	4	2
TX FRIO	3.32	PB1-5	1-Jun-57	4	2
TX GAINES	1.15	TS5-11	17-May-52	3	6
TX GAINES	16.75	UpK6-2	19-Apr-53	6	6
TX GAINES	1.51	UpK6-3	20-Apr-53	8	6
TX GAINES	2.50	UpK8-5	12-May-53	1	6
TX GAINES	5.38	TP7-2	30-Mar-55	10	6
TX GAINES	11.19	PB3-2	25-Jun-57	2	6
TX GALVESTON	79.98	UpK6-2	19-Apr-53	1	5
TX GALVESTON	2.72	UpK6-3	20-Apr-53	4	5
TX GALVESTON	1.34	UpK8-5	12-May-53	3	5
TX GALVESTON	6.40	TP6-3	24-Mar-55	3	5
TX GALVESTON	4.94	PB1-5	1-Jun-57	2	5
TX GARZA	5.43	UpK6-2	19-Apr-53	9	4
TX GARZA	3.18	TP6-2	23-Mar-55	6	4
TX GARZA	1.78	PB1-5	1-Jun-57	8	4
TX GARZA	3.21	PB18-6	12-Oct-57	1	4
TX GILLESPIE	0.87	UpK6-5	22-Apr-53	8	3
TX GILLESPIE	6.11	TP6-2	23-Mar-55	4	3

COUNTY	uCi/ sqm	SHOT DAY	DATE	RANK FOR DAY	TIMES IN TOP 10
TX GILLESPIE	4.94	PB1-5	1-Jun-57	2	3
TX GLASSCOCK	5.43	UpK6-2	19-Apr-53	9	3
TX GLASSCOCK	1.30	UpK6-5	22-Apr-53	4	3
TX GLASSCOCK	3.18	TP6-2	23-Mar-55	6	3
TX GOLIAD	4.68	UpK6-2	19-Apr-53	10	2
TX GOLIAD	3.32	PB1-5	1-Jun-57	4	2
TX GONZALES	4.79	BJ3-5	5-Nov-51	5	3
TX GONZALES	4.68	UpK6-2	19-Apr-53	10	3
TX GONZALES	4.94	PB1-5	1-Jun-57	2	3
TX GRAY	1.02	TS4-4	4-May-52	1	8
TX GRAY	1.02	TS4-4	4-May-52	1	8
TX GRAY	8.19	UpK3-5	4-Apr-53	2	8
TX GRAY	4.23	UpK6-3	20-Apr-53	2	8
TX GRAY	25.62	TP7-2	30-Mar-55	4	8
TX GRAY	31.48	TP10-2	6-May-55	6	8
TX GRAY	12.30	TP11-4	18-May-55	3	8
TX GRAY	1.29	PB18-6	12-Oct-57	6	8
TX GRAYSON	3.70	BJ5-6	24-Nov-51	2	11
TX GRAYSON	3.70	BJ5-6	24-Nov-51	2	11
TX GRAYSON	5.74	UpK1-9	25-Mar-53	8	11
TX GRAYSON	55.80	UpK7-3	27-Apr-53	4	11
TX GRAYSON	1.83	UpK8-4	11-May-53	4	11
TX GRAYSON	12.96	TP6-2	23-Mar-55	2	11
TX GRAYSON	1.78	PB1-5	1-Jun-57	8	11
TX GRAYSON	1.63	PB5-7	21-Jul-57	8	11
TX GRAYSON	0.83	PB13-8	13-Sep-57	9	11
TX GRAYSON	2.50	PB13-9	14-Sep-57	8	11
TX GRAYSON	1.56	PB18-7	13-Oct-57	4	11
TX GREGG	4.02	BJ3-5	5-Nov-51	8	8
TX GREGG	1.33	UpK5-4	14-Apr-53	7	8
TX GREGG	1.32	UpK8-3	10-May-53	9	8
TX GREGG	1.12	UpK8-4	11-May-53	9	8
TX GREGG	1.16	UpK8-5	12-May-53	5	8
TX GREGG	12.96	TP6-2	23-Mar-55	2	8
TX GREGG	2.42	PB1-6	2-Jun-57	4	8

COUNTY	uCi/ sqm	SHOT DAY	DATE	RANK FOR DAY	TIMES IN TOP 10
TX GREGG	2.60	PB18-7	13-Oct-57	1	8
TX GRIMES	4.79	BJ3-5	5-Nov-51	5	7
TX GRIMES	79.98	UpK6-2	19-Apr-53	1	7
TX GRIMES	2.72	UpK6-3	20-Apr-53	4	7
TX GRIMES	1.34	UpK8-5	12-May-53	3	7
TX GRIMES	6.40	TP6-3	24-Mar-55	3	7
TX GRIMES	3.80	PB5-4	18-Jul-57	10	7
TX GRIMES	2.60	PB18-7	13-Oct-57	1	7
TX GUADALUPE	4.68	UpK6-2	19-Apr-53	10	2
TX GUADALUPE	1.66	PB1-5	1-Jun-57	9	2
TX HALE	1.02	TS4-4	4-May-52	1	9
TX HALE	1.02	TS4-4	4-May-52	1	9
TX HALE	5.46	UpK3-5	4-Apr-53	4	9
TX HALE	4.23	UpK6-3	20-Apr-53	2	9
TX HALE	25.62	TP7-2	30-Mar-55	4	9
TX HALE	31.48	TP10-2	6-May-55	6	9
TX HALE	12.85	TP11-4	18-May-55	1	9
TX HALE	11.19	PB3-2	25-Jun-57	2	9
TX HALE	9.99	PB13-4	9-Sep-57	1	9
TX HALL	1.02	TS4-4	4-May-52	1	10
TX HALL	1.02	TS4-4	4-May-52	1	10
TX HALL	0.93	UpK3-3	2-Apr-53	8	10
TX HALL	8.19	UpK3-5	4-Apr-53	2	10
TX HALL	4.23	UpK6-3	20-Apr-53	2	10
TX HALL	6.07	UpK8-2	9-May-53	9	10
TX HALL	25.62	TP7-2	30-Mar-55	4	10
TX HALL	31.48	TP10-2	6-May-55	6	10
TX HALL	1.78	PB1-5	1-Jun-57	8	10
TX HALL	3.21	PB18-6	12-Oct-57	1	10
TX HAMILTON	3.04	UpK8-4	11-May-53	2	4
TX HAMILTON	12.96	TP6-2	23-Mar-55	2	4
TX HAMILTON	1.63	PB5-7	21-Jul-57	8	4
TX HAMILTON	2.60	PB18-7	13-Oct-57	1	4
TX HANSFORD	1.22	BJ5-6	24-Nov-51	6	13
TX HANSFORD	1.22	BJ5-6	24-Nov-51	6	13

COUNTY	uCi/ sqm	SHOT DAY	DATE	RANK FOR DAY	TIMES IN TOP 10
TX HANSFORD	1.02	TS4-4	4-May-52	1	13
TX HANSFORD	1.02	TS4-4	4-May-52	1	13
TX HANSFORD	1.74	TS4-6	6-May-52	5	13
TX HANSFORD	1.74	TS4-6	6-May-52	5	13
TX HANSFORD	5.46	UpK3-5	4-Apr-53	4	13
TX HANSFORD	4.23	UpK6-3	20-Apr-53	2	13
TX HANSFORD	102.46	TP7-2	30-Mar-55	2	13
TX HANSFORD	1.87	TP9-14	28-Apr-55	2	13
TX HANSFORD	31.48	TP10-2	6-May-55	6	13
TX HANSFORD	1.28	PB13-8	13-Sep-57	6	13
TX HANSFORD	1.92	PB18-6	12-Oct-57	3	13
TX HARDEMAN	5.43	UpK6-2	19-Apr-53	9	5
TX HARDEMAN	1.30	UpK6-5	22-Apr-53	4	5
TX HARDEMAN	3.18	TP6-2	23-Mar-55	6	5
TX HARDEMAN	1.78	PB1-5	1-Jun-57	8	5
TX HARDEMAN	3.21	PB18-6	12-Oct-57	1	5
TX HARDIN	79.98	UpK6-2	19-Apr-53	1	5
TX HARDIN	2.72	UpK6-3	20-Apr-53	4	5
TX HARDIN	0.81	UpK6-7	24-Apr-53	1	5
TX HARDIN	1.34	UpK8-5	12-May-53	3	5
TX HARDIN	6.40	TP6-3	24-Mar-55	3	5
TX HARRIS	79.98	UpK6-2	19-Apr-53	1	5
TX HARRIS	2.72	UpK6-3	20-Apr-53	4	5
TX HARRIS	1.34	UpK8-5	12-May-53	3	5
TX HARRIS	6.40	TP6-3	24-Mar-55	3	5
TX HARRIS	4.94	PB1-5	1-Jun-57	2	5
TX HARRISON	2.58	UpK5-4	14-Apr-53	2	8
TX HARRISON	1.32	UpK8-3	10-May-53	9	8
TX HARRISON	1.83	UpK8-4	11-May-53	4	8
TX HARRISON	1.16	UpK8-5	12-May-53	5	8
TX HARRISON	12.96	TP6-2	23-Mar-55	2	8
TX HARRISON	1.78	PB1-5	1-Jun-57	8	8
TX HARRISON	3.80	PB5-4	18-Jul-57	10	8
TX HARRISON	2.60	PB18-7	13-Oct-57	1	8
TX HARTLEY	1.02	TS4-4	4-May-52	1	13

COUNTY	uCi/ sqm	SHOT DAY	DATE	RANK FOR DAY	TIMES IN TOP 10
TX HARTLEY	1.02	TS4-4	4-May-52	1	13
TX HARTLEY	1.74	TS4-6	6-May-52	5	13
TX HARTLEY	1.74	TS4-6	6-May-52	5	13
TX HARTLEY	2.30	TS5-11	17-May-52	2	13
TX HARTLEY	2.05	UpK3-5	4-Apr-53	9	13
TX HARTLEY	4.23	UpK6-3	20-Apr-53	2	13
TX HARTLEY	51.23	TP7-2	30-Mar-55	3	13
TX HARTLEY	1.87	TP9-14	28-Apr-55	2	13
TX HARTLEY	31.48	TP10-2	6-May-55	6	13
TX HARTLEY	11.19	PB3-2	25-Jun-57	2	13
TX HARTLEY	9.99	PB13-4	9-Sep-57	1	13
TX HARTLEY	1.35	PB18-5	11-Oct-57	1	13
TX HASKELL	5.43	UpK6-2	19-Apr-53	9	4
TX HASKELL	1.30	UpK6-5	22-Apr-53	4	4
TX HASKELL	3.18	TP6-2	23-Mar-55	6	4
TX HASKELL	1.78	PB1-5	1-Jun-57	8	4
TX HAYS	4.68	UpK6-2	19-Apr-53	10	2
TX HAYS	4.94	PB1-5	1-Jun-57	2	2
TX HEMPHILL	1.02	TS4-4	4-May-52	1	9
TX HEMPHILL	1.02	TS4-4	4-May-52	1	9
TX HEMPHILL	8.19	UpK3-5	4-Apr-53	2	9
TX HEMPHILL	4.23	UpK6-3	20-Apr-53	2	9
TX HEMPHILL	1.83	UpK8-4	11-May-53	4	9
TX HEMPHILL	25.62	TP7-2	30-Mar-55	4	9
TX HEMPHILL	31.48	TP10-2	6-May-55	6	9
TX HEMPHILL	2.11	PB13-8	13-Sep-57	4	9
TX HEMPHILL	1.92	PB18-6	12-Oct-57	3	9
TX HENDERSON	3.04	UpK8-4	11-May-53	2	6
TX HENDERSON	1.16	UpK8-5	12-May-53	5	6
TX HENDERSON	12.96	TP6-2	23-Mar-55	2	6
TX HENDERSON	2.96	PB1-5	1-Jun-57	5	6
TX HENDERSON	1.63	PB5-7	21-Jul-57	8	6
TX HENDERSON	2.60	PB18-7	13-Oct-57	1	6
TX HIDALGO	4.68	UpK6-2	19-Apr-53	10	2
TX HIDALGO	3.32	PB1-5	1-Jun-57	4	2

COUNTY	uCi/ sqm	SHOT DAY	DATE	RANK FOR DAY	TIMES IN TOP 10
TX HILL	3.04	UpK8-4	11-May-53	2	5
TX HILL	1.16	UpK8-5	12-May-53	5	5
TX HILL	12.96	TP6-2	23-Mar-55	2	5
TX HILL	11.33	PB12-5	6-Sep-57	4	5
TX HILL	2.60	PB18-7	13-Oct-57	1	5
TX HOCKLEY	1.02	TS4-4	4-May-52	1	9
TX HOCKLEY	1.02	TS4-4	4-May-52	1	9
TX HOCKLEY	3.46	TS5-11	17-May-52	1	9
TX HOCKLEY	8.19	UpK3-5	4-Apr-53	2	9
TX HOCKLEY	4.23	UpK6-3	20-Apr-53	2	9
TX HOCKLEY	25.62	TP7-2	30-Mar-55	4	9
TX HOCKLEY	31.48	TP10-2	6-May-55	6	9
TX HOCKLEY	11.19	PB3-2	25-Jun-57	2	9
TX HOCKLEY	9.99	PB13-4	9-Sep-57	1	9
TX HOOD	1.83	UpK8-4	11-May-53	4	3
TX HOOD	12.96	TP6-2	23-Mar-55	2	3
TX HOOD	2.60	PB18-7	13-Oct-57	1	3
TX HOPKINS	1.22	UpK8-4	11-May-53	8	5
TX HOPKINS	1.16	UpK8-5	12-May-53	5	5
TX HOPKINS	12.96	TP6-2	23-Mar-55	2	5
TX HOPKINS	2.96	PB1-5	1-Jun-57	5	5
TX HOPKINS	2.60	PB18-7	13-Oct-57	1	5
TX HOUSTON	1.83	UpK8-4	11-May-53	4	6
TX HOUSTON	1.16	UpK8-5	12-May-53	5	6
TX HOUSTON	12.96	TP6-2	23-Mar-55	2	6
TX HOUSTON	2.96	PB1-5	1-Jun-57	5	6
TX HOUSTON	1.30	PB1-7	3-Jun-57	10	6
TX HOUSTON	2.60	PB18-7	13-Oct-57	1	6
TX HOWARD	5.43	UpK6-2	19-Apr-53	9	4
TX HOWARD	1.30	UpK6-5	22-Apr-53	4	4
TX HOWARD	3.18	TP6-2	23-Mar-55	6	4
TX HOWARD	1.78	PB1-5	1-Jun-57	8	4
TX HUDSPETH	1.15	TS5-11	17-May-52	3	9
TX HUDSPETH	16.75	UpK6-2	19-Apr-53	6	9
TX HUDSPETH	1.51	UpK6-3	20-Apr-53	8	9

COUNTY	uCi/ sqm	SHOT DAY	DATE	RANK FOR DAY	TIMES IN TOP 10
TX HUDSPETH	1.58	UpK6-4	21-Apr-53	3	9
TX HUDSPETH	1.25	UpK8-5	12-May-53	4	9
TX HUDSPETH	5.38	TP7-2	30-Mar-55	10	9
TX HUDSPETH	11.19	PB3-2	25-Jun-57	2	9
TX HUDSPETH	2.39	PB5-5	19-Jul-57	6	9
TX HUDSPETH	2.44	PB9-7	29-Aug-57	1	9
TX HUNT	1.83	UpK8-4	11-May-53	4	4
TX HUNT	12.96	TP6-2	23-Mar-55	2	4
TX HUNT	1.78	PB1-5	1-Jun-57	8	4
TX HUNT	2.60	PB18-7	13-Oct-57	1	4
TX HUTCHINSON	1.02	TS4-4	4-May-52	1	10
TX HUTCHINSON	1.02	TS4-4	4-May-52	1	10
TX HUTCHINSON	5.46	UpK3-5	4-Apr-53	4	10
TX HUTCHINSON	4.23	UpK6-3	20-Apr-53	2	10
TX HUTCHINSON	1.22	UpK8-4	11-May-53	8	10
TX HUTCHINSON	102.46	TP7-2	30-Mar-55	2	10
TX HUTCHINSON	31.48	TP10-2	6-May-55	6	10
TX HUTCHINSON	0.83	PB13-8	13-Sep-57	9	10
TX HUTCHINSON	0.88	PB18-5	11-Oct-57	4	10
TX HUTCHINSON	1.92	PB18-6	12-Oct-57	3	10
TX IRION	5.43	UpK6-2	19-Apr-53	9	6
TX IRION	1.58	UpK6-4	21-Apr-53	3	6
TX IRION	1.30	UpK6-5	22-Apr-53	4	6
TX IRION	3.18	TP6-2	23-Mar-55	6	6
TX IRION	1.78	PB1-5	1-Jun-57	8	6
TX IRION	1.04	PB18-7	13-Oct-57	7	6
TX JACK	1.83	UpK8-4	11-May-53	4	4
TX JACK	12.96	TP6-2	23-Mar-55	2	4
TX JACK	2.50	PB13-9	14-Sep-57	8	4
TX JACK	1.56	PB18-7	13-Oct-57	4	4
TX JACKSON	4.68	UpK6-2	19-Apr-53	10	2
TX JACKSON	8.26	PB1-5	1-Jun-57	1	2
TX JASPER	79.98	UpK6-2	19-Apr-53	1	7
TX JASPER	2.72	UpK6-3	20-Apr-53	4	7
TX JASPER	0.81	UpK6-7	24-Apr-53	1	7

COUNTY	uCi/ sqm	SHOT DAY	DATE	RANK FOR DAY	TIMES IN TOP 10
TX JASPER	1.34	UpK8-5	12-May-53	3	7
TX JASPER	6.40	TP6-3	24-Mar-55	3	7
TX JASPER	3.80	PB5-4	18-Jul-57	10	7
TX JASPER	0.80	PB13-7	12-Sep-57	7	7
TX JEFF DAVIS	16.75	UpK6-2	19-Apr-53	6	6
TX JEFF DAVIS	1.51	UpK6-3	20-Apr-53	8	6
TX JEFF DAVIS	1.25	UpK8-5	12-May-53	4	6
TX JEFF DAVIS	5.38	TP7-2	30-Mar-55	10	6
TX JEFF DAVIS	22.38	PB3-2	25-Jun-57	1	6
TX JEFF DAVIS	3.59	PB5-5	19-Jul-57	4	6
TX JEFFERSON	33.22	UpK6-2	19-Apr-53	3	5
TX JEFFERSON	1.13	UpK6-3	20-Apr-53	10	5
TX JEFFERSON	2.66	TP6-3	24-Mar-55	6	5
TX JEFFERSON	1.78	PB1-5	1-Jun-57	8	5
TX JEFFERSON	1.56	PB18-7	13-Oct-57	4	5
TX JIM HOGG	4.68	UpK6-2	19-Apr-53	10	2
TX JIM HOGG	8.26	PB1-5	1-Jun-57	1	2
TX JIM WELLS	4.68	UpK6-2	19-Apr-53	10	2
TX JIM WELLS	8.26	PB1-5	1-Jun-57	1	2
TX JOHNSON	1.08	UpK6-5	22-Apr-53	6	7
TX JOHNSON	1.83	UpK8-4	11-May-53	4	7
TX JOHNSON	12.96	TP6-2	23-Mar-55	2	7
TX JOHNSON	2.37	TP8-3	11-Apr-55	8	7
TX JOHNSON	1.78	PB1-5	1-Jun-57	8	7
TX JOHNSON	10.91	PB12-5	6-Sep-57	5	7
TX JOHNSON	2.60	PB18-7	13-Oct-57	1	7
TX JONES	5.43	UpK6-2	19-Apr-53	9	6
TX JONES	1.30	UpK6-5	22-Apr-53	4	6
TX JONES	1.83	UpK8-4	11-May-53	4	6
TX JONES	3.18	TP6-2	23-Mar-55	6	6
TX JONES	1.78	PB1-5	1-Jun-57	8	6
TX JONES	1.56	PB18-7	13-Oct-57	4	6
TX KARNES	4.68	UpK6-2	19-Apr-53	10	2
TX KARNES	4.94	PB1-5	1-Jun-57	2	2
TX KAUFMAN	5.38	TP6-2	23-Mar-55	5	3

COUNTY	uCi/ sqm	SHOT DAY	DATE	RANK FOR DAY	TIMES IN TOP 10
TX KAUFMAN	2.13	TP10-6	10-May-55	7	3
TX KAUFMAN	1.08	PB18-7	13-Oct-57	7	3
TX KENDALL	6.11	TP6-2	23-Mar-55	4	2
TX KENDALL	8.26	PB1-5	1-Jun-57	1	2
TX KENEDY	4.68	UpK6-2	19-Apr-53	10	2
TX KENEDY	4.94	PB1-5	1-Jun-57	2	2
TX KENT	5.43	UpK6-2	19-Apr-53	9	5
TX KENT	1.30	UpK6-5	22-Apr-53	4	5
TX KENT	3.18	TP6-2	23-Mar-55	6	5
TX KENT	1.78	PB1-5	1-Jun-57	8	5
TX KENT	1.92	PB18-6	12-Oct-57	3	5
TX KERR	6.11	TP6-2	23-Mar-55	4	2
TX KERR	3.32	PB1-5	1-Jun-57	4	2
TX KIMBLE	6.11	TP6-2	23-Mar-55	4	2
TX KIMBLE	3.32	PB1-5	1-Jun-57	4	2
TX KING	5.43	UpK6-2	19-Apr-53	9	5
TX KING	3.18	TP6-2	23-Mar-55	6	5
TX KING	1.78	PB1-5	1-Jun-57	8	5
TX KING	1.63	PB5-7	21-Jul-57	8	5
TX KING	1.92	PB18-6	12-Oct-57	3	5
TX KINNEY	6.11	TP6-2	23-Mar-55	4	1
TX KLEBERG	4.68	UpK6-2	19-Apr-53	10	2
TX KLEBERG	4.94	PB1-5	1-Jun-57	2	2
TX KNOX	5.43	UpK6-2	19-Apr-53	9	6
TX KNOX	1.30	UpK6-5	22-Apr-53	4	6
TX KNOX	1.83	UpK8-4	11-May-53	4	6
TX KNOX	3.18	TP6-2	23-Mar-55	6	6
TX KNOX	1.78	PB1-5	1-Jun-57	8	6
TX KNOX	1.92	PB18-6	12-Oct-57	3	6
TX LA SALLE	4.68	UpK6-2	19-Apr-53	10	2
TX LA SALLE	1.66	PB1-5	1-Jun-57	9	2
TX LAMAR	3.04	UpK8-4	11-May-53	2	5
TX LAMAR	12.96	TP6-2	23-Mar-55	2	5
TX LAMAR	1.70	PB1-7	3-Jun-57	3	5
TX LAMAR	4.15	PB13-9	14-Sep-57	7	5

COUNTY	uCi/ sqm	SHOT DAY	DATE	RANK FOR DAY	TIMES IN TOP 10
TX LAMAR	2.60	PB18-7	13-Oct-57	1	5
TX LAMB	1.02	TS4-4	4-May-52	1	10
TX LAMB	1.02	TS4-4	4-May-52	1	10
TX LAMB	2.30	TS5-11	17-May-52	2	10
TX LAMB	2.73	UpK3-5	4-Apr-53	8	10
TX LAMB	4.23	UpK6-3	20-Apr-53	2	10
TX LAMB	25.62	TP7-2	30-Mar-55	4	10
TX LAMB	31.48	TP10-2	6-May-55	6	10
TX LAMB	9.59	TP11-4	18-May-55	9	10
TX LAMB	11.19	PB3-2	25-Jun-57	2	10
TX LAMB	9.99	PB13-4	9-Sep-57	1	10
TX LAMPASAS	5.43	UpK6-2	19-Apr-53	9	8
TX LAMPASAS	2.37	UpK6-4	21-Apr-53	1	8
TX LAMPASAS	1.30	UpK6-5	22-Apr-53	4	8
TX LAMPASAS	3.04	UpK8-4	11-May-53	2	8
TX LAMPASAS	1.34	UpK8-5	12-May-53	3	8
TX LAMPASAS	3.18	TP6-2	23-Mar-55	6	8
TX LAMPASAS	1.78	PB1-5	1-Jun-57	8	8
TX LAMPASAS	2.60	PB18-7	13-Oct-57	1	8
TX LAVACA	4.68	UpK6-2	19-Apr-53	10	2
TX LAVACA	4.94	PB1-5	1-Jun-57	2	2
TX LEE	1.83	UpK8-4	11-May-53	4	4
TX LEE	1.16	UpK8-5	12-May-53	5	4
TX LEE	12.96	TP6-2	23-Mar-55	2	4
TX LEE	1.78	PB1-5	1-Jun-57	8	4
TX LEON	3.04	UpK8-4	11-May-53	2	5
TX LEON	1.16	UpK8-5	12-May-53	5	5
TX LEON	12.96	TP6-2	23-Mar-55	2	5
TX LEON	2.96	PB1-5	1-Jun-57	5	5
TX LEON	2.60	PB18-7	13-Oct-57	1	5
TX LIBERTY	79.98	UpK6-2	19-Apr-53	1	4
TX LIBERTY	2.72	UpK6-3	20-Apr-53	4	4
TX LIBERTY	1.34	UpK8-5	12-May-53	3	4
TX LIBERTY	6.40	TP6-3	24-Mar-55	3	4
TX LIMESTONE	3.04	UpK8-4	11-May-53	2	6

COUNTY	uCi/ sqm	SHOT DAY	DATE	RANK FOR DAY	TIMES IN TOP 10
TX LIMESTONE	1.16	UpK8-5	12-May-53	5	6
TX LIMESTONE	12.96	TP6-2	23-Mar-55	2	6
TX LIMESTONE	1.78	PB1-5	1-Jun-57	8	6
TX LIMESTONE	1.60	PB1-7	3-Jun-57	4	6
TX LIMESTONE	2.60	PB18-7	13-Oct-57	1	6
TX LIPSCOMB	2.48	BJ5-6	24-Nov-51	4	17
TX LIPSCOMB	2.48	BJ5-6	24-Nov-51	4	17
TX LIPSCOMB	1.02	TS4-4	4-May-52	1	17
TX LIPSCOMB	1.02	TS4-4	4-May-52	1	17
TX LIPSCOMB	1.74	TS4-6	6-May-52	5	17
TX LIPSCOMB	1.74	TS4-6	6-May-52	5	17
TX LIPSCOMB	3.81	TS6-7	31-May-52	8	17
TX LIPSCOMB	8.19	UpK3-5	4-Apr-53	2	17
TX LIPSCOMB	4.23	UpK6-3	20-Apr-53	2	17
TX LIPSCOMB	1.22	UpK8-4	11-May-53	8	17
TX LIPSCOMB	16.15	UpK9-4	22-May-53	9	17
TX LIPSCOMB	51.23	TP7-2	30-Mar-55	3	17
TX LIPSCOMB	31.48	TP10-2	6-May-55	6	17
TX LIPSCOMB	1.76	PB2-4	21-Jun-57	6	17
TX LIPSCOMB	1.63	PB5-7	21-Jul-57	8	17
TX LIPSCOMB	2.11	PB13-8	13-Sep-57	4	17
TX LIPSCOMB	1.92	PB18-6	12-Oct-57	3	17
TX LIVE OAK	4.68	UpK6-2	19-Apr-53	10	2
TX LIVE OAK	1.66	PB1-5	1-Jun-57	9	2
TX LLANO	5.43	UpK6-2	19-Apr-53	9	6
TX LLANO	1.58	UpK6-4	21-Apr-53	3	6
TX LLANO	1.34	UpK8-5	12-May-53	3	6
TX LLANO	3.18	TP6-2	23-Mar-55	6	6
TX LLANO	1.78	PB1-5	1-Jun-57	8	6
TX LLANO	2.60	PB18-7	13-Oct-57	1	6
TX LOVING	3.18	TS6-4	28-May-52	3	6
TX LOVING	16.75	UpK6-2	19-Apr-53	6	6
TX LOVING	1.51	UpK6-3	20-Apr-53	8	6
TX LOVING	1.25	UpK8-5	12-May-53	4	6
TX LOVING	5.38	TP7-2	30-Mar-55	10	6

COUNTY	uCi/ sqm	SHOT DAY	DATE	RANK FOR DAY	TIMES IN TOP 10
TX LOVING	11.19	PB3-2	25-Jun-57	2	6
TX LUBBOCK	1.02	TS4-4	4-May-52	1	10
TX LUBBOCK	1.02	TS4-4	4-May-52	1	10
TX LUBBOCK	5.46	UpK3-5	4-Apr-53	4	10
TX LUBBOCK	4.23	UpK6-3	20-Apr-53	2	10
TX LUBBOCK	25.62	TP7-2	30-Mar-55	4	10
TX LUBBOCK	31.48	TP10-2	6-May-55	6	10
TX LUBBOCK	11.19	PB3-2	25-Jun-57	2	10
TX LUBBOCK	1.63	PB5-7	21-Jul-57	8	10
TX LUBBOCK	9.99	PB13-4	9-Sep-57	1	10
TX LUBBOCK	1.35	PB18-5	11-Oct-57	1	10
TX LYNN	1.02	TS4-4	4-May-52	1	7
TX LYNN	1.02	TS4-4	4-May-52	1	7
TX LYNN	8.19	UpK3-5	4-Apr-53	2	7
TX LYNN	4.23	UpK6-3	20-Apr-53	2	7
TX LYNN	25.62	TP7-2	30-Mar-55	4	7
TX LYNN	31.48	TP10-2	6-May-55	6	7
TX LYNN	3.21	PB18-6	12-Oct-57	1	7
TX MADISON	4.02	BJ3-5	5-Nov-51	8	5
TX MADISON	3.04	UpK8-4	11-May-53	2	5
TX MADISON	12.96	TP6-2	23-Mar-55	2	5
TX MADISON	1.78	PB1-5	1-Jun-57	8	5
TX MADISON	2.60	PB18-7	13-Oct-57	1	5
TX MARION	2.58	UpK5-4	14-Apr-53	2	6
TX MARION	1.83	UpK8-4	11-May-53	4	6
TX MARION	1.16	UpK8-5	12-May-53	5	6
TX MARION	12.96	TP6-2	23-Mar-55	2	6
TX MARION	2.96	PB1-5	1-Jun-57	5	6
TX MARION	1.56	PB18-7	13-Oct-57	4	6
TX MARTIN	5.43	UpK6-2	19-Apr-53	9	4
TX MARTIN	1.30	UpK6-5	22-Apr-53	4	4
TX MARTIN	3.18	TP6-2	23-Mar-55	6	4
TX MARTIN	1.63	PB5-7	21-Jul-57	8	4
TX MASON	5.43	UpK6-2	19-Apr-53	9	5
TX MASON	0.80	UpK8-5	12-May-53	7	5

COUNTY	uCi/ sqm	SHOT DAY	DATE	RANK FOR DAY	TIMES IN TOP 10
TX MASON	3.18	TP6-2	23-Mar-55	6	5
TX MASON	1.78	PB1-5	1-Jun-57	8	5
TX MASON	2.60	PB18-7	13-Oct-57	1	5
TX MATAGORDA	4.68	UpK6-2	19-Apr-53	10	2
TX MATAGORDA	4.94	PB1-5	1-Jun-57	2	2
TX MAVERICK	6.11	TP6-2	23-Mar-55	4	1
TX MCCULLOCH	5.43	UpK6-2	19-Apr-53	9	5
TX MCCULLOCH	0.87	UpK6-5	22-Apr-53	8	5
TX MCCULLOCH	0.80	UpK8-5	12-May-53	7	5
TX MCCULLOCH	3.18	TP6-2	23-Mar-55	6	5
TX MCCULLOCH	2.60	PB18-7	13-Oct-57	1	5
TX MCLENNAN	3.04	UpK8-4	11-May-53	2	5
TX MCLENNAN	1.16	UpK8-5	12-May-53	5	5
TX MCLENNAN	12.96	TP6-2	23-Mar-55	2	5
TX MCLENNAN	1.63	PB5-7	21-Jul-57	8	5
TX MCLENNAN	2.60	PB18-7	13-Oct-57	1	5
TX MCMULLEN	4.68	UpK6-2	19-Apr-53	10	1
TX MEDINA	6.11	TP6-2	23-Mar-55	4	2
TX MEDINA	4.94	PB1-5	1-Jun-57	2	2
TX MENARD	5.43	UpK6-2	19-Apr-53	9	5
TX MENARD	1.30	UpK6-5	22-Apr-53	4	5
TX MENARD	0.80	UpK8-5	12-May-53	7	5
TX MENARD	3.18	TP6-2	23-Mar-55	6	5
TX MENARD	2.60	PB18-7	13-Oct-57	1	5
TX MIDLAND	5.43	UpK6-2	19-Apr-53	9	4
TX MIDLAND	2.37	UpK6-4	21-Apr-53	1	4
TX MIDLAND	0.87	UpK6-5	22-Apr-53	8	4
TX MIDLAND	3.18	TP6-2	23-Mar-55	6	4
TX MILAM	1.83	UpK8-4	11-May-53	4	5
TX MILAM	1.16	UpK8-5	12-May-53	5	5
TX MILAM	12.96	TP6-2	23-Mar-55	2	5
TX MILAM	1.78	PB1-5	1-Jun-57	8	5
TX MILAM	2.60	PB18-7	13-Oct-57	1	5
TX MILLS	5.43	UpK6-2	19-Apr-53	9	6
TX MILLS	3.04	UpK8-4	11-May-53	2	6

COUNTY	uCi/ sqm	SHOT DAY	DATE	RANK FOR DAY	TIMES IN TOP 10
TX MILLS	0.80	UpK8-5	12-May-53	7	6
TX MILLS	3.18	TP6-2	23-Mar-55	6	6
TX MILLS	1.63	PB5-7	21-Jul-57	8	6
TX MILLS	2.60	PB18-7	13-Oct-57	1	6
TX MITCHELL	5.43	UpK6-2	19-Apr-53	9	5
TX MITCHELL	2.16	UpK6-5	22-Apr-53	1	5
TX MITCHELL	3.18	TP6-2	23-Mar-55	6	5
TX MITCHELL	1.78	PB1-5	1-Jun-57	8	5
TX MITCHELL	1.63	PB5-7	21-Jul-57	8	5
TX MONTAGUE	1.22	BJ5-6	24-Nov-51	6	8
TX MONTAGUE	1.22	BJ5-6	24-Nov-51	6	8
TX MONTAGUE	5.74	UpK1-9	25-Mar-53	8	8
TX MONTAGUE	1.83	UpK8-4	11-May-53	4	8
TX MONTAGUE	12.96	TP6-2	23-Mar-55	2	8
TX MONTAGUE	1.78	PB1-5	1-Jun-57	8	8
TX MONTAGUE	0.83	PB13-8	13-Sep-57	9	8
TX MONTAGUE	2.50	PB13-9	14-Sep-57	8	8
TX MONTGOMERY	79.98	UpK6-2	19-Apr-53	1	6
TX MONTGOMERY	2.72	UpK6-3	20-Apr-53	4	6
TX MONTGOMERY	0.80	UpK8-5	12-May-53	7	6
TX MONTGOMERY	6.40	TP6-3	24-Mar-55	3	6
TX MONTGOMERY	1.78	PB1-5	1-Jun-57	8	6
TX MONTGOMERY	1.04	PB18-7	13-Oct-57	7	6
TX MOORE	1.02	TS4-4	4-May-52	1	11
TX MOORE	1.02	TS4-4	4-May-52	1	11
TX MOORE	3.48	TS4-6	6-May-52	1	11
TX MOORE	3.48	TS4-6	6-May-52	1	11
TX MOORE	4.23	UpK6-3	20-Apr-53	2	11
TX MOORE	25.62	TP7-2	30-Mar-55	4	11
TX MOORE	3.12	TP9-14	28-Apr-55	1	11
TX MOORE	31.48	TP10-2	6-May-55	6	11
TX MOORE	9.65	TP11-4	18-May-55	8	11
TX MOORE	11.19	PB3-2	25-Jun-57	2	11
TX MOORE	9.99	PB13-4	9-Sep-57	1	11
TX MORRIS	4.02	BJ3-5	5-Nov-51	8	8

COUNTY	uCi/ sqm	SHOT DAY	DATE	RANK FOR DAY	TIMES IN TOP 10
TX MORRIS	1.22	UpK3-6	5-Apr-53	6	8
TX MORRIS	2.58	UpK5-4	14-Apr-53	2	8
TX MORRIS	1.83	UpK8-4	11-May-53	4	8
TX MORRIS	1.16	UpK8-5	12-May-53	5	8
TX MORRIS	12.96	TP6-2	23-Mar-55	2	8
TX MORRIS	1.78	PB1-5	1-Jun-57	8	8
TX MORRIS	2.60	PB18-7	13-Oct-57	1	8
TX MOTLEY	1.02	TS4-4	4-May-52	1	9
TX MOTLEY	1.02	TS4-4	4-May-52	1	9
TX MOTLEY	5.46	UpK3-5	4-Apr-53	4	9
TX MOTLEY	4.23	UpK6-3	20-Apr-53	2	9
TX MOTLEY	6.07	UpK8-2	9-May-53	9	9
TX MOTLEY	25.62	TP7-2	30-Mar-55	4	9
TX MOTLEY	31.48	TP10-2	6-May-55	6	9
TX MOTLEY	2.96	PB1-5	1-Jun-57	5	9
TX MOTLEY	3.21	PB18-6	12-Oct-57	1	9
TX NACOGDOCHES	4.02	BJ3-5	5-Nov-51	8	5
TX NACOGDOCHES	1.83	UpK8-4	11-May-53	4	5
TX NACOGDOCHES	1.16	UpK8-5	12-May-53	5	5
TX NACOGDOCHES	6.40	TP6-3	24-Mar-55	3	5
TX NACOGDOCHES	2.60	PB18-7	13-Oct-57	1	5
TX NAVARRO	3.04	UpK8-4	11-May-53	2	5
TX NAVARRO	1.16	UpK8-5	12-May-53	5	5
TX NAVARRO	12.96	TP6-2	23-Mar-55	2	5
TX NAVARRO	1.78	PB1-5	1-Jun-57	8	5
TX NAVARRO	2.60	PB18-7	13-Oct-57	1	5
TX NEWTON	79.98	UpK6-2	19-Apr-53	1	6
TX NEWTON	2.72	UpK6-3	20-Apr-53	4	6
TX NEWTON	0.81	UpK6-7	24-Apr-53	1	6
TX NEWTON	1.34	UpK8-5	12-May-53	3	6
TX NEWTON	6.40	TP6-3	24-Mar-55	3	6
TX NEWTON	0.80	PB13-7	12-Sep-57	7	6
TX NOLAN	5.43	UpK6-2	19-Apr-53	9	4
TX NOLAN	1.30	UpK6-5	22-Apr-53	4	4
TX NOLAN	3.18	TP6-2	23-Mar-55	6	4

COUNTY	uCi/ sqm	SHOT DAY	DATE	RANK FOR DAY	TIMES IN TOP 10
TX NOLAN	2.96	PB1-5	1-Jun-57	5	4
TX NUECES	3.43	PB1-5	1-Jun-57	3	1
TX OCHILTREE	1.02	TS4-4	4-May-52	1	12
TX OCHILTREE	1.02	TS4-4	4-May-52	1	12
TX OCHILTREE	1.74	TS4-6	6-May-52	5	12
TX OCHILTREE	1.74	TS4-6	6-May-52	5	12
TX OCHILTREE	5.46	UpK3-5	4-Apr-53	4	12
TX OCHILTREE	4.23	UpK6-3	20-Apr-53	2	12
TX OCHILTREE	1.22	UpK8-4	11-May-53	8	12
TX OCHILTREE	102.46	TP7-2	30-Mar-55	2	12
TX OCHILTREE	1.25	TP9-14	28-Apr-55	3	12
TX OCHILTREE	31.48	TP10-2	6-May-55	6	12
TX OCHILTREE	1.28	PB13-8	13-Sep-57	6	12
TX OCHILTREE	1.92	PB18-6	12-Oct-57	3	12
TX OLDHAM	1.02	TS4-4	4-May-52	1	11
TX OLDHAM	1.02	TS4-4	4-May-52	1	11
TX OLDHAM	2.30	TS5-11	17-May-52	2	11
TX OLDHAM	5.46	UpK3-5	4-Apr-53	4	11
TX OLDHAM	4.23	UpK6-3	20-Apr-53	2	11
TX OLDHAM	153.69	TP7-2	30-Mar-55	1	11
TX OLDHAM	3.12	TP9-14	28-Apr-55	1	11
TX OLDHAM	31.48	TP10-2	6-May-55	6	11
TX OLDHAM	11.19	PB3-2	25-Jun-57	2	11
TX OLDHAM	9.99	PB13-4	9-Sep-57	1	11
TX OLDHAM	1.35	PB18-5	11-Oct-57	1	11
TX ORANGE	79.98	UpK6-2	19-Apr-53	1	6
TX ORANGE	2.72	UpK6-3	20-Apr-53	4	6
TX ORANGE	0.81	UpK6-7	24-Apr-53	1	6
TX ORANGE	1.34	UpK8-5	12-May-53	3	6
TX ORANGE	6.40	TP6-3	24-Mar-55	3	6
TX ORANGE	114.67	PB3-3	26-Jun-57	1	6
TX PALO PINTO	5.43	UpK6-2	19-Apr-53	9	7
TX PALO PINTO	0.87	UpK6-5	22-Apr-53	8	7
TX PALO PINTO	1.83	UpK8-4	11-May-53	4	7
TX PALO PINTO	3.18	TP6-2	23-Mar-55	6	7

COUNTY	uCi/ sqm	SHOT DAY	DATE	RANK FOR DAY	TIMES IN TOP 10
TX PALO PINTO	1.78	PB1-5	1-Jun-57	8	7
TX PALO PINTO	1.63	PB5-7	21-Jul-57	8	7
TX PALO PINTO	2.60	PB18-7	13-Oct-57	1	7
TX PANOLA	4.02	BJ3-5	5-Nov-51	8	6
TX PANOLA	1.33	UpK5-4	14-Apr-53	7	6
TX PANOLA	1.83	UpK8-4	11-May-53	4	6
TX PANOLA	6.40	TP6-3	24-Mar-55	3	6
TX PANOLA	2.96	PB1-5	1-Jun-57	5	6
TX PANOLA	2.60	PB18-7	13-Oct-57	1	6
TX PARKER	1.22	UpK8-4	11-May-53	8	4
TX PARKER	12.96	TP6-2	23-Mar-55	2	4
TX PARKER	11.55	PB12-5	6-Sep-57	3	4
TX PARKER	2.60	PB18-7	13-Oct-57	1	4
TX PARMER	1.02	TS4-4	4-May-52	1	10
TX PARMER	1.02	TS4-4	4-May-52	1	10
TX PARMER	2.30	TS5-11	17-May-52	2	10
TX PARMER	13.66	UpK3-5	4-Apr-53	1	10
TX PARMER	4.23	UpK6-3	20-Apr-53	2	10
TX PARMER	25.62	TP7-2	30-Mar-55	4	10
TX PARMER	31.48	TP10-2	6-May-55	6	10
TX PARMER	10.65	TP11-4	18-May-55	6	10
TX PARMER	11.19	PB3-2	25-Jun-57	2	10
TX PARMER	9.99	PB13-4	9-Sep-57	1	10
TX PECOS	6.11	TP6-2	23-Mar-55	4	2
TX PECOS	2.39	PB5-5	19-Jul-57	6	2
TX POLK	4.02	BJ3-5	5-Nov-51	8	7
TX POLK	79.98	UpK6-2	19-Apr-53	1	7
TX POLK	2.72	UpK6-3	20-Apr-53	4	7
TX POLK	1.34	UpK8-5	12-May-53	3	7
TX POLK	6.40	TP6-3	24-Mar-55	3	7
TX POLK	1.78	PB1-5	1-Jun-57	8	7
TX POLK	3.80	PB5-4	18-Jul-57	10	7
TX POTTER	1.02	TS4-4	4-May-52	1	9
TX POTTER	1.02	TS4-4	4-May-52	1	9
TX POTTER	5.46	UpK3-5	4-Apr-53	4	9

COUNTY	uCi/ sqm	SHOT DAY	DATE	RANK FOR DAY	TIMES IN TOP 10
TX POTTER	4.23	UpK6-3	20-Apr-53	2	9
TX POTTER	153.69	TP7-2	30-Mar-55	1	9
TX POTTER	31.48	TP10-2	6-May-55	6	9
TX POTTER	11.19	PB3-2	25-Jun-57	2	9
TX POTTER	9.99	PB13-4	9-Sep-57	1	9
TX POTTER	1.35	PB18-5	11-Oct-57	1	9
TX PRESIDIO	6.11	TP6-2	23-Mar-55	4	2
TX PRESIDIO	11.19	PB3-2	25-Jun-57	2	2
TX RAINS	1.83	UpK8-4	11-May-53	4	5
TX RAINS	1.16	UpK8-5	12-May-53	5	5
TX RAINS	12.96	TP6-2	23-Mar-55	2	5
TX RAINS	1.78	PB1-5	1-Jun-57	8	5
TX RAINS	2.60	PB18-7	13-Oct-57	1	5
TX RANDALL	1.15	TS5-11	17-May-52	3	9
TX RANDALL	2.76	TS6-4	28-May-52	5	9
TX RANDALL	3.40	UpK3-5	4-Apr-53	6	9
TX RANDALL	1.76	UpK6-3	20-Apr-53	6	9
TX RANDALL	10.64	TP7-2	30-Mar-55	5	9
TX RANDALL	11.52	TP11-4	18-May-55	4	9
TX RANDALL	11.19	PB3-2	25-Jun-57	2	9
TX RANDALL	9.99	PB13-4	9-Sep-57	1	9
TX RANDALL	0.88	PB18-5	11-Oct-57	4	9
TX REAGAN	0.87	UpK6-5	22-Apr-53	8	4
TX REAGAN	6.11	TP6-2	23-Mar-55	4	4
TX REAGAN	1.78	PB1-5	1-Jun-57	8	4
TX REAGAN	1.04	PB18-7	13-Oct-57	7	4
TX REAL	6.11	TP6-2	23-Mar-55	4	2
TX REAL	1.63	PB5-7	21-Jul-57	8	2
TX RED RIVER	3.91	UpK5-4	14-Apr-53	1	5
TX RED RIVER	1.83	UpK8-4	11-May-53	4	5
TX RED RIVER	12.96	TP6-2	23-Mar-55	2	5
TX RED RIVER	9.57	PB12-5	6-Sep-57	7	5
TX RED RIVER	2.60	PB18-7	13-Oct-57	1	5
TX REEVES	16.75	UpK6-2	19-Apr-53	6	7
TX REEVES	1.51	UpK6-3	20-Apr-53	8	7

COUNTY	uCi/ sqm	SHOT DAY	DATE	RANK FOR DAY	TIMES IN TOP 10
TX REEVES	1.03	UpK6-4	21-Apr-53	4	7
TX REEVES	1.25	UpK8-5	12-May-53	4	7
TX REEVES	5.38	TP7-2	30-Mar-55	10	7
TX REEVES	0.85	TP9-14	28-Apr-55	4	7
TX REEVES	11.19	PB3-2	25-Jun-57	2	7
TX REFUGIO	1066.98	R-2-2	3-Feb-51	1	3
TX REFUGIO	4.68	UpK6-2	19-Apr-53	10	3
TX REFUGIO	4.94	PB1-5	1-Jun-57	2	3
TX ROBERTS	2.48	BJ5-6	24-Nov-51	4	11
TX ROBERTS	2.48	BJ5-6	24-Nov-51	4	11
TX ROBERTS	1.02	TS4-4	4-May-52	1	11
TX ROBERTS	1.02	TS4-4	4-May-52	1	11
TX ROBERTS	8.19	UpK3-5	4-Apr-53	2	11
TX ROBERTS	4.23	UpK6-3	20-Apr-53	2	11
TX ROBERTS	1.83	UpK8-4	11-May-53	4	11
TX ROBERTS	25.62	TP7-2	30-Mar-55	4	11
TX ROBERTS	31.48	TP10-2	6-May-55	6	11
TX ROBERTS	1.28	PB13-8	13-Sep-57	6	11
TX ROBERTS	1.29	PB18-6	12-Oct-57	6	11
TX ROBERTSON	1.22	UpK8-4	11-May-53	8	5
TX ROBERTSON	1.16	UpK8-5	12-May-53	5	5
TX ROBERTSON	12.96	TP6-2	23-Mar-55	2	5
TX ROBERTSON	2.96	PB1-5	1-Jun-57	5	5
TX ROBERTSON	2.60	PB18-7	13-Oct-57	1	5
TX ROCKWALL	1.83	UpK8-4	11-May-53	4	7
TX ROCKWALL	6.04	UpK10-4	28-May-53	1	7
TX ROCKWALL	12.96	TP6-2	23-Mar-55	2	7
TX ROCKWALL	2.35	TP10-6	10-May-55	5	7
TX ROCKWALL	1.78	PB1-5	1-Jun-57	8	7
TX ROCKWALL	9.28	PB12-5	6-Sep-57	9	7
TX ROCKWALL	2.60	PB18-7	13-Oct-57	1	7
TX RUNNELS	5.43	UpK6-2	19-Apr-53	9	7
TX RUNNELS	1.58	UpK6-4	21-Apr-53	3	7
TX RUNNELS	1.30	UpK6-5	22-Apr-53	4	7
TX RUNNELS	3.18	TP6-2	23-Mar-55	6	7

COUNTY	uCi/ sqm	SHOT DAY	DATE	RANK FOR DAY	TIMES IN TOP 10
TX RUNNELS	2.96	PB1-5	1-Jun-57	5	7
TX RUNNELS	0.80	PB13-7	12-Sep-57	7	7
TX RUNNELS	1.56	PB18-7	13-Oct-57	4	7
TX RUSK	1.32	UpK8-3	10-May-53	9	7
TX RUSK	1.83	UpK8-4	11-May-53	4	7
TX RUSK	1.16	UpK8-5	12-May-53	5	7
TX RUSK	12.96	TP6-2	23-Mar-55	2	7
TX RUSK	2.96	PB1-5	1-Jun-57	5	7
TX RUSK	3.80	PB5-4	18-Jul-57	10	7
TX RUSK	2.60	PB18-7	13-Oct-57	1	7
TX SABINE	4.02	BJ3-5	5-Nov-51	8	5
TX SABINE	79.98	UpK6-2	19-Apr-53	1	5
TX SABINE	2.72	UpK6-3	20-Apr-53	4	5
TX SABINE	1.34	UpK8-5	12-May-53	3	5
TX SABINE	6.40	TP6-3	24-Mar-55	3	5
TX SAN AUGUSTIN	4.02	BJ3-5	5-Nov-51	8	8
TX SAN AUGUSTIN	79.98	UpK6-2	19-Apr-53	1	8
TX SAN AUGUSTIN	2.72	UpK6-3	20-Apr-53	4	8
TX SAN AUGUSTIN	0.81	UpK6-7	24-Apr-53	1	8
TX SAN AUGUSTIN	1.34	UpK8-5	12-May-53	3	8
TX SAN AUGUSTIN	6.40	TP6-3	24-Mar-55	3	8
TX SAN AUGUSTIN	2.96	PB1-5	1-Jun-57	5	8
TX SAN AUGUSTIN	0.94	PB5-6	20-Jul-57	9	8
TX SAN JACINTO	4.02	BJ3-5	5-Nov-51	8	4
TX SAN JACINTO	79.98	UpK6-2	19-Apr-53	1	4
TX SAN JACINTO	2.72	UpK6-3	20-Apr-53	4	4
TX SAN JACINTO	6.40	TP6-3	24-Mar-55	3	4
TX SAN PATRICIO	4.68	UpK6-2	19-Apr-53	10	2
TX SAN PATRICIO	4.94	PB1-5	1-Jun-57	2	2
TX SAN SABA	5.43	UpK6-2	19-Apr-53	9	7
TX SAN SABA	1.30	UpK6-5	22-Apr-53	4	7
TX SAN SABA	1.83	UpK8-4	11-May-53	4	7
TX SAN SABA	0.80	UpK8-5	12-May-53	7	7
TX SAN SABA	3.18	TP6-2	23-Mar-55	6	7
TX SAN SABA	1.78	PB1-5	1-Jun-57	8	7

COUNTY	uCi/ sqm	SHOT DAY	DATE	RANK FOR DAY	TIMES IN TOP 10
TX SAN SABA	2.60	PB18-7	13-Oct-57	1	7
TX SCHLEICHER	0.95	UpK6-4	21-Apr-53	5	4
TX SCHLEICHER	0.87	UpK6-5	22-Apr-53	8	4
TX SCHLEICHER	6.11	TP6-2	23-Mar-55	4	4
TX SCHLEICHER	2.60	PB18-7	13-Oct-57	1	4
TX SCURRY	5.43	UpK6-2	19-Apr-53	9	6
TX SCURRY	2.16	UpK6-5	22-Apr-53	1	6
TX SCURRY	3.18	TP6-2	23-Mar-55	6	6
TX SCURRY	1.78	PB1-5	1-Jun-57	8	6
TX SCURRY	1.63	PB5-7	21-Jul-57	8	6
TX SCURRY	0.83	PB13-8	13-Sep-57	9	6
TX SHACKELFORD	5.43	UpK6-2	19-Apr-53	9	7
TX SHACKELFORD	1.30	UpK6-5	22-Apr-53	4	7
TX SHACKELFORD	3.04	UpK8-4	11-May-53	2	7
TX SHACKELFORD	3.18	TP6-2	23-Mar-55	6	7
TX SHACKELFORD	1.63	PB5-7	21-Jul-57	8	7
TX SHACKELFORD	9.56	PB12-5	6-Sep-57	8	7
TX SHACKELFORD	1.04	PB18-7	13-Oct-57	7	7
TX SHELBY	1.83	UpK8-4	11-May-53	4	3
TX SHELBY	1.16	UpK8-5	12-May-53	5	3
TX SHELBY	6.40	TP6-3	24-Mar-55	3	3
TX SHERMAN	1.02	TS4-4	4-May-52	1	10
TX SHERMAN	1.02	TS4-4	4-May-52	1	10
TX SHERMAN	3.48	TS4-6	6-May-52	1	10
TX SHERMAN	3.48	TS4-6	6-May-52	1	10
TX SHERMAN	5.46	UpK3-5	4-Apr-53	4	10
TX SHERMAN	4.23	UpK6-3	20-Apr-53	2	10
TX SHERMAN	25.62	TP7-2	30-Mar-55	4	10
TX SHERMAN	31.48	TP10-2	6-May-55	6	10
TX SHERMAN	0.98	PB1-12	8-Jun-57	1	10
TX SHERMAN	0.83	PB3-8	1-Jul-57	5	10
TX SMITH	3.04	UpK8-4	11-May-53	2	5
TX SMITH	12.96	TP6-2	23-Mar-55	2	5
TX SMITH	2.96	PB1-5	1-Jun-57	5	5
TX SMITH	1.32	PB1-7	3-Jun-57	9	5

COUNTY	uCi/ sqm	SHOT DAY	DATE	RANK FOR DAY	TIMES IN TOP 10
TX SMITH	2.60	PB18-7	13-Oct-57	1	5
TX SOMERVELL	1.83	UpK8-4	11-May-53	4	3
TX SOMERVELL	12.96	TP6-2	23-Mar-55	2	3
TX SOMERVELL	2.60	PB18-7	13-Oct-57	1	3
TX STARR	4.68	UpK6-2	19-Apr-53	10	2
TX STARR	4.94	PB1-5	1-Jun-57	2	2
TX STEPHENS	5.43	UpK6-2	19-Apr-53	9	6
TX STEPHENS	1.30	UpK6-5	22-Apr-53	4	6
TX STEPHENS	1.83	UpK8-4	11-May-53	4	6
TX STEPHENS	3.18	TP6-2	23-Mar-55	6	6
TX STEPHENS	1.63	PB5-7	21-Jul-57	8	6
TX STEPHENS	1.04	PB18-7	13-Oct-57	7	6
TX STERLING	5.43	UpK6-2	19-Apr-53	9	5
TX STERLING	2.37	UpK6-4	21-Apr-53	1	5
TX STERLING	0.87	UpK6-5	22-Apr-53	8	5
TX STERLING	3.18	TP6-2	23-Mar-55	6	5
TX STERLING	1.78	PB1-5	1-Jun-57	8	5
TX STONEWALL	5.43	UpK6-2	19-Apr-53	9	4
TX STONEWALL	1.30	UpK6-5	22-Apr-53	4	4
TX STONEWALL	3.18	TP6-2	23-Mar-55	6	4
TX STONEWALL	1.78	PB1-5	1-Jun-57	8	4
TX SUTTON	1.44	UpK6-5	22-Apr-53	3	2
TX SUTTON	6.11	TP6-2	23-Mar-55	4	2
TX SWISHER	1.02	TS4-4	4-May-52	1	8
TX SWISHER	1.02	TS4-4	4-May-52	1	8
TX SWISHER	4.23	UpK6-3	20-Apr-53	2	8
TX SWISHER	25.62	TP7-2	30-Mar-55	4	8
TX SWISHER	31.48	TP10-2	6-May-55	6	8
TX SWISHER	11.19	PB3-2	25-Jun-57	2	8
TX SWISHER	9.99	PB13-4	9-Sep-57	1	8
TX SWISHER	3.21	PB18-6	12-Oct-57	1	8
TX TARRANT	1.83	UpK8-4	11-May-53	4	4
TX TARRANT	12.96	TP6-2	23-Mar-55	2	4
TX TARRANT	12.72	PB12-5	6-Sep-57	2	4
TX TARRANT	2.60	PB18-7	13-Oct-57	1	4

COUNTY	uCi/ sqm	SHOT DAY	DATE	RANK FOR DAY	TIMES IN TOP 10
TX TAYLOR	1.32	TP6-2	23-Mar-55	10	2
TX TAYLOR	1.78	PB1-5	1-Jun-57	8	2
TX TERRELL	6.11	TP6-2	23-Mar-55	4	1
TX TERRY	1.02	TS4-4	4-May-52	1	8
TX TERRY	1.02	TS4-4	4-May-52	1	8
TX TERRY	3.46	TS5-11	17-May-52	1	8
TX TERRY	8.19	UpK3-5	4-Apr-53	2	8
TX TERRY	4.23	UpK6-3	20-Apr-53	2	8
TX TERRY	25.62	TP7-2	30-Mar-55	4	8
TX TERRY	31.48	TP10-2	6-May-55	6	8
TX TERRY	11.19	PB3-2	25-Jun-57	2	8
TX THROCKMORTON	5.43	UpK6-2	19-Apr-53	9	4
TX THROCKMORTON	1.30	UpK6-5	22-Apr-53	4	4
TX THROCKMORTON	1.83	UpK8-4	11-May-53	4	4
TX THROCKMORTON	3.18	TP6-2	23-Mar-55	6	4
TX TITUS	2.58	UpK5-4	14-Apr-53	2	5
TX TITUS	1.12	UpK8-4	11-May-53	9	5
TX TITUS	1.16	UpK8-5	12-May-53	5	5
TX TITUS	12.96	TP6-2	23-Mar-55	2	5
TX TITUS	2.60	PB18-7	13-Oct-57	1	5
TX TOM GREEN	5.43	UpK6-2	19-Apr-53	9	6
TX TOM GREEN	1.58	UpK6-4	21-Apr-53	3	6
TX TOM GREEN	1.30	UpK6-5	22-Apr-53	4	6
TX TOM GREEN	3.18	TP6-2	23-Mar-55	6	6
TX TOM GREEN	1.78	PB1-5	1-Jun-57	8	6
TX TOM GREEN	1.56	PB18-7	13-Oct-57	4	6
TX TRAVIS	4.68	UpK6-2	19-Apr-53	10	3
TX TRAVIS	1.83	UpK8-4	11-May-53	4	3
TX TRAVIS	4.94	PB1-5	1-Jun-57	2	3
TX TRINITY	79.98	UpK6-2	19-Apr-53	1	6
TX TRINITY	2.72	UpK6-3	20-Apr-53	4	6
TX TRINITY	0.80	UpK8-5	12-May-53	7	6
TX TRINITY	6.40	TP6-3	24-Mar-55	3	6
TX TRINITY	2.96	PB1-5	1-Jun-57	5	6
TX TRINITY	2.60	PB18-7	13-Oct-57	1	6

COUNTY	uCi/ sqm	SHOT DAY	DATE	RANK FOR DAY	TIMES IN TOP 10
TX TYLER	79.98	UpK6-2	19-Apr-53	1	6
TX TYLER	2.72	UpK6-3	20-Apr-53	4	6
TX TYLER	1.34	UpK8-5	12-May-53	3	6
TX TYLER	6.40	TP6-3	24-Mar-55	3	6
TX TYLER	2.96	PB1-5	1-Jun-57	5	6
TX TYLER	1.07	PB9-8	30-Aug-57	2	6
TX UPSHUR	3.91	UpK5-4	14-Apr-53	1	4
TX UPSHUR	1.83	UpK8-4	11-May-53	4	4
TX UPSHUR	12.96	TP6-2	23-Mar-55	2	4
TX UPSHUR	2.60	PB18-7	13-Oct-57	1	4
TX UPTON	6.11	TP6-2	23-Mar-55	4	1
TX UVALDE	6.11	TP6-2	23-Mar-55	4	2
TX UVALDE	4.94	PB1-5	1-Jun-57	2	2
TX VAL VERDE	2.54	TP6-2	23-Mar-55	7	2
TX VAL VERDE	1.66	PB1-5	1-Jun-57	9	2
TX VAN ZANDT	3.04	UpK8-4	11-May-53	2	5
TX VAN ZANDT	12.96	TP6-2	23-Mar-55	2	5
TX VAN ZANDT	2.96	PB1-5	1-Jun-57	5	5
TX VAN ZANDT	1.63	PB5-7	21-Jul-57	8	5
TX VAN ZANDT	2.60	PB18-7	13-Oct-57	1	5
TX VICTORIA	4.68	UpK6-2	19-Apr-53	10	1
TX WALKER	4.02	BJ3-5	5-Nov-51	8	7
TX WALKER	79.98	UpK6-2	19-Apr-53	1	7
TX WALKER	2.72	UpK6-3	20-Apr-53	4	7
TX WALKER	1.34	UpK8-5	12-May-53	3	7
TX WALKER	6.40	TP6-3	24-Mar-55	3	7
TX WALKER	2.96	PB1-5	1-Jun-57	5	7
TX WALKER	2.60	PB18-7	13-Oct-57	1	7
TX WALLER	4.79	BJ3-5	5-Nov-51	5	6
TX WALLER	79.98	UpK6-2	19-Apr-53	1	6
TX WALLER	2.72	UpK6-3	20-Apr-53	4	6
TX WALLER	0.80	UpK8-5	12-May-53	7	6
TX WALLER	6.40	TP6-3	24-Mar-55	3	6
TX WALLER	8.26	PB1-5	1-Jun-57	1	6
TX WARD	16.75	UpK6-2	19-Apr-53	6	6

COUNTY	uCi/ sqm	SHOT DAY	DATE	RANK FOR DAY	TIMES IN TOP 10
TX WARD	1.51	UpK6-3	20-Apr-53	8	6
TX WARD	1.25	UpK8-5	12-May-53	4	6
TX WARD	5.38	TP7-2	30-Mar-55	10	6
TX WARD	11.19	PB3-2	25-Jun-57	2	6
TX WARD	2.39	PB5-5	19-Jul-57	6	6
TX WASHINGTON	4.79	BJ3-5	5-Nov-51	5	5
TX WASHINGTON	79.98	UpK6-2	19-Apr-53	1	5
TX WASHINGTON	2.72	UpK6-3	20-Apr-53	4	5
TX WASHINGTON	1.34	UpK8-5	12-May-53	3	5
TX WASHINGTON	6.40	TP6-3	24-Mar-55	3	5
TX WEBB	4.68	UpK6-2	19-Apr-53	10	1
TX WHARTON	4.68	UpK6-2	19-Apr-53	10	2
TX WHARTON	8.26	PB1-5	1-Jun-57	1	2
TX WHEELER	3.70	BJ5-6	24-Nov-51	2	11
TX WHEELER	3.70	BJ5-6	24-Nov-51	2	11
TX WHEELER	1.02	TS4-4	4-May-52	1	11
TX WHEELER	1.02	TS4-4	4-May-52	1	11
TX WHEELER	13.66	UpK3-5	4-Apr-53	1	11
TX WHEELER	4.23	UpK6-3	20-Apr-53	2	11
TX WHEELER	5.42	UpK9-5	23-May-53	2	11
TX WHEELER	51.23	TP7-2	30-Mar-55	3	11
TX WHEELER	31.48	TP10-2	6-May-55	6	11
TX WHEELER	0.81	PB14-7	20-Sep-57	9	11
TX WHEELER	3.21	PB18-6	12-Oct-57	1	11
TX WICHITA	5.43	UpK6-2	19-Apr-53	9	8
TX WICHITA	1.30	UpK6-5	22-Apr-53	4	8
TX WICHITA	1.83	UpK8-4	11-May-53	4	8
TX WICHITA	3.18	TP6-2	23-Mar-55	6	8
TX WICHITA	1.78	PB1-5	1-Jun-57	8	8
TX WICHITA	1.28	PB13-8	13-Sep-57	6	8
TX WICHITA	2.50	PB13-9	14-Sep-57	8	8
TX WICHITA	3.21	PB18-6	12-Oct-57	1	8
TX WILBARGER	5.43	UpK6-2	19-Apr-53	9	7
TX WILBARGER	2.16	UpK6-5	22-Apr-53	1	7
TX WILBARGER	3.18	TP6-2	23-Mar-55	6	7

COUNTY	uCi/ sqm	SHOT DAY	DATE	RANK FOR DAY	TIMES IN TOP 10
TX WILBARGER	2.96	PB1-5	1-Jun-57	5	7
TX WILBARGER	1.63	PB5-7	21-Jul-57	8	7
TX WILBARGER	2.50	PB13-9	14-Sep-57	8	7
TX WILBARGER	1.29	PB18-6	12-Oct-57	6	7
TX WILLACY	4.68	UpK6-2	19-Apr-53	10	2
TX WILLACY	4.94	PB1-5	1-Jun-57	2	2
TX WILLIAMSON	1.83	UpK8-4	11-May-53	4	4
TX WILLIAMSON	12.96	TP6-2	23-Mar-55	2	4
TX WILLIAMSON	2.96	PB1-5	1-Jun-57	5	4
TX WILLIAMSON	2.60	PB18-7	13-Oct-57	1	4
TX WILSON	4.68	UpK6-2	19-Apr-53	10	2
TX WILSON	3.32	PB1-5	1-Jun-57	4	2
TX WINKLER	1.15	TS5-11	17-May-52	3	6
TX WINKLER	16.75	UpK6-2	19-Apr-53	6	6
TX WINKLER	1.51	UpK6-3	20-Apr-53	8	6
TX WINKLER	2.50	UpK8-5	12-May-53	1	6
TX WINKLER	5.38	TP7-2	30-Mar-55	10	6
TX WINKLER	11.19	PB3-2	25-Jun-57	2	6
TX WISE	5.74	UpK1-9	25-Mar-53	8	6
TX WISE	1.83	UpK8-4	11-May-53	4	6
TX WISE	12.96	TP6-2	23-Mar-55	2	6
TX WISE	14.08	TP11-5	19-May-55	6	6
TX WISE	2.50	PB13-9	14-Sep-57	8	6
TX WISE	2.60	PB18-7	13-Oct-57	1	6
TX WOOD	3.04	UpK8-4	11-May-53	2	5
TX WOOD	1.16	UpK8-5	12-May-53	5	5
TX WOOD	12.96	TP6-2	23-Mar-55	2	5
TX WOOD	1.63	PB5-7	21-Jul-57	8	5
TX WOOD	2.60	PB18-7	13-Oct-57	1	5
TX YOAKUM	3.46	TS5-11	17-May-52	1	6
TX YOAKUM	16.75	UpK6-2	19-Apr-53	6	6
TX YOAKUM	1.51	UpK6-3	20-Apr-53	8	6
TX YOAKUM	2.50	UpK8-5	12-May-53	1	6
TX YOAKUM	5.38	TP7-2	30-Mar-55	10	6
TX YOAKUM	11.19	PB3-2	25-Jun-57	2	6

COUNTY	uCi/ sqm	SHOT DAY	DATE	RANK FOR DAY	TIMES IN TOP 10
TX YOUNG	5.43	UpK6-2	19-Apr-53	9	7
TX YOUNG	0.87	UpK6-5	22-Apr-53	8	7
TX YOUNG	3.04	UpK8-4	11-May-53	2	7
TX YOUNG	3.18	TP6-2	23-Mar-55	6	7
TX YOUNG	1.78	PB1-5	1-Jun-57	8	7
TX YOUNG	1.64	PB13-9	14-Sep-57	10	7
TX YOUNG	1.04	PB18-7	13-Oct-57	7	7
TX ZAPATA	4.68	UpK6-2	19-Apr-53	10	1
TX ZAVALA	6.11	TP6-2	23-Mar-55	4	2
TX ZAVALA	1.66	PB1-5	1-Jun-57	9	2

COUNTY	uCi/ sqm	SHOT DAY	DATE	RANK FOR DAY	TIMES IN TOP 10
UT BEAVER	5.86	TS4-1	1-May-52	5	10
UT BEAVER	5.86	TS4-1	1-May-52	5	10
UT BEAVER	119.89	TS6-1	25-May-52	6	10
UT BEAVER	5.51	UpK5-1	11-Apr-53	7	10
UT BEAVER	421.26	TP9-1	15-Apr-55	2	10
UT BEAVER	20.88	PB2-1	18-Jun-57	7	10
UT BEAVER	11.52	PB15-1	16-Sep-57	3	10
UT BEAVER	14.62	PB18-1	7-Oct-57	5	10
UT BEAVER	113.24	UE13-1	25-Apr-66	1	10
UT BEAVER	13.49	UE16-1	8-Dec-68	6	10
UT BOX ELDER1	2.72	BJ6-1	29-Nov-51	10	7
UT BOX ELDER1	2.72	BJ6-1	29-Nov-51	10	7
UT BOX ELDER1	138.46	TS7-1	1-Jun-52	4	7
UT BOX ELDER1	40.59	PB8-1	18-Aug-57	9	7
UT BOX ELDER1	11.30	PB15-1	16-Sep-57	4	7
UT BOX ELDER1	23.90	PB17-1	28-Sep-57	2	7
UT BOX ELDER1	0.99	BN-1	18-Dec-70	10	7
UT BOX ELDER2	417.17	TS5-1	7-May-52	2	8
UT BOX ELDER2	40.59	PB8-1	18-Aug-57	9	8
UT BOX ELDER2	43.10	PB9-1	23-Aug-57	10	8
UT BOX ELDER2	11.30	PB15-1	16-Sep-57	4	8
UT BOX ELDER2	23.90	PB17-1	28-Sep-57	2	8
UT BOX ELDER2	11.17	S-1	6-Jul-62	6	8
UT BOX ELDER2	3.89	UE13-1	25-Apr-66	10	8
UT BOX ELDER2	1.04	BN-1	18-Dec-70	9	8
UT CACHE	333.77	TS5-1	7-May-52	8	6
UT CACHE	536.48	UpK2-1	24-Mar-53	9	6
UT CACHE	11.30	PB15-1	16-Sep-57	4	6
UT CACHE	23.90	PB17-1	28-Sep-57	2	6
UT CACHE	3.89	UE13-1	25-Apr-66	10	6
UT CACHE	0.99	BN-1	18-Dec-70	10	6
UT CARBON	33.30	UpK10-1	25-May-53	6	2
UT CARBON	13.49	UE16-1	8-Dec-68	6	2
UT DAGGETT	240.97	TP10-1	5-May-55	7	2
UT DAGGETT	232.43	PB11-1	31-Aug-57	9	2

COUNTY	uCi/ sqm	SHOT DAY	DATE	RANK FOR DAY	TIMES IN TOP 10
UT DAVIS	375.47	TS5-1	7-May-52	3	10
UT DAVIS	723.08	UpK2-1	24-Mar-53	2	10
UT DAVIS	31.36	TP8-1	9-Apr-55	6	10
UT DAVIS	49.94	PB9-1	23-Aug-57	8	10
UT DAVIS	34.59	PB12-1	2-Sep-57	8	10
UT DAVIS	11.30	PB15-1	16-Sep-57	4	10
UT DAVIS	23.90	PB17-1	28-Sep-57	2	10
UT DAVIS	11.17	S-1	6-Jul-62	6	10
UT DAVIS	3.89	UE13-1	25-Apr-66	10	10
UT DAVIS	1.04	BN-1	18-Dec-70	9	10
UT DUCHESNE	33.45	UpK10-1	25-May-53	5	5
UT DUCHESNE	243.58	TP10-1	5-May-55	6	5
UT DUCHESNE	52.28	PB9-1	23-Aug-57	6	5
UT DUCHESNE	117.07	UE8-1	14-Jul-62	1	5
UT DUCHESNE	9.60	UE16-1	8-Dec-68	9	5
UT EMERY	91.50	TS6-1	25-May-52	8	3
UT EMERY	337.07	TP9-1	15-Apr-55	4	3
UT EMERY	10.98	UE8-1	14-Jul-62	8	3
UT GARFIELD	33.82	UpK11-1	4-Jun-53	8	7
UT GARFIELD	38.02	TP3-1	1-Mar-55	7	7
UT GARFIELD	75.38	TP7-1	29-Mar-55	4	7
UT GARFIELD	14.67	PB2-1	18-Jun-57	10	7
UT GARFIELD	26.86	PB3-1	24-Jun-57	9	7
UT GARFIELD	1238.14	PB11-1	31-Aug-57	1	7
UT GARFIELD	10.18	PB18-1	7-Oct-57	6	7
UT GRAND	82.10	TS6-1	25-May-52	10	2
UT GRAND	337.07	TP9-1	15-Apr-55	4	2
UT IRON1	13.73	TS4-1	1-May-52	1	11
UT IRON1	13.73	TS4-1	1-May-52	1	11
UT IRON1	509.71	TS6-1	25-May-52	2	11
UT IRON1	6.53	UpK5-1	11-Apr-53	4	11
UT IRON1	26.88	TP5-1	12-Mar-55	9	11
UT IRON1	72.13	TP7-1	29-Mar-55	6	11
UT IRON1	87.97	TP9-1	15-Apr-55	10	11
UT IRON1	43.12	PB2-1	18-Jun-57	1	11

COUNTY	uCi/ sqm	SHOT DAY	DATE	RANK FOR DAY	TIMES IN TOP 10
UT IRON1	59.51	PB3-1	24-Jun-57	3	11
UT IRON1	21.54	PB18-1	7-Oct-57	1	11
UT IRON1	13.11	UE8-1	14-Jul-62	6	11
UT IRON2	8.39	TS4-1	1-May-52	2	8
UT IRON2	8.39	TS4-1	1-May-52	2	8
UT IRON2	130.89	TS6-1	25-May-52	4	8
UT IRON2	5.20	UpK5-1	11-Apr-53	9	8
UT IRON2	67.79	TP7-1	29-Mar-55	7	8
UT IRON2	413.43	TP11-1	15-May-55	1	8
UT IRON2	21.00	PB2-1	18-Jun-57	6	8
UT IRON2	15.50	PB18-1	7-Oct-57	3	8
UT IRON3	4.86	TS4-1	1-May-52	6	8
UT IRON3	4.86	TS4-1	1-May-52	6	8
UT IRON3	6.23	UpK5-1	11-Apr-53	5	8
UT IRON3	36.62	TP3-1	1-Mar-55	8	8
UT IRON3	95.86	TP7-1	29-Mar-55	2	8
UT IRON3	180.05	TP11-1	15-May-55	4	8
UT IRON3	32.43	PB2-1	18-Jun-57	2	8
UT IRON3	15.50	PB18-1	7-Oct-57	3	8
UT JUAB	3.63	BJ6-1	29-Nov-51	9	11
UT JUAB	3.63	BJ6-1	29-Nov-51	9	11
UT JUAB	33.74	UpK10-1	25-May-53	4	11
UT JUAB	118.61	TP4-1	7-Mar-55	7	11
UT JUAB	164.19	TP10-1	5-May-55	10	11
UT JUAB	69.20	PB5-1	15-Jul-57	7	11
UT JUAB	33.93	PB8-1	18-Aug-57	10	11
UT JUAB	54.07	PB9-1	23-Aug-57	4	11
UT JUAB	21.81	PB15-1	16-Sep-57	1	11
UT JUAB	7.28	UE13-1	25-Apr-66	7	11
UT JUAB	16.96	UE16-1	8-Dec-68	5	11
UT KANE1	2334.05	UpK9-1	19-May-53	4	7
UT KANE1	47.69	UpK11-1	4-Jun-53	3	7
UT KANE1	64.42	TP3-1	1-Mar-55	4	7
UT KANE1	35.28	TP5-1	12-Mar-55	7	7
UT KANE1	31.36	TP6-1	22-Mar-55	6	7

COUNTY	uCi/ sqm	SHOT DAY	DATE	RANK FOR DAY	TIMES IN TOP 10
UT KANE1	33.75	TP7-1	29-Mar-55	10	7
UT KANE1	5.30	UE13-1	25-Apr-66	8	7
UT KANE2	212.16	UpK1-1	17-Mar-53	3	6
UT KANE2	2334.05	UpK9-1	19-May-53	5	6
UT KANE2	36.40	UpK11-1	4-Jun-53	6	6
UT KANE2	91.68	TP3-1	1-Mar-55	2	6
UT KANE2	47.74	TP7-1	29-Mar-55	8	6
UT KANE2	67.77	PB3-1	24-Jun-57	2	6
UT MILLARD	4.86	TS4-1	1-May-52	6	13
UT MILLARD	4.86	TS4-1	1-May-52	6	13
UT MILLARD	120.23	TS6-1	25-May-52	5	13
UT MILLARD	34.46	UpK10-1	25-May-53	2	13
UT MILLARD	119.47	TP4-1	7-Mar-55	6	13
UT MILLARD	26.40	TP8-1	9-Apr-55	10	13
UT MILLARD	87.66	TP11-1	15-May-55	9	13
UT MILLARD	43.27	PB5-1	15-Jul-57	9	13
UT MILLARD	45.16	PB12-1	2-Sep-57	4	13
UT MILLARD	11.52	PB15-1	16-Sep-57	3	13
UT MILLARD	14.49	UE8-1	14-Jul-62	5	13
UT MILLARD	18.87	UE13-1	25-Apr-66	6	13
UT MILLARD	25.45	UE16-1	8-Dec-68	4	13
UT MORGAN	375.47	TS5-1	7-May-52	3	10
UT MORGAN	725.72	UpK2-1	24-Mar-53	1	10
UT MORGAN	101.37	TP4-1	7-Mar-55	10	10
UT MORGAN	33.47	TP8-1	9-Apr-55	4	10
UT MORGAN	36.92	PB12-1	2-Sep-57	5	10
UT MORGAN	11.30	PB15-1	16-Sep-57	4	10
UT MORGAN	23.90	PB17-1	28-Sep-57	2	10
UT MORGAN	11.17	S-1	6-Jul-62	6	10
UT MORGAN	3.89	UE13-1	25-Apr-66	10	10
UT MORGAN	1.04	BN-1	18-Dec-70	9	10
UT PIUTE	92.16	TS6-1	25-May-52	7	9
UT PIUTE	41.63	TP7-1	29-Mar-55	9	9
UT PIUTE	337.07	TP9-1	15-Apr-55	4	9
UT PIUTE	82.13	TP11-1	15-May-55	10	9

COUNTY	uCi/ sqm	SHOT DAY	DATE	RANK FOR DAY	TIMES IN TOP 10
UT PIUTE	24.45	PB2-1	18-Jun-57	4	9
UT PIUTE	8.50	PB15-1	16-Sep-57	8	9
UT PIUTE	10.18	PB18-1	7-Oct-57	7	9
UT PIUTE	40.29	UE13-1	25-Apr-66	3	9
UT PIUTE	8.99	UE16-1	8-Dec-68	10	9
UT RICH	417.17	TS5-1	7-May-52	2	6
UT RICH	623.61	UpK2-1	24-Mar-53	5	6
UT RICH	11.30	PB15-1	16-Sep-57	4	6
UT RICH	23.90	PB17-1	28-Sep-57	2	6
UT RICH	3.89	UE13-1	25-Apr-66	10	6
UT RICH	1.04	BN-1	18-Dec-70	9	6
UT SALT LAKE	315.94	TS5-1	7-May-52	9	12
UT SALT LAKE	598.87	UpK2-1	24-Mar-53	6	12
UT SALT LAKE	118.48	TP4-1	7-Mar-55	7	12
UT SALT LAKE	37.24	TP8-1	9-Apr-55	2	12
UT SALT LAKE	42.33	PB5-1	15-Jul-57	10	12
UT SALT LAKE	47.10	PB9-1	23-Aug-57	9	12
UT SALT LAKE	32.62	PB12-1	2-Sep-57	10	12
UT SALT LAKE	13.29	PB15-1	16-Sep-57	2	12
UT SALT LAKE	24.38	PB17-1	28-Sep-57	1	12
UT SALT LAKE	11.17	S-1	6-Jul-62	6	12
UT SALT LAKE	3.89	UE13-1	25-Apr-66	10	12
UT SALT LAKE	1.91	BN-1	18-Dec-70	7	12
UT SAN JUAN	64.71	UpK1-1	17-Mar-53	5	3
UT SAN JUAN	32.98	UpK11-1	4-Jun-53	10	3
UT SAN JUAN	9.94	PB18-1	7-Oct-57	9	3
UT SANPETE	34.17	UpK10-1	25-May-53	3	5
UT SANPETE	8.50	PB15-1	16-Sep-57	8	5
UT SANPETE	11.71	UE8-1	14-Jul-62	7	5
UT SANPETE	39.58	UE13-1	25-Apr-66	4	5
UT SANPETE	26.98	UE16-1	8-Dec-68	3	5
UT SEVIER	92.16	TS6-1	25-May-52	7	8
UT SEVIER	34.17	UpK10-1	25-May-53	3	8
UT SEVIER	252.77	TP9-1	15-Apr-55	6	8
UT SEVIER	14.67	PB2-1	18-Jun-57	10	8

COUNTY	uCi/ sqm	SHOT DAY	DATE	RANK FOR DAY	TIMES IN TOP 10
UT SEVIER	9.40	PB15-1	16-Sep-57	7	8
UT SEVIER	9.51	UE8-1	14-Jul-62	10	8
UT SEVIER	74.83	UE13-1	25-Apr-66	2	8
UT SEVIER	13.49	UE16-1	8-Dec-68	6	8
UT SUMMIT	333.77	TS5-1	7-May-52	8	8
UT SUMMIT	116.91	TP4-1	7-Mar-55	9	8
UT SUMMIT	31.52	TP8-1	9-Apr-55	5	8
UT SUMMIT	33.85	PB12-1	2-Sep-57	9	8
UT SUMMIT	11.30	PB15-1	16-Sep-57	4	8
UT SUMMIT	15.85	PB17-1	28-Sep-57	6	8
UT SUMMIT	3.89	UE13-1	25-Apr-66	10	8
UT SUMMIT	0.99	BN-1	18-Dec-70	10	8
UT TOOELE1	5.89	BJ6-1	29-Nov-51	7	7
UT TOOELE1	5.89	BJ6-1	29-Nov-51	7	7
UT TOOELE1	868.71	TS5-1	7-May-52	1	7
UT TOOELE1	84.89	PB8-1	18-Aug-57	6	7
UT TOOELE1	36.17	PB12-1	2-Sep-57	6	7
UT TOOELE1	23.90	PB17-1	28-Sep-57	2	7
UT TOOELE1	0.99	BN-1	18-Dec-70	10	7
UT TOOELE2	2.72	BJ6-1	29-Nov-51	10	14
UT TOOELE2	2.72	BJ6-1	29-Nov-51	10	14
UT TOOELE2	418.68	TS5-1	7-May-52	2	14
UT TOOELE2	635.00	UpK2-1	24-Mar-53	4	14
UT TOOELE2	39.49	TP8-1	9-Apr-55	1	14
UT TOOELE2	42.39	PB8-1	18-Aug-57	8	14
UT TOOELE2	53.43	PB9-1	23-Aug-57	5	14
UT TOOELE2	34.59	PB12-1	2-Sep-57	8	14
UT TOOELE2	11.30	PB15-1	16-Sep-57	4	14
UT TOOELE2	23.90	PB17-1	28-Sep-57	2	14
UT TOOELE2	22.33	S-1	6-Jul-62	5	14
UT TOOELE2	3.89	UE13-1	25-Apr-66	10	14
UT TOOELE2	9.60	UE16-1	8-Dec-68	9	14
UT TOOELE2	1.04	BN-1	18-Dec-70	9	14
UT UINTAH	240.97	TP10-1	5-May-55	7	3
UT UINTAH	51.90	PB9-1	23-Aug-57	7	3

COUNTY	uCi/ sqm	SHOT DAY	DATE	RANK FOR DAY	TIMES IN TOP 10
UT UINTAH	235.73	PB11-1	31-Aug-57	8	3
UT UTAH	335.00	TS5-1	7-May-52	7	13
UT UTAH	33.74	UpK10-1	25-May-53	4	13
UT UTAH	157.55	TP4-1	7-Mar-55	4	13
UT UTAH	29.71	TP8-1	9-Apr-55	7	13
UT UTAH	246.19	TP10-1	5-May-55	3	13
UT UTAH	84.67	PB5-1	15-Jul-57	6	13
UT UTAH	40.59	PB8-1	18-Aug-57	9	13
UT UTAH	103.16	PB9-1	23-Aug-57	1	13
UT UTAH	46.17	PB12-1	2-Sep-57	3	13
UT UTAH	11.30	PB15-1	16-Sep-57	4	13
UT UTAH	15.85	PB17-1	28-Sep-57	6	13
UT UTAH	26.98	UE16-1	8-Dec-68	3	13
UT UTAH	1.04	BN-1	18-Dec-70	9	13
UT WASATCH	250.75	TS5-1	7-May-52	10	12
UT WASATCH	117.38	TP4-1	7-Mar-55	8	12
UT WASATCH	33.74	TP8-1	9-Apr-55	3	12
UT WASATCH	245.39	TP10-1	5-May-55	5	12
UT WASATCH	66.27	PB5-1	15-Jul-57	8	12
UT WASATCH	74.47	PB9-1	23-Aug-57	3	12
UT WASATCH	36.13	PB12-1	2-Sep-57	7	12
UT WASATCH	11.30	PB15-1	16-Sep-57	4	12
UT WASATCH	15.85	PB17-1	28-Sep-57	6	12
UT WASATCH	38.78	UE8-1	14-Jul-62	2	12
UT WASATCH	3.89	UE13-1	25-Apr-66	10	12
UT WASATCH	9.60	UE16-1	8-Dec-68	9	12
UT WASHINGTON1	2758.94	UpK9-1	19-May-53	3	6
UT WASHINGTON1	92.98	TP7-1	29-Mar-55	3	6
UT WASHINGTON1	188.87	PB3-1	24-Jun-57	1	6
UT WASHINGTON1	318.51	PB11-1	31-Aug-57	7	6
UT WASHINGTON1	20.12	PB18-1	7-Oct-57	2	6
UT WASHINGTON1	10.51	UE8-1	14-Jul-62	9	6
UT WASHINGTON2	1293.84	UpK1-1	17-Mar-53	1	8
UT WASHINGTON2	2.74	UpK4-1	6-Apr-53	1	8
UT WASHINGTON2	4392.11	UpK9-1	19-May-53	1	8

COUNTY	uCi/ sqm	SHOT DAY	DATE	RANK FOR DAY	TIMES IN TOP 10
UT WASHINGTON2	254.31	TP3-1	1-Mar-55	1	8
UT WASHINGTON2	1.73	TP6-2	23-Mar-55	8	8
UT WASHINGTON2	104.21	TP11-1	15-May-55	7	8
UT WASHINGTON2	27.52	PB3-1	24-Jun-57	8	8
UT WASHINGTON2	641.06	PB11-1	31-Aug-57	3	8
UT WASHINGTON3	591.85	UpK1-1	17-Mar-53	2	6
UT WASHINGTON3	3642.39	UpK9-1	19-May-53	2	6
UT WASHINGTON3	68.79	TP3-1	1-Mar-55	3	6
UT WASHINGTON3	52.16	PB3-1	24-Jun-57	5	6
UT WASHINGTON3	500.53	PB11-1	31-Aug-57	4	6
UT WASHINGTON3	15.36	PB18-1	7-Oct-57	4	6
UT WAYNE	82.99	TS6-1	25-May-52	9	3
UT WAYNE	19.44	PB2-1	18-Jun-57	9	3
UT WAYNE	10.18	PB18-1	7-Oct-57	7	3
UT WEBER	417.17	TS5-1	7-May-52	2	7
UT WEBER	679.45	UpK2-1	24-Mar-53	3	7
UT WEBER	11.30	PB15-1	16-Sep-57	4	7
UT WEBER	23.90	PB17-1	28-Sep-57	2	7
UT WEBER	11.17	S-1	6-Jul-62	6	7
UT WEBER	3.89	UE13-1	25-Apr-66	10	7
UT WEBER	1.04	BN-1	18-Dec-70	9	7

COUNTY	uCi/ sqm	SHOT DAY	DATE	RANK FOR DAY	TIMES IN TOP 10
VA ACCOMACK	11.58	BJ3-6	6-Nov-51	3	6
VA ACCOMACK	3.43	BJ4-7	11-Nov-51	1	6
VA ACCOMACK	1.51	PB5-8	22-Jul-57	7	6
VA ACCOMACK	4.72	PB5-9	23-Jul-57	1	6
VA ACCOMACK	1.51	PB8-7	24-Aug-57	10	6
VA ACCOMACK	66.88	S-6	11-Jul-62	1	6
VA ALBEMARLE	1.28	PB2-8	25-Jun-57	10	3
VA ALBEMARLE	1.51	PB5-8	22-Jul-57	7	3
VA ALBEMARLE	2.83	PB5-9	23-Jul-57	2	3
VA ALLEGHANY	1.18	UpK8-6	13-May-53	2	5
VA ALLEGHANY	2.99	PB5-8	22-Jul-57	4	5
VA ALLEGHANY	2.83	PB5-9	23-Jul-57	2	5
VA ALLEGHANY	85.56	S-4	9-Jul-62	9	5
VA ALLEGHANY	1.74	UE8-5	18-Jul-62	10	5
VA AMELIA	3.89	BJ2-2	31-Oct-51	10	4
VA AMELIA	8.96	PB5-8	22-Jul-57	2	4
VA AMELIA	2.54	PB8-7	24-Aug-57	7	4
VA AMELIA	149.13	S-4	9-Jul-62	5	4
VA AMHERST	1.23	UpK3-7	6-Apr-53	10	4
VA AMHERST	1.28	PB2-8	25-Jun-57	10	4
VA AMHERST	2.99	PB5-8	22-Jul-57	4	4
VA AMHERST	2.83	PB5-9	23-Jul-57	2	4
VA APPOMATTOX	6.42	BJ2-2	31-Oct-51	6	5
VA APPOMATTOX	1.79	TS5-4	10-May-52	2	5
VA APPOMATTOX	1.23	UpK3-7	6-Apr-53	10	5
VA APPOMATTOX	1.51	PB5-8	22-Jul-57	7	5
VA APPOMATTOX	2.83	PB5-9	23-Jul-57	2	5
VA ARLINGTON	21.07	BJ2-3	1-Nov-51	3	4
VA ARLINGTON	5.97	PB5-8	22-Jul-57	3	4
VA ARLINGTON	0.94	PB5-9	23-Jul-57	4	4
VA ARLINGTON	2.54	PB8-7	24-Aug-57	7	4
VA AUGUSTA	4.59	BJ3-5	5-Nov-51	6	4
VA AUGUSTA	1.18	UpK8-6	13-May-53	2	4
VA AUGUSTA	8.96	PB5-8	22-Jul-57	2	4
VA AUGUSTA	2.83	PB5-9	23-Jul-57	2	4

COUNTY	uCi/ sqm	SHOT DAY	DATE	RANK FOR DAY	TIMES IN TOP 10
VA BATH	1.55	TS5-4	10-May-52	6	4
VA BATH	8.96	PB5-8	22-Jul-57	2	4
VA BATH	2.83	PB5-9	23-Jul-57	2	4
VA BATH	208.78	S-4	9-Jul-62	3	4
VA BEDFORD	1.18	BJ1-2	29-Oct-51	4	7
VA BEDFORD	9.19	BJ3-5	5-Nov-51	2	7
VA BEDFORD	1.69	TS5-4	10-May-52	4	7
VA BEDFORD	1.62	PB2-9	26-Jun-57	10	7
VA BEDFORD	8.96	PB5-8	22-Jul-57	2	7
VA BEDFORD	2.83	PB5-9	23-Jul-57	2	7
VA BEDFORD	2.58	UE8-6	19-Jul-62	4	7
VA BLAND	2.78	PB2-5	22-Jun-57	4	3
VA BLAND	1.51	PB5-8	22-Jul-57	7	3
VA BLAND	1.89	PB5-9	23-Jul-57	3	3
VA BOTETOURT	1.23	UpK3-7	6-Apr-53	10	4
VA BOTETOURT	2.99	PB5-8	22-Jul-57	4	4
VA BOTETOURT	2.83	PB5-9	23-Jul-57	2	4
VA BOTETOURT	208.78	S-4	9-Jul-62	3	4
VA BRUNSWICK	6.42	BJ2-2	31-Oct-51	6	4
VA BRUNSWICK	1.79	TS5-4	10-May-52	2	4
VA BRUNSWICK	8.96	PB5-8	22-Jul-57	2	4
VA BRUNSWICK	2.54	PB8-7	24-Aug-57	7	4
VA BUCHANAN	1.04	UpK5-2	12-Apr-53	4	7
VA BUCHANAN	1.15	UpK6-5	22-Apr-53	5	7
VA BUCHANAN	2.03	TP6-4	25-Mar-55	8	7
VA BUCHANAN	0.97	TP6-5	26-Mar-55	8	7
VA BUCHANAN	2.78	PB2-5	22-Jun-57	4	7
VA BUCHANAN	5.97	PB5-8	22-Jul-57	3	7
VA BUCHANAN	149.13	S-4	9-Jul-62	5	7
VA BUCKINGHAM	6.42	BJ2-2	31-Oct-51	6	6
VA BUCKINGHAM	5.97	PB5-8	22-Jul-57	3	6
VA BUCKINGHAM	2.83	PB5-9	23-Jul-57	2	6
VA BUCKINGHAM	1.51	PB8-7	24-Aug-57	10	6
VA BUCKINGHAM	2.11	UE8-5	18-Jul-62	7	6
VA BUCKINGHAM	2.58	UE8-6	19-Jul-62	4	6

COUNTY	uCi/ sqm	SHOT DAY	DATE	RANK FOR DAY	TIMES IN TOP 10
VA CAMPBELL	6.42	BJ2-2	31-Oct-51	6	5
VA CAMPBELL	1.92	PB2-8	25-Jun-57	7	5
VA CAMPBELL	5.97	PB5-8	22-Jul-57	3	5
VA CAMPBELL	2.83	PB5-9	23-Jul-57	2	5
VA CAMPBELL	0.30	UE8-8	21-Jul-62	3	5
VA CAROLINE	1.36	TS5-4	10-May-52	10	4
VA CAROLINE	1.92	PB2-8	25-Jun-57	7	4
VA CAROLINE	1.51	PB5-8	22-Jul-57	7	4
VA CAROLINE	1.51	PB8-7	24-Aug-57	10	4
VA CARROLL	6.42	BJ2-2	31-Oct-51	6	5
VA CARROLL	1.62	PB2-9	26-Jun-57	10	5
VA CARROLL	8.96	PB5-8	22-Jul-57	2	5
VA CARROLL	2.83	PB5-9	23-Jul-57	2	5
VA CARROLL	94.45	S-4	9-Jul-62	7	5
VA CHARLES CITY	11.58	BJ3-6	6-Nov-51	3	5
VA CHARLES CITY	3.43	BJ4-7	11-Nov-51	1	5
VA CHARLES CITY	2.99	PB5-8	22-Jul-57	4	5
VA CHARLES CITY	2.54	PB8-7	24-Aug-57	7	5
VA CHARLES CITY	2.58	UE8-6	19-Jul-62	4	5
VA CHARLOTTE	6.42	BJ2-2	31-Oct-51	6	5
VA CHARLOTTE	1.51	PB5-8	22-Jul-57	7	5
VA CHARLOTTE	2.83	PB5-9	23-Jul-57	2	5
VA CHARLOTTE	2.54	PB8-7	24-Aug-57	7	5
VA CHARLOTTE	3.66	PB12-8	9-Sep-57	4	5
VA CHESTERFIELD	6.95	BJ3-6	6-Nov-51	6	5
VA CHESTERFIELD	3.43	BJ4-7	11-Nov-51	1	5
VA CHESTERFIELD	2.48	PB5-8	22-Jul-57	5	5
VA CHESTERFIELD	1.64	PB17-8	5-Oct-57	1	5
VA CHESTERFIELD	149.13	S-4	9-Jul-62	5	5
VA CLARKE	1.04	UpK4-4	9-Apr-53	5	4
VA CLARKE	8.96	PB5-8	22-Jul-57	2	4
VA CLARKE	0.94	PB5-9	23-Jul-57	4	4
VA CLARKE	1.51	PB8-7	24-Aug-57	10	4
VA CRAIG	1.55	TS5-4	10-May-52	6	4
VA CRAIG	1.51	PB5-8	22-Jul-57	7	4

COUNTY	uCi/ sqm	SHOT DAY	DATE	RANK FOR DAY	TIMES IN TOP 10
VA CRAIG	2.83	PB5-9	23-Jul-57	2	4
VA CRAIG	90.29	S-4	9-Jul-62	8	4
VA CULPEPER	8.96	PB5-8	22-Jul-57	2	4
VA CULPEPER	1.89	PB5-9	23-Jul-57	3	4
VA CULPEPER	2.54	PB8-7	24-Aug-57	7	4
VA CULPEPER	2.11	UE8-5	18-Jul-62	7	4
VA CUMBERLAND	6.42	BJ2-2	31-Oct-51	6	5
VA CUMBERLAND	5.97	PB5-8	22-Jul-57	3	5
VA CUMBERLAND	1.51	PB8-7	24-Aug-57	10	5
VA CUMBERLAND	2.11	UE8-5	18-Jul-62	7	5
VA CUMBERLAND	2.58	UE8-6	19-Jul-62	4	5
VA DICKENSON	1.55	UpK5-2	12-Apr-53	2	8
VA DICKENSON	1.15	UpK6-5	22-Apr-53	5	8
VA DICKENSON	2.03	TP6-4	25-Mar-55	8	8
VA DICKENSON	1.93	TP6-5	26-Mar-55	4	8
VA DICKENSON	1.39	PB2-5	22-Jun-57	8	8
VA DICKENSON	2.99	PB5-8	22-Jul-57	4	8
VA DICKENSON	3.79	PB17-3	30-Sep-57	10	8
VA DICKENSON	94.45	S-4	9-Jul-62	7	8
VA DINWIDDIE	6.95	BJ3-6	6-Nov-51	6	5
VA DINWIDDIE	3.43	BJ4-7	11-Nov-51	1	5
VA DINWIDDIE	1.84	TS5-4	10-May-52	1	5
VA DINWIDDIE	8.96	PB5-8	22-Jul-57	2	5
VA DINWIDDIE	2.54	PB8-7	24-Aug-57	7	5
VA ESSEX	1.92	PB2-8	25-Jun-57	7	3
VA ESSEX	2.99	PB5-8	22-Jul-57	4	3
VA ESSEX	2.54	PB8-7	24-Aug-57	7	3
VA FAIRFAX	5.97	PB5-8	22-Jul-57	3	2
VA FAIRFAX	2.54	PB8-7	24-Aug-57	7	2
VA FAUQUIER	2.75	TP11-9	23-May-55	9	4
VA FAUQUIER	2.99	PB5-8	22-Jul-57	4	4
VA FAUQUIER	2.83	PB5-9	23-Jul-57	2	4
VA FAUQUIER	2.70	UE8-6	19-Jul-62	3	4
VA FLOYD	9.19	BJ3-5	5-Nov-51	2	5
VA FLOYD	1.18	UpK8-6	13-May-53	2	5

COUNTY	uCi/ sqm	SHOT DAY	DATE	RANK FOR DAY	TIMES IN TOP 10
VA FLOYD	1.62	PB2-9	26-Jun-57	10	5
VA FLOYD	8.96	PB5-8	22-Jul-57	2	5
VA FLOYD	2.83	PB5-9	23-Jul-57	2	5
VA FLUVANNA	3.20	PB2-8	25-Jun-57	4	5
VA FLUVANNA	5.97	PB5-8	22-Jul-57	3	5
VA FLUVANNA	2.83	PB5-9	23-Jul-57	2	5
VA FLUVANNA	1.17	PB6-8	31-Jul-57	10	5
VA FLUVANNA	2.58	UE8-6	19-Jul-62	4	5
VA FRANKLIN	6.42	BJ2-2	31-Oct-51	6	4
VA FRANKLIN	1.51	PB5-8	22-Jul-57	7	4
VA FRANKLIN	4.72	PB5-9	23-Jul-57	1	4
VA FRANKLIN	149.13	S-4	9-Jul-62	5	4
VA FREDERICK	1.04	UpK4-4	9-Apr-53	5	5
VA FREDERICK	0.96	UpK8-6	13-May-53	3	5
VA FREDERICK	2.99	PB5-8	22-Jul-57	4	5
VA FREDERICK	2.83	PB5-9	23-Jul-57	2	5
VA FREDERICK	1.45	UE8-6	19-Jul-62	8	5
VA GILES	1.51	PB5-8	22-Jul-57	7	2
VA GILES	1.89	PB5-9	23-Jul-57	3	2
VA GLOUCESTER	11.58	BJ3-6	6-Nov-51	3	6
VA GLOUCESTER	3.43	BJ4-7	11-Nov-51	1	6
VA GLOUCESTER	5.97	PB5-8	22-Jul-57	3	6
VA GLOUCESTER	2.54	PB8-7	24-Aug-57	7	6
VA GLOUCESTER	1.21	PB17-8	5-Oct-57	6	6
VA GLOUCESTER	2.58	UE8-6	19-Jul-62	4	6
VA GOOCHLAND	1.92	PB2-8	25-Jun-57	7	5
VA GOOCHLAND	5.97	PB5-8	22-Jul-57	3	5
VA GOOCHLAND	2.83	PB5-9	23-Jul-57	2	5
VA GOOCHLAND	1.51	PB8-7	24-Aug-57	10	5
VA GOOCHLAND	149.13	S-4	9-Jul-62	5	5
VA GRAYSON	1.72	BJ1-2	29-Oct-51	2	4
VA GRAYSON	2.78	PB2-5	22-Jun-57	4	4
VA GRAYSON	1.51	PB5-8	22-Jul-57	7	4
VA GRAYSON	2.83	PB5-9	23-Jul-57	2	4
VA GREENE	1.74	TS5-4	10-May-52	3	5

COUNTY	uCi/ sqm	SHOT DAY	DATE	RANK FOR DAY	TIMES IN TOP 10
VA GREENE	1.92	PB2-8	25-Jun-57	7	5
VA GREENE	1.51	PB5-8	22-Jul-57	7	5
VA GREENE	2.83	PB5-9	23-Jul-57	2	5
VA GREENE	2.87	PB12-8	9-Sep-57	10	5
VA GREENSVILLE	11.58	BJ3-6	6-Nov-51	3	7
VA GREENSVILLE	3.43	BJ4-7	11-Nov-51	1	7
VA GREENSVILLE	1.74	TS5-4	10-May-52	3	7
VA GREENSVILLE	1.28	PB2-8	25-Jun-57	10	7
VA GREENSVILLE	1.51	PB5-8	22-Jul-57	7	7
VA GREENSVILLE	4.72	PB5-9	23-Jul-57	1	7
VA GREENSVILLE	4.62	UE8-6	19-Jul-62	1	7
VA HALIFAX	3.89	BJ2-2	31-Oct-51	10	5
VA HALIFAX	1.28	PB2-8	25-Jun-57	10	5
VA HALIFAX	1.51	PB5-8	22-Jul-57	7	5
VA HALIFAX	4.72	PB5-9	23-Jul-57	1	5
VA HALIFAX	2.54	PB8-7	24-Aug-57	7	5
VA HAMPTON	11.58	BJ3-6	6-Nov-51	3	6
VA HAMPTON	3.43	BJ4-7	11-Nov-51	1	6
VA HAMPTON	8.96	PB5-8	22-Jul-57	2	6
VA HAMPTON	1.51	PB8-7	24-Aug-57	10	6
VA HAMPTON	2.58	UE8-6	19-Jul-62	4	6
VA HAMPTON	0.16	UE8-8	21-Jul-62	6	6
VA HANOVER	1.36	TS5-4	10-May-52	10	6
VA HANOVER	1.06	TP9-10	24-Apr-55	1	6
VA HANOVER	8.96	PB5-8	22-Jul-57	2	6
VA HANOVER	2.54	PB8-7	24-Aug-57	7	6
VA HANOVER	1.59	PB17-8	5-Oct-57	3	6
VA HANOVER	149.13	S-4	9-Jul-62	5	6
VA HENRICO	6.95	BJ3-6	6-Nov-51	6	9
VA HENRICO	3.43	BJ4-7	11-Nov-51	1	9
VA HENRICO	1.33	TS4-5	5-May-52	7	9
VA HENRICO	1.33	TS4-5	5-May-52	7	9
VA HENRICO	1.40	TS5-4	10-May-52	9	9
VA HENRICO	8.96	PB5-8	22-Jul-57	2	9
VA HENRICO	2.54	PB8-7	24-Aug-57	7	9

COUNTY	uCi/ sqm	SHOT DAY	DATE	RANK FOR DAY	TIMES IN TOP 10
VA HENRICO	1.64	PB17-8	5-Oct-57	2	9
VA HENRICO	2.11	UE8-5	18-Jul-62	7	9
VA HENRY	6.42	BJ2-2	31-Oct-51	6	5
VA HENRY	1.32	PB2-8	25-Jun-57	9	5
VA HENRY	2.99	PB5-8	22-Jul-57	4	5
VA HENRY	2.83	PB5-9	23-Jul-57	2	5
VA HENRY	1.32	UE8-6	19-Jul-62	9	5
VA HIGHLAND	1.50	TS5-4	10-May-52	7	4
VA HIGHLAND	1.51	PB5-8	22-Jul-57	7	4
VA HIGHLAND	4.72	PB5-9	23-Jul-57	1	4
VA HIGHLAND	85.56	S-4	9-Jul-62	9	4
VA ISLE OF WIGH	11.58	BJ3-6	6-Nov-51	3	8
VA ISLE OF WIGH	3.43	BJ4-7	11-Nov-51	1	8
VA ISLE OF WIGH	1.04	UpK4-4	9-Apr-53	5	8
VA ISLE OF WIGH	8.96	PB5-8	22-Jul-57	2	8
VA ISLE OF WIGH	4.72	PB5-9	23-Jul-57	1	8
VA ISLE OF WIGH	1.51	PB8-7	24-Aug-57	10	8
VA ISLE OF WIGH	2.58	UE8-6	19-Jul-62	4	8
VA ISLE OF WIGH	0.16	UE8-8	21-Jul-62	6	8
VA JAMES CITY	11.58	BJ3-6	6-Nov-51	3	6
VA JAMES CITY	3.43	BJ4-7	11-Nov-51	1	6
VA JAMES CITY	2.99	PB5-8	22-Jul-57	4	6
VA JAMES CITY	1.51	PB8-7	24-Aug-57	10	6
VA JAMES CITY	3.39	PB12-8	9-Sep-57	6	6
VA JAMES CITY	2.58	UE8-6	19-Jul-62	4	6
VA KING AND QUE	4.31	BJ2-5	3-Nov-51	7	8
VA KING AND QUE	6.95	BJ3-6	6-Nov-51	6	8
VA KING AND QUE	3.43	BJ4-7	11-Nov-51	1	8
VA KING AND QUE	1.28	PB2-8	25-Jun-57	10	8
VA KING AND QUE	8.96	PB5-8	22-Jul-57	2	8
VA KING AND QUE	2.54	PB8-7	24-Aug-57	7	8
VA KING AND QUE	1.17	PB17-8	5-Oct-57	8	8
VA KING AND QUE	2.58	UE8-6	19-Jul-62	4	8
VA KING GEORGE	1.36	TS5-4	10-May-52	10	3
VA KING GEORGE	2.99	PB5-8	22-Jul-57	4	3

COUNTY	uCi/ sqm	SHOT DAY	DATE	RANK FOR DAY	TIMES IN TOP 10
VA KING GEORGE	2.54	PB8-7	24-Aug-57	7	3
VA KING WILLIAM	4.12	BJ2-5	3-Nov-51	8	9
VA KING WILLIAM	6.95	BJ3-6	6-Nov-51	6	9
VA KING WILLIAM	3.43	BJ4-7	11-Nov-51	1	9
VA KING WILLIAM	1.28	PB2-8	25-Jun-57	10	9
VA KING WILLIAM	8.96	PB5-8	22-Jul-57	2	9
VA KING WILLIAM	1.89	PB5-9	23-Jul-57	3	9
VA KING WILLIAM	2.54	PB8-7	24-Aug-57	7	9
VA KING WILLIAM	1.35	PB17-8	5-Oct-57	4	9
VA KING WILLIAM	2.58	UE8-6	19-Jul-62	4	9
VA LANCASTER	11.58	BJ3-6	6-Nov-51	3	7
VA LANCASTER	3.43	BJ4-7	11-Nov-51	1	7
VA LANCASTER	1.92	PB2-8	25-Jun-57	7	7
VA LANCASTER	2.99	PB5-8	22-Jul-57	4	7
VA LANCASTER	2.54	PB8-7	24-Aug-57	7	7
VA LANCASTER	1.12	PB17-8	5-Oct-57	9	7
VA LANCASTER	2.58	UE8-6	19-Jul-62	4	7
VA LEE	1.04	UpK5-2	12-Apr-53	4	6
VA LEE	3.11	TP6-4	25-Mar-55	3	6
VA LEE	0.97	TP6-5	26-Mar-55	8	6
VA LEE	4.18	PB2-5	22-Jun-57	3	6
VA LEE	1.51	PB5-8	22-Jul-57	7	6
VA LEE	3.79	PB17-3	30-Sep-57	10	6
VA LOUDOUN	8.96	PB5-8	22-Jul-57	2	4
VA LOUDOUN	1.89	PB5-9	23-Jul-57	3	4
VA LOUDOUN	1.51	PB8-7	24-Aug-57	10	4
VA LOUDOUN	2.11	UE8-5	18-Jul-62	7	4
VA LOUISA	8.96	PB5-8	22-Jul-57	2	4
VA LOUISA	1.51	PB8-7	24-Aug-57	10	4
VA LOUISA	1.29	PB17-8	5-Oct-57	5	4
VA LOUISA	1.32	UE8-6	19-Jul-62	9	4
VA LUNENBURG	6.42	BJ2-2	31-Oct-51	6	4
VA LUNENBURG	1.28	PB2-8	25-Jun-57	10	4
VA LUNENBURG	1.51	PB5-8	22-Jul-57	7	4
VA LUNENBURG	4.72	PB5-9	23-Jul-57	1	4

COUNTY	uCi/ sqm	SHOT DAY	DATE	RANK FOR DAY	TIMES IN TOP 10
VA MADISON	1.74	TS5-4	10-May-52	3	5
VA MADISON	1.77	UpK8-6	13-May-53	1	5
VA MADISON	8.96	PB5-8	22-Jul-57	2	5
VA MADISON	0.94	PB5-9	23-Jul-57	4	5
VA MADISON	1.95	PB6-8	31-Jul-57	5	5
VA MATHEWS	11.58	BJ3-6	6-Nov-51	3	7
VA MATHEWS	3.43	BJ4-7	11-Nov-51	1	7
VA MATHEWS	5.97	PB5-8	22-Jul-57	3	7
VA MATHEWS	2.83	PB5-9	23-Jul-57	2	7
VA MATHEWS	2.54	PB8-7	24-Aug-57	7	7
VA MATHEWS	1.09	PB17-8	5-Oct-57	10	7
VA MATHEWS	2.58	UE8-6	19-Jul-62	4	7
VA MECKLENBURG	6.42	BJ2-2	31-Oct-51	6	5
VA MECKLENBURG	1.28	PB2-8	25-Jun-57	10	5
VA MECKLENBURG	5.97	PB5-8	22-Jul-57	3	5
VA MECKLENBURG	2.83	PB5-9	23-Jul-57	2	5
VA MECKLENBURG	2.54	PB8-7	24-Aug-57	7	5
VA MIDDLESEX	11.58	BJ3-6	6-Nov-51	3	8
VA MIDDLESEX	3.43	BJ4-7	11-Nov-51	1	8
VA MIDDLESEX	1.92	PB2-8	25-Jun-57	7	8
VA MIDDLESEX	2.99	PB5-8	22-Jul-57	4	8
VA MIDDLESEX	4.72	PB5-9	23-Jul-57	1	8
VA MIDDLESEX	2.54	PB8-7	24-Aug-57	7	8
VA MIDDLESEX	1.18	PB17-8	5-Oct-57	7	8
VA MIDDLESEX	2.58	UE8-6	19-Jul-62	4	8
VA MONTGOMERY	1.51	PB5-8	22-Jul-57	7	2
VA MONTGOMERY	1.89	PB5-9	23-Jul-57	3	2
VA NELSON	1.23	UpK3-7	6-Apr-53	10	5
VA NELSON	1.28	PB2-8	25-Jun-57	10	5
VA NELSON	5.97	PB5-8	22-Jul-57	3	5
VA NELSON	2.83	PB5-9	23-Jul-57	2	5
VA NELSON	1.32	UE8-6	19-Jul-62	9	5
VA NEW KENT	5.01	BJ2-5	3-Nov-51	5	7
VA NEW KENT	6.95	BJ3-6	6-Nov-51	6	7
VA NEW KENT	3.43	BJ4-7	11-Nov-51	1	7

COUNTY	uCi/ sqm	SHOT DAY	DATE	RANK FOR DAY	TIMES IN TOP 10
VA NEW KENT	2.99	PB5-8	22-Jul-57	4	7
VA NEW KENT	2.83	PB5-9	23-Jul-57	2	7
VA NEW KENT	2.54	PB8-7	24-Aug-57	7	7
VA NEW KENT	2.58	UE8-6	19-Jul-62	4	7
VA NEWPORT NEWS	11.58	BJ3-6	6-Nov-51	3	7
VA NEWPORT NEWS	3.43	BJ4-7	11-Nov-51	1	7
VA NEWPORT NEWS	8.96	PB5-8	22-Jul-57	2	7
VA NEWPORT NEWS	4.72	PB5-9	23-Jul-57	1	7
VA NEWPORT NEWS	1.51	PB8-7	24-Aug-57	10	7
VA NEWPORT NEWS	2.58	UE8-6	19-Jul-62	4	7
VA NEWPORT NEWS	0.30	UE8-8	21-Jul-62	3	7
VA NORFOLK/CHES	11.58	BJ3-6	6-Nov-51	3	5
VA NORFOLK/CHES	3.43	BJ4-7	11-Nov-51	1	5
VA NORFOLK/CHES	1.28	PB2-8	25-Jun-57	10	5
VA NORFOLK/CHES	8.96	PB5-8	22-Jul-57	2	5
VA NORFOLK/CHES	4.72	PB5-9	23-Jul-57	1	5
VA NORTHAMPTON	11.58	BJ3-6	6-Nov-51	3	4
VA NORTHAMPTON	3.43	BJ4-7	11-Nov-51	1	4
VA NORTHAMPTON	5.97	PB5-8	22-Jul-57	3	4
VA NORTHAMPTON	2.54	PB8-7	24-Aug-57	7	4
VA NORTHUMBERLA	11.58	BJ3-6	6-Nov-51	3	5
VA NORTHUMBERLA	3.43	BJ4-7	11-Nov-51	1	5
VA NORTHUMBERLA	1.92	PB2-8	25-Jun-57	7	5
VA NORTHUMBERLA	2.99	PB5-8	22-Jul-57	4	5
VA NORTHUMBERLA	2.54	PB8-7	24-Aug-57	7	5
VA NOTTOWAY	3.89	BJ2-2	31-Oct-51	10	4
VA NOTTOWAY	8.96	PB5-8	22-Jul-57	2	4
VA NOTTOWAY	2.83	PB5-9	23-Jul-57	2	4
VA NOTTOWAY	2.54	PB8-7	24-Aug-57	7	4
VA ORANGE	2.75	TP11-9	23-May-55	9	5
VA ORANGE	1.28	PB2-8	25-Jun-57	10	5
VA ORANGE	2.99	PB5-8	22-Jul-57	4	5
VA ORANGE	2.83	PB5-9	23-Jul-57	2	5
VA ORANGE	2.15	UE8-6	19-Jul-62	5	5
VA PAGE	1.77	UpK8-6	13-May-53	1	4

COUNTY	uCi/ sqm	SHOT DAY	DATE	RANK FOR DAY	TIMES IN TOP 10
VA PAGE	1.28	PB2-8	25-Jun-57	10	4
VA PAGE	8.96	PB5-8	22-Jul-57	2	4
VA PAGE	1.89	PB5-9	23-Jul-57	3	4
VA PATRICK	6.42	BJ2-2	31-Oct-51	6	5
VA PATRICK	1.32	PB2-8	25-Jun-57	9	5
VA PATRICK	5.97	PB5-8	22-Jul-57	3	5
VA PATRICK	2.83	PB5-9	23-Jul-57	2	5
VA PATRICK	2.58	UE8-6	19-Jul-62	4	5
VA PITTSYLVANIA	3.89	BJ2-2	31-Oct-51	10	6
VA PITTSYLVANIA	1.92	PB2-8	25-Jun-57	7	6
VA PITTSYLVANIA	5.97	PB5-8	22-Jul-57	3	6
VA PITTSYLVANIA	2.83	PB5-9	23-Jul-57	2	6
VA PITTSYLVANIA	1.51	PB8-7	24-Aug-57	10	6
VA PITTSYLVANIA	1.32	UE8-6	19-Jul-62	9	6
VA POWHATAN	14.93	PB5-8	22-Jul-57	1	4
VA POWHATAN	1.89	PB5-9	23-Jul-57	3	4
VA POWHATAN	2.54	PB8-7	24-Aug-57	7	4
VA POWHATAN	149.13	S-4	9-Jul-62	5	4
VA PRINCE EDWAR	6.42	BJ2-2	31-Oct-51	6	4
VA PRINCE EDWAR	8.96	PB5-8	22-Jul-57	2	4
VA PRINCE EDWAR	2.83	PB5-9	23-Jul-57	2	4
VA PRINCE EDWAR	2.54	PB8-7	24-Aug-57	7	4
VA PRINCE GEORG	11.58	BJ3-6	6-Nov-51	3	7
VA PRINCE GEORG	3.43	BJ4-7	11-Nov-51	1	7
VA PRINCE GEORG	1.28	PB2-8	25-Jun-57	10	7
VA PRINCE GEORG	8.96	PB5-8	22-Jul-57	2	7
VA PRINCE GEORG	2.83	PB5-9	23-Jul-57	2	7
VA PRINCE GEORG	2.54	PB8-7	24-Aug-57	7	7
VA PRINCE GEORG	0.30	UE8-8	21-Jul-62	3	7
VA PRINCE WILLI	1.40	TS5-5	11-May-52	1	5
VA PRINCE WILLI	2.75	TP11-9	23-May-55	9	5
VA PRINCE WILLI	8.96	PB5-8	22-Jul-57	2	5
VA PRINCE WILLI	0.94	PB5-9	23-Jul-57	4	5
VA PRINCE WILLI	2.54	PB8-7	24-Aug-57	7	5
VA PULASKI	1.18	BJ1-2	29-Oct-51	4	4

COUNTY	uCi/ sqm	SHOT DAY	DATE	RANK FOR DAY	TIMES IN TOP 10
VA PULASKI	4.59	BJ3-5	5-Nov-51	6	4
VA PULASKI	2.99	PB5-8	22-Jul-57	4	4
VA PULASKI	1.89	PB5-9	23-Jul-57	3	4
VA RAPPAHANNOCK	1.04	UpK4-4	9-Apr-53	5	5
VA RAPPAHANNOCK	1.77	UpK8-6	13-May-53	1	5
VA RAPPAHANNOCK	2.75	TP11-9	23-May-55	9	5
VA RAPPAHANNOCK	5.97	PB5-8	22-Jul-57	3	5
VA RAPPAHANNOCK	2.83	PB5-9	23-Jul-57	2	5
VA RICHMOND	1.92	PB2-8	25-Jun-57	7	3
VA RICHMOND	2.99	PB5-8	22-Jul-57	4	3
VA RICHMOND	2.54	PB8-7	24-Aug-57	7	3
VA ROANOKE	1.18	BJ1-2	29-Oct-51	4	6
VA ROANOKE	9.19	BJ3-5	5-Nov-51	2	6
VA ROANOKE	1.60	TS5-4	10-May-52	5	6
VA ROANOKE	5.97	PB5-8	22-Jul-57	3	6
VA ROANOKE	1.89	PB5-9	23-Jul-57	3	6
VA ROANOKE	0.21	UE8-8	21-Jul-62	5	6
VA ROCKBRIDGE	5.97	PB5-8	22-Jul-57	3	2
VA ROCKBRIDGE	2.83	PB5-9	23-Jul-57	2	2
VA ROCKINGHAM	1.77	UpK8-6	13-May-53	1	4
VA ROCKINGHAM	8.96	PB5-8	22-Jul-57	2	4
VA ROCKINGHAM	1.89	PB5-9	23-Jul-57	3	4
VA ROCKINGHAM	25.03	S-6	11-Jul-62	2	4
VA RUSSELL	1.55	UpK5-2	12-Apr-53	2	5
VA RUSSELL	3.11	TP6-4	25-Mar-55	3	5
VA RUSSELL	1.45	TP6-5	26-Mar-55	5	5
VA RUSSELL	1.39	PB2-5	22-Jun-57	8	5
VA RUSSELL	2.99	PB5-8	22-Jul-57	4	5
VA SCOTT	1.55	UpK5-2	12-Apr-53	2	5
VA SCOTT	3.11	TP6-4	25-Mar-55	3	5
VA SCOTT	0.97	TP6-5	26-Mar-55	8	5
VA SCOTT	1.51	PB5-8	22-Jul-57	7	5
VA SCOTT	7.49	PB17-3	30-Sep-57	5	5
VA SHENANDOAH	1.77	UpK8-6	13-May-53	1	3
VA SHENANDOAH	1.51	PB5-8	22-Jul-57	7	3

COUNTY	uCi/ sqm	SHOT DAY	DATE	RANK FOR DAY	TIMES IN TOP 10
VA SHENANDOAH	2.83	PB5-9	23-Jul-57	2	3
VA SMYTH	1.39	PB2-5	22-Jun-57	8	4
VA SMYTH	2.23	PB5-8	22-Jul-57	6	4
VA SMYTH	2.83	PB5-9	23-Jul-57	2	4
VA SMYTH	77.90	S-4	9-Jul-62	10	4
VA SOUTHAMPTON	11.58	BJ3-6	6-Nov-51	3	5
VA SOUTHAMPTON	3.43	BJ4-7	11-Nov-51	1	5
VA SOUTHAMPTON	1.51	PB5-8	22-Jul-57	7	5
VA SOUTHAMPTON	1.51	PB8-7	24-Aug-57	10	5
VA SOUTHAMPTON	4.62	UE8-6	19-Jul-62	1	5
VA SPOTSYLVANIA	1.04	UpK4-4	9-Apr-53	5	6
VA SPOTSYLVANIA	2.75	TP11-9	23-May-55	9	6
VA SPOTSYLVANIA	1.92	PB2-8	25-Jun-57	7	6
VA SPOTSYLVANIA	1.51	PB5-8	22-Jul-57	7	6
VA SPOTSYLVANIA	1.89	PB5-9	23-Jul-57	3	6
VA SPOTSYLVANIA	1.51	PB8-7	24-Aug-57	10	6
VA STAFFORD	1.36	TS5-4	10-May-52	10	7
VA STAFFORD	2.75	TP11-9	23-May-55	9	7
VA STAFFORD	1.54	PB5-7	21-Jul-57	9	7
VA STAFFORD	8.96	PB5-8	22-Jul-57	2	7
VA STAFFORD	0.94	PB5-9	23-Jul-57	4	7
VA STAFFORD	2.54	PB8-7	24-Aug-57	7	7
VA STAFFORD	3.66	UE8-5	18-Jul-62	3	7
VA SUFFOLK/NANS	11.58	BJ3-6	6-Nov-51	3	7
VA SUFFOLK/NANS	3.43	BJ4-7	11-Nov-51	1	7
VA SUFFOLK/NANS	5.97	PB5-8	22-Jul-57	3	7
VA SUFFOLK/NANS	4.72	PB5-9	23-Jul-57	1	7
VA SUFFOLK/NANS	40.47	S-5	10-Jul-62	3	7
VA SUFFOLK/NANS	2.58	UE8-6	19-Jul-62	4	7
VA SUFFOLK/NANS	0.16	UE8-8	21-Jul-62	6	7
VA SURRY	11.58	BJ3-6	6-Nov-51	3	5
VA SURRY	3.43	BJ4-7	11-Nov-51	1	5
VA SURRY	8.96	PB5-8	22-Jul-57	2	5
VA SURRY	1.51	PB8-7	24-Aug-57	10	5
VA SURRY	0.30	UE8-8	21-Jul-62	3	5

COUNTY	uCi/ sqm	SHOT DAY	DATE	RANK FOR DAY	TIMES IN TOP 10
VA SUSSEX	11.58	BJ3-6	6-Nov-51	3	6
VA SUSSEX	3.43	BJ4-7	11-Nov-51	1	6
VA SUSSEX	1.51	PB5-8	22-Jul-57	7	6
VA SUSSEX	0.94	PB5-9	23-Jul-57	4	6
VA SUSSEX	1.51	PB8-7	24-Aug-57	10	6
VA SUSSEX	4.62	UE8-6	19-Jul-62	1	6
VA TAZEWELL	1.18	BJ1-2	29-Oct-51	4	6
VA TAZEWELL	2.78	PB2-5	22-Jun-57	4	6
VA TAZEWELL	5.97	PB5-8	22-Jul-57	3	6
VA TAZEWELL	2.83	PB5-9	23-Jul-57	2	6
VA TAZEWELL	15.70	PB6-3	26-Jul-57	5	6
VA TAZEWELL	1.32	UE8-6	19-Jul-62	9	6
VA VIRGINIA BEA	4.81	BJ3-6	6-Nov-51	8	7
VA VIRGINIA BEA	1.42	BJ4-7	11-Nov-51	2	7
VA VIRGINIA BEA	1.28	PB2-8	25-Jun-57	10	7
VA VIRGINIA BEA	8.96	PB5-8	22-Jul-57	2	7
VA VIRGINIA BEA	40.47	S-5	10-Jul-62	3	7
VA VIRGINIA BEA	2.58	UE8-6	19-Jul-62	4	7
VA VIRGINIA BEA	0.16	UE8-8	21-Jul-62	6	7
VA WARREN	2.75	TP11-9	23-May-55	9	3
VA WARREN	2.99	PB5-8	22-Jul-57	4	3
VA WARREN	2.83	PB5-9	23-Jul-57	2	3
VA WASHINGTON	1.55	UpK5-2	12-Apr-53	2	6
VA WASHINGTON	3.11	TP6-4	25-Mar-55	3	6
VA WASHINGTON	1.45	TP6-5	26-Mar-55	5	6
VA WASHINGTON	2.78	PB2-5	22-Jun-57	4	6
VA WASHINGTON	1.51	PB5-8	22-Jul-57	7	6
VA WASHINGTON	5.90	PB6-3	26-Jul-57	10	6
VA WESTMORELAND	1.92	PB2-8	25-Jun-57	7	4
VA WESTMORELAND	2.99	PB5-8	22-Jul-57	4	4
VA WESTMORELAND	0.94	PB5-9	23-Jul-57	4	4
VA WESTMORELAND	2.54	PB8-7	24-Aug-57	7	4
VA WISE	1.55	UpK5-2	12-Apr-53	2	6
VA WISE	3.11	TP6-4	25-Mar-55	3	6
VA WISE	0.97	TP6-5	26-Mar-55	8	6

COUNTY	uCi/ sqm	SHOT DAY	DATE	RANK FOR DAY	TIMES IN TOP 10
VA WISE	2.78	PB2-5	22-Jun-57	4	6
VA WISE	5.97	PB5-8	22-Jul-57	3	6
VA WISE	7.49	PB17-3	30-Sep-57	5	6
VA WYTHE	1.51	PB5-8	22-Jul-57	7	2
VA WYTHE	2.83	PB5-9	23-Jul-57	2	2
VA YORK	11.58	BJ3-6	6-Nov-51	3	7
VA YORK	3.43	BJ4-7	11-Nov-51	1	7
VA YORK	2.99	PB5-8	22-Jul-57	4	7
VA YORK	4.72	PB5-9	23-Jul-57	1	7
VA YORK	1.51	PB8-7	24-Aug-57	10	7
VA YORK	2.89	PB12-8	9-Sep-57	9	7
VA YORK	2.58	UE8-6	19-Jul-62	4	7

COUNTY	uCi/ sqm	SHOT DAY	DATE	RANK FOR DAY	TIMES IN TOP 10
VT ADDISON	1.42	R1-3	30-Jan-51	1	9
VT ADDISON	7.98	BJ2-2	31-Oct-51	4	9
VT ADDISON	1.37	TS4-6	6-May-52	9	9
VT ADDISON	1.37	TS4-6	6-May-52	9	9
VT ADDISON	170.38	UpK7-2	26-Apr-53	10	9
VT ADDISON	1.32	UpK8-3	10-May-53	9	9
VT ADDISON	1.36	PB5-6	20-Jul-57	6	9
VT ADDISON	5.97	PB5-8	22-Jul-57	3	9
VT ADDISON	0.24	UE7-5	15-Jul-62	5	9
VT BENNINGTON	1.42	R1-3	30-Jan-51	1	7
VT BENNINGTON	248.11	UpK7-2	26-Apr-53	6	7
VT BENNINGTON	2.63	UpK8-3	10-May-53	5	7
VT BENNINGTON	1.36	PB5-6	20-Jul-57	6	7
VT BENNINGTON	8.96	PB5-8	22-Jul-57	2	7
VT BENNINGTON	1.57	PB6-7	30-Jul-57	9	7
VT BENNINGTON	1.27	PB6-8	31-Jul-57	9	7
VT CALEDONIA	1.42	R1-3	30-Jan-51	1	10
VT CALEDONIA	7.98	BJ2-2	31-Oct-51	4	10
VT CALEDONIA	22.91	TS8-2	6-Jun-52	7	10
VT CALEDONIA	2.34	UpK4-3	8-Apr-53	6	10
VT CALEDONIA	3.03	UpK4-4	9-Apr-53	3	10
VT CALEDONIA	0.97	TP6-5	26-Mar-55	8	10
VT CALEDONIA	1.14	PB2-6	23-Jun-57	7	10
VT CALEDONIA	1.36	PB5-6	20-Jul-57	6	10
VT CALEDONIA	8.96	PB5-8	22-Jul-57	2	10
VT CALEDONIA	0.33	UE7-5	15-Jul-62	3	10
VT CHITTENDEN	1.42	R1-3	30-Jan-51	1	8
VT CHITTENDEN	5.01	BJ2-2	31-Oct-51	8	8
VT CHITTENDEN	1.32	UpK8-3	10-May-53	9	8
VT CHITTENDEN	1.72	UpK8-4	11-May-53	5	8
VT CHITTENDEN	1.14	PB2-6	23-Jun-57	7	8
VT CHITTENDEN	5.41	PB5-6	20-Jul-57	2	8
VT CHITTENDEN	8.96	PB5-8	22-Jul-57	2	8
VT CHITTENDEN	19.27	PB11-5	4-Sep-57	4	8
VT ESSEX	0.99	R1-3	30-Jan-51	3	9

COUNTY	uCi/ sqm	SHOT DAY	DATE	RANK FOR DAY	TIMES IN TOP 10
VT ESSEX	7.98	BJ2-2	31-Oct-51	4	9
VT ESSEX	2.34	UpK4-3	8-Apr-53	6	9
VT ESSEX	3.03	UpK4-4	9-Apr-53	3	9
VT ESSEX	0.97	TP6-5	26-Mar-55	8	9
VT ESSEX	1.14	PB2-6	23-Jun-57	7	9
VT ESSEX	8.08	PB5-6	20-Jul-57	1	9
VT ESSEX	14.93	PB5-8	22-Jul-57	1	9
VT ESSEX	0.12	UE7-5	15-Jul-62	9	9
VT FRANKLIN	1.42	R1-3	30-Jan-51	1	7
VT FRANKLIN	12.07	BJ2-2	31-Oct-51	1	7
VT FRANKLIN	1.32	UpK8-3	10-May-53	9	7
VT FRANKLIN	1.72	UpK8-4	11-May-53	5	7
VT FRANKLIN	0.97	TP6-5	26-Mar-55	8	7
VT FRANKLIN	2.68	PB5-6	20-Jul-57	3	7
VT FRANKLIN	1.51	PB5-8	22-Jul-57	7	7
VT GRAND ISLE	1.42	R1-3	30-Jan-51	1	7
VT GRAND ISLE	12.07	BJ2-2	31-Oct-51	1	7
VT GRAND ISLE	1.32	UpK8-3	10-May-53	9	7
VT GRAND ISLE	3.55	UpK8-4	11-May-53	1	7
VT GRAND ISLE	8.08	PB5-6	20-Jul-57	1	7
VT GRAND ISLE	1.51	PB5-8	22-Jul-57	7	7
VT GRAND ISLE	0.12	UE7-5	15-Jul-62	9	7
VT LAMOILLE	7.98	BJ2-2	31-Oct-51	4	7
VT LAMOILLE	1.32	UpK8-3	10-May-53	9	7
VT LAMOILLE	0.97	TP6-5	26-Mar-55	8	7
VT LAMOILLE	1.91	PB2-6	23-Jun-57	4	7
VT LAMOILLE	5.41	PB5-6	20-Jul-57	2	7
VT LAMOILLE	5.97	PB5-8	22-Jul-57	3	7
VT LAMOILLE	0.12	UE7-5	15-Jul-62	9	7
VT ORANGE	1.42	R1-3	30-Jan-51	1	8
VT ORANGE	4.09	BJ2-2	31-Oct-51	9	8
VT ORANGE	22.73	TS8-2	6-Jun-52	8	8
VT ORANGE	1.32	UpK8-3	10-May-53	9	8
VT ORANGE	0.97	TP6-5	26-Mar-55	8	8
VT ORANGE	5.41	PB5-6	20-Jul-57	2	8

COUNTY	uCi/ sqm	SHOT DAY	DATE	RANK FOR DAY	TIMES IN TOP 10
VT ORANGE	8.96	PB5-8	22-Jul-57	2	8
VT ORANGE	0.12	UE7-5	15-Jul-62	9	8
VT ORLEANS	12.07	BJ2-2	31-Oct-51	1	7
VT ORLEANS	1.32	UpK8-3	10-May-53	9	7
VT ORLEANS	1.72	UpK8-4	11-May-53	5	7
VT ORLEANS	0.97	TP6-5	26-Mar-55	8	7
VT ORLEANS	8.08	PB5-6	20-Jul-57	1	7
VT ORLEANS	2.99	PB5-8	22-Jul-57	4	7
VT ORLEANS	0.53	UE7-5	15-Jul-62	1	7
VT RUTLAND	1.42	R1-3	30-Jan-51	1	7
VT RUTLAND	7.98	BJ2-2	31-Oct-51	4	7
VT RUTLAND	21.74	TS8-2	6-Jun-52	10	7
VT RUTLAND	1.32	UpK8-3	10-May-53	9	7
VT RUTLAND	1.36	PB5-6	20-Jul-57	6	7
VT RUTLAND	5.97	PB5-8	22-Jul-57	3	7
VT RUTLAND	1.57	PB6-7	30-Jul-57	9	7
VT WASHINGTON	0.99	R1-3	30-Jan-51	3	9
VT WASHINGTON	7.98	BJ2-2	31-Oct-51	4	9
VT WASHINGTON	23.92	TS8-2	6-Jun-52	5	9
VT WASHINGTON	1.32	UpK8-3	10-May-53	9	9
VT WASHINGTON	1.45	TP6-5	26-Mar-55	5	9
VT WASHINGTON	1.14	PB2-6	23-Jun-57	7	9
VT WASHINGTON	5.41	PB5-6	20-Jul-57	2	9
VT WASHINGTON	8.96	PB5-8	22-Jul-57	2	9
VT WASHINGTON	0.12	UE7-5	15-Jul-62	9	9
VT WINDHAM	47.66	R1-2	29-Jan-51	2	7
VT WINDHAM	0.99	R1-3	30-Jan-51	3	7
VT WINDHAM	3.71	BJ3-7	7-Nov-51	10	7
VT WINDHAM	1.97	UpK8-3	10-May-53	7	7
VT WINDHAM	1.16	TP7-6	3-Apr-55	6	7
VT WINDHAM	1.36	PB5-6	20-Jul-57	6	7
VT WINDHAM	8.96	PB5-8	22-Jul-57	2	7
VT WINDSOR	1.17	R1-3	30-Jan-51	2	9
VT WINDSOR	7.98	BJ2-2	31-Oct-51	4	9
VT WINDSOR	23.88	TS8-2	6-Jun-52	6	9

COUNTY	uCi/ sqm	SHOT DAY	DATE	RANK FOR DAY	TIMES IN TOP 10
VT WINDSOR	1.32	UpK8-3	10-May-53	9	9
VT WINDSOR	1.32	UpK8-4	11-May-53	7	9
VT WINDSOR	0.97	TP6-5	26-Mar-55	8	9
VT WINDSOR	2.02	PB5-6	20-Jul-57	4	9
VT WINDSOR	2.99	PB5-8	22-Jul-57	4	9
VT WINDSOR	0.13	UE7-5	15-Jul-62	8	9

COUNTY	uCi/ sqm	SHOT DAY	DATE	RANK FOR DAY	TIMES IN TOP 10
WA ASOTIN	0.83	PB1-9	5-Jun-57	8	7
WA ASOTIN	12.28	PB6-3	26-Jul-57	6	7
WA ASOTIN	7.58	PB16-3	25-Sep-57	1	7
WA ASOTIN	2.58	PB16-4	26-Sep-57	7	7
WA ASOTIN	9.52	PB17-3	30-Sep-57	4	7
WA ASOTIN	0.85	PB18-4	10-Oct-57	7	7
WA ASOTIN	0.46	BN-4	21-Dec-70	2	7
WA BENTON	0.46	BN-4	21-Dec-70	2	1
WA CHELAN	0.23	BN-4	21-Dec-70	8	1
WA CLARK	1.83	TP6-3	24-Mar-55	7	1
WA COLUMBIA	0.83	PB1-9	5-Jun-57	8	7
WA COLUMBIA	12.28	PB6-3	26-Jul-57	6	7
WA COLUMBIA	7.58	PB16-3	25-Sep-57	1	7
WA COLUMBIA	7.74	PB16-4	26-Sep-57	5	7
WA COLUMBIA	9.52	PB17-3	30-Sep-57	4	7
WA COLUMBIA	0.85	PB18-4	10-Oct-57	7	7
WA COLUMBIA	0.46	BN-4	21-Dec-70	2	7
WA COWLITZ	1.83	TP6-3	24-Mar-55	7	1
WA DOUGLAS	13.81	UpK10-2	26-May-53	6	2
WA DOUGLAS	0.46	BN-4	21-Dec-70	2	2
WA FERRY	33.86	UpK10-2	26-May-53	1	2
WA FERRY	0.23	BN-4	21-Dec-70	8	2
WA GARFIELD	12.45	UpK10-2	26-May-53	10	7
WA GARFIELD	12.28	PB6-3	26-Jul-57	6	7
WA GARFIELD	7.58	PB16-3	25-Sep-57	1	7
WA GARFIELD	1.29	PB16-4	26-Sep-57	9	7
WA GARFIELD	6.34	PB17-3	30-Sep-57	6	7
WA GARFIELD	0.85	PB18-4	10-Oct-57	7	7
WA GARFIELD	0.46	BN-4	21-Dec-70	2	7
WA GRANT	0.38	BN-4	21-Dec-70	3	1
WA JEFFERSON	2.80	UpK2-12	4-Apr-53	7	1
WA KING	2.80	UpK2-12	4-Apr-53	7	2
WA KING	0.13	BN-4	21-Dec-70	10	2
WA KLICKITAT	0.26	BN-4	21-Dec-70	6	1
WA OKANOGAN	23.02	UpK10-2	26-May-53	4	1
WA PACIFIC	1.83	TP6-3	24-Mar-55	7	1

COUNTY	uCi/ sqm	SHOT DAY	DATE	RANK FOR DAY	TIMES IN TOP 10
WA PEND OREILLE	28.27	UpK10-2	26-May-53	2	6
WA PEND OREILLE	12.28	PB6-3	26-Jul-57	6	6
WA PEND OREILLE	7.58	PB16-3	25-Sep-57	1	6
WA PEND OREILLE	5.16	PB16-4	26-Sep-57	6	6
WA PEND OREILLE	6.34	PB17-3	30-Sep-57	6	6
WA PEND OREILLE	0.85	PB18-4	10-Oct-57	7	6
WA PIERCE	2.80	UpK2-12	4-Apr-53	7	1
WA SKAMANIA	1.83	TP6-3	24-Mar-55	7	2
WA SKAMANIA	0.64	BN-4	21-Dec-70	1	2
WA SNOHOMISH	2.80	UpK2-12	4-Apr-53	7	1
WA SPOKANE	0.83	PB1-9	5-Jun-57	8	6
WA SPOKANE	12.28	PB6-3	26-Jul-57	6	6
WA SPOKANE	7.58	PB16-3	25-Sep-57	1	6
WA SPOKANE	7.74	PB16-4	26-Sep-57	5	6
WA SPOKANE	6.34	PB17-3	30-Sep-57	6	6
WA SPOKANE	0.85	PB18-4	10-Oct-57	7	6
WA STEVENS	24.41	UpK10-2	26-May-53	3	1
WA WAHKIAKUM	1.83	TP6-3	24-Mar-55	7	1
WA WALLA WALLA	0.83	PB1-9	5-Jun-57	8	8
WA WALLA WALLA	12.28	PB6-3	26-Jul-57	6	8
WA WALLA WALLA	1.98	PB15-3	18-Sep-57	7	8
WA WALLA WALLA	7.58	PB16-3	25-Sep-57	1	8
WA WALLA WALLA	7.74	PB16-4	26-Sep-57	5	8
WA WALLA WALLA	9.52	PB17-3	30-Sep-57	4	8
WA WALLA WALLA	0.85	PB18-4	10-Oct-57	7	8
WA WALLA WALLA	0.46	BN-4	21-Dec-70	2	8
WA WHATCOM	2.80	UpK2-12	4-Apr-53	7	1
WA WHITMAN	0.83	PB1-9	5-Jun-57	8	7
WA WHITMAN	12.28	PB6-3	26-Jul-57	6	7
WA WHITMAN	7.58	PB16-3	25-Sep-57	1	7
WA WHITMAN	2.58	PB16-4	26-Sep-57	7	7
WA WHITMAN	6.34	PB17-3	30-Sep-57	6	7
WA WHITMAN	0.85	PB18-4	10-Oct-57	7	7
WA WHITMAN	0.46	BN-4	21-Dec-70	2	7
WA YAKIMA	0.26	BN-4	21-Dec-70	6	1

COUNTY	uCi/ sqm	SHOT DAY	DATE	RANK FOR DAY	TIMES IN TOP 10
WI ADAMS	1.06	TS5-8	14-May-52	4	8
WI ADAMS	28.08	TP10-2	6-May-55	7	8
WI ADAMS	2.48	PB5-7	21-Jul-57	6	8
WI ADAMS	13.72	PB9-3	25-Aug-57	6	8
WI ADAMS	8.37	PB9-4	26-Aug-57	7	8
WI ADAMS	2.65	PB18-8	14-Oct-57	8	8
WI ADAMS	0.13	UE7-5	15-Jul-62	8	8
WI ADAMS	14.56	UE12-4	17-Apr-65	2	8
WI ASHLAND	0.99	TS3-8	29-Apr-52	8	7
WI ASHLAND	0.99	TS3-8	29-Apr-52	8	7
WI ASHLAND	2.54	PB2-4	21-Jun-57	4	7
WI ASHLAND	10.11	PB5-5	19-Jul-57	1	7
WI ASHLAND	1.55	PB15-3	18-Sep-57	10	7
WI ASHLAND	1.16	PB15-5	20-Sep-57	7	7
WI ASHLAND	0.10	UE2-7	11-Mar-62	4	7
WI BARRON	0.99	TS3-8	29-Apr-52	8	7
WI BARRON	0.99	TS3-8	29-Apr-52	8	7
WI BARRON	2.54	PB2-4	21-Jun-57	4	7
WI BARRON	10.11	PB5-5	19-Jul-57	1	7
WI BARRON	1.97	PB8-5	22-Aug-57	9	7
WI BARRON	1.16	PB15-5	20-Sep-57	7	7
WI BARRON	0.13	UE7-5	15-Jul-62	8	7
WI BAYFIELD	0.99	TS3-8	29-Apr-52	8	7
WI BAYFIELD	0.99	TS3-8	29-Apr-52	8	7
WI BAYFIELD	2.37	UpK10-4	28-May-53	10	7
WI BAYFIELD	2.54	PB2-4	21-Jun-57	4	7
WI BAYFIELD	6.09	PB5-5	19-Jul-57	2	7
WI BAYFIELD	1.55	PB15-3	18-Sep-57	10	7
WI BAYFIELD	0.10	UE2-7	11-Mar-62	3	7
WI BROWN	0.85	PB4-3	7-Jul-57	10	3
WI BROWN	13.72	PB9-3	25-Aug-57	6	3
WI BROWN	8.37	PB9-4	26-Aug-57	7	3
WI BUFFALO	0.99	TS3-8	29-Apr-52	8	10
WI BUFFALO	0.99	TS3-8	29-Apr-52	8	10
WI BUFFALO	61.61	UpK9-2	20-May-53	10	10

COUNTY	uCi/ sqm	SHOT DAY	DATE	RANK FOR DAY	TIMES IN TOP 10
WI BUFFALO	2.54	PB2-4	21-Jun-57	4	10
WI BUFFALO	0.85	PB4-3	7-Jul-57	10	10
WI BUFFALO	4.02	PB5-5	19-Jul-57	3	10
WI BUFFALO	1.16	PB15-5	20-Sep-57	7	10
WI BUFFALO	0.17	UE2-7	11-Mar-62	2	10
WI BUFFALO	325.19	S-3	8-Jul-62	10	10
WI BUFFALO	0.22	UE7-5	15-Jul-62	6	10
WI BURNETT	2.54	PB2-4	21-Jun-57	4	6
WI BURNETT	10.11	PB5-5	19-Jul-57	1	6
WI BURNETT	1.97	PB8-5	22-Aug-57	9	6
WI BURNETT	1.11	PB9-8	30-Aug-57	1	6
WI BURNETT	1.55	PB15-3	18-Sep-57	10	6
WI BURNETT	0.10	UE2-7	11-Mar-62	3	6
WI CALUMET	0.94	TS5-9	15-May-52	5	6
WI CALUMET	1.10	PB4-3	7-Jul-57	7	6
WI CALUMET	13.72	PB9-3	25-Aug-57	6	6
WI CALUMET	8.37	PB9-4	26-Aug-57	7	6
WI CALUMET	0.91	PB18-9	15-Oct-57	5	6
WI CALUMET	14.56	UE12-4	17-Apr-65	2	6
WI CHIPPEWA	0.99	TS3-8	29-Apr-52	8	9
WI CHIPPEWA	0.99	TS3-8	29-Apr-52	8	9
WI CHIPPEWA	64.64	UpK9-2	20-May-53	7	9
WI CHIPPEWA	2.54	PB2-4	21-Jun-57	4	9
WI CHIPPEWA	1.02	PB4-3	7-Jul-57	8	9
WI CHIPPEWA	2.01	PB5-5	19-Jul-57	8	9
WI CHIPPEWA	1.16	PB15-5	20-Sep-57	7	9
WI CHIPPEWA	0.10	UE2-7	11-Mar-62	4	9
WI CHIPPEWA	0.13	UE7-5	15-Jul-62	8	9
WI CLARK	0.99	TS3-8	29-Apr-52	8	6
WI CLARK	0.99	TS3-8	29-Apr-52	8	6
WI CLARK	2.54	PB2-4	21-Jun-57	4	6
WI CLARK	1.36	PB4-3	7-Jul-57	4	6
WI CLARK	1.16	PB15-5	20-Sep-57	7	6
WI CLARK	0.13	UE7-5	15-Jul-62	8	6
WI COLUMBIA	13.72	PB9-3	25-Aug-57	6	5

COUNTY	uCi/ sqm	SHOT DAY	DATE	RANK FOR DAY	TIMES IN TOP 10
WI COLUMBIA	16.67	PB9-4	26-Aug-57	4	5
WI COLUMBIA	3.98	PB18-8	14-Oct-57	4	5
WI COLUMBIA	0.08	UE2-7	11-Mar-62	5	5
WI COLUMBIA	0.13	UE7-5	15-Jul-62	8	5
WI CRAWFORD	0.99	TS3-8	29-Apr-52	8	11
WI CRAWFORD	0.99	TS3-8	29-Apr-52	8	11
WI CRAWFORD	0.85	PB4-3	7-Jul-57	10	11
WI CRAWFORD	2.48	PB5-7	21-Jul-57	6	11
WI CRAWFORD	2.43	PB6-9	1-Aug-57	2	11
WI CRAWFORD	13.72	PB9-3	25-Aug-57	6	11
WI CRAWFORD	16.67	PB9-4	26-Aug-57	4	11
WI CRAWFORD	2.04	PB9-5	27-Aug-57	8	11
WI CRAWFORD	3.98	PB18-8	14-Oct-57	4	11
WI CRAWFORD	0.12	UE7-5	15-Jul-62	9	11
WI CRAWFORD	9.84	UE12-4	17-Apr-65	4	11
WI DANE	13.72	PB9-3	25-Aug-57	6	6
WI DANE	33.34	PB9-4	26-Aug-57	3	6
WI DANE	0.98	PB9-6	28-Aug-57	6	6
WI DANE	3.98	PB18-8	14-Oct-57	4	6
WI DANE	0.08	UE2-7	11-Mar-62	5	6
WI DANE	0.13	UE7-5	15-Jul-62	8	6
WI DODGE	0.93	PB4-3	7-Jul-57	9	4
WI DODGE	13.72	PB9-3	25-Aug-57	6	4
WI DODGE	8.37	PB9-4	26-Aug-57	7	4
WI DODGE	0.24	UE7-5	15-Jul-62	5	4
WI DOOR	1.06	TS5-8	14-May-52	4	3
WI DOOR	28.08	TP10-2	6-May-55	7	3
WI DOOR	3.01	PB9-5	27-Aug-57	6	3
WI DOUGLAS	2.69	UpK10-4	28-May-53	7	4
WI DOUGLAS	2.54	PB2-4	21-Jun-57	4	4
WI DOUGLAS	10.11	PB5-5	19-Jul-57	1	4
WI DOUGLAS	1.55	PB15-3	18-Sep-57	10	4
WI DUNN	0.99	TS3-8	29-Apr-52	8	9
WI DUNN	0.99	TS3-8	29-Apr-52	8	9
WI DUNN	67.78	UpK9-2	20-May-53	6	9

COUNTY	uCi/ sqm	SHOT DAY	DATE	RANK FOR DAY	TIMES IN TOP 10
WI DUNN	2.54	PB2-4	21-Jun-57	4	9
WI DUNN	1.44	PB4-3	7-Jul-57	3	9
WI DUNN	6.09	PB5-5	19-Jul-57	2	9
WI DUNN	1.16	PB15-5	20-Sep-57	7	9
WI DUNN	0.17	UE2-7	11-Mar-62	2	9
WI DUNN	0.12	UE7-5	15-Jul-62	9	9
WI EAU CLAIRE	0.99	TS3-8	29-Apr-52	8	8
WI EAU CLAIRE	0.99	TS3-8	29-Apr-52	8	8
WI EAU CLAIRE	2.54	PB2-4	21-Jun-57	4	8
WI EAU CLAIRE	1.02	PB4-3	7-Jul-57	8	8
WI EAU CLAIRE	2.01	PB5-5	19-Jul-57	8	8
WI EAU CLAIRE	1.16	PB15-5	20-Sep-57	7	8
WI EAU CLAIRE	0.10	UE2-7	11-Mar-62	4	8
WI EAU CLAIRE	0.22	UE7-5	15-Jul-62	6	8
WI FLORENCE	10.11	PB5-5	19-Jul-57	1	2
WI FLORENCE	1.97	PB8-5	22-Aug-57	9	2
WI FOND DU LAC	13.72	PB9-3	25-Aug-57	6	5
WI FOND DU LAC	8.37	PB9-4	26-Aug-57	7	5
WI FOND DU LAC	0.08	UE2-7	11-Mar-62	5	5
WI FOND DU LAC	0.24	UE7-5	15-Jul-62	5	5
WI FOND DU LAC	14.56	UE12-4	17-Apr-65	2	5
WI FOREST	1.06	TS5-8	14-May-52	4	3
WI FOREST	1.97	PB8-5	22-Aug-57	9	3
WI FOREST	0.91	PB18-9	15-Oct-57	5	3
WI GRANT	0.99	TS3-8	29-Apr-52	8	12
WI GRANT	0.99	TS3-8	29-Apr-52	8	12
WI GRANT	0.93	PB4-3	7-Jul-57	9	12
WI GRANT	4.15	PB5-7	21-Jul-57	2	12
WI GRANT	13.72	PB9-3	25-Aug-57	6	12
WI GRANT	33.34	PB9-4	26-Aug-57	3	12
WI GRANT	2.04	PB9-5	27-Aug-57	8	12
WI GRANT	0.89	PB9-7	29-Aug-57	6	12
WI GRANT	3.98	PB18-8	14-Oct-57	4	12
WI GRANT	0.08	UE2-7	11-Mar-62	5	12
WI GRANT	0.12	UE7-5	15-Jul-62	9	12

COUNTY	uCi/ sqm	SHOT DAY	DATE	RANK FOR DAY	TIMES IN TOP 10
WI GRANT	9.84	UE12-4	17-Apr-65	4	12
WI GREEN	2.25	TS8-7	11-Jun-52	3	9
WI GREEN	1.36	PB4-3	7-Jul-57	4	9
WI GREEN	13.72	PB9-3	25-Aug-57	6	9
WI GREEN	8.37	PB9-4	26-Aug-57	7	9
WI GREEN	2.04	PB9-5	27-Aug-57	8	9
WI GREEN	0.89	PB9-7	29-Aug-57	6	9
WI GREEN	2.65	PB18-8	14-Oct-57	8	9
WI GREEN	0.91	PB18-9	15-Oct-57	5	9
WI GREEN	0.33	UE7-5	15-Jul-62	3	9
WI GREEN LAKE	0.94	TS5-9	15-May-52	5	8
WI GREEN LAKE	0.93	PB4-3	7-Jul-57	9	8
WI GREEN LAKE	13.72	PB9-3	25-Aug-57	6	8
WI GREEN LAKE	8.37	PB9-4	26-Aug-57	7	8
WI GREEN LAKE	3.98	PB18-8	14-Oct-57	4	8
WI GREEN LAKE	0.08	UE2-7	11-Mar-62	5	8
WI GREEN LAKE	0.13	UE7-5	15-Jul-62	8	8
WI GREEN LAKE	14.56	UE12-4	17-Apr-65	2	8
WI IOWA	1.67	PB5-7	21-Jul-57	7	8
WI IOWA	13.72	PB9-3	25-Aug-57	6	8
WI IOWA	16.67	PB9-4	26-Aug-57	4	8
WI IOWA	2.04	PB9-5	27-Aug-57	8	8
WI IOWA	0.89	PB9-7	29-Aug-57	6	8
WI IOWA	3.98	PB18-8	14-Oct-57	4	8
WI IOWA	345.56	S-3	8-Jul-62	8	8
WI IOWA	0.33	UE7-5	15-Jul-62	3	8
WI IRON	2.54	PB2-4	21-Jun-57	4	6
WI IRON	6.09	PB5-5	19-Jul-57	2	6
WI IRON	1.97	PB8-5	22-Aug-57	9	6
WI IRON	1.55	PB15-3	18-Sep-57	10	6
WI IRON	1.16	PB15-5	20-Sep-57	7	6
WI IRON	0.10	UE2-7	11-Mar-62	3	6
WI JACKSON	0.99	TS3-8	29-Apr-52	8	7
WI JACKSON	0.99	TS3-8	29-Apr-52	8	7
WI JACKSON	1.50	PB2-4	21-Jun-57	8	7

COUNTY	uCi/ sqm	SHOT DAY	DATE	RANK FOR DAY	TIMES IN TOP 10
WI JACKSON	1.02	PB4-3	7-Jul-57	8	7
WI JACKSON	0.10	UE2-7	11-Mar-62	4	7
WI JACKSON	0.12	UE7-5	15-Jul-62	9	7
WI JACKSON	14.56	UE12-4	17-Apr-65	2	7
WI JEFFERSON	13.72	PB9-3	25-Aug-57	6	7
WI JEFFERSON	8.37	PB9-4	26-Aug-57	7	7
WI JEFFERSON	2.04	PB9-5	27-Aug-57	8	7
WI JEFFERSON	0.89	PB9-7	29-Aug-57	6	7
WI JEFFERSON	2.65	PB18-8	14-Oct-57	8	7
WI JEFFERSON	0.91	PB18-9	15-Oct-57	5	7
WI JEFFERSON	0.13	UE7-5	15-Jul-62	8	7
WI JUNEAU	1.06	TS5-8	14-May-52	4	10
WI JUNEAU	28.08	TP10-2	6-May-55	7	10
WI JUNEAU	1.02	PB4-3	7-Jul-57	8	10
WI JUNEAU	2.48	PB5-7	21-Jul-57	6	10
WI JUNEAU	13.72	PB9-3	25-Aug-57	6	10
WI JUNEAU	8.37	PB9-4	26-Aug-57	7	10
WI JUNEAU	2.65	PB18-8	14-Oct-57	8	10
WI JUNEAU	0.17	UE2-7	11-Mar-62	2	10
WI JUNEAU	0.12	UE7-5	15-Jul-62	9	10
WI JUNEAU	14.56	UE12-4	17-Apr-65	2	10
WI KENOSHA	1.65	PB5-6	20-Jul-57	5	3
WI KENOSHA	0.08	UE2-7	11-Mar-62	5	3
WI KENOSHA	0.12	UE7-5	15-Jul-62	9	3
WI KEWAUNEE	1.06	TS5-8	14-May-52	4	5
WI KEWAUNEE	1.10	PB4-3	7-Jul-57	7	5
WI KEWAUNEE	13.72	PB9-3	25-Aug-57	6	5
WI KEWAUNEE	8.37	PB9-4	26-Aug-57	7	5
WI KEWAUNEE	0.91	PB18-9	15-Oct-57	5	5
WI LA CROSSE	0.99	TS3-8	29-Apr-52	8	8
WI LA CROSSE	0.99	TS3-8	29-Apr-52	8	8
WI LA CROSSE	2.24	PB6-8	31-Jul-57	4	8
WI LA CROSSE	13.72	PB9-3	25-Aug-57	6	8
WI LA CROSSE	16.67	PB9-4	26-Aug-57	4	8
WI LA CROSSE	3.98	PB18-8	14-Oct-57	4	8

COUNTY	uCi/ sqm	SHOT DAY	DATE	RANK FOR DAY	TIMES IN TOP 10
WI LA CROSSE	0.17	UE2-7	11-Mar-62	2	8
WI LA CROSSE	0.24	UE7-5	15-Jul-62	5	8
WI LAFAYETTE	1.10	PB4-3	7-Jul-57	7	7
WI LAFAYETTE	4.15	PB5-7	21-Jul-57	2	7
WI LAFAYETTE	13.72	PB9-3	25-Aug-57	6	7
WI LAFAYETTE	8.37	PB9-4	26-Aug-57	7	7
WI LAFAYETTE	3.39	PB9-5	27-Aug-57	4	7
WI LAFAYETTE	3.98	PB18-8	14-Oct-57	4	7
WI LAFAYETTE	0.33	UE7-5	15-Jul-62	3	7
WI LANGLADE	0.85	PB4-3	7-Jul-57	10	4
WI LANGLADE	1.11	PB9-8	30-Aug-57	1	4
WI LANGLADE	1.16	PB15-5	20-Sep-57	7	4
WI LANGLADE	0.24	UE7-5	15-Jul-62	5	4
WI LINCOLN	0.94	TS5-9	15-May-52	5	5
WI LINCOLN	0.93	PB4-3	7-Jul-57	9	5
WI LINCOLN	2.01	PB5-5	19-Jul-57	8	5
WI LINCOLN	1.11	PB9-8	30-Aug-57	1	5
WI LINCOLN	1.16	PB15-5	20-Sep-57	7	5
WI MANITOWOC	1.06	TS5-8	14-May-52	4	6
WI MANITOWOC	1.10	PB4-3	7-Jul-57	7	6
WI MANITOWOC	13.72	PB9-3	25-Aug-57	6	6
WI MANITOWOC	8.37	PB9-4	26-Aug-57	7	6
WI MANITOWOC	0.91	PB18-9	15-Oct-57	5	6
WI MANITOWOC	7.97	UE12-4	17-Apr-65	8	6
WI MARATHON	1.06	TS5-8	14-May-52	4	7
WI MARATHON	0.93	PB4-3	7-Jul-57	9	7
WI MARATHON	6.09	PB5-5	19-Jul-57	2	7
WI MARATHON	1.18	PB5-6	20-Jul-57	7	7
WI MARATHON	1.16	PB15-5	20-Sep-57	7	7
WI MARATHON	0.10	UE2-7	11-Mar-62	3	7
WI MARATHON	0.13	UE7-5	15-Jul-62	8	7
WI MARINETTE	1.06	TS5-8	14-May-52	4	6
WI MARINETTE	0.85	PB4-3	7-Jul-57	10	6
WI MARINETTE	1.97	PB8-5	22-Aug-57	9	6
WI MARINETTE	13.72	PB9-3	25-Aug-57	6	6

COUNTY	uCi/ sqm	SHOT DAY	DATE	RANK FOR DAY	TIMES IN TOP 10
WI MARINETTE	8.37	PB9-4	26-Aug-57	7	6
WI MARINETTE	0.91	PB18-9	15-Oct-57	5	6
WI MARQUETTE	1.06	TS5-8	14-May-52	4	8
WI MARQUETTE	28.08	TP10-2	6-May-55	7	8
WI MARQUETTE	13.72	PB9-3	25-Aug-57	6	8
WI MARQUETTE	8.37	PB9-4	26-Aug-57	7	8
WI MARQUETTE	2.04	PB9-5	27-Aug-57	8	8
WI MARQUETTE	3.98	PB18-8	14-Oct-57	4	8
WI MARQUETTE	0.91	PB18-9	15-Oct-57	5	8
WI MARQUETTE	0.08	UE2-7	11-Mar-62	5	8
WI MENOMINEE	0.94	TS5-9	15-May-52	5	5
WI MENOMINEE	0.85	PB4-3	7-Jul-57	10	5
WI MENOMINEE	13.72	PB9-3	25-Aug-57	6	5
WI MENOMINEE	8.37	PB9-4	26-Aug-57	7	5
WI MENOMINEE	0.91	PB18-9	15-Oct-57	5	5
WI MILWAUKEE	0.08	UE2-7	11-Mar-62	5	2
WI MILWAUKEE	0.13	UE7-5	15-Jul-62	8	2
WI MONROE	0.99	TS3-8	29-Apr-52	8	10
WI MONROE	0.99	TS3-8	29-Apr-52	8	10
WI MONROE	1.10	PB4-3	7-Jul-57	7	10
WI MONROE	2.48	PB5-7	21-Jul-57	6	10
WI MONROE	13.72	PB9-3	25-Aug-57	6	10
WI MONROE	8.37	PB9-4	26-Aug-57	7	10
WI MONROE	2.65	PB18-8	14-Oct-57	8	10
WI MONROE	0.17	UE2-7	11-Mar-62	2	10
WI MONROE	0.12	UE7-5	15-Jul-62	9	10
WI MONROE	14.56	UE12-4	17-Apr-65	2	10
WI OCONTO	1.06	TS5-8	14-May-52	4	6
WI OCONTO	1.97	PB8-5	22-Aug-57	9	6
WI OCONTO	13.72	PB9-3	25-Aug-57	6	6
WI OCONTO	8.37	PB9-4	26-Aug-57	7	6
WI OCONTO	2.04	PB9-5	27-Aug-57	8	6
WI OCONTO	0.91	PB18-9	15-Oct-57	5	6
WI ONEIDA	2.01	PB5-5	19-Jul-57	8	3
WI ONEIDA	1.97	PB8-5	22-Aug-57	9	3

COUNTY	uCi/ sqm	SHOT DAY	DATE	RANK FOR DAY	TIMES IN TOP 10
WI ONEIDA	1.16	PB15-5	20-Sep-57	7	3
WI OUTAGAMIE	1.06	TS5-8	14-May-52	4	4
WI OUTAGAMIE	13.72	PB9-3	25-Aug-57	6	4
WI OUTAGAMIE	8.37	PB9-4	26-Aug-57	7	4
WI OUTAGAMIE	0.08	UE2-7	11-Mar-62	5	4
WI OZAUKEE	13.72	PB9-3	25-Aug-57	6	5
WI OZAUKEE	8.37	PB9-4	26-Aug-57	7	5
WI OZAUKEE	0.91	PB18-9	15-Oct-57	5	5
WI OZAUKEE	0.08	UE2-7	11-Mar-62	5	5
WI OZAUKEE	0.24	UE7-5	15-Jul-62	5	5
WI PEPIN	0.99	TS3-8	29-Apr-52	8	8
WI PEPIN	0.99	TS3-8	29-Apr-52	8	8
WI PEPIN	2.54	PB2-4	21-Jun-57	4	8
WI PEPIN	1.10	PB4-3	7-Jul-57	7	8
WI PEPIN	10.11	PB5-5	19-Jul-57	1	8
WI PEPIN	1.97	PB8-5	22-Aug-57	9	8
WI PEPIN	0.17	UE2-7	11-Mar-62	2	8
WI PEPIN	0.22	UE7-5	15-Jul-62	6	8
WI PIERCE	0.99	TS3-8	29-Apr-52	8	8
WI PIERCE	0.99	TS3-8	29-Apr-52	8	8
WI PIERCE	2.54	PB2-4	21-Jun-57	4	8
WI PIERCE	1.19	PB4-3	7-Jul-57	6	8
WI PIERCE	10.11	PB5-5	19-Jul-57	1	8
WI PIERCE	1.97	PB8-5	22-Aug-57	9	8
WI PIERCE	0.17	UE2-7	11-Mar-62	2	8
WI PIERCE	0.22	UE7-5	15-Jul-62	6	8
WI POLK	1.97	TS3-8	29-Apr-52	5	8
WI POLK	1.97	TS3-8	29-Apr-52	5	8
WI POLK	2.54	PB2-4	21-Jun-57	4	8
WI POLK	1.10	PB4-3	7-Jul-57	7	8
WI POLK	10.11	PB5-5	19-Jul-57	1	8
WI POLK	1.97	PB8-5	22-Aug-57	9	8
WI POLK	1.55	PB15-3	18-Sep-57	10	8
WI POLK	0.10	UE2-7	11-Mar-62	4	8
WI PORTAGE	1.06	TS5-8	14-May-52	4	7

COUNTY	uCi/ sqm	SHOT DAY	DATE	RANK FOR DAY	TIMES IN TOP 10
WI PORTAGE	0.93	PB4-3	7-Jul-57	9	7
WI PORTAGE	13.72	PB9-3	25-Aug-57	6	7
WI PORTAGE	8.37	PB9-4	26-Aug-57	7	7
WI PORTAGE	2.65	PB18-8	14-Oct-57	8	7
WI PORTAGE	0.10	UE2-7	11-Mar-62	4	7
WI PORTAGE	0.17	UE7-5	15-Jul-62	7	7
WI PRICE	1.50	PB2-4	21-Jun-57	8	6
WI PRICE	0.93	PB4-3	7-Jul-57	9	6
WI PRICE	4.02	PB5-5	19-Jul-57	3	6
WI PRICE	1.97	PB8-5	22-Aug-57	9	6
WI PRICE	0.10	UE2-7	11-Mar-62	4	6
WI PRICE	0.13	UE7-5	15-Jul-62	8	6
WI RACINE	0.99	PB5-6	20-Jul-57	9	4
WI RACINE	2.15	PB9-5	27-Aug-57	7	4
WI RACINE	0.91	PB18-9	15-Oct-57	5	4
WI RACINE	0.12	UE7-5	15-Jul-62	9	4
WI RICHLAND	0.99	TS3-8	29-Apr-52	8	11
WI RICHLAND	0.99	TS3-8	29-Apr-52	8	11
WI RICHLAND	1.10	PB4-3	7-Jul-57	7	11
WI RICHLAND	2.48	PB5-7	21-Jul-57	6	11
WI RICHLAND	3.38	PB6-8	31-Jul-57	1	11
WI RICHLAND	13.72	PB9-3	25-Aug-57	6	11
WI RICHLAND	16.67	PB9-4	26-Aug-57	4	11
WI RICHLAND	3.98	PB18-8	14-Oct-57	4	11
WI RICHLAND	0.08	UE2-7	11-Mar-62	5	11
WI RICHLAND	0.24	UE7-5	15-Jul-62	5	11
WI RICHLAND	9.84	UE12-4	17-Apr-65	4	11
WI ROCK	0.93	PB4-3	7-Jul-57	9	8
WI ROCK	1.65	PB5-6	20-Jul-57	5	8
WI ROCK	13.72	PB9-3	25-Aug-57	6	8
WI ROCK	16.67	PB9-4	26-Aug-57	4	8
WI ROCK	2.04	PB9-5	27-Aug-57	8	8
WI ROCK	2.65	PB18-8	14-Oct-57	8	8
WI ROCK	0.08	UE2-7	11-Mar-62	5	8
WI ROCK	0.12	UE7-5	15-Jul-62	9	8

COUNTY	uCi/ sqm	SHOT DAY	DATE	RANK FOR DAY	TIMES IN TOP 10
WI RUSK	3.95	TS3-8	29-Apr-52	3	9
WI RUSK	3.95	TS3-8	29-Apr-52	3	9
WI RUSK	69.47	UpK9-2	20-May-53	5	9
WI RUSK	2.54	PB2-4	21-Jun-57	4	9
WI RUSK	1.27	PB4-3	7-Jul-57	5	9
WI RUSK	10.11	PB5-5	19-Jul-57	1	9
WI RUSK	1.97	PB8-5	22-Aug-57	9	9
WI RUSK	1.16	PB15-5	20-Sep-57	7	9
WI RUSK	0.17	UE2-7	11-Mar-62	2	9
WI SAUK	1.67	PB5-7	21-Jul-57	7	7
WI SAUK	2.24	PB6-8	31-Jul-57	4	7
WI SAUK	13.72	PB9-3	25-Aug-57	6	7
WI SAUK	33.34	PB9-4	26-Aug-57	3	7
WI SAUK	3.98	PB18-8	14-Oct-57	4	7
WI SAUK	0.08	UE2-7	11-Mar-62	5	7
WI SAUK	0.13	UE7-5	15-Jul-62	8	7
WI SAWYER	0.99	TS3-8	29-Apr-52	8	8
WI SAWYER	0.99	TS3-8	29-Apr-52	8	8
WI SAWYER	72.96	UpK9-2	20-May-53	2	8
WI SAWYER	2.54	PB2-4	21-Jun-57	4	8
WI SAWYER	0.93	PB4-3	7-Jul-57	9	8
WI SAWYER	6.09	PB5-5	19-Jul-57	2	8
WI SAWYER	1.97	PB8-5	22-Aug-57	9	8
WI SAWYER	1.11	PB9-8	30-Aug-57	1	8
WI SHAWANO	0.94	TS5-9	15-May-52	5	6
WI SHAWANO	0.85	PB4-3	7-Jul-57	10	6
WI SHAWANO	13.72	PB9-3	25-Aug-57	6	6
WI SHAWANO	8.37	PB9-4	26-Aug-57	7	6
WI SHAWANO	0.91	PB18-9	15-Oct-57	5	6
WI SHAWANO	0.24	UE7-5	15-Jul-62	5	6
WI SHEBOYGAN	13.72	PB9-3	25-Aug-57	6	4
WI SHEBOYGAN	8.37	PB9-4	26-Aug-57	7	4
WI SHEBOYGAN	0.91	PB18-9	15-Oct-57	5	4
WI SHEBOYGAN	14.56	UE12-4	17-Apr-65	2	4
WI ST CROIX	0.99	TS3-8	29-Apr-52	8	9

COUNTY	uCi/ sqm	SHOT DAY	DATE	RANK FOR DAY	TIMES IN TOP 10
WI ST CROIX	0.99	TS3-8	29-Apr-52	8	9
WI ST CROIX	2.54	PB2-4	21-Jun-57	4	9
WI ST CROIX	1.10	PB4-3	7-Jul-57	7	9
WI ST CROIX	10.11	PB5-5	19-Jul-57	1	9
WI ST CROIX	1.97	PB8-5	22-Aug-57	9	9
WI ST CROIX	1.16	PB15-5	20-Sep-57	7	9
WI ST CROIX	0.10	UE2-7	11-Mar-62	3	9
WI ST CROIX	0.12	UE7-5	15-Jul-62	9	9
WI TAYLOR	0.99	TS3-8	29-Apr-52	8	7
WI TAYLOR	0.99	TS3-8	29-Apr-52	8	7
WI TAYLOR	2.54	PB2-4	21-Jun-57	4	7
WI TAYLOR	1.27	PB4-3	7-Jul-57	5	7
WI TAYLOR	1.16	PB15-5	20-Sep-57	7	7
WI TAYLOR	0.10	UE2-7	11-Mar-62	4	7
WI TAYLOR	0.13	UE7-5	15-Jul-62	8	7
WI TREMPEALEAU	0.99	TS3-8	29-Apr-52	8	10
WI TREMPEALEAU	0.99	TS3-8	29-Apr-52	8	10
WI TREMPEALEAU	70.05	UpK9-2	20-May-53	3	10
WI TREMPEALEAU	2.54	PB2-4	21-Jun-57	4	10
WI TREMPEALEAU	0.85	PB4-3	7-Jul-57	10	10
WI TREMPEALEAU	1.18	PB5-6	20-Jul-57	7	10
WI TREMPEALEAU	1.16	PB15-5	20-Sep-57	7	10
WI TREMPEALEAU	0.10	UE2-7	11-Mar-62	4	10
WI TREMPEALEAU	0.12	UE7-5	15-Jul-62	9	10
WI TREMPEALEAU	14.56	UE12-4	17-Apr-65	2	10
WI VERNON	0.99	TS3-8	29-Apr-52	8	11
WI VERNON	0.99	TS3-8	29-Apr-52	8	11
WI VERNON	1.10	PB4-3	7-Jul-57	7	11
WI VERNON	4.15	PB5-7	21-Jul-57	2	11
WI VERNON	1.63	PB6-9	1-Aug-57	3	11
WI VERNON	13.72	PB9-3	25-Aug-57	6	11
WI VERNON	8.37	PB9-4	26-Aug-57	7	11
WI VERNON	2.54	PB17-2	29-Sep-57	2	11
WI VERNON	3.98	PB18-8	14-Oct-57	4	11
WI VERNON	0.24	UE7-5	15-Jul-62	5	11

COUNTY	uCi/ sqm	SHOT DAY	DATE	RANK FOR DAY	TIMES IN TOP 10
WI VERNON	9.32	UE12-4	17-Apr-65	6	11
WI VILAS	4.02	PB5-5	19-Jul-57	3	3
WI VILAS	1.97	PB8-5	22-Aug-57	9	3
WI VILAS	1.16	PB15-5	20-Sep-57	7	3
WI WALWORTH	0.99	PB5-6	20-Jul-57	9	7
WI WALWORTH	13.72	PB9-3	25-Aug-57	6	7
WI WALWORTH	16.67	PB9-4	26-Aug-57	4	7
WI WALWORTH	2.15	PB9-5	27-Aug-57	7	7
WI WALWORTH	0.89	PB9-7	29-Aug-57	6	7
WI WALWORTH	0.08	UE2-7	11-Mar-62	5	7
WI WALWORTH	0.12	UE7-5	15-Jul-62	9	7
WI WASHBURN	0.99	TS3-8	29-Apr-52	8	9
WI WASHBURN	0.99	TS3-8	29-Apr-52	8	9
WI WASHBURN	73.09	UpK9-2	20-May-53	1	9
WI WASHBURN	2.54	PB2-4	21-Jun-57	4	9
WI WASHBURN	10.11	PB5-5	19-Jul-57	1	9
WI WASHBURN	1.97	PB8-5	22-Aug-57	9	9
WI WASHBURN	1.11	PB9-8	30-Aug-57	1	9
WI WASHBURN	1.55	PB15-3	18-Sep-57	10	9
WI WASHBURN	0.10	UE2-7	11-Mar-62	3	9
WI WASHINGTON	0.85	PB4-3	7-Jul-57	10	5
WI WASHINGTON	0.99	PB5-6	20-Jul-57	9	5
WI WASHINGTON	13.72	PB9-3	25-Aug-57	6	5
WI WASHINGTON	8.37	PB9-4	26-Aug-57	7	5
WI WASHINGTON	0.24	UE7-5	15-Jul-62	5	5
WI WAUKESHA	13.72	PB9-3	25-Aug-57	6	4
WI WAUKESHA	8.37	PB9-4	26-Aug-57	7	4
WI WAUKESHA	0.08	UE2-7	11-Mar-62	5	4
WI WAUKESHA	0.13	UE7-5	15-Jul-62	8	4
WI WAUPACA	1.06	TS5-8	14-May-52	4	8
WI WAUPACA	0.85	PB4-3	7-Jul-57	10	8
WI WAUPACA	20.62	PB9-3	25-Aug-57	4	8
WI WAUPACA	8.37	PB9-4	26-Aug-57	7	8
WI WAUPACA	2.65	PB18-8	14-Oct-57	8	8
WI WAUPACA	0.91	PB18-9	15-Oct-57	5	8

COUNTY	uCi/ sqm	SHOT DAY	DATE	RANK FOR DAY	TIMES IN TOP 10
WI WAUPACA	0.10	UE2-7	11-Mar-62	4	8
WI WAUPACA	7.97	UE12-4	17-Apr-65	8	8
WI WAUSHARA	1.06	TS5-8	14-May-52	4	7
WI WAUSHARA	28.08	TP10-2	6-May-55	7	7
WI WAUSHARA	13.72	PB9-3	25-Aug-57	6	7
WI WAUSHARA	8.37	PB9-4	26-Aug-57	7	7
WI WAUSHARA	2.65	PB18-8	14-Oct-57	8	7
WI WAUSHARA	0.13	UE7-5	15-Jul-62	8	7
WI WAUSHARA	14.56	UE12-4	17-Apr-65	2	7
WI WINNEBAGO	1.06	TS5-8	14-May-52	4	6
WI WINNEBAGO	0.85	PB4-3	7-Jul-57	10	6
WI WINNEBAGO	1.18	PB5-6	20-Jul-57	7	6
WI WINNEBAGO	13.72	PB9-3	25-Aug-57	6	6
WI WINNEBAGO	8.37	PB9-4	26-Aug-57	7	6
WI WINNEBAGO	14.56	UE12-4	17-Apr-65	2	6
WI WOOD	1.06	TS5-8	14-May-52	4	7
WI WOOD	0.93	PB4-3	7-Jul-57	9	7
WI WOOD	13.72	PB9-3	25-Aug-57	6	7
WI WOOD	12.52	PB9-4	26-Aug-57	5	7
WI WOOD	2.65	PB18-8	14-Oct-57	8	7
WI WOOD	0.17	UE2-7	11-Mar-62	2	7
WI WOOD	0.24	UE7-5	15-Jul-62	5	7

COUNTY	uCi/ sqm	SHOT DAY	DATE	RANK FOR DAY	TIMES IN TOP 10
WV BARBOUR	1.29	TP11-8	22-May-55	9	3
WV BARBOUR	5.97	PB5-8	22-Jul-57	3	3
WV BARBOUR	2.83	PB5-9	23-Jul-57	2	3
WV BERKELEY	1.04	UpK4-4	9-Apr-53	5	3
WV BERKELEY	8.96	PB5-8	22-Jul-57	2	3
WV BERKELEY	1.51	PB8-7	24-Aug-57	10	3
WV BOONE	1.51	PB5-8	22-Jul-57	7	3
WV BOONE	359.52	S-4	9-Jul-62	1	3
WV BOONE	2.15	UE8-6	19-Jul-62	5	3
WV BRAXTON	1.18	BJ1-2	29-Oct-51	4	5
WV BRAXTON	1.29	TP11-8	22-May-55	9	5
WV BRAXTON	5.97	PB5-8	22-Jul-57	3	5
WV BRAXTON	2.83	PB5-9	23-Jul-57	2	5
WV BRAXTON	85.56	S-4	9-Jul-62	9	5
WV BROOKE	10.89	UpK9-3	21-May-53	8	4
WV BROOKE	1.29	TP11-8	22-May-55	9	4
WV BROOKE	8.96	PB5-8	22-Jul-57	2	4
WV BROOKE	85.56	S-4	9-Jul-62	9	4
WV CABELL	2.03	TP6-4	25-Mar-55	8	5
WV CABELL	0.97	TP6-5	26-Mar-55	8	5
WV CABELL	8.96	PB5-8	22-Jul-57	2	5
WV CABELL	23.55	PB6-3	26-Jul-57	3	5
WV CABELL	2.23	PB13-7	12-Sep-57	1	5
WV CALHOUN	1.29	TP11-8	22-May-55	9	6
WV CALHOUN	8.96	PB5-8	22-Jul-57	2	6
WV CALHOUN	2.83	PB5-9	23-Jul-57	2	6
WV CALHOUN	0.81	PB15-7	22-Sep-57	9	6
WV CALHOUN	103.59	S-4	9-Jul-62	6	6
WV CALHOUN	1.74	UE8-5	18-Jul-62	10	6
WV CLAY	1.36	TS5-4	10-May-52	10	4
WV CLAY	1.51	PB5-8	22-Jul-57	7	4
WV CLAY	2.83	PB5-9	23-Jul-57	2	4
WV CLAY	90.29	S-4	9-Jul-62	8	4
WV DODDRIDGE	2.57	TP11-8	22-May-55	5	2
WV DODDRIDGE	8.96	PB5-8	22-Jul-57	2	2

COUNTY	uCi/ sqm	SHOT DAY	DATE	RANK FOR DAY	TIMES IN TOP 10
WV FAYETTE	1.36	TS5-4	10-May-52	10	5
WV FAYETTE	2.99	PB5-8	22-Jul-57	4	5
WV FAYETTE	4.72	PB5-9	23-Jul-57	1	5
WV FAYETTE	85.56	S-4	9-Jul-62	9	5
WV FAYETTE	6.62	S-6	11-Jul-62	7	5
WV GILMER	8.96	PB5-8	22-Jul-57	2	5
WV GILMER	2.83	PB5-9	23-Jul-57	2	5
WV GILMER	0.81	PB15-7	22-Sep-57	9	5
WV GILMER	103.59	S-4	9-Jul-62	6	5
WV GILMER	1.74	UE8-5	18-Jul-62	10	5
WV GRANT	1.01	UpK6-5	22-Apr-53	7	5
WV GRANT	1.29	TP11-8	22-May-55	9	5
WV GRANT	5.97	PB5-8	22-Jul-57	3	5
WV GRANT	2.83	PB5-9	23-Jul-57	2	5
WV GRANT	85.56	S-4	9-Jul-62	9	5
WV GREENBRIER	1.45	TS5-4	10-May-52	8	5
WV GREENBRIER	5.97	PB5-8	22-Jul-57	3	5
WV GREENBRIER	2.83	PB5-9	23-Jul-57	2	5
WV GREENBRIER	103.59	S-4	9-Jul-62	6	5
WV GREENBRIER	1.74	UE8-5	18-Jul-62	10	5
WV HAMPSHIRE	1.04	UpK4-4	9-Apr-53	5	3
WV HAMPSHIRE	8.96	PB5-8	22-Jul-57	2	3
WV HAMPSHIRE	1.89	PB5-9	23-Jul-57	3	3
WV HANCOCK	7.20	BJ2-2	31-Oct-51	5	6
WV HANCOCK	11.32	UpK9-3	21-May-53	5	6
WV HANCOCK	1.72	TP3-3	3-Mar-55	8	6
WV HANCOCK	3.86	TP11-8	22-May-55	3	6
WV HANCOCK	4.68	TP11-9	23-May-55	5	6
WV HANCOCK	8.96	PB5-8	22-Jul-57	2	6
WV HARDY	1.18	UpK8-6	13-May-53	2	3
WV HARDY	8.96	PB5-8	22-Jul-57	2	3
WV HARDY	2.83	PB5-9	23-Jul-57	2	3
WV HARRISON	2.57	TP11-8	22-May-55	5	6
WV HARRISON	3.12	TP11-9	23-May-55	8	6
WV HARRISON	8.96	PB5-8	22-Jul-57	2	6

COUNTY	uCi/ sqm	SHOT DAY	DATE	RANK FOR DAY	TIMES IN TOP 10
WV HARRISON	0.94	PB5-9	23-Jul-57	4	6
WV HARRISON	103.59	S-4	9-Jul-62	6	6
WV HARRISON	1.74	UE8-5	18-Jul-62	10	6
WV JACKSON	3.86	TP11-8	22-May-55	3	3
WV JACKSON	4.68	TP11-9	23-May-55	5	3
WV JACKSON	8.96	PB5-8	22-Jul-57	2	3
WV JEFFERSON	1.04	UpK4-4	9-Apr-53	5	3
WV JEFFERSON	8.96	PB5-8	22-Jul-57	2	3
WV JEFFERSON	0.94	PB5-9	23-Jul-57	4	3
WV KANAWHA	1.18	BJ1-2	29-Oct-51	4	5
WV KANAWHA	1.18	UpK8-6	13-May-53	2	5
WV KANAWHA	5.97	PB5-8	22-Jul-57	3	5
WV KANAWHA	2.83	PB5-9	23-Jul-57	2	5
WV KANAWHA	208.78	S-4	9-Jul-62	3	5
WV LEWIS	2.57	TP11-8	22-May-55	5	6
WV LEWIS	3.12	TP11-9	23-May-55	8	6
WV LEWIS	8.96	PB5-8	22-Jul-57	2	6
WV LEWIS	1.89	PB5-9	23-Jul-57	3	6
WV LEWIS	85.56	S-4	9-Jul-62	9	6
WV LEWIS	1.74	UE8-5	18-Jul-62	10	6
WV LINCOLN	1.04	UpK5-2	12-Apr-53	4	8
WV LINCOLN	2.03	TP6-4	25-Mar-55	8	8
WV LINCOLN	0.97	TP6-5	26-Mar-55	8	8
WV LINCOLN	5.97	PB5-8	22-Jul-57	3	8
WV LINCOLN	15.70	PB6-3	26-Jul-57	5	8
WV LINCOLN	2.23	PB13-7	12-Sep-57	1	8
WV LINCOLN	1.08	PB15-7	22-Sep-57	5	8
WV LINCOLN	208.78	S-4	9-Jul-62	3	8
WV LOGAN	1.04	UpK5-2	12-Apr-53	4	8
WV LOGAN	2.03	TP6-4	25-Mar-55	8	8
WV LOGAN	0.97	TP6-5	26-Mar-55	8	8
WV LOGAN	13.06	PB4-5	9-Jul-57	7	8
WV LOGAN	1.51	PB5-8	22-Jul-57	7	8
WV LOGAN	3.43	PB6-7	30-Jul-57	1	8
WV LOGAN	1.62	PB6-8	31-Jul-57	6	8

COUNTY	uCi/ sqm	SHOT DAY	DATE	RANK FOR DAY	TIMES IN TOP 10
WV LOGAN	359.52	S-4	9-Jul-62	1	8
WV MARION	3.86	TP11-8	22-May-55	3	5
WV MARION	4.68	TP11-9	23-May-55	5	5
WV MARION	14.93	PB5-8	22-Jul-57	1	5
WV MARION	2.83	PB5-9	23-Jul-57	2	5
WV MARION	1.74	UE8-5	18-Jul-62	10	5
WV MARSHALL	1.72	TP3-3	3-Mar-55	8	6
WV MARSHALL	2.57	TP11-8	22-May-55	5	6
WV MARSHALL	3.12	TP11-9	23-May-55	8	6
WV MARSHALL	8.96	PB5-8	22-Jul-57	2	6
WV MARSHALL	2.38	PB6-7	30-Jul-57	4	6
WV MARSHALL	103.59	S-4	9-Jul-62	6	6
WV MASON	3.45	BJ3-5	5-Nov-51	10	8
WV MASON	3.86	TP11-8	22-May-55	3	8
WV MASON	4.68	TP11-9	23-May-55	5	8
WV MASON	5.97	PB5-8	22-Jul-57	3	8
WV MASON	23.55	PB6-3	26-Jul-57	3	8
WV MASON	2.23	PB13-7	12-Sep-57	1	8
WV MASON	1.08	PB15-7	22-Sep-57	5	8
WV MASON	208.78	S-4	9-Jul-62	3	8
WV MCDOWELL	33.77	UpK9-4	22-May-53	1	7
WV MCDOWELL	4.18	PB2-5	22-Jun-57	3	7
WV MCDOWELL	13.06	PB4-5	9-Jul-57	7	7
WV MCDOWELL	2.99	PB5-8	22-Jul-57	4	7
WV MCDOWELL	2.83	PB5-9	23-Jul-57	2	7
WV MCDOWELL	149.13	S-4	9-Jul-62	5	7
WV MCDOWELL	1.32	UE8-6	19-Jul-62	9	7
WV MERCER	1.18	BJ1-2	29-Oct-51	4	7
WV MERCER	4.59	BJ3-5	5-Nov-51	6	7
WV MERCER	1.40	TS5-4	10-May-52	9	7
WV MERCER	2.78	PB2-5	22-Jun-57	4	7
WV MERCER	5.97	PB5-8	22-Jul-57	3	7
WV MERCER	2.83	PB5-9	23-Jul-57	2	7
WV MERCER	149.13	S-4	9-Jul-62	5	7
WV MINERAL	1.29	TP11-8	22-May-55	9	3

COUNTY	uCi/ sqm	SHOT DAY	DATE	RANK FOR DAY	TIMES IN TOP 10
WV MINERAL	5.97	PB5-8	22-Jul-57	3	3
WV MINERAL	2.83	PB5-9	23-Jul-57	2	3
WV MINGO	1.04	UpK5-2	12-Apr-53	4	8
WV MINGO	2.03	TP6-4	25-Mar-55	8	8
WV MINGO	0.97	TP6-5	26-Mar-55	8	8
WV MINGO	2.78	PB2-5	22-Jun-57	4	8
WV MINGO	1.51	PB5-8	22-Jul-57	7	8
WV MINGO	2.06	PB6-7	30-Jul-57	5	8
WV MINGO	7.49	PB17-3	30-Sep-57	5	8
WV MINGO	149.13	S-4	9-Jul-62	5	8
WV MONONGALIA	1.29	TP11-8	22-May-55	9	4
WV MONONGALIA	8.96	PB5-8	22-Jul-57	2	4
WV MONONGALIA	2.83	PB5-9	23-Jul-57	2	4
WV MONONGALIA	85.56	S-4	9-Jul-62	9	4
WV MONROE	1.51	PB5-8	22-Jul-57	7	5
WV MONROE	2.83	PB5-9	23-Jul-57	2	5
WV MONROE	208.78	S-4	9-Jul-62	3	5
WV MONROE	1.74	UE8-5	18-Jul-62	10	5
WV MONROE	1.32	UE8-6	19-Jul-62	9	5
WV MORGAN	1.04	UpK4-4	9-Apr-53	5	4
WV MORGAN	5.39	UpK10-6	30-May-53	4	4
WV MORGAN	8.96	PB5-8	22-Jul-57	2	4
WV MORGAN	2.83	PB5-9	23-Jul-57	2	4
WV NICHOLAS	5.97	PB5-8	22-Jul-57	3	3
WV NICHOLAS	2.83	PB5-9	23-Jul-57	2	3
WV NICHOLAS	85.56	S-4	9-Jul-62	9	3
WV OHIO	48.19	R1-2	29-Jan-51	1	6
WV OHIO	4.59	BJ3-5	5-Nov-51	6	6
WV OHIO	1.72	TP3-3	3-Mar-55	8	6
WV OHIO	2.57	TP11-8	22-May-55	5	6
WV OHIO	3.12	TP11-9	23-May-55	8	6
WV OHIO	8.96	PB5-8	22-Jul-57	2	6
WV PENDLETON	1.77	UpK8-6	13-May-53	1	4
WV PENDLETON	1.51	PB5-8	22-Jul-57	7	4
WV PENDLETON	2.83	PB5-9	23-Jul-57	2	4

COUNTY	uCi/ sqm	SHOT DAY	DATE	RANK FOR DAY	TIMES IN TOP 10
WV PENDLETON	208.78	S-4	9-Jul-62	3	4
WV PLEASANTS	41.49	R1-2	29-Jan-51	3	4
WV PLEASANTS	1.29	TP11-8	22-May-55	9	4
WV PLEASANTS	1.54	PB5-7	21-Jul-57	9	4
WV PLEASANTS	8.96	PB5-8	22-Jul-57	2	4
WV POCAHONTAS	1.45	TS5-4	10-May-52	8	4
WV POCAHONTAS	8.96	PB5-8	22-Jul-57	2	4
WV POCAHONTAS	2.83	PB5-9	23-Jul-57	2	4
WV POCAHONTAS	90.29	S-4	9-Jul-62	8	4
WV PRESTON	1.29	TP11-8	22-May-55	9	3
WV PRESTON	8.96	PB5-8	22-Jul-57	2	3
WV PRESTON	4.72	PB5-9	23-Jul-57	1	3
WV PUTNAM	1.72	BJ1-2	29-Oct-51	2	7
WV PUTNAM	2.57	TP11-8	22-May-55	5	7
WV PUTNAM	3.12	TP11-9	23-May-55	8	7
WV PUTNAM	2.99	PB5-8	22-Jul-57	4	7
WV PUTNAM	1.34	PB13-7	12-Sep-57	2	7
WV PUTNAM	1.08	PB15-7	22-Sep-57	5	7
WV PUTNAM	85.56	S-4	9-Jul-62	9	7
WV RALEIGH	1.72	BJ1-2	29-Oct-51	2	7
WV RALEIGH	1.18	UpK8-6	13-May-53	2	7
WV RALEIGH	1.51	PB5-8	22-Jul-57	7	7
WV RALEIGH	2.83	PB5-9	23-Jul-57	2	7
WV RALEIGH	208.78	S-4	9-Jul-62	3	7
WV RALEIGH	1.74	UE8-5	18-Jul-62	10	7
WV RALEIGH	1.32	UE8-6	19-Jul-62	9	7
WV RANDOLPH	5.97	PB5-8	22-Jul-57	3	4
WV RANDOLPH	2.83	PB5-9	23-Jul-57	2	4
WV RANDOLPH	85.56	S-4	9-Jul-62	9	4
WV RANDOLPH	1.74	UE8-5	18-Jul-62	10	4
WV RITCHIE	2.57	TP11-8	22-May-55	5	3
WV RITCHIE	3.12	TP11-9	23-May-55	8	3
WV RITCHIE	8.96	PB5-8	22-Jul-57	2	3
WV ROANE	1.29	TP11-8	22-May-55	9	4
WV ROANE	8.96	PB5-8	22-Jul-57	2	4

COUNTY	uCi/ sqm	SHOT DAY	DATE	RANK FOR DAY	TIMES IN TOP 10
WV ROANE	0.81	PB15-7	22-Sep-57	9	4
WV ROANE	3.35	UE8-5	18-Jul-62	4	4
WV SUMMERS	1.72	BJ1-2	29-Oct-51	2	6
WV SUMMERS	4.59	BJ3-5	5-Nov-51	6	6
WV SUMMERS	1.18	UpK8-6	13-May-53	2	6
WV SUMMERS	8.96	PB5-8	22-Jul-57	2	6
WV SUMMERS	1.89	PB5-9	23-Jul-57	3	6
WV SUMMERS	208.78	S-4	9-Jul-62	3	6
WV TAYLOR	1.18	BJ1-2	29-Oct-51	4	2
WV TAYLOR	8.96	PB5-8	22-Jul-57	2	2
WV TUCKER	2.15	TP6-4	25-Mar-55	7	5
WV TUCKER	1.29	TP11-8	22-May-55	9	5
WV TUCKER	5.97	PB5-8	22-Jul-57	3	5
WV TUCKER	4.72	PB5-9	23-Jul-57	1	5
WV TUCKER	85.56	S-4	9-Jul-62	9	5
WV TYLER	4.59	BJ3-5	5-Nov-51	6	4
WV TYLER	1.29	TP11-8	22-May-55	9	4
WV TYLER	14.93	PB5-8	22-Jul-57	1	4
WV TYLER	90.29	S-4	9-Jul-62	8	4
WV UPSHUR	3.86	TP11-8	22-May-55	3	5
WV UPSHUR	4.68	TP11-9	23-May-55	5	5
WV UPSHUR	8.96	PB5-8	22-Jul-57	2	5
WV UPSHUR	0.94	PB5-9	23-Jul-57	4	5
WV UPSHUR	103.59	S-4	9-Jul-62	6	5
WV WAYNE	1.04	UpK5-2	12-Apr-53	4	7
WV WAYNE	3.11	TP6-4	25-Mar-55	3	7
WV WAYNE	0.97	TP6-5	26-Mar-55	8	7
WV WAYNE	8.96	PB5-8	22-Jul-57	2	7
WV WAYNE	1.34	PB13-7	12-Sep-57	2	7
WV WAYNE	1.08	PB15-7	22-Sep-57	5	7
WV WAYNE	208.78	S-4	9-Jul-62	3	7
WV WEBSTER	1.51	PB5-8	22-Jul-57	7	3
WV WEBSTER	4.72	PB5-9	23-Jul-57	1	3
WV WEBSTER	90.29	S-4	9-Jul-62	8	3
WV WETZEL	2.57	TP11-8	22-May-55	5	5

COUNTY	uCi/ sqm	SHOT DAY	DATE	RANK FOR DAY	TIMES IN TOP 10
WV WETZEL	3.12	TP11-9	23-May-55	8	5
WV WETZEL	14.93	PB5-8	22-Jul-57	1	5
WV WETZEL	85.56	S-4	9-Jul-62	9	5
WV WETZEL	1.74	UE8-5	18-Jul-62	10	5
WV WIRT	3.86	TP11-8	22-May-55	3	7
WV WIRT	4.68	TP11-9	23-May-55	5	7
WV WIRT	5.97	PB5-8	22-Jul-57	3	7
WV WIRT	0.81	PB15-7	22-Sep-57	9	7
WV WIRT	85.56	S-4	9-Jul-62	9	7
WV WIRT	8.10	S-6	11-Jul-62	5	7
WV WIRT	3.35	UE8-5	18-Jul-62	4	7
WV WOOD	4.59	BJ3-5	5-Nov-51	6	6
WV WOOD	1.29	TP11-8	22-May-55	9	6
WV WOOD	1.17	PB2-4	21-Jun-57	10	6
WV WOOD	8.96	PB5-8	22-Jul-57	2	6
WV WOOD	85.56	S-4	9-Jul-62	9	6
WV WOOD	1.74	UE8-5	18-Jul-62	10	6
WV WYOMING	13.06	PB4-5	9-Jul-57	7	4
WV WYOMING	1.51	PB5-8	22-Jul-57	7	4
WV WYOMING	4.72	PB5-9	23-Jul-57	1	4
WV WYOMING	149.13	S-4	9-Jul-62	5	4

COUNTY	uCi/ sqm	SHOT DAY	DATE	RANK FOR DAY	TIMES IN TOP 10
WY ALBANY	9.53	UpK11-2	5-Jun-53	4	14
WY ALBANY	3.16	TP3-2	2-Mar-55	7	14
WY ALBANY	1.47	TP8-2	10-Apr-55	9	14
WY ALBANY	24.89	TP10-2	6-May-55	9	14
WY ALBANY	7.30	PB2-2	19-Jun-57	3	14
WY ALBANY	0.84	PB5-8	22-Jul-57	8	14
WY ALBANY	0.83	PB8-2	19-Aug-57	8	14
WY ALBANY	3.68	PB9-4	26-Aug-57	9	14
WY ALBANY	2.37	PB18-3	9-Oct-57	3	14
WY ALBANY	1.44	PB18-4	10-Oct-57	6	14
WY ALBANY	0.44	UE2-5	9-Mar-62	1	14
WY ALBANY	0.11	UE2-6	10-Mar-62	6	14
WY ALBANY	5.91	S-1	6-Jul-62	7	14
WY ALBANY	19.93	UE8-3	16-Jul-62	2	14
WY BIG HORN	16.38	UpK8-2	9-May-53	3	16
WY BIG HORN	2.63	UpK8-3	10-May-53	5	16
WY BIG HORN	1.22	UpK8-4	11-May-53	8	16
WY BIG HORN	2.07	UpK10-7	31-May-53	3	16
WY BIG HORN	1.32	PB2-2	19-Jun-57	9	16
WY BIG HORN	1.65	PB8-2	19-Aug-57	6	16
WY BIG HORN	3.07	PB8-4	21-Aug-57	10	16
WY BIG HORN	3.07	PB8-4	21-Aug-57	10	16
WY BIG HORN	1.86	PB9-2	24-Aug-57	9	16
WY BIG HORN	3.68	PB9-4	26-Aug-57	9	16
WY BIG HORN	0.83	PB9-6	28-Aug-57	8	16
WY BIG HORN	3.47	PB15-2	17-Sep-57	10	16
WY BIG HORN	0.44	UE2-5	9-Mar-62	1	16
WY BIG HORN	0.21	UE2-6	10-Mar-62	3	16
WY BIG HORN	4.20	UE7-3	13-Jul-62	1	16
WY BIG HORN	0.20	BN-5	22-Dec-70	3	16
WY CAMPBELL	23.57	UpK2-2	25-Mar-53	5	16
WY CAMPBELL	4.73	UpK8-1	8-May-53	6	16
WY CAMPBELL	1.89	TP3-2	2-Mar-55	9	16
WY CAMPBELL	1.25	TP8-2	10-Apr-55	10	16
WY CAMPBELL	20.60	TP11-2	16-May-55	5	16

COUNTY	uCi/ sqm	SHOT DAY	DATE	RANK FOR DAY	TIMES IN TOP 10
WY CAMPBELL	10.10	PB5-3	17-Jul-57	4	16
WY CAMPBELL	0.83	PB8-2	19-Aug-57	8	16
WY CAMPBELL	3.07	PB8-4	21-Aug-57	10	16
WY CAMPBELL	3.07	PB8-4	21-Aug-57	10	16
WY CAMPBELL	7.76	PB9-3	25-Aug-57	9	16
WY CAMPBELL	7.36	PB9-4	26-Aug-57	7	16
WY CAMPBELL	6.81	PB11-2	1-Sep-57	6	16
WY CAMPBELL	1.44	PB18-4	10-Oct-57	6	16
WY CAMPBELL	0.22	UE2-6	10-Mar-62	2	16
WY CAMPBELL	2.66	S-2	7-Jul-62	10	16
WY CAMPBELL	9.84	UE12-4	17-Apr-65	4	16
WY CARBON	31.57	UpK10-1	25-May-53	7	2
WY CARBON	15.24	PB17-1	28-Sep-57	6	2
WY CONVERSE	23.57	UpK2-2	25-Mar-53	5	15
WY CONVERSE	1.58	UpK8-1	8-May-53	8	15
WY CONVERSE	2.61	TP8-2	10-Apr-55	4	15
WY CONVERSE	13.73	TP11-2	16-May-55	8	15
WY CONVERSE	14.59	PB2-2	19-Jun-57	2	15
WY CONVERSE	13.07	PB2-3	20-Jun-57	7	15
WY CONVERSE	0.83	PB8-2	19-Aug-57	8	15
WY CONVERSE	7.76	PB9-3	25-Aug-57	9	15
WY CONVERSE	11.05	PB9-4	26-Aug-57	6	15
WY CONVERSE	2.37	PB18-3	9-Oct-57	3	15
WY CONVERSE	1.44	PB18-4	10-Oct-57	6	15
WY CONVERSE	0.44	UE2-5	9-Mar-62	1	15
WY CONVERSE	0.22	UE2-6	10-Mar-62	2	15
WY CONVERSE	0.10	UE2-7	11-Mar-62	3	15
WY CONVERSE	4.66	S-2	7-Jul-62	8	15
WY CROOK	45.45	UpK2-2	25-Mar-53	4	15
WY CROOK	3.16	UpK8-1	8-May-53	7	15
WY CROOK	9.68	PB5-2	16-Jul-57	6	15
WY CROOK	10.10	PB5-3	17-Jul-57	4	15
WY CROOK	0.83	PB8-2	19-Aug-57	8	15
WY CROOK	4.60	PB8-4	21-Aug-57	8	15
WY CROOK	4.60	PB8-4	21-Aug-57	8	15

COUNTY	uCi/ sqm	SHOT DAY	DATE	RANK FOR DAY	TIMES IN TOP 10
WY CROOK	15.53	PB9-3	25-Aug-57	5	15
WY CROOK	7.36	PB9-4	26-Aug-57	7	15
WY CROOK	6.81	PB11-2	1-Sep-57	6	15
WY CROOK	1.44	PB18-4	10-Oct-57	6	15
WY CROOK	0.31	UE2-5	9-Mar-62	4	15
WY CROOK	0.22	UE2-6	10-Mar-62	2	15
WY CROOK	2.48	UE7-3	13-Jul-62	4	15
WY CROOK	9.84	UE12-4	17-Apr-65	4	15
WY FREMONT	564.41	UpK2-1	24-Mar-53	8	2
WY FREMONT	15.24	PB17-1	28-Sep-57	6	2
WY GOSHEN	3.71	UpK2-7	30-Mar-53	8	17
WY GOSHEN	4.00	TP3-2	2-Mar-55	5	17
WY GOSHEN	10.18	PB5-3	17-Jul-57	3	17
WY GOSHEN	4.87	PB5-4	18-Jul-57	8	17
WY GOSHEN	1.65	PB8-2	19-Aug-57	6	17
WY GOSHEN	6.85	PB8-3	20-Aug-57	10	17
WY GOSHEN	3.68	PB9-4	26-Aug-57	9	17
WY GOSHEN	0.83	PB9-6	28-Aug-57	8	17
WY GOSHEN	2.37	PB18-3	9-Oct-57	3	17
WY GOSHEN	1.44	PB18-4	10-Oct-57	6	17
WY GOSHEN	0.86	PB18-6	12-Oct-57	7	17
WY GOSHEN	1.18	UE2-4	8-Mar-62	1	17
WY GOSHEN	0.44	UE2-5	9-Mar-62	1	17
WY GOSHEN	0.11	UE2-6	10-Mar-62	6	17
WY GOSHEN	0.10	UE2-7	11-Mar-62	3	17
WY GOSHEN	3.33	S-2	7-Jul-62	9	17
WY GOSHEN	10.77	UE8-3	16-Jul-62	7	17
WY HOT SPRINGS	41.70	TS5-2	8-May-52	1	19
WY HOT SPRINGS	16.38	UpK8-2	9-May-53	3	19
WY HOT SPRINGS	1.32	UpK8-3	10-May-53	9	19
WY HOT SPRINGS	1.22	UpK8-4	11-May-53	8	19
WY HOT SPRINGS	2.07	UpK10-7	31-May-53	3	19
WY HOT SPRINGS	1.56	PB2-4	21-Jun-57	7	19
WY HOT SPRINGS	2.63	PB4-2	6-Jul-57	5	19
WY HOT SPRINGS	1.65	PB8-2	19-Aug-57	6	19

COUNTY	uCi/ sqm	SHOT DAY	DATE	RANK FOR DAY	TIMES IN TOP 10
WY HOT SPRINGS	3.07	PB8-4	21-Aug-57	10	19
WY HOT SPRINGS	3.07	PB8-4	21-Aug-57	10	19
WY HOT SPRINGS	3.60	PB9-2	24-Aug-57	5	19
WY HOT SPRINGS	3.68	PB9-4	26-Aug-57	9	19
WY HOT SPRINGS	0.83	PB9-6	28-Aug-57	8	19
WY HOT SPRINGS	0.80	PB13-7	12-Sep-57	7	19
WY HOT SPRINGS	1.72	PB15-3	18-Sep-57	8	19
WY HOT SPRINGS	0.12	UE2-6	10-Mar-62	5	19
WY HOT SPRINGS	4.73	S-1	6-Jul-62	9	19
WY HOT SPRINGS	2.86	UE7-3	13-Jul-62	3	19
WY HOT SPRINGS	0.14	BN-5	22-Dec-70	4	19
WY JOHNSON	23.77	TS5-2	8-May-52	10	17
WY JOHNSON	23.57	UpK2-2	25-Mar-53	5	17
WY JOHNSON	3.16	UpK8-1	8-May-53	7	17
WY JOHNSON	2.26	UpK10-3	27-May-53	4	17
WY JOHNSON	7.33	UpK11-2	5-Jun-53	7	17
WY JOHNSON	13.73	TP11-2	16-May-55	8	17
WY JOHNSON	2.63	PB4-2	6-Jul-57	5	17
WY JOHNSON	5.09	PB5-3	17-Jul-57	8	17
WY JOHNSON	0.83	PB8-2	19-Aug-57	8	17
WY JOHNSON	3.07	PB8-4	21-Aug-57	10	17
WY JOHNSON	3.07	PB8-4	21-Aug-57	10	17
WY JOHNSON	2.73	PB9-2	24-Aug-57	6	17
WY JOHNSON	3.68	PB9-4	26-Aug-57	9	17
WY JOHNSON	0.80	PB13-7	12-Sep-57	7	17
WY JOHNSON	0.21	UE2-6	10-Mar-62	3	17
WY JOHNSON	4.73	S-1	6-Jul-62	9	17
WY JOHNSON	4.20	UE7-3	13-Jul-62	1	17
WY LARAMIE	2.11	TP8-2	10-Apr-55	7	11
WY LARAMIE	3.03	PB2-2	19-Jun-57	5	11
WY LARAMIE	3.18	PB8-4	21-Aug-57	9	11
WY LARAMIE	3.18	PB8-4	21-Aug-57	9	11
WY LARAMIE	1.02	PB9-6	28-Aug-57	5	11
WY LARAMIE	3.55	PB18-3	9-Oct-57	2	11
WY LARAMIE	0.65	UE2-4	8-Mar-62	5	11

COUNTY	uCi/ sqm	SHOT DAY	DATE	RANK FOR DAY	TIMES IN TOP 10
WY LARAMIE	0.44	UE2-5	9-Mar-62	1	11
WY LARAMIE	0.11	UE2-6	10-Mar-62	6	11
WY LARAMIE	0.10	UE2-7	11-Mar-62	3	11
WY LARAMIE	9.26	UE8-3	16-Jul-62	8	11
WY LINCOLN	568.58	UpK2-1	24-Mar-53	7	4
WY LINCOLN	19.02	PB17-1	28-Sep-57	4	4
WY LINCOLN	5.45	S-1	6-Jul-62	8	4
WY LINCOLN	3.89	UE13-1	25-Apr-66	10	4
WY NATRONA	35.36	TS5-2	8-May-52	2	21
WY NATRONA	1.95	TS5-8	14-May-52	2	21
WY NATRONA	9.79	UpK2-2	25-Mar-53	7	21
WY NATRONA	3.25	UpK10-3	27-May-53	1	21
WY NATRONA	5.29	TP8-2	10-Apr-55	2	21
WY NATRONA	29.18	PB2-2	19-Jun-57	1	21
WY NATRONA	1.30	PB6-8	31-Jul-57	9	21
WY NATRONA	1.65	PB8-2	19-Aug-57	6	21
WY NATRONA	3.07	PB8-4	21-Aug-57	10	21
WY NATRONA	3.07	PB8-4	21-Aug-57	10	21
WY NATRONA	7.76	PB9-3	25-Aug-57	9	21
WY NATRONA	7.36	PB9-4	26-Aug-57	7	21
WY NATRONA	0.83	PB9-6	28-Aug-57	8	21
WY NATRONA	1.44	PB18-4	10-Oct-57	6	21
WY NATRONA	0.86	PB18-6	12-Oct-57	7	21
WY NATRONA	3.77	UE2-3	7-Mar-62	1	21
WY NATRONA	0.83	UE2-4	8-Mar-62	4	21
WY NATRONA	0.31	UE2-5	9-Mar-62	4	21
WY NATRONA	0.22	UE2-6	10-Mar-62	2	21
WY NATRONA	4.73	S-1	6-Jul-62	9	21
WY NATRONA	7.97	UE12-4	17-Apr-65	9	21
WY NIOBRARA	30.75	TS5-2	8-May-52	3	17
WY NIOBRARA	195.97	TS7-3	3-Jun-52	3	17
WY NIOBRARA	2.49	UpK2-7	30-Mar-53	10	17
WY NIOBRARA	4.00	TP3-2	2-Mar-55	5	17
WY NIOBRARA	14.24	TP11-2	16-May-55	7	17
WY NIOBRARA	21.11	PB2-3	20-Jun-57	4	17

COUNTY	uCi/ sqm	SHOT DAY	DATE	RANK FOR DAY	TIMES IN TOP 10
WY NIOBRARA	10.18	PB5-3	17-Jul-57	3	17
WY NIOBRARA	4.87	PB5-4	18-Jul-57	8	17
WY NIOBRARA	1.65	PB8-2	19-Aug-57	6	17
WY NIOBRARA	11.05	PB9-4	26-Aug-57	6	17
WY NIOBRARA	2.37	PB18-3	9-Oct-57	3	17
WY NIOBRARA	1.44	PB18-4	10-Oct-57	6	17
WY NIOBRARA	1.51	UE2-3	7-Mar-62	4	17
WY NIOBRARA	0.22	UE2-6	10-Mar-62	2	17
WY NIOBRARA	0.10	UE2-7	11-Mar-62	3	17
WY NIOBRARA	3.33	S-2	7-Jul-62	9	17
WY NIOBRARA	11.43	UE12-4	17-Apr-65	3	17
WY PARK	12.33	UpK8-2	9-May-53	4	17
WY PARK	2.63	UpK8-3	10-May-53	5	17
WY PARK	1.22	UpK8-4	11-May-53	8	17
WY PARK	4.13	UpK10-7	31-May-53	2	17
WY PARK	1.32	PB2-2	19-Jun-57	9	17
WY PARK	1.97	PB4-2	6-Jul-57	8	17
WY PARK	1.30	PB6-8	31-Jul-57	9	17
WY PARK	44.80	PB8-2	19-Aug-57	1	17
WY PARK	2.73	PB9-2	24-Aug-57	6	17
WY PARK	1.72	PB15-3	18-Sep-57	8	17
WY PARK	2.63	UE2-3	7-Mar-62	2	17
WY PARK	0.37	UE2-5	9-Mar-62	2	17
WY PARK	0.23	UE2-6	10-Mar-62	1	17
WY PARK	2.32	UE7-2	12-Jul-62	7	17
WY PARK	33.02	UE12-3	16-Apr-65	9	17
WY PARK	0.16	BN-4	21-Dec-70	9	17
WY PARK	0.22	BN-5	22-Dec-70	1	17
WY PLATTE	3.16	TP3-2	2-Mar-55	7	13
WY PLATTE	24.89	TP10-2	6-May-55	9	13
WY PLATTE	34.74	TP11-2	16-May-55	2	13
WY PLATTE	7.30	PB2-2	19-Jun-57	3	13
WY PLATTE	0.83	PB8-2	19-Aug-57	8	13
WY PLATTE	7.36	PB9-4	26-Aug-57	7	13
WY PLATTE	0.83	PB9-6	28-Aug-57	8	13

COUNTY	uCi/ sqm	SHOT DAY	DATE	RANK FOR DAY	TIMES IN TOP 10
WY PLATTE	2.37	PB18-3	9-Oct-57	3	13
WY PLATTE	1.44	PB18-4	10-Oct-57	6	13
WY PLATTE	0.22	UE2-6	10-Mar-62	2	13
WY PLATTE	0.10	UE2-7	11-Mar-62	3	13
WY PLATTE	3.33	S-2	7-Jul-62	9	13
WY PLATTE	11.37	UE8-3	16-Jul-62	6	13
WY SHERIDAN	32.75	UpK8-2	9-May-53	2	17
WY SHERIDAN	5.26	UpK8-3	10-May-53	3	17
WY SHERIDAN	1.22	UpK8-4	11-May-53	8	17
WY SHERIDAN	2.07	UpK10-7	31-May-53	3	17
WY SHERIDAN	1.32	PB2-2	19-Jun-57	9	17
WY SHERIDAN	2.63	PB4-2	6-Jul-57	5	17
WY SHERIDAN	0.83	PB8-2	19-Aug-57	8	17
WY SHERIDAN	3.07	PB8-4	21-Aug-57	10	17
WY SHERIDAN	3.07	PB8-4	21-Aug-57	10	17
WY SHERIDAN	1.86	PB9-2	24-Aug-57	9	17
WY SHERIDAN	7.36	PB9-4	26-Aug-57	7	17
WY SHERIDAN	3.47	PB15-2	17-Sep-57	10	17
WY SHERIDAN	0.31	UE2-5	9-Mar-62	4	17
WY SHERIDAN	0.21	UE2-6	10-Mar-62	3	17
WY SHERIDAN	4.20	UE7-3	13-Jul-62	1	17
WY SHERIDAN	7.73	UE12-4	17-Apr-65	10	17
WY SHERIDAN	0.20	BN-5	22-Dec-70	3	17
WY SUBLETTE	564.41	UpK2-1	24-Mar-53	8	3
WY SUBLETTE	8.05	PB15-1	16-Sep-57	9	3
WY SUBLETTE	15.24	PB17-1	28-Sep-57	6	3
WY SWEETWATER	384.54	PB11-1	31-Aug-57	6	2
WY SWEETWATER	22.80	PB17-1	28-Sep-57	2	2
WY TETON	28.94	TS5-2	8-May-52	6	16
WY TETON	1.15	TS5-8	14-May-52	3	16
WY TETON	1.31	TS5-9	15-May-52	4	16
WY TETON	1.32	PB2-2	19-Jun-57	9	16
WY TETON	29.83	PB8-2	19-Aug-57	2	16
WY TETON	1.86	PB9-2	24-Aug-57	9	16
WY TETON	6.73	PB9-3	25-Aug-57	10	16

COUNTY	uCi/ sqm	SHOT DAY	DATE	RANK FOR DAY	TIMES IN TOP 10
WY TETON	1.72	PB15-3	18-Sep-57	8	16
WY TETON	10.96	PB17-5	2-Oct-57	3	16
WY TETON	2.17	UE2-3	7-Mar-62	3	16
WY TETON	0.44	UE2-5	9-Mar-62	1	16
WY TETON	0.21	UE2-6	10-Mar-62	3	16
WY TETON	47.38	UE12-3	16-Apr-65	4	16
WY TETON	9.32	UE12-4	17-Apr-65	6	16
WY TETON	0.88	BN-3	20-Dec-70	5	16
WY TETON	0.21	BN-5	22-Dec-70	2	16
WY UINTA	29.08	TP8-1	9-Apr-55	9	4
WY UINTA	8.05	PB15-1	16-Sep-57	9	4
WY UINTA	22.80	PB17-1	28-Sep-57	2	4
WY UINTA	3.89	UE13-1	25-Apr-66	10	4
WY WASHAKIE	23.57	UpK2-2	25-Mar-53	5	11
WY WASHAKIE	4.73	UpK8-1	8-May-53	6	11
WY WASHAKIE	0.83	PB8-2	19-Aug-57	8	11
WY WASHAKIE	4.60	PB8-4	21-Aug-57	8	11
WY WASHAKIE	4.60	PB8-4	21-Aug-57	8	11
WY WASHAKIE	3.60	PB9-2	24-Aug-57	5	11
WY WASHAKIE	3.68	PB9-4	26-Aug-57	9	11
WY WASHAKIE	0.80	PB13-7	12-Sep-57	7	11
WY WASHAKIE	0.23	UE2-6	10-Mar-62	1	11
WY WASHAKIE	4.73	S-1	6-Jul-62	9	11
WY WASHAKIE	1.75	UE7-3	13-Jul-62	7	11
WY WESTON	45.45	UpK2-2	25-Mar-53	4	14
WY WESTON	1.58	UpK8-1	8-May-53	8	14
WY WESTON	9.68	PB5-2	16-Jul-57	6	14
WY WESTON	15.19	PB5-3	17-Jul-57	1	14
WY WESTON	0.85	PB5-6	20-Jul-57	10	14
WY WESTON	0.83	PB8-2	19-Aug-57	8	14
WY WESTON	7.76	PB9-3	25-Aug-57	9	14
WY WESTON	3.68	PB9-4	26-Aug-57	9	14
WY WESTON	2.37	PB18-3	9-Oct-57	3	14
WY WESTON	1.44	PB18-4	10-Oct-57	6	14
WY WESTON	0.22	UE2-6	10-Mar-62	2	14

COUNTY	uCi/ sqm	SHOT DAY	DATE	RANK FOR DAY	TIMES IN TOP 10
WY WESTON	0.10	UE2-7	11-Mar-62	3	14
WY WESTON	3.33	S-2	7-Jul-62	9	14
WY WESTON	11.43	UE12-4	17-Apr-65	3	14

ESTIMATED TOTAL FALLOUT FROM NEVADA NUCLEAR TESTS 1951-1970 MICROCURIES/SQUARE METER COUNTIES RANKED WHERE No. 1=Highest Fallout ; No. 3094 = Lowest Fallout					
COUNTY	**uCi/sq meter**	**RANK**	**COUNTY**	**uCi/sq meter**	**RANK**
AL AUTAUGA	67.00	2726	AL JACKSON	75.11	2515
AL BALDWIN	97.20	2066	AL JEFFERSON	74.23	2540
AL BARBOUR	80.25	2387	AL LAMAR	62.89	2818
AL BIBB	76.56	2480	AL LAUDERDALE	78.67	2426
AL BLOUNT	68.06	2693	AL LAWRENCE	65.33	2756
AL BULLOCK	68.35	2685	AL LEE	76.76	2469
AL BUTLER	66.43	2737	AL LIMESTONE	68.82	2669
AL CALHOUN	69.31	2658	AL LOWNDES	82.10	2342
AL CHAMBERS	77.16	2458	AL MACON	64.12	2783
AL CHEROKEE	66.79	2731	AL MADISON	81.11	2371
AL CHILTON	67.28	2715	AL MARENGO	153.13	1425
AL CHOCTAW	66.51	2736	AL MARION	64.78	2764
AL CLARKE	92.15	2143	AL MARSHALL	75.45	2504
AL CLAY	74.89	2519	AL MOBILE	161.77	1331
AL CLEBURNE	69.68	2648	AL MONROE	109.78	1915
AL COFFEE	69.24	2660	AL MONTGOMERY	58.20	2872
AL COLBERT	85.67	2262	AL MORGAN	75.29	2510
AL CONECUH	126.83	1721	AL PERRY	80.53	2381
AL COOSA	65.01	2761	AL PICKENS	82.05	2343
AL COVINGTON	79.21	2410	AL PIKE	81.04	2374
AL CRENSHAW	62.93	2817	AL RANDOLPH	75.17	2512
AL CULLMAN	76.66	2478	AL RUSSELL	69.22	2661
AL DALE	81.53	2359	AL SHELBY	62.72	2821
AL DALLAS	74.23	2541	AL ST CLAIR	68.02	2696
AL DE KALB	78.25	2437	AL SUMTER	78.34	2433
AL ELMORE	73.82	2554	AL TALLADEGA	76.02	2488
AL ESCAMBIA	88.51	2212	AL TALLAPOOSA	73.24	2566
AL ETOWAH	79.30	2404	AL TUSCALOOSA	88.45	2215
AL FAYETTE	64.47	2773	AL WALKER	66.82	2730
AL FRANKLIN	69.36	2656	AL WASHINGTON	86.64	2246
AL GENEVA	98.64	2048	AL WILCOX	111.28	1886
AL GREENE	98.75	2046	AL WINSTON	63.23	2810
AL HALE	84.50	2291	AR ARKANSAS	178.26	1176
AL HENRY	55.99	2894	AR ASHLEY	101.39	2010
AL HOUSTON	73.84	2553	AR BAXTER	200.56	978

ESTIMATED TOTAL FALLOUT FROM NEVADA NUCLEAR TESTS 1951-1970 MICROCURIES/SQUARE METER COUNTIES RANKED WHERE No. 1=Highest Fallout ; No. 3094 = Lowest Fallout					
COUNTY	**uCi/sq meter**	**RANK**	**COUNTY**	**uCi/sq meter**	**RANK**
AR BENTON	213.49	875	AR LEE	145.82	1498
AR BOONE	169.60	1267	AR LINCOLN	136.33	1616
AR BRADLEY	107.04	1948	AR LITTLE RIVER	157.85	1374
AR CALHOUN	84.09	2301	AR LOGAN	172.28	1232
AR CARROLL	169.11	1270	AR LONOKE	172.49	1230
AR CHICOT	96.31	2076	AR MADISON	206.89	925
AR CLARK	150.33	1451	AR MARION	191.56	1054
AR CLAY	168.96	1273	AR MILLER	140.49	1566
AR CLEBURNE	198.43	1000	AR MISSISSIPPI	158.00	1370
AR CLEVELAND	121.64	1772	AR MONROE	159.00	1359
AR COLUMBIA	125.79	1734	AR MONTGOMERY	170.78	1251
AR CONWAY	172.37	1231	AR NEVADA	138.77	1590
AR CRAIGHEAD	199.46	991	AR NEWTON	200.71	976
AR CRAWFORD	182.17	1144	AR OUACHITA	121.31	1777
AR CRITTENDEN	151.20	1446	AR PERRY	204.15	946
AR CROSS	149.50	1457	AR PHILLIPS	139.18	1584
AR DALLAS	127.34	1713	AR PIKE	147.91	1477
AR DESHA	143.17	1529	AR POINSETT	146.32	1494
AR DREW	106.13	1963	AR POLK	182.22	1141
AR FAULKNER	214.41	870	AR POPE	199.88	987
AR FRANKLIN	195.88	1024	AR PRAIRIE	208.86	913
AR FULTON	220.35	828	AR PULASKI	173.11	1228
AR GARLAND	140.88	1560	AR RANDOLPH	178.51	1170
AR GRANT	154.21	1412	AR SALINE	182.66	1135
AR GREENE	177.56	1188	AR SCOTT	163.75	1315
AR HEMPSTEAD	134.01	1653	AR SEARCY	179.10	1162
AR HOT SPRING	141.04	1558	AR SEBASTIAN	170.18	1258
AR HOWARD	153.48	1419	AR SEVIER	169.83	1262
AR INDEPENDENCE	191.48	1056	AR SHARP	212.11	885
AR IZARD	197.52	1013	AR ST FRANCIS	165.32	1304
AR JACKSON	206.84	926	AR STONE	166.92	1294
AR JEFFERSON	176.73	1192	AR UNION	116.60	1836
AR JOHNSON	195.50	1029	AR VAN BUREN	197.73	1007
AR LAFAYETTE	142.21	1544	AR WASHINGTON	204.23	944
AR LAWRENCE	209.90	900	AR WHITE	190.48	1059

ESTIMATED TOTAL FALLOUT FROM NEVADA NUCLEAR TESTS 1951-1970
MICROCURIES/SQUARE METER
COUNTIES RANKED
WHERE No. 1=Highest Fallout ; No. 3094 = Lowest Fallout

COUNTY	uCi/sq meter	RANK	COUNTY	uCi/sq meter	RANK
AR WOODRUFF	184.69	1116	CA INYO2	275.94	579
AR YELL	176.21	1200	CA INYO3	1750.08	18
AZ APACHE	1424.32	33	CA KERN	62.29	2829
AZ COCHISE	156.55	1382	CA KINGS	45.69	2955
AZ COCONINO1	2057.89	14	CA LAKE	40.23	2982
AZ COCONINO2	1591.21	26	CA LASSEN	66.76	2732
AZ COCONINO3	401.28	315	CA LOS ANGELES	114.58	1856
AZ GILA	177.81	1184	CA MADERA	35.95	2996
AZ GRAHAM	173.51	1225	CA MARIN	41.75	2974
AZ GREENLEE	159.50	1356	CA MARIPOSA	42.34	2972
AZ MARICOPA	173.90	1220	CA MENDOCINO	25.74	3059
AZ MOHAVE1	1593.40	25	CA MERCED	30.51	3032
AZ MOHAVE2	2997.49	8	CA MODOC	46.82	2946
AZ MOHAVE3	598.72	174	CA MONO	260.93	644
AZ MOHAVE4	598.72	175	CA MONTEREY	49.72	2932
AZ NAVAJO	932.91	72	CA NAPA	33.78	3009
AZ PIMA	265.35	623	CA NEVADA	89.33	2197
AZ PINAL	178.28	1174	CA ORANGE	43.20	2968
AZ SANTA CRUZ	222.89	804	CA PLACER	63.38	2802
AZ YAVAPAI	220.60	826	CA PLUMAS	62.51	2826
AZ YUMA	151.35	1442	CA RIVERSIDE	25.13	3063
CA ALAMEDA	31.39	3027	CA SACRAMENTO	36.29	2994
CA ALPINE	111.18	1889	CA SAN BENITO	44.58	2961
CA AMADOR	66.94	2728	CA SAN BERNADIN	313.57	443
CA BUTTE	52.30	2919	CA SAN DIEGO	46.88	2944
CA CALAVERAS	57.81	2875	CA SAN FRANCISC	26.35	3051
CA COLUSA	39.65	2984	CA SAN JOAQUIN	30.91	3031
CA CONTRA COSTA	31.53	3024	CA SAN LUIS OBI	46.84	2945
CA DEL NORTE	32.61	3016	CA SAN MATEO	34.79	3002
CA EL DORADO	52.88	2916	CA SANTA BARBAR	52.88	2915
CA FRESNO	41.38	2976	CA SANTA CLARA	34.42	3005
CA GLENN	67.53	2709	CA SANTA CRUZ	46.22	2952
CA HUMBOLDT	28.15	3045	CA SHASTA	25.58	3062
CA IMPERIAL	25.78	3058	CA SIERRA	87.53	2230
CA INYO1	275.94	578	CA SISKIYOU	29.46	3037

ESTIMATED TOTAL FALLOUT FROM NEVADA NUCLEAR TESTS 1951-1970
MICROCURIES/SQUARE METER
COUNTIES RANKED
WHERE No. 1=Highest Fallout ; No. 3094 = Lowest Fallout

COUNTY	uCi/sq meter	RANK	COUNTY	uCi/sq meter	RANK
CA SOLANO	29.28	3038	CO GRAND	596.92	176
CA SONOMA	52.32	2918	CO GUNNISON	1946.42	15
CA STANISLAUS	32.37	3020	CO HINSDALE	1146.12	49
CA SUTTER	33.36	3011	CO HUERFANO	377.27	339
CA TEHAMA	54.96	2899	CO JACKSON	319.89	428
CA TRINITY	29.20	3039	CO JEFFERSON	485.81	227
CA TULARE	50.29	2929	CO KIOWA	337.59	400
CA TUOLUMNE	52.64	2917	CO KIT CARSON	290.13	514
CA VENTURA	62.96	2816	CO LA PLATA	1111.60	55
CA YOLO	32.52	3017	CO LAKE	825.34	87
CA YUBA	57.77	2876	CO LARIMER	363.36	360
CO ADAMS	448.57	266	CO LAS ANIMAS	374.45	344
CO ALAMOSA	683.08	142	CO LINCOLN	290.25	513
CO ARAPAHOE	434.52	278	CO LOGAN	250.13	682
CO ARCHULETA	1438.43	31	CO MESA	1267.33	40
CO BACA	344.02	385	CO MINERAL	1156.07	47
CO BENT	282.15	546	CO MOFFAT	708.22	120
CO BOULDER	479.57	234	CO MONTEZUMA	801.36	93
CO CHAFFEE	414.92	303	CO MONTROSE	1050.29	62
CO CHEYENNE	266.48	616	CO MORGAN	463.49	253
CO CLEAR CREEK	503.31	218	CO OTERO	285.78	531
CO CONEJOS	1831.89	16	CO OURAY	802.94	90
CO COSTILLA	698.66	126	CO PARK	313.63	442
CO CROWLEY	273.02	592	CO PHILLIPS	363.22	361
CO CUSTER	424.59	289	CO PITKIN	1018.63	65
CO DELTA	1104.38	57	CO PROWERS	274.58	586
CO DENVER	289.95	516	CO PUEBLO	245.45	697
CO DOLORES	889.93	78	CO RIO BLANCO	973.92	69
CO DOUGLAS	472.96	243	CO RIO GRANDE	905.59	75
CO EAGLE	717.55	115	CO ROUTT	447.71	269
CO EL PASO	276.54	570	CO SAGUACHE	715.97	116
CO ELBERT	271.32	601	CO SAN JUAN	793.25	94
CO FREMONT	376.02	342	CO SAN MIGUEL	808.07	89
CO GARFIELD	1199.96	43	CO SEDGWICK	279.14	559
CO GILPIN	476.81	236	CO SUMMIT	641.44	158

ESTIMATED TOTAL FALLOUT FROM NEVADA NUCLEAR TESTS 1951-1970
MICROCURIES/SQUARE METER
COUNTIES RANKED
WHERE No. 1=Highest Fallout ; No. 3094 = Lowest Fallout

COUNTY	uCi/sq meter	RANK	COUNTY	uCi/sq meter	RANK
CO TELLER	315.04	439	FL GADSDEN	66.98	2727
CO WASHINGTON	267.24	613	FL GILCHRIST	74.53	2527
CO WELD	319.24	429	FL GLADES	48.99	2937
CO YUMA	280.74	553	FL GULF	72.95	2572
CT FAIRFIELD	163.91	1314	FL HAMILTON	67.98	2697
CT HARTFORD	143.46	1524	FL HARDEE	40.25	2981
CT LITCHFIELD	160.04	1350	FL HENDRY	54.44	2902
CT MIDDLESEX	209.66	903	FL HERNANDO	78.16	2441
CT NEW HAVEN	178.70	1168	FL HIGHLANDS	49.30	2934
CT NEW LONDON	159.86	1353	FL HILLSBOROUGH	87.59	2226
CT TOLLAND	178.07	1180	FL HOLMES	96.34	2074
CT WINDHAM	157.52	1378	FL INDIAN RIVER	47.34	2943
DC WASHINGTON	81.36	2364	FL JACKSON	63.53	2798
DE KENT	98.60	2051	FL JEFFERSON	68.30	2689
DE NEW CASTLE	99.25	2035	FL LAFAYETTE	62.72	2822
DE SUSSEX	99.50	2031	FL LAKE	98.13	2055
FL ALACHUA	76.14	2486	FL LEE	49.85	2931
FL BAKER	62.97	2815	FL LEON	83.94	2305
FL BAY	65.20	2758	FL LEVY	46.60	2949
FL BRADFORD	58.35	2868	FL LIBERTY	72.35	2590
FL BREVARD	79.15	2412	FL MADISON	75.80	2498
FL BROWARD	50.65	2926	FL MANATEE	51.71	2921
FL CALHOUN	75.93	2494	FL MARION	73.20	2567
FL CHARLOTTE	44.87	2959	FL MARTIN	49.01	2936
FL CITRUS	71.84	2608	FL MONROE	53.50	2911
FL CLAY	65.43	2755	FL NASSAU	75.08	2517
FL COLLIER	57.16	2880	FL OKALOOSA	136.68	1612
FL COLUMBIA	61.72	2834	FL OKEECHOBEE	48.54	2940
FL DADE	50.34	2928	FL ORANGE	73.98	2547
FL DE SOTO	43.68	2965	FL OSCEOLA	63.83	2793
FL DIXIE	72.24	2595	FL PALM BEACH	55.81	2897
FL DUVAL	44.65	2960	FL PASCO	58.55	2866
FL ESCAMBIA	85.11	2276	FL PINELLAS	72.12	2599
FL FLAGLER	78.60	2427	FL POLK	75.12	2514
FL FRANKLIN	70.31	2635	FL PUTNAM	70.93	2622

ESTIMATED TOTAL FALLOUT FROM NEVADA NUCLEAR TESTS 1951-1970
MICROCURIES/SQUARE METER
COUNTIES RANKED
WHERE No. 1=Highest Fallout ; No. 3094 = Lowest Fallout

COUNTY	uCi/sq meter	RANK	COUNTY	uCi/sq meter	RANK
FL SANTA ROSA	91.40	2155	GA CATOOSA	68.37	2683
FL SARASOTA	57.18	2879	GA CHARLTON	68.52	2676
FL SEMINOLE	73.56	2559	GA CHATHAM	83.37	2323
FL ST JOHNS	81.30	2366	GA CHATTAHOOCHE	72.20	2598
FL ST LUCIE	53.46	2913	GA CHATTOOGA	69.58	2651
FL SUMTER	81.74	2353	GA CHEROKEE	68.42	2680
FL SUWANNEE	70.77	2626	GA CLARKE	73.67	2555
FL TAYLOR	62.70	2824	GA CLAY	85.66	2263
FL UNION	59.75	2857	GA CLAYTON	81.11	2370
FL VOLUSIA	84.99	2281	GA CLINCH	74.31	2537
FL WAKULLA	71.82	2610	GA COBB	63.62	2796
FL WALTON	62.78	2820	GA COFFEE	64.99	2762
FL WASHINGTON	63.95	2788	GA COLQUITT	83.47	2321
GA APPLING	79.26	2406	GA COLUMBIA	61.73	2833
GA ATKINSON	74.39	2532	GA COLUMBUS	75.96	2491
GA BACON	77.87	2447	GA COOK	73.54	2560
GA BAKER	81.34	2365	GA COWETA	67.59	2706
GA BALDWIN	74.52	2528	GA CRAWFORD	81.94	2348
GA BANKS	72.40	2586	GA CRISP	75.37	2505
GA BARROW	63.84	2792	GA DADE	63.85	2791
GA BARTOW	66.37	2739	GA DAWSON	64.19	2782
GA BEN HILL	71.98	2605	GA DE KALB	68.37	2684
GA BERRIEN	76.72	2473	GA DECATUR	68.46	2677
GA BIBB	73.42	2564	GA DODGE	69.86	2643
GA BLECKLEY	77.84	2448	GA DOOLY	71.79	2611
GA BRANTLEY	73.63	2556	GA DOUGHERTY	80.67	2378
GA BROOKS	66.72	2734	GA DOUGLAS	76.55	2481
GA BRYAN	92.65	2133	GA EARLY	66.31	2743
GA BULLOCH	86.03	2253	GA ECHOLS	64.67	2767
GA BURKE	82.04	2344	GA EFFINGHAM	90.82	2165
GA BUTTS	68.26	2690	GA ELBERT	62.79	2819
GA CALHOUN	84.50	2292	GA EMANUEL	75.35	2508
GA CAMDEN	75.96	2490	GA EVANS	86.28	2251
GA CANDLER	73.10	2570	GA FANNIN	85.45	2266
GA CARROLL	81.72	2355	GA FAYETTE	82.01	2345

ESTIMATED TOTAL FALLOUT FROM NEVADA NUCLEAR TESTS 1951-1970
MICROCURIES/SQUARE METER
COUNTIES RANKED
WHERE No. 1=Highest Fallout ; No. 3094 = Lowest Fallout

COUNTY	uCi/sq meter	RANK	COUNTY	uCi/sq meter	RANK
GA FLOYD	75.70	2499	GA LOWNDES	71.06	2621
GA FORSYTH	64.57	2770	GA LUMPKIN	66.17	2747
GA FRANKLIN	65.05	2759	GA MACON	59.56	2858
GA FULTON	62.70	2823	GA MADISON	72.38	2587
GA GILMER	72.20	2597	GA MARION	60.01	2855
GA GLASCOCK	74.38	2533	GA MCDUFFIE	74.36	2534
GA GLYNN	83.17	2325	GA MCINTOSH	82.59	2334
GA GORDON	78.79	2424	GA MERIWETHER	69.44	2654
GA GRADY	65.44	2754	GA MILLER	56.46	2889
GA GREENE	80.01	2391	GA MITCHELL	78.92	2418
GA GWINNETT	72.58	2583	GA MONROE	76.91	2464
GA HABERSHAM	72.27	2592	GA MONTGOMERY	90.07	2181
GA HALL	68.74	2672	GA MORGAN	78.84	2421
GA HANCOCK	70.56	2631	GA MURRAY	71.43	2618
GA HARALSON	69.90	2642	GA NEWTON	66.33	2741
GA HARRIS	66.33	2742	GA OCONEE	69.64	2649
GA HART	66.24	2746	GA OGLETHORPE	68.06	2694
GA HEARD	56.36	2890	GA PAULDING	76.95	2463
GA HENRY	67.27	2716	GA PEACH	81.95	2347
GA HOUSTON	75.46	2503	GA PICKENS	64.44	2775
GA IRWIN	71.49	2615	GA PIERCE	68.73	2673
GA JACKSON	70.89	2623	GA PIKE	70.80	2625
GA JASPER	67.70	2703	GA POLK	60.05	2854
GA JEFF DAVIS	79.92	2392	GA PULASKI	85.78	2257
GA JEFFERSON	80.30	2384	GA PUTNAM	68.31	2688
GA JENKINS	85.09	2277	GA QUITMAN	85.90	2255
GA JOHNSON	78.80	2423	GA RABUN	107.40	1945
GA JONES	69.57	2652	GA RANDOLPH	86.42	2249
GA LAMAR	67.10	2724	GA RICHMOND	67.20	2720
GA LANIER	77.90	2446	GA ROCKDALE	62.46	2827
GA LAURENS	60.24	2852	GA SCHLEY	83.98	2304
GA LEE	78.28	2436	GA SCREVEN	90.60	2168
GA LIBERTY	105.78	1967	GA SEMINOLE	54.74	2901
GA LINCOLN	63.42	2801	GA SPALDING	71.59	2613
GA LONG	102.65	2000	GA STEPHENS	70.25	2636

ESTIMATED TOTAL FALLOUT FROM NEVADA NUCLEAR TESTS 1951-1970
MICROCURIES/SQUARE METER
COUNTIES RANKED
WHERE No. 1=Highest Fallout ; No. 3094 = Lowest Fallout

COUNTY	uCi/sq meter	RANK	COUNTY	uCi/sq meter	RANK
GA STEWART	75.98	2489	IA APPANOOSE	662.58	154
GA SUMTER	77.01	2461	IA AUDUBON	589.86	183
GA TALBOT	63.78	2794	IA BENTON	765.06	99
GA TALIAFERRO	74.94	2518	IA BLACK HAWK	463.64	252
GA TATTNALL	83.93	2306	IA BOONE	766.49	98
GA TAYLOR	67.24	2719	IA BREMER	464.18	251
GA TELFAIR	76.95	2462	IA BUCHANAN	409.10	308
GA TERRELL	88.53	2211	IA BUENA VISTA	697.46	127
GA THOMAS	76.76	2468	IA BUTLER	391.42	325
GA TIFT	73.91	2550	IA CALHOUN	691.41	132
GA TOOMBS	78.13	2442	IA CARROLL	786.42	95
GA TOWNS	92.74	2131	IA CASS	626.99	163
GA TREUTLEN	59.88	2856	IA CEDAR	332.60	410
GA TROUP	74.27	2539	IA CERRO GORDO	368.00	350
GA TURNER	72.70	2578	IA CHEROKEE	684.50	141
GA TWIGGS	78.85	2420	IA CHICKASAW	196.56	1022
GA UNION	80.27	2386	IA CLARKE	564.31	196
GA UPSON	62.20	2830	IA CLAY	687.38	138
GA WALKER	72.46	2585	IA CLAYTON	343.16	389
GA WALTON	67.93	2698	IA CLINTON	334.31	408
GA WARE	64.38	2776	IA CRAWFORD	672.60	150
GA WARREN	74.49	2530	IA DALLAS	376.32	340
GA WASHINGTON	72.82	2574	IA DAVIS	638.73	160
GA WAYNE	96.01	2088	IA DECATUR	569.18	195
GA WEBSTER	84.44	2296	IA DELAWARE	557.21	198
GA WHEELER	93.35	2124	IA DES MOINES	294.80	496
GA WHITE	83.42	2322	IA DICKINSON	676.03	147
GA WHITFIELD	67.27	2717	IA DUBUQUE	359.66	362
GA WILCOX	75.90	2495	IA EMMET	657.61	156
GA WILKES	67.39	2713	IA FAYETTE	415.77	301
GA WILKINSON	74.20	2542	IA FLOYD	689.63	135
GA WORTH	74.32	2536	IA FRANKLIN	447.05	270
IA ADAIR	595.29	178	IA FREMONT	306.47	458
IA ADAMS	595.04	179	IA GREENE	715.09	117
IA ALLAMAKEE	321.59	425	IA GRUNDY	176.93	1191

ESTIMATED TOTAL FALLOUT FROM NEVADA NUCLEAR TESTS 1951-1970 MICROCURIES/SQUARE METER COUNTIES RANKED WHERE No. 1=Highest Fallout ; No. 3094 = Lowest Fallout					
COUNTY	uCi/sq meter	RANK	COUNTY	uCi/sq meter	RANK
IA GUTHRIE	526.32	208	IA PALO ALTO	677.76	146
IA HAMILTON	734.10	110	IA PLYMOUTH	692.97	131
IA HANCOCK	684.75	140	IA POCAHONTAS	694.93	129
IA HARDIN	404.51	312	IA POLK	335.33	405
IA HARRISON	316.77	434	IA POTTAWATTAMI	374.69	343
IA HENRY	505.05	216	IA POWESHIEK	439.89	276
IA HOWARD	1133.00	51	IA RINGGOLD	613.99	167
IA HUMBOLDT	702.65	122	IA SAC	688.68	136
IA IDA	717.75	114	IA SCOTT	257.67	658
IA IOWA	691.02	133	IA SHELBY	623.95	165
IA JACKSON	352.36	371	IA SIOUX	699.28	125
IA JASPER	767.92	97	IA STORY	754.09	103
IA JEFFERSON	756.13	102	IA TAMA	498.16	221
IA JOHNSON	477.17	235	IA TAYLOR	417.13	299
IA JONES	295.04	495	IA UNION	591.19	182
IA KEOKUK	538.84	205	IA VAN BUREN	813.71	88
IA KOSSUTH	685.66	139	IA WAPELLO	608.54	170
IA LEE	802.44	91	IA WARREN	351.68	373
IA LINN	419.87	293	IA WASHINGTON	756.97	101
IA LOUISA	484.20	230	IA WAYNE	580.62	185
IA LUCAS	199.92	986	IA WEBSTER	349.03	377
IA LYON	687.43	137	IA WINNEBAGO	679.33	145
IA MADISON	601.13	172	IA WINNESHIEK	693.28	130
IA MAHASKA	570.21	194	IA WOODBURY	842.74	82
IA MARION	594.37	180	IA WORTH	1113.05	54
IA MARSHALL	763.97	100	IA WRIGHT	343.74	387
IA MILLS	339.17	398	ID ADA	419.43	296
IA MITCHELL	1116.94	53	ID ADAMS	236.86	737
IA MONONA	700.07	124	ID BANNOCK	287.31	526
IA MONROE	661.74	155	ID BEAR LAKE	864.58	81
IA MONTGOMERY	415.13	302	ID BENEWAH	131.43	1679
IA MUSCATINE	276.43	573	ID BINGHAM	186.83	1089
IA O BRIEN	712.89	119	ID BLAINE	908.33	74
IA OSCEOLA	701.76	123	ID BOISE	472.78	244
IA PAGE	352.38	370	ID BONNER	106.14	1962

ESTIMATED TOTAL FALLOUT FROM NEVADA NUCLEAR TESTS 1951-1970
MICROCURIES/SQUARE METER
COUNTIES RANKED
WHERE No. 1=Highest Fallout ; No. 3094 = Lowest Fallout

COUNTY	uCi/sq meter	RANK	COUNTY	uCi/sq meter	RANK
ID BONNEVILLE	249.64	683	IL ADAMS	452.94	263
ID BOUNDARY	85.01	2280	IL ALEXANDER	114.59	1855
ID BUTTE	328.23	415	IL BOND	195.56	1027
ID CAMAS	501.99	219	IL BOONE	265.21	625
ID CANYON	233.53	751	IL BROWN	346.19	381
ID CARIBOU	462.68	255	IL BUREAU	110.07	1905
ID CASSIA	428.02	284	IL CALHOUN	218.14	842
ID CLARK	262.09	637	IL CARROLL	123.50	1755
ID CLEARWATER	304.95	461	IL CASS	358.02	363
ID CUSTER	1015.84	66	IL CHAMPAIGN	238.17	729
ID ELMORE	421.16	291	IL CHRISTIAN	326.02	420
ID FRANKLIN	690.01	134	IL CLARK	206.29	933
ID FREMONT	240.54	715	IL CLAY	239.80	720
ID GEM	1030.75	63	IL CLINTON	102.64	2001
ID GOODING	465.33	250	IL COLES	343.44	388
ID IDAHO	572.08	193	IL COOK	72.27	2594
ID JEFFERSON	372.64	347	IL CRAWFORD	150.35	1450
ID JEROME	347.26	380	IL CUMBERLAND	177.82	1183
ID KOOTENAI	96.10	2086	IL DE KALB	131.26	1684
ID LATAH	171.31	1245	IL DE WITT	264.13	631
ID LEMHI	718.73	113	IL DOUGLAS	276.62	569
ID LEWIS	327.89	417	IL DU PAGE	81.19	2368
ID LINCOLN	391.10	327	IL EDGAR	183.52	1130
ID MADISON	288.44	521	IL EDWARDS	116.56	1837
ID MINIDOKA	339.50	397	IL EFFINGHAM	276.34	574
ID NEZ PERCE	184.19	1122	IL FAYETTE	188.47	1076
ID ONEIDA	524.09	209	IL FORD	199.75	989
ID OWYHEE	475.67	239	IL FRANKLIN	89.10	2205
ID PAYETTE	323.74	422	IL FULTON	483.86	231
ID POWER	280.32	555	IL GALLATIN	116.41	1840
ID SHOSHONE	228.01	774	IL GREENE	334.93	407
ID TETON	299.85	478	IL GRUNDY	131.27	1683
ID TWIN FALLS	401.92	314	IL HAMILTON	83.83	2308
ID VALLEY	579.41	188	IL HANCOCK	801.68	92
ID WASHINGTON	396.33	321	IL HARDIN	79.05	2414

ESTIMATED TOTAL FALLOUT FROM NEVADA NUCLEAR TESTS 1951-1970
MICROCURIES/SQUARE METER
COUNTIES RANKED
WHERE No. 1=Highest Fallout ; No. 3094 = Lowest Fallout

COUNTY	uCi/sq meter	RANK	COUNTY	uCi/sq meter	RANK
IL HENDERSON	295.91	492	IL OGLE	151.25	1445
IL HENRY	276.87	567	IL PEORIA	298.99	480
IL IROQUOIS	141.65	1551	IL PERRY	104.65	1981
IL JACKSON	117.94	1810	IL PIATT	278.23	563
IL JASPER	307.99	454	IL PIKE	408.83	309
IL JEFFERSON	141.30	1554	IL POPE	104.00	1987
IL JERSEY	234.66	745	IL PULASKI	114.02	1862
IL JO DAVIESS	162.73	1321	IL PUTNAM	145.04	1507
IL JOHNSON	89.30	2198	IL RANDOLPH	135.99	1625
IL KANE	120.02	1789	IL RICHLAND	200.94	970
IL KANKAKEE	138.61	1594	IL ROCK ISLAND	278.35	562
IL KENDALL	81.52	2361	IL SALINE	88.46	2214
IL KNOX	287.84	524	IL SANGAMON	342.19	390
IL LA SALLE	153.06	1430	IL SCHUYLER	316.66	435
IL LAKE	100.76	2017	IL SCOTT	237.82	734
IL LAWRENCE	159.85	1354	IL SHELBY	399.15	318
IL LEE	154.39	1408	IL ST CLAIR	120.40	1784
IL LIVINGSTON	124.55	1744	IL STARK	284.64	537
IL LOGAN	251.47	673	IL STEPHENSON	154.05	1413
IL MACON	294.08	499	IL TAZEWELL	265.64	621
IL MACOUPIN	324.48	421	IL UNION	112.14	1878
IL MADISON	110.91	1893	IL VERMILION	215.72	859
IL MARION	94.28	2108	IL WABASH	129.47	1697
IL MARSHALL	307.29	456	IL WARREN	492.77	224
IL MASON	279.40	558	IL WASHINGTON	98.91	2042
IL MASSAC	87.54	2228	IL WAYNE	123.36	1758
IL MCDONOUGH	496.30	223	IL WHITE	90.69	2167
IL MCHENRY	119.31	1797	IL WHITESIDE	168.31	1280
IL MCLEAN	267.04	615	IL WILL	134.58	1642
IL MENARD	204.83	940	IL WILLIAMSON	91.54	2152
IL MERCER	265.10	627	IL WINNEBAGO	291.52	511
IL MONROE	116.34	1841	IL WOODFORD	296.35	489
IL MONTGOMERY	331.57	412	IN ADAMS	195.71	1026
IL MORGAN	238.26	728	IN ALLEN	389.97	328
IL MOULTRIE	402.59	313	IN BARTHOLOMEW	176.52	1194

ESTIMATED TOTAL FALLOUT FROM NEVADA NUCLEAR TESTS 1951-1970
MICROCURIES/SQUARE METER
COUNTIES RANKED
WHERE No. 1=Highest Fallout ; No. 3094 = Lowest Fallout

COUNTY	uCi/sq meter	RANK	COUNTY	uCi/sq meter	RANK
IN BENTON	109.69	1918	IN JEFFERSON	228.95	770
IN BLACKFORD	67.86	2701	IN JENNINGS	74.65	2522
IN BOONE	117.49	1819	IN JOHNSON	101.73	2009
IN BROWN	109.20	1922	IN KNOX	188.70	1073
IN CARROLL	208.04	919	IN KOSCIUSKO	100.52	2019
IN CASS	240.10	718	IN LA PORTE	96.10	2085
IN CLARK	182.28	1139	IN LAGRANGE	83.60	2319
IN CLAY	117.54	1817	IN LAKE	123.13	1761
IN CLINTON	195.46	1030	IN LAWRENCE	338.63	399
IN CRAWFORD	239.42	722	IN MADISON	197.12	1016
IN DAVIESS	140.39	1570	IN MARION	268.81	608
IN DE KALB	63.21	2811	IN MARSHALL	92.35	2139
IN DEARBORN	104.11	1985	IN MARTIN	221.04	823
IN DECATUR	130.84	1687	IN MIAMI	148.98	1468
IN DELAWARE	250.72	678	IN MONROE	215.73	858
IN DUBOIS	193.79	1040	IN MONTGOMERY	126.31	1729
IN ELKHART	89.17	2203	IN MORGAN	113.07	1867
IN FAYETTE	72.80	2575	IN NEWTON	122.43	1766
IN FLOYD	180.96	1154	IN NOBLE	75.95	2492
IN FOUNTAIN	116.61	1835	IN OHIO	77.66	2450
IN FRANKLIN	209.64	904	IN ORANGE	209.13	910
IN FULTON	194.99	1035	IN OWEN	229.91	767
IN GIBSON	102.98	1996	IN PARKE	112.42	1873
IN GRANT	296.61	488	IN PERRY	146.60	1488
IN GREENE	240.78	712	IN PIKE	100.85	2016
IN HAMILTON	297.84	487	IN PORTER	123.18	1759
IN HANCOCK	89.13	2204	IN POSEY	132.79	1665
IN HARRISON	169.16	1269	IN PULASKI	139.30	1583
IN HENDRICKS	248.96	688	IN PUTNAM	111.78	1884
IN HENRY	231.98	757	IN RANDOLPH	169.54	1268
IN HOWARD	292.78	506	IN RIPLEY	109.10	1926
IN HUNTINGTON	154.57	1405	IN RUSH	110.33	1901
IN JACKSON	264.34	629	IN SCOTT	166.59	1298
IN JASPER	134.16	1649	IN SHELBY	185.80	1101
IN JAY	89.62	2193	IN SPENCER	146.39	1493

ESTIMATED TOTAL FALLOUT FROM NEVADA NUCLEAR TESTS 1951-1970
MICROCURIES/SQUARE METER
COUNTIES RANKED
WHERE No. 1=Highest Fallout ; No. 3094 = Lowest Fallout

COUNTY	uCi/sq meter	RANK	COUNTY	uCi/sq meter	RANK
IN ST JOSEPH	186.12	1096	KS COMANCHE	134.80	1640
IN STARKE	92.81	2128	KS COWLEY	211.79	889
IN STEUBEN	84.63	2287	KS CRAWFORD	206.50	931
IN SULLIVAN	144.10	1516	KS DECATUR	265.98	618
IN SWITZERLAND	241.57	708	KS DICKINSON	227.38	777
IN TIPPECANOE	154.61	1404	KS DONIPHAN	134.77	1641
IN TIPTON	289.95	517	KS DOUGLAS	161.07	1340
IN UNION	196.61	1021	KS EDWARDS	183.58	1128
IN VANDERBURGH	149.21	1463	KS ELK	190.03	1063
IN VERMILLION	188.01	1084	KS ELLIS	216.16	854
IN VIGO	139.98	1575	KS ELLSWORTH	220.03	830
IN WABASH	158.87	1362	KS FINNEY	269.76	604
IN WARREN	153.32	1422	KS FORD	295.76	493
IN WARRICK	155.96	1386	KS FRANKLIN	161.67	1335
IN WASHINGTON	168.18	1282	KS GEARY	200.84	974
IN WAYNE	383.88	332	KS GOVE	269.47	606
IN WELLS	289.66	518	KS GRAHAM	282.11	547
IN WHITE	130.76	1688	KS GRANT	279.58	556
IN WHITLEY	72.07	2600	KS GRAY	252.87	668
KS ALLEN	166.05	1301	KS GREELEY	249.20	686
KS ANDERSON	147.89	1479	KS GREENWOOD	200.31	980
KS ATCHISON	167.67	1287	KS HAMILTON	264.21	630
KS BARBER	161.67	1336	KS HARPER	189.77	1065
KS BARTON	235.30	743	KS HARVEY	209.43	906
KS BOURBON	144.28	1513	KS HASKELL	249.31	685
KS BROWN	158.14	1369	KS HODGEMAN	275.98	577
KS BUTLER	203.82	949	KS JACKSON	165.86	1302
KS CHASE	202.69	956	KS JEFFERSON	162.16	1327
KS CHAUTAUQUA	204.61	941	KS JEWELL	168.75	1275
KS CHEROKEE	184.35	1120	KS JOHNSON	170.51	1255
KS CHEYENNE	284.15	538	KS KEARNY	277.54	566
KS CLARK	165.68	1303	KS KINGMAN	206.59	930
KS CLAY	176.29	1198	KS KIOWA	166.55	1299
KS CLOUD	165.20	1306	KS LABETTE	188.47	1077
KS COFFEY	168.14	1283	KS LANE	272.51	593

ESTIMATED TOTAL FALLOUT FROM NEVADA NUCLEAR TESTS 1951-1970
MICROCURIES/SQUARE METER
COUNTIES RANKED
WHERE No. 1=Highest Fallout ; No. 3094 = Lowest Fallout

COUNTY	uCi/sq meter	RANK
KS LEAVENWORTH	198.20	1003
KS LINCOLN	234.41	747
KS LINN	149.75	1454
KS LOGAN	265.89	620
KS LYON	203.06	953
KS MARION	228.94	771
KS MARSHALL	168.48	1277
KS MCPHERSON	222.68	806
KS MEADE	223.61	797
KS MIAMI	161.75	1332
KS MITCHELL	220.97	825
KS MONTGOMERY	174.40	1218
KS MORRIS	190.34	1060
KS MORTON	237.66	735
KS NEMAHA	155.14	1398
KS NEOSHO	187.31	1086
KS NESS	304.84	462
KS NORTON	287.81	525
KS OSAGE	159.23	1358
KS OSBORNE	229.57	769
KS OTTAWA	220.20	829
KS PAWNEE	185.54	1104
KS PHILLIPS	195.46	1031
KS POTTAWATOMIE	184.08	1125
KS PRATT	185.16	1111
KS RAWLINS	263.92	632
KS RENO	210.80	895
KS REPUBLIC	207.57	923
KS RICE	207.14	924
KS RILEY	194.53	1039
KS ROOKS	222.28	810
KS RUSH	209.14	909
KS RUSSELL	274.53	587
KS SALINE	258.72	652
KS SCOTT	268.81	609
KS SEDGWICK	170.95	1249
KS SEWARD	212.65	883
KS SHAWNEE	206.62	929
KS SHERIDAN	249.00	687
KS SHERMAN	182.19	1142
KS SMITH	195.17	1032
KS STAFFORD	188.98	1068
KS STANTON	285.09	534
KS STEVENS	391.16	326
KS SUMNER	203.46	951
KS THOMAS	248.87	689
KS TREGO	285.09	535
KS WABAUNSEE	221.06	822
KS WALLACE	273.76	590
KS WASHINGTON	178.90	1164
KS WICHITA	260.37	646
KS WILSON	184.60	1117
KS WOODSON	173.67	1222
KS WYANDOTTE	179.62	1159
KY ADAIR	70.59	2629
KY ALLEN	65.47	2753
KY ANDERSON	230.02	766
KY BALLARD	107.93	1939
KY BARREN	77.49	2455
KY BATH	298.51	486
KY BELL	123.89	1750
KY BOONE	181.34	1150
KY BOURBON	162.06	1328
KY BOYD	290.65	512
KY BOYLE	118.36	1804
KY BRACKEN	197.71	1008
KY BREATHITT	281.97	548
KY BRECKINRIDGE	80.79	2377
KY BULLITT	109.66	1919
KY BUTLER	96.34	2075

ESTIMATED TOTAL FALLOUT FROM NEVADA NUCLEAR TESTS 1951-1970
MICROCURIES/SQUARE METER
COUNTIES RANKED
WHERE No. 1=Highest Fallout ; No. 3094 = Lowest Fallout

COUNTY	uCi/sq meter	RANK	COUNTY	uCi/sq meter	RANK
KY CALDWELL	85.39	2269	KY HENRY	302.12	468
KY CALLOWAY	85.31	2272	KY HICKMAN	143.07	1530
KY CAMPBELL	384.98	330	KY HOPKINS	73.43	2562
KY CARLISLE	92.88	2127	KY JACKSON	241.60	707
KY CARROLL	365.47	356	KY JEFFERSON	96.73	2071
KY CARTER	271.91	598	KY JESSAMINE	181.89	1145
KY CASEY	91.15	2158	KY JOHNSON	265.28	624
KY CHRISTIAN	96.15	2083	KY KENTON	341.43	392
KY CLARK	310.87	446	KY KNOTT	186.28	1095
KY CLAY	120.33	1785	KY KNOX	115.44	1850
KY CLINTON	69.84	2644	KY LARUE	70.59	2630
KY CRITTENDEN	74.63	2523	KY LAUREL	85.48	2265
KY CUMBERLAND	128.36	1706	KY LAWRENCE	430.44	281
KY DAVIESS	131.64	1678	KY LEE	74.02	2546
KY EDMONSON	83.67	2316	KY LESLIE	214.64	867
KY ELLIOTT	273.37	591	KY LETCHER	176.11	1201
KY ESTILL	176.69	1193	KY LEWIS	423.58	290
KY FAYETTE	155.29	1395	KY LINCOLN	99.52	2030
KY FLEMING	275.51	581	KY LIVINGSTON	99.05	2038
KY FLOYD	260.65	645	KY LOGAN	74.60	2526
KY FRANKLIN	307.06	457	KY LYON	76.89	2465
KY FULTON	138.66	1593	KY MADISON	168.99	1272
KY GALLATIN	279.44	557	KY MAGOFFIN	276.07	576
KY GARRARD	109.85	1914	KY MARION	63.30	2807
KY GRANT	261.54	639	KY MARSHALL	116.90	1827
KY GRAVES	132.70	1671	KY MARTIN	142.50	1539
KY GRAYSON	60.49	2848	KY MASON	272.40	594
KY GREEN	68.44	2678	KY MCCRACKEN	109.99	1907
KY GREENUP	446.90	271	KY MCCREARY	85.56	2264
KY HANCOCK	143.94	1519	KY MCLEAN	84.66	2286
KY HARDIN	68.40	2681	KY MEADE	79.25	2407
KY HARLAN	162.37	1324	KY MENIFEE	302.92	465
KY HARRISON	336.08	403	KY MERCER	115.55	1847
KY HART	61.41	2839	KY METCALFE	81.38	2362
KY HENDERSON	100.07	2025	KY MONROE	93.60	2118

ESTIMATED TOTAL FALLOUT FROM NEVADA NUCLEAR TESTS 1951-1970
MICROCURIES/SQUARE METER
COUNTIES RANKED
WHERE No. 1=Highest Fallout ; No. 3094 = Lowest Fallout

COUNTY	uCi/sq meter	RANK	COUNTY	uCi/sq meter	RANK
KY MONTGOMERY	215.87	857	LA ALLEN	222.35	807
KY MORGAN	249.46	684	LA ASCENSION	178.38	1173
KY MUHLENBERG	79.19	2411	LA ASSUMPTION	200.05	984
KY NELSON	56.56	2887	LA AVOYELLES	184.92	1113
KY NICHOLAS	330.17	413	LA BEAUREGARD	273.81	589
KY OHIO	109.91	1911	LA BIENVILLE	97.91	2056
KY OLDHAM	218.52	839	LA BOSSIER	103.98	1988
KY OWEN	379.35	337	LA CADDO	105.71	1969
KY OWSLEY	241.50	710	LA CALCASIEU	253.35	667
KY PENDLETON	215.09	862	LA CALDWELL	91.03	2163
KY PERRY	192.11	1049	LA CAMERON	250.94	677
KY PIKE	244.73	701	LA CATAHOULA	163.15	1318
KY POWELL	438.24	277	LA CLAIBORNE	83.55	2320
KY PULASKI	115.93	1843	LA CONCORDIA	168.79	1274
KY ROBERTSON	286.09	530	LA DE SOTO	89.19	2202
KY ROCKCASTLE	168.28	1281	LA EAST BATON R	208.78	914
KY ROWAN	378.91	338	LA EAST CARROLL	97.08	2067
KY RUSSELL	115.46	1849	LA EAST FELICIA	190.99	1057
KY SCOTT	318.02	431	LA EVANGELINE	257.84	657
KY SHELBY	252.62	669	LA FRANKLIN	94.00	2112
KY SIMPSON	72.67	2579	LA GRANT	298.56	485
KY SPENCER	187.71	1085	LA IBERIA	219.69	832
KY TAYLOR	66.30	2744	LA IBERVILLE	209.84	901
KY TODD	116.86	1828	LA JACKSON	98.26	2054
KY TRIGG	89.91	2185	LA JEFFERSON	119.38	1793
KY TRIMBLE	428.98	283	LA JEFFERSON DA	217.90	844
KY UNION	119.67	1792	LA LA SALLE	222.24	811
KY WARREN	95.34	2096	LA LAFAYETTE	270.68	602
KY WASHINGTON	97.73	2057	LA LAFOURCHE	181.37	1149
KY WAYNE	90.38	2169	LA LINCOLN	95.75	2091
KY WEBSTER	84.63	2288	LA LIVINGSTON	185.76	1103
KY WHITLEY	88.04	2220	LA MADISON	116.82	1832
KY WOLFE	284.06	539	LA MOREHOUSE	98.63	2050
KY WOODFORD	315.17	438	LA NATCHITOCHES	93.47	2122
LA ACADIA	253.86	666	LA ORLEANS	80.38	2383

ESTIMATED TOTAL FALLOUT FROM NEVADA NUCLEAR TESTS 1951-1970
MICROCURIES/SQUARE METER
COUNTIES RANKED
WHERE No. 1=Highest Fallout ; No. 3094 = Lowest Fallout

COUNTY	uCi/sq meter	RANK	COUNTY	uCi/sq meter	RANK
LA OUACHITA	110.37	1900	MA HAMPSHIRE	147.89	1478
LA PLAQUEMINES	161.84	1330	MA MIDDLESEX	245.10	699
LA POINTE COUPE	213.38	876	MA NANTUCKET	120.20	1787
LA RAPIDES	291.94	510	MA NORFOLK	299.85	477
LA RED RIVER	84.02	2303	MA PLYMOUTH	292.95	505
LA RICHLAND	113.16	1866	MA SUFFOLK	223.89	794
LA SABINE	173.15	1226	MA WORCESTER	158.84	1363
LA ST BERNARD	151.32	1443	MD ALLEGANY	127.72	1712
LA ST CHARLES	181.70	1148	MD ANNE ARUNDEL	81.55	2358
LA ST HELENA	236.29	739	MD BALTIMORE	145.29	1503
LA ST JAMES	148.83	1470	MD CALVERT	84.68	2285
LA ST JOHN THE	153.59	1418	MD CAROLINE	74.35	2535
LA ST LANDRY	161.72	1333	MD CARROLL	64.94	2763
LA ST MARTIN	199.87	988	MD CECIL	83.83	2309
LA ST MARY	201.52	964	MD CHARLES	80.96	2376
LA ST TAMMANY	161.68	1334	MD DORCHESTER	78.87	2419
LA TANGIPAHOA	230.30	761	MD FREDERICK	78.20	2439
LA TENSAS	223.49	799	MD GARRETT	175.44	1206
LA TERREBONNE	210.21	899	MD HARFORD	147.77	1480
LA UNION	99.34	2033	MD HOWARD	82.42	2335
LA VERMILION	217.26	846	MD KENT	81.21	2367
LA VERNON	171.58	1241	MD MONTGOMERY	88.72	2210
LA WASHINGTON	162.57	1322	MD PRINCE GEORG	87.26	2236
LA WEBSTER	108.86	1930	MD QUEEN ANNES	73.59	2558
LA WEST BATON R	211.55	891	MD SOMERSET	98.87	2043
LA WEST CARROLL	92.93	2126	MD ST MARYS	87.12	2239
LA WEST FELICIA	211.86	887	MD TALBOT	80.27	2385
LA WINN	99.98	2027	MD WASHINGTON	78.83	2422
MA BARNSTABLE	171.73	1238	MD WICOMICO	84.55	2290
MA BERKSHIRE	170.66	1252	MD WORCESTER	93.86	2116
MA BRISTOL	243.10	704	ME ANDROSCOGGIN	103.00	1995
MA DUKES	172.14	1233	ME AROOSTOOK	184.10	1124
MA ESSEX	195.04	1033	ME CUMBERLAND	108.08	1936
MA FRANKLIN	131.67	1677	ME FRANKLIN	98.81	2045
MA HAMPDEN	161.11	1339	ME HANCOCK	122.02	1769

ESTIMATED TOTAL FALLOUT FROM NEVADA NUCLEAR TESTS 1951-1970
MICROCURIES/SQUARE METER
COUNTIES RANKED
WHERE No. 1=Highest Fallout ; No. 3094 = Lowest Fallout

COUNTY	uCi/sq meter	RANK	COUNTY	uCi/sq meter	RANK
ME KENNEBEC	104.37	1983	MI GENESEE	159.25	1357
ME KNOX	112.88	1869	MI GLADWIN	61.81	2831
ME LINCOLN	108.87	1929	MI GOGEBIC	114.28	1859
ME OXFORD	109.74	1916	MI GRAND TRAVER	54.38	2904
ME PENOBSCOT	146.55	1491	MI GRATIOT	103.53	1990
ME PISCATAQUIS	137.55	1605	MI HILLSDALE	145.16	1504
ME SAGADAHOC	105.09	1978	MI HOUGHTON	113.33	1864
ME SOMERSET	92.10	2144	MI HURON	111.71	1885
ME WALDO	108.99	1927	MI INGHAM	216.88	849
ME WASHINGTON	136.13	1621	MI IONIA	64.09	2784
ME YORK	105.65	1972	MI IOSCO	71.45	2617
MI ALCONA	65.90	2751	MI IRON	94.57	2105
MI ALGER	81.75	2352	MI ISABELLA	68.75	2670
MI ALLEGAN	91.08	2160	MI JACKSON	96.26	2077
MI ALPENA	43.91	2963	MI KALAMAZOO	134.39	1647
MI ANTRIM	64.32	2778	MI KALKASKA	49.25	2935
MI ARENAC	70.71	2627	MI KENT	141.98	1547
MI BARAGA	91.29	2156	MI KEWEENAW	110.52	1899
MI BARRY	116.93	1826	MI LAKE	51.63	2922
MI BAY	67.47	2711	MI LAPEER	148.38	1472
MI BENZIE	57.31	2878	MI LEELANAU	47.93	2941
MI BERRIEN	94.75	2103	MI LENAWEE	155.80	1389
MI BRANCH	109.73	1917	MI LIVINGSTON	151.69	1437
MI CALHOUN	133.55	1656	MI LUCE	79.24	2408
MI CASS	80.17	2388	MI MACKINAC	64.29	2779
MI CHARLEVOIX	68.04	2695	MI MACOMB	130.54	1690
MI CHEBOYGAN	63.52	2799	MI MANISTEE	58.39	2867
MI CHIPPEWA	67.75	2702	MI MARQUETTE	72.58	2582
MI CLARE	61.46	2837	MI MASON	58.63	2865
MI CLINTON	133.31	1659	MI MECOSTA	60.78	2846
MI CRAWFORD	62.38	2828	MI MENOMINEE	93.51	2119
MI DELTA	76.80	2467	MI MIDLAND	90.71	2166
MI DICKINSON	79.50	2397	MI MISSAUKEE	57.09	2882
MI EATON	221.79	816	MI MONROE	147.43	1484
MI EMMET	57.08	2883	MI MONTCALM	89.80	2189

ESTIMATED TOTAL FALLOUT FROM NEVADA NUCLEAR TESTS 1951-1970 MICROCURIES/SQUARE METER COUNTIES RANKED WHERE No. 1=Highest Fallout ; No. 3094 = Lowest Fallout					
COUNTY	**uCi/sq meter**	**RANK**	**COUNTY**	**uCi/sq meter**	**RANK**
MI MONTMORENCY	47.55	2942	MN CHIPPEWA	93.93	2114
MI MUSKEGON	87.16	2238	MN CHISAGO	109.33	1921
MI NEWAYGO	83.85	2307	MN CLAY	53.49	2912
MI OAKLAND	166.96	1292	MN CLEARWATER	66.75	2733
MI OCEANA	84.36	2298	MN COOK	92.53	2136
MI OGEMAW	58.09	2874	MN COTTONWOOD	258.55	653
MI ONTONAGON	108.94	1928	MN CROW WING	99.54	2029
MI OSCEOLA	56.85	2885	MN DAKOTA	140.82	1562
MI OSCODA	48.56	2939	MN DODGE	348.90	379
MI OTSEGO	54.13	2905	MN DOUGLAS	69.49	2653
MI OTTAWA	121.28	1779	MN FARIBAULT	729.28	111
MI PRESQUE ISLE	50.41	2927	MN FILLMORE	492.01	225
MI ROSCOMMON	63.46	2800	MN FREEBORN	129.19	1703
MI SAGINAW	110.23	1903	MN GOODHUE	340.66	394
MI SANILAC	120.11	1788	MN GRANT	63.35	2805
MI SCHOOLCRAFT	92.59	2134	MN HENNEPIN	97.32	2064
MI SHIAWASSEE	146.61	1487	MN HOUSTON	322.07	424
MI ST CLAIR	97.60	2061	MN HUBBARD	71.83	2609
MI ST JOSEPH	78.58	2428	MN ISANTI	104.03	1986
MI TUSCOLA	96.22	2079	MN ITASCA	87.88	2222
MI VAN BUREN	132.87	1663	MN JACKSON	299.96	476
MI WASHTENAW	78.31	2435	MN KANABEC	102.52	2002
MI WAYNE	143.62	1520	MN KANDIYOHI	102.26	2005
MI WEXFORD	58.33	2869	MN KITTSON	65.68	2752
MN AITKIN	107.18	1947	MN KOOCHICHING	89.98	2183
MN ANOKA	114.97	1852	MN LAC QUI PARL	94.85	2102
MN BECKER	74.62	2524	MN LAKE	99.03	2039
MN BELTRAMI	78.02	2443	MN LAKE OF THE	58.31	2870
MN BENTON	105.81	1965	MN LE SUEUR	470.75	245
MN BIG STONE	74.52	2529	MN LINCOLN	300.62	474
MN BLUE EARTH	444.10	274	MN LYON	303.09	464
MN BROWN	446.85	272	MN MAHNOMEN	61.42	2838
MN CARLTON	134.57	1643	MN MARSHALL	55.72	2898
MN CARVER	123.47	1756	MN MARTIN	461.98	257
MN CASS	91.62	2150	MN MCLEOD	115.54	1848

ESTIMATED TOTAL FALLOUT FROM NEVADA NUCLEAR TESTS 1951-1970
MICROCURIES/SQUARE METER
COUNTIES RANKED
WHERE No. 1=Highest Fallout ; No. 3094 = Lowest Fallout

COUNTY	uCi/sq meter	RANK	COUNTY	uCi/sq meter	RANK
MN MEEKER	111.10	1890	MN WASHINGTON	127.84	1711
MN MILLE LACS	98.66	2047	MN WATONWAN	721.07	112
MN MORRISON	91.88	2149	MN WILKIN	55.96	2895
MN MOWER	752.61	105	MN WINONA	351.25	374
MN MURRAY	451.02	265	MN WRIGHT	109.12	1925
MN NICOLLET	454.61	262	MN YELLOW MEDIC	92.79	2129
MN NOBLES	354.17	369	MO ADAIR	317.01	432
MN NORMAN	54.42	2903	MO ANDREW	128.26	1708
MN OLMSTED	138.76	1591	MO ATCHISON	233.25	752
MN OTTER TAIL	74.60	2525	MO AUDRAIN	368.39	349
MN PENNINGTON	67.32	2714	MO BARRY	206.50	932
MN PINE	121.04	1780	MO BARTON	174.46	1217
MN PIPESTONE	462.27	256	MO BATES	164.25	1312
MN POLK	68.91	2666	MO BENTON	222.08	814
MN POPE	76.74	2472	MO BOLLINGER	104.16	1984
MN RAMSEY	128.08	1710	MO BOONE	173.71	1221
MN RED LAKE	61.23	2842	MO BUCHANAN	125.98	1732
MN REDWOOD	184.71	1115	MO BUTLER	149.27	1460
MN RENVILLE	111.03	1891	MO CALDWELL	265.60	622
MN RICE	485.73	228	MO CALLAWAY	227.26	779
MN ROCK	666.03	153	MO CAMDEN	245.57	696
MN ROSEAU	63.04	2814	MO CAPE GIRARDE	114.92	1853
MN SCOTT	127.01	1719	MO CARROLL	185.33	1108
MN SHERBURNE	117.09	1824	MO CARTER	154.27	1409
MN SIBLEY	298.64	483	MO CASS	190.10	1061
MN ST LOUIS	111.94	1882	MO CEDAR	182.26	1140
MN STEARNS	95.03	2098	MO CHARITON	275.82	580
MN STEELE	348.96	378	MO CHRISTIAN	200.00	985
MN STEVENS	71.99	2603	MO CLARK	580.35	186
MN SWIFT	84.48	2294	MO CLAY	165.17	1307
MN TODD	96.12	2084	MO CLINTON	177.46	1189
MN TRAVERSE	68.37	2682	MO COLE	201.12	969
MN WABASHA	484.91	229	MO COOPER	189.98	1064
MN WADENA	89.29	2199	MO CRAWFORD	167.25	1291
MN WASECA	465.92	248	MO DADE	175.22	1208

ESTIMATED TOTAL FALLOUT FROM NEVADA NUCLEAR TESTS 1951-1970
MICROCURIES/SQUARE METER
COUNTIES RANKED
WHERE No. 1=Highest Fallout ; No. 3094 = Lowest Fallout

COUNTY	uCi/sq meter	RANK	COUNTY	uCi/sq meter	RANK
MO DALLAS	178.80	1166	MO MERCER	187.00	1088
MO DAVIESS	366.44	355	MO MILLER	213.30	877
MO DE KALB	230.24	763	MO MISSISSIPPI	142.73	1535
MO DENT	191.99	1050	MO MONITEAU	171.75	1237
MO DOUGLAS	202.38	957	MO MONROE	267.15	614
MO DUNKLIN	157.78	1375	MO MONTGOMERY	143.19	1528
MO FRANKLIN	218.88	837	MO MORGAN	240.03	719
MO GASCONADE	178.86	1165	MO NEW MADRID	162.78	1320
MO GENTRY	166.68	1297	MO NEWTON	222.19	813
MO GREENE	195.52	1028	MO NODAWAY	182.47	1137
MO GRUNDY	268.72	610	MO OREGON	201.24	967
MO HARRISON	255.44	664	MO OSAGE	230.87	759
MO HENRY	219.86	831	MO OZARK	197.59	1011
MO HICKORY	149.32	1458	MO PEMISCOT	149.83	1453
MO HOLT	123.63	1754	MO PERRY	117.74	1814
MO HOWARD	179.18	1161	MO PETTIS	240.58	714
MO HOWELL	185.83	1100	MO PHELPS	190.82	1058
MO IRON	134.33	1648	MO PIKE	393.60	322
MO JACKSON	142.98	1532	MO PLATTE	153.70	1417
MO JASPER	198.36	1001	MO POLK	204.13	948
MO JEFFERSON	162.88	1319	MO PULASKI	208.03	920
MO JOHNSON	178.00	1181	MO PUTNAM	483.39	232
MO KNOX	752.10	106	MO RALLS	519.93	212
MO LACLEDE	208.36	917	MO RANDOLPH	251.59	672
MO LAFAYETTE	205.70	937	MO RAY	216.24	852
MO LAWRENCE	221.40	817	MO REYNOLDS	149.11	1466
MO LEWIS	577.00	189	MO RIPLEY	167.83	1286
MO LINCOLN	189.76	1066	MO SALINE	191.74	1053
MO LINN	336.83	401	MO SCHUYLER	458.11	260
MO LIVINGSTON	227.05	782	MO SCOTLAND	680.01	144
MO MACON	344.05	384	MO SCOTT	109.15	1924
MO MADISON	108.56	1932	MO SHANNON	153.44	1420
MO MARIES	188.17	1081	MO SHELBY	440.85	275
MO MARION	367.02	352	MO ST CHARLES	144.23	1514
MO MCDONALD	215.60	861	MO ST CLAIR	138.19	1598

ESTIMATED TOTAL FALLOUT FROM NEVADA NUCLEAR TESTS 1951-1970 MICROCURIES/SQUARE METER COUNTIES RANKED WHERE No. 1=Highest Fallout ; No. 3094 = Lowest Fallout					
COUNTY	uCi/sq meter	RANK	COUNTY	uCi/sq meter	RANK
MO ST FRANCOIS	138.33	1597	MS GREENE	129.37	1699
MO ST LOUIS	119.34	1794	MS GRENADA	122.02	1770
MO STE GENEVIEV	102.28	2004	MS HANCOCK	160.33	1347
MO STODDARD	136.31	1618	MS HARRISON	135.50	1633
MO STONE	171.57	1242	MS HINDS	170.17	1259
MO SULLIVAN	343.93	386	MS HOLMES	74.66	2521
MO TANEY	130.02	1692	MS HUMPHREYS	106.24	1960
MO TEXAS	199.34	992	MS ISSAQUENA	112.00	1881
MO VERNON	188.93	1069	MS ITAWAMBA	86.73	2245
MO WARREN	134.94	1638	MS JACKSON	135.17	1637
MO WASHINGTON	133.40	1658	MS JASPER	95.76	2090
MO WAYNE	160.05	1349	MS JEFFERSON	153.40	1421
MO WEBSTER	194.85	1036	MS JEFFERSON DA	69.36	2655
MO WORTH	419.80	294	MS JONES	76.87	2466
MO WRIGHT	212.77	881	MS KEMPER	74.09	2543
MS ADAMS	217.81	845	MS LAFAYETTE	112.14	1879
MS ALCORN	108.51	1933	MS LAMAR	121.50	1774
MS AMITE	122.32	1767	MS LAUDERDALE	143.20	1527
MS ATTALA	83.68	2315	MS LAWRENCE	75.53	2502
MS BENTON	126.13	1731	MS LEAKE	81.92	2349
MS BOLIVAR	124.04	1748	MS LEE	116.04	1842
MS CALHOUN	105.61	1974	MS LEFLORE	108.02	1937
MS CARROLL	118.87	1801	MS LINCOLN	100.19	2023
MS CHICKASAW	135.23	1634	MS LOWNDES	67.12	2723
MS CHOCTAW	83.25	2324	MS MADISON	100.44	2022
MS CLAIBORNE	156.23	1384	MS MARION	149.09	1467
MS CLARKE	107.88	1941	MS MARSHALL	140.46	1568
MS CLAY	89.84	2188	MS MONROE	89.91	2186
MS COAHOMA	134.11	1650	MS MONTGOMERY	121.29	1778
MS COPIAH	117.20	1822	MS NESHOBA	68.67	2674
MS COVINGTON	88.19	2219	MS NEWTON	106.27	1958
MS DE SOTO	142.45	1540	MS NOXUBEE	68.91	2667
MS FORREST	144.58	1511	MS OKTIBBEHA	100.44	2021
MS FRANKLIN	197.42	1015	MS PANOLA	132.70	1670
MS GEORGE	155.74	1390	MS PEARL RIVER	164.73	1309

ESTIMATED TOTAL FALLOUT FROM NEVADA NUCLEAR TESTS 1951-1970
MICROCURIES/SQUARE METER
COUNTIES RANKED
WHERE No. 1=Highest Fallout ; No. 3094 = Lowest Fallout

COUNTY	uCi/sq meter	RANK
MS PERRY	112.43	1872
MS PIKE	158.59	1367
MS PONTOTOC	118.17	1809
MS PRENTISS	110.75	1894
MS QUITMAN	141.00	1559
MS RANKIN	78.41	2432
MS SCOTT	92.01	2147
MS SHARKEY	110.02	1906
MS SIMPSON	72.51	2584
MS SMITH	85.70	2261
MS STONE	140.50	1565
MS SUNFLOWER	102.65	1999
MS TALLAHATCHIE	131.78	1675
MS TATE	136.02	1623
MS TIPPAH	134.04	1652
MS TISHOMINGO	112.37	1874
MS TUNICA	126.44	1727
MS UNION	114.56	1858
MS WALTHALL	130.07	1691
MS WARREN	223.93	793
MS WASHINGTON	105.39	1977
MS WAYNE	105.61	1973
MS WEBSTER	96.42	2073
MS WILKINSON	205.95	935
MS WINSTON	75.17	2513
MS YALOBUSHA	129.67	1696
MS YAZOO	114.63	1854
MT BEAVERHEAD	491.65	226
MT BIG HORN	221.36	818
MT BLAINE	251.26	674
MT BROADWATER	398.63	319
MT CARBON	258.42	655
MT CARTER	231.88	758
MT CASCADE	316.85	433
MT CHOUTEAU	451.47	264
MT CUSTER	356.19	366
MT DANIELS	201.30	965
MT DAWSON	333.33	409
MT DEER LODGE	639.75	159
MT FALLON	272.15	597
MT FERGUS	302.46	466
MT FLATHEAD	212.79	879
MT GALLATIN	399.89	316
MT GARFIELD	206.65	928
MT GLACIER	140.42	1569
MT GOLDEN VALLE	201.73	963
MT GRANITE	244.72	702
MT HILL	238.29	727
MT JEFFERSON	417.89	298
MT JUDITH BASIN	504.76	217
MT LAKE	200.74	975
MT LEWIS AND CL	398.03	320
MT LIBERTY	193.74	1041
MT LINCOLN	126.79	1722
MT MADISON	426.45	285
MT MCCONE	269.59	605
MT MEAGHER	496.96	222
MT MINERAL	271.46	600
MT MISSOULA	351.68	372
MT MUSSELSHELL	228.53	773
MT PARK	309.04	451
MT PETROLEUM	199.49	990
MT PHILLIPS	182.51	1136
MT PONDERA	190.06	1062
MT POWDER RIVER	373.25	346
MT POWELL	419.89	292
MT PRAIRIE	252.44	670
MT RAVALLI	232.76	756
MT RICHLAND	301.52	473
MT ROOSEVELT	260.23	647

ESTIMATED TOTAL FALLOUT FROM NEVADA NUCLEAR TESTS 1951-1970
MICROCURIES/SQUARE METER
COUNTIES RANKED
WHERE No. 1=Highest Fallout ; No. 3094 = Lowest Fallout

COUNTY	uCi/sq meter	RANK	COUNTY	uCi/sq meter	RANK
MT ROSEBUD	200.85	973	NC CLEVELAND	81.73	2354
MT SANDERS	183.01	1134	NC COLUMBUS	76.09	2487
MT SHERIDAN	237.86	732	NC CRAVEN	87.62	2225
MT SILVER BOW	444.52	273	NC CUMBERLAND	80.65	2380
MT STILLWATER	233.82	750	NC CURRITUCK	277.62	565
MT SWEET GRASS	223.04	802	NC DARE	89.65	2192
MT TETON	293.78	500	NC DAVIDSON	85.81	2256
MT TOOLE	205.78	936	NC DAVIE	66.39	2738
MT TREASURE	176.09	1202	NC DUPLIN	77.15	2460
MT VALLEY	203.08	952	NC DURHAM	91.59	2151
MT WHEATLAND	381.33	336	NC EDGECOMBE	69.61	2650
MT WIBAUX	287.92	523	NC FORSYTH	76.76	2470
MT YELLOWSTONE	140.22	1571	NC FRANKLIN	58.85	2863
NC ALAMANCE	83.62	2318	NC GASTON	69.75	2646
NC ALEXANDER	87.55	2227	NC GATES	81.97	2346
NC ALLEGHANY	90.09	2180	NC GRAHAM	92.06	2146
NC ANSON	64.24	2780	NC GRANVILLE	124.51	1746
NC ASHE	117.82	1813	NC GREENE	69.09	2664
NC AVERY	90.06	2182	NC GUILFORD	87.26	2237
NC BEAUFORT	143.23	1526	NC HALIFAX	72.90	2573
NC BERTIE	86.99	2241	NC HARNETT	73.95	2548
NC BLADEN	71.98	2604	NC HAYWOOD	104.82	1979
NC BRUNSWICK	67.15	2722	NC HENDERSON	100.95	2014
NC BUNCOMBE	91.89	2148	NC HERTFORD	84.49	2293
NC BURKE	94.91	2101	NC HOKE	68.35	2686
NC CABARRUS	60.93	2845	NC HYDE	105.78	1966
NC CALDWELL	90.25	2172	NC IREDELL	86.18	2252
NC CAMDEN	276.48	571	NC JACKSON	109.86	1913
NC CARTERET	87.47	2232	NC JOHNSTON	66.25	2745
NC CASWELL	63.13	2812	NC JONES	112.35	1875
NC CATAWBA	83.63	2317	NC LEE	74.09	2544
NC CHATHAM	123.39	1757	NC LENOIR	80.66	2379
NC CHEROKEE	100.10	2024	NC LINCOLN	75.55	2501
NC CHOWAN	172.58	1229	NC MACON	109.95	1910
NC CLAY	106.48	1956	NC MADISON	90.90	2164

ESTIMATED TOTAL FALLOUT FROM NEVADA NUCLEAR TESTS 1951-1970
MICROCURIES/SQUARE METER
COUNTIES RANKED
WHERE No. 1=Highest Fallout ; No. 3094 = Lowest Fallout

COUNTY	uCi/sq meter	RANK	COUNTY	uCi/sq meter	RANK
NC MARTIN	121.41	1775	NC WARREN	79.09	2413
NC MCDOWELL	92.30	2140	NC WASHINGTON	176.46	1197
NC MECKLENBURG	64.03	2787	NC WATAUGA	96.21	2080
NC MITCHELL	96.88	2069	NC WAYNE	79.31	2403
NC MONTGOMERY	70.25	2637	NC WILKES	122.07	1768
NC MOORE	88.35	2216	NC WILSON	76.67	2477
NC NASH	70.67	2628	NC YADKIN	92.53	2137
NC NEW HANOVER	88.99	2207	NC YANCEY	96.04	2087
NC NORTHAMPTON	146.21	1496	ND ADAMS	225.25	789
NC ONSLOW	89.65	2191	ND BARNES	181.81	1147
NC ORANGE	102.76	1997	ND BENSON	165.29	1305
NC PAMLICO	92.07	2145	ND BILLINGS	384.09	331
NC PASQUOTANK	276.73	568	ND BOTTINEAU	271.89	599
NC PENDER	81.53	2360	ND BOWMAN	222.04	815
NC PERQUIMANS	278.75	561	ND BURKE	277.80	564
NC PERSON	135.79	1631	ND BURLEIGH	226.72	784
NC PITT	85.40	2268	ND CASS	200.14	982
NC POLK	86.73	2244	ND CAVALIER	137.38	1607
NC RANDOLPH	84.56	2289	ND DICKEY	209.06	912
NC RICHMOND	65.94	2750	ND DIVIDE	281.26	549
NC ROBESON	73.87	2552	ND DUNN	310.82	447
NC ROCKINGHAM	59.23	2860	ND EDDY	167.47	1290
NC ROWAN	73.59	2557	ND EMMONS	223.70	795
NC RUTHERFORD	79.55	2396	ND FOSTER	199.01	997
NC SAMPSON	73.16	2568	ND GOLDEN VALLE	306.28	459
NC SCOTLAND	67.19	2721	ND GRAND FORKS	174.75	1211
NC STANLY	68.23	2691	ND GRANT	227.38	778
NC STOKES	67.51	2710	ND GRIGGS	174.70	1213
NC SURRY	89.23	2201	ND HETTINGER	240.36	716
NC SWAIN	105.71	1968	ND KIDDER	154.02	1414
NC TRANSYLVANIA	105.70	1970	ND LA MOURE	184.37	1119
NC TYRRELL	63.38	2803	ND LOGAN	177.79	1185
NC UNION	69.71	2647	ND MCHENRY	272.28	595
NC VANCE	132.86	1664	ND MCINTOSH	183.56	1129
NC WAKE	91.07	2161	ND MCKENZIE	298.57	484

ESTIMATED TOTAL FALLOUT FROM NEVADA NUCLEAR TESTS 1951-1970
MICROCURIES/SQUARE METER
COUNTIES RANKED
WHERE No. 1=Highest Fallout ; No. 3094 = Lowest Fallout

COUNTY	uCi/sq meter	RANK
ND MCLEAN	255.79	663
ND MERCER	282.55	545
ND MORTON	240.70	713
ND MOUNTRAIL	257.51	659
ND NELSON	171.20	1246
ND OLIVER	280.87	551
ND PEMBINA	161.36	1337
ND PIERCE	117.55	1816
ND RAMSEY	155.22	1396
ND RANSOM	221.34	820
ND RENVILLE	250.37	679
ND RICHLAND	261.48	640
ND ROLETTE	120.98	1781
ND SARGENT	212.77	880
ND SHERIDAN	212.87	878
ND SIOUX	174.55	1216
ND SLOPE	265.02	628
ND STARK	332.40	411
ND STEELE	163.69	1316
ND STUTSMAN	177.69	1187
ND TOWNER	125.65	1739
ND TRAILL	181.29	1151
ND WALSH	180.09	1158
ND WARD	286.69	528
ND WELLS	136.52	1614
ND WILLIAMS	180.53	1156
NE ADAMS	258.47	654
NE ANTELOPE	426.39	286
NE ARTHUR	230.04	765
NE BANNER	207.95	922
NE BLAINE	205.99	934
NE BOONE	203.52	950
NE BOX BUTTE	171.60	1239
NE BOYD	187.17	1087
NE BROWN	200.85	972
NE BUFFALO	295.92	491
NE BURT	169.73	1266
NE BUTLER	322.46	423
NE CASS	285.64	533
NE CEDAR	373.69	345
NE CHASE	294.60	497
NE CHERRY	310.14	449
NE CHEYENNE	251.00	675
NE CLAY	282.72	544
NE COLFAX	341.55	391
NE CUMING	412.08	304
NE CUSTER	328.23	416
NE DAKOTA	681.23	143
NE DAWES	202.92	954
NE DAWSON	283.22	541
NE DEUEL	225.70	787
NE DIXON	746.75	107
NE DODGE	169.83	1263
NE DOUGLAS	295.98	490
NE DUNDY	259.36	649
NE FILLMORE	238.83	724
NE FRANKLIN	180.79	1155
NE FRONTIER	288.81	519
NE FURNAS	281.18	550
NE GAGE	250.33	680
NE GARDEN	209.21	908
NE GARFIELD	424.61	288
NE GOSPER	276.45	572
NE GRANT	201.22	968
NE GREELEY	354.57	368
NE HALL	208.39	916
NE HAMILTON	250.28	681
NE HARLAN	196.76	1019
NE HAYES	260.96	643
NE HITCHCOCK	252.01	671

ESTIMATED TOTAL FALLOUT FROM NEVADA NUCLEAR TESTS 1951-1970
MICROCURIES/SQUARE METER
COUNTIES RANKED
WHERE No. 1=Highest Fallout ; No. 3094 = Lowest Fallout

COUNTY	uCi/sq meter	RANK	COUNTY	uCi/sq meter	RANK
NE HOLT	591.41	181	NE SEWARD	257.20	660
NE HOOKER	241.52	709	NE SHERIDAN	200.08	983
NE HOWARD	552.42	200	NE SHERMAN	540.41	204
NE JEFFERSON	192.68	1046	NE SIOUX	300.12	475
NE JOHNSON	241.64	706	NE STANTON	704.90	121
NE KEARNEY	204.15	947	NE THAYER	212.68	882
NE KEITH	292.52	508	NE THOMAS	259.31	650
NE KEYA PAHA	171.59	1240	NE THURSTON	696.54	128
NE KIMBALL	193.74	1042	NE VALLEY	335.99	404
NE KNOX	174.69	1214	NE WASHINGTON	476.43	237
NE LANCASTER	230.16	764	NE WAYNE	383.32	333
NE LINCOLN	314.86	440	NE WEBSTER	211.34	892
NE LOGAN	293.35	502	NE WHEELER	191.49	1055
NE LOUP	600.80	173	NE YORK	248.60	690
NE MADISON	458.60	259	NH BELKNAP	118.83	1802
NE MCPHERSON	236.20	740	NH CARROLL	141.90	1549
NE MERRICK	351.04	375	NH CHESHIRE	133.70	1654
NE MORRILL	227.49	776	NH COOS	110.75	1895
NE NANCE	349.17	376	NH GRAFTON	123.15	1760
NE NEMAHA	221.34	819	NH HILLSBOROUGH	128.34	1707
NE NUCKOLLS	237.14	736	NH MERRIMACK	118.92	1799
NE OTOE	214.21	871	NH ROCKINGHAM	155.02	1400
NE PAWNEE	157.69	1376	NH STRAFFORD	126.96	1720
NE PERKINS	290.04	515	NH SULLIVAN	121.63	1773
NE PHELPS	285.70	532	NJ ATLANTIC	176.48	1196
NE PIERCE	178.21	1178	NJ BERGEN	141.56	1552
NE PLATTE	335.17	406	NJ BURLINGTON	215.08	863
NE POLK	326.46	419	NJ CAMDEN	114.07	1861
NE RED WILLOW	269.36	607	NJ CAPE MAY	132.71	1668
NE RICHARDSON	160.59	1345	NJ CUMBERLAND	109.87	1912
NE ROCK	144.42	1512	NJ ESSEX	149.16	1464
NE SALINE	276.31	575	NJ GLOUCESTER	112.18	1876
NE SARPY	149.74	1455	NJ HUDSON	152.72	1431
NE SAUNDERS	301.76	471	NJ HUNTERDON	101.00	2013
NE SCOTTS BLUFF	137.52	1606	NJ MERCER	131.77	1676

ESTIMATED TOTAL FALLOUT FROM NEVADA NUCLEAR TESTS 1951-1970
MICROCURIES/SQUARE METER
COUNTIES RANKED
WHERE No. 1=Highest Fallout ; No. 3094 = Lowest Fallout

COUNTY	uCi/sq meter	RANK
NJ MIDDLESEX	138.37	1595
NJ MONMOUTH	174.65	1215
NJ MORRIS	135.90	1627
NJ OCEAN	196.25	1023
NJ PASSAIC	136.71	1611
NJ SALEM	102.74	1998
NJ SOMERSET	120.89	1782
NJ SUSSEX	118.30	1806
NJ UNION	143.52	1522
NJ WARREN	122.76	1763
NM BERNALILLO	834.31	84
NM CATRON	341.26	393
NM CHAVES	579.58	187
NM COLFAX	549.53	202
NM CURRY	483.12	233
NM DE BACA	510.97	215
NM DONA ANA	128.49	1705
NM EDDY	234.51	746
NM GRANT	150.44	1449
NM GUADALUPE	613.32	168
NM HARDING	523.15	210
NM HIDALGO	133.13	1661
NM LEA	223.61	796
NM LINCOLN	576.40	190
NM LOS ALAMOS	672.40	151
NM LUNA	139.55	1580
NM MCKINLEY	1166.85	46
NM MORA	552.54	199
NM OTERO	148.19	1473
NM QUAY	537.89	206
NM RIO ARRIBA	573.74	191
NM ROOSEVELT	463.20	254
NM SAN JUAN	715.04	118
NM SAN MIGUEL	582.80	184
NM SANDOVAL	1013.70	67
NM SANTA FE	835.30	83
NM SIERRA	214.95	865
NM SOCORRO	418.16	297
NM TAOS	517.26	214
NM TORRANCE	827.51	86
NM UNION	527.59	207
NM VALENCIA	903.87	76
NV CARSON CITY	469.25	247
NV CHURCHILL	612.70	169
NV CLARK1	3315.80	6
NV CLARK2	448.45	267
NV CLARK3	171.84	1236
NV DOUGLAS	457.11	261
NV ELKO	752.66	104
NV ESMERALDA1	366.58	354
NV ESMERALDA2	674.28	149
NV EUREKA	1196.26	44
NV HUMBOLDT	321.05	427
NV LANDER1	475.40	240
NV LANDER2	735.91	109
NV LINCOLN1	3420.75	5
NV LINCOLN2	2191.79	12
NV LYON	475.84	238
NV MINERAL	672.09	152
NV NYE1	624.58	164
NV NYE2	3822.15	3
NV NYE3	118.32	1805
NV PERSHING	366.83	353
NV STOREY	461.29	258
NV WASHOE	473.32	242
NV WHITE PINE1	1599.72	24
NV WHITE PINE2	2451.01	10
NV WHITE PINE3	2170.94	13
NY ALBANY	275.00	584
NY ALLEGANY	153.11	1427

ESTIMATED TOTAL FALLOUT FROM NEVADA NUCLEAR TESTS 1951-1970 MICROCURIES/SQUARE METER COUNTIES RANKED WHERE No. 1=Highest Fallout ; No. 3094 = Lowest Fallout					
COUNTY	**uCi/sq meter**	**RANK**	**COUNTY**	**uCi/sq meter**	**RANK**
NY BRONX	133.23	1660	NY OSWEGO	136.31	1617
NY BROOME	73.42	2563	NY OTSEGO	147.71	1482
NY CATTARAUGUS	146.59	1489	NY PUTNAM	146.26	1495
NY CAYUGA	99.60	2028	NY QUEENS	157.91	1371
NY CHAUTAUQUA	177.97	1182	NY RENSSELAER	606.82	171
NY CHEMUNG	100.60	2018	NY RICHMOND	138.18	1600
NY CHENANGO	98.57	2052	NY ROCKLAND	118.68	1803
NY CLINTON	154.56	1406	NY SARATOGA	411.91	305
NY COLUMBIA	303.39	463	NY SCHENECTADY	315.70	437
NY CORTLAND	104.71	1980	NY SCHOHARIE	178.26	1177
NY DELAWARE	142.68	1536	NY SCHUYLER	84.85	2282
NY DUTCHESS	134.11	1651	NY SENECA	127.13	1716
NY ERIE	186.65	1091	NY ST LAWRENCE	160.96	1341
NY ESSEX	201.81	961	NY STEUBEN	147.03	1485
NY FRANKLIN	140.46	1567	NY SUFFOLK	159.96	1351
NY FULTON	406.06	310	NY SULLIVAN	148.02	1475
NY GENESEE	178.27	1175	NY TIOGA	82.11	2341
NY GREENE	201.74	962	NY TOMPKINS	122.71	1764
NY HAMILTON	183.93	1126	NY ULSTER	170.88	1250
NY HERKIMER	137.29	1608	NY WARREN	614.55	166
NY JEFFERSON	139.75	1577	NY WASHINGTON	636.95	161
NY KINGS	155.17	1397	NY WAYNE	153.12	1426
NY LEWIS	107.53	1944	NY WESTCHESTER	155.54	1394
NY LIVINGSTON	164.20	1313	NY WYOMING	175.81	1204
NY MADISON	102.47	2003	NY YATES	106.69	1952
NY MONROE	104.46	1982	OH ADAMS	280.85	552
NY MONTGOMERY	188.81	1072	OH ALLEN	302.13	467
NY NASSAU	170.19	1257	OH ASHLAND	209.30	907
NY NEW YORK	141.49	1553	OH ASHTABULA	107.95	1938
NY NIAGARA	166.76	1296	OH ATHENS	243.67	703
NY ONEIDA	106.62	1954	OH AUGLAIZE	174.86	1209
NY ONONDAGA	86.99	2240	OH BELMONT	197.63	1010
NY ONTARIO	152.05	1434	OH BROWN	222.75	805
NY ORANGE	130.85	1686	OH BUTLER	405.21	311
NY ORLEANS	143.48	1523	OH CARROLL	88.87	2209

ESTIMATED TOTAL FALLOUT FROM NEVADA NUCLEAR TESTS 1951-1970
MICROCURIES/SQUARE METER
COUNTIES RANKED
WHERE No. 1=Highest Fallout ; No. 3094 = Lowest Fallout

COUNTY	uCi/sq meter	RANK	COUNTY	uCi/sq meter	RANK
OH CHAMPAIGN	313.47	444	OH LOGAN	224.51	792
OH CLARK	345.23	383	OH LORAIN	151.45	1439
OH CLERMONT	224.75	791	OH LUCAS	183.46	1131
OH CLINTON	336.37	402	OH MADISON	238.86	723
OH COLUMBIANA	195.02	1034	OH MAHONING	188.57	1075
OH COSHOCTON	227.16	781	OH MARION	183.82	1127
OH CRAWFORD	173.14	1227	OH MEDINA	381.46	335
OH CUYAHOGA	175.41	1207	OH MEIGS	186.45	1094
OH DARKE	209.77	902	OH MERCER	106.84	1950
OH DEFIANCE	70.46	2633	OH MIAMI	310.64	448
OH DELAWARE	184.12	1123	OH MONROE	185.22	1110
OH ERIE	178.71	1167	OH MONTGOMERY	376.08	341
OH FAIRFIELD	166.86	1295	OH MORGAN	293.18	503
OH FAYETTE	170.27	1256	OH MORROW	179.37	1160
OH FRANKLIN	107.66	1943	OH MUSKINGUM	216.81	851
OH FULTON	282.76	543	OH NOBLE	157.46	1379
OH GALLIA	230.78	760	OH OTTAWA	114.09	1860
OH GEAUGA	156.65	1381	OH PAULDING	214.96	864
OH GREENE	162.17	1326	OH PERRY	301.86	470
OH GUERNSEY	217.20	848	OH PICKAWAY	77.56	2453
OH HAMILTON	195.85	1025	OH PIKE	292.40	509
OH HANCOCK	292.78	507	OH PORTAGE	237.84	733
OH HARDIN	111.20	1888	OH PREBLE	283.02	542
OH HARRISON	184.39	1118	OH PUTNAM	139.50	1581
OH HENRY	93.49	2120	OH RICHLAND	307.34	455
OH HIGHLAND	308.83	453	OH ROSS	177.37	1190
OH HOCKING	181.01	1153	OH SANDUSKY	192.16	1048
OH HOLMES	102.25	2006	OH SCIOTO	432.52	280
OH HURON	148.13	1474	OH SENECA	367.41	351
OH JACKSON	295.19	494	OH SHELBY	75.34	2509
OH JEFFERSON	85.01	2279	OH STARK	189.37	1067
OH KNOX	302.02	469	OH SUMMIT	126.40	1728
OH LAKE	198.27	1002	OH TRUMBULL	204.18	945
OH LAWRENCE	268.69	611	OH TUSCARAWAS	199.16	993
OH LICKING	311.08	445	OH UNION	219.50	835

ESTIMATED TOTAL FALLOUT FROM NEVADA NUCLEAR TESTS 1951-1970
MICROCURIES/SQUARE METER
COUNTIES RANKED
WHERE No. 1=Highest Fallout ; No. 3094 = Lowest Fallout

COUNTY	uCi/sq meter	RANK	COUNTY	uCi/sq meter	RANK
OH VAN WERT	159.88	1352	OK GREER	178.99	1163
OH VINTON	162.26	1325	OK HARMON	164.30	1311
OH WARREN	339.51	396	OK HARPER	139.12	1585
OH WASHINGTON	186.80	1090	OK HASKELL	183.18	1133
OH WAYNE	106.21	1961	OK HUGHES	204.92	939
OH WILLIAMS	62.58	2825	OK JACKSON	117.93	1811
OH WOOD	196.96	1017	OK JEFFERSON	126.52	1726
OH WYANDOT	278.97	560	OK JOHNSTON	153.92	1416
OK ADAIR	185.07	1112	OK KAY	210.73	896
OK ALFALFA	186.09	1097	OK KINGFISHER	158.66	1365
OK ATOKA	221.30	821	OK KIOWA	120.73	1783
OK BEAVER	197.52	1014	OK LATIMER	185.79	1102
OK BECKHAM	183.23	1132	OK LE FLORE	167.53	1288
OK BLAINE	134.82	1639	OK LINCOLN	168.99	1271
OK BRYAN	185.31	1109	OK LOGAN	151.46	1438
OK CADDO	132.75	1667	OK LOVE	155.86	1388
OK CANADIAN	143.59	1521	OK MAJOR	136.56	1613
OK CARTER	144.13	1515	OK MARSHALL	140.70	1563
OK CHEROKEE	240.18	717	OK MAYES	220.97	824
OK CHOCTAW	198.82	998	OK MCCLAIN	129.09	1704
OK CIMARRON	305.57	460	OK MCCURTAIN	191.92	1052
OK CLEVELAND	126.62	1724	OK MCINTOSH	202.36	959
OK COAL	145.06	1506	OK MURRAY	125.33	1740
OK COMANCHE	141.10	1557	OK MUSKOGEE	216.06	855
OK COTTON	118.91	1800	OK NOBLE	188.16	1082
OK CRAIG	256.47	661	OK NOWATA	214.13	872
OK CREEK	248.25	692	OK OKFUSKEE	171.54	1243
OK CUSTER	117.53	1818	OK OKLAHOMA	124.65	1743
OK DELAWARE	227.67	775	OK OKMULGEE	185.46	1106
OK DEWEY	137.71	1603	OK OSAGE	209.12	911
OK ELLIS	214.43	869	OK OTTAWA	236.33	738
OK GARFIELD	174.83	1210	OK PAWNEE	188.44	1078
OK GARVIN	124.31	1747	OK PAYNE	156.08	1385
OK GRADY	124.53	1745	OK PITTSBURG	220.59	827
OK GRANT	186.02	1099	OK PONTOTOC	178.45	1171

ESTIMATED TOTAL FALLOUT FROM NEVADA NUCLEAR TESTS 1951-1970
MICROCURIES/SQUARE METER
COUNTIES RANKED
WHERE No. 1=Highest Fallout ; No. 3094 = Lowest Fallout

COUNTY	uCi/sq meter	RANK	COUNTY	uCi/sq meter	RANK
OK POTTAWATOMIE	153.31	1423	OR LINCOLN	29.66	3034
OK PUSHMATAHA	186.08	1098	OR LINN	26.24	3052
OK ROGER MILLS	223.49	800	OR MALHEUR	135.88	1629
OK ROGERS	214.77	866	OR MARION	28.94	3041
OK SEMINOLE	191.93	1051	OR MORROW	35.28	3001
OK SEQUOYAH	210.89	893	OR MULTNOMAH	21.83	3081
OK STEPHENS	127.03	1717	OR POLK	20.32	3085
OK TEXAS	355.90	367	OR SHERMAN	23.60	3069
OK TILLMAN	103.73	1989	OR TILLAMOOK	27.70	3048
OK TULSA	274.19	588	OR UMATILLA	105.49	1975
OK WAGONER	216.01	856	OR UNION	171.97	1235
OK WASHINGTON	211.87	886	OR WALLOWA	330.15	414
OK WASHITA	177.74	1186	OR WASCO	22.13	3079
OK WOODS	150.44	1448	OR WASHINGTON	20.75	3084
OK WOODWARD	124.73	1742	OR WHEELER	29.89	3033
OR BAKER	309.42	450	OR YAMHILL	19.15	3090
OR BENTON	19.91	3087	PA ADAMS	136.00	1624
OR CLACKAMAS	31.33	3029	PA ALLEGHENY	158.99	1360
OR CLATSOP	27.72	3047	PA ARMSTRONG	198.49	999
OR COLUMBIA	24.01	3068	PA BEAVER	202.16	960
OR COOS	25.58	3061	PA BEDFORD	135.90	1628
OR CROOK	36.25	2995	PA BERKS	93.26	2125
OR CURRY	28.94	3042	PA BLAIR	103.25	1994
OR DESCHUTES	22.77	3075	PA BRADFORD	84.48	2295
OR DOUGLAS	27.82	3046	PA BUCKS	95.66	2093
OR GILLIAM	25.67	3060	PA BUTLER	172.02	1234
OR GRANT	193.57	1044	PA CAMBRIA	194.73	1037
OR HARNEY	109.99	1908	PA CAMERON	143.05	1531
OR HOOD RIVER	23.31	3070	PA CARBON	94.97	2099
OR JACKSON	19.87	3088	PA CENTRE	148.83	1469
OR JEFFERSON	25.87	3056	PA CHESTER	173.60	1224
OR JOSEPHINE	26.16	3053	PA CLARION	197.53	1012
OR KLAMATH	29.58	3035	PA CLEARFIELD	208.49	915
OR LAKE	40.32	2979	PA CLINTON	154.62	1403
OR LANE	29.11	3040	PA COLUMBIA	76.37	2484

ESTIMATED TOTAL FALLOUT FROM NEVADA NUCLEAR TESTS 1951-1970
MICROCURIES/SQUARE METER
COUNTIES RANKED
WHERE No. 1=Highest Fallout ; No. 3094 = Lowest Fallout

COUNTY	uCi/sq meter	RANK
PA CRAWFORD	202.37	958
PA CUMBERLAND	63.69	2795
PA DAUPHIN	61.52	2835
PA DELAWARE	157.87	1373
PA ELK	258.76	651
PA ERIE	157.65	1377
PA FAYETTE	215.72	860
PA FOREST	222.23	812
PA FRANKLIN	75.88	2497
PA FULTON	223.57	798
PA GREENE	176.23	1199
PA HUNTINGDON	148.74	1471
PA INDIANA	199.07	995
PA JEFFERSON	185.47	1105
PA JUNIATA	66.16	2748
PA LACKAWANNA	140.86	1561
PA LANCASTER	135.81	1630
PA LAWRENCE	182.37	1138
PA LEBANON	61.27	2841
PA LEHIGH	96.91	2068
PA LUZERNE	117.92	1812
PA LYCOMING	71.97	2606
PA MCKEAN	125.94	1733
PA MERCER	204.25	943
PA MIFFLIN	70.05	2641
PA MONROE	112.89	1868
PA MONTGOMERY	97.23	2065
PA MONTOUR	82.70	2330
PA NORTHAMPTON	99.27	2034
PA NORTHUMBERLA	142.30	1542
PA PERRY	60.44	2849
PA PHILADELPHIA	81.64	2356
PA PIKE	123.85	1751
PA POTTER	131.42	1680
PA SCHUYLKILL	84.26	2299
PA SNYDER	234.87	744
PA SOMERSET	167.86	1285
PA SULLIVAN	86.35	2250
PA SUSQUEHANNA	94.53	2106
PA TIOGA	98.85	2044
PA UNION	82.19	2337
PA VENANGO	327.78	418
PA WARREN	158.65	1366
PA WASHINGTON	188.17	1080
PA WAYNE	138.17	1601
PA WESTMORELAND	193.73	1043
PA WYOMING	115.89	1844
PA YORK	145.76	1499
RI BRISTOL	181.88	1146
RI KENT	176.49	1195
RI NEWPORT	138.35	1596
RI PROVIDENCE	149.24	1462
RI WASHINGTON	192.32	1047
SC ABBEVILLE	67.58	2708
SC AIKEN	59.08	2861
SC ALLENDALE	67.58	2707
SC ANDERSON	75.89	2496
SC BAMBERG	75.10	2516
SC BARNWELL	68.75	2671
SC BEAUFORT	81.64	2357
SC BERKELEY	84.14	2300
SC CALHOUN	89.03	2206
SC CHARLESTON	82.12	2339
SC CHEROKEE	72.65	2580
SC CHESTER	68.42	2679
SC CHESTERFIELD	79.39	2399
SC CLARENDON	92.25	2141
SC COLLETON	82.92	2326
SC DARLINGTON	82.60	2332
SC DILLON	77.40	2457

ESTIMATED TOTAL FALLOUT FROM NEVADA NUCLEAR TESTS 1951-1970
MICROCURIES/SQUARE METER
COUNTIES RANKED
WHERE No. 1=Highest Fallout ; No. 3094 = Lowest Fallout

COUNTY	uCi/sq meter	RANK	COUNTY	uCi/sq meter	RANK
SC DORCHESTER	83.76	2311	SD BRULE	261.41	641
SC EDGEFIELD	68.94	2665	SD BUFFALO	214.06	873
SC FAIRFIELD	66.33	2740	SD BUTTE	219.59	833
SC FLORENCE	112.70	1871	SD CAMPBELL	196.85	1018
SC GEORGETOWN	72.38	2588	SD CHARLES MIX	209.53	905
SC GREENVILLE	88.31	2218	SD CLARK	199.10	994
SC GREENWOOD	70.89	2624	SD CLAY	356.72	365
SC HAMPTON	71.08	2620	SD CODINGTON	206.67	927
SC HORRY	87.36	2233	SD CORSON	233.18	753
SC JASPER	93.45	2123	SD CUSTER	783.73	96
SC KERSHAW	77.43	2456	SD DAVISON	411.81	306
SC LANCASTER	67.26	2718	SD DAY	200.18	981
SC LAURENS	71.23	2619	SD DEUEL	90.19	2177
SC LEE	94.10	2110	SD DEWEY	272.15	596
SC LEXINGTON	63.85	2790	SD DOUGLAS	208.25	918
SC MARION	89.24	2200	SD EDMUNDS	184.79	1114
SC MARLBORO	81.17	2369	SD FALL RIVER	552.01	201
SC MCCORMICK	70.39	2634	SD FAULK	245.38	698
SC NEWBERRY	69.10	2663	SD GRANT	72.23	2596
SC OCONEE	106.65	1953	SD GREGORY	197.67	1009
SC ORANGEBURG	86.44	2248	SD HAAKON	315.99	436
SC PICKENS	94.35	2107	SD HAMLIN	652.91	157
SC RICHLAND	59.43	2859	SD HAND	200.44	979
SC SALUDA	67.92	2699	SD HANSON	415.77	300
SC SPARTANBURG	81.01	2375	SD HARDING	256.05	662
SC SUMTER	90.14	2178	SD HUGHES	166.94	1293
SC UNION	68.16	2692	SD HUTCHINSON	572.09	192
SC WILLIAMSBURG	78.23	2438	SD HYDE	227.26	780
SC YORK	72.75	2577	SD JACKSON	283.53	540
SD AURORA	254.20	665	SD JERAULD	356.73	364
SD BEADLE	211.77	890	SD JONES	216.17	853
SD BENNETT	233.14	754	SD KINGSBURY	596.37	177
SD BON HOMME	158.72	1364	SD LAKE	547.18	203
SD BROOKINGS	399.83	317	SD LAWRENCE	219.50	834
SD BROWN	169.78	1264	SD LINCOLN	392.72	323

ESTIMATED TOTAL FALLOUT FROM NEVADA NUCLEAR TESTS 1951-1970
MICROCURIES/SQUARE METER
COUNTIES RANKED
WHERE No. 1=Highest Fallout ; No. 3094 = Lowest Fallout

COUNTY	uCi/sq meter	RANK	COUNTY	uCi/sq meter	RANK
SD LYMAN	275.23	583	TN CARTER	121.78	1771
SD MARSHALL	188.05	1083	TN CHEATHAM	90.37	2170
SD MCCOOK	557.97	197	TN CHESTER	103.36	1992
SD MCPHERSON	167.52	1289	TN CLAIBORNE	126.25	1730
SD MEADE	261.79	638	TN CLAY	112.80	1870
SD MELLETTE	274.71	585	TN COCKE	99.23	2036
SD MINER	517.85	213	TN COFFEE	96.25	2078
SD MINNEHAHA	364.60	358	TN CROCKETT	135.22	1635
SD MOODY	473.91	241	TN CUMBERLAND	92.75	2130
SD PENNINGTON	218.16	841	TN DAVIDSON	86.63	2247
SD PERKINS	223.01	803	TN DE KALB	77.64	2452
SD POTTER	236.13	741	TN DECATUR	90.21	2176
SD ROBERTS	223.44	801	TN DICKSON	95.59	2094
SD SANBORN	214.57	868	TN DYER	147.51	1483
SD SHANNON	523.04	211	TN FAYETTE	133.49	1657
SD SPINK	225.92	786	TN FENTRESS	91.06	2162
SD STANLEY	631.65	162	TN FRANKLIN	68.84	2668
SD SULLY	238.73	725	TN GIBSON	163.28	1317
SD TODD	234.05	749	TN GILES	72.27	2593
SD TRIPP	205.21	938	TN GRAINGER	73.13	2569
SD TURNER	739.52	108	TN GREENE	90.36	2171
SD UNION	385.78	329	TN GRUNDY	95.70	2092
SD WALWORTH	219.41	836	TN HAMBLEN	80.51	2382
SD WASHABAUGH	1071.25	58	TN HAMILTON	83.68	2314
SD YANKTON	411.77	307	TN HANCOCK	123.94	1749
SD ZIEBACH	265.94	619	TN HARDEMAN	128.21	1709
TN ANDERSON	105.68	1971	TN HARDIN	85.74	2258
TN BEDFORD	81.84	2350	TN HAWKINS	92.70	2132
TN BENTON	72.01	2602	TN HAYWOOD	145.76	1500
TN BLEDSOE	95.78	2089	TN HENDERSON	87.71	2224
TN BLOUNT	97.52	2062	TN HENRY	95.07	2097
TN BRADLEY	89.87	2187	TN HICKMAN	84.76	2283
TN CAMPBELL	125.07	1741	TN HOUSTON	75.94	2493
TN CANNON	67.03	2725	TN HUMPHREYS	90.23	2175
TN CARROLL	96.16	2082	TN JACKSON	71.48	2616

ESTIMATED TOTAL FALLOUT FROM NEVADA NUCLEAR TESTS 1951-1970
MICROCURIES/SQUARE METER
COUNTIES RANKED
WHERE No. 1=Highest Fallout ; No. 3094 = Lowest Fallout

COUNTY	uCi/sq meter	RANK
TN JEFFERSON	84.74	2284
TN JOHNSON	85.28	2273
TN KNOX	81.10	2372
TN LAKE	154.93	1402
TN LAUDERDALE	151.74	1436
TN LAWRENCE	73.32	2565
TN LEWIS	63.62	2797
TN LINCOLN	71.78	2612
TN LOUDON	86.74	2243
TN MACON	58.83	2864
TN MADISON	149.11	1465
TN MARION	70.23	2638
TN MARSHALL	82.60	2333
TN MAURY	81.81	2351
TN MCMINN	93.87	2115
TN MCNAIRY	110.95	1892
TN MEIGS	78.94	2416
TN MONROE	94.01	2111
TN MONTGOMERY	98.53	2053
TN MOORE	67.44	2712
TN MORGAN	88.96	2208
TN OBION	151.32	1444
TN OVERTON	78.55	2430
TN PERRY	81.04	2373
TN PICKETT	108.30	1935
TN POLK	89.34	2196
TN PUTNAM	85.33	2270
TN RHEA	90.24	2173
TN ROANE	98.99	2040
TN ROBERTSON	79.78	2394
TN RUTHERFORD	61.35	2840
TN SCOTT	109.49	1920
TN SEQUATCHIE	90.23	2174
TN SEVIER	112.12	1880
TN SHELBY	116.83	1830
TN SMITH	71.87	2607
TN STEWART	80.02	2390
TN SULLIVAN	115.78	1846
TN SUMNER	78.56	2429
TN TIPTON	155.71	1391
TN TROUSDALE	58.13	2873
TN UNICOI	85.72	2259
TN UNION	75.18	2511
TN VAN BUREN	84.37	2297
TN WARREN	85.23	2274
TN WASHINGTON	158.29	1368
TN WAYNE	74.08	2545
TN WEAKLEY	156.74	1380
TN WHITE	78.96	2415
TN WILLIAMSON	77.51	2454
TN WILSON	72.37	2589
TX ANDERSON	82.65	2331
TX ANDREWS	169.76	1265
TX ANGELINA	155.89	1387
TX ARANSAS	36.63	2992
TX ARCHER	92.20	2142
TX ARMSTRONG	210.48	897
TX ATASCOSA	35.79	2997
TX AUSTIN	132.69	1672
TX BAILEY	288.73	520
TX BANDERA	58.87	2862
TX BASTROP	41.79	2973
TX BAYLOR	83.81	2310
TX BEE	38.93	2987
TX BELL	69.81	2645
TX BEXAR	46.34	2951
TX BLANCO	61.02	2844
TX BORDEN	93.49	2121
TX BOSQUE	82.11	2340
TX BOWIE	144.72	1508

ESTIMATED TOTAL FALLOUT FROM NEVADA NUCLEAR TESTS 1951-1970
MICROCURIES/SQUARE METER
COUNTIES RANKED
WHERE No. 1=Highest Fallout ; No. 3094 = Lowest Fallout

COUNTY	uCi/sq meter	RANK	COUNTY	uCi/sq meter	RANK
TX BRAZORIA	141.14	1555	TX CULBERSON	160.64	1344
TX BRAZOS	66.64	2735	TX DALLAM	425.79	287
TX BREWSTER	46.76	2947	TX DALLAS	160.40	1346
TX BRISCOE	243.08	705	TX DAWSON	106.60	1955
TX BROOKS	35.47	2999	TX DE WITT	41.32	2977
TX BROWN	74.29	2538	TX DEAF SMITH	298.81	482
TX BURLESON	61.79	2832	TX DELTA	132.59	1673
TX BURNET	54.77	2900	TX DENTON	171.44	1244
TX CALDWELL	42.49	2970	TX DICKENS	106.28	1957
TX CALHOUN	39.98	2983	TX DIMMIT	42.40	2971
TX CALLAHAN	88.31	2217	TX DONLEY	213.89	874
TX CAMERON	38.73	2988	TX DUVAL	31.78	3022
TX CAMP	139.70	1579	TX EASTLAND	79.56	2395
TX CARSON	211.85	888	TX ECTOR	116.44	1839
TX CASS	137.63	1604	TX EDWARDS	33.93	3008
TX CASTRO	270.56	603	TX EL PASO	216.86	850
TX CHAMBERS	147.77	1481	TX ELLIS	127.02	1718
TX CHEROKEE	100.47	2020	TX ERATH	76.64	2479
TX CHILDRESS	178.17	1179	TX FALLS	68.34	2687
TX CLAY	116.67	1834	TX FANNIN	210.85	894
TX COCHRAN	246.80	695	TX FAYETTE	53.71	2909
TX COKE	56.71	2886	TX FISHER	96.84	2070
TX COLEMAN	65.01	2760	TX FLOYD	199.03	996
TX COLLIN	178.52	1169	TX FOARD	97.38	2063
TX COLLINGSWORT	188.68	1074	TX FORT BEND	132.89	1662
TX COLORADO	50.19	2930	TX FRANKLIN	117.20	1823
TX COMAL	49.56	2933	TX FREESTONE	74.86	2520
TX COMANCHE	67.65	2704	TX FRIO	53.69	2910
TX CONCHO	70.53	2632	TX GAINES	200.86	971
TX COOKE	174.71	1212	TX GALVESTON	129.22	1702
TX CORYELL	63.26	2808	TX GARZA	103.28	1993
TX COTTLE	92.57	2135	TX GILLESPIE	56.51	2888
TX CRANE	53.73	2908	TX GLASSCOCK	53.99	2907
TX CROCKETT	45.13	2958	TX GOLIAD	32.39	3019
TX CROSBY	171.10	1247	TX GONZALES	43.42	2966

ESTIMATED TOTAL FALLOUT FROM NEVADA NUCLEAR TESTS 1951-1970
MICROCURIES/SQUARE METER
COUNTIES RANKED
WHERE No. 1=Highest Fallout ; No. 3094 = Lowest Fallout

COUNTY	uCi/sq meter	RANK	COUNTY	uCi/sq meter	RANK
TX GRAY	184.20	1121	TX JIM WELLS	42.97	2969
TX GRAYSON	201.27	966	TX JOHNSON	106.27	1959
TX GREGG	99.43	2032	TX JONES	69.16	2662
TX GRIMES	144.02	1518	TX KARNES	35.66	2998
TX GUADALUPE	31.51	3025	TX KAUFMAN	110.70	1896
TX HALE	241.08	711	TX KENDALL	48.88	2938
TX HALL	197.99	1005	TX KENEDY	33.01	3014
TX HAMILTON	75.36	2507	TX KENT	92.36	2138
TX HANSFORD	288.07	522	TX KERR	51.93	2920
TX HARDEMAN	116.46	1838	TX KIMBLE	45.90	2953
TX HARDIN	138.94	1587	TX KING	78.55	2431
TX HARRIS	142.59	1537	TX KINNEY	31.60	3023
TX HARRISON	98.95	2041	TX KLEBERG	36.54	2993
TX HARTLEY	340.26	395	TX KNOX	94.93	2100
TX HASKELL	85.20	2275	TX LA SALLE	32.65	3015
TX HAYS	40.30	2980	TX LAMAR	194.60	1038
TX HEMPHILL	176.06	1203	TX LAMB	265.16	626
TX HENDERSON	99.13	2037	TX LAMPASAS	64.58	2769
TX HIDALGO	33.07	3013	TX LAVACA	41.67	2975
TX HILL	88.01	2221	TX LEE	60.06	2853
TX HOCKLEY	248.43	691	TX LEON	72.99	2571
TX HOOD	89.97	2184	TX LIBERTY	142.59	1538
TX HOPKINS	149.27	1461	TX LIMESTONE	87.87	2223
TX HOUSTON	77.66	2451	TX LIPSCOMB	263.33	634
TX HOWARD	74.41	2531	TX LIVE OAK	34.46	3003
TX HUDSPETH	186.48	1093	TX LLANO	61.22	2843
TX HUNT	142.77	1534	TX LOVING	168.47	1278
TX HUTCHINSON	287.17	527	TX LUBBOCK	218.33	840
TX IRION	56.03	2892	TX LYNN	197.85	1006
TX JACK	97.67	2058	TX MADISON	107.90	1940
TX JACKSON	43.33	2967	TX MARION	103.38	1991
TX JASPER	161.14	1338	TX MARTIN	89.46	2194
TX JEFF DAVIS	146.70	1486	TX MASON	63.24	2809
TX JEFFERSON	85.08	2278	TX MATAGORDA	46.75	2948
TX JIM HOGG	39.53	2985	TX MAVERICK	34.23	3007

ESTIMATED TOTAL FALLOUT FROM NEVADA NUCLEAR TESTS 1951-1970 MICROCURIES/SQUARE METER COUNTIES RANKED WHERE No. 1=Highest Fallout ; No. 3094 = Lowest Fallout					
COUNTY	uCi/sq meter	RANK	COUNTY	uCi/sq meter	RANK
TX MCCULLOCH	72.62	2581	TX REEVES	138.74	1592
TX MCLENNAN	84.03	2302	TX REFUGIO	1106.74	56
TX MCMULLEN	45.35	2956	TX ROBERTS	188.87	1070
TX MEDINA	56.94	2884	TX ROBERTSON	67.63	2705
TX MENARD	53.45	2914	TX ROCKWALL	134.40	1646
TX MIDLAND	64.72	2765	TX RUNNELS	64.72	2766
TX MILAM	64.24	2781	TX RUSK	109.16	1923
TX MILLS	56.14	2891	TX SABINE	155.12	1399
TX MITCHELL	64.53	2772	TX SAN AUGUSTIN	155.00	1401
TX MONTAGUE	129.79	1694	TX SAN JACINTO	133.59	1655
TX MONTGOMERY	139.45	1582	TX SAN PATRICIO	34.37	3006
TX MOORE	247.50	693	TX SAN SABA	63.06	2813
TX MORRIS	107.82	1942	TX SCHLEICHER	45.71	2954
TX MOTLEY	186.60	1092	TX SCURRY	82.78	2327
TX NACOGDOCHES	79.24	2409	TX SHACKELFORD	73.91	2549
TX NAVARRO	91.20	2157	TX SHELBY	70.17	2639
TX NEWTON	160.87	1343	TX SHERMAN	226.66	785
TX NOLAN	72.31	2591	TX SMITH	122.97	1762
TX NUECES	31.44	3026	TX SOMERVELL	79.33	2402
TX OCHILTREE	280.36	554	TX STARR	37.24	2991
TX OLDHAM	429.87	282	TX STEPHENS	76.71	2474
TX ORANGE	250.94	676	TX STERLING	64.34	2777
TX PALO PINTO	91.11	2159	TX STONEWALL	93.98	2113
TX PANOLA	89.66	2190	TX SUTTON	37.53	2990
TX PARKER	109.96	1909	TX SWISHER	235.61	742
TX PARMER	301.61	472	TX TARRANT	146.10	1497
TX PECOS	51.20	2924	TX TAYLOR	45.26	2957
TX POLK	150.93	1447	TX TERRELL	34.42	3004
TX POTTER	364.51	359	TX TERRY	229.84	768
TX PRESIDIO	58.24	2871	TX THROCKMORTON	76.41	2483
TX RAINS	147.92	1476	TX TITUS	125.68	1738
TX RANDALL	200.63	977	TX TOM GREEN	55.96	2896
TX REAGAN	55.99	2893	TX TRAVIS	46.39	2950
TX REAL	35.29	3000	TX TRINITY	143.39	1525
TX RED RIVER	141.95	1548	TX TYLER	141.98	1546

ESTIMATED TOTAL FALLOUT FROM NEVADA NUCLEAR TESTS 1951-1970
MICROCURIES/SQUARE METER
COUNTIES RANKED
WHERE No. 1=Highest Fallout ; No. 3094 = Lowest Fallout

COUNTY	uCi/sq meter	RANK	COUNTY	uCi/sq meter	RANK
TX UPSHUR	110.57	1898	UT GRAND	1133.04	50
TX UPTON	44.36	2962	UT IRON1	1468.87	28
TX UVALDE	39.21	2986	UT IRON2	1220.93	42
TX VAL VERDE	24.90	3065	UT IRON3	1297.81	37
TX VAN ZANDT	142.89	1533	UT JUAB	1267.54	39
TX VICTORIA	33.75	3010	UT KANE1	2832.65	9
TX WALKER	151.35	1441	UT KANE2	3114.65	7
TX WALLER	136.37	1615	UT MILLARD	1438.13	32
TX WARD	123.63	1753	UT MORGAN	1718.89	19
TX WASHINGTON	129.31	1700	UT PIUTE	1148.28	48
TX WEBB	31.08	3030	UT RICH	1377.29	34
TX WHARTON	51.52	2923	UT SALT LAKE	1601.58	23
TX WHEELER	222.30	809	UT SAN JUAN	957.07	71
TX WICHITA	98.63	2049	UT SANPETE	1070.27	59
TX WILBARGER	87.48	2231	UT SEVIER	1058.64	60
TX WILLACY	37.74	2989	UT SUMMIT	1452.86	30
TX WILLIAMSON	69.32	2657	UT TOOELE1	1519.07	27
TX WILSON	32.52	3018	UT TOOELE2	1678.75	21
TX WINKLER	145.63	1502	UT UINTAH	986.33	68
TX WISE	131.40	1681	UT UTAH	1760.93	17
TX WOOD	140.05	1572	UT WASATCH	1452.94	29
TX YOAKUM	218.67	838	UT WASHINGTON1	3547.02	4
TX YOUNG	77.70	2449	UT WASHINGTON2	6788.45	1
TX ZAPATA	33.19	3012	UT WASHINGTON3	4952.76	2
TX ZAVALA	40.58	2978	UT WAYNE	874.74	80
UT BEAVER	1376.48	35	UT WEBER	1644.96	22
UT BOX ELDER1	675.37	148	VA ACCOMACK	181.29	1152
UT BOX ELDER2	1280.59	38	VA ALBEMARLE	54.10	2906
UT CACHE	1241.60	41	VA ALLEGHANY	132.75	1666
UT CARBON	886.62	79	VA AMELIA	225.09	790
UT DAGGETT	914.18	73	VA AMHERST	122.52	1765
UT DAVIS	1708.92	20	VA APPOMATTOX	119.32	1796
UT DUCHESNE	893.73	77	VA ARLINGTON	87.28	2235
UT EMERY	1167.87	45	VA AUGUSTA	120.30	1786
UT GARFIELD	2304.25	11	VA BATH	261.17	642

ESTIMATED TOTAL FALLOUT FROM NEVADA NUCLEAR TESTS 1951-1970
MICROCURIES/SQUARE METER
COUNTIES RANKED
WHERE No. 1=Highest Fallout ; No. 3094 = Lowest Fallout

COUNTY	uCi/sq meter	RANK	COUNTY	uCi/sq meter	RANK
VA BEDFORD	132.34	1674	VA HENRY	75.63	2500
VA BLAND	57.12	2881	VA HIGHLAND	136.03	1622
VA BOTETOURT	258.06	656	VA ISLE OF WIGH	83.70	2313
VA BRUNSWICK	78.31	2434	VA JAMES CITY	76.69	2475
VA BUCHANAN	228.60	772	VA KING AND QUE	152.08	1433
VA BUCKINGHAM	64.47	2774	VA KING GEORGE	79.86	2393
VA CAMPBELL	118.20	1808	VA KING WILLIAM	158.97	1361
VA CAROLINE	127.15	1714	VA LANCASTER	86.76	2242
VA CARROLL	166.41	1300	VA LEE	119.74	1791
VA CHARLES CITY	144.04	1517	VA LOUDOUN	140.00	1574
VA CHARLOTTE	60.35	2850	VA LOUISA	70.16	2640
VA CHESTERFIELD	217.97	843	VA LUNENBURG	125.78	1735
VA CLARKE	64.55	2771	VA MADISON	125.75	1736
VA CRAIG	136.90	1610	VA MATHEWS	82.28	2336
VA CULPEPER	85.97	2254	VA MECKLENBURG	131.14	1685
VA CUMBERLAND	73.88	2551	VA MIDDLESEX	91.47	2154
VA DICKENSON	174.22	1219	VA MONTGOMERY	116.85	1829
VA DINWIDDIE	82.72	2328	VA NELSON	117.57	1815
VA ESSEX	72.05	2601	VA NEW KENT	145.75	1501
VA FAIRFAX	142.11	1545	VA NEWPORT NEWS	85.43	2267
VA FAUQUIER	136.26	1619	VA NORFOLK/CHES	82.15	2338
VA FLOYD	135.67	1632	VA NORTHAMPTON	78.00	2444
VA FLUVANNA	66.88	2729	VA NORTHUMBERLA	87.28	2234
VA FRANKLIN	208.00	921	VA NOTTOWAY	78.19	2440
VA FREDERICK	115.87	1845	VA ORANGE	135.21	1636
VA GILES	111.22	1887	VA PAGE	60.56	2847
VA GLOUCESTER	76.76	2471	VA PATRICK	117.38	1820
VA GOOCHLAND	225.62	788	VA PITTSYLVANIA	131.38	1682
VA GRAYSON	108.51	1934	VA POWHATAN	230.28	762
VA GREENE	110.57	1897	VA PRINCE EDWAR	126.54	1725
VA GREENSVILLE	88.48	2213	VA PRINCE GEORG	95.44	2095
VA HALIFAX	67.87	2700	VA PRINCE WILLI	78.79	2425
VA HAMPTON	78.94	2417	VA PULASKI	119.32	1795
VA HANOVER	226.93	783	VA RAPPAHANNOCK	126.67	1723
VA HENRICO	94.70	2104	VA RICHMOND	72.75	2576

ESTIMATED TOTAL FALLOUT FROM NEVADA NUCLEAR TESTS 1951-1970
MICROCURIES/SQUARE METER
COUNTIES RANKED
WHERE No. 1=Highest Fallout ; No. 3094 = Lowest Fallout

COUNTY	uCi/sq meter	RANK
VA ROANOKE	130.01	1693
VA ROCKBRIDGE	113.21	1865
VA ROCKINGHAM	134.43	1645
VA RUSSELL	118.22	1807
VA SCOTT	116.81	1833
VA SHENANDOAH	108.58	1931
VA SMYTH	146.57	1490
VA SOUTHAMPTON	79.48	2398
VA SPOTSYLVANIA	129.74	1695
VA STAFFORD	85.32	2271
VA SUFFOLK/NANS	125.74	1737
VA SURRY	79.39	2400
VA SUSSEX	79.27	2405
VA TAZEWELL	153.18	1424
VA VIRGINIA BEA	107.00	1949
VA WARREN	141.13	1556
VA WASHINGTON	105.44	1976
VA WESTMORELAND	77.91	2445
VA WISE	100.85	2015
VA WYTHE	64.08	2785
VA YORK	76.43	2482
VT ADDISON	299.82	479
VT BENNINGTON	391.69	324
VT CALEDONIA	113.65	1863
VT CHITTENDEN	217.23	847
VT ESSEX	119.87	1790
VT FRANKLIN	153.10	1428
VT GRAND ISLE	168.63	1276
VT LAMOILLE	136.20	1620
VT ORANGE	129.44	1698
VT ORLEANS	123.70	1752
VT RUTLAND	193.38	1045
VT WASHINGTON	138.19	1599
VT WINDHAM	198.12	1004
VT WINDSOR	142.26	1543
WA ADAMS	27.63	3049
WA ASOTIN	160.23	1348
WA BENTON	26.02	3054
WA CHELAN	26.78	3050
WA CLALLAM	20.01	3086
WA CLARK	22.90	3074
WA COLUMBIA	111.87	1883
WA COWLITZ	25.97	3055
WA DOUGLAS	31.91	3021
WA FERRY	57.70	2877
WA FRANKLIN	22.94	3073
WA GARFIELD	202.80	955
WA GRANT	24.19	3067
WA GRAYS HARBOR	23.15	3071
WA ISLAND	12.37	3094
WA JEFFERSON	21.71	3082
WA KING	24.55	3066
WA KITSAP	18.54	3091
WA KITTITAS	25.87	3057
WA KLICKITAT	21.18	3083
WA LEWIS	22.55	3076
WA LINCOLN	29.53	3036
WA MASON	19.27	3089
WA OKANOGAN	43.74	2964
WA PACIFIC	22.27	3078
WA PEND OREILLE	112.18	1877
WA PIERCE	28.74	3044
WA SAN JUAN	13.49	3093
WA SKAGIT	21.85	3080
WA SKAMANIA	31.37	3028
WA SNOHOMISH	28.88	3043
WA SPOKANE	76.68	2476
WA STEVENS	50.84	2925
WA THURSTON	17.53	3092
WA WAHKIAKUM	24.94	3064

ESTIMATED TOTAL FALLOUT FROM NEVADA NUCLEAR TESTS 1951-1970
MICROCURIES/SQUARE METER
COUNTIES RANKED
WHERE No. 1=Highest Fallout ; No. 3094 = Lowest Fallout

COUNTY	uCi/sq meter	RANK
WA WALLA WALLA	90.09	2179
WA WHATCOM	23.06	3072
WA WHITMAN	89.44	2195
WA YAKIMA	22.41	3077
WI ADAMS	151.37	1440
WI ASHLAND	138.99	1586
WI BARRON	138.94	1588
WI BAYFIELD	130.68	1689
WI BROWN	69.25	2659
WI BUFFALO	470.52	246
WI BURNETT	135.93	1626
WI CALUMET	101.09	2012
WI CHIPPEWA	138.06	1602
WI CLARK	134.48	1644
WI COLUMBIA	142.36	1541
WI CRAWFORD	244.85	700
WI DANE	210.26	898
WI DODGE	97.64	2060
WI DOOR	79.38	2401
WI DOUGLAS	140.01	1573
WI DUNN	154.45	1407
WI EAU CLAIRE	141.72	1550
WI FLORENCE	77.16	2459
WI FOND DU LAC	106.77	1951
WI FOREST	87.53	2229
WI GRANT	364.66	357
WI GREEN	262.55	636
WI GREEN LAKE	110.31	1902
WI IOWA	498.94	220
WI IRON	132.70	1669
WI JACKSON	154.21	1411
WI JEFFERSON	180.27	1157
WI JUNEAU	170.08	1261
WI KENOSHA	83.72	2312
WI KEWAUNEE	81.38	2363
WI LA CROSSE	160.88	1342
WI LAFAYETTE	286.41	529
WI LANGLADE	91.53	2153
WI LINCOLN	115.25	1851
WI MANITOWOC	82.71	2329
WI MARATHON	110.14	1904
WI MARINETTE	101.98	2007
WI MARQUETTE	137.23	1609
WI MENOMINEE	96.53	2072
WI MILWAUKEE	63.37	2804
WI MONROE	162.01	1329
WI OCONTO	101.22	2011
WI ONEIDA	117.00	1825
WI OUTAGAMIE	76.31	2485
WI OZAUKEE	85.70	2260
WI PEPIN	175.56	1205
WI PIERCE	139.92	1576
WI POLK	138.85	1589
WI PORTAGE	119.09	1798
WI PRICE	127.14	1715
WI RACINE	75.36	2506
WI RICHLAND	238.04	731
WI ROCK	259.55	648
WI RUSK	144.68	1510
WI SAUK	234.20	748
WI SAWYER	157.91	1372
WI SHAWANO	97.66	2059
WI SHEBOYGAN	99.99	2026
WI ST CROIX	140.58	1564
WI TAYLOR	129.31	1701
WI TREMPEALEAU	168.46	1279
WI VERNON	321.13	426
WI VILAS	114.58	1857
WI WALWORTH	293.18	504
WI WASHBURN	153.10	1429

ESTIMATED TOTAL FALLOUT FROM NEVADA NUCLEAR TESTS 1951-1970
MICROCURIES/SQUARE METER
COUNTIES RANKED
WHERE No. 1=Highest Fallout ; No. 3094 = Lowest Fallout

COUNTY	uCi/sq meter	RANK
WI WASHINGTON	93.82	2117
WI WAUKESHA	106.01	1964
WI WAUPACA	121.41	1776
WI WAUSHARA	139.70	1578
WI WINNEBAGO	101.93	2008
WI WOOD	145.15	1505
WV BARBOUR	65.28	2757
WV BERKELEY	63.34	2806
WV BOONE	419.44	295
WV BRAXTON	159.64	1355
WV BROOKE	156.49	1383
WV CABELL	94.27	2109
WV CALHOUN	171.07	1248
WV CLAY	149.31	1459
WV DODDRIDGE	64.61	2768
WV FAYETTE	155.64	1393
WV GILMER	170.58	1254
WV GRANT	150.18	1452
WV GREENBRIER	170.62	1253
WV HAMPSHIRE	61.51	2836
WV HANCOCK	96.17	2081
WV HARDY	116.83	1831
WV HARRISON	173.63	1223
WV JACKSON	80.10	2389
WV JEFFERSON	68.65	2675
WV KANAWHA	275.26	582
WV LEWIS	164.36	1310
WV LINCOLN	298.83	481
WV LOGAN	447.71	268
WV MARION	152.48	1432
WV MARSHALL	188.85	1071
WV MASON	308.84	452
WV MCDOWELL	263.91	633
WV MERCER	233.01	755
WV MINERAL	60.31	2851
WV MINGO	239.78	721
WV MONONGALIA	149.60	1456
WV MONROE	268.56	612
WV MORGAN	71.57	2614
WV NICHOLAS	152.02	1435
WV OHIO	117.24	1821
WV PENDLETON	262.98	635
WV PLEASANTS	107.27	1946
WV POCAHONTAS	146.41	1492
WV PRESTON	73.48	2561
WV PUTNAM	154.26	1410
WV RALEIGH	266.26	617
WV RANDOLPH	155.70	1392
WV RITCHIE	66.14	2749
WV ROANE	63.92	2789
WV SUMMERS	284.76	536
WV TAYLOR	64.05	2786
WV TUCKER	162.50	1323
WV TYLER	185.35	1107
WV UPSHUR	182.17	1143
WV WAYNE	293.62	501
WV WEBSTER	144.69	1509
WV WETZEL	164.79	1308
WV WIRT	170.12	1260
WV WOOD	178.44	1172
WV WYOMING	222.33	808
WY ALBANY	345.87	382
WY BIG HORN	153.97	1415
WY CAMPBELL	370.39	348
WY CARBON	829.62	85
WY CONVERSE	465.71	249
WY CROOK	238.10	730
WY FREMONT	1374.23	36
WY GOSHEN	212.60	884
WY HOT SPRINGS	168.03	1284

ESTIMATED TOTAL FALLOUT FROM NEVADA NUCLEAR TESTS 1951-1970
MICROCURIES/SQUARE METER
COUNTIES RANKED
WHERE No. 1=Highest Fallout ; No. 3094 = Lowest Fallout

COUNTY	uCi/sq meter	RANK
WY JOHNSON	381.57	334
WY LARAMIE	204.46	942
WY LINCOLN	1053.35	61
WY NATRONA	294.32	498
WY NIOBRARA	434.45	279
WY PARK	238.71	726
WY PLATTE	319.02	430
WY SHERIDAN	188.26	1079
WY SUBLETTE	1132.07	52
WY SWEETWATER	963.72	70
WY TETON	247.18	694
WY UINTA	1030.29	64
WY WASHAKIE	314.19	441
WY WESTON	196.73	1020

ESTIMATED TOTAL FALLOUT FROM NEVADA NUCLEAR TESTS 1951-1970
MICROCURIES/SQUARE METER
COUNTIES RANKED
WHERE No. 1=Highest Fallout ; No. 3094 = Lowest Fallout
SORTED B RANK

COUNTY	uCi/sq meter	RANK	COUNTY	uCi/sq meter	RANK
UT WASHINGTON2	6788.45	1	WY FREMONT	1374.23	36
UT WASHINGTON3	4952.76	2	UT IRON3	1297.81	37
NV NYE2	3822.15	3	UT BOX ELDER2	1280.59	38
UT WASHINGTON1	3547.02	4	UT JUAB	1267.54	39
NV LINCOLN1	3420.75	5	CO MESA	1267.33	40
NV CLARK1	3315.80	6	UT CACHE	1241.60	41
UT KANE2	3114.65	7	UT IRON2	1220.93	42
AZ MOHAVE2	2997.49	8	CO GARFIELD	1199.96	43
UT KANE1	2832.65	9	NV EUREKA	1196.26	44
NV WHITE PINE2	2451.01	10	UT EMERY	1167.87	45
UT GARFIELD	2304.25	11	NM MCKINLEY	1166.85	46
NV LINCOLN2	2191.79	12	CO MINERAL	1156.07	47
NV WHITE PINE3	2170.94	13	UT PIUTE	1148.28	48
AZ COCONINO1	2057.89	14	CO HINSDALE	1146.12	49
CO GUNNISON	1946.42	15	UT GRAND	1133.04	50
CO CONEJOS	1831.89	16	IA HOWARD	1133.00	51
UT UTAH	1760.93	17	WY SUBLETTE	1132.07	52
CA INYO3	1750.08	18	IA MITCHELL	1116.94	53
UT MORGAN	1718.89	19	IA WORTH	1113.05	54
UT DAVIS	1708.92	20	CO LA PLATA	1111.60	55
UT TOOELE2	1678.75	21	TX REFUGIO	1106.74	56
UT WEBER	1644.96	22	CO DELTA	1104.38	57
UT SALT LAKE	1601.58	23	SD WASHABAUGH	1071.25	58
NV WHITE PINE1	1599.72	24	UT SANPETE	1070.27	59
AZ MOHAVE1	1593.40	25	UT SEVIER	1058.64	60
AZ COCONINO2	1591.21	26	WY LINCOLN	1053.35	61
UT TOOELE1	1519.07	27	CO MONTROSE	1050.29	62
UT IRON1	1468.87	28	ID GEM	1030.75	63
UT WASATCH	1452.94	29	WY UINTA	1030.29	64
UT SUMMIT	1452.86	30	CO PITKIN	1018.63	65
CO ARCHULETA	1438.43	31	ID CUSTER	1015.84	66
UT MILLARD	1438.13	32	NM SANDOVAL	1013.70	67
AZ APACHE	1424.32	33	UT UINTAH	986.33	68
UT RICH	1377.29	34	CO RIO BLANCO	973.92	69
UT BEAVER	1376.48	35	WY SWEETWATER	963.72	70

ESTIMATED TOTAL FALLOUT FROM NEVADA NUCLEAR TESTS 1951-1970
MICROCURIES/SQUARE METER
COUNTIES RANKED
WHERE No. 1=Highest Fallout ; No. 3094 = Lowest Fallout
SORTED B RANK

COUNTY	uCi/sq meter	RANK
UT SAN JUAN	957.07	71
AZ NAVAJO	932.91	72
UT DAGGETT	914.18	73
ID BLAINE	908.33	74
CO RIO GRANDE	905.59	75
NM VALENCIA	903.87	76
UT DUCHESNE	893.73	77
CO DOLORES	889.93	78
UT CARBON	886.62	79
UT WAYNE	874.74	80
ID BEAR LAKE	864.58	81
IA WOODBURY	842.74	82
NM SANTA FE	835.30	83
NM BERNALILLO	834.31	84
WY CARBON	829.62	85
NM TORRANCE	827.51	86
CO LAKE	825.34	87
IA VAN BUREN	813.71	88
CO SAN MIGUEL	808.07	89
CO OURAY	802.94	90
IA LEE	802.44	91
IL HANCOCK	801.68	92
CO MONTEZUMA	801.36	93
CO SAN JUAN	793.25	94
IA CARROLL	786.42	95
SD CUSTER	783.73	96
IA JASPER	767.92	97
IA BOONE	766.49	98
IA BENTON	765.06	99
IA MARSHALL	763.97	100
IA WASHINGTON	756.97	101
IA JEFFERSON	756.13	102
IA STORY	754.09	103
NV ELKO	752.66	104
MN MOWER	752.61	105
MO KNOX	752.10	106
NE DIXON	746.75	107
SD TURNER	739.52	108
NV LANDER2	735.91	109
IA HAMILTON	734.10	110
MN FARIBAULT	729.28	111
MN WATONWAN	721.07	112
ID LEMHI	718.73	113
IA IDA	717.75	114
CO EAGLE	717.55	115
CO SAGUACHE	715.97	116
IA GREENE	715.09	117
NM SAN JUAN	715.04	118
IA O BRIEN	712.89	119
CO MOFFAT	708.22	120
NE STANTON	704.90	121
IA HUMBOLDT	702.65	122
IA OSCEOLA	701.76	123
IA MONONA	700.07	124
IA SIOUX	699.28	125
CO COSTILLA	698.66	126
IA BUENA VISTA	697.46	127
NE THURSTON	696.54	128
IA POCAHONTAS	694.93	129
IA WINNESHIEK	693.28	130
IA PLYMOUTH	692.97	131
IA CALHOUN	691.41	132
IA IOWA	691.02	133
ID FRANKLIN	690.01	134
IA FLOYD	689.63	135
IA SAC	688.68	136
IA LYON	687.43	137
IA CLAY	687.38	138
IA KOSSUTH	685.66	139
IA HANCOCK	684.75	140

ESTIMATED TOTAL FALLOUT FROM NEVADA NUCLEAR TESTS 1951-1970 MICROCURIES/SQUARE METER COUNTIES RANKED WHERE No. 1=Highest Fallout ; No. 3094 = Lowest Fallout SORTED B RANK					
COUNTY	uCi/sq meter	RANK	COUNTY	uCi/sq meter	RANK
IA CHEROKEE	684.50	141	CO GRAND	596.92	176
CO ALAMOSA	683.08	142	SD KINGSBURY	596.37	177
NE DAKOTA	681.23	143	IA ADAIR	595.29	178
MO SCOTLAND	680.01	144	IA ADAMS	595.04	179
IA WINNEBAGO	679.33	145	IA MARION	594.37	180
IA PALO ALTO	677.76	146	NE HOLT	591.41	181
IA DICKINSON	676.03	147	IA UNION	591.19	182
UT BOX ELDER1	675.37	148	IA AUDUBON	589.86	183
NV ESMERALDA2	674.28	149	NM SAN MIGUEL	582.80	184
IA CRAWFORD	672.60	150	IA WAYNE	580.62	185
NM LOS ALAMOS	672.40	151	MO CLARK	580.35	186
NV MINERAL	672.09	152	NM CHAVES	579.58	187
MN ROCK	666.03	153	ID VALLEY	579.41	188
IA APPANOOSE	662.58	154	MO LEWIS	577.00	189
IA MONROE	661.74	155	NM LINCOLN	576.40	190
IA EMMET	657.61	156	NM RIO ARRIBA	573.74	191
SD HAMLIN	652.91	157	SD HUTCHINSON	572.09	192
CO SUMMIT	641.44	158	ID IDAHO	572.08	193
MT DEER LODGE	639.75	159	IA MAHASKA	570.21	194
IA DAVIS	638.73	160	IA DECATUR	569.18	195
NY WASHINGTON	636.95	161	IA CLARKE	564.31	196
SD STANLEY	631.65	162	SD MCCOOK	557.97	197
IA CASS	626.99	163	IA DELAWARE	557.21	198
NV NYE1	624.58	164	NM MORA	552.54	199
IA SHELBY	623.95	165	NE HOWARD	552.42	200
NY WARREN	614.55	166	SD FALL RIVER	552.01	201
IA RINGGOLD	613.99	167	NM COLFAX	549.53	202
NM GUADALUPE	613.32	168	SD LAKE	547.18	203
NV CHURCHILL	612.70	169	NE SHERMAN	540.41	204
IA WAPELLO	608.54	170	IA KEOKUK	538.84	205
NY RENSSELAER	606.82	171	NM QUAY	537.89	206
IA MADISON	601.13	172	NM UNION	527.59	207
NE LOUP	600.80	173	IA GUTHRIE	526.32	208
AZ MOHAVE3	598.72	174	ID ONEIDA	524.09	209
AZ MOHAVE4	598.72	175	NM HARDING	523.15	210

ESTIMATED TOTAL FALLOUT FROM NEVADA NUCLEAR TESTS 1951-1970 MICROCURIES/SQUARE METER COUNTIES RANKED WHERE No. 1=Highest Fallout ; No. 3094 = Lowest Fallout SORTED B RANK					
COUNTY	uCi/sq meter	RANK	COUNTY	uCi/sq meter	RANK
SD SHANNON	523.04	211	WI BUFFALO	470.52	246
MO RALLS	519.93	212	NV CARSON CITY	469.25	247
SD MINER	517.85	213	MN WASECA	465.92	248
NM TAOS	517.26	214	WY CONVERSE	465.71	249
NM DE BACA	510.97	215	ID GOODING	465.33	250
IA HENRY	505.05	216	IA BREMER	464.18	251
MT JUDITH BASIN	504.76	217	IA BLACK HAWK	463.64	252
CO CLEAR CREEK	503.31	218	CO MORGAN	463.49	253
ID CAMAS	501.99	219	NM ROOSEVELT	463.20	254
WI IOWA	498.94	220	ID CARIBOU	462.68	255
IA TAMA	498.16	221	MN PIPESTONE	462.27	256
MT MEAGHER	496.96	222	MN MARTIN	461.98	257
IL MCDONOUGH	496.30	223	NV STOREY	461.29	258
IL WARREN	492.77	224	NE MADISON	458.60	259
MN FILLMORE	492.01	225	MO SCHUYLER	458.11	260
MT BEAVERHEAD	491.65	226	NV DOUGLAS	457.11	261
CO JEFFERSON	485.81	227	MN NICOLLET	454.61	262
MN RICE	485.73	228	IL ADAMS	452.94	263
MN WABASHA	484.91	229	MT CHOUTEAU	451.47	264
IA LOUISA	484.20	230	MN MURRAY	451.02	265
IL FULTON	483.86	231	CO ADAMS	448.57	266
MO PUTNAM	483.39	232	NV CLARK2	448.45	267
NM CURRY	483.12	233	WV LOGAN	447.71	268
CO BOULDER	479.57	234	CO ROUTT	447.71	269
IA JOHNSON	477.17	235	IA FRANKLIN	447.05	270
CO GILPIN	476.81	236	KY GREENUP	446.90	271
NE WASHINGTON	476.43	237	MN BROWN	446.85	272
NV LYON	475.84	238	MT SILVER BOW	444.52	273
ID OWYHEE	475.67	239	MN BLUE EARTH	444.10	274
NV LANDER1	475.40	240	MO SHELBY	440.85	275
SD MOODY	473.91	241	IA POWESHIEK	439.89	276
NV WASHOE	473.32	242	KY POWELL	438.24	277
CO DOUGLAS	472.96	243	CO ARAPAHOE	434.52	278
ID BOISE	472.78	244	WY NIOBRARA	434.45	279
MN LE SUEUR	470.75	245	OH SCIOTO	432.52	280

ESTIMATED TOTAL FALLOUT FROM NEVADA NUCLEAR TESTS 1951-1970
MICROCURIES/SQUARE METER
COUNTIES RANKED
WHERE No. 1=Highest Fallout ; No. 3094 = Lowest Fallout
SORTED B RANK

COUNTY	uCi/sq meter	RANK	COUNTY	uCi/sq meter	RANK
KY LAWRENCE	430.44	281	MT GALLATIN	399.89	316
TX OLDHAM	429.87	282	SD BROOKINGS	399.83	317
KY TRIMBLE	428.98	283	IL SHELBY	399.15	318
ID CASSIA	428.02	284	MT BROADWATER	398.63	319
MT MADISON	426.45	285	MT LEWIS AND CL	398.03	320
NE ANTELOPE	426.39	286	ID WASHINGTON	396.33	321
TX DALLAM	425.79	287	MO PIKE	393.60	322
NE GARFIELD	424.61	288	SD LINCOLN	392.72	323
CO CUSTER	424.59	289	VT BENNINGTON	391.69	324
KY LEWIS	423.58	290	IA BUTLER	391.42	325
ID ELMORE	421.16	291	KS STEVENS	391.16	326
MT POWELL	419.89	292	ID LINCOLN	391.10	327
IA LINN	419.87	293	IN ALLEN	389.97	328
MO WORTH	419.80	294	SD UNION	385.78	329
WV BOONE	419.44	295	KY CAMPBELL	384.98	330
ID ADA	419.43	296	ND BILLINGS	384.09	331
NM SOCORRO	418.16	297	IN WAYNE	383.88	332
MT JEFFERSON	417.89	298	NE WAYNE	383.32	333
IA TAYLOR	417.13	299	WY JOHNSON	381.57	334
SD HANSON	415.77	300	OH MEDINA	381.46	335
IA FAYETTE	415.77	301	MT WHEATLAND	381.33	336
IA MONTGOMERY	415.13	302	KY OWEN	379.35	337
CO CHAFFEE	414.92	303	KY ROWAN	378.91	338
NE CUMING	412.08	304	CO HUERFANO	377.27	339
NY SARATOGA	411.91	305	IA DALLAS	376.32	340
SD DAVISON	411.81	306	OH MONTGOMERY	376.08	341
SD YANKTON	411.77	307	CO FREMONT	376.02	342
IA BUCHANAN	409.10	308	IA POTTAWATTAMI	374.69	343
IL PIKE	408.83	309	CO LAS ANIMAS	374.45	344
NY FULTON	406.06	310	NE CEDAR	373.69	345
OH BUTLER	405.21	311	MT POWDER RIVER	373.25	346
IA HARDIN	404.51	312	ID JEFFERSON	372.64	347
IL MOULTRIE	402.59	313	WY CAMPBELL	370.39	348
ID TWIN FALLS	401.92	314	MO AUDRAIN	368.39	349
AZ COCONINO3	401.28	315	IA CERRO GORDO	368.00	350

ESTIMATED TOTAL FALLOUT FROM NEVADA NUCLEAR TESTS 1951-1970
MICROCURIES/SQUARE METER
COUNTIES RANKED
WHERE No. 1=Highest Fallout ; No. 3094 = Lowest Fallout
SORTED BY RANK

COUNTY	uCi/sq meter	RANK	COUNTY	uCi/sq meter	RANK
OH SENECA	367.41	351	MO SULLIVAN	343.93	386
MO MARION	367.02	352	IA WRIGHT	343.74	387
NV PERSHING	366.83	353	IL COLES	343.44	388
NV ESMERALDA1	366.58	354	IA CLAYTON	343.16	389
MO DAVIESS	366.44	355	IL SANGAMON	342.19	390
KY CARROLL	365.47	356	NE COLFAX	341.55	391
WI GRANT	364.66	357	KY KENTON	341.43	392
SD MINNEHAHA	364.60	358	NM CATRON	341.26	393
TX POTTER	364.51	359	MN GOODHUE	340.66	394
CO LARIMER	363.36	360	TX HARTLEY	340.26	395
CO PHILLIPS	363.22	361	OH WARREN	339.51	396
IA DUBUQUE	359.66	362	ID MINIDOKA	339.50	397
IL CASS	358.02	363	IA MILLS	339.17	398
SD JERAULD	356.73	364	IN LAWRENCE	338.63	399
SD CLAY	356.72	365	CO KIOWA	337.59	400
MT CUSTER	356.19	366	MO LINN	336.83	401
OK TEXAS	355.90	367	OH CLINTON	336.37	402
NE GREELEY	354.57	368	KY HARRISON	336.08	403
MN NOBLES	354.17	369	NE VALLEY	335.99	404
IA PAGE	352.38	370	IA POLK	335.33	405
IA JACKSON	352.36	371	NE PLATTE	335.17	406
MT MISSOULA	351.68	372	IL GREENE	334.93	407
IA WARREN	351.68	373	IA CLINTON	334.31	408
MN WINONA	351.25	374	MT DAWSON	333.33	409
NE MERRICK	351.04	375	IA CEDAR	332.60	410
NE NANCE	349.17	376	ND STARK	332.40	411
IA WEBSTER	349.03	377	IL MONTGOMERY	331.57	412
MN STEELE	348.96	378	KY NICHOLAS	330.17	413
MN DODGE	348.90	379	OR WALLOWA	330.15	414
ID JEROME	347.26	380	ID BUTTE	328.23	415
IL BROWN	346.19	381	NE CUSTER	328.23	416
WY ALBANY	345.87	382	ID LEWIS	327.89	417
OH CLARK	345.23	383	PA VENANGO	327.78	418
MO MACON	344.05	384	NE POLK	326.46	419
CO BACA	344.02	385	IL CHRISTIAN	326.02	420

ESTIMATED TOTAL FALLOUT FROM NEVADA NUCLEAR TESTS 1951-1970
MICROCURIES/SQUARE METER
COUNTIES RANKED
WHERE No. 1=Highest Fallout ; No. 3094 = Lowest Fallout
SORTED BY RANK

COUNTY	uCi/sq meter	RANK	COUNTY	uCi/sq meter	RANK
IL MACOUPIN	324.48	421	IL MARSHALL	307.29	456
ID PAYETTE	323.74	422	KY FRANKLIN	307.06	457
NE BUTLER	322.46	423	IA FREMONT	306.47	458
MN HOUSTON	322.07	424	ND GOLDEN VALLE	306.28	459
IA ALLAMAKEE	321.59	425	OK CIMARRON	305.57	460
WI VERNON	321.13	426	ID CLEARWATER	304.95	461
NV HUMBOLDT	321.05	427	KS NESS	304.84	462
CO JACKSON	319.89	428	NY COLUMBIA	303.39	463
CO WELD	319.24	429	MN LYON	303.09	464
WY PLATTE	319.02	430	KY MENIFEE	302.92	465
KY SCOTT	318.02	431	MT FERGUS	302.46	466
MO ADAIR	317.01	432	OH ALLEN	302.13	467
MT CASCADE	316.85	433	KY HENRY	302.12	468
IA HARRISON	316.77	434	OH KNOX	302.02	469
IL SCHUYLER	316.66	435	OH PERRY	301.86	470
SD HAAKON	315.99	436	NE SAUNDERS	301.76	471
NY SCHENECTADY	315.70	437	TX PARMER	301.61	472
KY WOODFORD	315.17	438	MT RICHLAND	301.52	473
CO TELLER	315.04	439	MN LINCOLN	300.62	474
NE LINCOLN	314.86	440	NE SIOUX	300.12	475
WY WASHAKIE	314.19	441	MN JACKSON	299.96	476
CO PARK	313.63	442	MA NORFOLK	299.85	477
CA SAN BERNADIN	313.57	443	ID TETON	299.85	478
OH CHAMPAIGN	313.47	444	VT ADDISON	299.82	479
OH LICKING	311.08	445	IL PEORIA	298.99	480
KY CLARK	310.87	446	WV LINCOLN	298.83	481
ND DUNN	310.82	447	TX DEAF SMITH	298.81	482
OH MIAMI	310.64	448	MN SIBLEY	298.64	483
NE CHERRY	310.14	449	ND MCKENZIE	298.57	484
OR BAKER	309.42	450	LA GRANT	298.56	485
MT PARK	309.04	451	KY BATH	298.51	486
WV MASON	308.84	452	IN HAMILTON	297.84	487
OH HIGHLAND	308.83	453	IN GRANT	296.61	488
IL JASPER	307.99	454	IL WOODFORD	296.35	489
OH RICHLAND	307.34	455	NE DOUGLAS	295.98	490

ESTIMATED TOTAL FALLOUT FROM NEVADA NUCLEAR TESTS 1951-1970 MICROCURIES/SQUARE METER COUNTIES RANKED WHERE No. 1=Highest Fallout ; No. 3094 = Lowest Fallout SORTED B RANK					
COUNTY	uCi/sq meter	RANK	COUNTY	uCi/sq meter	RANK
NE BUFFALO	295.92	491	ID BANNOCK	287.31	526
IL HENDERSON	295.91	492	TX HUTCHINSON	287.17	527
KS FORD	295.76	493	ND WARD	286.69	528
OH JACKSON	295.19	494	WI LAFAYETTE	286.41	529
IA JONES	295.04	495	KY ROBERTSON	286.09	530
IA DES MOINES	294.80	496	CO OTERO	285.78	531
NE CHASE	294.60	497	NE PHELPS	285.70	532
WY NATRONA	294.32	498	NE CASS	285.64	533
IL MACON	294.08	499	KS STANTON	285.09	534
MT TETON	293.78	500	KS TREGO	285.09	535
WV WAYNE	293.62	501	WV SUMMERS	284.76	536
NE LOGAN	293.35	502	IL STARK	284.64	537
OH MORGAN	293.18	503	KS CHEYENNE	284.15	538
WI WALWORTH	293.18	504	KY WOLFE	284.06	539
MA PLYMOUTH	292.95	505	SD JACKSON	283.53	540
IN HOWARD	292.78	506	NE DAWSON	283.22	541
OH HANCOCK	292.78	507	OH PREBLE	283.02	542
NE KEITH	292.52	508	OH FULTON	282.76	543
OH PIKE	292.40	509	NE CLAY	282.72	544
LA RAPIDES	291.94	510	ND MERCER	282.55	545
IL WINNEBAGO	291.52	511	CO BENT	282.15	546
KY BOYD	290.65	512	KS GRAHAM	282.11	547
CO LINCOLN	290.25	513	KY BREATHITT	281.97	548
CO KIT CARSON	290.13	514	ND DIVIDE	281.26	549
NE PERKINS	290.04	515	NE FURNAS	281.18	550
CO DENVER	289.95	516	ND OLIVER	280.87	551
IN TIPTON	289.95	517	OH ADAMS	280.85	552
IN WELLS	289.66	518	CO YUMA	280.74	553
NE FRONTIER	288.81	519	TX OCHILTREE	280.36	554
TX BAILEY	288.73	520	ID POWER	280.32	555
ID MADISON	288.44	521	KS GRANT	279.58	556
TX HANSFORD	288.07	522	KY GALLATIN	279.44	557
MT WIBAUX	287.92	523	IL MASON	279.40	558
IL KNOX	287.84	524	CO SEDGWICK	279.14	559
KS NORTON	287.81	525	OH WYANDOT	278.97	560

ESTIMATED TOTAL FALLOUT FROM NEVADA NUCLEAR TESTS 1951-1970 MICROCURIES/SQUARE METER COUNTIES RANKED WHERE No. 1=Highest Fallout ; No. 3094 = Lowest Fallout SORTED B RANK					
COUNTY	uCi/sq meter	RANK	COUNTY	uCi/sq meter	RANK
NC PERQUIMANS	278.75	561	SD DEWEY	272.15	596
IL ROCK ISLAND	278.35	562	MT FALLON	272.15	597
IL PIATT	278.23	563	KY CARTER	271.91	598
ND BURKE	277.80	564	ND BOTTINEAU	271.89	599
NC CURRITUCK	277.62	565	MT MINERAL	271.46	600
KS KEARNY	277.54	566	CO ELBERT	271.32	601
IL HENRY	276.87	567	LA LAFAYETTE	270.68	602
NC PASQUOTANK	276.73	568	TX CASTRO	270.56	603
IL DOUGLAS	276.62	569	KS FINNEY	269.76	604
CO EL PASO	276.54	570	MT MCCONE	269.59	605
NC CAMDEN	276.48	571	KS GOVE	269.47	606
NE GOSPER	276.45	572	NE RED WILLOW	269.36	607
IA MUSCATINE	276.43	573	IN MARION	268.81	608
IL EFFINGHAM	276.34	574	KS SCOTT	268.81	609
NE SALINE	276.31	575	MO GRUNDY	268.72	610
KY MAGOFFIN	276.07	576	OH LAWRENCE	268.69	611
KS HODGEMAN	275.98	577	WV MONROE	268.56	612
CA INYO1	275.94	578	CO WASHINGTON	267.24	613
CA INYO2	275.94	579	MO MONROE	267.15	614
MO CHARITON	275.82	580	IL MCLEAN	267.04	615
KY FLEMING	275.51	581	CO CHEYENNE	266.48	616
WV KANAWHA	275.26	582	WV RALEIGH	266.26	617
SD LYMAN	275.23	583	KS DECATUR	265.98	618
NY ALBANY	275.00	584	SD ZIEBACH	265.94	619
SD MELLETTE	274.71	585	KS LOGAN	265.89	620
CO PROWERS	274.58	586	IL TAZEWELL	265.64	621
KS RUSSELL	274.53	587	MO CALDWELL	265.60	622
OK TULSA	274.19	588	AZ PIMA	265.35	623
LA BEAUREGARD	273.81	589	KY JOHNSON	265.28	624
KS WALLACE	273.76	590	IL BOONE	265.21	625
KY ELLIOTT	273.37	591	TX LAMB	265.16	626
CO CROWLEY	273.02	592	IL MERCER	265.10	627
KS LANE	272.51	593	ND SLOPE	265.02	628
KY MASON	272.40	594	IN JACKSON	264.34	629
ND MCHENRY	272.28	595	KS HAMILTON	264.21	630

ESTIMATED TOTAL FALLOUT FROM NEVADA NUCLEAR TESTS 1951-1970
MICROCURIES/SQUARE METER
COUNTIES RANKED
WHERE No. 1=Highest Fallout ; No. 3094 = Lowest Fallout
SORTED B RANK

COUNTY	uCi/sq meter	RANK
IL DE WITT	264.13	631
KS RAWLINS	263.92	632
WV MCDOWELL	263.91	633
TX LIPSCOMB	263.33	634
WV PENDLETON	262.98	635
WI GREEN	262.55	636
ID CLARK	262.09	637
SD MEADE	261.79	638
KY GRANT	261.54	639
ND RICHLAND	261.48	640
SD BRULE	261.41	641
VA BATH	261.17	642
NE HAYES	260.96	643
CA MONO	260.93	644
KY FLOYD	260.65	645
KS WICHITA	260.37	646
MT ROOSEVELT	260.23	647
WI ROCK	259.55	648
NE DUNDY	259.36	649
NE THOMAS	259.31	650
PA ELK	258.76	651
KS SALINE	258.72	652
MN COTTONWOOD	258.55	653
NE ADAMS	258.47	654
MT CARBON	258.42	655
VA BOTETOURT	258.06	656
LA EVANGELINE	257.84	657
IA SCOTT	257.67	658
ND MOUNTRAIL	257.51	659
NE SEWARD	257.20	660
OK CRAIG	256.47	661
SD HARDING	256.05	662
ND MCLEAN	255.79	663
MO HARRISON	255.44	664
SD AURORA	254.20	665
LA ACADIA	253.86	666
LA CALCASIEU	253.35	667
KS GRAY	252.87	668
KY SHELBY	252.62	669
MT PRAIRIE	252.44	670
NE HITCHCOCK	252.01	671
MO RANDOLPH	251.59	672
IL LOGAN	251.47	673
MT BLAINE	251.26	674
NE CHEYENNE	251.00	675
TX ORANGE	250.94	676
LA CAMERON	250.94	677
IN DELAWARE	250.72	678
ND RENVILLE	250.37	679
NE GAGE	250.33	680
NE HAMILTON	250.28	681
CO LOGAN	250.13	682
ID BONNEVILLE	249.64	683
KY MORGAN	249.46	684
KS HASKELL	249.31	685
KS GREELEY	249.20	686
KS SHERIDAN	249.00	687
IN HENDRICKS	248.96	688
KS THOMAS	248.87	689
NE YORK	248.60	690
TX HOCKLEY	248.43	691
OK CREEK	248.25	692
TX MOORE	247.50	693
WY TETON	247.18	694
TX COCHRAN	246.80	695
MO CAMDEN	245.57	696
CO PUEBLO	245.45	697
SD FAULK	245.38	698
MA MIDDLESEX	245.10	699
WI CRAWFORD	244.85	700

ESTIMATED TOTAL FALLOUT FROM NEVADA NUCLEAR TESTS 1951-1970
MICROCURIES/SQUARE METER
COUNTIES RANKED
WHERE No. 1=Highest Fallout ; No. 3094 = Lowest Fallout
SORTED B RANK

COUNTY	uCi/sq meter	RANK	COUNTY	uCi/sq meter	RANK
KY PIKE	244.73	701	NE NUCKOLLS	237.14	736
MT GRANITE	244.72	702	ID ADAMS	236.86	737
OH ATHENS	243.67	703	OK OTTAWA	236.33	738
MA BRISTOL	243.10	704	LA ST HELENA	236.29	739
TX BRISCOE	243.08	705	NE MCPHERSON	236.20	740
NE JOHNSON	241.64	706	SD POTTER	236.13	741
KY JACKSON	241.60	707	TX SWISHER	235.61	742
IN SWITZERLAND	241.57	708	KS BARTON	235.30	743
NE HOOKER	241.52	709	PA SNYDER	234.87	744
KY OWSLEY	241.50	710	IL JERSEY	234.66	745
TX HALE	241.08	711	NM EDDY	234.51	746
IN GREENE	240.78	712	KS LINCOLN	234.41	747
ND MORTON	240.70	713	WI SAUK	234.20	748
MO PETTIS	240.58	714	SD TODD	234.05	749
ID FREMONT	240.54	715	MT STILLWATER	233.82	750
ND HETTINGER	240.36	716	ID CANYON	233.53	751
OK CHEROKEE	240.18	717	MO ATCHISON	233.25	752
IN CASS	240.10	718	SD CORSON	233.18	753
MO MORGAN	240.03	719	SD BENNETT	233.14	754
IL CLAY	239.80	720	WV MERCER	233.01	755
WV MINGO	239.78	721	MT RAVALLI	232.76	756
IN CRAWFORD	239.42	722	IN HENRY	231.98	757
OH MADISON	238.86	723	MT CARTER	231.88	758
NE FILLMORE	238.83	724	MO OSAGE	230.87	759
SD SULLY	238.73	725	OH GALLIA	230.78	760
WY PARK	238.71	726	LA TANGIPAHOA	230.30	761
MT HILL	238.29	727	VA POWHATAN	230.28	762
IL MORGAN	238.26	728	MO DE KALB	230.24	763
IL CHAMPAIGN	238.17	729	NE LANCASTER	230.16	764
WY CROOK	238.10	730	NE ARTHUR	230.04	765
WI RICHLAND	238.04	731	KY ANDERSON	230.02	766
MT SHERIDAN	237.86	732	IN OWEN	229.91	767
OH PORTAGE	237.84	733	TX TERRY	229.84	768
IL SCOTT	237.82	734	KS OSBORNE	229.57	769
KS MORTON	237.66	735	IN JEFFERSON	228.95	770

ESTIMATED TOTAL FALLOUT FROM NEVADA NUCLEAR TESTS 1951-1970
MICROCURIES/SQUARE METER
COUNTIES RANKED
WHERE No. 1=Highest Fallout ; No. 3094 = Lowest Fallout
SORTED B RANK

COUNTY	uCi/sq meter	RANK
KS MARION	228.94	771
VA BUCHANAN	228.60	772
MT MUSSELSHELL	228.53	773
ID SHOSHONE	228.01	774
OK DELAWARE	227.67	775
NE MORRILL	227.49	776
KS DICKINSON	227.38	777
ND GRANT	227.38	778
MO CALLAWAY	227.26	779
SD HYDE	227.26	780
OH COSHOCTON	227.16	781
MO LIVINGSTON	227.05	782
VA HANOVER	226.93	783
ND BURLEIGH	226.72	784
TX SHERMAN	226.66	785
SD SPINK	225.92	786
NE DEUEL	225.70	787
VA GOOCHLAND	225.62	788
ND ADAMS	225.25	789
VA AMELIA	225.09	790
OH CLERMONT	224.75	791
OH LOGAN	224.51	792
MS WARREN	223.93	793
MA SUFFOLK	223.89	794
ND EMMONS	223.70	795
NM LEA	223.61	796
KS MEADE	223.61	797
PA FULTON	223.57	798
LA TENSAS	223.49	799
OK ROGER MILLS	223.49	800
SD ROBERTS	223.44	801
MT SWEET GRASS	223.04	802
SD PERKINS	223.01	803
AZ SANTA CRUZ	222.89	804
OH BROWN	222.75	805

COUNTY	uCi/sq meter	RANK
KS MCPHERSON	222.68	806
LA ALLEN	222.35	807
WV WYOMING	222.33	808
TX WHEELER	222.30	809
KS ROOKS	222.28	810
LA LA SALLE	222.24	811
PA FOREST	222.23	812
MO NEWTON	222.19	813
MO BENTON	222.08	814
ND BOWMAN	222.04	815
MI EATON	221.79	816
MO LAWRENCE	221.40	817
MT BIG HORN	221.36	818
NE NEMAHA	221.34	819
ND RANSOM	221.34	820
OK ATOKA	221.30	821
KS WABAUNSEE	221.06	822
IN MARTIN	221.04	823
OK MAYES	220.97	824
KS MITCHELL	220.97	825
AZ YAVAPAI	220.60	826
OK PITTSBURG	220.59	827
AR FULTON	220.35	828
KS OTTAWA	220.20	829
KS ELLSWORTH	220.03	830
MO HENRY	219.86	831
LA IBERIA	219.69	832
SD BUTTE	219.59	833
SD LAWRENCE	219.50	834
OH UNION	219.50	835
SD WALWORTH	219.41	836
MO FRANKLIN	218.88	837
TX YOAKUM	218.67	838
KY OLDHAM	218.52	839
TX LUBBOCK	218.33	840

ESTIMATED TOTAL FALLOUT FROM NEVADA NUCLEAR TESTS 1951-1970
MICROCURIES/SQUARE METER
COUNTIES RANKED
WHERE No. 1=Highest Fallout ; No. 3094 = Lowest Fallout
SORTED B RANK

COUNTY	uCi/sq meter	RANK	COUNTY	uCi/sq meter	RANK
SD PENNINGTON	218.16	841	LA POINTE COUPE	213.38	876
IL CALHOUN	218.14	842	MO MILLER	213.30	877
VA CHESTERFIELD	217.97	843	ND SHERIDAN	212.87	878
LA JEFFERSON DA	217.90	844	MT FLATHEAD	212.79	879
MS ADAMS	217.81	845	ND SARGENT	212.77	880
LA VERMILION	217.26	846	MO WRIGHT	212.77	881
VT CHITTENDEN	217.23	847	NE THAYER	212.68	882
OH GUERNSEY	217.20	848	KS SEWARD	212.65	883
MI INGHAM	216.88	849	WY GOSHEN	212.60	884
TX EL PASO	216.86	850	AR SHARP	212.11	885
OH MUSKINGUM	216.81	851	OK WASHINGTON	211.87	886
MO RAY	216.24	852	LA WEST FELICIA	211.86	887
SD JONES	216.17	853	TX CARSON	211.85	888
KS ELLIS	216.16	854	KS COWLEY	211.79	889
OK MUSKOGEE	216.06	855	SD BEADLE	211.77	890
OK WAGONER	216.01	856	LA WEST BATON R	211.55	891
KY MONTGOMERY	215.87	857	NE WEBSTER	211.34	892
IN MONROE	215.73	858	OK SEQUOYAH	210.89	893
IL VERMILION	215.72	859	TX FANNIN	210.85	894
PA FAYETTE	215.72	860	KS RENO	210.80	895
MO MCDONALD	215.60	861	OK KAY	210.73	896
KY PENDLETON	215.09	862	TX ARMSTRONG	210.48	897
NJ BURLINGTON	215.08	863	WI DANE	210.26	898
OH PAULDING	214.96	864	LA TERREBONNE	210.21	899
NM SIERRA	214.95	865	AR LAWRENCE	209.90	900
OK ROGERS	214.77	866	LA IBERVILLE	209.84	901
KY LESLIE	214.64	867	OH DARKE	209.77	902
SD SANBORN	214.57	868	CT MIDDLESEX	209.66	903
OK ELLIS	214.43	869	IN FRANKLIN	209.64	904
AR FAULKNER	214.41	870	SD CHARLES MIX	209.53	905
NE OTOE	214.21	871	KS HARVEY	209.43	906
OK NOWATA	214.13	872	OH ASHLAND	209.30	907
SD BUFFALO	214.06	873	NE GARDEN	209.21	908
TX DONLEY	213.89	874	KS RUSH	209.14	909
AR BENTON	213.49	875	IN ORANGE	209.13	910

ESTIMATED TOTAL FALLOUT FROM NEVADA NUCLEAR TESTS 1951-1970
MICROCURIES/SQUARE METER
COUNTIES RANKED
WHERE No. 1=Highest Fallout ; No. 3094 = Lowest Fallout
SORTED B RANK

COUNTY	uCi/sq meter	RANK	COUNTY	uCi/sq meter	RANK
OK OSAGE	209.12	911	AR PERRY	204.15	946
ND DICKEY	209.06	912	NE KEARNEY	204.15	947
AR PRAIRIE	208.86	913	MO POLK	204.13	948
LA EAST BATON R	208.78	914	KS BUTLER	203.82	949
PA CLEARFIELD	208.49	915	NE BOONE	203.52	950
NE HALL	208.39	916	KS SUMNER	203.46	951
MO LACLEDE	208.36	917	MT VALLEY	203.08	952
SD DOUGLAS	208.25	918	KS LYON	203.06	953
IN CARROLL	208.04	919	NE DAWES	202.92	954
MO PULASKI	208.03	920	WA GARFIELD	202.80	955
VA FRANKLIN	208.00	921	KS CHASE	202.69	956
NE BANNER	207.95	922	MO DOUGLAS	202.38	957
KS REPUBLIC	207.57	923	PA CRAWFORD	202.37	958
KS RICE	207.14	924	OK MCINTOSH	202.36	959
AR MADISON	206.89	925	PA BEAVER	202.16	960
AR JACKSON	206.84	926	NY ESSEX	201.81	961
SD CODINGTON	206.67	927	NY GREENE	201.74	962
MT GARFIELD	206.65	928	MT GOLDEN VALLE	201.73	963
KS SHAWNEE	206.62	929	LA ST MARY	201.52	964
KS KINGMAN	206.59	930	MT DANIELS	201.30	965
KS CRAWFORD	206.50	931	TX GRAYSON	201.27	966
MO BARRY	206.50	932	MO OREGON	201.24	967
IL CLARK	206.29	933	NE GRANT	201.22	968
NE BLAINE	205.99	934	MO COLE	201.12	969
MS WILKINSON	205.95	935	IL RICHLAND	200.94	970
MT TOOLE	205.78	936	TX GAINES	200.86	971
MO LAFAYETTE	205.70	937	NE BROWN	200.85	972
SD TRIPP	205.21	938	MT ROSEBUD	200.85	973
OK HUGHES	204.92	939	KS GEARY	200.84	974
IL MENARD	204.83	940	MT LAKE	200.74	975
KS CHAUTAUQUA	204.61	941	AR NEWTON	200.71	976
WY LARAMIE	204.46	942	TX RANDALL	200.63	977
PA MERCER	204.25	943	AR BAXTER	200.56	978
AR WASHINGTON	204.23	944	SD HAND	200.44	979
OH TRUMBULL	204.18	945	KS GREENWOOD	200.31	980

ESTIMATED TOTAL FALLOUT FROM NEVADA NUCLEAR TESTS 1951-1970 MICROCURIES/SQUARE METER COUNTIES RANKED WHERE No. 1=Highest Fallout ; No. 3094 = Lowest Fallout SORTED B RANK					
COUNTY	**uCi/sq meter**	**RANK**	**COUNTY**	**uCi/sq meter**	**RANK**
SD DAY	200.18	981	IN MADISON	197.12	1016
ND CASS	200.14	982	OH WOOD	196.96	1017
NE SHERIDAN	200.08	983	SD CAMPBELL	196.85	1018
LA ASSUMPTION	200.05	984	NE HARLAN	196.76	1019
MO CHRISTIAN	200.00	985	WY WESTON	196.73	1020
IA LUCAS	199.92	986	IN UNION	196.61	1021
AR POPE	199.88	987	IA CHICKASAW	196.56	1022
LA ST MARTIN	199.87	988	NJ OCEAN	196.25	1023
IL FORD	199.75	989	AR FRANKLIN	195.88	1024
MT PETROLEUM	199.49	990	OH HAMILTON	195.85	1025
AR CRAIGHEAD	199.46	991	IN ADAMS	195.71	1026
MO TEXAS	199.34	992	IL BOND	195.56	1027
OH TUSCARAWAS	199.16	993	MO GREENE	195.52	1028
SD CLARK	199.10	994	AR JOHNSON	195.50	1029
PA INDIANA	199.07	995	IN CLINTON	195.46	1030
TX FLOYD	199.03	996	KS PHILLIPS	195.46	1031
ND FOSTER	199.01	997	KS SMITH	195.17	1032
OK CHOCTAW	198.82	998	MA ESSEX	195.04	1033
PA ARMSTRONG	198.49	999	OH COLUMBIANA	195.02	1034
AR CLEBURNE	198.43	1000	IN FULTON	194.99	1035
MO JASPER	198.36	1001	MO WEBSTER	194.85	1036
OH LAKE	198.27	1002	PA CAMBRIA	194.73	1037
KS LEAVENWORTH	198.20	1003	TX LAMAR	194.60	1038
VT WINDHAM	198.12	1004	KS RILEY	194.53	1039
TX HALL	197.99	1005	IN DUBOIS	193.79	1040
TX LYNN	197.85	1006	MT LIBERTY	193.74	1041
AR VAN BUREN	197.73	1007	NE KIMBALL	193.74	1042
KY BRACKEN	197.71	1008	PA WESTMORELAND	193.73	1043
SD GREGORY	197.67	1009	OR GRANT	193.57	1044
OH BELMONT	197.63	1010	VT RUTLAND	193.38	1045
MO OZARK	197.59	1011	NE JEFFERSON	192.68	1046
PA CLARION	197.53	1012	RI WASHINGTON	192.32	1047
AR IZARD	197.52	1013	OH SANDUSKY	192.16	1048
OK BEAVER	197.52	1014	KY PERRY	192.11	1049
MS FRANKLIN	197.42	1015	MO DENT	191.99	1050

ESTIMATED TOTAL FALLOUT FROM NEVADA NUCLEAR TESTS 1951-1970 MICROCURIES/SQUARE METER COUNTIES RANKED WHERE No. 1=Highest Fallout ; No. 3094 = Lowest Fallout SORTED BY RANK					
COUNTY	**uCi/sq meter**	**RANK**	**COUNTY**	**uCi/sq meter**	**RANK**
OK SEMINOLE	191.93	1051	KS NEOSHO	187.31	1086
OK MCCURTAIN	191.92	1052	NE BOYD	187.17	1087
MO SALINE	191.74	1053	MO MERCER	187.00	1088
AR MARION	191.56	1054	ID BINGHAM	186.83	1089
NE WHEELER	191.49	1055	OH WASHINGTON	186.80	1090
AR INDEPENDENCE	191.48	1056	NY ERIE	186.65	1091
LA EAST FELICIA	190.99	1057	TX MOTLEY	186.60	1092
MO PHELPS	190.82	1058	TX HUDSPETH	186.48	1093
AR WHITE	190.48	1059	OH MEIGS	186.45	1094
KS MORRIS	190.34	1060	KY KNOTT	186.28	1095
MO CASS	190.10	1061	IN ST JOSEPH	186.12	1096
MT PONDERA	190.06	1062	OK ALFALFA	186.09	1097
KS ELK	190.03	1063	OK PUSHMATAHA	186.08	1098
MO COOPER	189.98	1064	OK GRANT	186.02	1099
KS HARPER	189.77	1065	MO HOWELL	185.83	1100
MO LINCOLN	189.76	1066	IN SHELBY	185.80	1101
OH STARK	189.37	1067	OK LATIMER	185.79	1102
KS STAFFORD	188.98	1068	LA LIVINGSTON	185.76	1103
MO VERNON	188.93	1069	KS PAWNEE	185.54	1104
TX ROBERTS	188.87	1070	PA JEFFERSON	185.47	1105
WV MARSHALL	188.85	1071	OK OKMULGEE	185.46	1106
NY MONTGOMERY	188.81	1072	WV TYLER	185.35	1107
IN KNOX	188.70	1073	MO CARROLL	185.33	1108
TX COLLINGSWORT	188.68	1074	OK BRYAN	185.31	1109
OH MAHONING	188.57	1075	OH MONROE	185.22	1110
IL FAYETTE	188.47	1076	KS PRATT	185.16	1111
KS LABETTE	188.47	1077	OK ADAIR	185.07	1112
OK PAWNEE	188.44	1078	LA AVOYELLES	184.92	1113
WY SHERIDAN	188.26	1079	SD EDMUNDS	184.79	1114
PA WASHINGTON	188.17	1080	MN REDWOOD	184.71	1115
MO MARIES	188.17	1081	AR WOODRUFF	184.69	1116
OK NOBLE	188.16	1082	KS WILSON	184.60	1117
SD MARSHALL	188.05	1083	OH HARRISON	184.39	1118
IN VERMILLION	188.01	1084	ND LA MOURE	184.37	1119
KY SPENCER	187.71	1085	KS CHEROKEE	184.35	1120

ESTIMATED TOTAL FALLOUT FROM NEVADA NUCLEAR TESTS 1951-1970
MICROCURIES/SQUARE METER
COUNTIES RANKED
WHERE No. 1=Highest Fallout ; No. 3094 = Lowest Fallout
SORTED B RANK

COUNTY	uCi/sq meter	RANK	COUNTY	uCi/sq meter	RANK
TX GRAY	184.20	1121	ND WILLIAMS	180.53	1156
ID NEZ PERCE	184.19	1122	WI JEFFERSON	180.27	1157
OH DELAWARE	184.12	1123	ND WALSH	180.09	1158
ME AROOSTOOK	184.10	1124	KS WYANDOTTE	179.62	1159
KS POTTAWATOMIE	184.08	1125	OH MORROW	179.37	1160
NY HAMILTON	183.93	1126	MO HOWARD	179.18	1161
OH MARION	183.82	1127	AR SEARCY	179.10	1162
KS EDWARDS	183.58	1128	OK GREER	178.99	1163
ND MCINTOSH	183.56	1129	KS WASHINGTON	178.90	1164
IL EDGAR	183.52	1130	MO GASCONADE	178.86	1165
OH LUCAS	183.46	1131	MO DALLAS	178.80	1166
OK BECKHAM	183.23	1132	OH ERIE	178.71	1167
OK HASKELL	183.18	1133	CT NEW HAVEN	178.70	1168
MT SANDERS	183.01	1134	TX COLLIN	178.52	1169
AR SALINE	182.66	1135	AR RANDOLPH	178.51	1170
MT PHILLIPS	182.51	1136	OK PONTOTOC	178.45	1171
MO NODAWAY	182.47	1137	WV WOOD	178.44	1172
PA LAWRENCE	182.37	1138	LA ASCENSION	178.38	1173
IN CLARK	182.28	1139	AZ PINAL	178.28	1174
MO CEDAR	182.26	1140	NY GENESEE	178.27	1175
AR POLK	182.22	1141	AR ARKANSAS	178.26	1176
KS SHERMAN	182.19	1142	NY SCHOHARIE	178.26	1177
WV UPSHUR	182.17	1143	NE PIERCE	178.21	1178
AR CRAWFORD	182.17	1144	TX CHILDRESS	178.17	1179
KY JESSAMINE	181.89	1145	CT TOLLAND	178.07	1180
RI BRISTOL	181.88	1146	MO JOHNSON	178.00	1181
ND BARNES	181.81	1147	NY CHAUTAUQUA	177.97	1182
LA ST CHARLES	181.70	1148	IL CUMBERLAND	177.82	1183
LA LAFOURCHE	181.37	1149	AZ GILA	177.81	1184
KY BOONE	181.34	1150	ND LOGAN	177.79	1185
ND TRAILL	181.29	1151	OK WASHITA	177.74	1186
VA ACCOMACK	181.29	1152	ND STUTSMAN	177.69	1187
OH HOCKING	181.01	1153	AR GREENE	177.56	1188
IN FLOYD	180.96	1154	MO CLINTON	177.46	1189
NE FRANKLIN	180.79	1155	OH ROSS	177.37	1190

ESTIMATED TOTAL FALLOUT FROM NEVADA NUCLEAR TESTS 1951-1970
MICROCURIES/SQUARE METER
COUNTIES RANKED
WHERE No. 1=Highest Fallout ; No. 3094 = Lowest Fallout
SORTED BY RANK

COUNTY	uCi/sq meter	RANK	COUNTY	uCi/sq meter	RANK
IA GRUNDY	176.93	1191	LA SABINE	173.15	1226
AR JEFFERSON	176.73	1192	OH CRAWFORD	173.14	1227
KY ESTILL	176.69	1193	AR PULASKI	173.11	1228
IN BARTHOLOMEW	176.52	1194	NC CHOWAN	172.58	1229
RI KENT	176.49	1195	AR LONOKE	172.49	1230
NJ ATLANTIC	176.48	1196	AR CONWAY	172.37	1231
NC WASHINGTON	176.46	1197	AR LOGAN	172.28	1232
KS CLAY	176.29	1198	MA DUKES	172.14	1233
PA GREENE	176.23	1199	PA BUTLER	172.02	1234
AR YELL	176.21	1200	OR UNION	171.97	1235
KY LETCHER	176.11	1201	NV CLARK3	171.84	1236
MT TREASURE	176.09	1202	MO MONITEAU	171.75	1237
TX HEMPHILL	176.06	1203	MA BARNSTABLE	171.73	1238
NY WYOMING	175.81	1204	NE BOX BUTTE	171.60	1239
WI PEPIN	175.56	1205	NE KEYA PAHA	171.59	1240
MD GARRETT	175.44	1206	LA VERNON	171.58	1241
OH CUYAHOGA	175.41	1207	MO STONE	171.57	1242
MO DADE	175.22	1208	OK OKFUSKEE	171.54	1243
OH AUGLAIZE	174.86	1209	TX DENTON	171.44	1244
OK GARFIELD	174.83	1210	ID LATAH	171.31	1245
ND GRAND FORKS	174.75	1211	ND NELSON	171.20	1246
TX COOKE	174.71	1212	TX CROSBY	171.10	1247
ND GRIGGS	174.70	1213	WV CALHOUN	171.07	1248
NE KNOX	174.69	1214	KS SEDGWICK	170.95	1249
NJ MONMOUTH	174.65	1215	NY ULSTER	170.88	1250
ND SIOUX	174.55	1216	AR MONTGOMERY	170.78	1251
MO BARTON	174.46	1217	MA BERKSHIRE	170.66	1252
KS MONTGOMERY	174.40	1218	WV GREENBRIER	170.62	1253
VA DICKENSON	174.22	1219	WV GILMER	170.58	1254
AZ MARICOPA	173.90	1220	KS JOHNSON	170.51	1255
MO BOONE	173.71	1221	OH FAYETTE	170.27	1256
KS WOODSON	173.67	1222	NY NASSAU	170.19	1257
WV HARRISON	173.63	1223	AR SEBASTIAN	170.18	1258
PA CHESTER	173.60	1224	MS HINDS	170.17	1259
AZ GRAHAM	173.51	1225	WV WIRT	170.12	1260

ESTIMATED TOTAL FALLOUT FROM NEVADA NUCLEAR TESTS 1951-1970 MICROCURIES/SQUARE METER COUNTIES RANKED WHERE No. 1=Highest Fallout ; No. 3094 = Lowest Fallout SORTED B RANK					
COUNTY	uCi/sq meter	RANK	COUNTY	uCi/sq meter	RANK
WI JUNEAU	170.08	1261	NY NIAGARA	166.76	1296
AR SEVIER	169.83	1262	MO GENTRY	166.68	1297
NE DODGE	169.83	1263	IN SCOTT	166.59	1298
SD BROWN	169.78	1264	KS KIOWA	166.55	1299
TX ANDREWS	169.76	1265	VA CARROLL	166.41	1300
NE BURT	169.73	1266	KS ALLEN	166.05	1301
AR BOONE	169.60	1267	KS JACKSON	165.86	1302
IN RANDOLPH	169.54	1268	KS CLARK	165.68	1303
IN HARRISON	169.16	1269	AR ST FRANCIS	165.32	1304
AR CARROLL	169.11	1270	ND BENSON	165.29	1305
OK LINCOLN	168.99	1271	KS CLOUD	165.20	1306
KY MADISON	168.99	1272	MO CLAY	165.17	1307
AR CLAY	168.96	1273	WV WETZEL	164.79	1308
LA CONCORDIA	168.79	1274	MS PEARL RIVER	164.73	1309
KS JEWELL	168.75	1275	WV LEWIS	164.36	1310
VT GRAND ISLE	168.63	1276	OK HARMON	164.30	1311
KS MARSHALL	168.48	1277	MO BATES	164.25	1312
TX LOVING	168.47	1278	NY LIVINGSTON	164.20	1313
WI TREMPEALEAU	168.46	1279	CT FAIRFIELD	163.91	1314
IL WHITESIDE	168.31	1280	AR SCOTT	163.75	1315
KY ROCKCASTLE	168.28	1281	ND STEELE	163.69	1316
IN WASHINGTON	168.18	1282	TN GIBSON	163.28	1317
KS COFFEY	168.14	1283	LA CATAHOULA	163.15	1318
WY HOT SPRINGS	168.03	1284	MO JEFFERSON	162.88	1319
PA SOMERSET	167.86	1285	MO NEW MADRID	162.78	1320
MO RIPLEY	167.83	1286	IL JO DAVIESS	162.73	1321
KS ATCHISON	167.67	1287	LA WASHINGTON	162.57	1322
OK LE FLORE	167.53	1288	WV TUCKER	162.50	1323
SD MCPHERSON	167.52	1289	KY HARLAN	162.37	1324
ND EDDY	167.47	1290	OH VINTON	162.26	1325
MO CRAWFORD	167.25	1291	OH GREENE	162.17	1326
MI OAKLAND	166.96	1292	KS JEFFERSON	162.16	1327
SD HUGHES	166.94	1293	KY BOURBON	162.06	1328
AR STONE	166.92	1294	WI MONROE	162.01	1329
OH FAIRFIELD	166.86	1295	LA PLAQUEMINES	161.84	1330

ESTIMATED TOTAL FALLOUT FROM NEVADA NUCLEAR TESTS 1951-1970
MICROCURIES/SQUARE METER
COUNTIES RANKED
WHERE No. 1=Highest Fallout ; No. 3094 = Lowest Fallout
SORTED B RANK

COUNTY	uCi/sq meter	RANK
AL MOBILE	161.77	1331
KS MIAMI	161.75	1332
LA ST LANDRY	161.72	1333
LA ST TAMMANY	161.68	1334
KS FRANKLIN	161.67	1335
KS BARBER	161.67	1336
ND PEMBINA	161.36	1337
TX JASPER	161.14	1338
MA HAMPDEN	161.11	1339
KS DOUGLAS	161.07	1340
NY ST LAWRENCE	160.96	1341
WI LA CROSSE	160.88	1342
TX NEWTON	160.87	1343
TX CULBERSON	160.64	1344
NE RICHARDSON	160.59	1345
TX DALLAS	160.40	1346
MS HANCOCK	160.33	1347
WA ASOTIN	160.23	1348
MO WAYNE	160.05	1349
CT LITCHFIELD	160.04	1350
NY SUFFOLK	159.96	1351
OH VAN WERT	159.88	1352
CT NEW LONDON	159.86	1353
IL LAWRENCE	159.85	1354
WV BRAXTON	159.64	1355
AZ GREENLEE	159.50	1356
MI GENESEE	159.25	1357
KS OSAGE	159.23	1358
AR MONROE	159.00	1359
PA ALLEGHENY	158.99	1360
VA KING WILLIAM	158.97	1361
IN WABASH	158.87	1362
MA WORCESTER	158.84	1363
SD BON HOMME	158.72	1364
OK KINGFISHER	158.66	1365
PA WARREN	158.65	1366
MS PIKE	158.59	1367
TN WASHINGTON	158.29	1368
KS BROWN	158.14	1369
AR MISSISSIPPI	158.00	1370
NY QUEENS	157.91	1371
WI SAWYER	157.91	1372
PA DELAWARE	157.87	1373
AR LITTLE RIVER	157.85	1374
MO DUNKLIN	157.78	1375
NE PAWNEE	157.69	1376
PA ERIE	157.65	1377
CT WINDHAM	157.52	1378
OH NOBLE	157.46	1379
TN WEAKLEY	156.74	1380
OH GEAUGA	156.65	1381
AZ COCHISE	156.55	1382
WV BROOKE	156.49	1383
MS CLAIBORNE	156.23	1384
OK PAYNE	156.08	1385
IN WARRICK	155.96	1386
TX ANGELINA	155.89	1387
OK LOVE	155.86	1388
MI LENAWEE	155.80	1389
MS GEORGE	155.74	1390
TN TIPTON	155.71	1391
WV RANDOLPH	155.70	1392
WV FAYETTE	155.64	1393
NY WESTCHESTER	155.54	1394
KY FAYETTE	155.29	1395
ND RAMSEY	155.22	1396
NY KINGS	155.17	1397
KS NEMAHA	155.14	1398
TX SABINE	155.12	1399
NH ROCKINGHAM	155.02	1400

ESTIMATED TOTAL FALLOUT FROM NEVADA NUCLEAR TESTS 1951-1970
MICROCURIES/SQUARE METER
COUNTIES RANKED
WHERE No. 1=Highest Fallout ; No. 3094 = Lowest Fallout
SORTED B RANK

COUNTY	uCi/sq meter	RANK	COUNTY	uCi/sq meter	RANK
TX SAN AUGUSTIN	155.00	1401	TN LAUDERDALE	151.74	1436
TN LAKE	154.93	1402	MI LIVINGSTON	151.69	1437
PA CLINTON	154.62	1403	OK LOGAN	151.46	1438
IN TIPPECANOE	154.61	1404	OH LORAIN	151.45	1439
IN HUNTINGTON	154.57	1405	WI ADAMS	151.37	1440
NY CLINTON	154.56	1406	TX WALKER	151.35	1441
WI DUNN	154.45	1407	AZ YUMA	151.35	1442
IL LEE	154.39	1408	LA ST BERNARD	151.32	1443
MO CARTER	154.27	1409	TN OBION	151.32	1444
WV PUTNAM	154.26	1410	IL OGLE	151.25	1445
WI JACKSON	154.21	1411	AR CRITTENDEN	151.20	1446
AR GRANT	154.21	1412	TX POLK	150.93	1447
IL STEPHENSON	154.05	1413	OK WOODS	150.44	1448
ND KIDDER	154.02	1414	NM GRANT	150.44	1449
WY BIG HORN	153.97	1415	IL CRAWFORD	150.35	1450
OK JOHNSTON	153.92	1416	AR CLARK	150.33	1451
MO PLATTE	153.70	1417	WV GRANT	150.18	1452
LA ST JOHN THE	153.59	1418	MO PEMISCOT	149.83	1453
AR HOWARD	153.48	1419	KS LINN	149.75	1454
MO SHANNON	153.44	1420	NE SARPY	149.74	1455
MS JEFFERSON	153.40	1421	WV MONONGALIA	149.60	1456
IN WARREN	153.32	1422	AR CROSS	149.50	1457
OK POTTAWATOMIE	153.31	1423	MO HICKORY	149.32	1458
VA TAZEWELL	153.18	1424	WV CLAY	149.31	1459
AL MARENGO	153.13	1425	MO BUTLER	149.27	1460
NY WAYNE	153.12	1426	TX HOPKINS	149.27	1461
NY ALLEGANY	153.11	1427	RI PROVIDENCE	149.24	1462
VT FRANKLIN	153.10	1428	IN VANDERBURGH	149.21	1463
WI WASHBURN	153.10	1429	NJ ESSEX	149.16	1464
IL LA SALLE	153.06	1430	TN MADISON	149.11	1465
NJ HUDSON	152.72	1431	MO REYNOLDS	149.11	1466
WV MARION	152.48	1432	MS MARION	149.09	1467
VA KING AND QUE	152.08	1433	IN MIAMI	148.98	1468
NY ONTARIO	152.05	1434	PA CENTRE	148.83	1469
WV NICHOLAS	152.02	1435	LA ST JAMES	148.83	1470

ESTIMATED TOTAL FALLOUT FROM NEVADA NUCLEAR TESTS 1951-1970
MICROCURIES/SQUARE METER
COUNTIES RANKED
WHERE No. 1=Highest Fallout ; No. 3094 = Lowest Fallout
SORTED BY RANK

COUNTY	uCi/sq meter	RANK
PA HUNTINGDON	148.74	1471
MI LAPEER	148.38	1472
NM OTERO	148.19	1473
OH HURON	148.13	1474
NY SULLIVAN	148.02	1475
TX RAINS	147.92	1476
AR PIKE	147.91	1477
MA HAMPSHIRE	147.89	1478
KS ANDERSON	147.89	1479
MD HARFORD	147.77	1480
TX CHAMBERS	147.77	1481
NY OTSEGO	147.71	1482
TN DYER	147.51	1483
MI MONROE	147.43	1484
NY STEUBEN	147.03	1485
TX JEFF DAVIS	146.70	1486
MI SHIAWASSEE	146.61	1487
IN PERRY	146.60	1488
NY CATTARAUGUS	146.59	1489
VA SMYTH	146.57	1490
ME PENOBSCOT	146.55	1491
WV POCAHONTAS	146.41	1492
IN SPENCER	146.39	1493
AR POINSETT	146.32	1494
NY PUTNAM	146.26	1495
NC NORTHAMPTON	146.21	1496
TX TARRANT	146.10	1497
AR LEE	145.82	1498
PA YORK	145.76	1499
TN HAYWOOD	145.76	1500
VA NEW KENT	145.75	1501
TX WINKLER	145.63	1502
MD BALTIMORE	145.29	1503
MI HILLSDALE	145.16	1504
WI WOOD	145.15	1505
OK COAL	145.06	1506
IL PUTNAM	145.04	1507
TX BOWIE	144.72	1508
WV WEBSTER	144.69	1509
WI RUSK	144.68	1510
MS FORREST	144.58	1511
NE ROCK	144.42	1512
KS BOURBON	144.28	1513
MO ST CHARLES	144.23	1514
OK CARTER	144.13	1515
IN SULLIVAN	144.10	1516
VA CHARLES CITY	144.04	1517
TX GRIMES	144.02	1518
KY HANCOCK	143.94	1519
MI WAYNE	143.62	1520
OK CANADIAN	143.59	1521
NJ UNION	143.52	1522
NY ORLEANS	143.48	1523
CT HARTFORD	143.46	1524
TX TRINITY	143.39	1525
NC BEAUFORT	143.23	1526
MS LAUDERDALE	143.20	1527
MO MONTGOMERY	143.19	1528
AR DESHA	143.17	1529
KY HICKMAN	143.07	1530
PA CAMERON	143.05	1531
MO JACKSON	142.98	1532
TX VAN ZANDT	142.89	1533
TX HUNT	142.77	1534
MO MISSISSIPPI	142.73	1535
NY DELAWARE	142.68	1536
TX HARRIS	142.59	1537
TX LIBERTY	142.59	1538
KY MARTIN	142.50	1539
MS DE SOTO	142.45	1540

ESTIMATED TOTAL FALLOUT FROM NEVADA NUCLEAR TESTS 1951-1970
MICROCURIES/SQUARE METER
COUNTIES RANKED
WHERE No. 1=Highest Fallout ; No. 3094 = Lowest Fallout
SORTED B RANK

COUNTY	uCi/sq meter	RANK
WI COLUMBIA	142.36	1541
PA NORTHUMBERLA	142.30	1542
VT WINDSOR	142.26	1543
AR LAFAYETTE	142.21	1544
VA FAIRFAX	142.11	1545
TX TYLER	141.98	1546
MI KENT	141.98	1547
TX RED RIVER	141.95	1548
NH CARROLL	141.90	1549
WI EAU CLAIRE	141.72	1550
IL IROQUOIS	141.65	1551
NJ BERGEN	141.56	1552
NY NEW YORK	141.49	1553
IL JEFFERSON	141.30	1554
TX BRAZORIA	141.14	1555
VA WARREN	141.13	1556
OK COMANCHE	141.10	1557
AR HOT SPRING	141.04	1558
MS QUITMAN	141.00	1559
AR GARLAND	140.88	1560
PA LACKAWANNA	140.86	1561
MN DAKOTA	140.82	1562
OK MARSHALL	140.70	1563
WI ST CROIX	140.58	1564
MS STONE	140.50	1565
AR MILLER	140.49	1566
NY FRANKLIN	140.46	1567
MS MARSHALL	140.46	1568
MT GLACIER	140.42	1569
IN DAVIESS	140.39	1570
MT YELLOWSTONE	140.22	1571
TX WOOD	140.05	1572
WI DOUGLAS	140.01	1573
VA LOUDOUN	140.00	1574
IN VIGO	139.98	1575
WI PIERCE	139.92	1576
NY JEFFERSON	139.75	1577
WI WAUSHARA	139.70	1578
TX CAMP	139.70	1579
NM LUNA	139.55	1580
OH PUTNAM	139.50	1581
TX MONTGOMERY	139.45	1582
IN PULASKI	139.30	1583
AR PHILLIPS	139.18	1584
OK HARPER	139.12	1585
WI ASHLAND	138.99	1586
TX HARDIN	138.94	1587
WI BARRON	138.94	1588
WI POLK	138.85	1589
AR NEVADA	138.77	1590
MN OLMSTED	138.76	1591
TX REEVES	138.74	1592
KY FULTON	138.66	1593
IL KANKAKEE	138.61	1594
NJ MIDDLESEX	138.37	1595
RI NEWPORT	138.35	1596
MO ST FRANCOIS	138.33	1597
MO ST CLAIR	138.19	1598
VT WASHINGTON	138.19	1599
NY RICHMOND	138.18	1600
PA WAYNE	138.17	1601
WI CHIPPEWA	138.06	1602
OK DEWEY	137.71	1603
TX CASS	137.63	1604
ME PISCATAQUIS	137.55	1605
NE SCOTTS BLUFF	137.52	1606
ND CAVALIER	137.38	1607
NY HERKIMER	137.29	1608
WI MARQUETTE	137.23	1609
VA CRAIG	136.90	1610

ESTIMATED TOTAL FALLOUT FROM NEVADA NUCLEAR TESTS 1951-1970 MICROCURIES/SQUARE METER COUNTIES RANKED WHERE No. 1=Highest Fallout ; No. 3094 = Lowest Fallout SORTED B RANK					
COUNTY	uCi/sq meter	RANK	COUNTY	uCi/sq meter	RANK
NJ PASSAIC	136.71	1611	TX ROCKWALL	134.40	1646
FL OKALOOSA	136.68	1612	MI KALAMAZOO	134.39	1647
OK MAJOR	136.56	1613	MO IRON	134.33	1648
ND WELLS	136.52	1614	IN JASPER	134.16	1649
TX WALLER	136.37	1615	MS COAHOMA	134.11	1650
AR LINCOLN	136.33	1616	NY DUTCHESS	134.11	1651
NY OSWEGO	136.31	1617	MS TIPPAH	134.04	1652
MO STODDARD	136.31	1618	AR HEMPSTEAD	134.01	1653
VA FAUQUIER	136.26	1619	NH CHESHIRE	133.70	1654
VT LAMOILLE	136.20	1620	TX SAN JACINTO	133.59	1655
ME WASHINGTON	136.13	1621	MI CALHOUN	133.55	1656
VA HIGHLAND	136.03	1622	TN FAYETTE	133.49	1657
MS TATE	136.02	1623	MO WASHINGTON	133.40	1658
PA ADAMS	136.00	1624	MI CLINTON	133.31	1659
IL RANDOLPH	135.99	1625	NY BRONX	133.23	1660
WI BURNETT	135.93	1626	NM HIDALGO	133.13	1661
NJ MORRIS	135.90	1627	TX FORT BEND	132.89	1662
PA BEDFORD	135.90	1628	MI VAN BUREN	132.87	1663
OR MALHEUR	135.88	1629	NC VANCE	132.86	1664
PA LANCASTER	135.81	1630	IN POSEY	132.79	1665
NC PERSON	135.79	1631	VA ALLEGHANY	132.75	1666
VA FLOYD	135.67	1632	OK CADDO	132.75	1667
MS HARRISON	135.50	1633	NJ CAPE MAY	132.71	1668
MS CHICKASAW	135.23	1634	WI IRON	132.70	1669
TN CROCKETT	135.22	1635	MS PANOLA	132.70	1670
VA ORANGE	135.21	1636	KY GRAVES	132.70	1671
MS JACKSON	135.17	1637	TX AUSTIN	132.69	1672
MO WARREN	134.94	1638	TX DELTA	132.59	1673
OK BLAINE	134.82	1639	VA BEDFORD	132.34	1674
KS COMANCHE	134.80	1640	MS TALLAHATCHIE	131.78	1675
KS DONIPHAN	134.77	1641	NJ MERCER	131.77	1676
IL WILL	134.58	1642	MA FRANKLIN	131.67	1677
MN CARLTON	134.57	1643	KY DAVIESS	131.64	1678
WI CLARK	134.48	1644	ID BENEWAH	131.43	1679
VA ROCKINGHAM	134.43	1645	PA POTTER	131.42	1680

ESTIMATED TOTAL FALLOUT FROM NEVADA NUCLEAR TESTS 1951-1970
MICROCURIES/SQUARE METER
COUNTIES RANKED
WHERE No. 1=Highest Fallout ; No. 3094 = Lowest Fallout
SORTED BY RANK

COUNTY	uCi/sq meter	RANK	COUNTY	uCi/sq meter	RANK
TX WISE	131.40	1681	NY SENECA	127.13	1716
VA PITTSYLVANIA	131.38	1682	OK STEPHENS	127.03	1717
IL GRUNDY	131.27	1683	TX ELLIS	127.02	1718
IL DE KALB	131.26	1684	MN SCOTT	127.01	1719
VA MECKLENBURG	131.14	1685	NH STRAFFORD	126.96	1720
NY ORANGE	130.85	1686	AL CONECUH	126.83	1721
IN DECATUR	130.84	1687	MT LINCOLN	126.79	1722
IN WHITE	130.76	1688	VA RAPPAHANNOCK	126.67	1723
WI BAYFIELD	130.68	1689	OK CLEVELAND	126.62	1724
MI MACOMB	130.54	1690	VA PRINCE EDWAR	126.54	1725
MS WALTHALL	130.07	1691	OK JEFFERSON	126.52	1726
MO TANEY	130.02	1692	MS TUNICA	126.44	1727
VA ROANOKE	130.01	1693	OH SUMMIT	126.40	1728
TX MONTAGUE	129.79	1694	IN MONTGOMERY	126.31	1729
VA SPOTSYLVANIA	129.74	1695	TN CLAIBORNE	126.25	1730
MS YALOBUSHA	129.67	1696	MS BENTON	126.13	1731
IL WABASH	129.47	1697	MO BUCHANAN	125.98	1732
VT ORANGE	129.44	1698	PA MCKEAN	125.94	1733
MS GREENE	129.37	1699	AR COLUMBIA	125.79	1734
TX WASHINGTON	129.31	1700	VA LUNENBURG	125.78	1735
WI TAYLOR	129.31	1701	VA MADISON	125.75	1736
TX GALVESTON	129.22	1702	VA SUFFOLK/NANS	125.74	1737
MN FREEBORN	129.19	1703	TX TITUS	125.68	1738
OK MCCLAIN	129.09	1704	ND TOWNER	125.65	1739
NM DONA ANA	128.49	1705	OK MURRAY	125.33	1740
KY CUMBERLAND	128.36	1706	TN CAMPBELL	125.07	1741
NH HILLSBOROUGH	128.34	1707	OK WOODWARD	124.73	1742
MO ANDREW	128.26	1708	OK OKLAHOMA	124.65	1743
TN HARDEMAN	128.21	1709	IL LIVINGSTON	124.55	1744
MN RAMSEY	128.08	1710	OK GRADY	124.53	1745
MN WASHINGTON	127.84	1711	NC GRANVILLE	124.51	1746
MD ALLEGANY	127.72	1712	OK GARVIN	124.31	1747
AR DALLAS	127.34	1713	MS BOLIVAR	124.04	1748
VA CAROLINE	127.15	1714	TN HANCOCK	123.94	1749
WI PRICE	127.14	1715	KY BELL	123.89	1750

ESTIMATED TOTAL FALLOUT FROM NEVADA NUCLEAR TESTS 1951-1970
MICROCURIES/SQUARE METER
COUNTIES RANKED
WHERE No. 1=Highest Fallout ; No. 3094 = Lowest Fallout
SORTED B RANK

COUNTY	uCi/sq meter	RANK	COUNTY	uCi/sq meter	RANK
PA PIKE	123.85	1751	VA AUGUSTA	120.30	1786
VT ORLEANS	123.70	1752	MA NANTUCKET	120.20	1787
TX WARD	123.63	1753	MI SANILAC	120.11	1788
MO HOLT	123.63	1754	IL KANE	120.02	1789
IL CARROLL	123.50	1755	VT ESSEX	119.87	1790
MN CARVER	123.47	1756	VA LEE	119.74	1791
NC CHATHAM	123.39	1757	KY UNION	119.67	1792
IL WAYNE	123.36	1758	LA JEFFERSON	119.38	1793
IN PORTER	123.18	1759	MO ST LOUIS	119.34	1794
NH GRAFTON	123.15	1760	VA PULASKI	119.32	1795
IN LAKE	123.13	1761	VA APPOMATTOX	119.32	1796
TX SMITH	122.97	1762	IL MCHENRY	119.31	1797
NJ WARREN	122.76	1763	WI PORTAGE	119.09	1798
NY TOMPKINS	122.71	1764	NH MERRIMACK	118.92	1799
VA AMHERST	122.52	1765	OK COTTON	118.91	1800
IN NEWTON	122.43	1766	MS CARROLL	118.87	1801
MS AMITE	122.32	1767	NH BELKNAP	118.83	1802
NC WILKES	122.07	1768	NY ROCKLAND	118.68	1803
ME HANCOCK	122.02	1769	KY BOYLE	118.36	1804
MS GRENADA	122.02	1770	NV NYE3	118.32	1805
TN CARTER	121.78	1771	NJ SUSSEX	118.30	1806
AR CLEVELAND	121.64	1772	VA RUSSELL	118.22	1807
NH SULLIVAN	121.63	1773	VA CAMPBELL	118.20	1808
MS LAMAR	121.50	1774	MS PONTOTOC	118.17	1809
NC MARTIN	121.41	1775	IL JACKSON	117.94	1810
WI WAUPACA	121.41	1776	OK JACKSON	117.93	1811
AR OUACHITA	121.31	1777	PA LUZERNE	117.92	1812
MS MONTGOMERY	121.29	1778	NC ASHE	117.82	1813
MI OTTAWA	121.28	1779	MO PERRY	117.74	1814
MN PINE	121.04	1780	VA NELSON	117.57	1815
ND ROLETTE	120.98	1781	ND PIERCE	117.55	1816
NJ SOMERSET	120.89	1782	IN CLAY	117.54	1817
OK KIOWA	120.73	1783	OK CUSTER	117.53	1818
IL ST CLAIR	120.40	1784	IN BOONE	117.49	1819
KY CLAY	120.33	1785	VA PATRICK	117.38	1820

ESTIMATED TOTAL FALLOUT FROM NEVADA NUCLEAR TESTS 1951-1970
MICROCURIES/SQUARE METER
COUNTIES RANKED
WHERE No. 1=Highest Fallout ; No. 3094 = Lowest Fallout
SORTED BY RANK

COUNTY	uCi/sq meter	RANK	COUNTY	uCi/sq meter	RANK
WV OHIO	117.24	1821	CA LOS ANGELES	114.58	1856
MS COPIAH	117.20	1822	WI VILAS	114.58	1857
TX FRANKLIN	117.20	1823	MS UNION	114.56	1858
MN SHERBURNE	117.09	1824	MI GOGEBIC	114.28	1859
WI ONEIDA	117.00	1825	OH OTTAWA	114.09	1860
MI BARRY	116.93	1826	NJ CAMDEN	114.07	1861
KY MARSHALL	116.90	1827	IL PULASKI	114.02	1862
KY TODD	116.86	1828	VT CALEDONIA	113.65	1863
VA MONTGOMERY	116.85	1829	MI HOUGHTON	113.33	1864
TN SHELBY	116.83	1830	VA ROCKBRIDGE	113.21	1865
WV HARDY	116.83	1831	LA RICHLAND	113.16	1866
LA MADISON	116.82	1832	IN MORGAN	113.07	1867
VA SCOTT	116.81	1833	PA MONROE	112.89	1868
TX CLAY	116.67	1834	ME KNOX	112.88	1869
IN FOUNTAIN	116.61	1835	TN CLAY	112.80	1870
AR UNION	116.60	1836	SC FLORENCE	112.70	1871
IL EDWARDS	116.56	1837	MS PERRY	112.43	1872
TX HARDEMAN	116.46	1838	IN PARKE	112.42	1873
TX ECTOR	116.44	1839	MS TISHOMINGO	112.37	1874
IL GALLATIN	116.41	1840	NC JONES	112.35	1875
IL MONROE	116.34	1841	NJ GLOUCESTER	112.18	1876
MS LEE	116.04	1842	WA PEND OREILLE	112.18	1877
KY PULASKI	115.93	1843	IL UNION	112.14	1878
PA WYOMING	115.89	1844	MS LAFAYETTE	112.14	1879
VA FREDERICK	115.87	1845	TN SEVIER	112.12	1880
TN SULLIVAN	115.78	1846	MS ISSAQUENA	112.00	1881
KY MERCER	115.55	1847	MN ST LOUIS	111.94	1882
MN MCLEOD	115.54	1848	WA COLUMBIA	111.87	1883
KY RUSSELL	115.46	1849	IN PUTNAM	111.78	1884
KY KNOX	115.44	1850	MI HURON	111.71	1885
WI LINCOLN	115.25	1851	AL WILCOX	111.28	1886
MN ANOKA	114.97	1852	VA GILES	111.22	1887
MO CAPE GIRARDE	114.92	1853	OH HARDIN	111.20	1888
MS YAZOO	114.63	1854	CA ALPINE	111.18	1889
IL ALEXANDER	114.59	1855	MN MEEKER	111.10	1890

ESTIMATED TOTAL FALLOUT FROM NEVADA NUCLEAR TESTS 1951-1970
MICROCURIES/SQUARE METER
COUNTIES RANKED
WHERE No. 1=Highest Fallout ; No. 3094 = Lowest Fallout
SORTED BY RANK

COUNTY	uCi/sq meter	RANK
MN RENVILLE	111.03	1891
TN MCNAIRY	110.95	1892
IL MADISON	110.91	1893
MS PRENTISS	110.75	1894
NH COOS	110.75	1895
TX KAUFMAN	110.70	1896
VA GREENE	110.57	1897
TX UPSHUR	110.57	1898
MI KEWEENAW	110.52	1899
LA OUACHITA	110.37	1900
IN RUSH	110.33	1901
WI GREEN LAKE	110.31	1902
MI SAGINAW	110.23	1903
WI MARATHON	110.14	1904
IL BUREAU	110.07	1905
MS SHARKEY	110.02	1906
KY MCCRACKEN	109.99	1907
OR HARNEY	109.99	1908
TX PARKER	109.96	1909
NC MACON	109.95	1910
KY OHIO	109.91	1911
NJ CUMBERLAND	109.87	1912
NC JACKSON	109.86	1913
KY GARRARD	109.85	1914
AL MONROE	109.78	1915
ME OXFORD	109.74	1916
MI BRANCH	109.73	1917
IN BENTON	109.69	1918
KY BULLITT	109.66	1919
TN SCOTT	109.49	1920
MN CHISAGO	109.33	1921
IN BROWN	109.20	1922
TX RUSK	109.16	1923
MO SCOTT	109.15	1924
MN WRIGHT	109.12	1925
IN RIPLEY	109.10	1926
ME WALDO	108.99	1927
MI ONTONAGON	108.94	1928
ME LINCOLN	108.87	1929
LA WEBSTER	108.86	1930
VA SHENANDOAH	108.58	1931
MO MADISON	108.56	1932
MS ALCORN	108.51	1933
VA GRAYSON	108.51	1934
TN PICKETT	108.30	1935
ME CUMBERLAND	108.08	1936
MS LEFLORE	108.02	1937
OH ASHTABULA	107.95	1938
KY BALLARD	107.93	1939
TX MADISON	107.90	1940
MS CLARKE	107.88	1941
TX MORRIS	107.82	1942
OH FRANKLIN	107.66	1943
NY LEWIS	107.53	1944
GA RABUN	107.40	1945
WV PLEASANTS	107.27	1946
MN AITKIN	107.18	1947
AR BRADLEY	107.04	1948
VA VIRGINIA BEA	107.00	1949
OH MERCER	106.84	1950
WI FOND DU LAC	106.77	1951
NY YATES	106.69	1952
SC OCONEE	106.65	1953
NY ONEIDA	106.62	1954
TX DAWSON	106.60	1955
NC CLAY	106.48	1956
TX DICKENS	106.28	1957
MS NEWTON	106.27	1958
TX JOHNSON	106.27	1959
MS HUMPHREYS	106.24	1960

ESTIMATED TOTAL FALLOUT FROM NEVADA NUCLEAR TESTS 1951-1970
MICROCURIES/SQUARE METER
COUNTIES RANKED
WHERE No. 1=Highest Fallout ; No. 3094 = Lowest Fallout
SORTED BY RANK

COUNTY	uCi/sq meter	RANK	COUNTY	uCi/sq meter	RANK
OH WAYNE	106.21	1961	IN GIBSON	102.98	1996
ID BONNER	106.14	1962	NC ORANGE	102.76	1997
AR DREW	106.13	1963	NJ SALEM	102.74	1998
WI WAUKESHA	106.01	1964	MS SUNFLOWER	102.65	1999
MN BENTON	105.81	1965	GA LONG	102.65	2000
NC HYDE	105.78	1966	IL CLINTON	102.64	2001
GA LIBERTY	105.78	1967	MN KANABEC	102.52	2002
NC SWAIN	105.71	1968	NY MADISON	102.47	2003
LA CADDO	105.71	1969	MO STE GENEVIEV	102.28	2004
NC TRANSYLVANIA	105.70	1970	MN KANDIYOHI	102.26	2005
TN ANDERSON	105.68	1971	OH HOLMES	102.25	2006
ME YORK	105.65	1972	WI MARINETTE	101.98	2007
MS WAYNE	105.61	1973	WI WINNEBAGO	101.93	2008
MS CALHOUN	105.61	1974	IN JOHNSON	101.73	2009
OR UMATILLA	105.49	1975	AR ASHLEY	101.39	2010
VA WASHINGTON	105.44	1976	WI OCONTO	101.22	2011
MS WASHINGTON	105.39	1977	WI CALUMET	101.09	2012
ME SAGADAHOC	105.09	1978	NJ HUNTERDON	101.00	2013
NC HAYWOOD	104.82	1979	NC HENDERSON	100.95	2014
NY CORTLAND	104.71	1980	VA WISE	100.85	2015
IL PERRY	104.65	1981	IN PIKE	100.85	2016
NY MONROE	104.46	1982	IL LAKE	100.76	2017
ME KENNEBEC	104.37	1983	NY CHEMUNG	100.60	2018
MO BOLLINGER	104.16	1984	IN KOSCIUSKO	100.52	2019
IN DEARBORN	104.11	1985	TX CHEROKEE	100.47	2020
MN ISANTI	104.03	1986	MS OKTIBBEHA	100.44	2021
IL POPE	104.00	1987	MS MADISON	100.44	2022
LA BOSSIER	103.98	1988	MS LINCOLN	100.19	2023
OK TILLMAN	103.73	1989	NC CHEROKEE	100.10	2024
MI GRATIOT	103.53	1990	KY HENDERSON	100.07	2025
TX MARION	103.38	1991	WI SHEBOYGAN	99.99	2026
TN CHESTER	103.36	1992	LA WINN	99.98	2027
TX GARZA	103.28	1993	NY CAYUGA	99.60	2028
PA BLAIR	103.25	1994	MN CROW WING	99.54	2029
ME ANDROSCOGGIN	103.00	1995	KY LINCOLN	99.52	2030

ESTIMATED TOTAL FALLOUT FROM NEVADA NUCLEAR TESTS 1951-1970 MICROCURIES/SQUARE METER COUNTIES RANKED WHERE No. 1=Highest Fallout ; No. 3094 = Lowest Fallout SORTED B RANK					
COUNTY	**uCi/sq meter**	**RANK**	**COUNTY**	**uCi/sq meter**	**RANK**
DE SUSSEX	99.50	2031	AL BALDWIN	97.20	2066
TX GREGG	99.43	2032	LA EAST CARROLL	97.08	2067
LA UNION	99.34	2033	PA LEHIGH	96.91	2068
PA NORTHAMPTON	99.27	2034	NC MITCHELL	96.88	2069
DE NEW CASTLE	99.25	2035	TX FISHER	96.84	2070
TN COCKE	99.23	2036	KY JEFFERSON	96.73	2071
TX HENDERSON	99.13	2037	WI MENOMINEE	96.53	2072
KY LIVINGSTON	99.05	2038	MS WEBSTER	96.42	2073
MN LAKE	99.03	2039	FL HOLMES	96.34	2074
TN ROANE	98.99	2040	KY BUTLER	96.34	2075
TX HARRISON	98.95	2041	AR CHICOT	96.31	2076
IL WASHINGTON	98.91	2042	MI JACKSON	96.26	2077
MD SOMERSET	98.87	2043	TN COFFEE	96.25	2078
PA TIOGA	98.85	2044	MI TUSCOLA	96.22	2079
ME FRANKLIN	98.81	2045	NC WATAUGA	96.21	2080
AL GREENE	98.75	2046	WV HANCOCK	96.17	2081
MN MILLE LACS	98.66	2047	TN CARROLL	96.16	2082
AL GENEVA	98.64	2048	KY CHRISTIAN	96.15	2083
TX WICHITA	98.63	2049	MN TODD	96.12	2084
LA MOREHOUSE	98.63	2050	IN LA PORTE	96.10	2085
DE KENT	98.60	2051	ID KOOTENAI	96.10	2086
NY CHENANGO	98.57	2052	NC YANCEY	96.04	2087
TN MONTGOMERY	98.53	2053	GA WAYNE	96.01	2088
LA JACKSON	98.26	2054	TN BLEDSOE	95.78	2089
FL LAKE	98.13	2055	MS JASPER	95.76	2090
LA BIENVILLE	97.91	2056	LA LINCOLN	95.75	2091
KY WASHINGTON	97.73	2057	TN GRUNDY	95.70	2092
TX JACK	97.67	2058	PA BUCKS	95.66	2093
WI SHAWANO	97.66	2059	TN DICKSON	95.59	2094
WI DODGE	97.64	2060	VA PRINCE GEORG	95.44	2095
MI ST CLAIR	97.60	2061	KY WARREN	95.34	2096
TN BLOUNT	97.52	2062	TN HENRY	95.07	2097
TX FOARD	97.38	2063	MN STEARNS	95.03	2098
MN HENNEPIN	97.32	2064	PA CARBON	94.97	2099
PA MONTGOMERY	97.23	2065	TX KNOX	94.93	2100

ESTIMATED TOTAL FALLOUT FROM NEVADA NUCLEAR TESTS 1951-1970
MICROCURIES/SQUARE METER
COUNTIES RANKED
WHERE No. 1=Highest Fallout ; No. 3094 = Lowest Fallout
SORTED BY RANK

COUNTY	uCi/sq meter	RANK	COUNTY	uCi/sq meter	RANK
NC BURKE	94.91	2101	MN COOK	92.53	2136
MN LAC QUI PARL	94.85	2102	NC YADKIN	92.53	2137
MI BERRIEN	94.75	2103	TX KENT	92.36	2138
VA HENRICO	94.70	2104	IN MARSHALL	92.35	2139
MI IRON	94.57	2105	NC MCDOWELL	92.30	2140
PA SUSQUEHANNA	94.53	2106	SC CLARENDON	92.25	2141
SC PICKENS	94.35	2107	TX ARCHER	92.20	2142
IL MARION	94.28	2108	AL CLARKE	92.15	2143
WV CABELL	94.27	2109	ME SOMERSET	92.10	2144
SC LEE	94.10	2110	NC PAMLICO	92.07	2145
TN MONROE	94.01	2111	NC GRAHAM	92.06	2146
LA FRANKLIN	94.00	2112	MS SCOTT	92.01	2147
TX STONEWALL	93.98	2113	NC BUNCOMBE	91.89	2148
MN CHIPPEWA	93.93	2114	MN MORRISON	91.88	2149
TN MCMINN	93.87	2115	MN CASS	91.62	2150
MD WORCESTER	93.86	2116	NC DURHAM	91.59	2151
WI WASHINGTON	93.82	2117	IL WILLIAMSON	91.54	2152
KY MONROE	93.60	2118	WI LANGLADE	91.53	2153
MI MENOMINEE	93.51	2119	VA MIDDLESEX	91.47	2154
OH HENRY	93.49	2120	FL SANTA ROSA	91.40	2155
TX BORDEN	93.49	2121	MI BARAGA	91.29	2156
LA NATCHITOCHES	93.47	2122	TX NAVARRO	91.20	2157
SC JASPER	93.45	2123	KY CASEY	91.15	2158
GA WHEELER	93.35	2124	TX PALO PINTO	91.11	2159
PA BERKS	93.26	2125	MI ALLEGAN	91.08	2160
LA WEST CARROLL	92.93	2126	NC WAKE	91.07	2161
KY CARLISLE	92.88	2127	TN FENTRESS	91.06	2162
IN STARKE	92.81	2128	LA CALDWELL	91.03	2163
MN YELLOW MEDIC	92.79	2129	NC MADISON	90.90	2164
TN CUMBERLAND	92.75	2130	GA EFFINGHAM	90.82	2165
GA TOWNS	92.74	2131	MI MIDLAND	90.71	2166
TN HAWKINS	92.70	2132	IL WHITE	90.69	2167
GA BRYAN	92.65	2133	GA SCREVEN	90.60	2168
MI SCHOOLCRAFT	92.59	2134	KY WAYNE	90.38	2169
TX COTTLE	92.57	2135	TN CHEATHAM	90.37	2170

ESTIMATED TOTAL FALLOUT FROM NEVADA NUCLEAR TESTS 1951-1970 MICROCURIES/SQUARE METER COUNTIES RANKED WHERE No. 1=Highest Fallout ; No. 3094 = Lowest Fallout SORTED B RANK					
COUNTY	uCi/sq meter	RANK	COUNTY	uCi/sq meter	RANK
TN GREENE	90.36	2171	SC CALHOUN	89.03	2206
NC CALDWELL	90.25	2172	NC NEW HANOVER	88.99	2207
TN RHEA	90.24	2173	TN MORGAN	88.96	2208
TN SEQUATCHIE	90.23	2174	OH CARROLL	88.87	2209
TN HUMPHREYS	90.23	2175	MD MONTGOMERY	88.72	2210
TN DECATUR	90.21	2176	GA TERRELL	88.53	2211
SD DEUEL	90.19	2177	AL ESCAMBIA	88.51	2212
SC SUMTER	90.14	2178	VA GREENSVILLE	88.48	2213
WA WALLA WALLA	90.09	2179	IL SALINE	88.46	2214
NC ALLEGHANY	90.09	2180	AL TUSCALOOSA	88.45	2215
GA MONTGOMERY	90.07	2181	NC MOORE	88.35	2216
NC AVERY	90.06	2182	TX CALLAHAN	88.31	2217
MN KOOCHICHING	89.98	2183	SC GREENVILLE	88.31	2218
TX HOOD	89.97	2184	MS COVINGTON	88.19	2219
KY TRIGG	89.91	2185	KY WHITLEY	88.04	2220
MS MONROE	89.91	2186	TX HILL	88.01	2221
TN BRADLEY	89.87	2187	MN ITASCA	87.88	2222
MS CLAY	89.84	2188	TX LIMESTONE	87.87	2223
MI MONTCALM	89.80	2189	TN HENDERSON	87.71	2224
TX PANOLA	89.66	2190	NC CRAVEN	87.62	2225
NC ONSLOW	89.65	2191	FL HILLSBOROUGH	87.59	2226
NC DARE	89.65	2192	NC ALEXANDER	87.55	2227
IN JAY	89.62	2193	IL MASSAC	87.54	2228
TX MARTIN	89.46	2194	WI FOREST	87.53	2229
WA WHITMAN	89.44	2195	CA SIERRA	87.53	2230
TN POLK	89.34	2196	TX WILBARGER	87.48	2231
CA NEVADA	89.33	2197	NC CARTERET	87.47	2232
IL JOHNSON	89.30	2198	SC HORRY	87.36	2233
MN WADENA	89.29	2199	VA NORTHUMBERLA	87.28	2234
SC MARION	89.24	2200	VA ARLINGTON	87.28	2235
NC SURRY	89.23	2201	MD PRINCE GEORG	87.26	2236
LA DE SOTO	89.19	2202	NC GUILFORD	87.26	2237
IN ELKHART	89.17	2203	MI MUSKEGON	87.16	2238
IN HANCOCK	89.13	2204	MD ST MARYS	87.12	2239
IL FRANKLIN	89.10	2205	NY ONONDAGA	86.99	2240

ESTIMATED TOTAL FALLOUT FROM NEVADA NUCLEAR TESTS 1951-1970 MICROCURIES/SQUARE METER COUNTIES RANKED WHERE No. 1=Highest Fallout ; No. 3094 = Lowest Fallout SORTED B RANK					
COUNTY	**uCi/sq meter**	**RANK**	**COUNTY**	**uCi/sq meter**	**RANK**
NC BERTIE	86.99	2241	FL ESCAMBIA	85.11	2276
VA LANCASTER	86.76	2242	GA JENKINS	85.09	2277
TN LOUDON	86.74	2243	TX JEFFERSON	85.08	2278
NC POLK	86.73	2244	OH JEFFERSON	85.01	2279
MS ITAWAMBA	86.73	2245	ID BOUNDARY	85.01	2280
AL WASHINGTON	86.64	2246	FL VOLUSIA	84.99	2281
TN DAVIDSON	86.63	2247	NY SCHUYLER	84.85	2282
SC ORANGEBURG	86.44	2248	TN HICKMAN	84.76	2283
GA RANDOLPH	86.42	2249	TN JEFFERSON	84.74	2284
PA SULLIVAN	86.35	2250	MD CALVERT	84.68	2285
GA EVANS	86.28	2251	KY MCLEAN	84.66	2286
NC IREDELL	86.18	2252	IN STEUBEN	84.63	2287
GA BULLOCH	86.03	2253	KY WEBSTER	84.63	2288
VA CULPEPER	85.97	2254	NC RANDOLPH	84.56	2289
GA QUITMAN	85.90	2255	MD WICOMICO	84.55	2290
NC DAVIDSON	85.81	2256	AL HALE	84.50	2291
GA PULASKI	85.78	2257	GA CALHOUN	84.50	2292
TN HARDIN	85.74	2258	NC HERTFORD	84.49	2293
TN UNICOI	85.72	2259	MN SWIFT	84.48	2294
WI OZAUKEE	85.70	2260	PA BRADFORD	84.48	2295
MS SMITH	85.70	2261	GA WEBSTER	84.44	2296
AL COLBERT	85.67	2262	TN VAN BUREN	84.37	2297
GA CLAY	85.66	2263	MI OCEANA	84.36	2298
KY MCCREARY	85.56	2264	PA SCHUYLKILL	84.26	2299
KY LAUREL	85.48	2265	SC BERKELEY	84.14	2300
GA FANNIN	85.45	2266	AR CALHOUN	84.09	2301
VA NEWPORT NEWS	85.43	2267	TX MCLENNAN	84.03	2302
NC PITT	85.40	2268	LA RED RIVER	84.02	2303
KY CALDWELL	85.39	2269	GA SCHLEY	83.98	2304
TN PUTNAM	85.33	2270	FL LEON	83.94	2305
VA STAFFORD	85.32	2271	GA TATTNALL	83.93	2306
KY CALLOWAY	85.31	2272	MI NEWAYGO	83.85	2307
TN JOHNSON	85.28	2273	IL HAMILTON	83.83	2308
TN WARREN	85.23	2274	MD CECIL	83.83	2309
TX HASKELL	85.20	2275	TX BAYLOR	83.81	2310

ESTIMATED TOTAL FALLOUT FROM NEVADA NUCLEAR TESTS 1951-1970
MICROCURIES/SQUARE METER
COUNTIES RANKED
WHERE No. 1=Highest Fallout ; No. 3094 = Lowest Fallout
SORTED BY RANK

COUNTY	uCi/sq meter	RANK	COUNTY	uCi/sq meter	RANK
SC DORCHESTER	83.76	2311	NC GATES	81.97	2346
WI KENOSHA	83.72	2312	GA PEACH	81.95	2347
VA ISLE OF WIGH	83.70	2313	GA CRAWFORD	81.94	2348
TN HAMILTON	83.68	2314	MS LEAKE	81.92	2349
MS ATTALA	83.68	2315	TN BEDFORD	81.84	2350
KY EDMONSON	83.67	2316	TN MAURY	81.81	2351
NC CATAWBA	83.63	2317	MI ALGER	81.75	2352
NC ALAMANCE	83.62	2318	FL SUMTER	81.74	2353
IN LAGRANGE	83.60	2319	NC CLEVELAND	81.73	2354
LA CLAIBORNE	83.55	2320	GA CARROLL	81.72	2355
GA COLQUITT	83.47	2321	PA PHILADELPHIA	81.64	2356
GA WHITE	83.42	2322	SC BEAUFORT	81.64	2357
GA CHATHAM	83.37	2323	MD ANNE ARUNDEL	81.55	2358
MS CHOCTAW	83.25	2324	AL DALE	81.53	2359
GA GLYNN	83.17	2325	NC PENDER	81.53	2360
SC COLLETON	82.92	2326	IL KENDALL	81.52	2361
TX SCURRY	82.78	2327	KY METCALFE	81.38	2362
VA DINWIDDIE	82.72	2328	WI KEWAUNEE	81.38	2363
WI MANITOWOC	82.71	2329	DC WASHINGTON	81.36	2364
PA MONTOUR	82.70	2330	GA BAKER	81.34	2365
TX ANDERSON	82.65	2331	FL ST JOHNS	81.30	2366
SC DARLINGTON	82.60	2332	MD KENT	81.21	2367
TN MARSHALL	82.60	2333	IL DU PAGE	81.19	2368
GA MCINTOSH	82.59	2334	SC MARLBORO	81.17	2369
MD HOWARD	82.42	2335	GA CLAYTON	81.11	2370
VA MATHEWS	82.28	2336	AL MADISON	81.11	2371
PA UNION	82.19	2337	TN KNOX	81.10	2372
VA NORFOLK/CHES	82.15	2338	TN PERRY	81.04	2373
SC CHARLESTON	82.12	2339	AL PIKE	81.04	2374
TX BOSQUE	82.11	2340	SC SPARTANBURG	81.01	2375
NY TIOGA	82.11	2341	MD CHARLES	80.96	2376
AL LOWNDES	82.10	2342	KY BRECKINRIDGE	80.79	2377
AL PICKENS	82.05	2343	GA DOUGHERTY	80.67	2378
GA BURKE	82.04	2344	NC LENOIR	80.66	2379
GA FAYETTE	82.01	2345	NC CUMBERLAND	80.65	2380

ESTIMATED TOTAL FALLOUT FROM NEVADA NUCLEAR TESTS 1951-1970
MICROCURIES/SQUARE METER
COUNTIES RANKED
WHERE No. 1=Highest Fallout ; No. 3094 = Lowest Fallout
SORTED B RANK

COUNTY	uCi/sq meter	RANK	COUNTY	uCi/sq meter	RANK
AL PERRY	80.53	2381	TN MEIGS	78.94	2416
TN HAMBLEN	80.51	2382	VA HAMPTON	78.94	2417
LA ORLEANS	80.38	2383	GA MITCHELL	78.92	2418
GA JEFFERSON	80.30	2384	MD DORCHESTER	78.87	2419
MD TALBOT	80.27	2385	GA TWIGGS	78.85	2420
GA UNION	80.27	2386	GA MORGAN	78.84	2421
AL BARBOUR	80.25	2387	MD WASHINGTON	78.83	2422
MI CASS	80.17	2388	GA JOHNSON	78.80	2423
WV JACKSON	80.10	2389	GA GORDON	78.79	2424
TN STEWART	80.02	2390	VA PRINCE WILLI	78.79	2425
GA GREENE	80.01	2391	AL LAUDERDALE	78.67	2426
GA JEFF DAVIS	79.92	2392	FL FLAGLER	78.60	2427
VA KING GEORGE	79.86	2393	MI ST JOSEPH	78.58	2428
TN ROBERTSON	79.78	2394	TN SUMNER	78.56	2429
TX EASTLAND	79.56	2395	TN OVERTON	78.55	2430
NC RUTHERFORD	79.55	2396	TX KING	78.55	2431
MI DICKINSON	79.50	2397	MS RANKIN	78.41	2432
VA SOUTHAMPTON	79.48	2398	AL SUMTER	78.34	2433
SC CHESTERFIELD	79.39	2399	VA BRUNSWICK	78.31	2434
VA SURRY	79.39	2400	MI WASHTENAW	78.31	2435
WI DOOR	79.38	2401	GA LEE	78.28	2436
TX SOMERVELL	79.33	2402	AL DE KALB	78.25	2437
NC WAYNE	79.31	2403	SC WILLIAMSBURG	78.23	2438
AL ETOWAH	79.30	2404	MD FREDERICK	78.20	2439
VA SUSSEX	79.27	2405	VA NOTTOWAY	78.19	2440
GA APPLING	79.26	2406	FL HERNANDO	78.16	2441
KY MEADE	79.25	2407	GA TOOMBS	78.13	2442
MI LUCE	79.24	2408	MN BELTRAMI	78.02	2443
TX NACOGDOCHES	79.24	2409	VA NORTHAMPTON	78.00	2444
AL COVINGTON	79.21	2410	VA WESTMORELAND	77.91	2445
KY MUHLENBERG	79.19	2411	GA LANIER	77.90	2446
FL BREVARD	79.15	2412	GA BACON	77.87	2447
NC WARREN	79.09	2413	GA BLECKLEY	77.84	2448
IL HARDIN	79.05	2414	TX YOUNG	77.70	2449
TN WHITE	78.96	2415	IN OHIO	77.66	2450

ESTIMATED TOTAL FALLOUT FROM NEVADA NUCLEAR TESTS 1951-1970
MICROCURIES/SQUARE METER
COUNTIES RANKED
WHERE No. 1=Highest Fallout ; No. 3094 = Lowest Fallout
SORTED B RANK

COUNTY	uCi/sq meter	RANK
TX HOUSTON	77.66	2451
TN DE KALB	77.64	2452
OH PICKAWAY	77.56	2453
TN WILLIAMSON	77.51	2454
KY BARREN	77.49	2455
SC KERSHAW	77.43	2456
SC DILLON	77.40	2457
AL CHAMBERS	77.16	2458
WI FLORENCE	77.16	2459
NC DUPLIN	77.15	2460
GA SUMTER	77.01	2461
GA TELFAIR	76.95	2462
GA PAULDING	76.95	2463
GA MONROE	76.91	2464
KY LYON	76.89	2465
MS JONES	76.87	2466
MI DELTA	76.80	2467
GA THOMAS	76.76	2468
AL LEE	76.76	2469
NC FORSYTH	76.76	2470
VA GLOUCESTER	76.76	2471
MN POPE	76.74	2472
GA BERRIEN	76.72	2473
TX STEPHENS	76.71	2474
VA JAMES CITY	76.69	2475
WA SPOKANE	76.68	2476
NC WILSON	76.67	2477
AL CULLMAN	76.66	2478
TX ERATH	76.64	2479
AL BIBB	76.56	2480
GA DOUGLAS	76.55	2481
VA YORK	76.43	2482
TX THROCKMORTON	76.41	2483
PA COLUMBIA	76.37	2484
WI OUTAGAMIE	76.31	2485
FL ALACHUA	76.14	2486
NC COLUMBUS	76.09	2487
AL TALLADEGA	76.02	2488
GA STEWART	75.98	2489
GA CAMDEN	75.96	2490
GA COLUMBUS	75.96	2491
IN NOBLE	75.95	2492
TN HOUSTON	75.94	2493
FL CALHOUN	75.93	2494
GA WILCOX	75.90	2495
SC ANDERSON	75.89	2496
PA FRANKLIN	75.88	2497
FL MADISON	75.80	2498
GA FLOYD	75.70	2499
VA HENRY	75.63	2500
NC LINCOLN	75.55	2501
MS LAWRENCE	75.53	2502
GA HOUSTON	75.46	2503
AL MARSHALL	75.45	2504
GA CRISP	75.37	2505
WI RACINE	75.36	2506
TX HAMILTON	75.36	2507
GA EMANUEL	75.35	2508
OH SHELBY	75.34	2509
AL MORGAN	75.29	2510
TN UNION	75.18	2511
AL RANDOLPH	75.17	2512
MS WINSTON	75.17	2513
FL POLK	75.12	2514
AL JACKSON	75.11	2515
SC BAMBERG	75.10	2516
FL NASSAU	75.08	2517
GA TALIAFERRO	74.94	2518
AL CLAY	74.89	2519
TX FREESTONE	74.86	2520

ESTIMATED TOTAL FALLOUT FROM NEVADA NUCLEAR TESTS 1951-1970 MICROCURIES/SQUARE METER COUNTIES RANKED WHERE No. 1=Highest Fallout ; No. 3094 = Lowest Fallout SORTED B RANK					
COUNTY	uCi/sq meter	RANK	COUNTY	uCi/sq meter	RANK
MS HOLMES	74.66	2521	GA BRANTLEY	73.63	2556
IN JENNINGS	74.65	2522	NC ROWAN	73.59	2557
KY CRITTENDEN	74.63	2523	MD QUEEN ANNES	73.59	2558
MN BECKER	74.62	2524	FL SEMINOLE	73.56	2559
MN OTTER TAIL	74.60	2525	GA COOK	73.54	2560
KY LOGAN	74.60	2526	WV PRESTON	73.48	2561
FL GILCHRIST	74.53	2527	KY HOPKINS	73.43	2562
GA BALDWIN	74.52	2528	NY BROOME	73.42	2563
MN BIG STONE	74.52	2529	GA BIBB	73.42	2564
GA WARREN	74.49	2530	TN LAWRENCE	73.32	2565
TX HOWARD	74.41	2531	AL TALLAPOOSA	73.24	2566
GA ATKINSON	74.39	2532	FL MARION	73.20	2567
GA GLASCOCK	74.38	2533	NC SAMPSON	73.16	2568
GA MCDUFFIE	74.36	2534	TN GRAINGER	73.13	2569
MD CAROLINE	74.35	2535	GA CANDLER	73.10	2570
GA WORTH	74.32	2536	TX LEON	72.99	2571
GA CLINCH	74.31	2537	FL GULF	72.95	2572
TX BROWN	74.29	2538	NC HALIFAX	72.90	2573
GA TROUP	74.27	2539	GA WASHINGTON	72.82	2574
AL JEFFERSON	74.23	2540	IN FAYETTE	72.80	2575
AL DALLAS	74.23	2541	VA RICHMOND	72.75	2576
GA WILKINSON	74.20	2542	SC YORK	72.75	2577
MS KEMPER	74.09	2543	GA TURNER	72.70	2578
NC LEE	74.09	2544	KY SIMPSON	72.67	2579
TN WAYNE	74.08	2545	SC CHEROKEE	72.65	2580
KY LEE	74.02	2546	TX MCCULLOCH	72.62	2581
FL ORANGE	73.98	2547	MI MARQUETTE	72.58	2582
NC HARNETT	73.95	2548	GA GWINNETT	72.58	2583
TX SHACKELFORD	73.91	2549	MS SIMPSON	72.51	2584
GA TIFT	73.91	2550	GA WALKER	72.46	2585
VA CUMBERLAND	73.88	2551	GA BANKS	72.40	2586
NC ROBESON	73.87	2552	GA MADISON	72.38	2587
AL HOUSTON	73.84	2553	SC GEORGETOWN	72.38	2588
AL ELMORE	73.82	2554	TN WILSON	72.37	2589
GA CLARKE	73.67	2555	FL LIBERTY	72.35	2590

ESTIMATED TOTAL FALLOUT FROM NEVADA NUCLEAR TESTS 1951-1970 MICROCURIES/SQUARE METER COUNTIES RANKED WHERE No. 1=Highest Fallout ; No. 3094 = Lowest Fallout SORTED BY RANK					
COUNTY	**uCi/sq meter**	**RANK**	**COUNTY**	**uCi/sq meter**	**RANK**
TX NOLAN	72.31	2591	FL SUWANNEE	70.77	2626
GA HABERSHAM	72.27	2592	MI ARENAC	70.71	2627
TN GILES	72.27	2593	NC NASH	70.67	2628
IL COOK	72.27	2594	KY ADAIR	70.59	2629
FL DIXIE	72.24	2595	KY LARUE	70.59	2630
SD GRANT	72.23	2596	GA HANCOCK	70.56	2631
GA GILMER	72.20	2597	TX CONCHO	70.53	2632
GA CHATTAHOOCHE	72.20	2598	OH DEFIANCE	70.46	2633
FL PINELLAS	72.12	2599	SC MCCORMICK	70.39	2634
IN WHITLEY	72.07	2600	FL FRANKLIN	70.31	2635
VA ESSEX	72.05	2601	GA STEPHENS	70.25	2636
TN BENTON	72.01	2602	NC MONTGOMERY	70.25	2637
MN STEVENS	71.99	2603	TN MARION	70.23	2638
NC BLADEN	71.98	2604	TX SHELBY	70.17	2639
GA BEN HILL	71.98	2605	VA LOUISA	70.16	2640
PA LYCOMING	71.97	2606	PA MIFFLIN	70.05	2641
TN SMITH	71.87	2607	GA HARALSON	69.90	2642
FL CITRUS	71.84	2608	GA DODGE	69.86	2643
MN HUBBARD	71.83	2609	KY CLINTON	69.84	2644
FL WAKULLA	71.82	2610	TX BELL	69.81	2645
GA DOOLY	71.79	2611	NC GASTON	69.75	2646
TN LINCOLN	71.78	2612	NC UNION	69.71	2647
GA SPALDING	71.59	2613	AL CLEBURNE	69.68	2648
WV MORGAN	71.57	2614	GA OCONEE	69.64	2649
GA IRWIN	71.49	2615	NC EDGECOMBE	69.61	2650
TN JACKSON	71.48	2616	GA CHATTOOGA	69.58	2651
MI IOSCO	71.45	2617	GA JONES	69.57	2652
GA MURRAY	71.43	2618	MN DOUGLAS	69.49	2653
SC LAURENS	71.23	2619	GA MERIWETHER	69.44	2654
SC HAMPTON	71.08	2620	MS JEFFERSON DA	69.36	2655
GA LOWNDES	71.06	2621	AL FRANKLIN	69.36	2656
FL PUTNAM	70.93	2622	TX WILLIAMSON	69.32	2657
GA JACKSON	70.89	2623	AL CALHOUN	69.31	2658
SC GREENWOOD	70.89	2624	WI BROWN	69.25	2659
GA PIKE	70.80	2625	AL COFFEE	69.24	2660

ESTIMATED TOTAL FALLOUT FROM NEVADA NUCLEAR TESTS 1951-1970
MICROCURIES/SQUARE METER
COUNTIES RANKED
WHERE No. 1=Highest Fallout ; No. 3094 = Lowest Fallout
SORTED BY RANK

COUNTY	uCi/sq meter	RANK	COUNTY	uCi/sq meter	RANK
AL RUSSELL	69.22	2661	AL ST CLAIR	68.02	2696
TX JONES	69.16	2662	FL HAMILTON	67.98	2697
SC NEWBERRY	69.10	2663	GA WALTON	67.93	2698
NC GREENE	69.09	2664	SC SALUDA	67.92	2699
SC EDGEFIELD	68.94	2665	VA HALIFAX	67.87	2700
MN POLK	68.91	2666	IN BLACKFORD	67.86	2701
MS NOXUBEE	68.91	2667	MI CHIPPEWA	67.75	2702
TN FRANKLIN	68.84	2668	GA JASPER	67.70	2703
AL LIMESTONE	68.82	2669	TX COMANCHE	67.65	2704
MI ISABELLA	68.75	2670	TX ROBERTSON	67.63	2705
SC BARNWELL	68.75	2671	GA COWETA	67.59	2706
GA HALL	68.74	2672	SC ALLENDALE	67.58	2707
GA PIERCE	68.73	2673	SC ABBEVILLE	67.58	2708
MS NESHOBA	68.67	2674	CA GLENN	67.53	2709
WV JEFFERSON	68.65	2675	NC STOKES	67.51	2710
GA CHARLTON	68.52	2676	MI BAY	67.47	2711
GA DECATUR	68.46	2677	TN MOORE	67.44	2712
KY GREEN	68.44	2678	GA WILKES	67.39	2713
SC CHESTER	68.42	2679	MN PENNINGTON	67.32	2714
GA CHEROKEE	68.42	2680	AL CHILTON	67.28	2715
KY HARDIN	68.40	2681	GA HENRY	67.27	2716
MN TRAVERSE	68.37	2682	GA WHITFIELD	67.27	2717
GA CATOOSA	68.37	2683	SC LANCASTER	67.26	2718
GA DE KALB	68.37	2684	GA TAYLOR	67.24	2719
AL BULLOCK	68.35	2685	GA RICHMOND	67.20	2720
NC HOKE	68.35	2686	NC SCOTLAND	67.19	2721
TX FALLS	68.34	2687	NC BRUNSWICK	67.15	2722
GA PUTNAM	68.31	2688	MS LOWNDES	67.12	2723
FL JEFFERSON	68.30	2689	GA LAMAR	67.10	2724
GA BUTTS	68.26	2690	TN CANNON	67.03	2725
NC STANLY	68.23	2691	AL AUTAUGA	67.00	2726
SC UNION	68.16	2692	FL GADSDEN	66.98	2727
AL BLOUNT	68.06	2693	CA AMADOR	66.94	2728
GA OGLETHORPE	68.06	2694	VA FLUVANNA	66.88	2729
MI CHARLEVOIX	68.04	2695	AL WALKER	66.82	2730

ESTIMATED TOTAL FALLOUT FROM NEVADA NUCLEAR TESTS 1951-1970
MICROCURIES/SQUARE METER
COUNTIES RANKED
WHERE No. 1=Highest Fallout ; No. 3094 = Lowest Fallout
SORTED BY RANK

COUNTY	uCi/sq meter	RANK	COUNTY	uCi/sq meter	RANK
AL CHEROKEE	66.79	2731	TX RUNNELS	64.72	2766
CA LASSEN	66.76	2732	GA ECHOLS	64.67	2767
MN CLEARWATER	66.75	2733	WV DODDRIDGE	64.61	2768
GA BROOKS	66.72	2734	TX LAMPASAS	64.58	2769
TX BRAZOS	66.64	2735	GA FORSYTH	64.57	2770
AL CHOCTAW	66.51	2736	VA CLARKE	64.55	2771
AL BUTLER	66.43	2737	TX MITCHELL	64.53	2772
NC DAVIE	66.39	2738	AL FAYETTE	64.47	2773
GA BARTOW	66.37	2739	VA BUCKINGHAM	64.47	2774
SC FAIRFIELD	66.33	2740	GA PICKENS	64.44	2775
GA NEWTON	66.33	2741	GA WARE	64.38	2776
GA HARRIS	66.33	2742	TX STERLING	64.34	2777
GA EARLY	66.31	2743	MI ANTRIM	64.32	2778
KY TAYLOR	66.30	2744	MI MACKINAC	64.29	2779
NC JOHNSTON	66.25	2745	NC ANSON	64.24	2780
GA HART	66.24	2746	TX MILAM	64.24	2781
GA LUMPKIN	66.17	2747	GA DAWSON	64.19	2782
PA JUNIATA	66.16	2748	AL MACON	64.12	2783
WV RITCHIE	66.14	2749	MI IONIA	64.09	2784
NC RICHMOND	65.94	2750	VA WYTHE	64.08	2785
MI ALCONA	65.90	2751	WV TAYLOR	64.05	2786
MN KITTSON	65.68	2752	NC MECKLENBURG	64.03	2787
KY ALLEN	65.47	2753	FL WASHINGTON	63.95	2788
GA GRADY	65.44	2754	WV ROANE	63.92	2789
FL CLAY	65.43	2755	SC LEXINGTON	63.85	2790
AL LAWRENCE	65.33	2756	GA DADE	63.85	2791
WV BARBOUR	65.28	2757	GA BARROW	63.84	2792
FL BAY	65.20	2758	FL OSCEOLA	63.83	2793
GA FRANKLIN	65.05	2759	GA TALBOT	63.78	2794
TX COLEMAN	65.01	2760	PA CUMBERLAND	63.69	2795
AL COOSA	65.01	2761	GA COBB	63.62	2796
GA COFFEE	64.99	2762	TN LEWIS	63.62	2797
MD CARROLL	64.94	2763	FL JACKSON	63.53	2798
AL MARION	64.78	2764	MI CHEBOYGAN	63.52	2799
TX MIDLAND	64.72	2765	MI ROSCOMMON	63.46	2800

ESTIMATED TOTAL FALLOUT FROM NEVADA NUCLEAR TESTS 1951-1970 MICROCURIES/SQUARE METER COUNTIES RANKED WHERE No. 1=Highest Fallout ; No. 3094 = Lowest Fallout SORTED BY RANK					
COUNTY	**uCi/sq meter**	**RANK**	**COUNTY**	**uCi/sq meter**	**RANK**
GA LINCOLN	63.42	2801	WV HAMPSHIRE	61.51	2836
CA PLACER	63.38	2802	MI CLARE	61.46	2837
NC TYRRELL	63.38	2803	MN MAHNOMEN	61.42	2838
WI MILWAUKEE	63.37	2804	KY HART	61.41	2839
MN GRANT	63.35	2805	TN RUTHERFORD	61.35	2840
WV BERKELEY	63.34	2806	PA LEBANON	61.27	2841
KY MARION	63.30	2807	MN RED LAKE	61.23	2842
TX CORYELL	63.26	2808	TX LLANO	61.22	2843
TX MASON	63.24	2809	TX BLANCO	61.02	2844
AL WINSTON	63.23	2810	NC CABARRUS	60.93	2845
IN DE KALB	63.21	2811	MI MECOSTA	60.78	2846
NC CASWELL	63.13	2812	VA PAGE	60.56	2847
TX SAN SABA	63.06	2813	KY GRAYSON	60.49	2848
MN ROSEAU	63.04	2814	PA PERRY	60.44	2849
FL BAKER	62.97	2815	VA CHARLOTTE	60.35	2850
CA VENTURA	62.96	2816	WV MINERAL	60.31	2851
AL CRENSHAW	62.93	2817	GA LAURENS	60.24	2852
AL LAMAR	62.89	2818	TX LEE	60.06	2853
GA ELBERT	62.79	2819	GA POLK	60.05	2854
FL WALTON	62.78	2820	GA MARION	60.01	2855
AL SHELBY	62.72	2821	GA TREUTLEN	59.88	2856
FL LAFAYETTE	62.72	2822	FL UNION	59.75	2857
GA FULTON	62.70	2823	GA MACON	59.56	2858
FL TAYLOR	62.70	2824	SC RICHLAND	59.43	2859
OH WILLIAMS	62.58	2825	NC ROCKINGHAM	59.23	2860
CA PLUMAS	62.51	2826	SC AIKEN	59.08	2861
GA ROCKDALE	62.46	2827	TX BANDERA	58.87	2862
MI CRAWFORD	62.38	2828	NC FRANKLIN	58.85	2863
CA KERN	62.29	2829	TN MACON	58.83	2864
GA UPSON	62.20	2830	MI MASON	58.63	2865
MI GLADWIN	61.81	2831	FL PASCO	58.55	2866
TX BURLESON	61.79	2832	MI MANISTEE	58.39	2867
GA COLUMBIA	61.73	2833	FL BRADFORD	58.35	2868
FL COLUMBIA	61.72	2834	MI WEXFORD	58.33	2869
PA DAUPHIN	61.52	2835	MN LAKE OF THE	58.31	2870

ESTIMATED TOTAL FALLOUT FROM NEVADA NUCLEAR TESTS 1951-1970
MICROCURIES/SQUARE METER
COUNTIES RANKED
WHERE No. 1=Highest Fallout ; No. 3094 = Lowest Fallout
SORTED BY RANK

COUNTY	uCi/sq meter	RANK	COUNTY	uCi/sq meter	RANK
TX PRESIDIO	58.24	2871	VA ALBEMARLE	54.10	2906
AL MONTGOMERY	58.20	2872	TX GLASSCOCK	53.99	2907
TN TROUSDALE	58.13	2873	TX CRANE	53.73	2908
MI OGEMAW	58.09	2874	TX FAYETTE	53.71	2909
CA CALAVERAS	57.81	2875	TX FRIO	53.69	2910
CA YUBA	57.77	2876	FL MONROE	53.50	2911
WA FERRY	57.70	2877	MN CLAY	53.49	2912
MI BENZIE	57.31	2878	FL ST LUCIE	53.46	2913
FL SARASOTA	57.18	2879	TX MENARD	53.45	2914
FL COLLIER	57.16	2880	CA SANTA BARBAR	52.88	2915
VA BLAND	57.12	2881	CA EL DORADO	52.88	2916
MI MISSAUKEE	57.09	2882	CA TUOLUMNE	52.64	2917
MI EMMET	57.08	2883	CA SONOMA	52.32	2918
TX MEDINA	56.94	2884	CA BUTTE	52.30	2919
MI OSCEOLA	56.85	2885	TX KERR	51.93	2920
TX COKE	56.71	2886	FL MANATEE	51.71	2921
KY NELSON	56.56	2887	MI LAKE	51.63	2922
TX GILLESPIE	56.51	2888	TX WHARTON	51.52	2923
GA MILLER	56.46	2889	TX PECOS	51.20	2924
GA HEARD	56.36	2890	WA STEVENS	50.84	2925
TX MILLS	56.14	2891	FL BROWARD	50.65	2926
TX IRION	56.03	2892	MI PRESQUE ISLE	50.41	2927
TX REAGAN	55.99	2893	FL DADE	50.34	2928
AL HENRY	55.99	2894	CA TULARE	50.29	2929
MN WILKIN	55.96	2895	TX COLORADO	50.19	2930
TX TOM GREEN	55.96	2896	FL LEE	49.85	2931
FL PALM BEACH	55.81	2897	CA MONTEREY	49.72	2932
MN MARSHALL	55.72	2898	TX COMAL	49.56	2933
CA TEHAMA	54.96	2899	FL HIGHLANDS	49.30	2934
TX BURNET	54.77	2900	MI KALKASKA	49.25	2935
GA SEMINOLE	54.74	2901	FL MARTIN	49.01	2936
FL HENDRY	54.44	2902	FL GLADES	48.99	2937
MN NORMAN	54.42	2903	TX KENDALL	48.88	2938
MI GRAND TRAVER	54.38	2904	MI OSCODA	48.56	2939
MI OTSEGO	54.13	2905	FL OKEECHOBEE	48.54	2940

ESTIMATED TOTAL FALLOUT FROM NEVADA NUCLEAR TESTS 1951-1970 MICROCURIES/SQUARE METER COUNTIES RANKED WHERE No. 1=Highest Fallout ; No. 3094 = Lowest Fallout SORTED BY RANK					
COUNTY	**uCi/sq meter**	**RANK**	**COUNTY**	**uCi/sq meter**	**RANK**
MI LEELANAU	47.93	2941	CA FRESNO	41.38	2976
MI MONTMORENCY	47.55	2942	TX DE WITT	41.32	2977
FL INDIAN RIVER	47.34	2943	TX ZAVALA	40.58	2978
CA SAN DIEGO	46.88	2944	OR LAKE	40.32	2979
CA SAN LUIS OBI	46.84	2945	TX HAYS	40.30	2980
CA MODOC	46.82	2946	FL HARDEE	40.25	2981
TX BREWSTER	46.76	2947	CA LAKE	40.23	2982
TX MATAGORDA	46.75	2948	TX CALHOUN	39.98	2983
FL LEVY	46.60	2949	CA COLUSA	39.65	2984
TX TRAVIS	46.39	2950	TX JIM HOGG	39.53	2985
TX BEXAR	46.34	2951	TX UVALDE	39.21	2986
CA SANTA CRUZ	46.22	2952	TX BEE	38.93	2987
TX KIMBLE	45.90	2953	TX CAMERON	38.73	2988
TX SCHLEICHER	45.71	2954	TX WILLACY	37.74	2989
CA KINGS	45.69	2955	TX SUTTON	37.53	2990
TX MCMULLEN	45.35	2956	TX STARR	37.24	2991
TX TAYLOR	45.26	2957	TX ARANSAS	36.63	2992
TX CROCKETT	45.13	2958	TX KLEBERG	36.54	2993
FL CHARLOTTE	44.87	2959	CA SACRAMENTO	36.29	2994
FL DUVAL	44.65	2960	OR CROOK	36.25	2995
CA SAN BENITO	44.58	2961	CA MADERA	35.95	2996
TX UPTON	44.36	2962	TX ATASCOSA	35.79	2997
MI ALPENA	43.91	2963	TX KARNES	35.66	2998
WA OKANOGAN	43.74	2964	TX BROOKS	35.47	2999
FL DE SOTO	43.68	2965	TX REAL	35.29	3000
TX GONZALES	43.42	2966	OR MORROW	35.28	3001
TX JACKSON	43.33	2967	CA SAN MATEO	34.79	3002
CA ORANGE	43.20	2968	TX LIVE OAK	34.46	3003
TX JIM WELLS	42.97	2969	TX TERRELL	34.42	3004
TX CALDWELL	42.49	2970	CA SANTA CLARA	34.42	3005
TX DIMMIT	42.40	2971	TX SAN PATRICIO	34.37	3006
CA MARIPOSA	42.34	2972	TX MAVERICK	34.23	3007
TX BASTROP	41.79	2973	TX EDWARDS	33.93	3008
CA MARIN	41.75	2974	CA NAPA	33.78	3009
TX LAVACA	41.67	2975	TX VICTORIA	33.75	3010

www.ingramcontent.com/pod-product-compliance
Lightning Source LLC
LaVergne TN
LVHW052204200926
848481LV00001B/24

* 9 7 8 1 8 8 1 0 4 3 2 9 4 *